Collins
POSTCODE ATLAS | BRITAIN AND NORTHERN IRELAND

CONTENTS

Key to map symbols	ii
Structure of postcodes	iii
Postcode areas	iv-v
Britain maps (4.1 miles to 1 inch)	2-107
Northern Ireland map (9 miles to 1 inch)	108-109
Urban maps (1.6 miles to 1 inch)	110-122
London (1.3 miles to 1 inch)	110-113
West Midlands	114-115
Manchester	116-117
Leeds and Bradford	118-119
Liverpool	120
Newcastle upon Tyne	121
Glasgow	122
London key to map pages and symbols	123
London mapping (3.2 inches to 1 mile)	124-131
Administrative areas information	132-145
Administrative areas map (19.7 miles to inch)	146-151
London boroughs map (2.6 miles to 1 inch)	152-153
Index to London street names	154-166
Index to Great Britain place names	167-223
Index to Northern Ireland place names	224

Published by Collins
An imprint of HarperCollins Publishers
Westerhill Road, Bishopbriggs, Glasgow, G64 2QT

www.harpercollins.co.uk

Copyright © HarperCollins Publishers Ltd 2016

Collins® is a registered trademark of HarperCollins Publishers Limited

Mapping generated from Collins Bartholomew digital databases

Postcode boundaries and codes copyright © Royal Mail Group plc

The postcode boundary information published in this atlas is compiled from the Postcode Address File (PAF) and reproduced with the permission of Royal Mail Group plc. The copyright and database rights in PAF are owned by Royal Mail Group plc. Details included in this atlas are subject to change without notice.

The grid on this map is the National Grid taken from the Ordnance Survey map with the permission of the Controller of Her Majesty's Stationery Office. British population figures are derived from the 2011 census. Source: Office for National Statistics website: www.ons.gov.uk Crown copyright material is reproduced with the permission of the Controller of HMSO. Northern Ireland populations derived from the 2011 Census. Source: Office for National Statistics website: www.ons.gov.uk

All rights reserved. No part of this publication may be reproduced, stored in a retrieval system, or transmitted, in any form or by any means, electronic, mechanical, photocopying, recording or otherwise, without the prior written permission of the publisher and copyright owners.

The contents of this publication are believed correct at the time of printing. Nevertheless, the publisher can accept no responsibility for errors or omissions, changes in the detail given, or for any expense or loss thereby caused.

The representation of a road, track or footpath is no evidence of a right of way.

Printed in China

ISBN 978 0 00 821154 7   10 9 8 7 6 5 4 3 2 1

e-mail: roadcheck@harpercollins.co.uk
facebook.com/collinsmaps
@Collinsmaps

MIX
Paper from responsible sources
FSC C007454

FSC™ is a non-profit international organisation established to promote the responsible management of the world's forests. Products carrying the FSC label are independently certified to assure consumers that they come from forests that are managed to meet the social, economic and ecological needs of present and future generations, and other controlled sources.

Find out more about HarperCollins and the environment at
www.harpercollins.co.uk/green

Key to map symbols

Postcode information

PL	Area code
▬▬▬	Area boundary
35	District code
▬▬▬	District boundary

Britain map symbols (pages 2-107)

M4	Motorway
M6 Toll	Toll Motorway
8 — 9	Motorway junction with full / limited access
Maidstone / Birch / Sarn	Motorway service areas (off road, full, limited access)
A48	Primary route dual / single carriageway
A5	'A' road dual / single carriageway
B1403	'B' road dual / single carriageway
▬▬▬	Minor road
▭▭▭	Restricted access due to road condition or private ownership
▬▬▬	Roads with passing places
▭▭▭	Roads proposed or under construction
32b	Multi-level junction (occasionally with junction number)
○─○	Roundabout
)====(	Road tunnel
»»	Steep hill (arrows point downhill)
××	Level crossing
Toll	Toll
▬▬▬	Railway line and station
▬▬▬	Preserved railway line and station
▬▬▬	Railway tunnel
✈ ✈	Airport with / without scheduled services
Ⓗ	Heliport
	Built up area
▫ ▫ ▫	Towns, villages and other settlements
▬▬▬	National boundary
▬ ▬ ▬	County / Unitary Authority boundary
468 ▲941	Spot / Summit height in metres
	Lake, dam and river
	Canal / Dry canal / Canal tunnel
	Beach

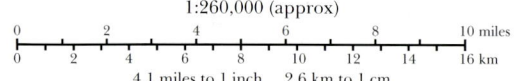

1:260,000 (approx)

0 2 4 6 8 10 miles
0 2 4 6 8 10 12 14 16 km
4.1 miles to 1 inch   2.6 km to 1 cm

Northern Ireland map symbols (pages 108-109)

M1	Motorway
5 — 9	Motorway junction with full / limited access
A28 — N4	Primary / National primary route
A29 — N52	'A' road / National secondary route
B113 R408	'B' road / Regional road
▬▬▬	Minor road
▭▭▭	Road under construction
○─○	Multi-level junction / roundabout
»»	Steep hill (arrows point downhill)
⊕ ✈	International / domestic airport
▬▬▬	Canal
▬ ▬ ▬	International boundary
▬ ▬ ▬	District / County boundary
▲754	Summit height (in metres)
	Built up areas
	Beach

Conurbation map symbols (pages 110-122)

M73	Motorway
M6 Toll	Toll motorway
5 — 4 FRANKLEY SERVICES	Motorway junctions with full / limited access
	Motorway service area
A725	Primary route dual / single carriageway
A4054	'A' road dual / single carriageway
B7078	'B' road dual / single carriageway
▬▬▬	Minor road
○ ○ ○	Roundabout
▬▬▬	Railway line and station
⊖ Ⓢ ▪	London Underground / Subway / Metro / Light rail station
▬▬▬	Railway tunnel
▭▭▭	Airport with scheduled services
	Built up areas
▬	Public building
▬ ▬ ▬	County / Unitary Authority boundary
	Woodland / Park
▲266	Spot height in metres
▬▬▬	Congestion charging zone

Structure of postcodes

Postcodes operate at five levels.

Level 1. Areas are denoted by the first one or two letters of the code, eg GL. These areas are then divided into districts.

Level 2. Districts are denoted by the number or numbers in the first part of the postcode, eg GL52. Districts are further subdivided into sectors.

Level 3. Subdistricts are a further special division of districts and only occur in London, eg EC1A.

Level 4. Sectors are denoted by the number in the second part of the postcode, eg GL52 5.

Level 5. The final two letters of the code denote a group of houses or an individual building, eg GL52 5HH.

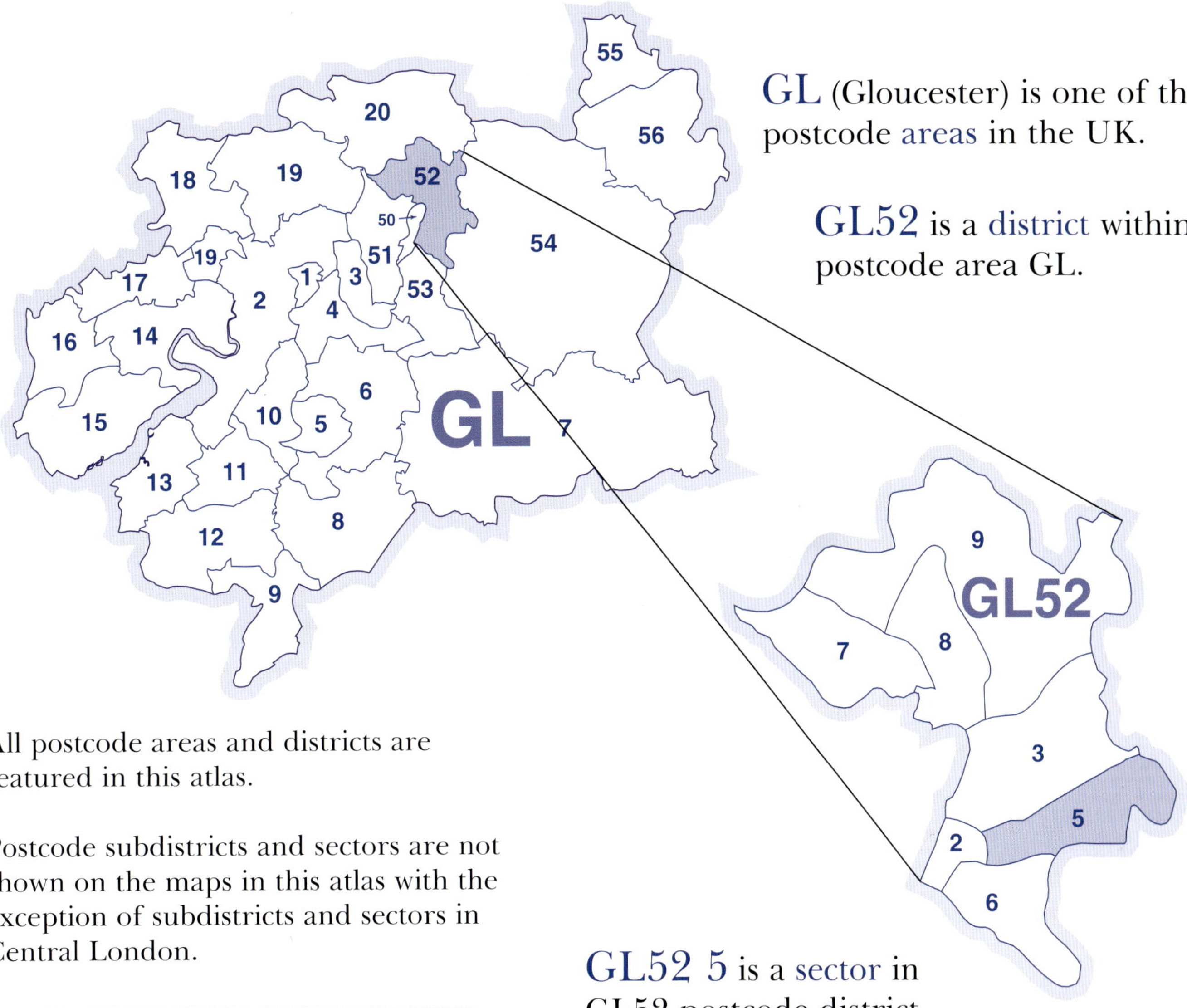

GL (Gloucester) is one of the postcode areas in the UK.

GL52 is a district within postcode area GL.

All postcode areas and districts are featured in this atlas.

Postcode subdistricts and sectors are not shown on the maps in this atlas with the exception of subdistricts and sectors in Central London.

GL52 5 is a sector in GL52 postcode district.

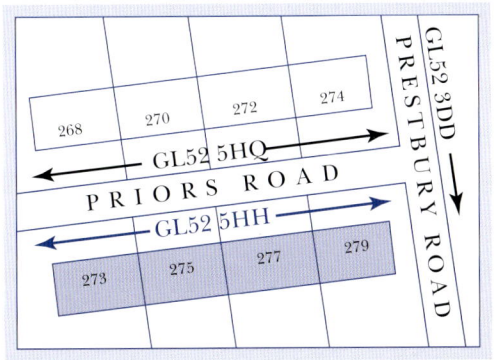

GL52 5HH is the postcode. It pinpoints a group of houses and in some cases individual business premises.

Postcode areas

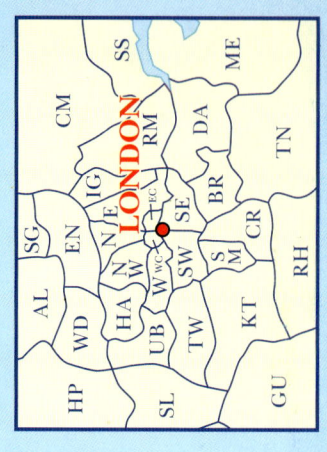

Postcode	& Area
SA	Swansea
SE	London SE
SG	Stevenage
SK	Stockport
SL	Slough
SM	Sutton
SN	Swindon
SO	Southampton
SP	Salisbury
SR	Sunderland
SS	Southend-on-Sea
ST	Stoke-on-Trent
SW	London SW
SY	Shrewsbury
TA	Taunton
TD	Galashiels
TF	Telford
TN	Royal Tunbridge Wells
TQ	Torquay
TR	Truro
TS	Teesside
TW	Twickenham
UB	Southall
W	London W
WA	Warrington
WC	London WC
WD	Watford
WF	Wakefield
WN	Wigan
WR	Worcester
WS	Walsall
WV	Wolverhampton
YO	York
ZE	Shetland

Postcode	& Area
AB	Aberdeen
AL	St Albans
B	Birmingham
BA	Bath
BB	Blackburn
BD	Bradford
BH	Bournemouth
BL	Bolton
BN	Brighton
BR	Bromley
BS	Bristol
BT	Northern Ireland
CA	Carlisle
CB	Cambridge
CF	Cardiff
CH	Chester
CM	Chelmsford
CO	Colchester
CR	Croydon
CT	Canterbury
CV	Coventry
CW	Crewe
DA	Dartford
DD	Dundee
DE	Derby
DG	Dumfries
DH	Durham
DL	Darlington
DN	Doncaster
DT	Dorchester
DY	Dudley
E	London E
EC	London EC
EH	Edinburgh
EN	Enfield
EX	Exeter
FK	Falkirk
FY	Blackpool
G	Glasgow
GL	Gloucester
GU	Guildford
GY	Guernsey
HA	Harrow
HD	Huddersfield
HG	Harrogate
HP	Hemel Hempstead
HR	Hereford
HS	Hebrides
HU	Kingston upon Hull
HX	Halifax
IG	Ilford
IM	Isle of Man
IP	Ipswich
IV	Inverness
JE	Jersey
KA	Kilmarnock
KT	Kingston-upon-Thames
KW	Kirkwall
KY	Kirkcaldy
L	Liverpool
LA	Lancaster
LD	Llandrindod Wells
LE	Leicester
LL	Llandudno
LN	Lincoln
LS	Leeds
LU	Luton
M	Manchester
ME	Medway
MK	Milton Keynes
ML	Motherwell
N	London N
NE	Newcastle upon Tyne
NG	Nottingham
NN	Northampton
NP	Newport
NR	Norwich
NW	London NW
OL	Oldham
OX	Oxford
PA	Paisley
PE	Peterborough
PH	Perth
PL	Plymouth
PO	Portsmouth
PR	Preston
RG	Reading
RH	Redhill
RM	Romford
S	Sheffield

v

Map of UK Postcode Areas

France

English Channel

Channel Islands — GY, JE

Irish Sea

Ireland

Isles of Scilly — TR

Isle of Man — IM

Cities and Postcode Areas

- Stranraer
- Belfast — BT
- Carlisle — CA
- Middlesbrough — TS, SR, DH, DL
- Lancaster — LA, FY
- Leeds — LS, HG, BD, YO, HU, DN
- Liverpool — L, PR, BB, HX, HD, WF
- Manchester — M, BL, OL, WN, WA, SK
- Lincoln — LN, NG
- Nottingham — NG, DE, S
- Birmingham — B, CV, WS, DY, WV, ST, TF, CW, CH
- Norwich — NR
- Cambridge — CB, PE, IP
- Ipswich — IP, CO, CM, SG
- London — EN, HA, WD, AL, LU, MK, NN, LE, HP, SL, UB, TW, KT, BR, DA, RM, SS, ME, CT
- Oxford — OX, RG, SN, GL, WR, HR, LD, SY, LL
- Cardiff — CF, NP, SA
- Swansea — SA
- Aberystwyth
- Bristol — BS, BA, SP, DT, BH
- Southampton — SO, PO, GU, RH, TN
- Brighton — BN
- Folkestone
- Exeter — EX, TA, TQ, PL, TR
- Plymouth — PL

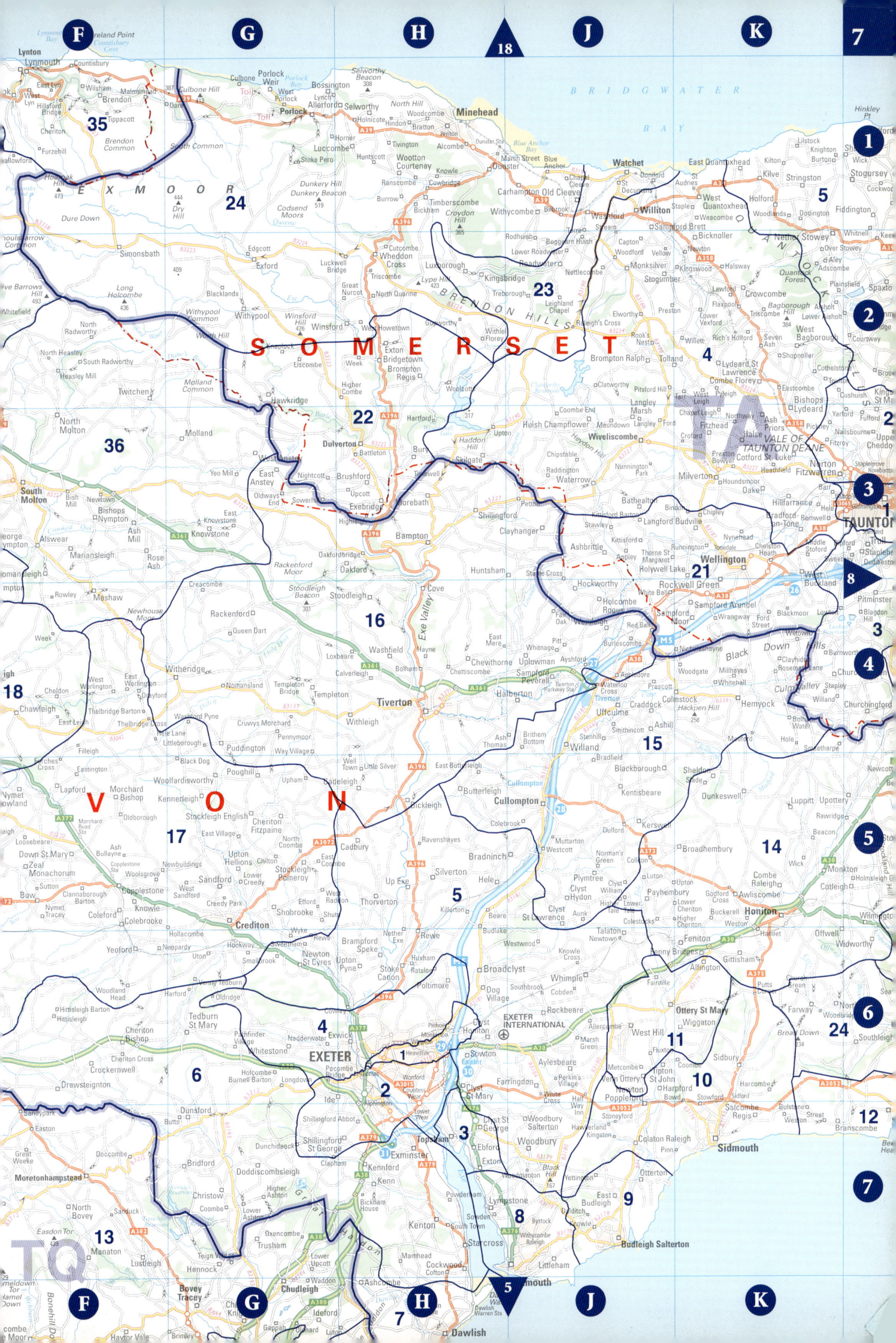

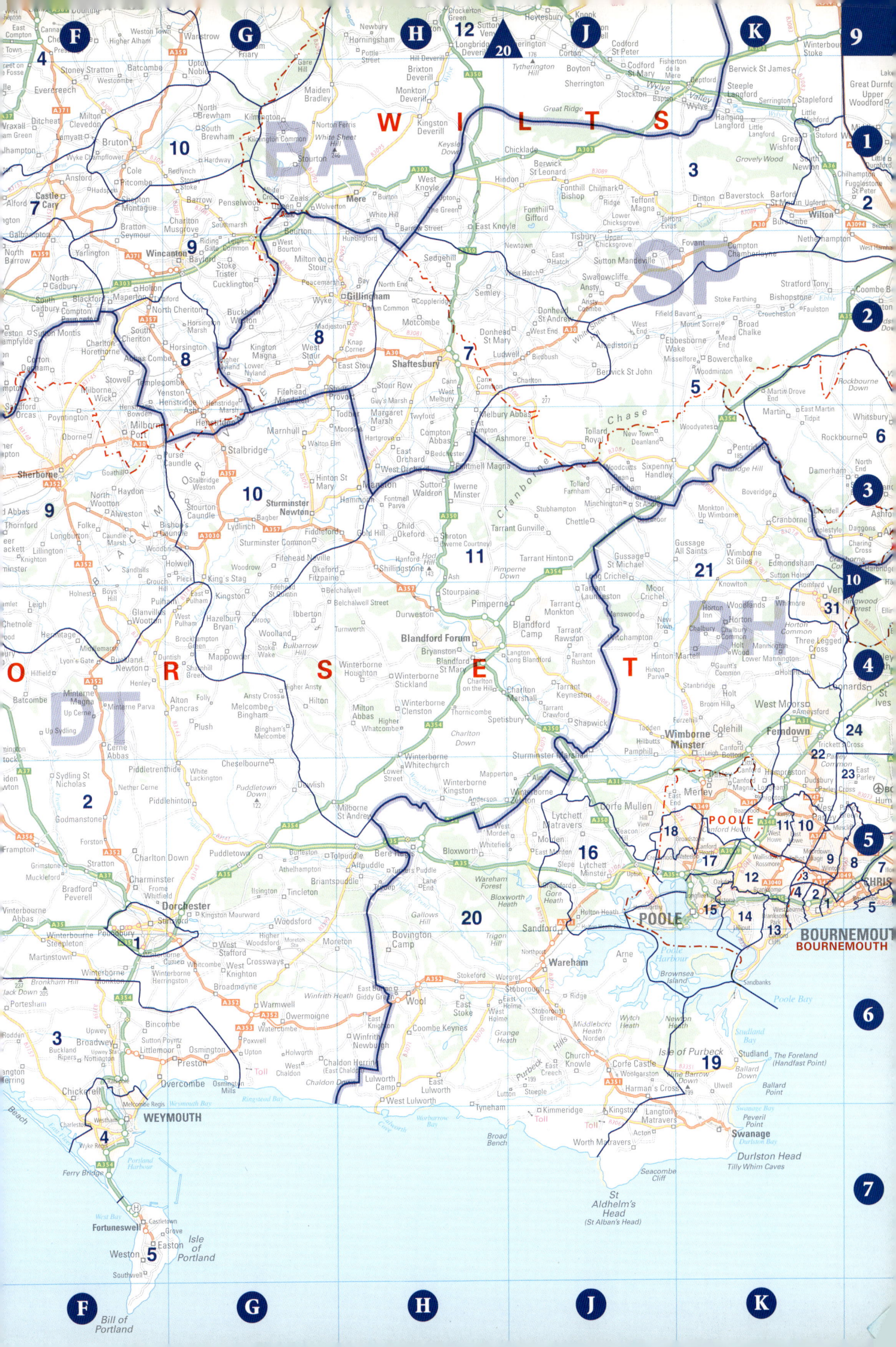

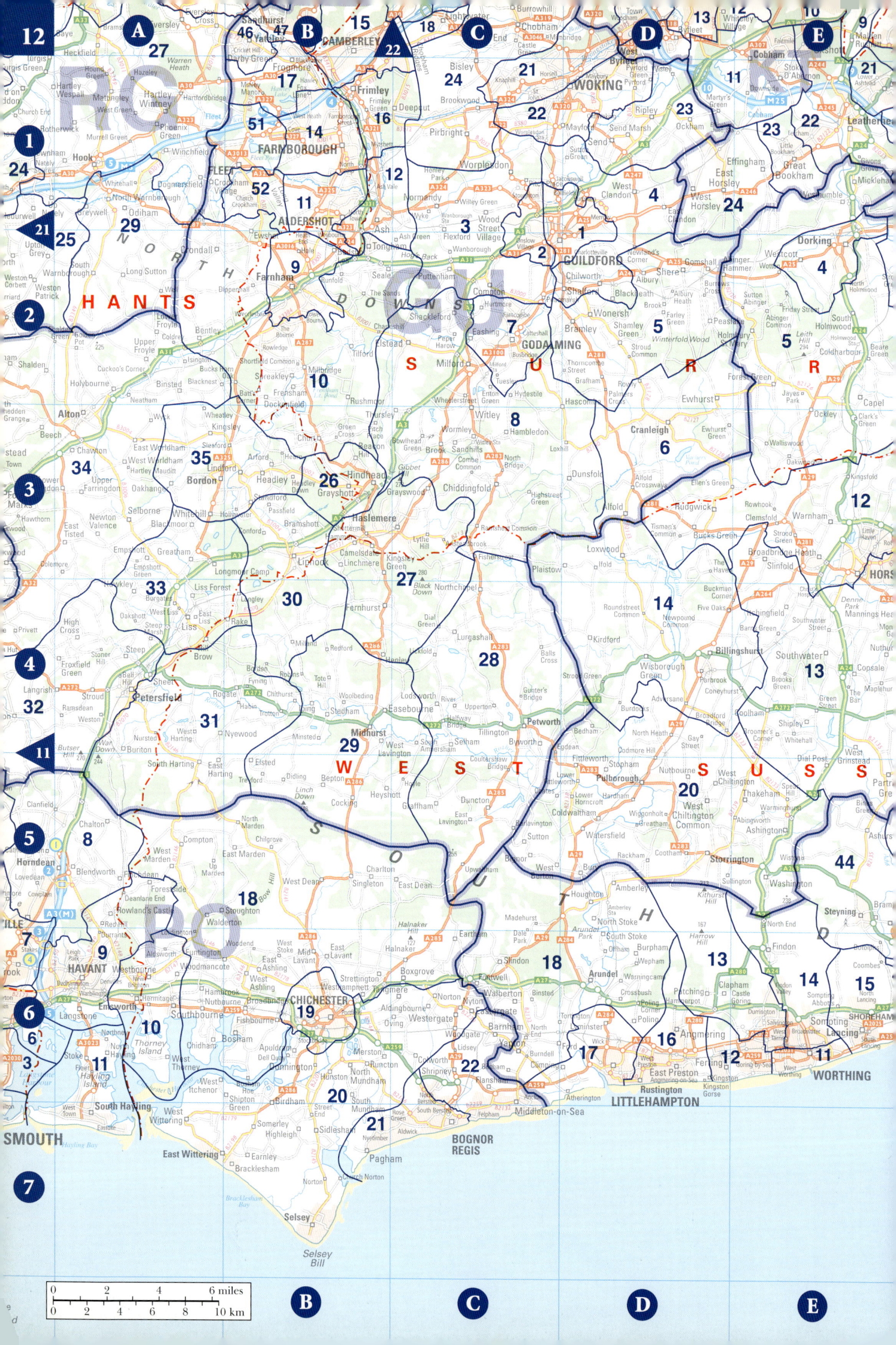

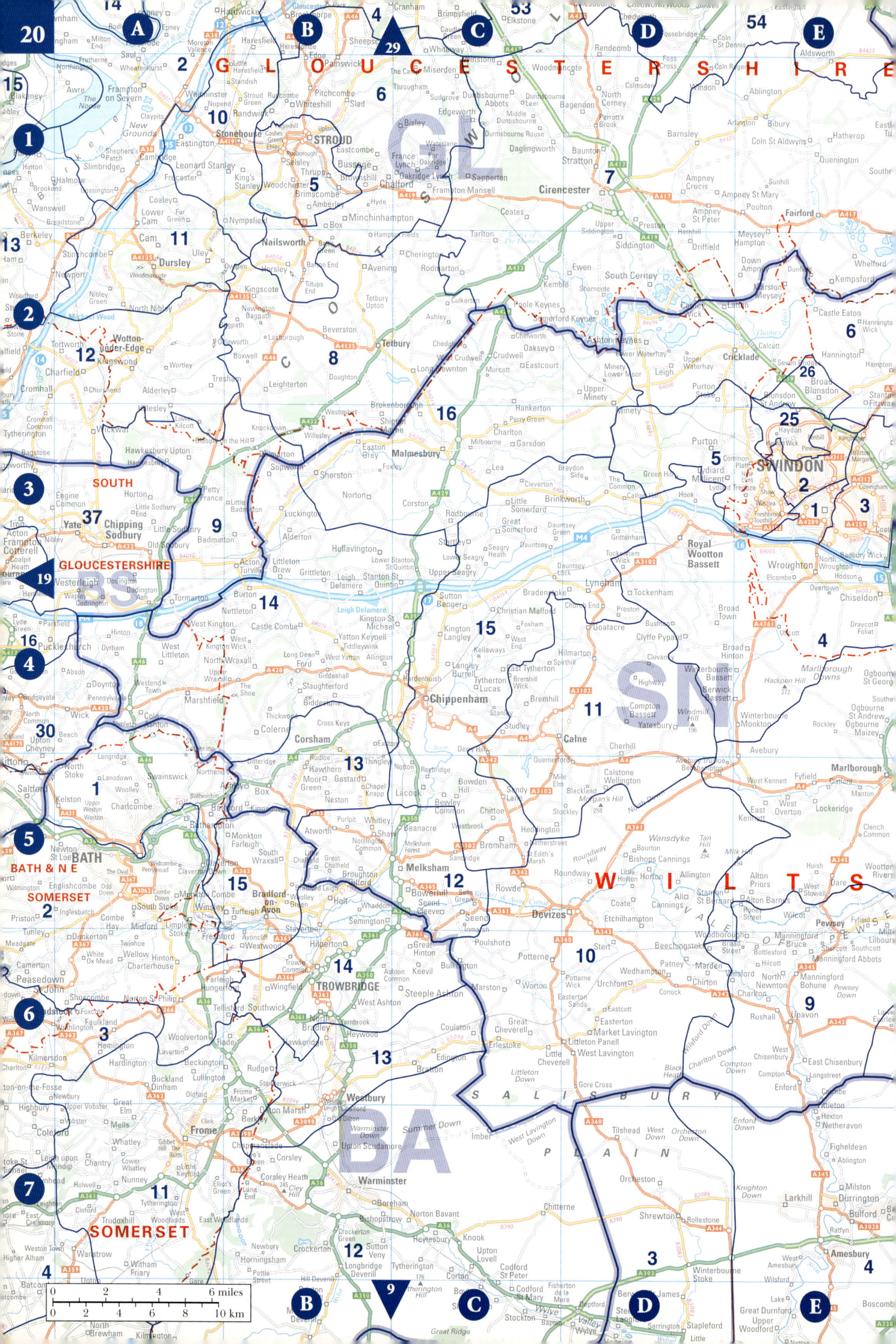

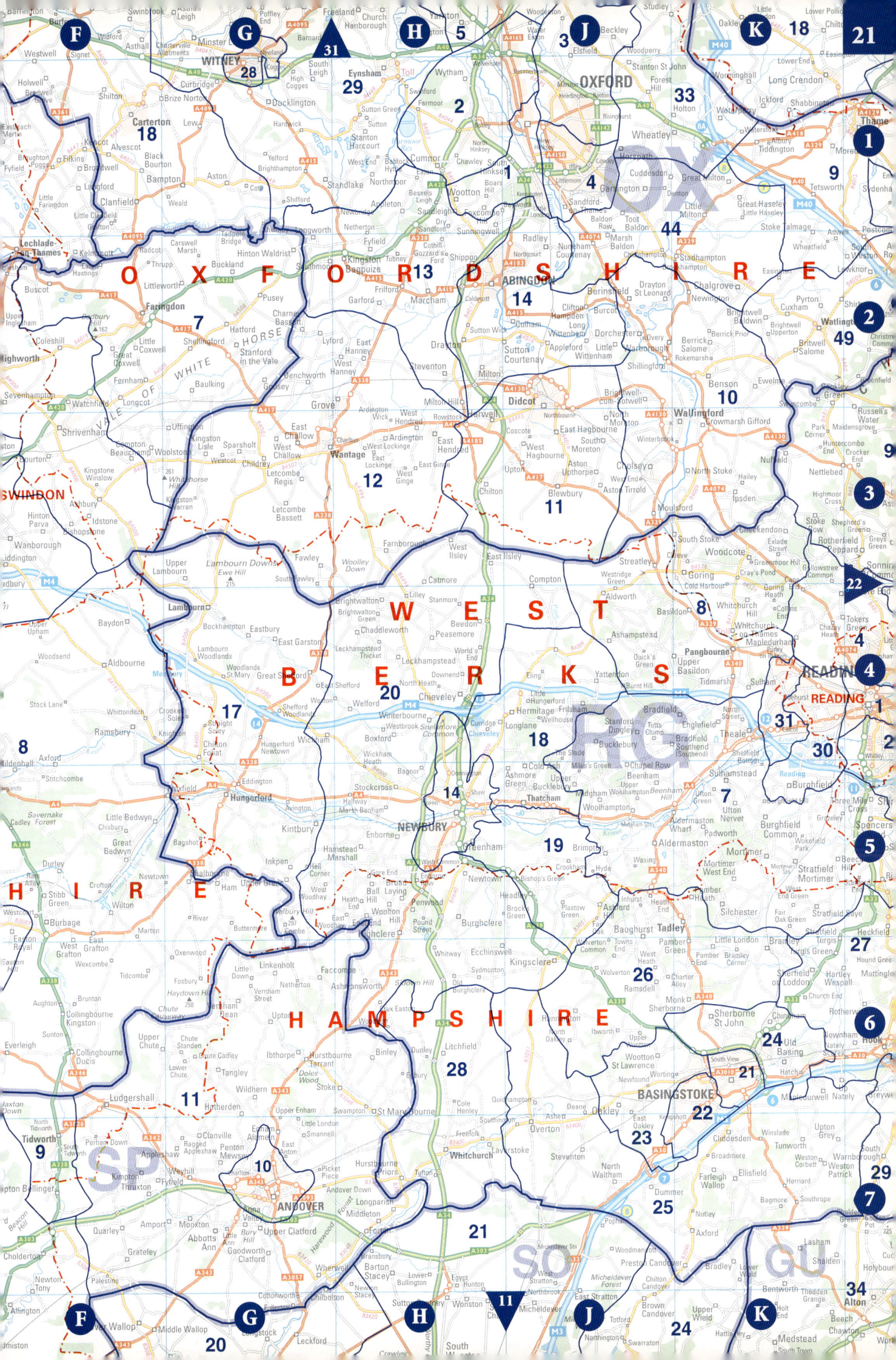

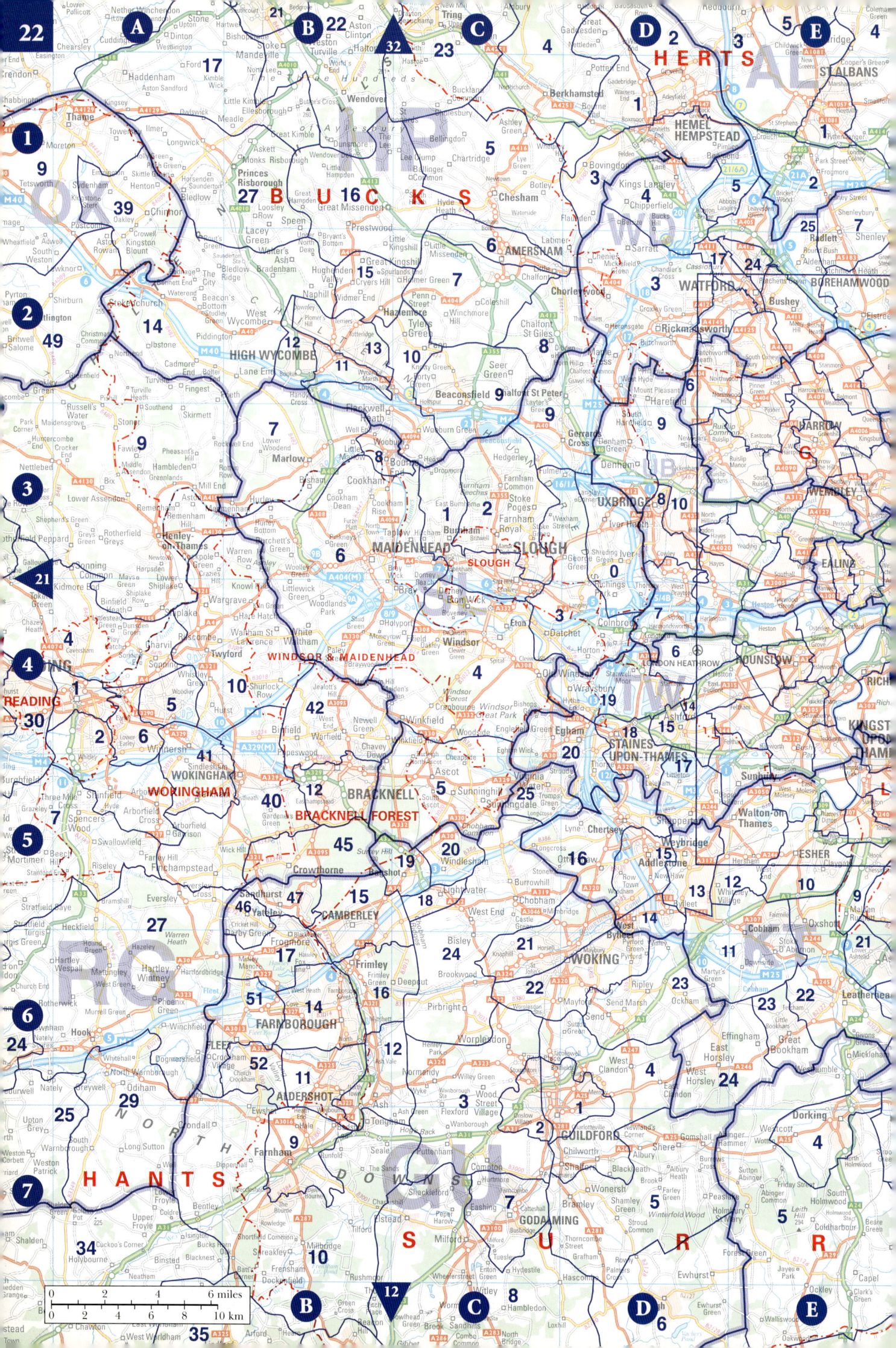

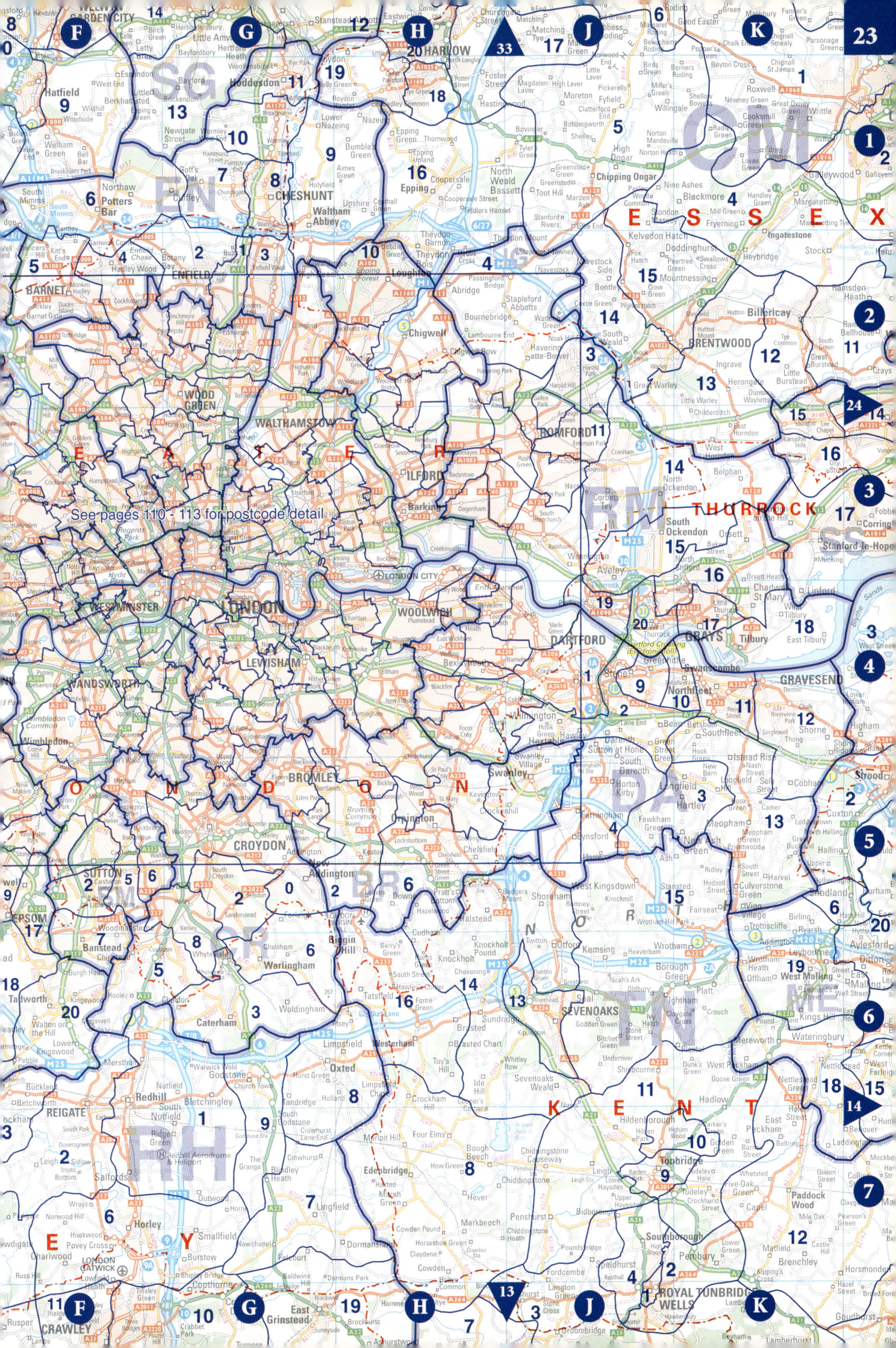

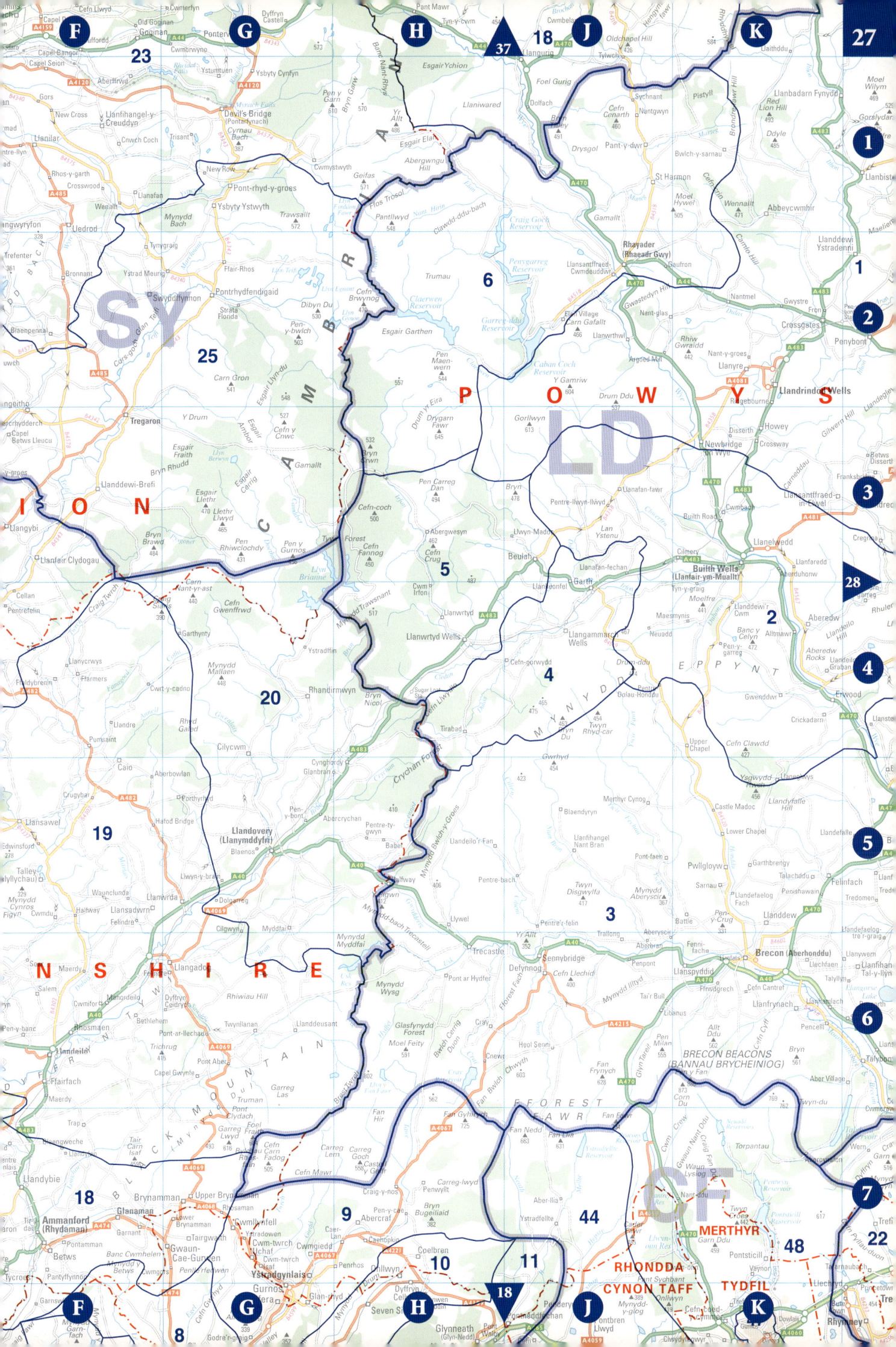

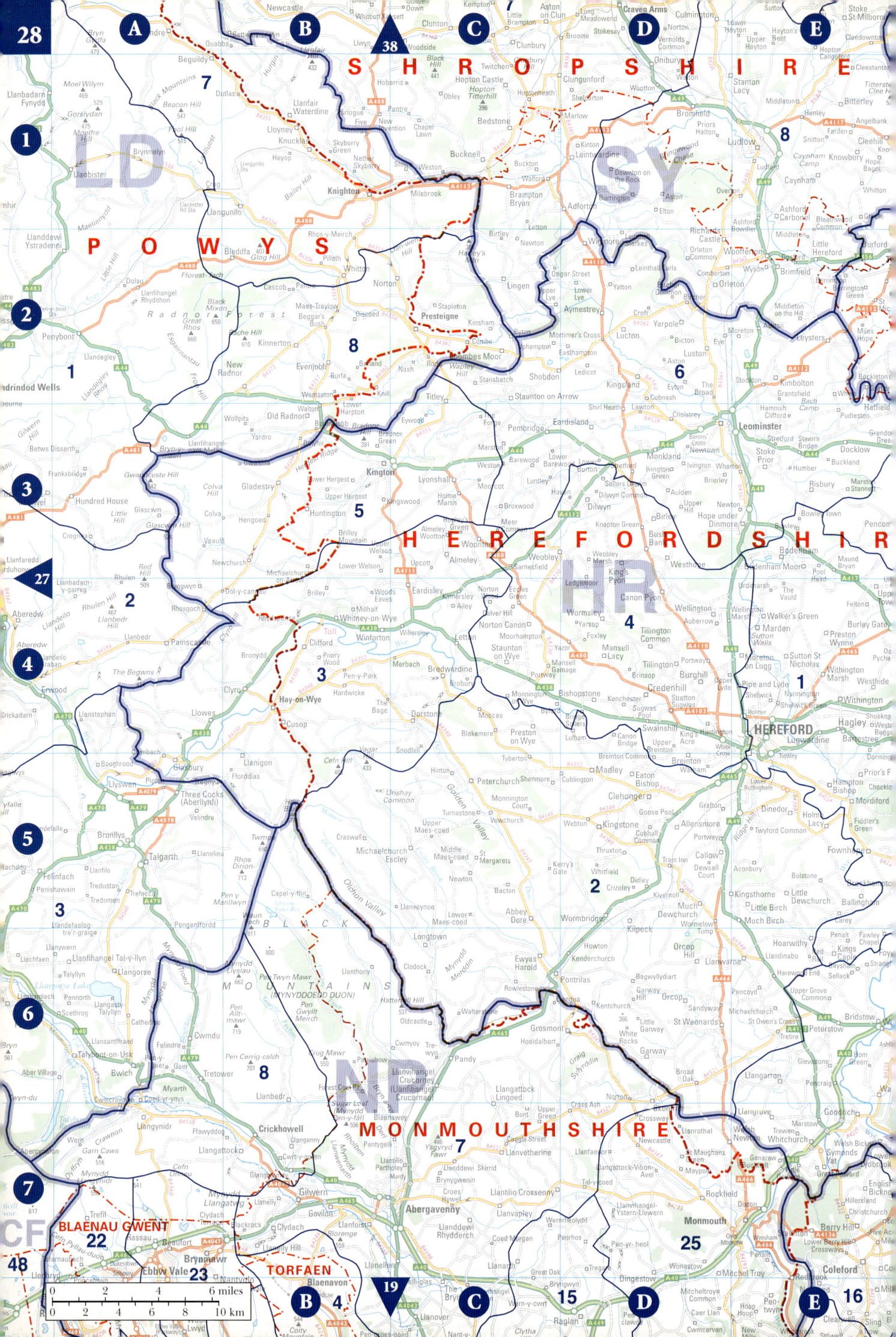

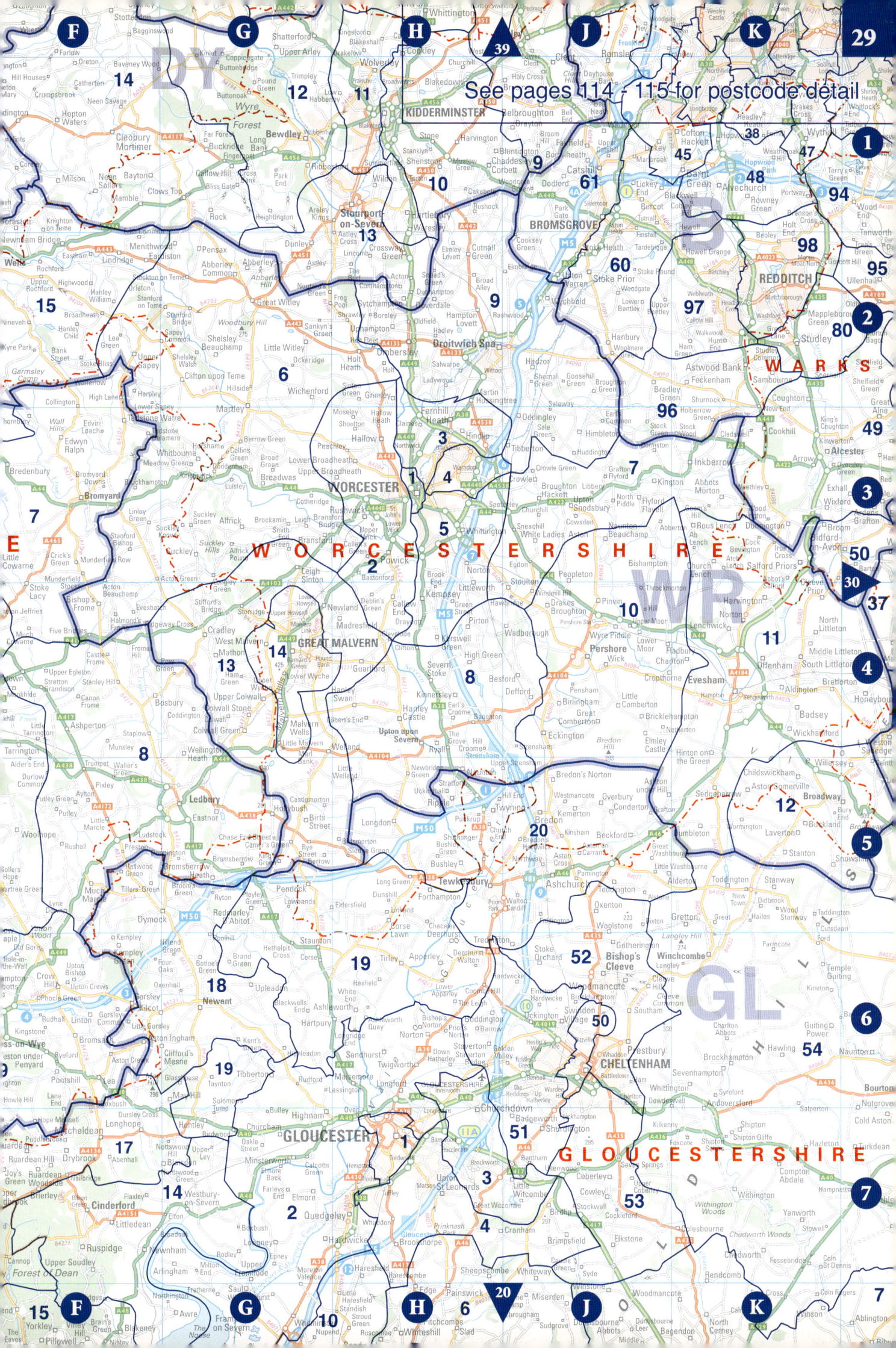

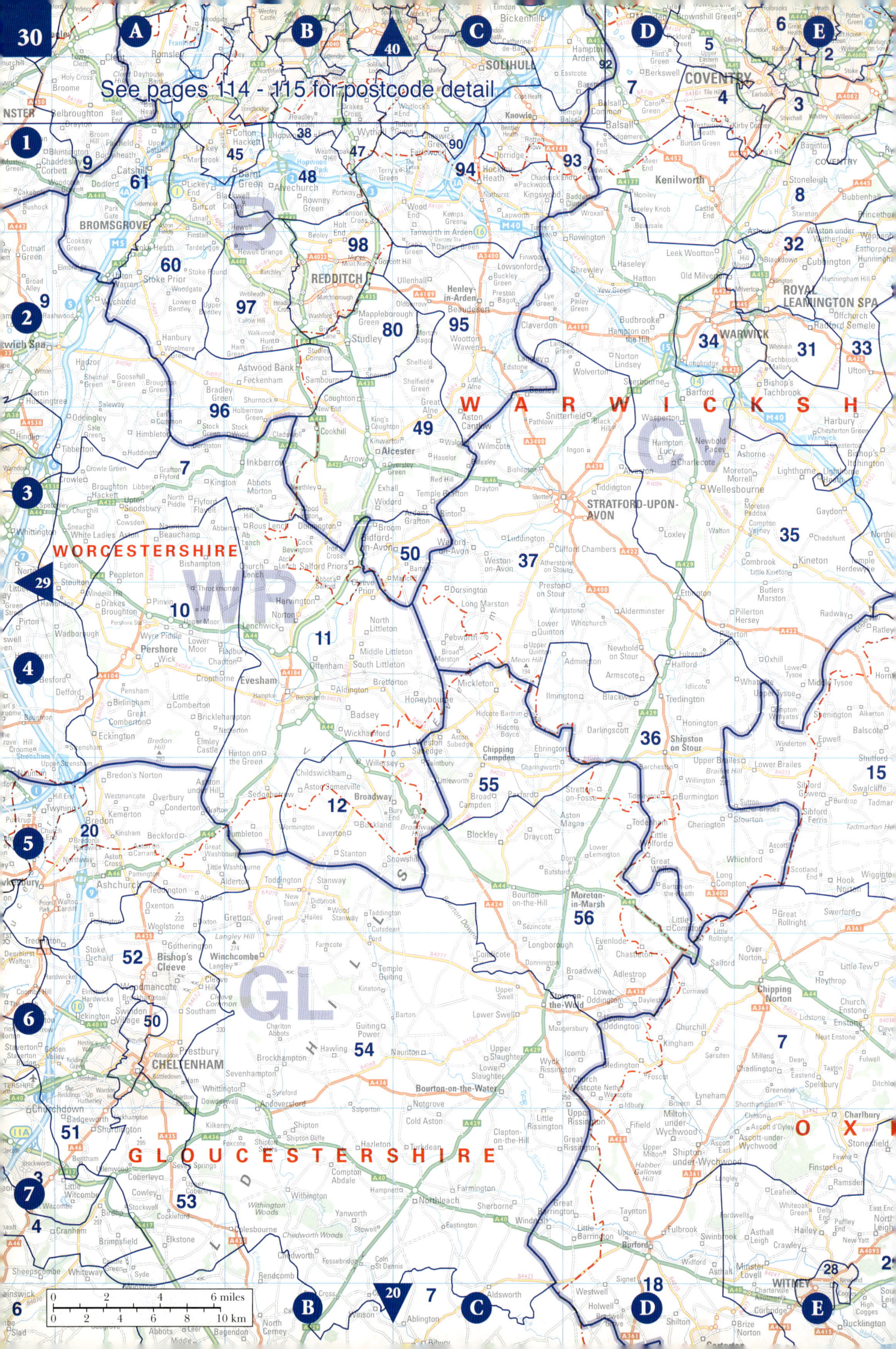

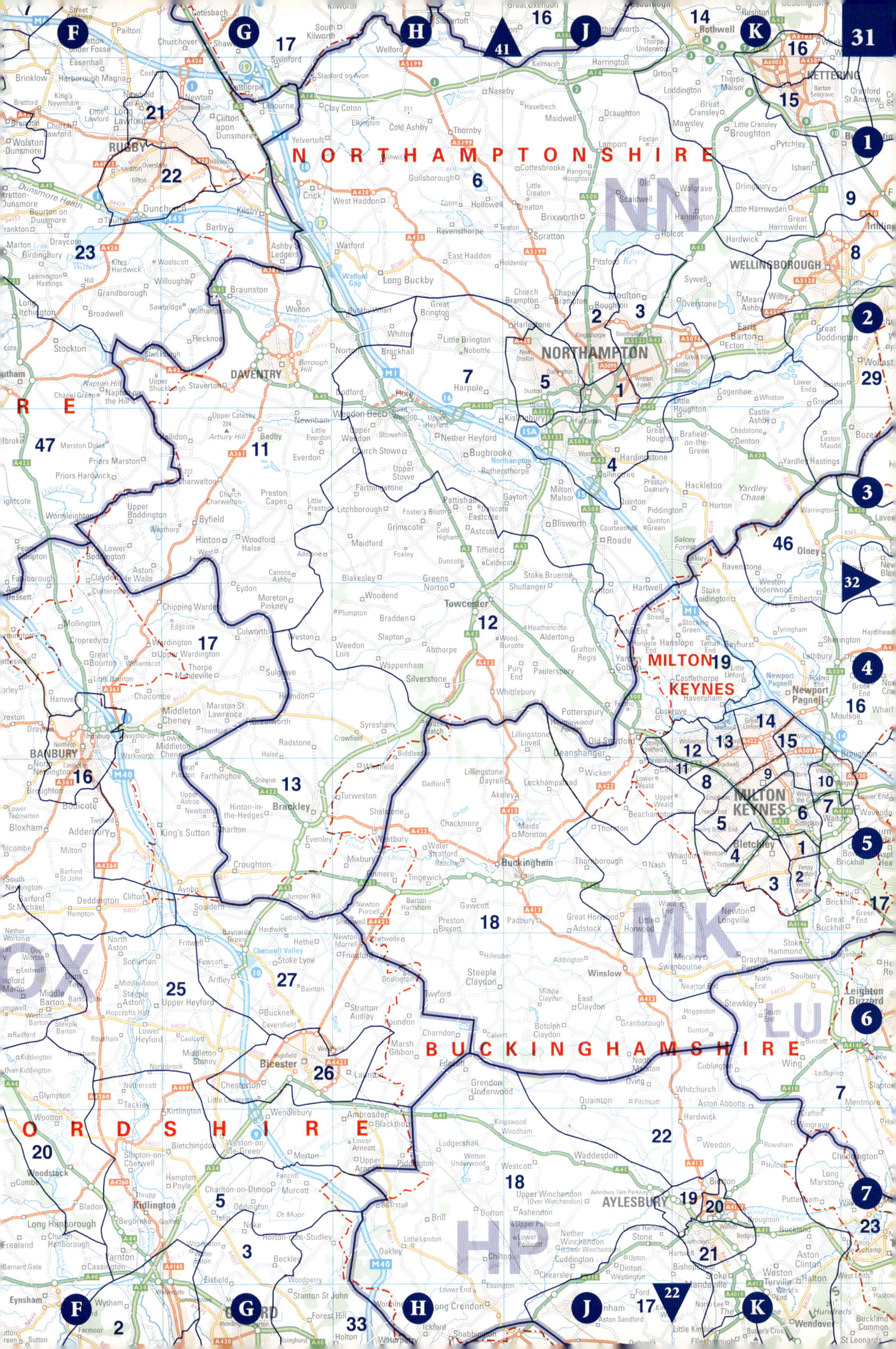

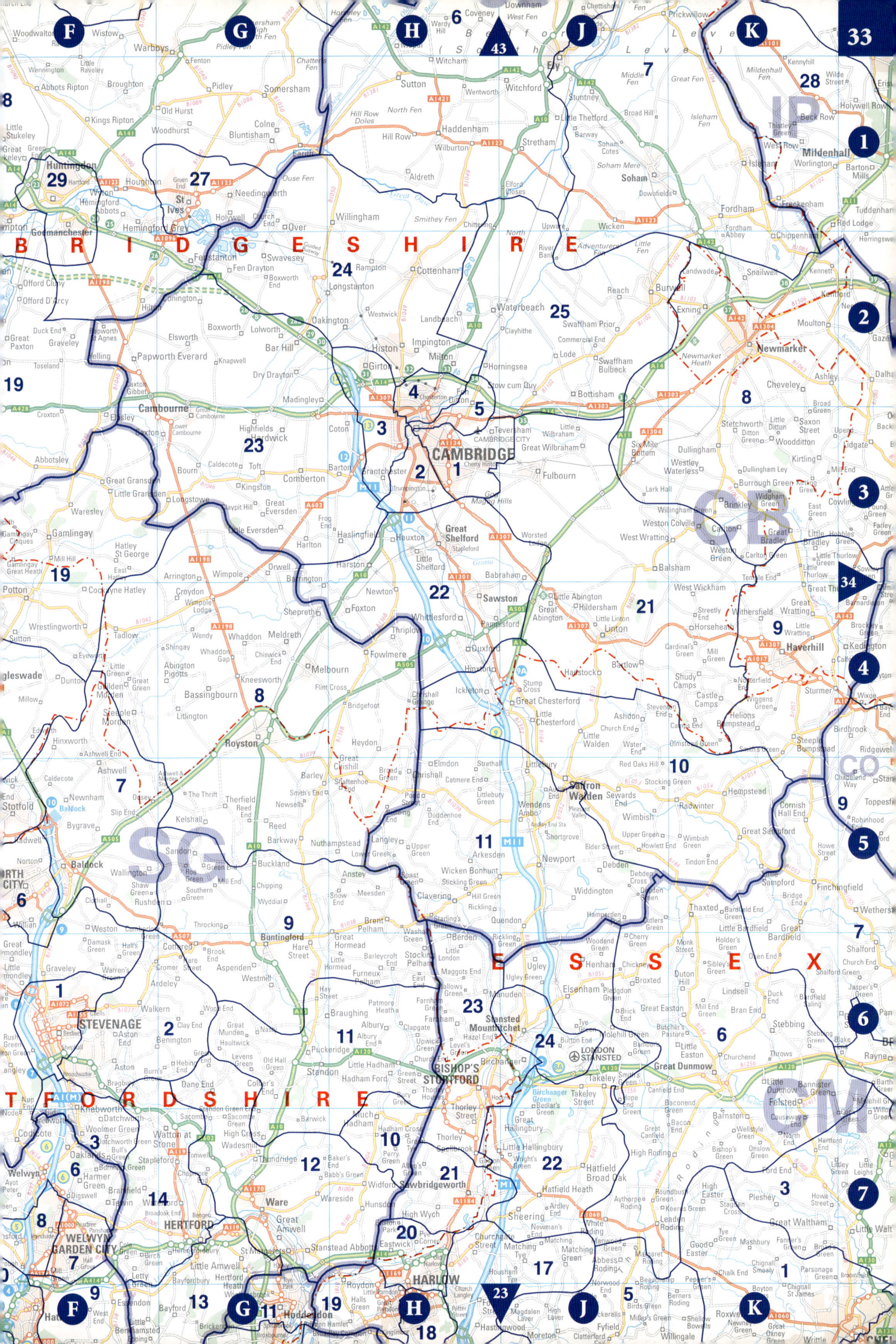

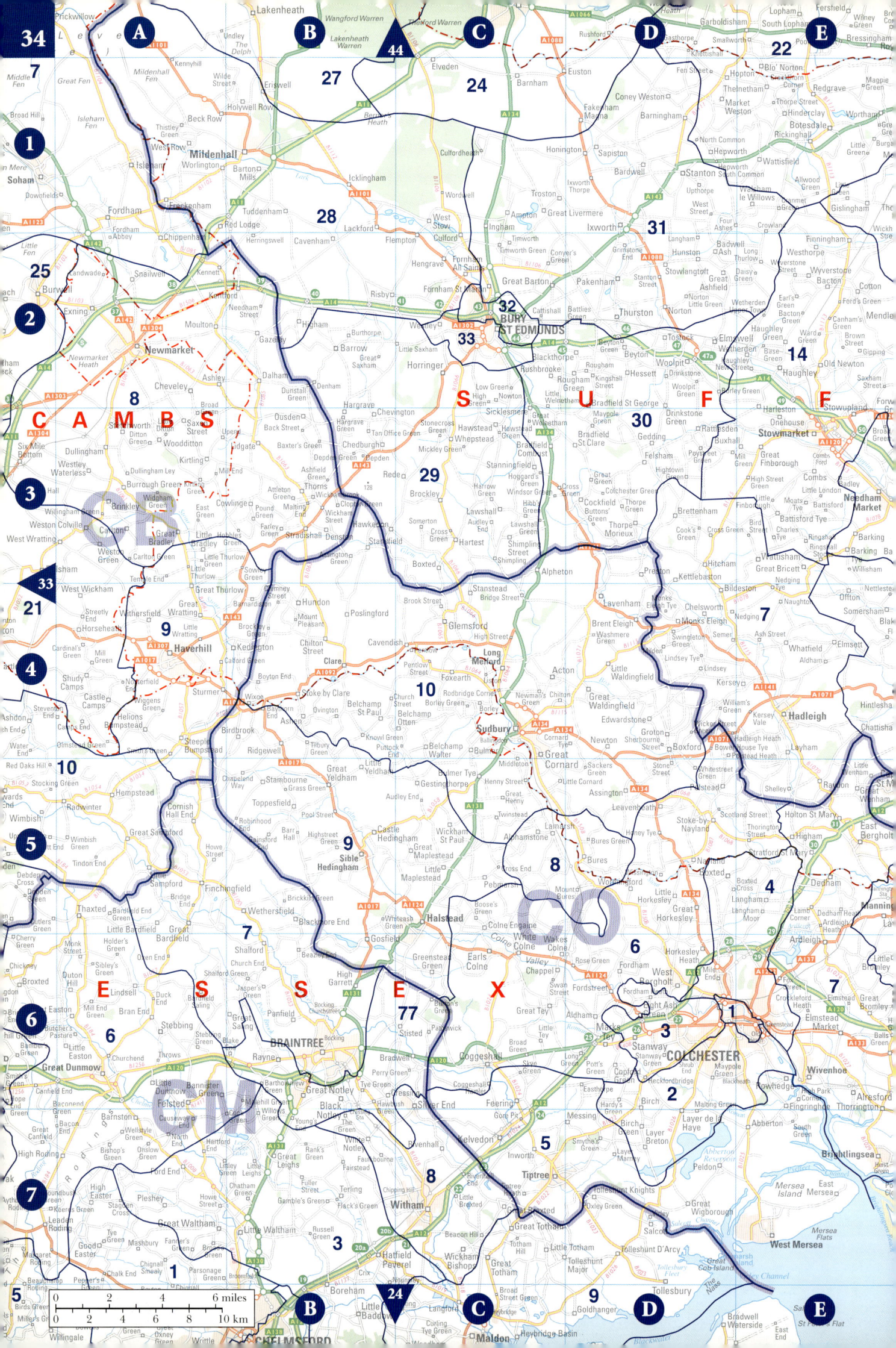

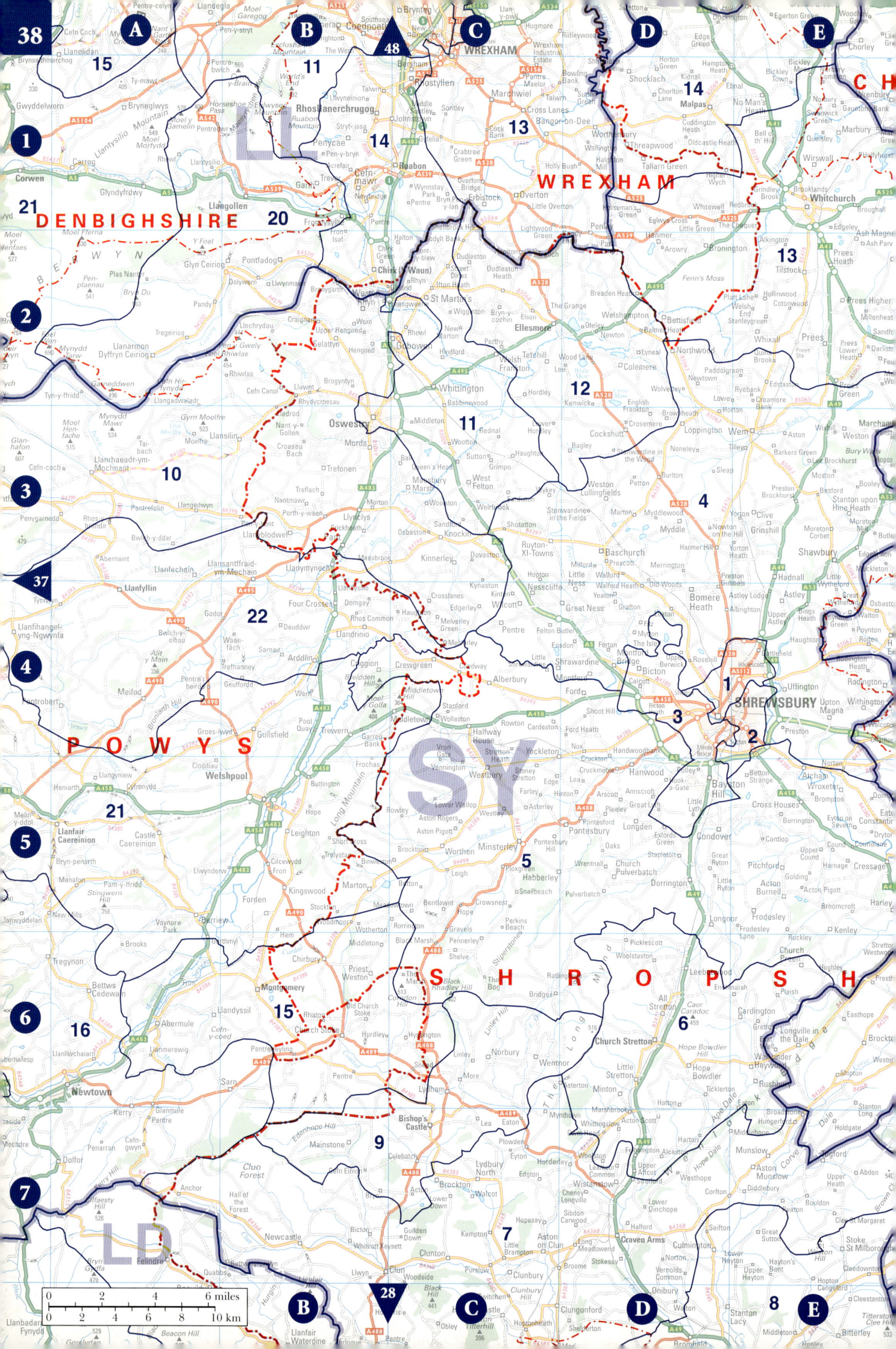

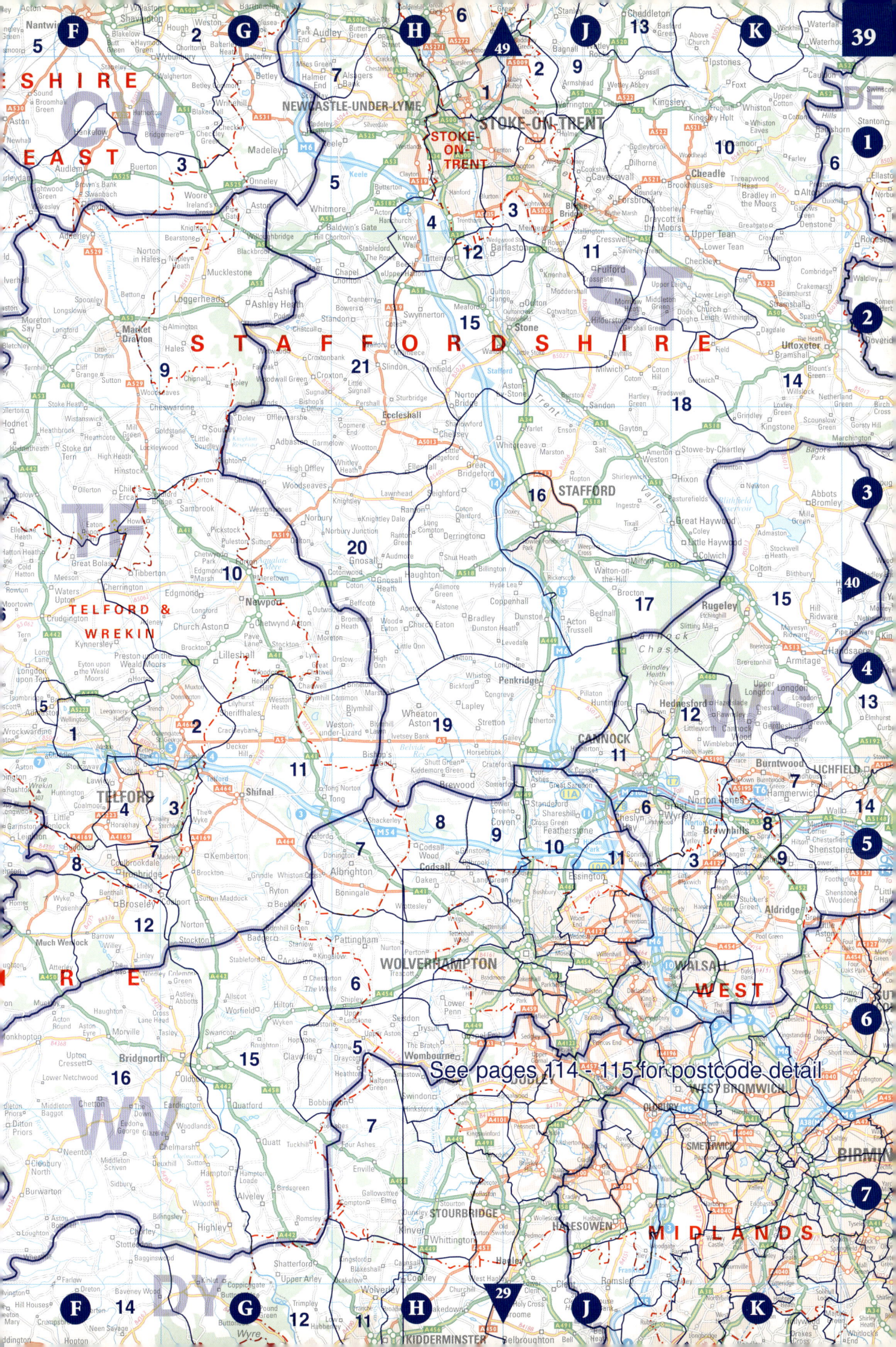

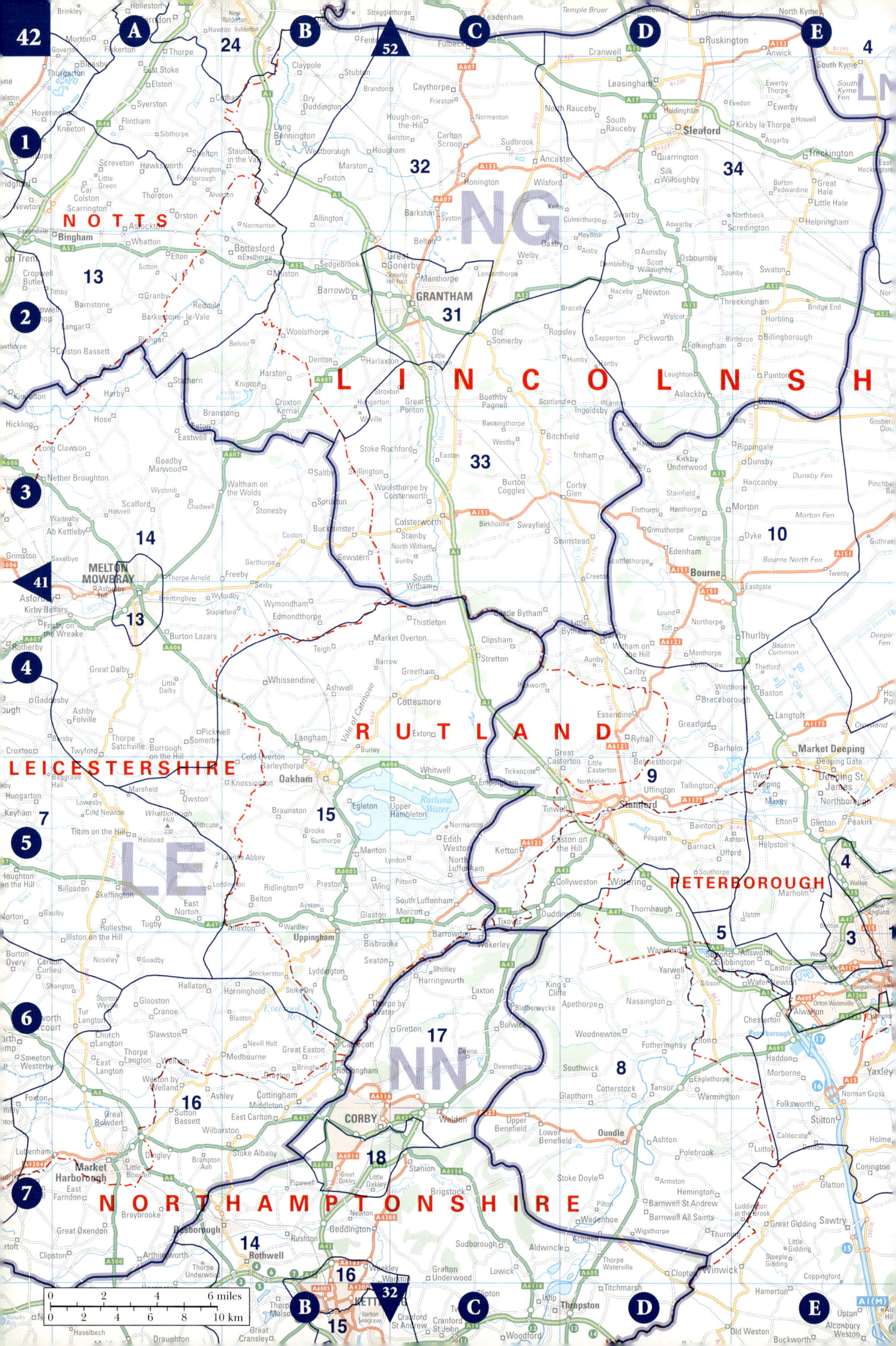

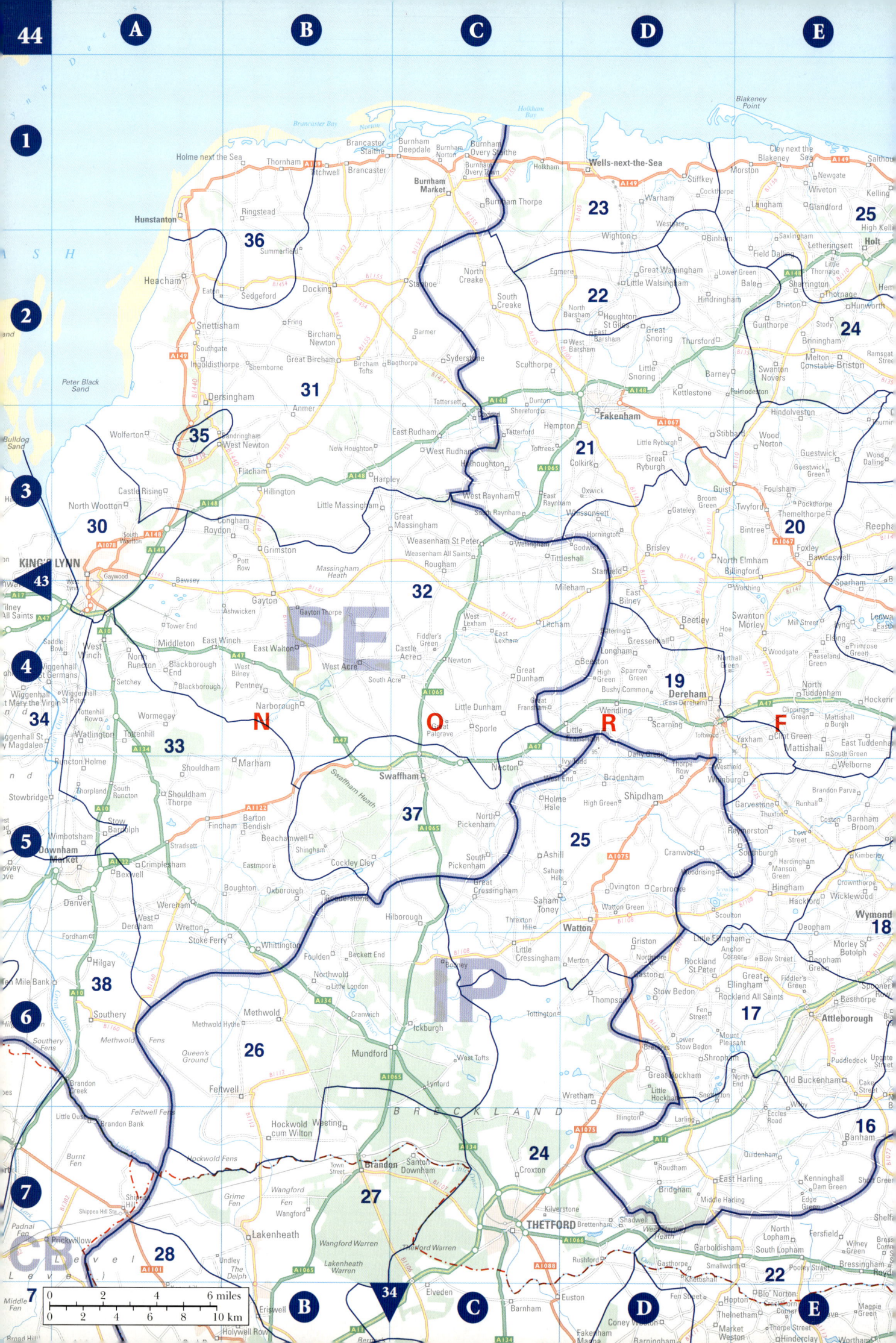

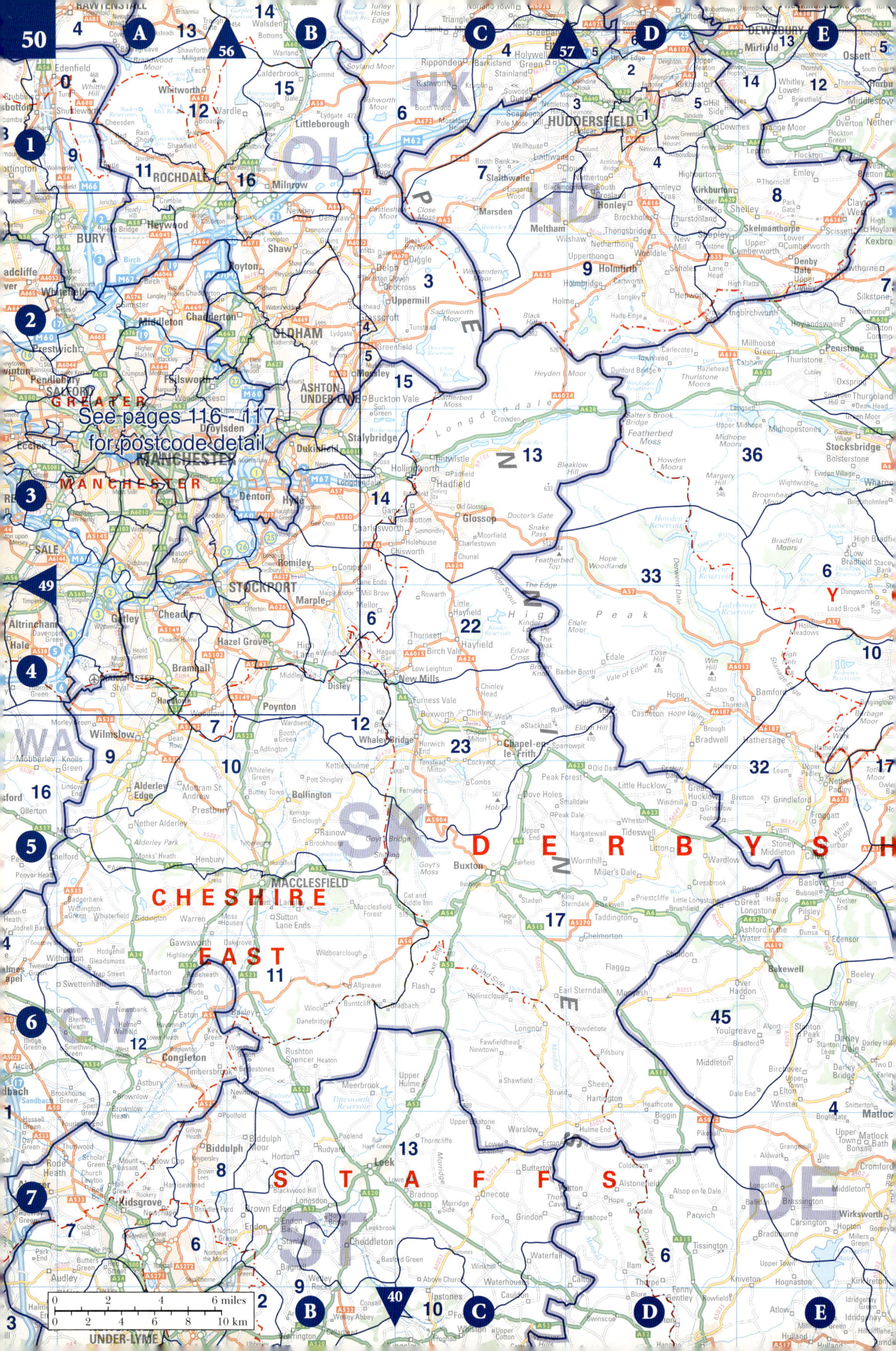

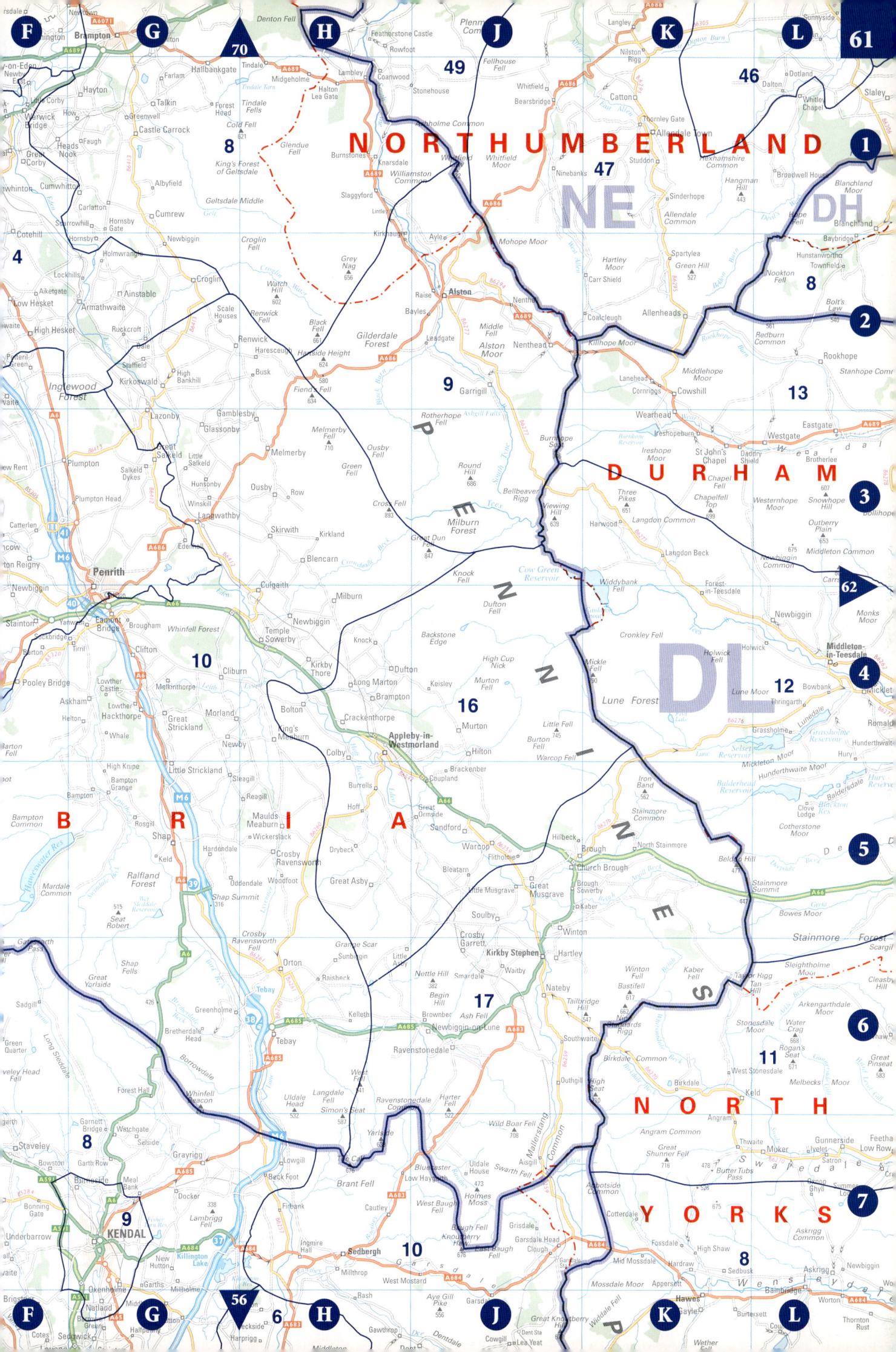

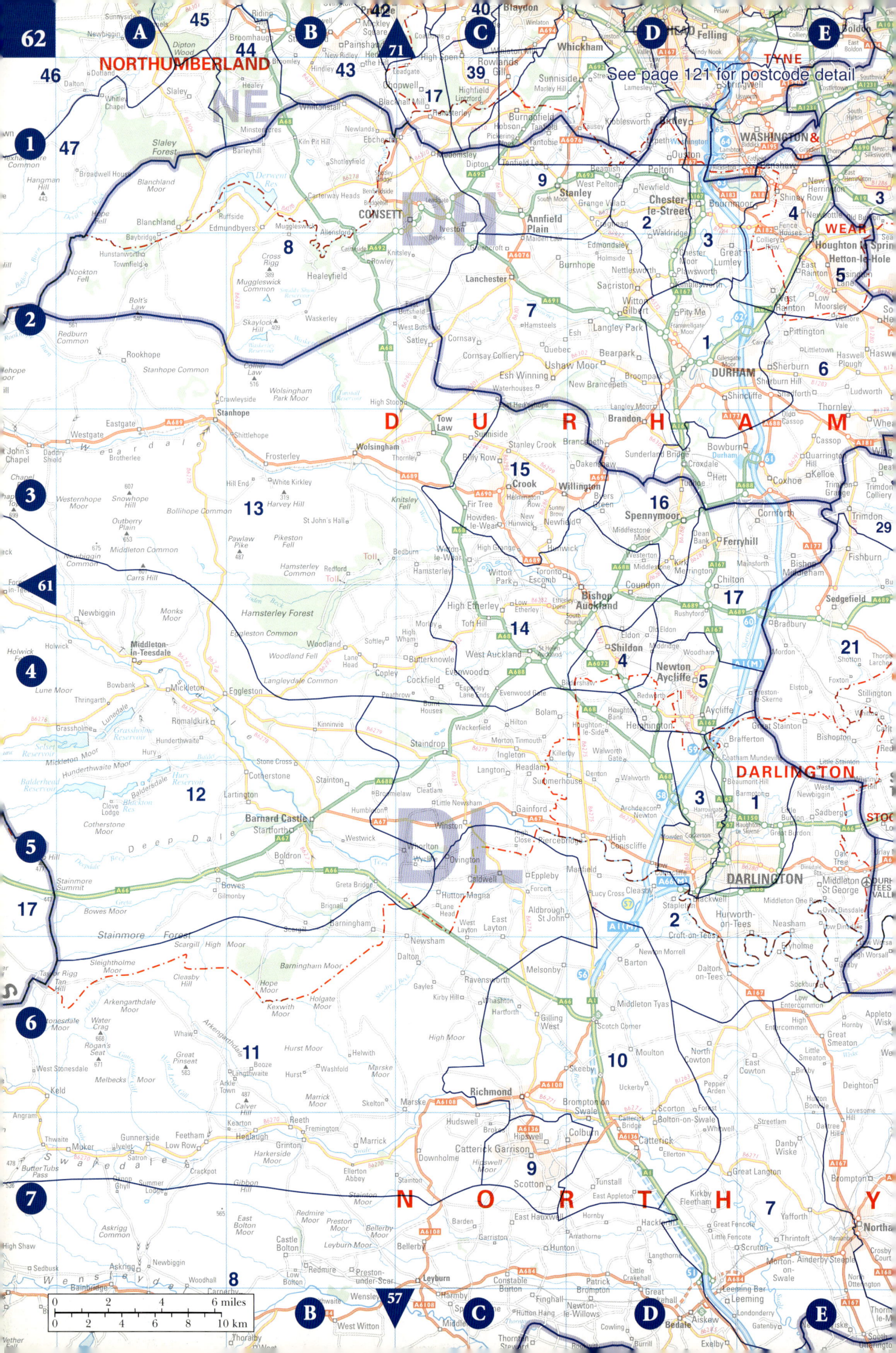

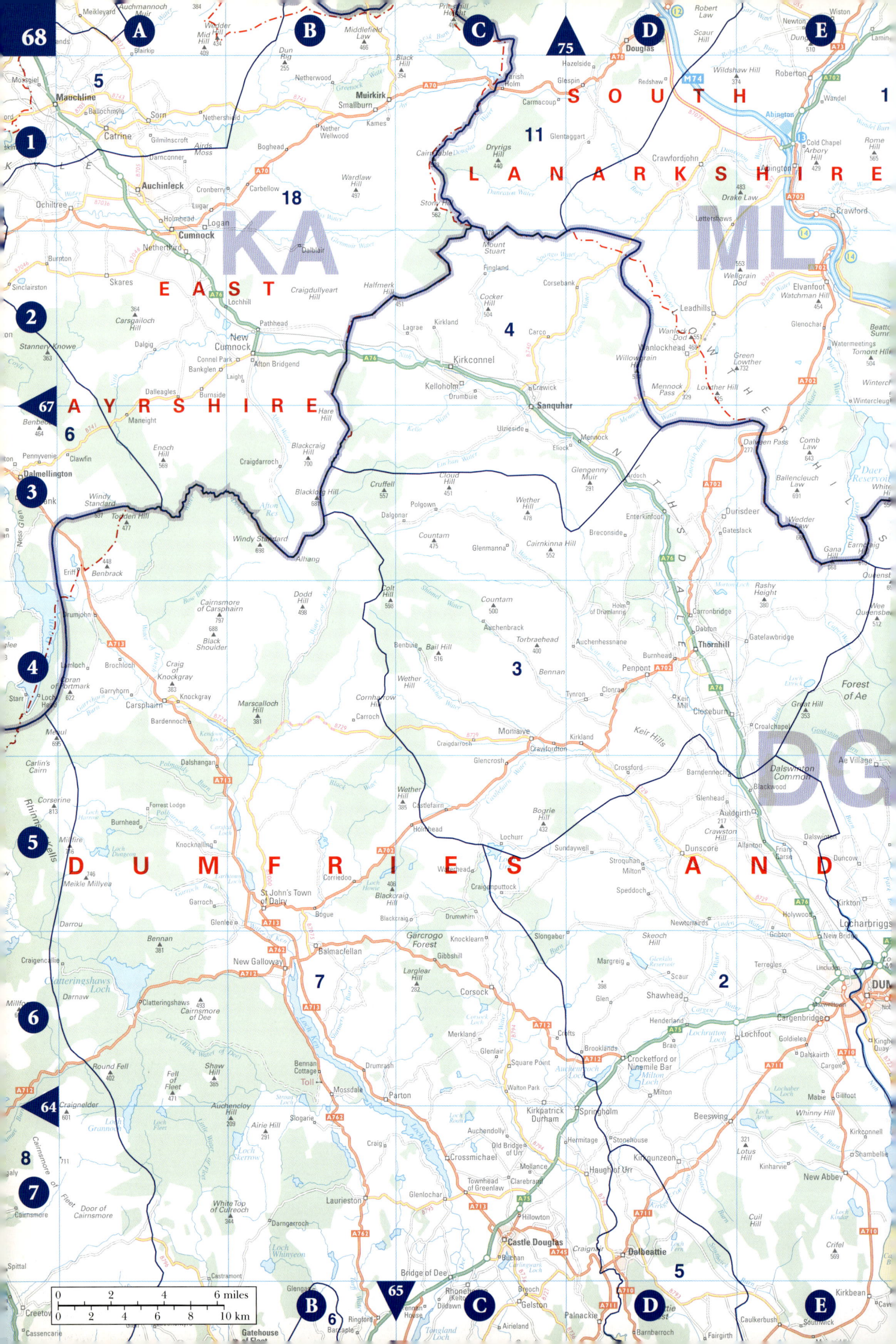

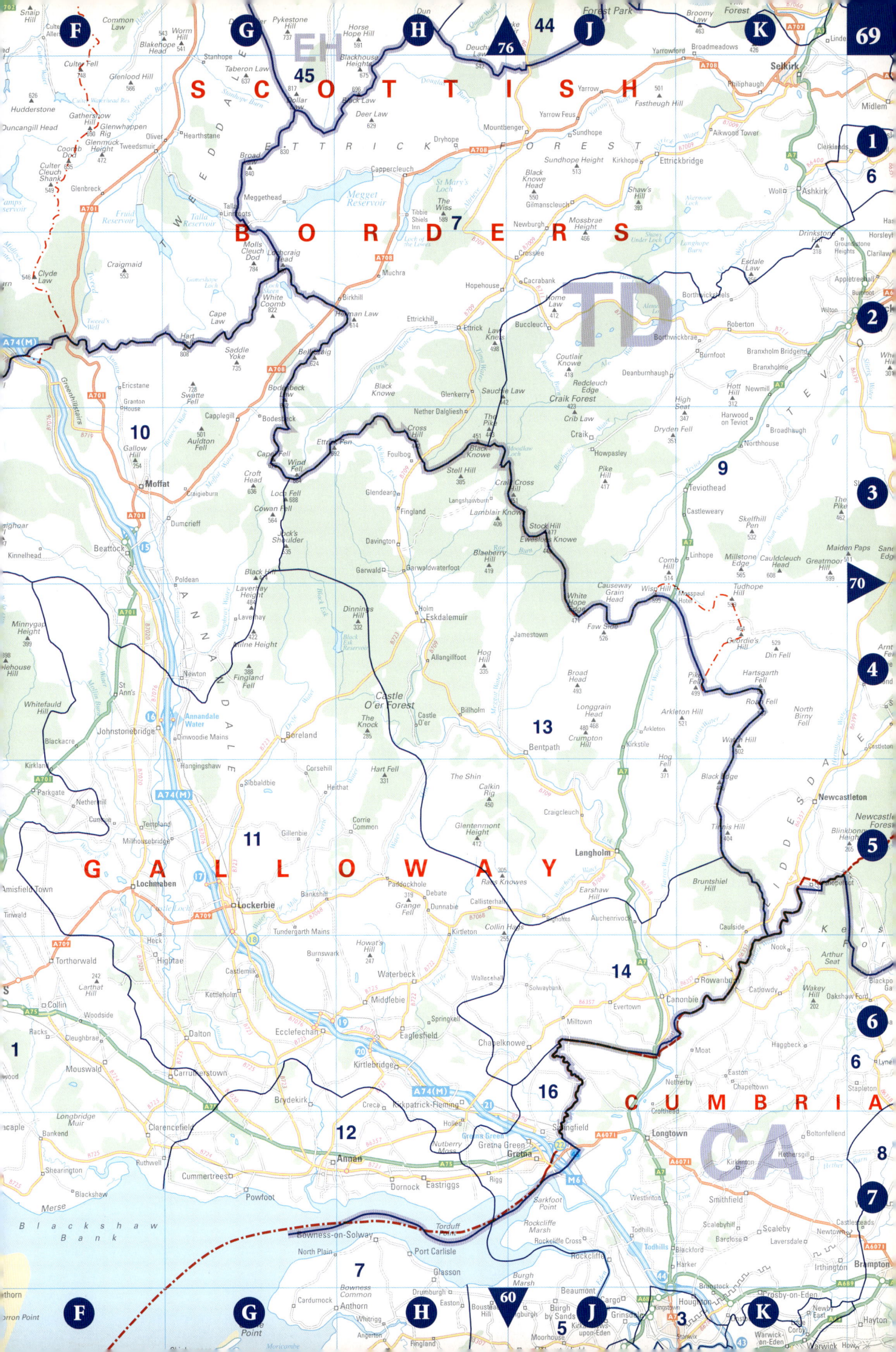

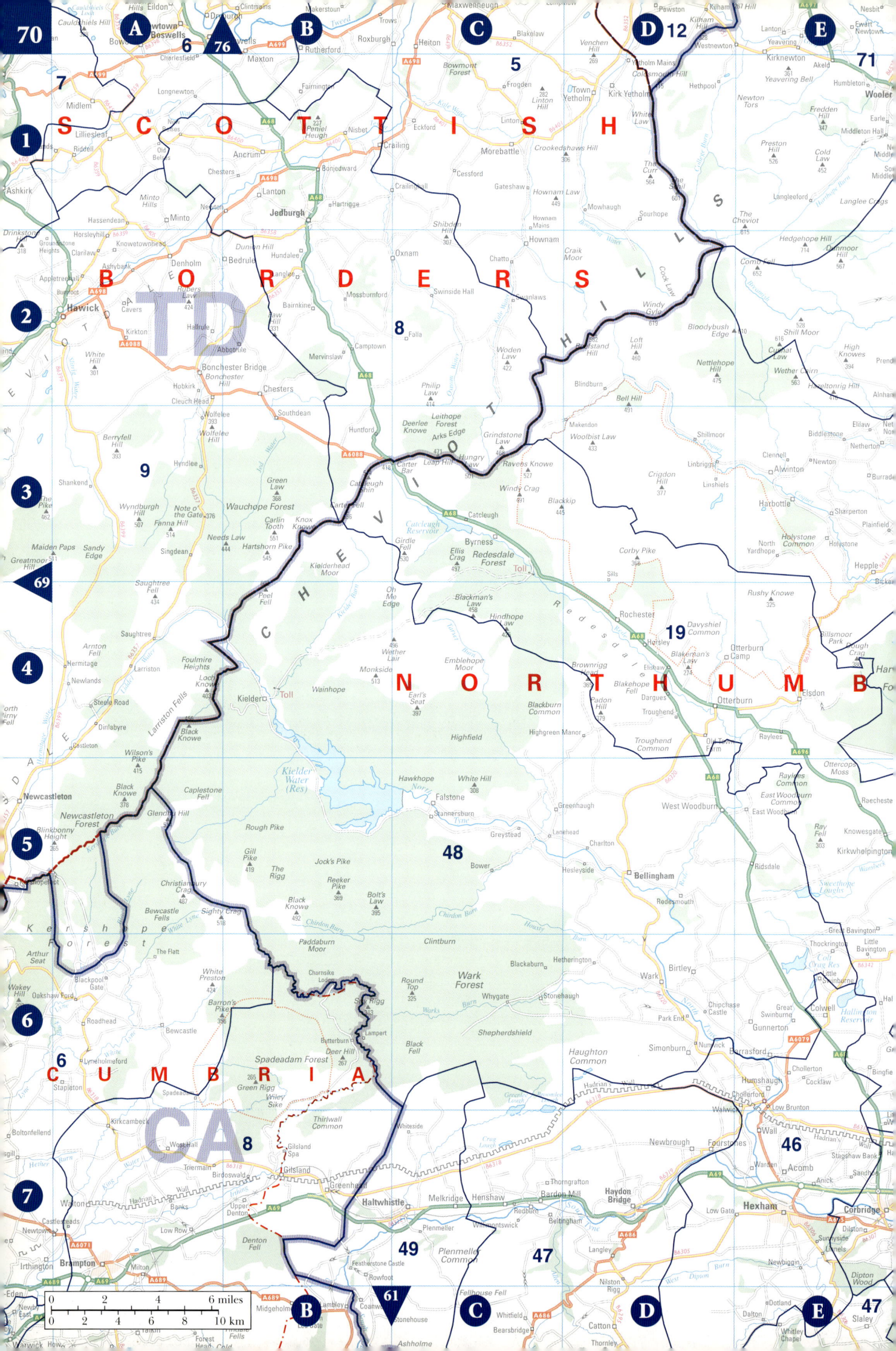

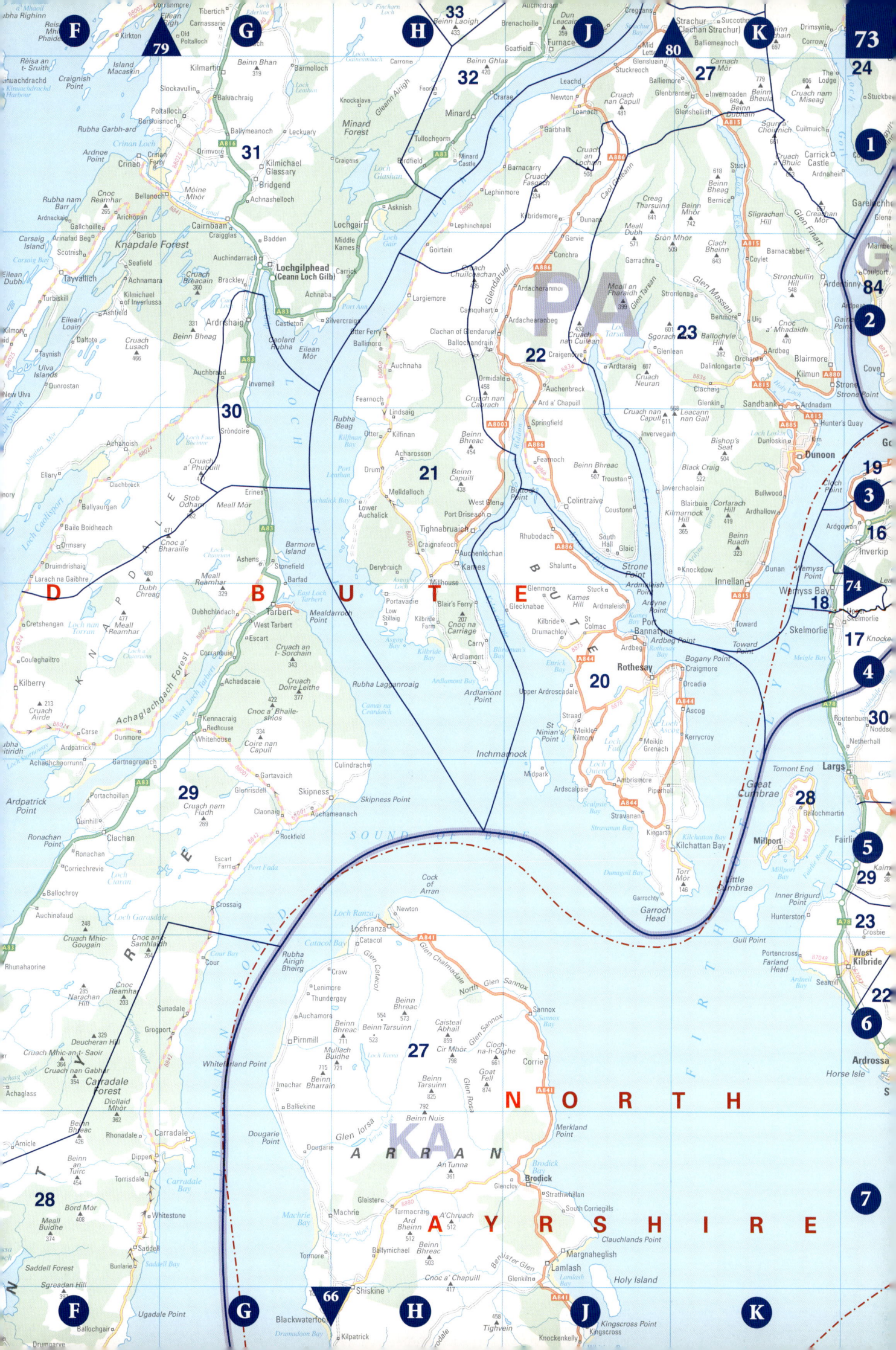

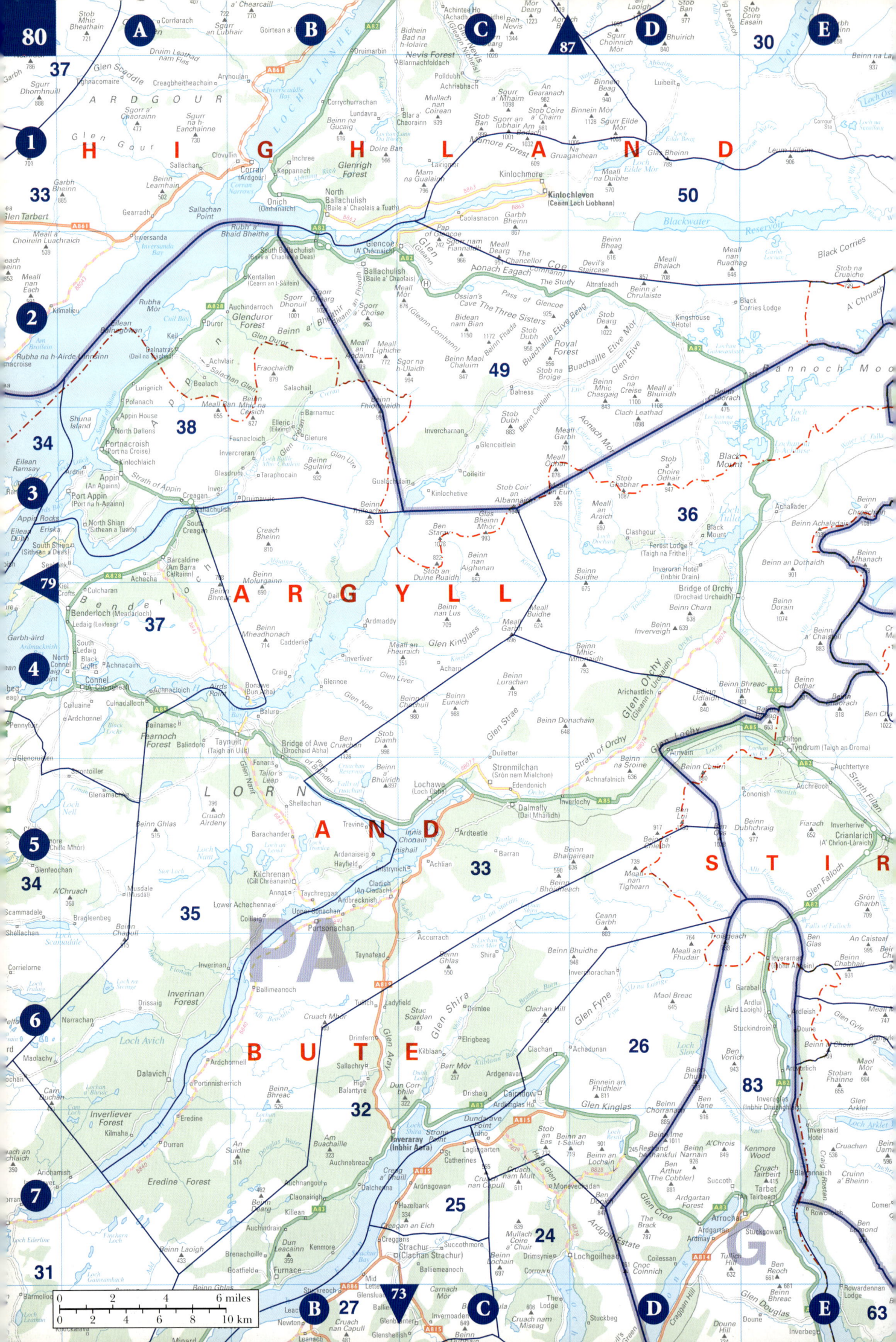

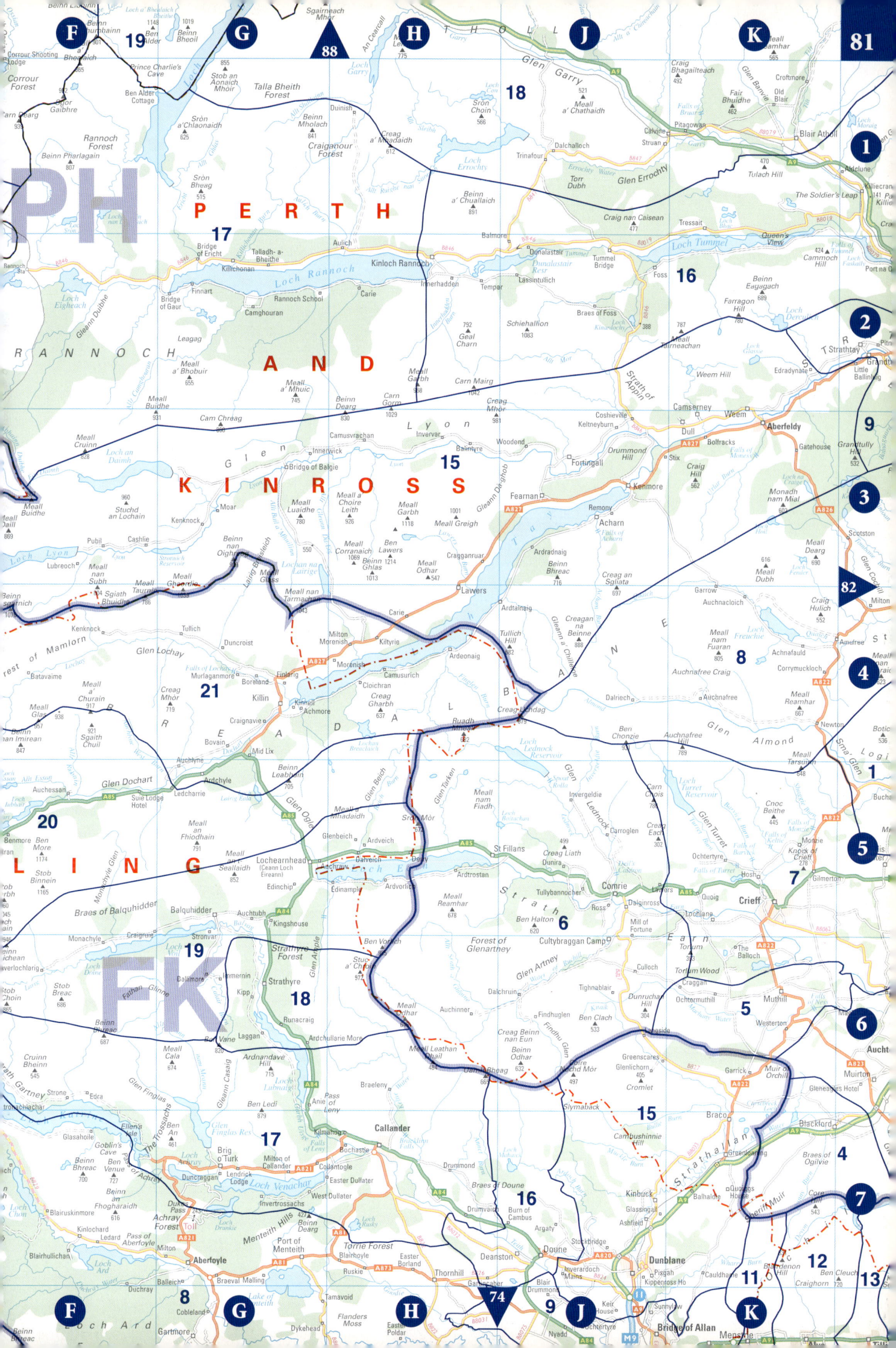

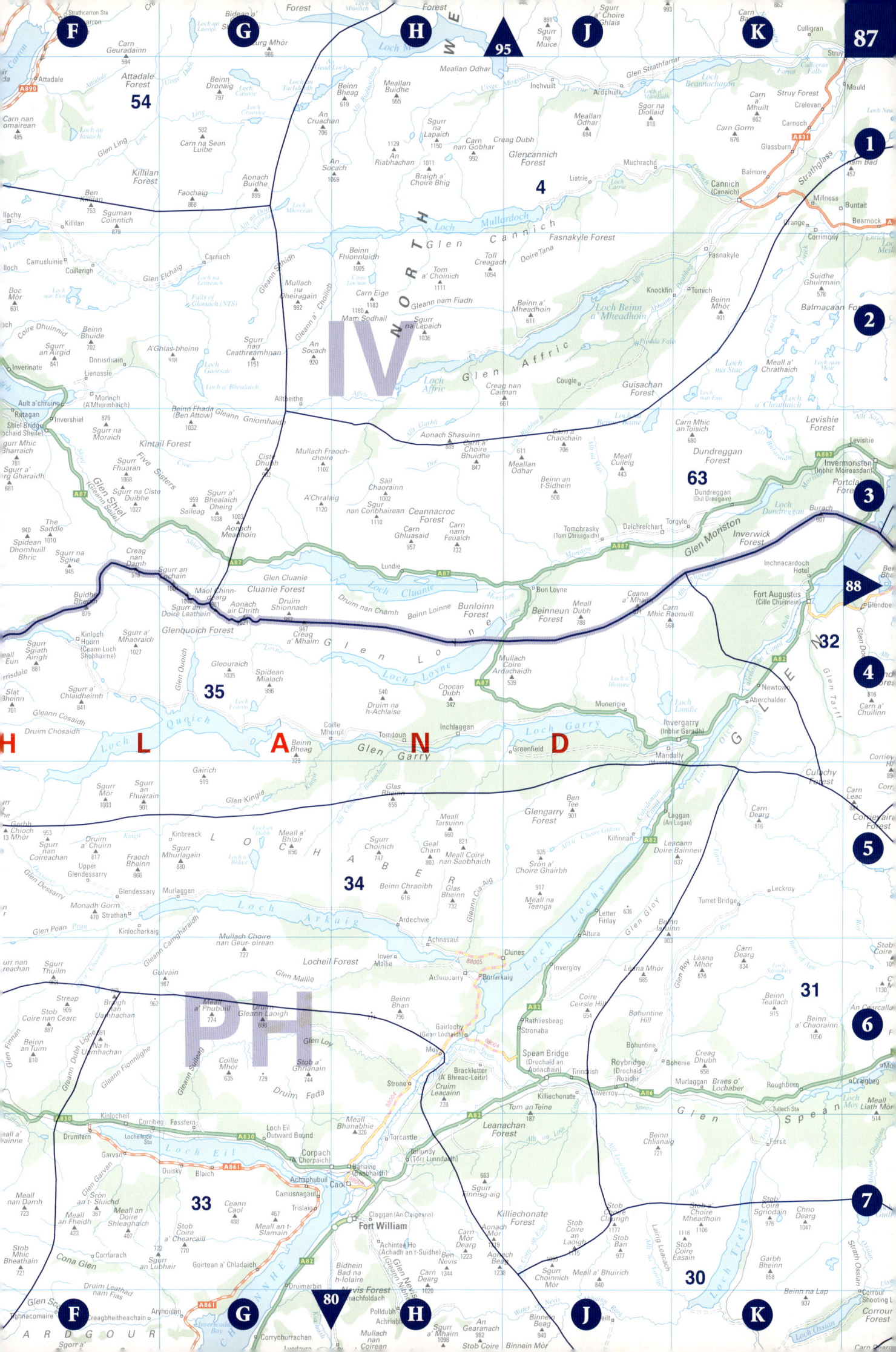

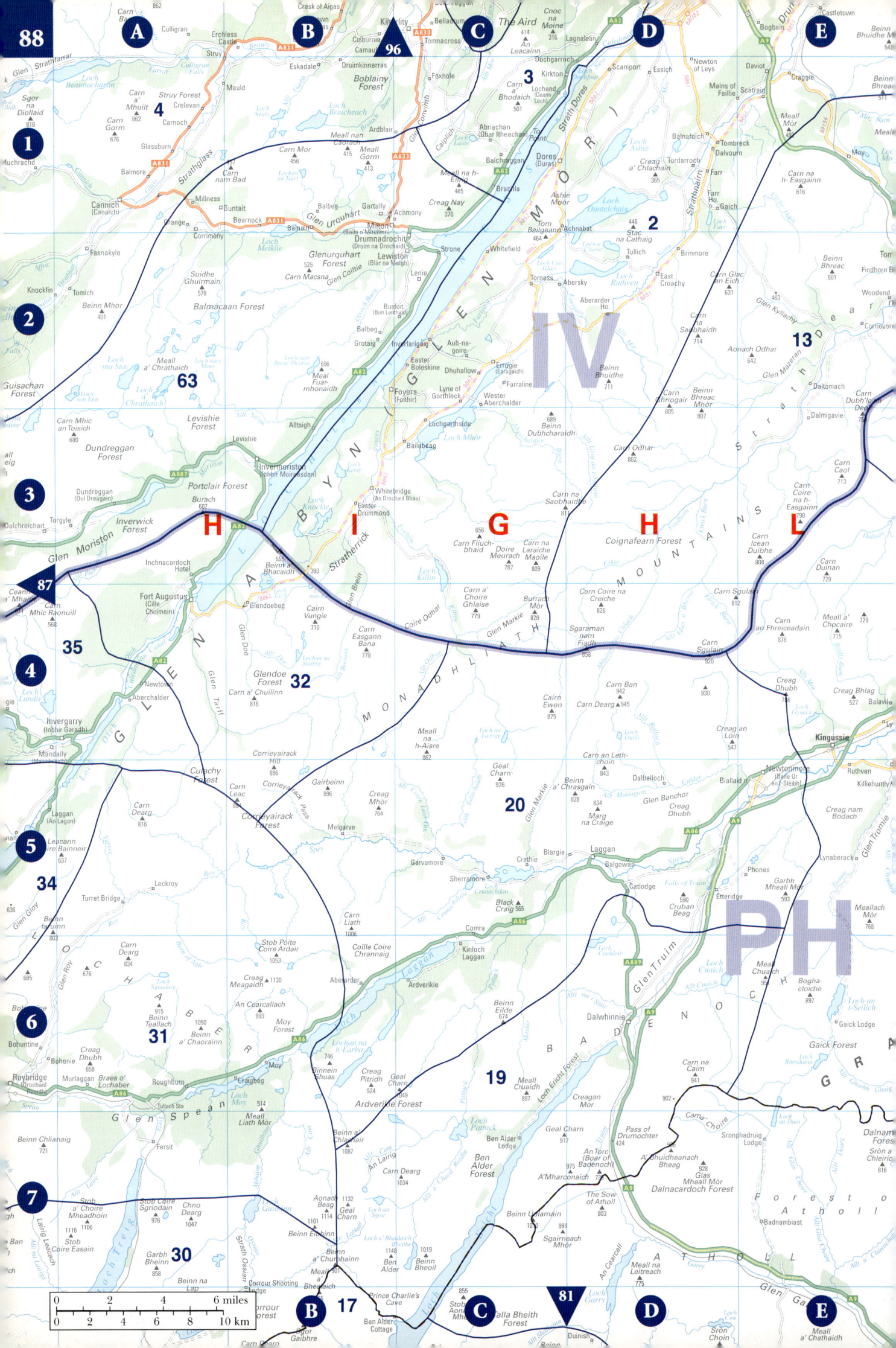

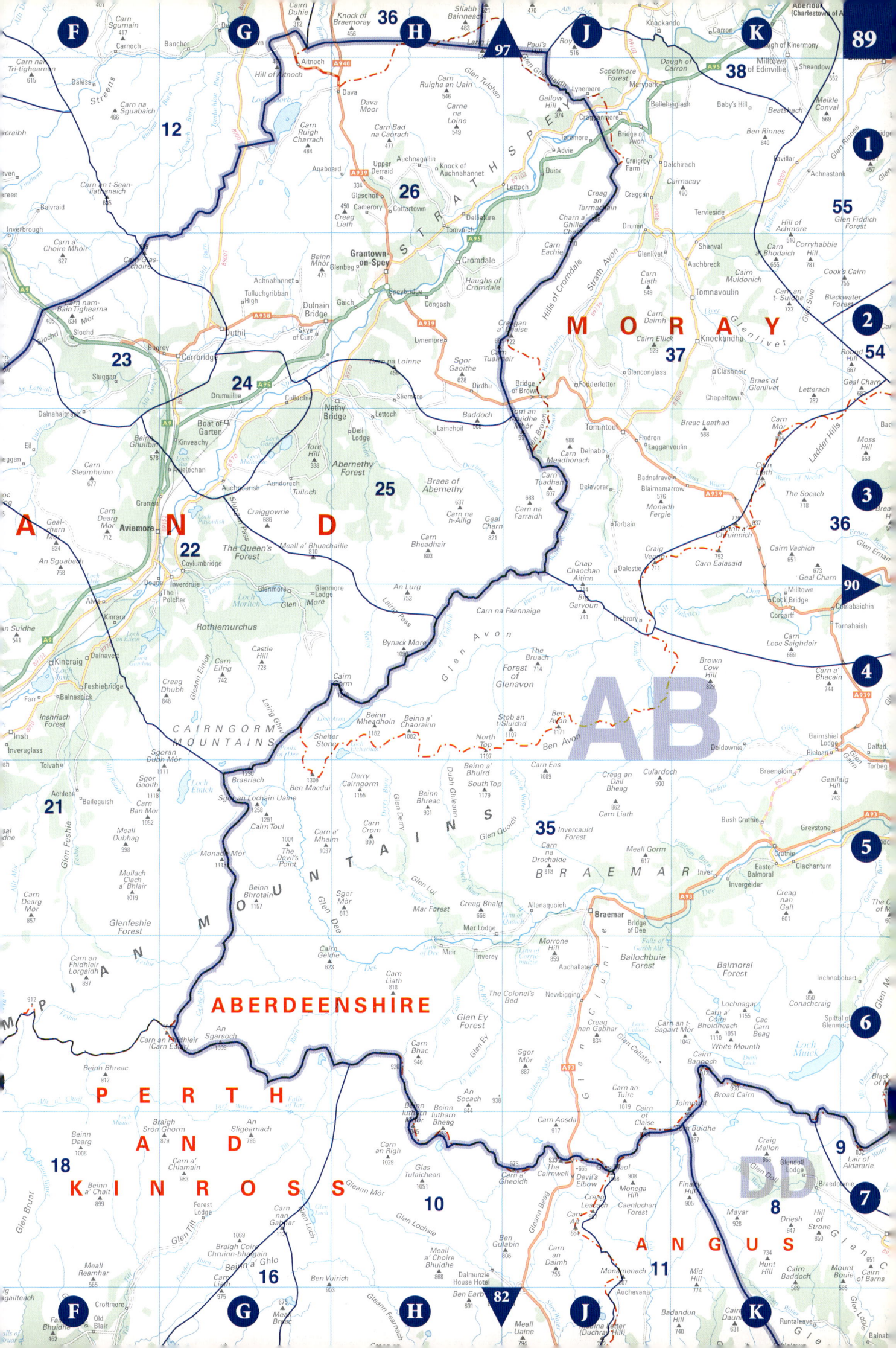

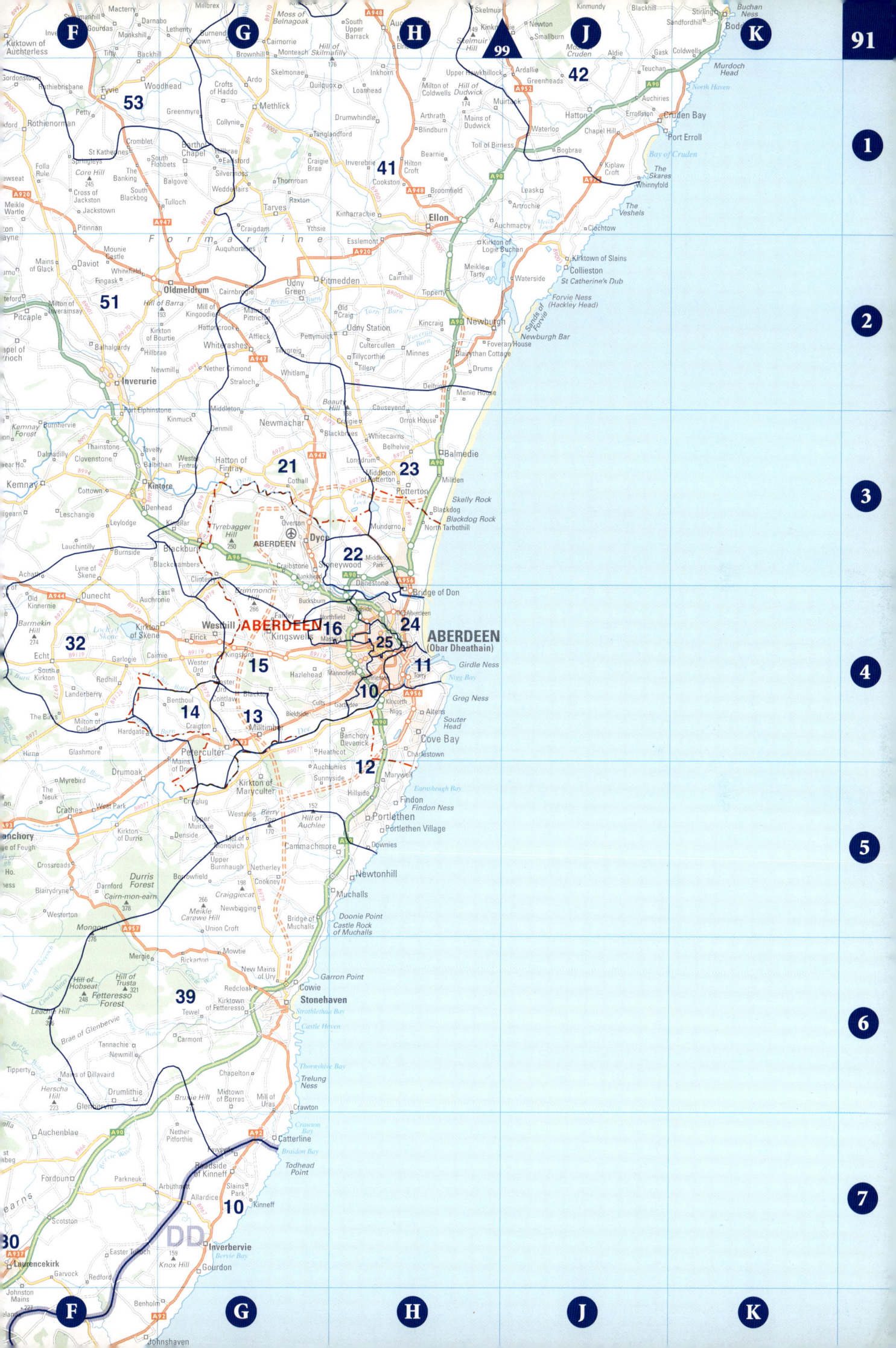

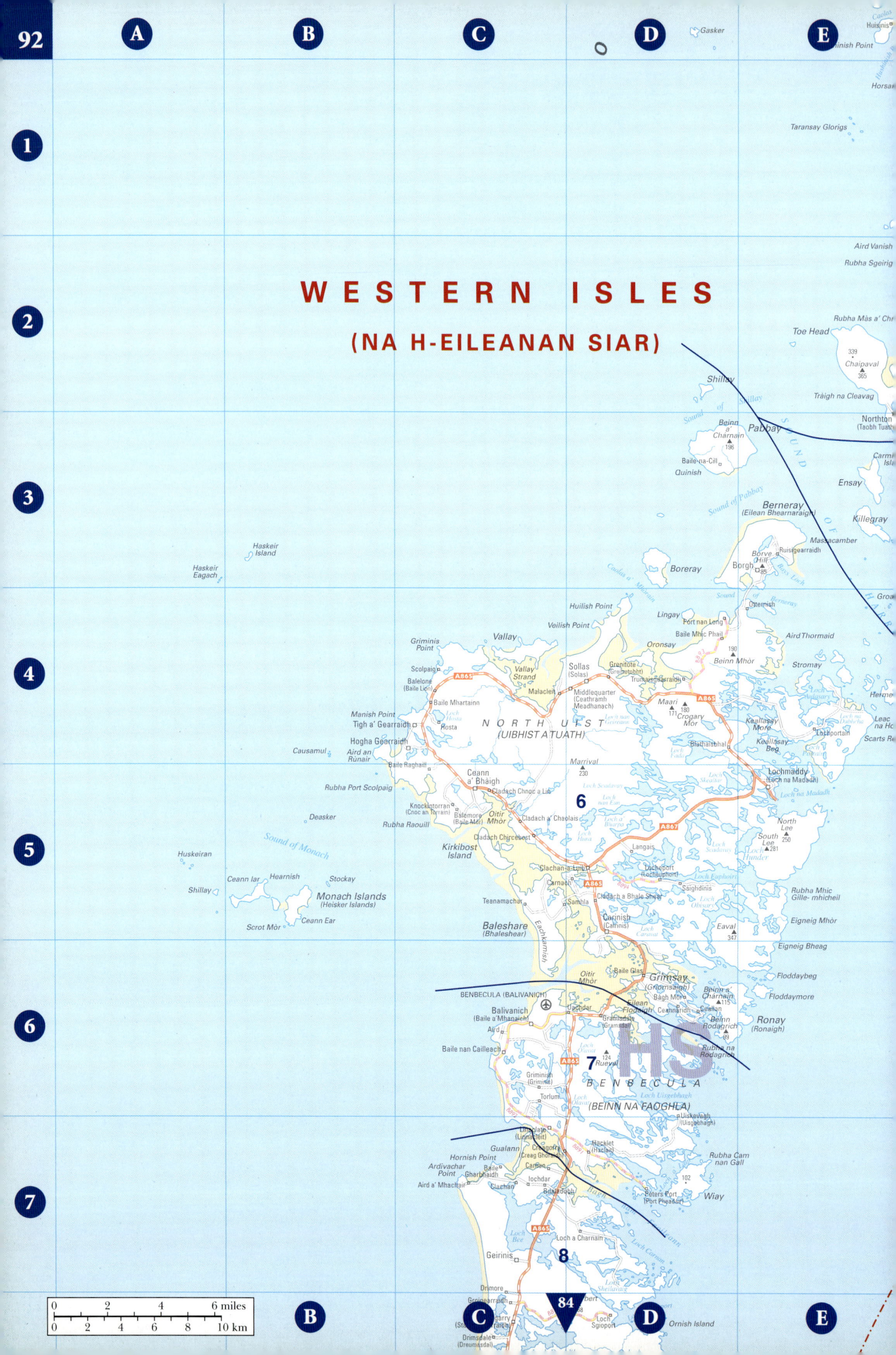

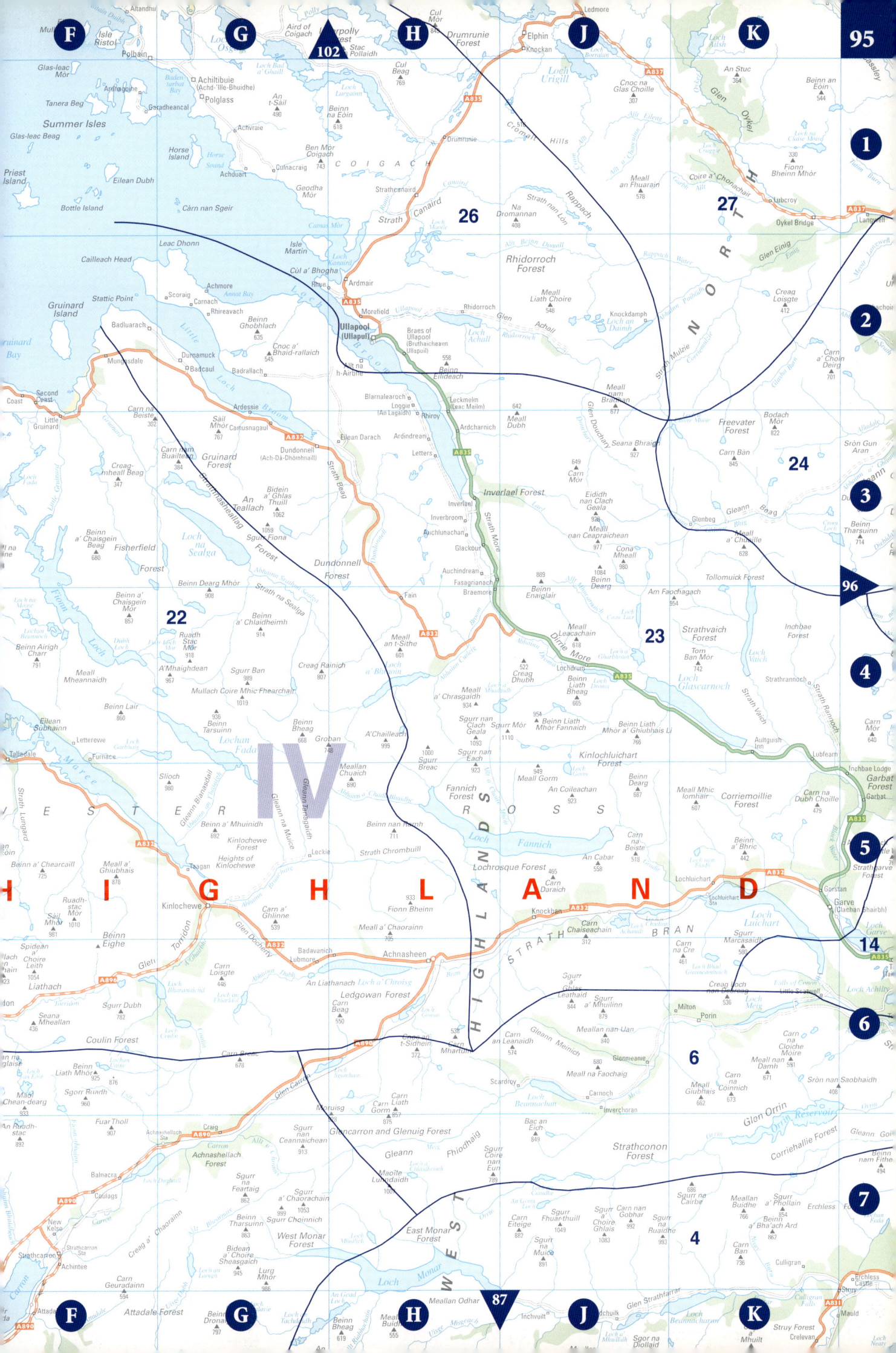

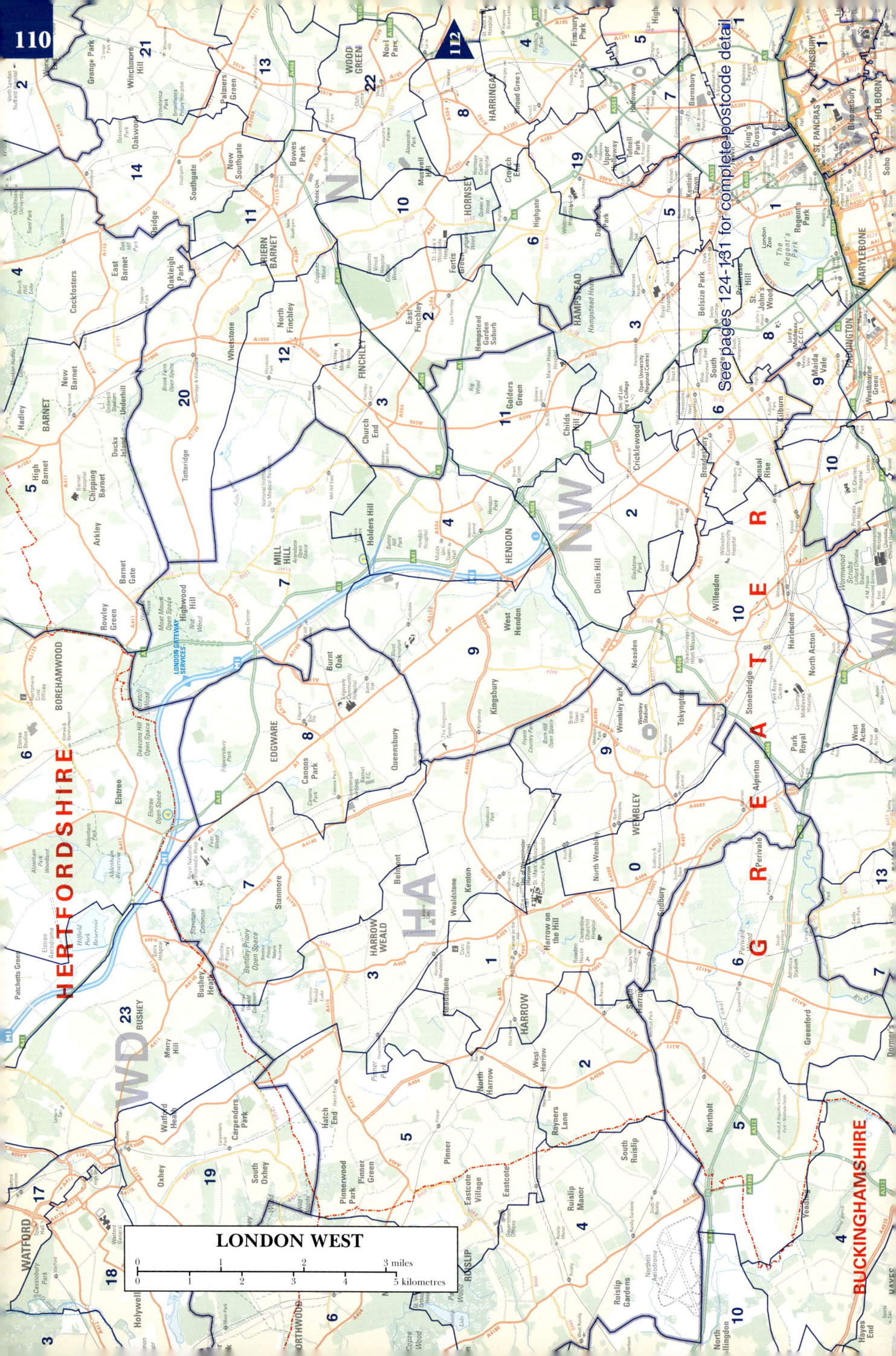

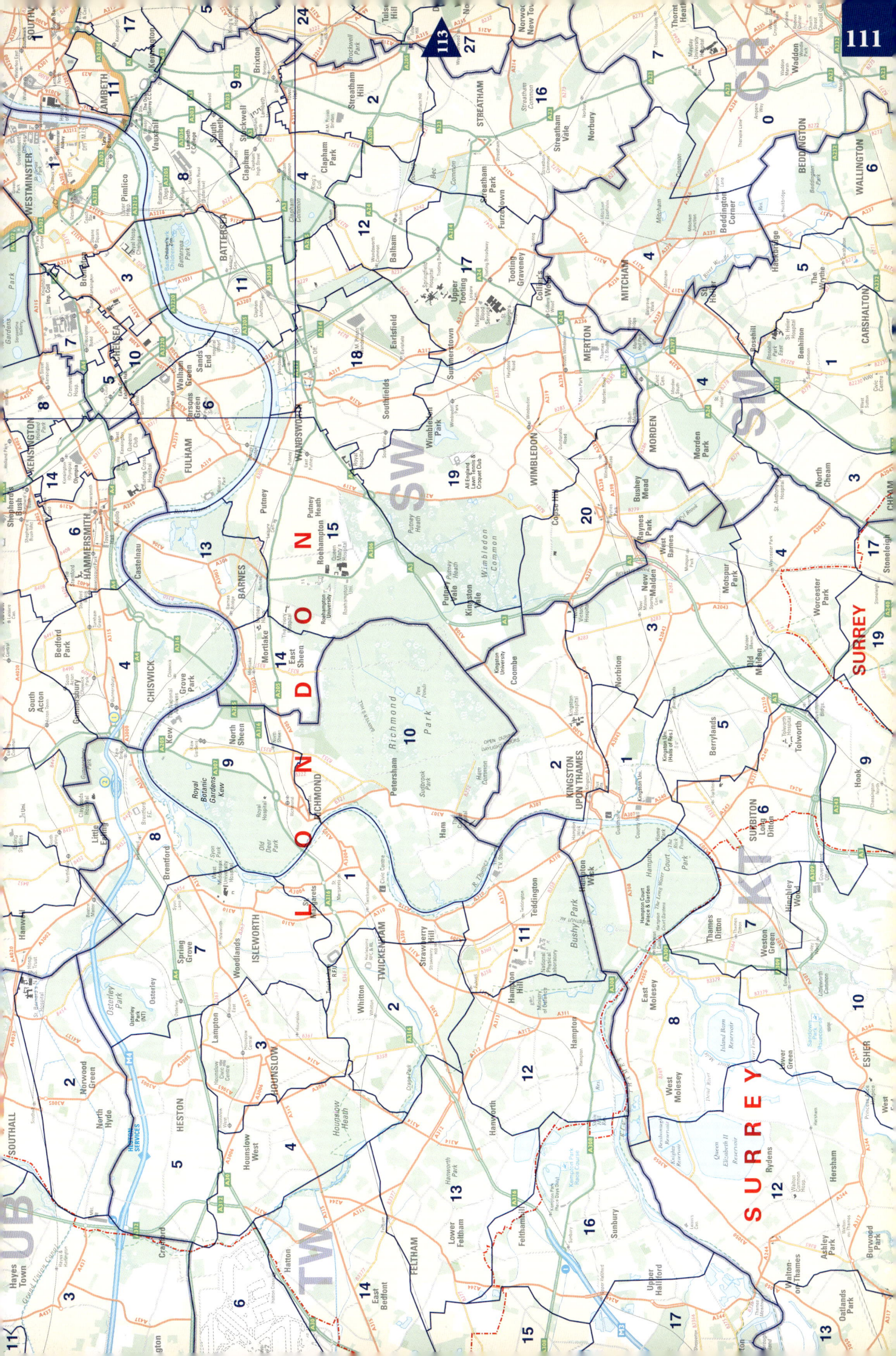

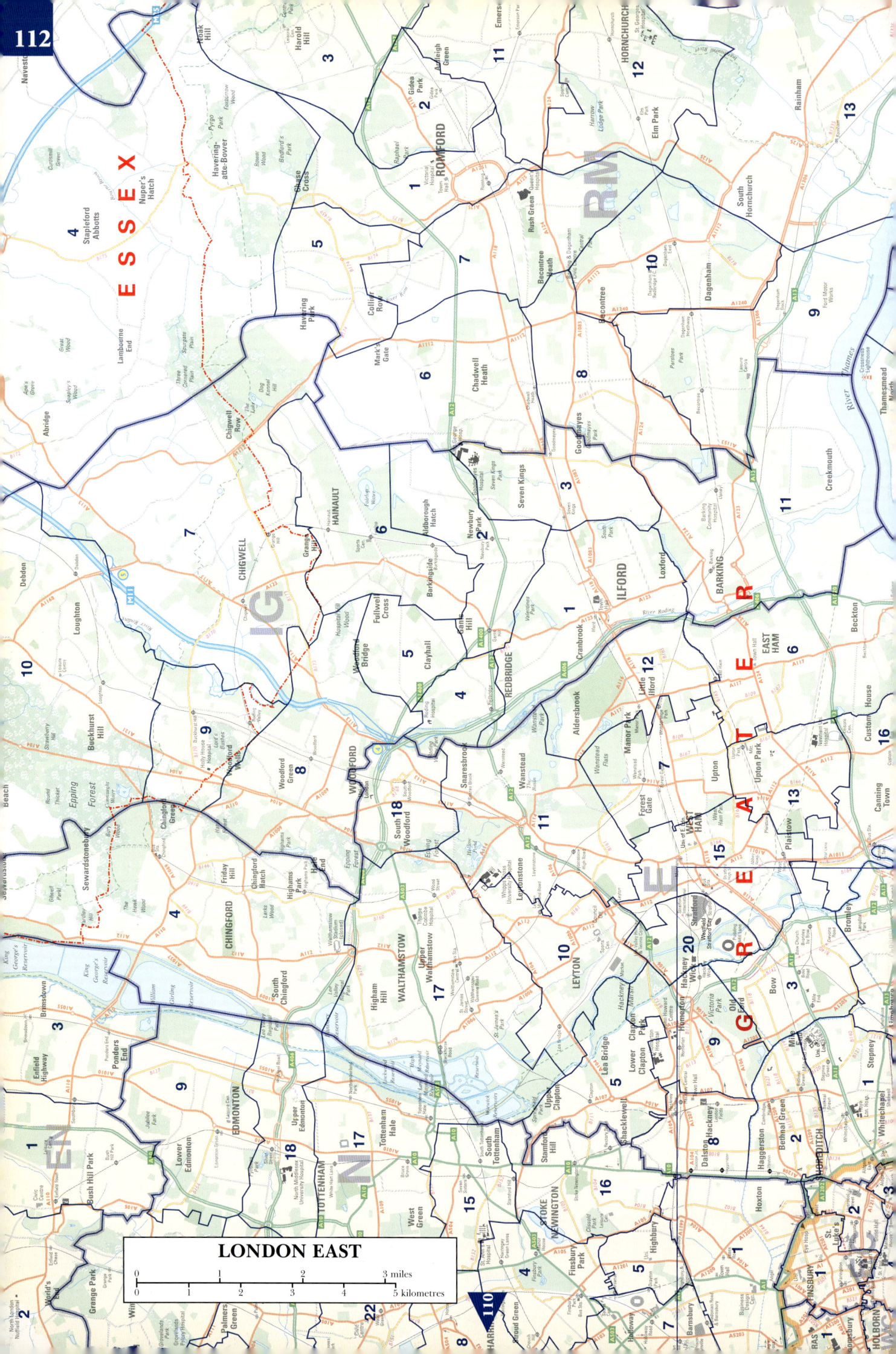

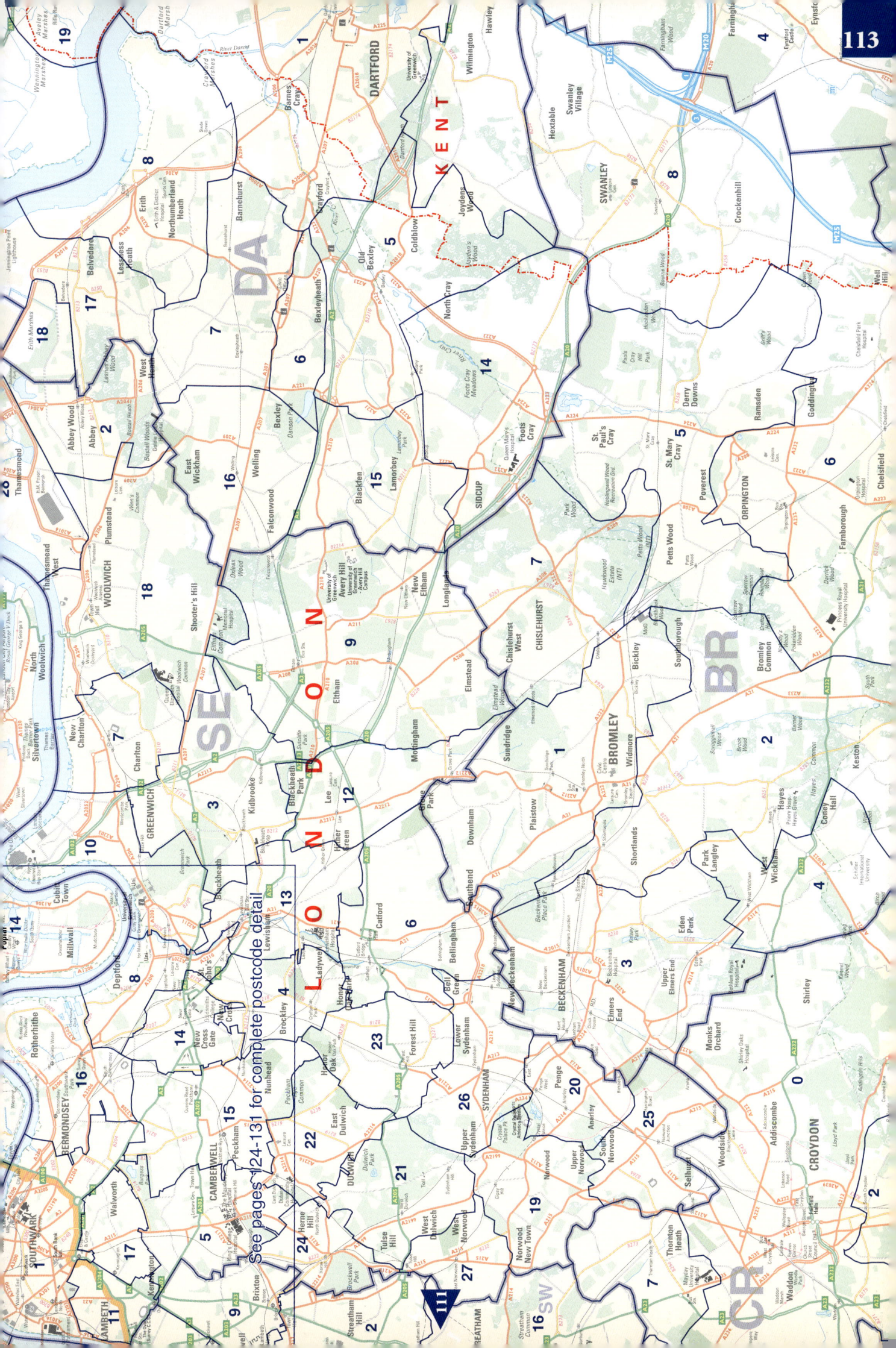

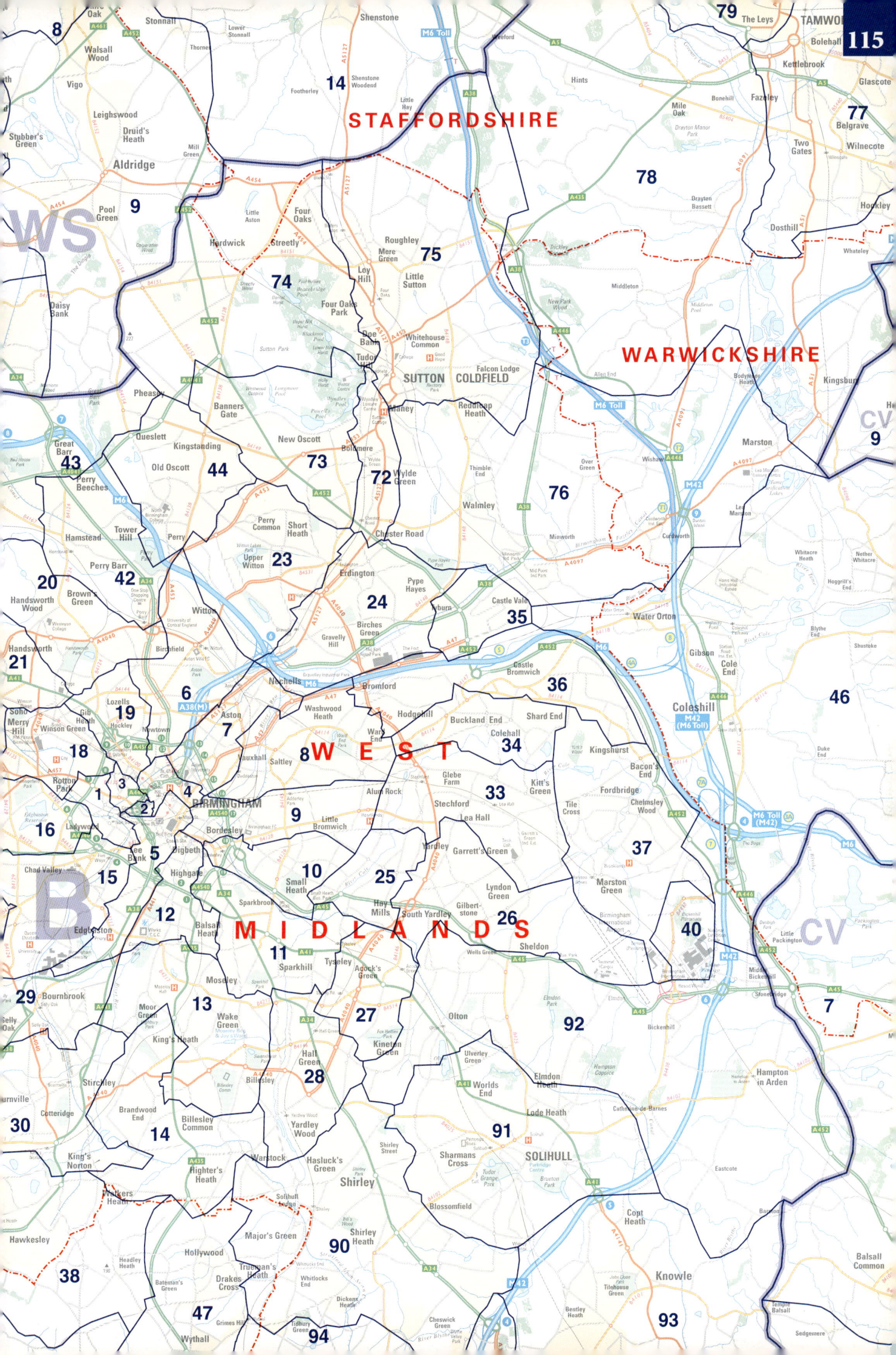

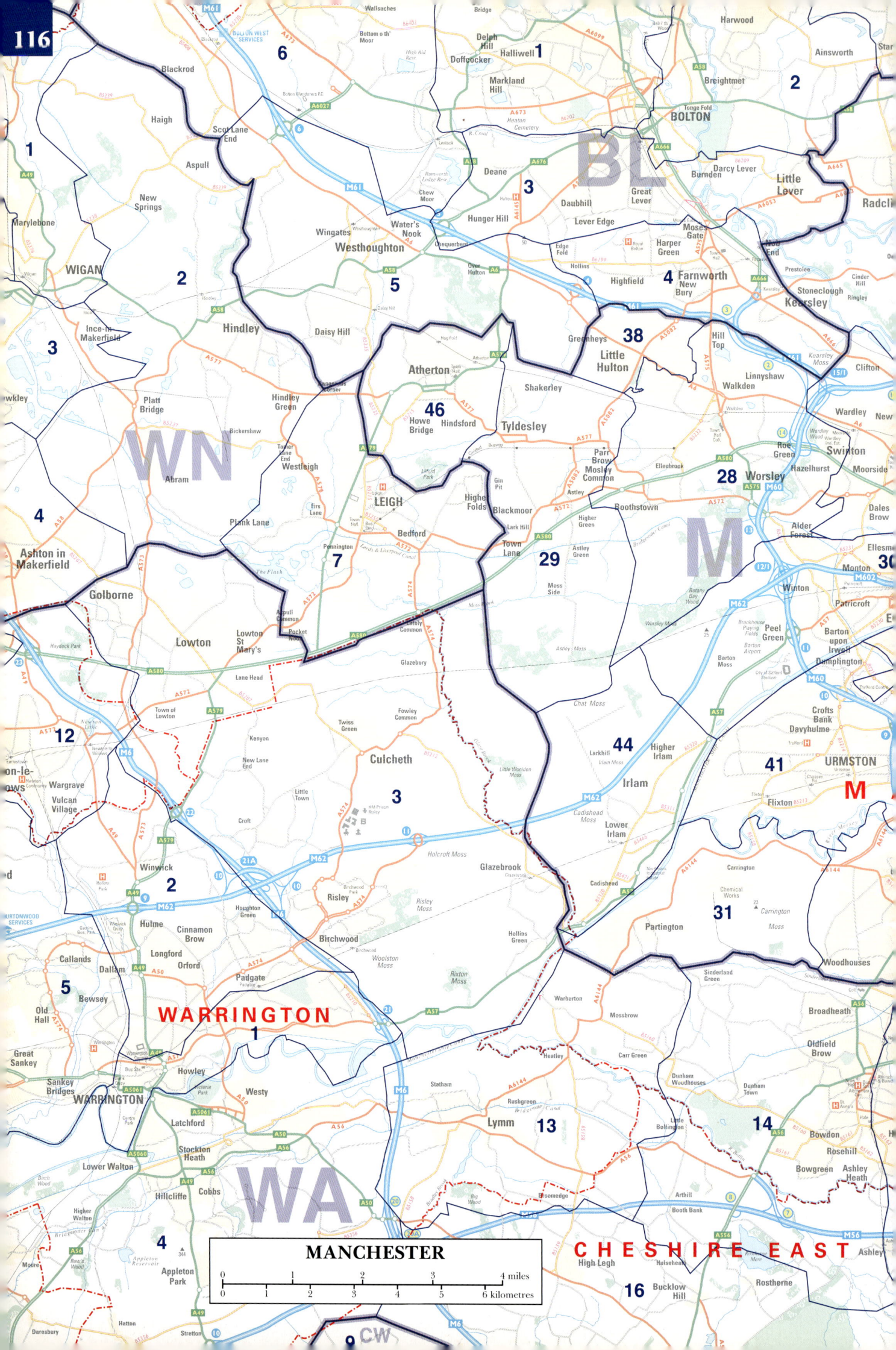

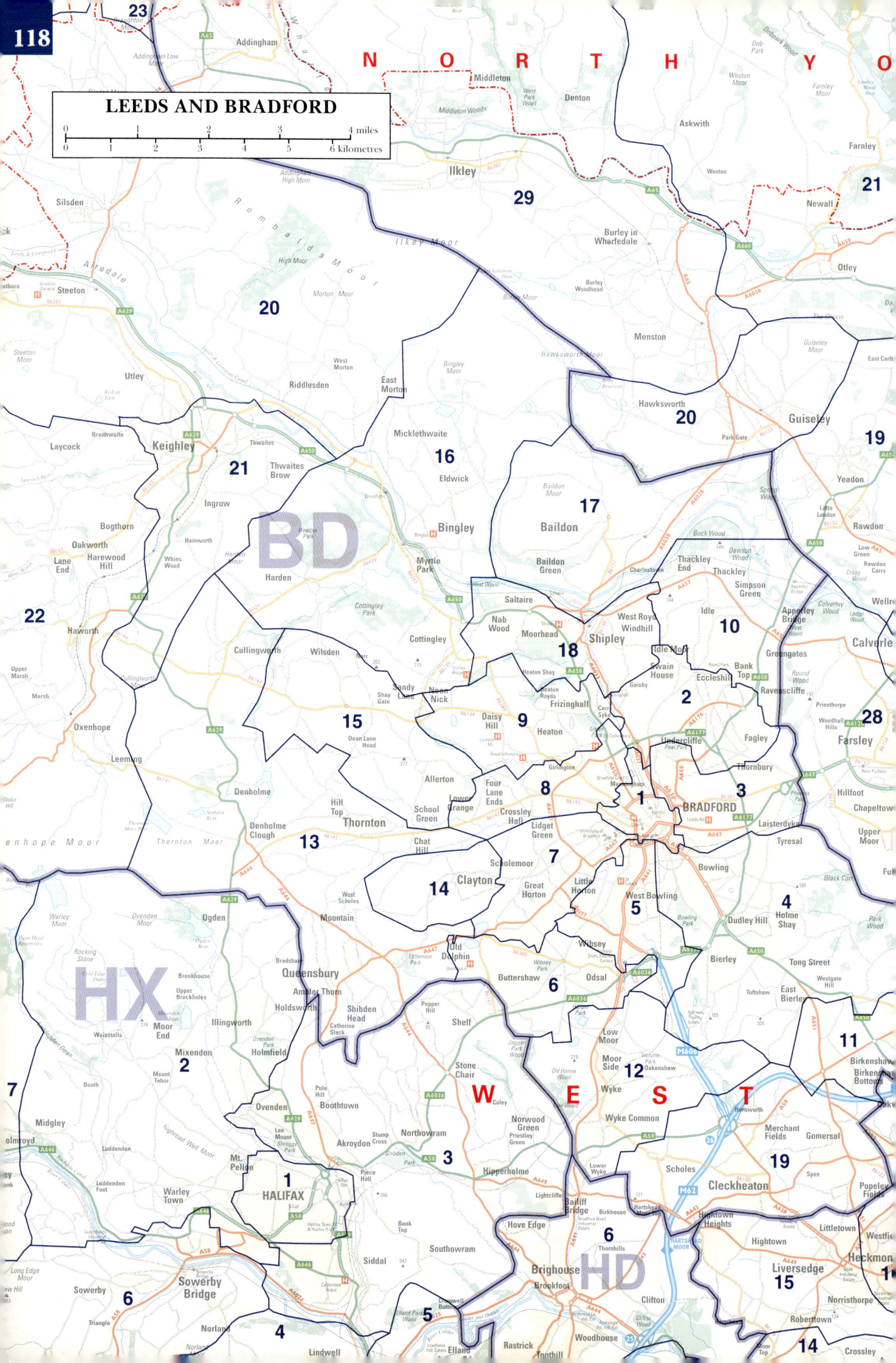

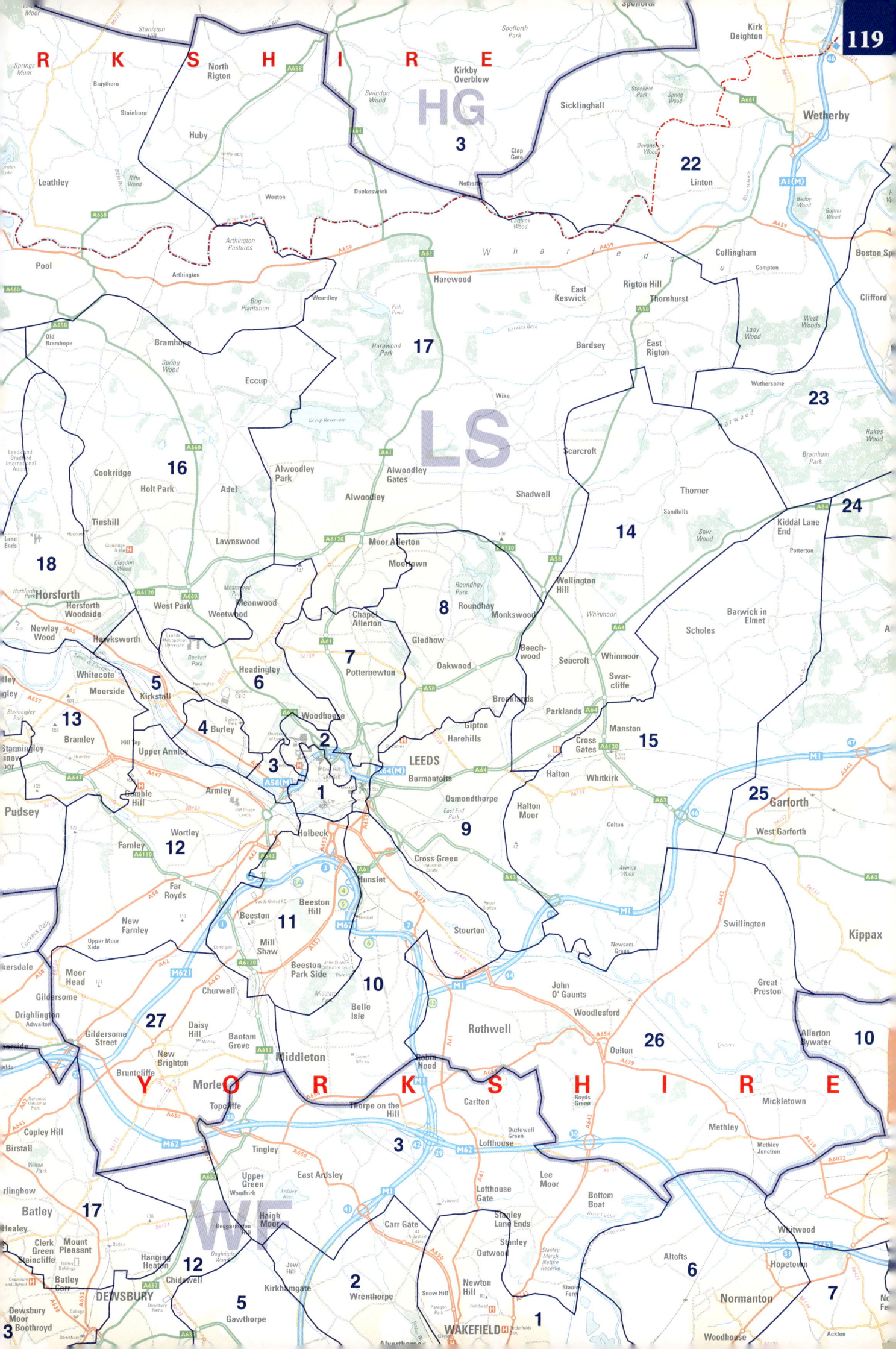

London key to map pages

123

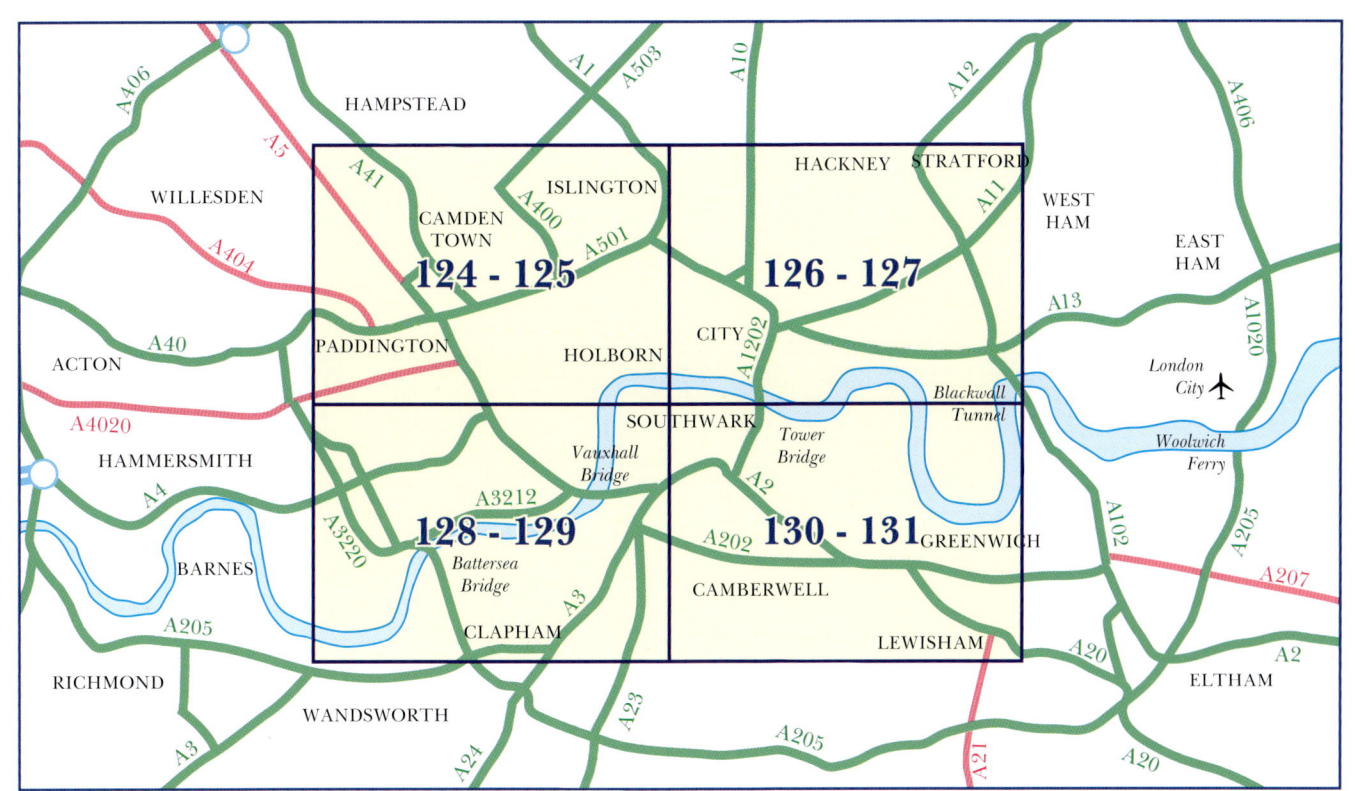

London key to symbols

	Postcode area boundary		Cycle path	▲	Youth hostel
E	Postcode area		Track/Footpath	m	Historic site
	Postcode district boundary		Long distance footpath	+	Church
4	Postcode district	P	Pedestrian ferry	☾	Mosque
3N	Postcode sub-district (central area)		Borough boundary	✡	Synagogue
	Postcode sector boundary		Main National Rail station	✶	Windmill
5	Postcode sector		Other National Rail station		Leisure & tourism
	Extent of congestion charging zone		London Underground station		Shopping
M4	Motorway		Docklands Light Railway station		Administration & law
Dual A4	Primary route		Pedestrian ferry landing stage		Health & welfare
Dual A40	'A' road	P	Car park		Education
B504	'B' road		Bus/Coach station		Industry & commerce
	Other road/One way street	H	Heliport		Cemetery
	Toll	USA	Embassy		Golf course
	Street market	Pol	Police station		Public open space/Allotments
	Restricted access road	Fire Sta	Fire station		Park/Garden/Sports ground
	Pedestrian street	PO	Post Office		Wood/Forest
		Lib	Library		Orchard
		i	Information centre for visitors		Built-up area

SCALE

0 — 1/4 — 1/2 — 3/4 — 1 mile
0 — 0.25 — 0.5 — 0.75 — 1 — 1.25 — 1.5 kilometres

1:20,000   3.2 inches to 1 mile / 5 cm to 1 km

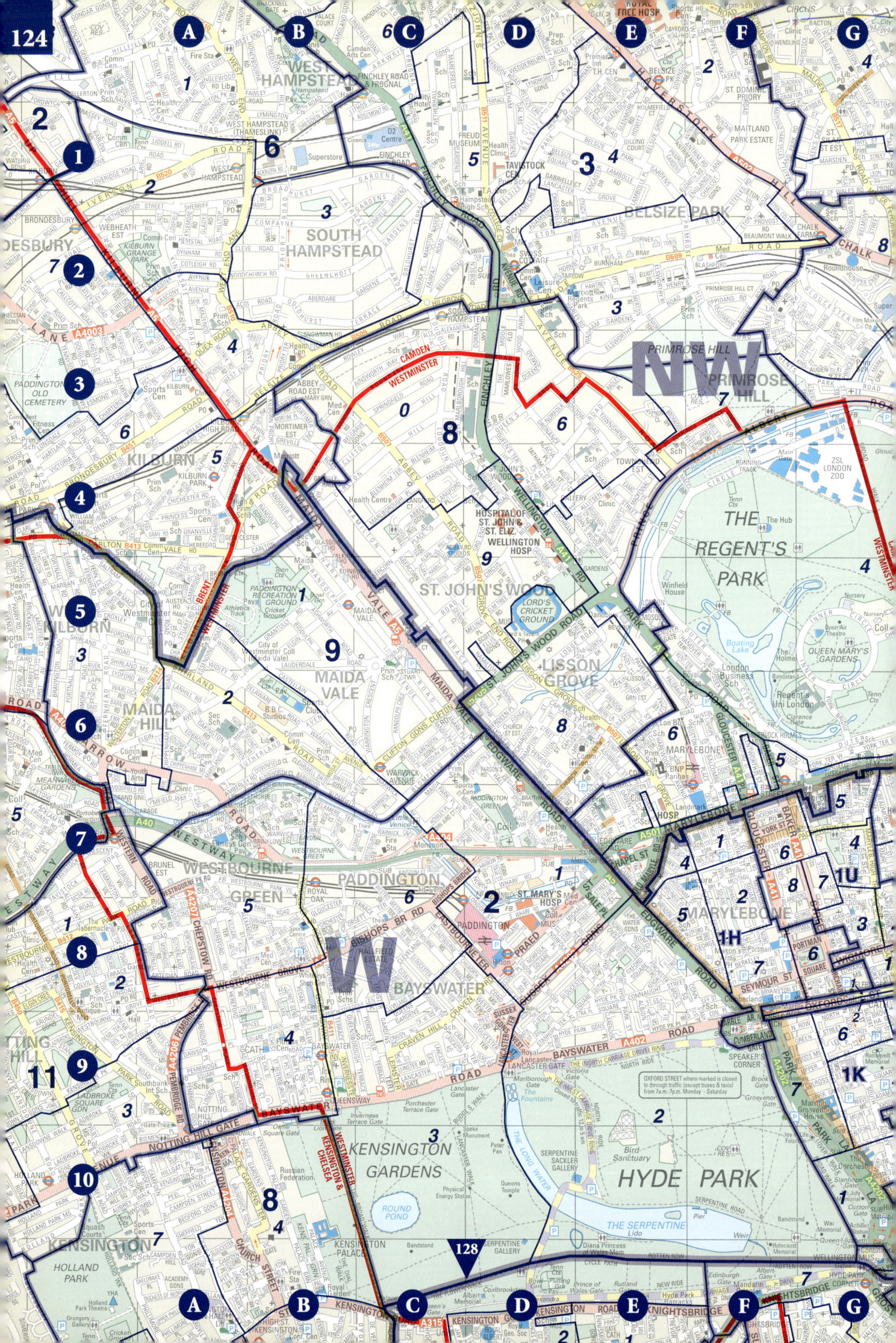

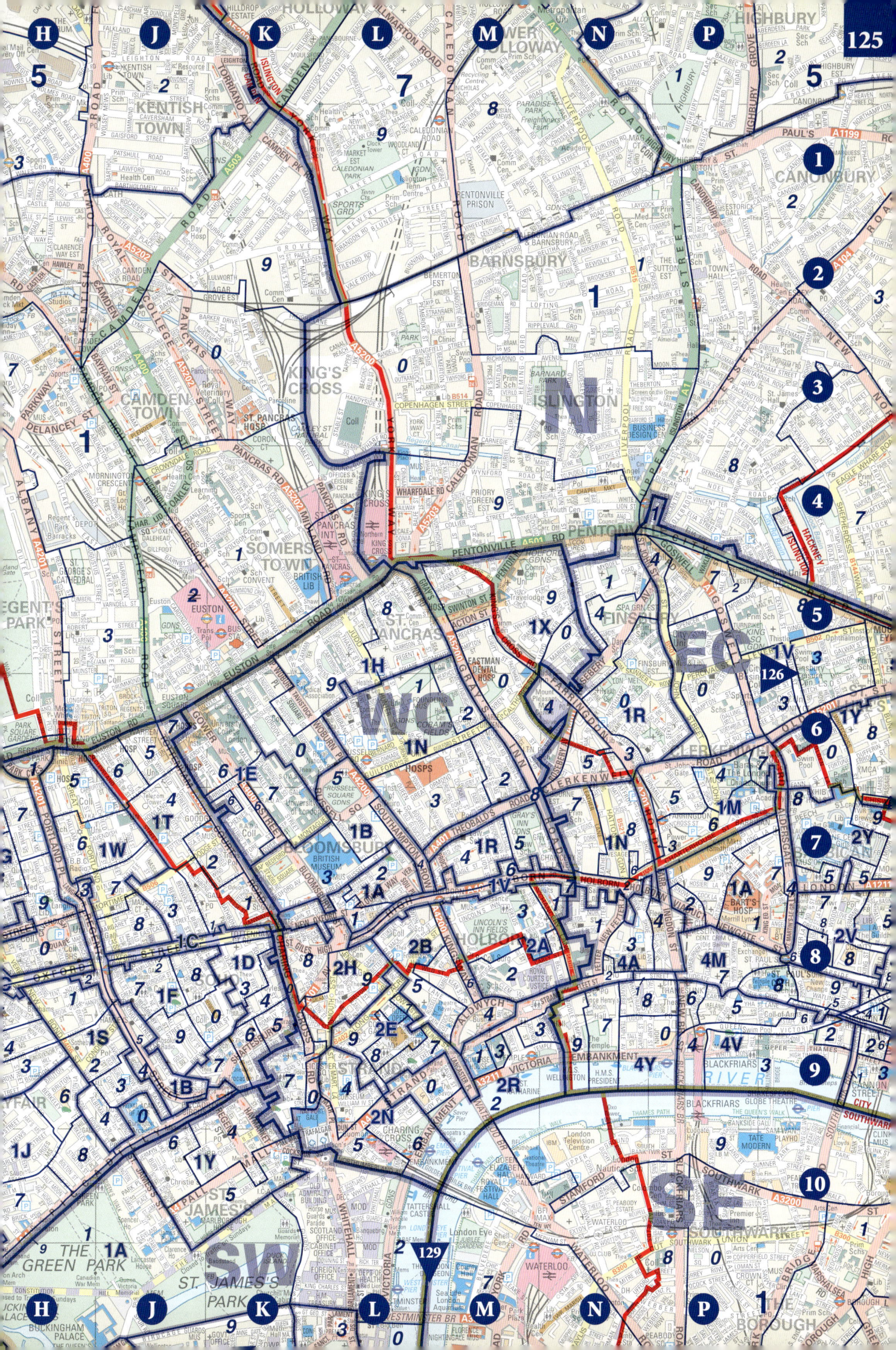

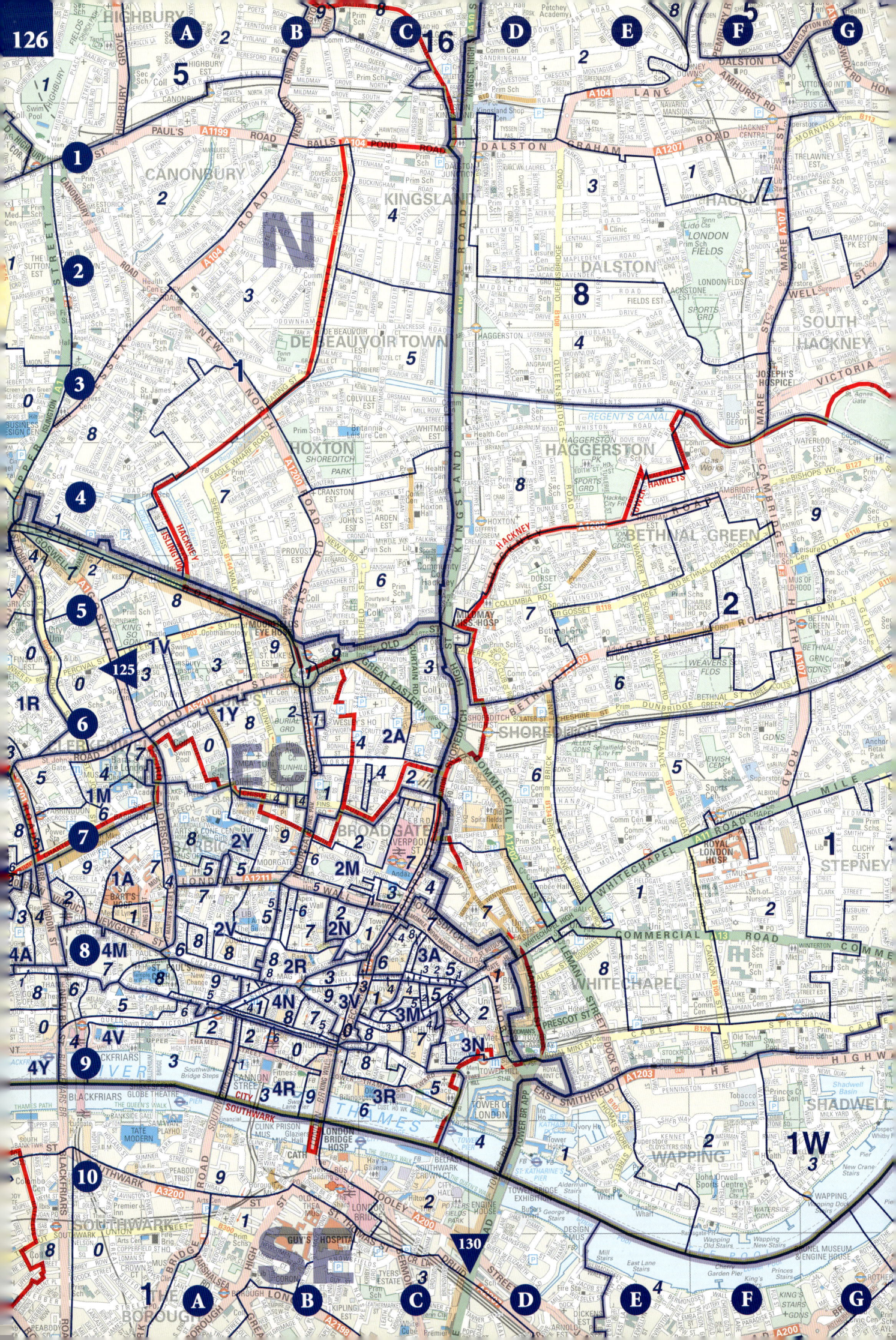

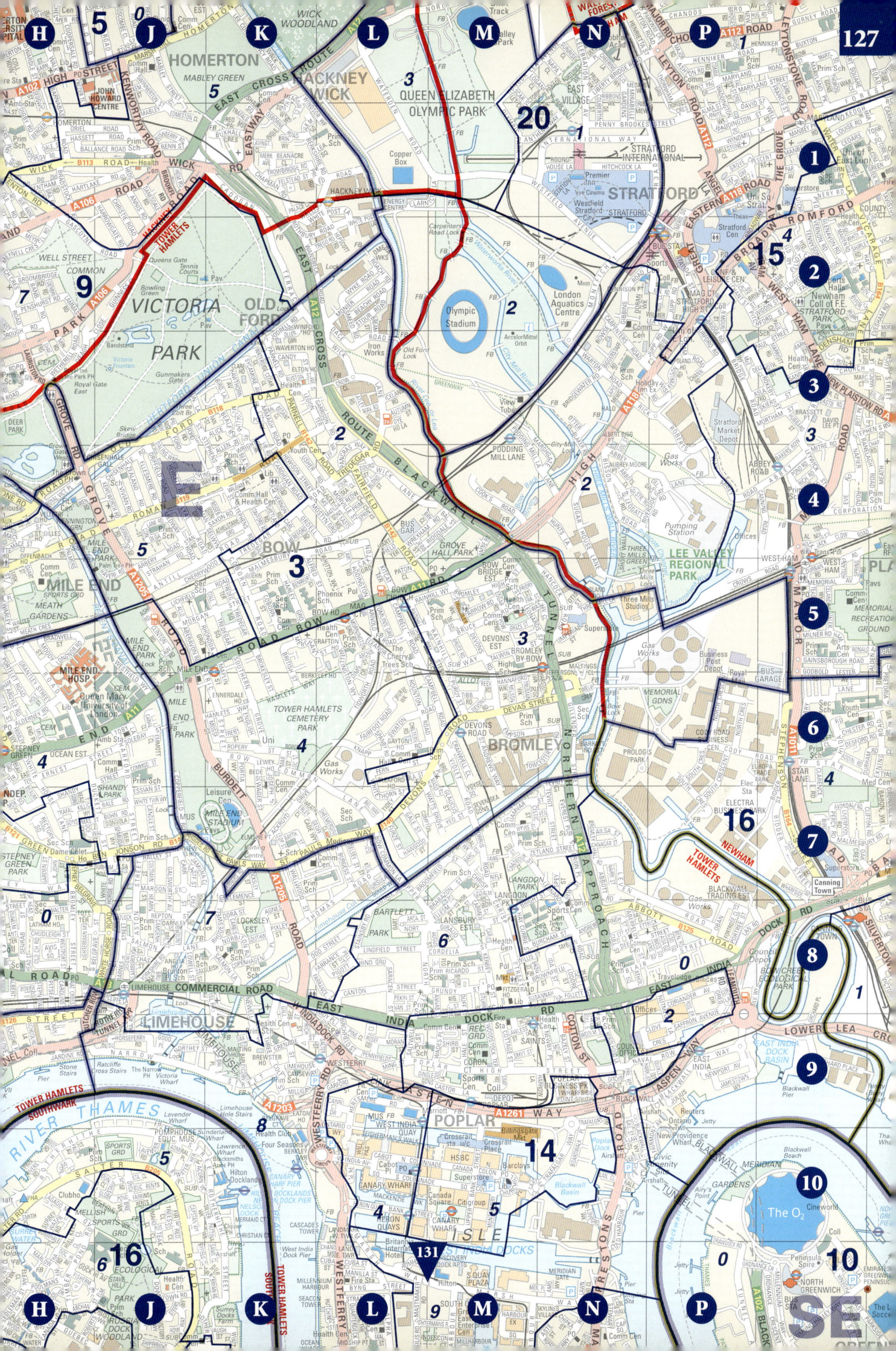

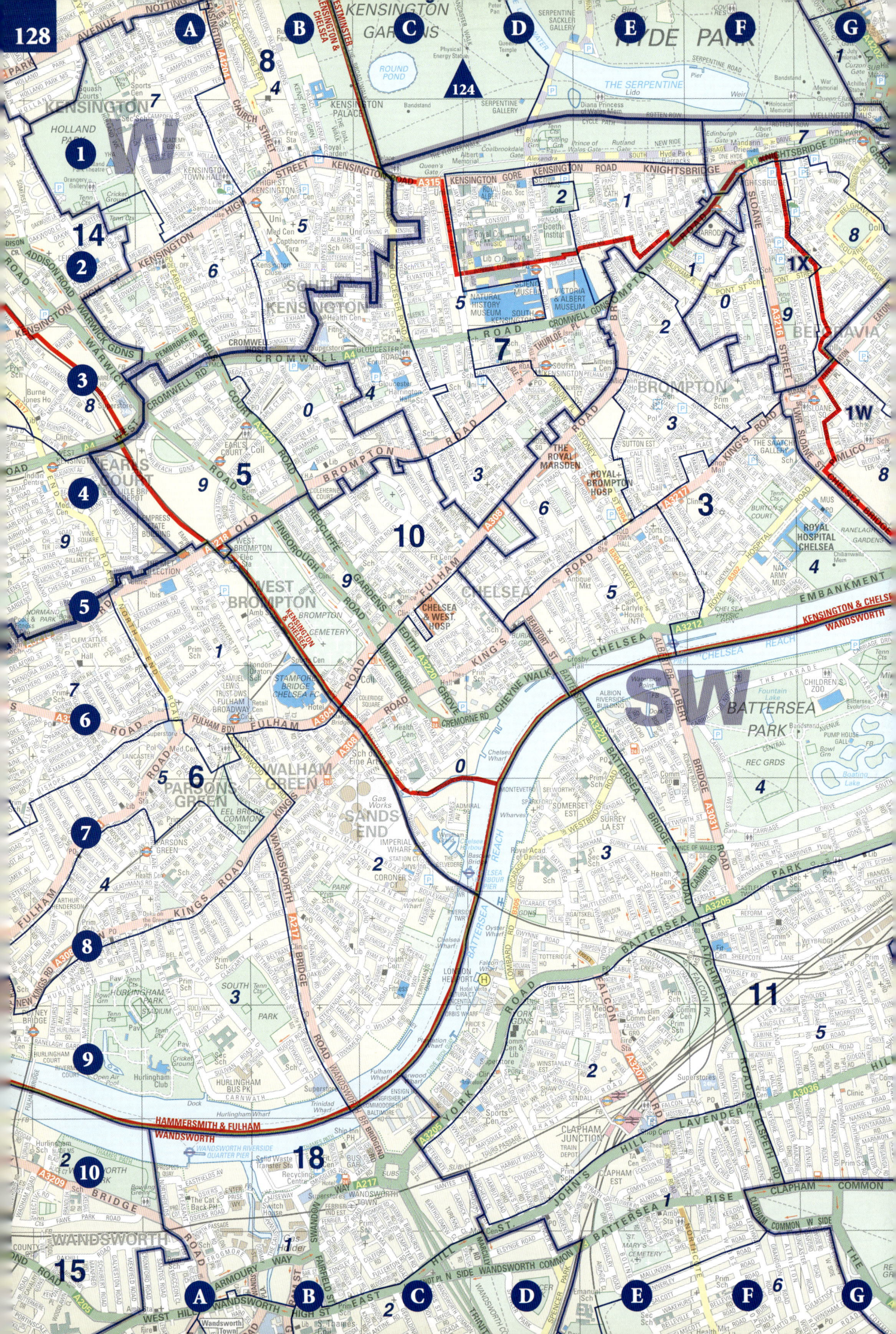

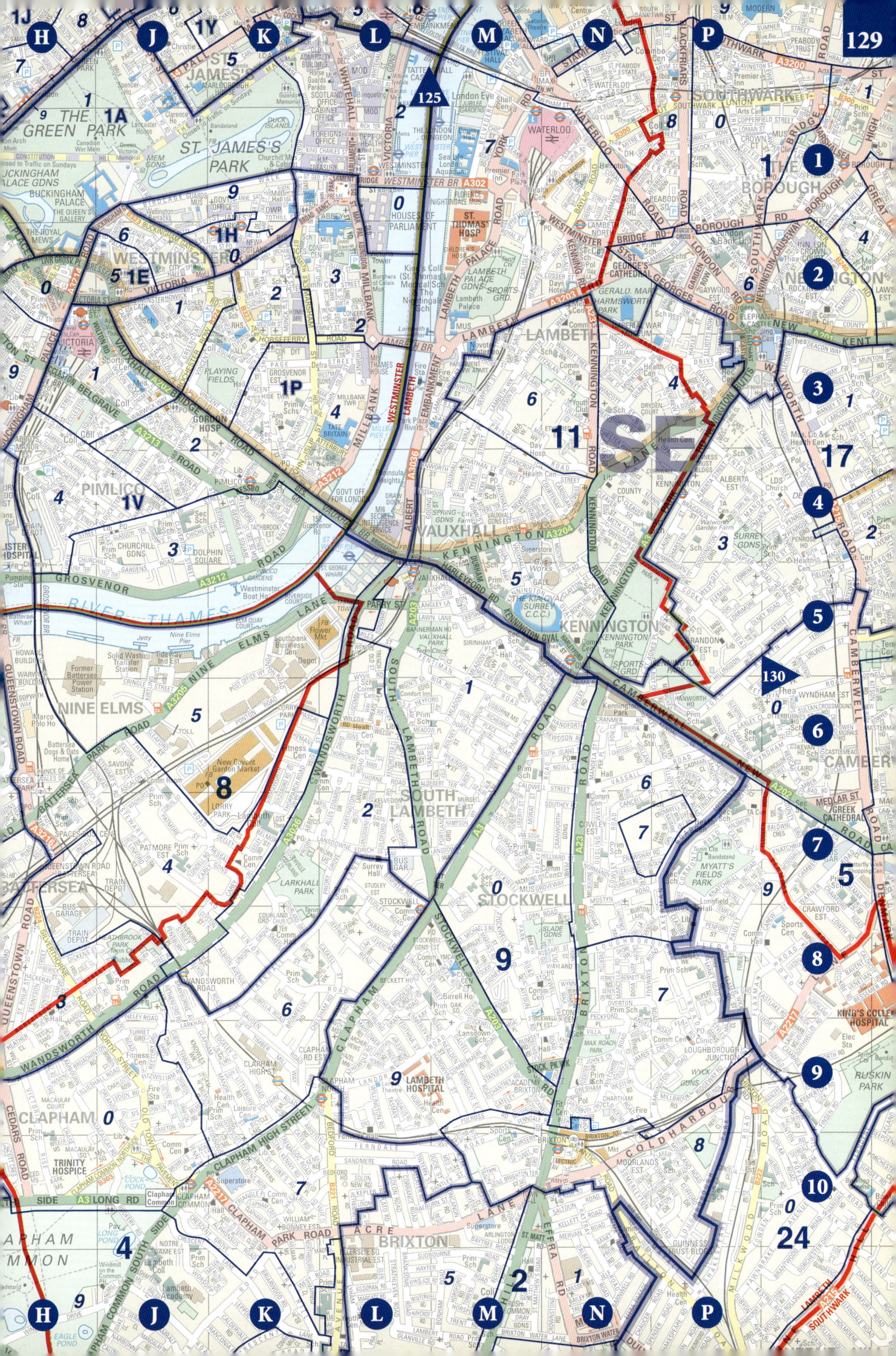

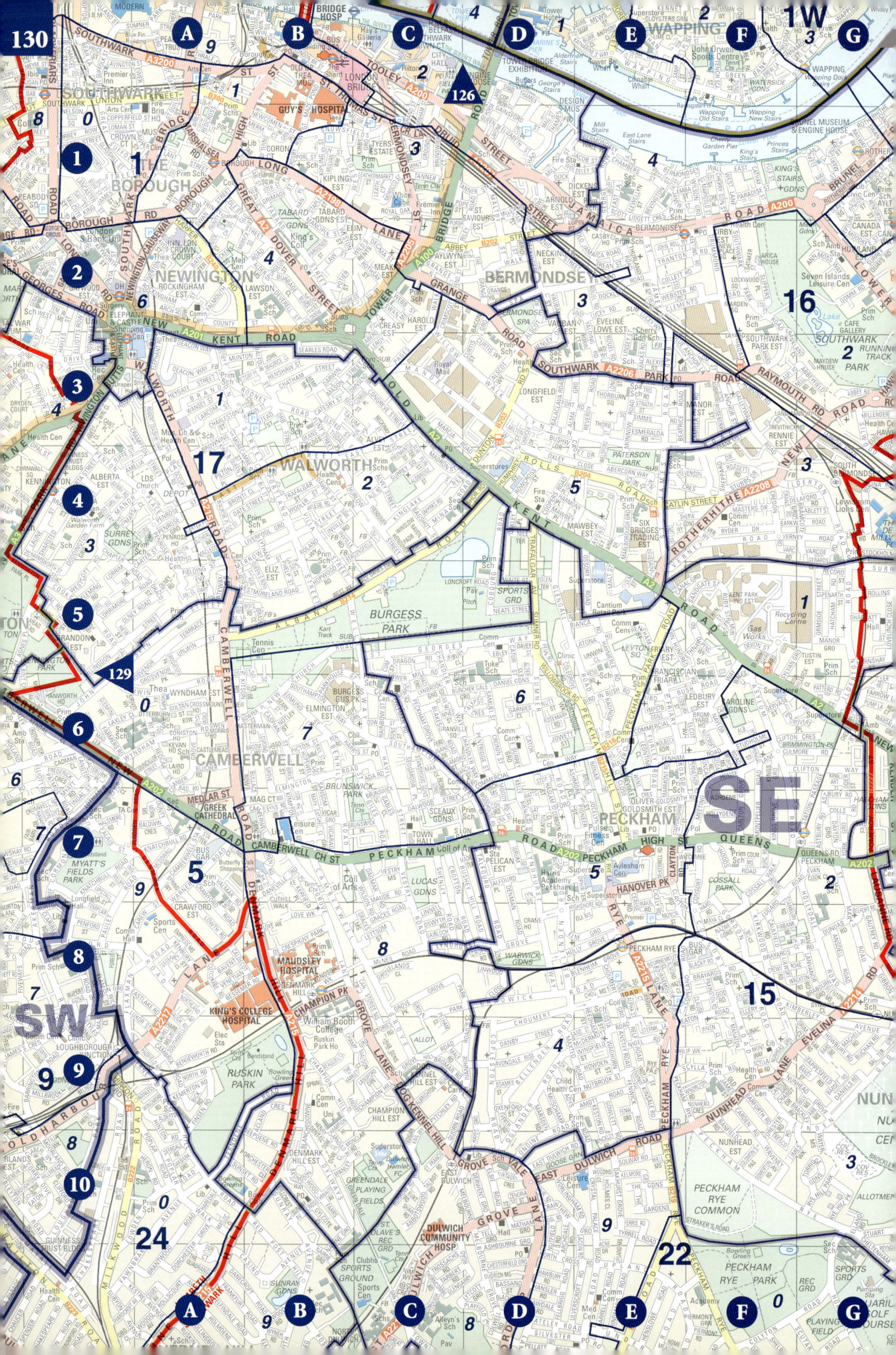

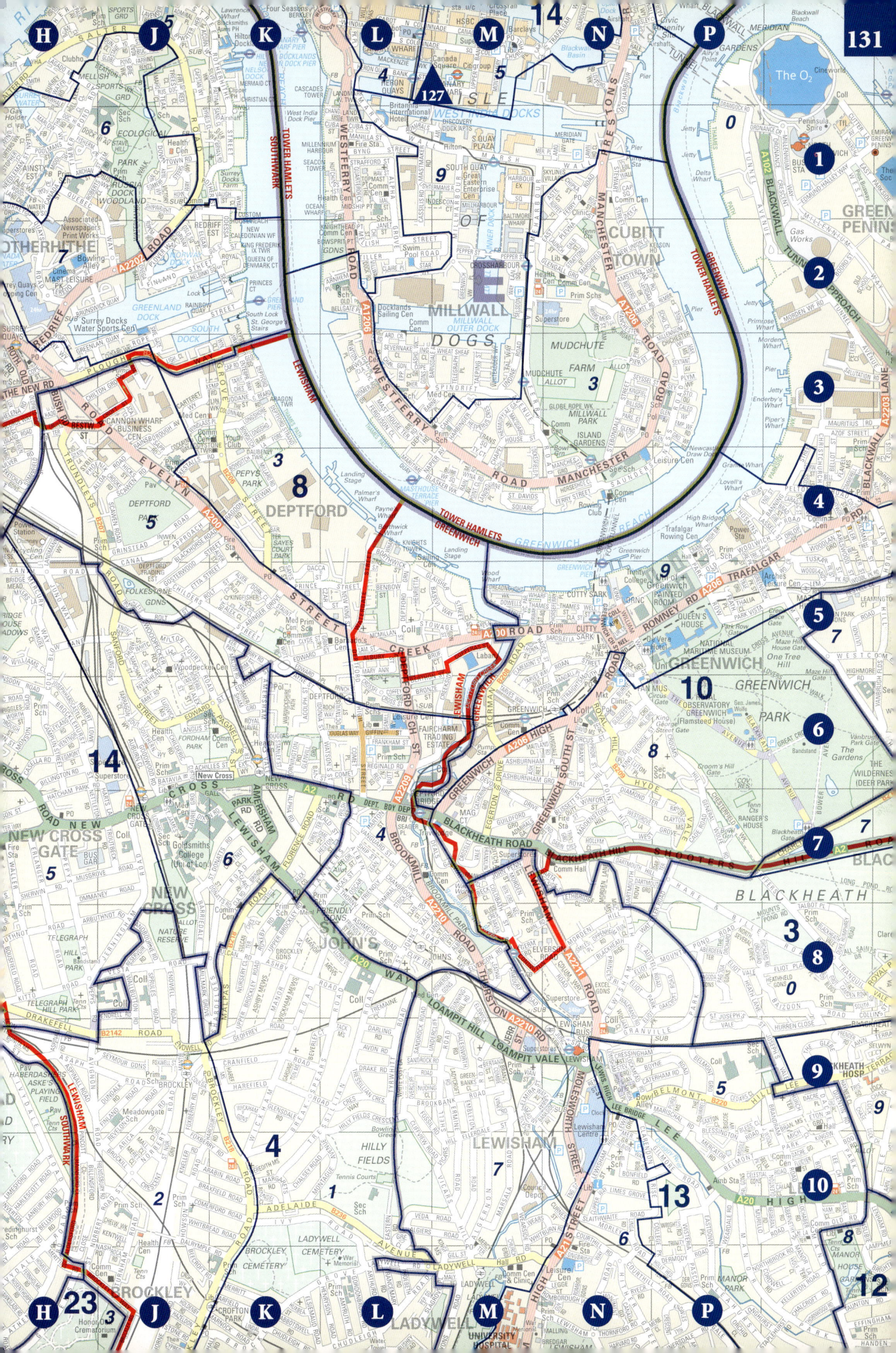

Administrative areas

Notes: Listed below are the administrative areas for Great Britain, Northern Ireland and Isle of Man used in this Postcode Atlas. Where an area is dual language, the English form is given first, followed by the alternative in parenthesis. Each entry includes its standard abbreviation in *italics* which will appear in the index. Population figures are derived from 2011 Census information. A brief description of the area then follows, which includes: adjoining administrative areas; main centres (based on descending order of population); historical, physical and economic characteristics. For English counties or former Metropolitan counties, each district, city or borough authority is listed under the heading, **Districts.**

Aberdeen *Aberdeen* Population: 228,990.
Unitary authority surrounding Aberdeen, Scotland's third largest city, on the NE coast and neighbouring Aberdeenshire. Aberdeen is the major commercial and administrative centre for N Scotland. It is a major fishing port in Scotland, with docks at the mouth of the River Dee, and is the oil and gas capital of Europe.

Aberdeenshire *Aber.* Population: 260,500.
Unitary authority on the NE coast of Scotland neighbouring Aberdeen, Angus, Highland, Moray and Perth & Kinross. Main centres are Peterhead, Fraserburgh, Inverurie, Stonehaven, Ellon, Banchory, Portlethan and Huntly. Aberdeenshire is split geographically into two main areas. The W is dominated by the Grampian Mountains, now part of the Cairngorms National Park and is largely unpopulated. The undulating lowlands of the E are mainly rural and are populated by farming and fishing communities. The major rivers are the Dee, which flows through Royal Deeside, and the Don.

Angus *Angus* Population: 116,660.
Unitary authority on the E coast of Scotland neighbouring Aberdeenshire, Dundee and Perth & Kinross. The chief centres are Arbroath, Forfar, Montrose, Carnoustie, the ancient cathedral city of Brechin, Kirriemuir and Monifieth. Angus occupies an area of 2200 square km and is an important agricultural area. It combines ancient relics and castles with highland terrain and market towns. Rivers include the North Esk, Isla and Prosen Water.

Antrim & Newtownabbey *A. & N.* Population: 139,966
Borough Council combining the former councils of Antrim and Newtownabbey. The area stretches from the NW shores of Belfast Lough to the eastern shores of Lough Neagh, the largest freshwater lake in the United Kingdom. Bounded by Belfast City, Lisburn & Castlereagh, Mid & East Antrim, Mid Ulster and Armagh City, Banbridge & Craigavon Councils. Main towns are Antrim and Newtownabbey, also Ballyclare which holds one of the oldest horse fairs in Ireland, Crumlin and Templepatrick. Small businesses and agriculture feature as well as heavy engineering, construction, transport and distribution. Industry and tourism is served by Belfast International airport.

Ards & North Down *Ards & N.D.* Population: 157,931
Borough Council combining the former councils of Ards and North Down with the administrative centre at Bangor. Bounded by Belfast City, Lisburn & Castlereagh, Newry, Mourne & Down councils. The northern boundary is Belfast Lough and this area is densely populated but further S the Ards Peninsula lies between the Irish Sea and Strangford Lough, an Area of Outstanding Natural Beauty (AONB). Other towns include Newtownards, Helen's Bay, Holywood, Donaghdee and Portaferry where a ferry travels across to the side of the lough on the southern tip of the peninsula. Main industries are light engineering, retail, agriculture and tourism. The Copeland Islands lie about a mile northeast off the coast.

Argyll & Bute *Arg. & B.* Population: 87,660.
Unitary authority on the W coast of Scotland combining mainland and island life and neighbouring Highland, Inverclyde, North Ayrshire, Perth & Kinross, Stirling and West Dunbartonshire. The main towns are Helensburgh, Dunoon, Oban, Campbeltown, Rothesay and Lochgilphead. It includes the former districts of Argyll and Bute as well as the islands of Islay, Jura, Colonsay and Mull. The main industries are fishing, agriculture, whisky production and tourism.

Armagh City, Banbridge & Craigavon *A., B. & C.* Population: 205,711.
Borough Council created from the three former councils, stretches from the southern shores of Lough Neagh to the Border with Ireland, with Craigavon designated the administrative centre. It is also bordered by Mid Ulster, Antrim & Newtownabbey, Lisburn & Castlereagh, Newry, Mourne & Down councils. The city of Armagh is the ecclesiastical capital of Ireland, home to residences for the Archbishops of both the Church of Ireland and the Roman Catholic Church. Other settlements include the cathedral town of Dromore, also Lurgan which houses the Carnegie library, Portadown, Keady and Gilford. Agriculture, especially orchard fruits and their processing is an important industry, also textiles, small scale manufacturing and a growing IT sector.

Bath & North East Somerset *B. & N.E.Som.* Population: 182,021.
Unitary authority in SW England neighbouring Bristol, North Somerset, Somerset, South Gloucestershire and Wiltshire. It surrounds the city of Bath, and includes the towns of Keynsham, Radstock and Midsomer Norton. The Georgian spa of Bath is considered to be one of the most beautiful cities in Britain, and is an important commercial and ecclesiastical centre popular with tourists. The River Avon flows through the area.

Bedford *Bed.* Population: 163,924
Unitary authority in England formed in April 2009 from the northern part of Bedfordshire county. Bounded by Milton Keynes, Northamptonshire, Cambridgeshire and Central Bedfordshire. The main centres are Bedford and Kempston. The main river is the Great Ouse.

Belfast City *Belfast.* Population: 336,830
Capital of Northern Ireland, it sits at the mouth of Belfast Lough which is fed by the river Lagan. Shipbuilding is still important to this busy port where the Titanic was built. The Titanic Quarter is now part of a redevelopment programme for the older dock area. There are ferry routes to Scotland, England and the Isle of Man. Business and tourism is also served by the city airport. Aircraft manufacturing, textiles, construction, oil refining and brewing are also important industries. It is bordered by Antrim & Newtownabbey, Lisburn & Castlereagh, Ards & North Down councils.

Blackburn with Darwen *B'burn.* Population: 146,743
Unitary authority in NW England surrounding Blackburn and Darwen and neighbouring Greater Manchester and Lancashire. Blackburn is a market and retail centre with a wide spread of industry including textiles, brewing and electronic engineering.

Blackpool *B'pool* Population: 140,501
Unitary authority on the NW coast of England surrounding Blackpool and neighbouring Lancashire. Blackpool receives around 10 million visitors each year, making it the most popular seaside resort in Europe. Attractions including the Tower, Pleasure Beach, Winter Gardens and Illuminations.

Blaenau Gwent *B.Gwent* Population: 69,674
Unitary authority in S Wales bounded by Caerphilly, Monmouthshire, Powys and Torfaen. The chief towns are Ebbw Vale, Tredegar, Bryn-mawr and Abertillery. The area was previously dependent upon coal, iron and steel industries but has since developed a broader industrial base. Part of the Brecon Beacons are in the N of the area.

Bournemouth *Bourne.* Population: 191,390
Unitary authority on the S coast of England surrounding Bournemouth and neighbouring Dorset and Poole. Bournemouth is a major resort, conference and commercial centre.

Bracknell Forest *Brack.F.* Population: 118,025
Unitary authority to the W of Greater London and bounded by Hampshire, Surrey, Windsor & Maidenhead and Wokingham. Bracknell is the chief town, while to the N of the area there are the villages of Winkfield and Binfield. To the S lies forest and heathland, and the towns of Crowthorne and Sandhurst. Bracknell has many hi-tech industries, and is a shopping and leisure centre.

Bridgend (Pen-y-Bont ar Ogwr). *Bridgend* Population: 141,214
Unitary authority in S Wales bounded by Neath Port Talbot, Rhondda Cynon Taff, Vale of Glamorgan and the sea. Main centres are Bridgend, Maesteg and Porthcawl. The area is mountainous to the N, having ribbon development along river valleys; there is greater urbanisation in the S.

Brighton & Hove *B. & H.* Population: 281,076
Unitary authority on the S coast of England neighbouring East Sussex and West Sussex. It encompasses the seaside resort of Brighton, which is a major commercial and conference centre, and the surrounding area which includes Hove, Portslade-by-Sea, Portslade, Rottingdean, Saltdean and part of the South Downs.

Bristol *Bristol* Population: 442,474
Unitary authority in SW England neighbouring Bath & North East Somerset, North Somerset, South Gloucestershire and the Bristol Channel. The area includes the city of Bristol and surrounding urban area, including Avonmouth. Bristol is an important industrial and commercial centre of W England. A former major port, its character varies from docks and a busy city centre, to parks and gardens and Georgian terracing. The city hosts the Balloon Fiesta and Harbour Regatta. River Avon forms part of the W border of the area.

Buckinghamshire *Bucks.* Population: 521,922
S midland county of England bounded by Central Bedfordshire, Greater London, Hertfordshire, Northamptonshire, Oxfordshire, Surrey, Windsor & Maidenhead and Wokingham. Chief towns are High Wycombe, the county town of Aylesbury, Amersham, Chesham, Marlow and Beaconsfield, around which, and other smaller towns, is a variety of light industry, as well as extensive residential areas. The chalk downs of the Chiltern Hills traverse the S part of the county, which is otherwise mostly flat. The River Thames flows along its S border.
Districts: Aylesbury Vale; Chiltern; South Bucks; Wycombe.

Caerphilly (Caerffili). *Caerp.* Population: 179,941
Unitary authority in S Wales bordered by Blaenau Gwent, Cardiff, Merthyr Tydfil, Rhondda Cynon Taff and Torfaen. The chief centres are Caerphilly, Ystrad Mynach, Risca, Bargoed, Blackwood and Bedwas. The geography of the area varies from open moorland to busy market towns. The former mining industry has been replaced by electronics and automotive companies, with tourism also being important to the local economy. Rivers include the Rhymney and Sirhowy.

Cambridgeshire *Cambs.* Population: 639,818
County of E England bounded by Bedford, Central Bedfordshire, Essex, Hertfordshire, Lincolnshire, Norfolk, Northamptonshire, Peterborough and Suffolk. Cambridgeshire is mostly flat, with fenland to N and E, although there are low chalk hills in the S and SE. Chief centres are the city and county town of Cambridge, Wisbech, St. Ives, March, Huntingdon, St. Neots and the cathedral city of Ely. Agriculture is a major industry with sugar beet, potatoes and corn all important crops; soft fruit and vegetable cultivation and canning are also significant rural industries. There has been recent growth of medical, pharmaceutical and hi-tech industries around Cambridge. Rivers include the Cam, Nene, and Great Ouse.
Districts: Cambridge; East Cambridgeshire; Fenland; Huntingdonshire; South Cambridgeshire.

Cardiff (Caerdydd). *Cardiff* Population: 354,294
Unitary authority in S Wales surrounding the city of Cardiff and bordered by Caerphilly, Newport, Rhondda Cynon Taff, Vale of Glamorgan and the Bristol Channel. Cardiff, the capital of Wales, is a major administrative, commercial, cultural and tourism centre. It contains the Welsh Office, Welsh National Stadium (Principality Stadium), remains of medieval castle, cathedral at Llandarff and university. Cardiff docks, which were formerly used to export Welsh coal, are part of an ongoing major redevelopment. The city has excellent shopping facilities, notably at the St. David's Centre. The birthplace of Roald Dahl.

Carmarthenshire (Sir Gaerfyrddin). *Carmar.* Population: 184,898
Unitary authority in S Wales bounded by Ceredigion, Neath Port Talbot, Pembrokeshire, Powys, Swansea and the sea. The chief towns are Llanelli, Carmarthen and Ammanford. The geography varies from the Brecon Beacons in the E, to the river valleys in the N, and the fishing villages, beaches and coastal towns in the S. The 50m coastline runs along the S of the area. Rivers include the Tywi, Cothi, Gwendaeth Fach and Gwendaeth Fawr.

Causeway Coast & Glens *C. C. & G.* Population: 142,303
District Council created from the amalgamation of Balleymoney, Coleraine, Limavady and Moyle councils with Coleraine as its centre. Tourists are drawn to seaside towns such as Portrush but this area is dominated by the rural beauty of the Antrim Coast & Glens AONB and the amazing landscape of basalt columns now an UNESCO World Heritage Site of the Giant's Causeway in the N and the Sperrin Mountains (an AONB) in the S. The inhabited Rathlin Island lies off the N coast but is home to Northern Ireland's biggest seabird breeding colony, managed by the RSPB. Lough Foyle in the W forms a boundary with Ireland but the area is also bounded by Derry City & Strabane, Mid & East Antrim and Mid Ulster councils. Agriculture and tourism are the main industries. Bushmills is home to the world's oldest licensed distillery.

Central Bedfordshire *Cen Beds.* Population: 269,076
Unitary authority in England formed in April 2009 from the southern part of Bedfordshire county (the former districts of Mid and South Bedfordshire). Bounded by Milton Keynes, Bedford, Cambridgeshire, Hertfordshire, Luton and Buckinghamshire. The main centres are Dunstable, Leighton Buzzard and Biggleswade. It includes the N end of the Chiltern Hills but is otherwise flat.

Ceredigion *Cere.* Population: 75,425
Unitary authority in W Wales bounded by Carmarthenshire, Gwynedd, Pembrokeshire, Powys and the sea at Cardigan Bay. The main towns are Aberystwyth, Cardigan, Aberaeron, Lampeter, Tregaron and Llandysul. Part of the Cambrian Mountains lie in the E of the area and the 50m coast has many sandy beaches. Tourism and agriculture are the most important industries. The main river is the Teifi.

Cheshire East *Ches.E.* Population: 374,179
Unitary authority of NW England formed in April 2009 from three former districts of the county of Cheshire. Bounded by Warrington, Greater Manchester, Derbyshire, Staffordshire, Shropshire and Cheshire West & Chester. Chief centres are Crewe, Macclesfield, Nantwich, Wilmslow and Congleton. The foothills of the Pennines enter the NW of the area.

Cheshire West & Chester *Ches.W & C.* Population: 332,210
Unitary authority of NW England formed in April 2009 from three former districts of the county of Cheshire. Bounded by Merseyside, Halton, Warrington, Cheshire East, Shropshire and the Welsh authorities of Wrexham and Flintshire. Chief centres are the cathedral city of Chester and the towns of Ellesmere Port, Northwich and Winsford. The country is mainly flat with the rural areas of the S and W noted for dairy products. To the N and W are the estuaries of the River Dee and River Mersey.

Clackmannanshire *Clack.* Population: 51,190
Unitary authority in central Scotland neighbouring Fife, Perth & Kinross and Stirling. The N includes the Ochil Hills,

while the lowland surrounding the Forth estuary contains the chief towns which are Alloa, Tullibody, Tillicoultry and Alva. Clackmannanshire has over 50 sites of nature conservation and five historic castles and towers. The main rivers are the Devon and the Forth.

Conwy *Conwy* Population: 116,287
Unitary authority in N Wales bordered by Denbighshire, Gwynedd and the sea. The chief towns are Colwyn Bay, Llandudno, Abergele, Rhôs-on-Sea and Conwy. Around 40 per cent of Conwy is within Snowdonia National Park and there are 29m of coastline. The coastal resorts attract tourism which is a key industry, but agriculture and light manufacturing are also important to the local economy. The main river is the Conwy.

Cornwall *Cornw.* Population: 545,335
Unitary authority of SW England bounded by Devon and the sea. Chief centres are St. Austell, Falmouth, Penzance, the cathedral city and administrative centre of Truro, Redruth, Camborne and Newquay. The coastline is wild and rocky; headlands and cliffs are interspersed with large sandy beaches in the N, and deeply indented with river estuaries in the S. The interior is dominated by areas of moorland, notably the granite mass of Bodmin Moor in the NE. There are also farmlands providing rich cattle-grazing, and deep river valleys. The climate is mild, and flower cultivation is carried on extensively. The many derelict tin mines are witness to the former importance of this industry; there has recently been a partial revival. The chief industry is tourism. China clay is produced in large quantities in the St. Austell area, and there is some fishing. Rivers include the Tamar, forming the boundary with Devon; Fowey, East and West Looe, Fal, Camel, and Lynher.

Cumbria *Cumb.* Population: 497,874
County of NW England bounded by Durham, Lancashire, Northumberland and North Yorkshire; the Scottish authorities of Dumfries & Galloway and Scottish Borders; and the Solway Firth and Irish Sea. Chief centres are the city of Carlisle and the towns of Barrow-in-Furness, Whitehaven, Workington, Kendal, Penrith and Ulverston. A narrow strip of flat country along the coast widens to a plain in the N and around Carlisle. Otherwise the county is composed of mountains, moorland and lakes, and includes the scenically famous Lake District. Cumbria is mostly rural and uncultivated, with industry centred on Carlisle and the urban centres. Whitehaven, Workington, and Maryport all once relied on coal, while Barrow-in-Furness developed due to shipbuilding and heavy industry. There are links with nuclear technology: Calder Hall, N of Seascale, was Britain's first atomic power station, Sellafield is the site of a nuclear reprocessing plant and Trident submarines were built at Barrow-in-Furness. Tourism in the Lake District and sheep farming are also important industries. The area is noted for its radial drainage, with Windermere and Ullswater being the largest of the lakes and the River Eden being the chief of many rivers.
Districts: Allerdale; Barrow-in-Furness; Carlisle; Copeland; Eden; South Lakeland.

Darlington *Darl.* Population: 105,367
Unitary authority in NE England surrounding Darlington and neighbouring Durham, North Yorkshire and Stockton-on-Tees. Darlington has a variety of industries, including iron, steel and textiles. The River Tees forms the S border.

Denbighshire (Sir Ddinbych). *Denb.* Population: 94,791
Unitary authority in N Wales neighbouring Conwy, Flintshire, Gwynedd, Powys, Wrexham and the sea. The chief towns are Rhyl, Prestatyn, Denbigh, Ruthin, the ancient city of St. Asaph, and Llangollen. Main industries are tourism, centred on the coastal resorts of Rhyl and Prestatyn, and agriculture. Rivers include the Morwynion.

Derby *Derby* Population: 252,463
Unitary authority in central England surrounding the city of Derby and bordered by Derbyshire. Derby has a history dating back to Roman times and is now important in the rail industry; other key industries are manufacturing and aerospace engineering. The River Derwent passes through the area.

Derbyshire *Derbys.* Population: 779,804
Midland county of England bounded by Cheshire East, Derby, Greater Manchester, Leicestershire, Nottinghamshire, South Yorkshire, Staffordshire and West Yorkshire. Chief towns are Chesterfield, Long Eaton, Swadlincote, Ilkeston, Staveley, Dronfield, Alfreton, Heanor and Buxton. The high steep hills in the N, which include the dramatic scenery of The Peak, are the S extremity of The Pennines, and provide grazing for sheep and cattle. There is some textile industry in the towns of the N and W, while the S of the county is dominated by heavy industry, mining, and quarrying. Tourism is based on the scenic Peak District National Park, most of which falls in the county. Principal rivers are the Dove, forming much of the boundary with Staffordshire and noted for its scenery and fishing, and the Derwent; the Trent flows through the S corner of the county.
Districts: Amber Valley; Bolsover; Chesterfield; Derbyshire Dales; Erewash; High Peak; North East Derbyshire; South Derbyshire.

Derry City & Strabane *Derry & Str.* Population: 149,198
District Council with a long border with Ireland in the west and the Sperrin Mountains (an AONB) in the east. City of Londonderry/Derry is an historic seaport on the estuary of the river Foyle with city walls, still complete, built in 1619 and recognised as some of the finest in Europe. Other towns include the historic market centre of Strabane on the river Mourne, Newtownstewart, New Buildings and the model linen village of Sion Mills. Along with Ireland, Fermanagh & Omagh, Mid Ulster and Causeway Coast & Glens councils border this area. Agriculture is the main industry as well as textiles and manufacturing.

Devon *Devon* Population: 765,302
Large county in SW peninsula of England bounded by Cornwall, Dorset, Plymouth, Somerset, Torbay and the Bristol and English Channels. The chief centres are the city of Exeter, Exmouth, Barnstaple, Newton Abbot, Tiverton, Bideford and Teignmouth. The county includes the W end of Exmoor and the whole of the granite mass of Dartmoor, whose summit, High Willhays, is the highest point in S England. Moorland areas apart, the county is largely given over to agriculture, and on the coast, to fishing and tourism. On Dartmoor there are quarries and a military training area; there are china clay workings in the S. Daffodils are grown commercially in River Tamar valley. Chief rivers are Exe, Teign, Dart, Avon, Erme, Tamar and Tavy in the S; and Taw and Torridge in the N. The granite island of Lundy is included in the county for administrative purposes.
Districts: East Devon; Exeter; Mid Devon; North Devon; South Hams; Teignbridge; Torridge; West Devon.

Dorset *Dorset* Population: 418,269
County in SW England bounded by Bournemouth, Devon, Hampshire, Poole, Somerset, Wiltshire and the English Channel. The chief towns are Weymouth, Christchurch, Wimborne Minster, the county town of Dorchester, Bridport, Swanage and Blandford Forum. The county is hilly, with chalk downs and impressive geological formations along the coastline. Sand, gravel, stone and oil extraction takes place around the Isle of Portland and the Isle of Purbeck. Dorset is also noted for its agricultural and dairy produce. Tourism is an important industry due to the beautiful scenery, the proliferation of prehistoric and Roman remains, and the connection with Thomas Hardy's Wessex. Among numerous minor rivers are the Stour, Frome, and Piddle or Trent.
Districts: Christchurch; East Dorset; North Dorset; Purbeck; West Dorset; Weymouth & Portland.

Dumfries & Galloway *D. & G.* Population: 149,940
Unitary authority in SW Scotland neighbouring East Ayrshire, Scottish Borders, South Ayrshire, South Lanarkshire, the English county of Cumbria and the sea. It comprises the former counties of Dumfries, Kirkcudbright and Wigtown. Chief towns are Dumfries, Stranraer, Annan, Dalbeattie, Lockerbie, Castle Douglas, Newton Stewart and Kirkcudbright. The hilly area to the N is largely given over to sheep-grazing and afforestation, while farther S there is some good-quality arable farmland. At the extreme W of the area is the peninsula known as the Rinns of Galloway, and the port of Cairnryan, which provides passenger and car ferry services to Northern Ireland. Main rivers are the Esk, Annan, Nith,

Dee and Cree which descend S to the Solway Firth from the Tweedsmuir Hills, Lowther Hills and the Rhinns of Kells in the N.

Dundee *Dundee* Population: 148,260
Unitary authority on the E coast of Scotland surrounding the city of Dundee and neighbouring Angus and Perth & Kinross. Dundee is Scotland's fourth largest city and is a centre of excellence in a variety of areas from telecommunications to medical research. The Firth of Tay borders Dundee to the S.

Durham *Dur.* Population: 517,773
Unitary authority in NE England bounded by Cumbria, Darlington, Hartlepool, Northumberland, North Yorkshire, Stockton-on-Tees, Tyne & Wear and the North Sea. Chief centres are the cathedral city of Durham; and the towns of Chester-le-Street, Peterlee, Newton Aycliffe, Bishop Auckland, Seaham and Consett. The W part includes The Peninnes and consists mostly of open moorlands which provide rough sheep-grazing and water for the urban areas from a number of large reservoirs. Economic activity is concentrated on the lowland in the E which is more heavily populated, and was formerly a centre for coal-mining and heavy industry. Diversification has since provided a broad industrial base. The principal rivers are the Tees and the Wear.

East Ayrshire *E.Ayr.* Population: 122,150
Unitary authority in SW Scotland bounded by Dumfries & Galloway, East Renfrewshire, North Ayrshire, South Ayrshire and South Lanarkshire. The principal towns are Kilmarnock, Cumnock, Stewarton, Galston and Auchinleck. Traditional industries centred on textiles and lace in the Irvine valley, coal mining and engineering. Dairy farming is also an important industry, particularly beef and sheep production. The area is a popular tourist destination, with several castles, battle sites and associations with Robert Burns and Keir Hardie. Rivers include the Irvine, Annick and Cessnock.

East Dunbartonshire *E.Dun.* Population: 106,730
Unitary authority in central Scotland bounded by Glasgow, North Lanarkshire, Stirling and West Dunbartonshire. The chief centres are Bearsden, Bishopbriggs, Kirkintilloch and Milngavie. Much of the urban and industrial development occurs on the N periphery of Greater Glasgow. The Campsie Fells lie in the N of the area.

East Lothian *E.Loth.* Population: 102,050
Unitary authority in central Scotland neighbouring Edinburgh, Midlothian, Scottish Borders and the North Sea. The main towns are Musselburgh, Haddington, Tranent, Prestonpans, Dunbar, North Berwick and Cockenzie and Port Seton. There are 43m of varied coastline and the topography includes the Lammermuir Hills in the S, and the ancient volcanoes at North Berwick and Traprain. Much of the urban and industrial development is in the NW and N of the area. Rivers include Whitehead Water, the Tyne, Peffer Burn and Gifford Water.

East Renfrewshire *E.Renf.* Population: 92,380
Unitary authority in SW Scotland bounded by East Ayrshire, Glasgow, Inverclyde, North Ayrshire, Renfrewshire and South Lanarkshire. The principal centres are Newton Mearns, Clarkston, Barrhead and Giffnock, which lie on the S periphery of Greater Glasgow. Over two-thirds of East Renfrewshire is farmland; the rest being mostly residential, with some light industry.

East Riding of Yorkshire *E.Riding* Population: 337,115
Unitary authority on the E coast of England neighbouring Kingston upon Hull, North Lincolnshire, North Yorkshire, South Yorkshire and York. The chief centres are Bridlington, Beverley, Goole, Great Driffield, Hornsea, Brough, Hedon and Withernsea. The area is mostly low-lying, except for the central ridge which forms part of The Wolds. The coastline is subject to much erosion, with material being moved from Flamborough Head to the large spit of Spurn Head, at the mouth of the River Humber. Key industries in the area include agriculture, aerospace, gas and oil industries.

East Sussex *E.Suss.* Population: 539,766
County of SE England bounded by Brighton & Hove, Kent, Surrey, West Sussex and the English Channel. Main towns are Eastbourne, Hastings, Bexhill, Seaford, Crowborough, Hailsham, Peacehaven and the county town of Lewes; Rye is a small historic town in the E of the county. In the W, the coast is backed by the chalk ridge of the South Downs, ending with the white cliffs of the Seven Sisters and Beachy Head, just W of Eastbourne. E of this point, there are extensive areas of reclaimed marshland, which provide good sheep-grazing. Inland is the heavily wooded Weald, a former centre of the iron industry, interspersed with hill ridges, the largest being the open heathland of Ashdown Forest. Rivers, none large, include the Cuckmere, Ouse, Rother, and upper reaches of the Medway.
Districts: Eastbourne; Hastings; Lewes; Rother; Wealden.

Edinburgh *Edin.* Population: 492,680
Unitary authority on the E coast of central Scotland surrounding the city of Edinburgh and neighbouring East Lothian, Midlothian, West Lothian and the sea at the Firth of Forth. Edinburgh as the capital of Scotland, is a major administrative, cultural, commercial and tourist centre. It contains most of Scotland's national and cultural institutions. Its historic core is centred around Edinburgh Castle and the Royal Mile, attracting much tourism. The city is also a centre for education and scientific research; other important industries are electronics and food and drink production. The river Water of Leith runs through the city to the docks at Leith.

Essex *Essex* Population: 1,431,953
County of SE England bounded by Cambridgeshire, Greater London, Hertfordshire, Southend, Suffolk, Thurrock and the sea at the Thames estuary and North Sea. Chief towns are Basildon, the county town of Chelmsford, Colchester, Harlow, Brentwood, Clacton-on-Sea, Loughton, Canvey Island, Billericay and Braintree. The landscape is mostly flat or gently undulating, and the low-lying coast is deeply indented with river estuaries. Along the county's S and W sides, there is a concentration of urban development, with a mixture of light engineering and service industries. In the N and central parts are farmlands, orchards, market and nursery gardens. The NE coast has the busy passenger and container port of Harwich, and the popular seaside resort of Clacton-on-Sea. Rivers include the Stour, forming part of the boundary with Suffolk, the Lea, forming part of the boundary with Hertfordshire, and the Blackwater.
Districts: Basildon; Braintree; Brentwood; Castle Point; Chelmsford; Colchester; Epping Forest; Harlow; Maldon; Rochford; Tendring; Uttlesford.

Falkirk *Falk.* Population: 157,640
Unitary authority in central Scotland surrounding Falkirk and neighbouring Clackmannanshire, Fife, North Lanarkshire, Stirling and West Lothian. Main towns are Falkirk, Grangemouth, Polmont, Stenhousemuir and Bo'ness. Petrochemical and chemical industries are important to the local economy, as well as bus manufacturing, toffees and paper-making. The Firth of Forth borders Falkirk to the N. Other rivers include the Carron and Pow Burn.

Fermanagh & Omagh *F. & O.* Population: 114,992
District Council combining the two former councils, with the administrative centres at Enniskillen and Omagh. In the southwest of Northern Ireland, this rural area forms a long boundary with Ireland and includes the large bodies of water - Upper and Lower Lough Erne which have about 150 small islands within them. This makes the area popular for outdoor and water-based activities. It is also bounded by Mid Ulster and Derry City & Strabane councils. Enniskillen, with its ancient castle, lies between the two loughs on the river Erne. Omagh is surrounded by hills and lies in a basin at the confluence of several rivers. This sparsely populated area is rural with scattered villages. Wooded areas have been expanded with conifer plantations for timber production.

Fife *Fife* Population: 367,260
Unitary authority in E Scotland neighbouring Clackmannanshire and Perth & Kinross, and lying between the Firth of Tay and Firth of Forth. Main towns are Dunfermline, Kirkcaldy, Glenrothes, Buckhaven,

Cowdenbeath and St. Andrews. Fife comprises the former county of the same name, known since ancient times as the Kingdom of Fife, and is noted for its fine coastline with many distinctive small towns and fishing ports. The historic town of St. Andrews, on the coast between the two firths, is a university town, and the home of the world's premier golf club. Inland, the area is outstandingly fertile, with agriculture being an important industry. The SW of the area is a former coal-mining area.

Flintshire (Sir y Fflint). *Flints.* Population: 153,804
Unitary authority in N Wales neighbouring Conwy, Denbighshire, Wrexham, Cheshire West & Chester and the mouth of the River Dee. Main towns are Buckley, Connah's Quay, Flint, Hawarden, Shotton, Queensferry, Mold and Holywell. Known as the Gateway to N Wales, the landscape varies from the mountains which form the Clwydian Range, to small villages and woodlands.

Glasgow *Glas.* Population: 599,650
Unitary authority in SW Scotland surrounding Glasgow and bounded by East Dunbartonshire, East Renfrewshire, North Lanarkshire, Renfrewshire, South Lanarkshire and West Dunbartonshire. Glasgow is Scotland's largest city and its principal industrial and shopping centre. The city developed significantly due to heavy industry, notably shipbuilding, being centred on the Clyde. While such industry has declined, Glasgow has emerged as a major cultural centre of Europe, due to its impressive arts and cultural scene. The River Clyde runs through the city.

Gloucestershire *Glos.* Population: 611,332
County of W England bounded by Herefordshire, Oxfordshire, South Gloucestershire, Swindon, Warwickshire, Wiltshire, Worcestershire and the Welsh authority of Monmouthshire. Main centres are the cathedral city and county town of Gloucester and the towns of Cheltenham, Stroud, Cirencester and Dursley. The limestone mass of the Cotswold Hills dominates the centre of the county, and provides the characteristic pale golden stone of many of its buildings. The River Severn forms a wide valley to the W, ending in a long tidal estuary, beyond which are the hills of the Forest of Dean. Industry is centred on the fertile Severn Vale, with aerospace, light engineering, food production, and service industries in and around the towns; in rural areas market gardening and orchards dominate. The River Thames rises in the county, and forms part of its S boundary in the vicinity of Lechlade. Apart from the Severn and the Thames, there is the River Wye, which forms part of the boundary with Monmouthshire, and many smaller rivers, among them the Chelt, Coln, Evenlode, Leach, Leadon, and Windrush.
Districts: Cheltenham; Cotswold; Forest of Dean; Gloucester; Stroud; Tewkesbury.

Greater London *Gt.Lon.* Population: 8,538,689
Former metropolitan county of 32 boroughs and the City of London which together form the conurbation of London, the capital of the UK. Greater London is the largest financial, commercial, cultural, distribution and communications centre in the country, including all but primary industrial sectors. London developed from the City of London, a walled Roman settlement on the Thames, and Westminster, which was a Saxon religious settlement and later a Norman seat of government. The Great Fire of 1666 destroyed most of the medieval city, and was followed by a period of rebuilding and rapid, unplanned expansion. Industrialisation and improved public transport over the last two centuries have caused major suburban growth, and the absorption of most of the surrounding settlements and countryside. Tourism is a major industry, with most attractions situated in and around the historic core, and along the Thames bankside. Other notable tourist areas include Greenwich, Hampstead, Kew and Richmond. Industrial activity is widespread, with major concentrations in the E along the Thames. Leisure facilities include national and major sports stadiums, and many big parks and gardens. Airports at Heathrow and docklands. The main river is the Thames.
Districts: Barking & Dagenham; Barnet; Bexley; Brent; Bromley; Camden; City of London; City of Westminster; Croydon; Ealing; Enfield; Greenwich; Hackney; Hammersmith & Fulham; Haringey; Harrow; Havering; Hillingdon; Hounslow; Islington; Kensington & Chelsea; Kingston upon Thames; Lambeth; Lewisham; Merton; Newham; Redbridge; Richmond upon Thames; Southwark; Sutton; Tower Hamlets; Waltham Forest; Wandsworth.

Greater Manchester *Gt.Man.* Population: 2,732,854
Former metropolitan county of NW England neighbouring Blackburn with Darwen, Cheshire East, Derbyshire, Lancashire, Merseyside, Warrington and West Yorkshire. It comprises the near-continuous urban complex which includes the adjoining cities of Manchester and Salford; and towns including Bolton, Stockport, Oldham, Rochdale, Wigan, Bury and Sale. The conurbation is framed by the wild moorland of The Pennines to the N and the Peak District and Cheshire Plain to the S. Development occurred during the 18c and 19c, creating a series of cotton producing textile towns, while Manchester established itself as the commercial and trading hub, later becoming an inland port linked to Liverpool via the canal network. As textile production declined, the industrial base of the area broadened to include brewing, food production, electronics, plastics, printing, light engineering, financial, leisure and service sectors. Retail is based on town shopping centres and malls such as the Arndale and Trafford Centres. There are many major sporting venues in the area, and cultural facilities include the Manchester Central Convention Complex, numerous universities, museums and galleries and a diverse nightlife. The area is served by Manchester Airport. Main rivers are Irwell and Mersey.
Districts: Bolton; Bury; Manchester; Oldham; Rochdale; Salford; Stockport; Tameside; Trafford; Wigan.

Gwynedd *Gwyn.* Population: 122,273
Unitary authority in NW Wales bounded by Ceredigion, Conwy, Denbighshire, Isle of Anglesey, Powys and the sea. Main centres are the cathedral city of Bangor, Caernarfon, Ffestiniog, Blaenau Ffestiniog, Llanddeiniolen, Pwllheli, Llanllynfi, Bethesda and Porthmadog. The whole mainland area, except the Lleyn Peninsula in the NW, is extremely mountainous and contains the scenically famous Snowdonia National Park. There is slate-quarrying in the Ffestiniog valley, otherwise sheep-farming and tourism are the principal occupations; the coastline has been much developed for the holiday trade. The area contains many lakes and reservoirs, among them are Llyn Trawsfynedd, Llyn Celyn and Llyn Tegid. Of the many rivers, the Wnion and the Dyfi, which flows through part of the area, are most significant.

Halton *Halton* Population: 126,354
Unitary authority in NW England neighbouring Cheshire West & Chester, Merseyside and Warrington. The principal towns are Runcorn and Widnes, separated by the River Mersey. The area is industrialised, being dominated by petro-chemicals and chemicals industries due to the nearby salt mines and port facilities.

Hampshire *Hants.* Population: 1,346,136
County of S England bounded by Bracknell Forest, Dorset, Portsmouth, Southampton, Surrey, West Berkshire, West Sussex, Wiltshire, Wokingham and the English Channel. Main towns are Basingstoke, Gosport, Waterlooville, Farnborough, Aldershot, Eastleigh, Havant, the ancient city and county town of Winchester, Andover and Fleet. The centre of the county consists largely of chalk downs interspersed with fertile valleys. In the SW is the New Forest, while in the NE is the military area centred on Aldershot. The much indented coastline borders The Solent and looks across to the Isle of Wight. Main industries are in the service sector, with chemicals and pharmaceuticals also important. The chief rivers are the Itchen and Test, both chalk streams flowing into Southampton Water, and the Meon flowing into The Solent.
Districts: Basingstoke & Deane; East Hampshire; Eastleigh; Fareham; Gosport; Hart; Havant; New Forest; Rushmoor; Test Valley; Winchester.

Hartlepool *Hart.* Population: 92,590
Unitary authority on the NE coast of England surrounding Hartlepool and bordering Darlington, Durham, Stockton-on-Tees and the North Sea. Fishing is a major industry and a marina has been created from part of the old docks. The mouth of the River Tees forms part of the E border.

Herefordshire *Here.* Population: 187,160
Unitary authority in W England bounded by Gloucestershire, Shropshire, Worcestershire and the Welsh authorities of Monmouthshire and Powys. Main centres are the cathedral city of Hereford and the towns of Ross-on-Wye, Leominster and Ledbury. Herefordshire lies between the Malvern Hills to the E and the Black Mountains to the W. It is mainly rural, with dairy farming, orchards and market gardening in evidence. The main river is the Wye, which provides excellent fishing.

Hertfordshire *Herts.* Population: 1,154,766
S midland county of England bounded by Central Bedfordshire, Buckinghamshire, Cambridgeshire, Essex, Greater London and Luton. Chief centres are Watford, the cathedral city of St. Albans, Hemel Hempstead, Stevenage, Cheshunt, Welwyn Garden City, Hoddesdon, Hitchin, Letchworth and Hatfield; the county town is Hertford. The Chilterns rise along the W border, and there are chalk hills in the N around Royston; otherwise the landscape is mostly flat or gently undulating. There is a mixture of rural and urban life, with agricultural and hi-tech industries represented. While the urban centres in the S lie on the N periphery of the Greater London conurbation, there are many villages with the traditional large green or common. The more urban S part of the county includes a dense network of major roads bypassing, and leading N from London. Rivers include the Colne, Ivel, and Lee.
Districts: Broxbourne; Dacorum; East Herts; Hertsmere; North Hertfordshire; St. Albans; Stevenage; Three Rivers; Watford; Welwyn Hatfield.

Highland *High.* Population: 233,100
Unitary authority covering a large part of N Scotland and neighbouring Aberdeenshire, Argyll & Bute, Moray and Perth & Kinross. It contains a mixture of mainland and island life, comprising the former districts of Badenoch and Strathspey, Caithness, Inverness, Lochaber, Nairn, Ross and Cromarty, Skye and Lochalsh and Sutherland. Main towns are Inverness, Fort William, Thurso, Nairn, Wick, Alness and Dingwall. Overall, Highland is very sparsely inhabited, being wild and remote in character. It is scenically outstanding, containing as it does part of the Cairngorm Mountains, Ben Nevis, and the North West Highlands. Many of the finest sea and inland lochs in Scotland are also here, such as Loch Ness, Loch Linnhe, Loch Torridon and Loch Broom. The discovery of North Sea oil has made an impact on the towns and villages around the Moray Firth. Elsewhere, tourism, crofting, fishing and skiing are important locally.

Inverclyde *Inclyde* Population: 79,860
Unitary authority on the W coast of central Scotland, on the S bank of the River Clyde. It is bordered by North Ayrshire, Renfrewshire and the Firth of Clyde. The chief towns are Greenock, Port Glasgow, Gourock and Kilmacolm.

Isle of Anglesey (Sir Ynys Môn). *I.o.A.* Population: 70,169
Unitary authority island of NW Wales divided from Gwynedd and the mainland by the Menai Strait, and with Holy Island lying to the W. Main towns are Holyhead, Llangefni, Amlwch and Menai Bridge. Anglesey has 125m of coastline and 16 beaches. Agriculture is an important industry to the island, with other industries including aluminium smelting and food processing. Holyhead is an important port terminus for Dublin, Ireland. Rivers include the Braint and Cefni.

Isle of Man *I.o.M.* Population: 87,127
Self-governing island in the Irish Sea, situated in the centre of the British Isles. The chief towns are Douglas, Ramsey, Peel, Castletown, Port St. Mary, Port Erin and Laxey. Apart from the N tip, the topography is generally mountainous, rising to a peak at Snaefell. The main industries are agriculture, fishing and tourism as well as financial services and manufacturing. The island is synonymous with motorsport, being the home of the internationally renowned Tourist Trophy (TT) Circuit. Rivers include the Glen Auldyn and Neb.

Isle of Wight *I.o.W.* Population: 139,105
County and island with an area of 147 square miles or 381 square km, separated from the S coast of England by The Solent. Chief towns are the capital, Newport, Ryde, Cowes, Shanklin, Sandown, Ventnor and Yarmouth. The island is geologically diverse, composed of sedimentary rocks and contains many important fossil remains. Tourism flourishes owing to the mild climate and the natural beauty of the island. There are Royal associations as Queen Victoria lived and died at Osborne House in the N of the island. There is a strong naval tradition, with the island historically acting as a defence for Portsmouth. Cowes is internationally famous for yachting. There are ferry and hovercraft connections at Cowes, Ryde, Fishbourne and Yarmouth (ferry to Lymington). Chief river is the Medina.

Isles of Scilly *I.o.S.* Population: 2,280
Group of some 140 islands 48m/45km SW of Land's End, Cornwall, of which five are inhabited: Bryher, St. Agnes, St. Martin's, St. Mary's and Tresco. Chief industries are fishing, and the growing of early flowers and vegetables due to the exceptionally mild climate.

Kent *Kent* Population: 1,510,354
South-easternmost county of England bounded by East Sussex, Greater London, Medway, Surrey and the sea at the Thames estuary and the Strait of Dover. Chief centres are the county town of Maidstone, Royal Tunbridge Wells, Dartford, Margate, Ashford, Gravesend, Folkestone, Sittingbourne, Ramsgate, the cathedral city of Canterbury, Tonbridge and Dover. The chalk ridge of the North Downs runs along the N side, then SE to Folkestone and Dover. The River Medway cuts through the chalk in the vicinity of Maidstone, and there are low lying areas to the E of Canterbury and of Tonbridge, on Romney Marsh in the S, and bordering the Thames estuary in the N. Chief industrial areas are around Maidstone, Ashford and Tonbridge; Dover and Folkestone are major ports, with the Channel Tunnel terminus to the N of Folkestone; Sheerness is a port of growing importance. Industrial activity includes mineral extraction, cement manufacture and papermaking. On the highly productive agricultural land, Kent's reputation as the Garden of England is earned, with market gardening, fruit and hop production. Romney Marsh is used for extensive sheep-grazing. Rivers include the Medway, Stour, and Beult.
Districts: Ashford; Canterbury; Dartford; Dover; Gravesham; Maidstone; Sevenoaks; Shepway; Swale; Thanet; Tonbridge & Malling; Tunbridge Wells.

Kingston upon Hull *Hull* Population: 257,710
Unitary authority on the E coast of England surrounding the city of Kingston upon Hull and bounded by East Riding of Yorkshire and the mouth of the River Humber. Kingston upon Hull is a major sea port and a great industrial city, with key industries including chemicals, food processing, pharmaceuticals and engineering. The River Hull passes through the area, and the River Humber forms the S border.

Lancashire *Lancs.* Population: 1,184,735
County of NW England bounded by Blackburn with Darwen, Cumbria, Greater Manchester, Merseyside, North Yorkshire, West Yorkshire and the Irish Sea. Chief centres are the administrative city of Preston, Burnley, Morecambe, the historic county town of Lancaster, Skelmersdale, Lytham St. Anne's, Leyland, Accrington and Chorley; Fleetwood and Heysham are ports. The inland side of the county is hilly and includes the wild and impressive Forest of Bowland. The W side contains the coastal plain, where vegetables are extensively cultivated. The S is largely urban; industries include cotton spinning and weaving, chemicals, glass, rubber, electrical goods, and motor vehicles. The principal rivers are the Lune and the Ribble.
Districts: Burnley; Chorley; Fylde; Hyndburn; Lancaster; Pendle; Preston; Ribble Valley; Rossendale; South Ribble; West Lancashire; Wyre.

Leicester *Leic.* Population: 337,653
Unitary authority in central England surrounding Leicester and bounded by Leicestershire. It is one of the leading shopping regions in the Midlands. Traditional industries such as hosiery and footwear, as well as hi-tech industries, are important to the local economy. Leicester is aiming to be one of the most environmentally-friendly cities in Europe. It is involved in pioneering electronic toll road schemes in order to encourage the use of public transport. The Rivers Sence and Soar run through the area.

Leicestershire *Leics.* Population: 667,905
Midland county of England bounded by Derbyshire, Leicester, Lincolnshire, Northamptonshire, Nottinghamshire, Rutland, Staffordshire and Warwickshire. Chief towns are Loughborough, Hinckley, Wigston, Coalville, Melton Mowbray, Oadby, Market Harborough, Shepshed and Ashby de la Zouch. The landscape is mostly of low, rolling hills. E and W of Leicester are areas of higher ground, notably Charnwood Forest. The W is largely industrial; industries include light engineering, hosiery, and footwear. The E is rural, with large fields and scattered woods, and is noted for field sports and food production. Part of the legacy left by the Roman occupation of Leicestershire are the Great North Road, Watling Street and Fosse Way which dissect the county. River Soar traverses the county from S to N, while River Welland forms part of the boundary with Northamptonshire to the S.
Districts: Blaby; Charnwood; Harborough; Hinckley & Bosworth; Melton; North West Leicestershire; Oadby & Wigston.

Lincolnshire *Lincs.* Population: 731,516
County of E England bounded by Cambridgeshire, Leicestershire, Norfolk, Northamptonshire, North East Lincolnshire, North Lincolnshire, Nottinghamshire, Peterborough, Rutland and the North Sea. Main towns are the cathedral city and county town of Lincoln and the towns of Boston, Grantham, Gainsborough, Spalding, Stamford, Skegness and Louth. Much of the county is flat and includes a large area of The Fens in the S. This reclaimed marshland is richly fertile, producing large crops of peas (for canning), sugar beet, potatoes, corn, and around Spalding, flower bulbs. Two ranges of hills traverse the county N and S: the narrow limestone ridge, a continuation of the Cotswold Hills, running from Grantham to Scunthorpe, and the chalk Wolds, about 12m/20km wide, running N from Spilsby and Horncastle. Apart from agriculture, industries include manufacture of agricultural machinery and tourism, which is centred on historic Lincoln, and the coastal resorts of Skegness and Mablethorpe. The rivers, of which the chief are the Witham and Welland, are largely incorporated into the extensive land-drainage system, and scarcely distinguishable from man-made channels.
Districts: Boston; East Lindsey; Lincoln; North Kesteven; South Holland; South Kesteven; West Lindsey.

Lisburn & Castlereagh *L. & C.* Population: 138,627
City Council lying to the S of Belfast. Lisburn is the administrative centre and sits in the valley of the river Lagan which flows through to the capital. Historically it was the centre of the linen industry. Agriculture and textiles are the main industries but most people will commute to the capital for work. Close to the village of Drumbo is the Giant's Ring, a massive 656 feet (200m) diameter earthwork. The area is surrounded by Antrim & Newtownabbey, Armagh City, Banbridge & Craigavon, Newry, Mourne & Down, Ards & North Down and Belfast City councils. As well as Castlereagh, other towns include Carryduff, Dundonald and Dunmurry.

Luton *Luton* Population: 210,962
Unitary authority in SE England surrounding Luton and bounded by Central Bedfordshire and Hertfordshire. Luton is one of the major centres of employment and manufacturing in SE England, with automotive, electrical and retail industries among the most important. The production and export of high fashion and straw hats remains a feature of the local economy. London Luton Airport is situated in the SE of the area, and the River Lea rises nearby.

Medway *Med.* Population: 274,015
Unitary authority on SE coast of England S of the River Thames estuary and neighbouring Kent. The chief centres are Gillingham, the historic naval base of Chatham, Strood and the cathedral city of Rochester. The S part of the area, surrounding the River Medway, is largely urban and industrialised. The marshland to the N includes a power station on the Isle of Grain, but is mostly rural, and contains Northward Hill Nature Reserve which is a haven for birds.

Merseyside *Mersey.* Population: 1,391,113
Former metropolitan county of NW England. It neighbours Cheshire West & Chester, Greater Manchester, Halton, Lancashire, Warrington and the sea. It comprises the near-continuous urban complex which includes the city of Liverpool and the towns of St. Helens, Birkenhead, Southport, Bootle, Wallasey, Bebington, Huyton and Crosby. The county straddles the long, wide estuary of the River Mersey, which accounts for the development of the area. During the 18c, growing Imperial trade of goods and slaves, led to the explosion of urban development surrounding the docks at Liverpool, Birkenhead and Bootle. Liverpool went on to become Britain's premier transatlantic port and a significant terminus during the migration flows of the 19c, leading to an ethnically diverse city culture. Over the last century the docks have declined, leaving behind an impressive waterfront and cityscape as testament to a mercantile and maritime heritage. Inland, the urban spread has reached the industrial town of St. Helens which is famed for glass production. To the N are the residential areas of Crosby, Formby and the coastal resort of Southport. The area includes race courses at Aintree and Haydock, and Liverpool John Lennon Airport at Speke.
Districts: Knowsley; Liverpool; St. Helens; Sefton; Wirral.

Merthyr Tydfil *M.Tyd.* Population: 59,065
Unitary authority in S Wales bounded by Caerphilly, Powys and Rhondda Cynon Taff. Main centres are the town of Merthyr Tydfil and the villages of Treharris, Abercanaid and Troedyrhiw. The area stretches from the Brecon Beacons, along the Taff Valley, to the centre of the former Welsh coal mining district. The local economy has diversified from primary industry, with Merthyr Tydfil being an important centre for public administration, shopping and employment for the region. The River Taff flows through the area.

Mid & East Antrim *M. & E. Ant.* Population: 136,642
Borough Council formed by the joining of Ballymena, Larne and Carrickfergus councils with the town of Ballymena now the administrative centre. The western boundary is the river Bann, famous for its coarse fishing and the area stretches across the Antrim Hills to the E Antrim coast. The coastline displays an amazing range of geological features from the most ancient rocks to remains of the last ice age. Larne is a busy port for freight and ferry passengers. The surrounding councils are Causeway Coast & Glens, Antrim & Newtownabbey and Mid Ulster. Other towns include Broughshane, Carnlough, Cullybuckey and Ballygalley. Main industries are based on freight, manufacturing and agriculture.

Middlesbrough *Middbro.* Population: 139,119
Unitary authority in NE England surrounding Middlesbrough and bounded by North Yorkshire, Redcar & Cleveland and Stockton-on-Tees. Middlesbrough is an industrial town, with chemical and petro-chemical industries in evidence. It is also an important sub-regional shopping and entertainment centre between Leeds and Newcastle upon Tyne.

Midlothian *Midloth.* Population: 86,210
Unitary authority in central Scotland neighbouring East Lothian, Edinburgh and Scottish Borders. Main towns are Penicuik, Bonnyrigg, Dalkeith, Gorebridge and Loanhead. The area is mostly rural, including the rolling moorland of the Pentland Hills and Moorfoot Hills in the S. To the N, the urban area is comprised of satellite towns to the SE of Edinburgh. Rivers include Tyne Water and South Esk.

Mid Ulster *M. Ulster* Population: 142,895
District Council made from the joining the former councils of Cookstown, Dungannon and Magherafelt. Surrounded by six other councils Mid Ulster also has a border with Ireland. Lough Neagh and the river Bann forms part of the eastern boundary with the Sperrin Mountains (an AONB) in the N. Cookstown boasts the longest main street in Northern Ireland with over one mile of shops. Wellbrook Mill nearby has the last working linen mill which used to be part of a major industry in the area. This large council area is mainly rural with the emphasis on agriculture with small businesses and manufacturing.

Milton Keynes *M.K.* Population: 259,245
S midland unitary authority of England bounded by Bedford, Central Bedfordshire, Buckinghamshire and Northamptonshire. The area includes the city of Milton

Keynes, Bletchley, Newport Pagnell, Great Linford, Stony Stratford and Wolverton. Over the past 30 years, the area has undergone the fastest rate of growth in the country, attracting numerous industries. The Great Ouse and Ouzel rivers pass through the area.

Monmouthshire (Sir Fynwy). *Mon.* Population: 92,336
Unitary authority in SE Wales bounded by Blaenau Gwent, Newport, Powys, Torfaen, the English areas of Gloucestershire, Herefordshire and the Bristol Channel. The main towns are Abergavenny, Caldicot, Chepstow and Monmouth. Part of the Brecon Beacons are found in NW Monmouthshire, whereas the SW area is mainly flat. Agriculture, mineral extraction and the service sector are important to the local economy. Rivers include the Wye, which forms part of E border, and the Usk, Trothy and Monnow.

Moray *Moray* Population: 94,750
Unitary authority in N Scotland neighboured by Aberdeenshire, Highland and the sea. Main towns are Elgin, Forres, Buckie, Lossiemouth and Keith. The area is mainly mountainous, including part of the Cairngorm Mountains in the S. It is dissected by many deep river valleys, most notably that of the River Spey. Along with the local grain and peat, the abundant waters provide the raw materials for half of Scotland's malt whisky distilleries, leading to the Whisky Trail and much tourism through Speyside.

Neath Port Talbot (Castell-nedd Port Talbot). *N.P.T.* Population: 140,490
Unitary authority in S Wales neighbouring Bridgend, Powys, Rhondda Cynon Taff, Swansea and the sea. The chief centres are Neath, Port Talbot, Pontardawe, Baglan, Glyncorrwg and Briton Ferry. The area is mostly mountainous, divided up by the river valleys of the Tawe, Neath, Afan and Dulais, which all flow out to sea at Swansea Bay. The lower valley of the River Neath is heavily industrialised.

Newport (Casnewydd). *Newport* Population: 146,841
Unitary authority on the S coast of Wales, N of the mouth of the River Severn, and bounded by Caerphilly, Cardiff, Monmouthshire and Torfaen. Main centres are Newport, Liswerry, Malpas and Caerleon. Steel manufacturing and hi-tech industries are important to the local economy. The rivers Ebbw and Usk run through the area.

Newry, Mourne & Down *N., M. & D.* Population: 175,403
District Council created from joining two former councils. Newry is now the administrative centre for the area which occupies the southeastern corner of Northern Ireland. As well as boundary with Ireland, it borders Ards & North Down, Lisburn & Castlereagh, and Armagh City, Banbridge & Craigavon. This scenic area includes the Mourne Mountains which have two designated AONB's and the long coastline has a notable sandy beach at Dundrum Bay. The area also includes the western shores of Strangford Lough. Newry has long developed cross-border trade and other important towns include the port of Warrenpoint, the fishing port of Killkeel, Downpatrick, Rostrevor and Ballynahinch. As well as fishing, other industries include engineering and agriculture.

Norfolk *Norf.* Population: 877,710
County of E England bounded by Cambridgeshire, Lincolnshire, Suffolk and the North Sea. Chief centres are the cathedral city and county town of Norwich, Great Yarmouth on the E coast, the expanding port of King's Lynn near the mouth of the Great Ouse and The Wash, Thetford, which is known as the Breckland 'capital', Dereham and Wymondham. Norfolk is mainly flat or gently undulating, with fenland in the W characterised by large drainage channels emptying into The Wash. In the SW is Breckland, an expanse of heath and conifer forest used for military training; other afforested areas are near King's Lynn and North Walsham. NE of Norwich are The Broads, an area of meres and rivers popular for boating; reeds for thatching are grown here. The N Norfolk coastline is an Area of Outstanding Natural Beauty and Heritage Coast, and includes the popular resorts of Cromer and Sheringham. Otherwise the county is almost entirely agricultural, with farming an important activity; service and manufacturing industries are also significant. Rivers include the Great Ouse, Bure, Nar, Wensum, Wissey, and Yare; the Little Ouse and Waveney both enter the county briefly, but mainly form the boundary with Suffolk.
Districts: Breckland; Broadland; Great Yarmouth; King's Lynn & West Norfolk; North Norfolk; Norwich; South Norfolk.

North Ayrshire *N.Ayr.* Population: 136,450
Unitary authority in central Scotland including the islands of Arran, Great Cumbrae and Little Cumbrae. It is bounded by East Ayrshire, East Renfrewshire, Inverclyde, Renfrewshire, South Ayrshire and the sea. The principal towns are Irvine, Kilwinning, Saltcoats, Largs, Ardrossan, Stevenston and Kirbirnie. The area includes mountains and part of Clyde Muirshiel Regional Park in the N, and the lower lands of Cunninghame in the S. There is a maritime heritage to the area; ferry routes operate from Largs and Ardrossan. Rivers include the Garnock, Dusk Water and Noddsdale Water.

North East Lincolnshire *N.E.Lincs.* Population: 159,804
Unitary authority in NE England, S of the mouth of the River Humber and bounded by Lincolnshire, North Lincolnshire and the North Sea. Chief towns are Grimsby, Cleethorpes and Immingham. Grimsby and Cleethorpes together are the shopping and commercial centres of the area. Fishing, food, tourism, chemical and port industries are all important to the local economy. The main rivers are the Humber and Freshney.

North Lanarkshire *N.Lan.* Population: 337,950
Unitary authority in central Scotland neighbouring East Dunbartonshire, Falkirk, Glasgow, South Lanarkshire, Stirling and West Lothian. The chief centres are Cumbernauld, Coatbridge, Airdrie, Motherwell, Wishaw and Bellshill. North Lanarkshire contains a mixture of urban and rural areas, and formerly depended heavily upon the coal, engineering and steel industries. Regeneration and diversification have occurred in recent years.

North Lincolnshire *N.Lincs.* Population: 169,247
Unitary authority in NE England neighbouring East Riding of Yorkshire, Leicestershire, Norfolk, North East Lincolnshire, Nottinghamshire, Peterborough, Rutland, South Yorkshire and the River Humber. The main centres are Scunthorpe, Bottesford, Barton-upon-Humber and Brigg. The area is mainly rural, but does include oil refineries, steel and manufacturing industries; the River Humber provides pool and wharf facilities. Rivers include the Humber, Trent and the Old Ancholme.

North Somerset *N.Som.* Population: 208,154
Unitary authority in W England, S of the mouth of the River Severn, and neighbouring Bath & North East Somerset, Bristol, Somerset and the Bristol Channel. Chief towns are Weston-super-Mare, Clevedon, Nailsea and Portishead. The area is largely rural with tourism, centred on the coastal resort of Weston-super-Mare, being a major industry. Bristol International Airport is located in the E of the area.

North Yorkshire *N.Yorks.* Population: 601,536
Large county of N England bounded by Cumbria, Darlington, Durham, East Riding of Yorkshire, Lancashire, Middlesbrough, Redcar & Cleveland, South Yorkshire, Stockton-on-Tees, West Yorkshire, York and the North Sea. Main centres are Harrogate, Scarborough, Hetton, Selby, the cathedral city of Ripon, the county town of Northallerton, Whitby, Skipton and Knaresborough. Apart from the wide plain around York, through which flow River Ouse and its tributaries, and the smaller Vale of Pickering, watered by the Derwent and its tributary the Rye, the county is dominated by two ranges of hills; The Pennines in the W and the Cleveland Hills in the NE. The plains are pastoral and agricultural, while the hills provide rough sheep-grazing. The county includes the popular resorts of Scarborough and Whitby, and the majority of the North York Moors and Yorkshire Dales National Parks which promote tourism. Other economic activities include light engineering, service and hi-tech industries. Principal rivers are the Ouse, fed by the Derwent, Swale, Ure, Nidd and Wharfe, and draining into the Humber; the Esk, flowing into the North Sea at Whitby; and in the W, the Ribble, passing out into Lancashire and the Irish Sea.
Districts: Craven; Hambleton; Harrogate; Richmondshire; Ryedale; Scarborough; Selby.

Northamptonshire *Northants.* Population: 714,392
Midland county of England bounded by Bedford, Buckinghamshire, Cambridgeshire, Leicestershire, Lincolnshire, Milton Keynes, Oxfordshire, Peterborough, Rutland and Warwickshire. Chief towns are Northampton, Corby, Kettering, Wellingborough, Rushden and Daventry. The county consists largely of undulating agricultural country rising locally to low hills, especially along the W border. Large fields and scattered woods provide terrain for field sports. Northamptonshire still retains its rural and agricultural charm, despite undergoing rapid population growth recently. There are many villages of architectural, scenic and historic interest. Industrial development is modest, concentrating on the traditional footwear manufacture. Corby is undergoing regeneration following the decline of its steel industry. Tourism is set to increase due to the county's natural Middle England ambience, and the seasonal opening of the Althorp Estate, the family home and resting place of Diana, Princess of Wales. The principal rivers are the Nene and Welland.
Districts: Corby; Daventry; East Northamptonshire; Kettering; Northampton; South Northamptonshire; Wellingborough.

Northumberland *Northumb.* Population: 315,987
Northernmost unitary authority of England bounded by Cumbria, Durham and Tyne & Wear, the Scottish authority of Scottish Borders and the North Sea. The principal towns are Blyth, Ashington, Cramlington, Bedlington, Morpeth, Berwick-upon-Tweed, Prudhoe and Hexham. There is some industry in the SE coastal area, otherwise it is almost entirely rural, the greater part being high moorland, culminating in the Cheviot Hills along the Scottish border. The most spectacular stretches of Hadrian's Wall traverse the area to the N of Haltwhistle and Hexham. There is extensive afforestation, including Kielder Forest Park and part of the Northumberland National Park in the NW; parts of these forests are used for military training. The large reservoir, Kielder Water, also occurs in the NW of the area. Rivers include the Aln, Blyth, Breamish, Coquet, East and West Allen, North and South Tyne, Till, and Wansbeck. The Tweed forms part of the Scottish border and flows out to sea at Berwick-upon-Tweed.

Nottingham *Nott.* Population: 314,268
Unitary authority in central England surrounding the city of Nottingham and bounded by Nottinghamshire. The city of Nottingham has a long history, having been granted many Royal Charters; Nottingham Castle and Wollaton Hall are among its many historical buildings. It is also an industrial and engineering centre, and a university city. Its main industries include the manufacture of chemicals, tobacco, cycles, lace and hosiery. The River Trent flows through the city.

Nottinghamshire *Notts.* Population: 801,390
Midland county of England bounded by Derbyshire, Leicestershire, Lincolnshire, North Lincolnshire, Nottingham and South Yorkshire. Principal towns are Mansfield, Carlton, Sutton in Ashfield, Arnold, Worksop, Newark-on-Trent, West Bridgford, Beeston, Stapleford, Hucknall and Kirkby in Ashfield. Much of the county is rural, with extensive woodlands in the central area of The Dukeries, part of the larger Sherwood Forest. Cattle-grazing is the chief farming activity. Around the large towns there is much industry, including iron and steel, engineering, knitwear, pharmaceuticals, and coal-mining. The county has associations with Robin Hood, at Sherwood Forest, and D.H. Lawrence, at Eastwood. The most important river is the Trent.
Districts: Ashfield; Bassetlaw; Broxtowe; Gedling; Mansfield; Newark & Sherwood; Rushcliffe.

Orkney *Ork.* Population: 21,590
Group of some fifteen main islands and numerous smaller islands, islets and rocks. Designated an Islands Area for administrative purposes, and lying N of the NE end of the Scottish mainland across the Pentland Firth. Kirkwall is the capital, situated on the island Mainland, 24m/38km N of Duncansby Head. Stromness is the only other town. About twenty of the islands are inhabited. In general the islands are low-lying but have steep, high cliffs on W side. The climate is generally mild for the latitude but storms are frequent.

Fishing and farming (mainly cattle-rearing) are the chief industries. The oil industry is also represented, with an oil terminal on the island of Flotta, and oil service bases at Car Ness and Stromness, Mainland and at Lyness, Hoy. Lesser industries include whisky distilling, knitwear and tourism. The islands are noted for their unique prehistoric and archaeological remains. The main airport is at Grimsetter, near Kirkwall, with most of the populated islands being served by airstrips. Ferries also operate from the Scottish mainland, and between islands in the group.

Oxfordshire *Oxon.* Population: 672,516
S midland county of England bounded by Buckinghamshire, Gloucestershire, Northamptonshire, Reading, Swindon, Warwickshire, West Berkshire, Wiltshire and Wokingham. Chief centres are the county town, cathedral and university city of Oxford and towns of Banbury, Abingdon, Bicester, Witney, Didcot, Thame and Henley-on-Thames. Burford and Chipping Norton are small Cotswold towns in the W and NW respectively. The landscape is predominantly flat or gently undulating, forming part of the Thames Valley. High ground occurs where the Chiltern Hills enter the county in the SE and the Cotswold Hills in the NW. The county is largely agricultural, with industries centred on the towns. Scientific, medical and research establishments are attracted by the proximity of Oxford's universities. Printing and publishing industries have their greatest concentration outside London. The motor industry is well represented with car manufacture at Cowley, Oxford, and the county has the world's largest concentration of performance car development and manufacturing. Tourism, attracted to stately homes, notably Blenheim Palace, and Oxford city centre, is also important. Chief rivers are the Thames (or Isis), Cherwell, Ock, Thame, and Windrush.
Districts: Cherwell; Oxford; South Oxfordshire; Vale of White Horse; West Oxfordshire.

Pembrokeshire (Sir Benfro). *Pembs.* Population: 123,666
Unitary authority in the SW corner of Wales neighbouring Carmarthenshire, Ceredigion and the sea. The chief centres are Haverfordwest, Pembroke Dock, Pembroke, Tenby, Saundersfoot, Neyland, Fishguard and the ancient cathedral city of St. David's. Key industries are tourism, agriculture and oil refining. The deep estuarial waters of Milford Haven provide a berth for oil tankers. A large part of Pembrokeshire's coastline forms Britain's only coastal National Park. Ferries sail from Fishguard and Pembroke Dock to Rosslare, Ireland.

Perth & Kinross *P. & K.* Population: 148,880
Unitary authority in Scotland bounded by Aberdeenshire, Angus, Argyll & Bute, Clackmannanshire, Fife, Highland and Stirling. Chief centres are the city of Perth, Blairgowrie, Crieff, Kinross, Auchterader and Pitlochry. The area is mountainous, containing large areas of remote open moorland, especially in the N and W; the vast upland expanses of Breadalbane, Rannoch and Atholl, form the S edge of the Grampian Mountains. The lower land of the S and E is more heavily populated and is dominated by the ancient city of Perth. The area is rich in history as it links the Highlands to the N with the central belt and lowlands to the S via important mountain passes, most notably the Pass of Dromochter. The area has many castles, and Scottish Kings were traditionally enthroned at Scone Abbey, to the N of Perth. There are many lochs, including Loch Rannoch and Loch Tay. Main industries are tourism and whisky production. The world famous Gleneagles golf course is in the S of the area. Rivers include the Tay, Almond and Earn.

Peterborough *Peter.* Population: 190,461
Unitary authority in E England neighbouring Cambridgeshire, Lincolnshire, Northamptonshire and Rutland. The area includes the city of Peterborough, which lies at the heart of an important agricultural area. Developing as a railway hub, it has become a major industrial, distribution and shopping centre. The River Nene passes through Peterborough.

Plymouth *Plym.* Population: 261,546
Unitary authority on the SW coast of England surrounding the city of Plymouth and neighbouring Cornwall and Devon. Plymouth stands at the mouth of the River Tamar and is the largest city on the S coast of England. It has strong mercantile

and naval traditions; it is closely linked with Sir Francis Drake, and has maintained a Royal Naval Dockyard for 300 years. Plymouth is a regional shopping centre and a popular resort.

Poole *Poole* Population: 150,109
Unitary authority on S coast of England surrounding Poole and bordered by Bournemouth and Dorset. Poole Harbour is the second largest natural harbour in the world, which enabled Poole to prosper through trading, especially with Newfoundland. Poole has now attracted a variety of industries including boat-building, fishing, pottery, engineering and electronics. Ferries run to the Channel Islands and France.

Portsmouth *Ports.* Population: 209,085
Unitary authority on the S coast of England surrounding the city of Portsmouth and bordered by Hampshire. Portsmouth developed as a strategic port around Portsmouth Harbour, and it is still the home of the Royal Navy. It has become a culturally diverse centre, attracting a wide range of industries which include leisure, tourism, financial services, distribution, manufacturing and hi-tech industries.

Powys *Powys* Population: 132,675
Large unitary authority in central Wales bordering Blaenau Gwent, Caerphilly, Carmarthenshire, Ceredigion, Denbighshire, Gwynedd, Merthyr Tydfil, Monmouthshire, Neath Port Talbot, Rhondda Cynon Taff, Wrexham and the English areas of Herefordshire and Shropshire. Main centres are Newtown, Gurnos, Brecon, Welshpool, Ystradgynlais, Llanllwchaiarn, Llandrindod Wells, Knighton, Llanidloes, Builth Wells and Machynlleth. Powys is almost entirely rural, with mountainous terrain; most of the Brecon Beacons National Park falls within the S part of the area, while the Cambrian Mountains are in the W. There is considerable afforestation, and a number of large reservoirs, including Lake Vyrnwy. To the N of Brecon, on Mynydd Eppynt, is an extensive military training area. Main economic activities are agriculture, which is predominantly based around hill farming. Tourism is significant, owing to the natural beauty of the area, and innovative attractions such as the Centre for Alternative Technology. Industrial development is gradually increasing. Among the many rivers, the largest are the Severn, Usk, and Wye.

Reading *Read.* Population: 160,825
Unitary authority in S England to W of Greater London, surrounding Reading and bordered by Oxfordshire, West Berkshire, Windsor & Maidenhead and Wokingham. Reading developed as a crossing point of the River Thames and River Kennet. Traditional industries include brewing and food production, notably biscuits. These are accompanied by an increasing sector of hi-tech and computer-based companies, attracted by Reading's location in the M4 corridor. Reading has also established itself as a major entertainments centre.

Redcar & Cleveland *R. & C.* Population: 135,042
Unitary authority on the NE coast of England neighbouring Hartlepool, Middlesbrough and North Yorkshire. The main centres are Redcar, South Bank, Eston, Guisborough, Marske-by-the-Sea, Saltburn-by-the-Sea, Loftus and Skelton. The area is one of great contrasts. It combines rural villages, market towns and coastal resorts, along with heavily populated urban areas and industrialised port facilities. Industries include steel-making, due to the local ironstone, and chemicals, based around the River Tees to the NW of the area. The coastal towns attract some tourism. The River Tees forms part of the border to the W.

Renfrewshire *Renf.* Population: 174,230
Unitary authority in central Scotland bordering East Renfrewshire, Glasgow, Inverclyde, North Ayrshire, West Dunbartonshire and the Firth of Clyde. Main centres are Paisley, Renfrew, Johnstone, Erskine and Linwood. The area emerges W from the Greater Glasgow periphery into a contrasting countryside of highlands, lochs and glens. Industry is centred on the urban area and includes electronics, engineering, food and drink production and service sectors; in rural areas to the W, agriculture is still important. The W part of the area includes some of Clyde Muirshiels Regional Park; Glasgow Airport is in the E.

Rhondda Cynon Taff (Rhondda Cynon Taf). *R.C.T.* Population: 236,888
Unitary authority in S Wales bounded by Bridgend, Caerphilly, Cardiff, Merthyr Tydfil, Neath Port Talbot, Powys and Vale of Glamorgan. The principal towns are Treorchy, Aberdare, Pontypridd, Ferndale and Mountain Ash. Rhondda Cynon Taff is a mountainous area, dissected by deep narrow valleys, with urbanisation typified by ribbon development. The area was the former heart of the Welsh coal mining industry, and has experienced a sharp economic decline as pits closed. Diversification into light engineering and service sectors are gradually improving the industrial base. Main rivers are the Rhondda and Cynon.

Rutland *Rut.* Population: 38,022
Unitary authority in E England neighbouring Leicestershire, Lincolnshire, Northamptonshire and Peterborough. The main town is Oakham. Agriculture is the main industry; other important industries are engineering, cement-making, plastics, clothing and tourism. The area includes the large reservoir, Rutland Water, which is an important feature for leisure, tourism and wildlife.

Scottish Borders *Sc.Bord.* Population: 114,030
Administrative region of SE Scotland bordering Dumfries & Galloway, East Lothian, Midlothian, South Lanarkshire, West Lothian, the English counties of Cumbria and Northumberland and the North Sea. It comprises the former counties of Berwick, Peebles, Roxburgh and Selkirk. Main towns are Hawick, Galashiels, Peebles, Kelso, Selkirk and Jedburgh. It extends from the Tweedsmuir Hills in the W to the North Sea on either side of St. Abb's Head in the E, and from the Pentland, Moorfoot and Lammermuir Hills in the N to the Cheviot Hills and the English border in the S. The fertile area of rich farmland between the hills to N and S is known as The Merse. The area around Peebles and Galashiels is noted for woollen manufacture. Elsewhere, the electronics industry is of growing importance. The River Tweed rises in the extreme W and flows between Kelso and Coldstream, finally passing into England, 4m/6km W of Berwick-upon-Tweed.

Shetland *Shet.* Population: 23,230
Group of over 100 islands, lying beyond Orkney to the NE of the Scottish mainland; Sumburgh Head being about 100m/160km from Duncansby Head. Designated an Islands Area for administrative purposes, the chief islands are Mainland, on which the capital and chief port of Lerwick is situated, Unst and Yell. Some twenty of the islands are inhabited. The islands are mainly low-lying, the highest point being Ronas Hill, on Mainland. The oil industry has made an impact on Shetland, with oil service bases at Lerwick and Sandwick, and a large terminal at Sullom Voe. Other industries include cattle and sheep-rearing, knitwear and fishing. The climate is mild, considering the latitude, but severe storms are frequent. The islands are famous for the small Shetland breed of pony, which is renowned for its strength and hardiness. There is an airport at Sumburgh, on S part of Mainland.

Shropshire *Shrop.* Population: 310,121
W midland unitary authority of England bounded by Cheshire East, Cheshire West & Chester, Herefordshire, Staffordshire, Telford & Wrekin, Worcestershire and the Welsh authorities of Powys and Wrexham. Main towns are Shrewsbury, Oswestry, Bridgnorth, Market Drayton, Ludlow and Whitchurch. The S and W borders are hilly, with large areas of open moorland, including The Long Mynd and Wenlock Edge, which provide good sheep-grazing. Elsewhere the county undulates towards the Severn Valley, which provides fertile agricultural land served by prosperous market towns. Agricultural output includes dairy, poultry and pig farming, along with corn crops. As the former heart of the Marches of Wales, Shropshire contains the remains of numerous border defences. There are also the remains of several monasteries, for instance, at Much Wenlock and Buildwas. The most important river is the Severn, which flows across the county from W to SE; others include the Clun, Corve, Perry, Rea Brook, and Teme.

Slough *Slo.* Population: 144,575
Unitary authority in SE England to the W of London, surrounding Slough and bordering Buckinghamshire, Greater London, Surrey and Windsor & Maidenhead. Slough has grown significantly over the past 30 years, and is a major regional shopping centre. Industry is centred on the large Slough Trading Estate, which was planned after World War I. Numerous sectors are represented in Slough, among them is confectionery.

Somerset *Som.* Population: 541,609
County in SW England bounded by Bath & North East Somerset, Devon, Dorset, North Somerset, Wiltshire and the Bristol Channel. The chief centres are the county town of Taunton, Yeovil, Bridgwater, Frome, Chard, Street, Burnham-on-Sea, Highbridge, the small cathedral city of Wells, Wellington and Minehead. Somerset consists of several hill ranges, including the Mendip, Polden, Quantock, Brendon Hills, along with most of Exmoor. These uplands are separated by valleys, or, on either side of the River Parrett, by the extensive marshy flats of Sedgemoor. Economic activity is mainly based on agriculture in the fertile vales, with manufacturing, distribution and service industries centred on the urban areas. Tourism is important with attractions including Exmoor National Park, a holiday complex at Minehead and the county's natural rural charm. Somerset also holds one of Europe's largest music festivals at Glastonbury. The chief rivers are Axe, Brue, Parrett, and Tone, draining into the Bristol Channel; and Barle and Exe, rising on Exmoor and flowing into Devon and the English Channel.
Districts: Mendip; Sedgemoor; South Somerset; Taunton Deane; West Somerset.

South Ayrshire *S.Ayr.* Population: 112,510
Unitary authority in SW Scotland bounded by Dumfries & Galloway, East Ayrshire, North Ayrshire and the sea. The chief towns are Ayr, Troon, Prestwick, Girvan and Maybole. The area consists of a long coastline, with lowlands surrounding Ayr Bay and higher ground to the S. Agriculture is a major economic activity on the uplands. To the N, aerospace and hi-tech industries are located near Prestwick International Airport and Ayr, the main retail centre. Notable sporting venues include a race course at Ayr and open championship golf courses at Troon and Turnberry. Tourism is a major feature of the local economy. The area was the birthplace of Robert the Bruce and Robert Burns; it contains Scotland's first country park at Culzean Castle; and it has a holiday camp on the coast near Ayr. Rivers include the Ayr, Water of Girvan and Stinchar.

South Gloucestershire *S.Glos.* Population: 271,556
Unitary authority in SW England neighbouring Bath & North East Somerset, Bristol, Gloucestershire and Wiltshire. The chief centres are Kingswood, Chipping Sodbury, Mangotsfield, Frampton Cotterell, Yate, Thornbury, Patchway and Filton. The S part of the area lies on the N and E fringes of Bristol. The Cotswold Hills are in the E, and the Severn Vale in the W. Main industries are in the S, and include aerospace engineering; the N is mainly agricultural. South Gloucestershire includes the English side of both Severn road bridges. Badminton Park in the E of the area, is the location for the Badminton Horse Trials. The River Severn borders the area to the NW.

South Lanarkshire *S.Lan.* Population: 315,360
Unitary authority in central Scotland bordering Dumfries & Galloway, East Ayrshire, East Renfrewshire, Glasgow, North Lanarkshire, Scottish Borders and West Lothian. The main towns are East Kilbride, Hamilton, Blantyre, Larkhall, Carluke, Lanark and Bothwell. Urban development is mainly in the N, merging with the SE periphery of Greater Glasgow. The S part is mostly farmland and not highly populated. Tourism is mainly centred on the picturesque valley of the upper Clyde; there is a race course at Hamilton. The area has associations with the industrial philanthropist, Robert Owen, who built New Lanark, now a UNESCO World Heritage Site. Rivers include the Clyde, Avon and Dippool Water.

South Yorkshire *S.Yorks.* Population: 1,365,847
Former metropolitan county of N England bordered by Derbyshire, East Riding of Yorkshire, North Lincolnshire, North Yorkshire, Nottinghamshire and West Yorkshire. It comprises the industrial and urban area around the city of Sheffield and the towns of Rotherham, Barnsley and Doncaster. Located at the heart of a major coalfield, South Yorkshire prospered through the development of heavy industry. Barnsley and Rotherham were coal mining towns, with steel and fine cutlery centred on Sheffield. The decline of these industries has led to the area redefining itself. Sheffield has become a centre of learning, tourism and conferences, aided by its environmental improvements. Barnsley, Rotherham and Doncaster have increased their industrial base, especially via light industries. Leisure and recreation are an important feature of the area, with venues including Barnsley's Metrodome, Doncaster's race course and Dome, and Sheffield's Arena. Retail has increased with city and town centre redevelopment, and the Meadowhall complex. The surrounding countryside includes country parks at Rother Valley and Thrybergh, with part of the Peak District National Park W and NW of Sheffield. The chief river is the Don.
Districts: Barnsley; Doncaster; Rotherham; Sheffield.

Southampton *S'ham.* Population: 245,290
Unitary authority on the S coast of England surrounding the city of Southampton, and bordered by Hampshire. Southampton owes much to the deep waters of Southampton Water, which have enabled the development of Europe's busiest cruise port. Water and the waterfront remain very important to the local economy, with marine technology, oceanography, boat shows and yacht races all prominent. The city is also a leading media, recreational, entertainment and retail centre. The chief river is the Itchen.

Southend *S'end* Population: 177,931
Unitary authority in SE England, N of the mouth of the River Thames, surrounding Southend-on-Sea and bordering Essex. Southend is a commerical, residential, shopping and holiday centre, with tourism among its main industries. It includes a 7m shoreline from Leigh-on-Sea to Shoeburyness, a famous pier and a sea life centre.

Staffordshire *Staffs.* Population: 860,165
Midland county of England bounded by Cheshire East, Derbyshire, Leicestershire, Shropshire, Stoke-on-Trent, Telford & Wrekin, Warwickshire, West Midlands and Worcestershire. Chief centres are Newcastle-under-Lyme, Tamworth, the county town of Stafford, Burton upon Trent, Cannock, Burntwood, the cathedral city of Lichfield, Kidsgrove, Rugeley and Leek. The urban development occurs around the West Midlands conurbation in the S, where main industries include engineering, iron and steel, rubber goods and leather production, while to the N, there is an urban concentration around Stoke-on-Trent. Burton upon Trent is noted for brewing. The ancient hunting forest and former mining district of Cannock Chase is in the centre of the county and contains preserved tracts of moorland. In the NE lies part of the Peak District National Park. The rest of the county is predominantly agricultural, with milk, wheat and sugar beet produced. To the E of Leek, moorland broken up by limestone walls extends across the Manifold valley to the Derbyshire border. In additon to the Trent, which dominates much of the county, rivers include the Blithe, Manifold, Sow and Tame. River Dove forms the boundary with Derbyshire.
Districts: Cannock Chase; East Staffordshire; Lichfield; Newcastle-under-Lyme; South Staffordshire; Stafford; Staffordshire Moorlands; Tamworth.

Stirling *Stir.* Population: 91,580
Unitary authority in central Scotland neighbouring Argyll & Bute, Clackmannanshire, East Dunbartonshire, Falkirk, North Lanarkshire, Perth & Kinross and West Dunbartonshire. The chief centres are Stirling, the ancient cathedral city of Dunblane, Bannockburn, Bridge of Allan and Callander. The fertile agricultural lands of the Forth valley are in the centre of the area, bounded by mountains: The Trossachs and the mountain peaks of Ben Lomond, Ben More and Ben Lui in the N, while in the S are the Campsie Fells. Tourism is an important industry with Stirling including many sites

of historical significance to Scotland, particularly during the struggle to retain independence. There are associations with Rob Roy, and the battle site of Bannockburn. Other features include the Loch Lomond and the Trossachs National Park and the Queen Elizabeth Forest Park. There are several lochs, including Loch Lomond, which forms part of the W border, and Loch Katrine. Scotland's only lake named as such, Lake of Menteith, is also in Stirling. The main river is the Forth.

Stockton-on-Tees *Stock.* Population: 194,119
Unitary authority in NE England neighbouring Darlington, Durham, Hartlepool, Middlesbrough, North Yorkshire and Redcar & Cleveland. The main centres are Stockton-on-Tees, Billingham, Thornaby-on-Tees, Eaglescliffe, Egglescliffe and Yarm. The area has a diverse mix of picturesque villages, large-scale urbanisation and heavy industry. The area has recently undergone major renewal and regeneration, with industries now including electronics, food technology and chemical production. Stockton is the main shopping centre for the area, and includes the Teesside Retail Park. The main river is the Tees, which is controlled by the Tees Barrage. This has created Britain's largest purpose-built whitewater canoeing course.

Stoke-on-Trent *Stoke* Population: 251,027
Unitary authority in England surrounding the city of Stoke-on-Trent and neighbouring Staffordshire. The city has six town centres: Burslem, Fenton, Hanley, Longton, Stoke-upon-Trent and Tunstall. Hanley is where most current city centre activities are located. The area forms The Potteries, and is the largest claywear producer in the world, although now it is largely a finishing centre for imported pottery. There are a wide variety of other industries, including steel, engineering, paper, glass and furniture. Stoke-on-Trent is a centre of employment, leisure and shopping for the surrounding areas. It is noted for its environmental approach, particularly with land reclamation which accounts for around 10 per cent of the city area; sites include Festival Park, Central Forest Park and Westport Lake. The River Trent flows through the area.

Suffolk *Suff.* Population: 738,512
Easternmost county of England bounded by Cambridgeshire, Essex, Norfolk and the North Sea. Main towns are the county town of Ipswich, Lowestoft, Bury St. Edmunds, Felixstowe, Sudbury, Haverhill, Newmarket, Stowmarket and Woodbridge. The county is low-lying and gently undulating. It is almost entirely agricultural, with cereal crops and oil seed rape in abundance. The low coastline, behind which are areas of heath and marsh, afforested in places, is subject to much erosion; it is deeply indented with long river estuaries which provide good sailing. The NW corner of the county forms part of Breckland. The central region includes many notable historic Wool Towns, for instance, Lavenham. Apart from agriculture, industries include electronics, telecommunications, printing and port facilities. Lowestoft is a prominent fishing port and Felixstowe is a container port of growing importance. River Stour forms the S boundary with Essex, and the Little Ouse and Waveney form most of the N boundary with Norfolk. The many other small rivers include the Alde with its estuary the Ore, Deben, and Gipping with its estuary the Orwell, in the E and Lark in the W.
Districts: Babergh; Forest Heath; Ipswich; Mid Suffolk; St. Edmundsbury; Suffolk Coastal; Waveney.

Surrey *Surr.* Population: 1,161,256
County of SE England bounded by Bracknell Forest, East Sussex, Greater London, Hampshire, Kent, Slough, West Sussex and Windsor & Maidenhead. The principal towns are Woking, the cathedral and university town of Guildford, Staines-upon-Thames, Leatherhead, Farnham, Epsom, Ewell, Sunbury, Walton-on-Thames, Weybridge, Egham, Redhill, Reigate, Esher, Camberley, Frimley and Godalming. The chalk ridge of the North Downs, gently sloping on the N side but forming a steep escarpment on the S, traverses the county from E to W. Extensive sandy heaths in the W are much used for military training. The county is heavily wooded, and contains many traces of the former iron industry in the predominantly rural S. Much of the urbanised E and N areas include commuter or dormitory towns which form the residential outskirts of the Greater London conurbation.

Industries include the agricultural activites of dairy farming and horticulture. Tourism and recreation are also important, with Surrey including numerous stately homes, Wentworth golf course, four race courses, and a theme park at Thorpe Park. The chief river is the Thames, into which flow the Wey and the Mole.
Districts: Elmbridge; Epsom & Ewell; Guildford; Mole Valley; Reigate & Banstead; Runnymede; Spelthorne; Surrey Heath; Tandridge; Waverley; Woking.

Swansea (Abertawe). *Swan.* Population: 241,297
Unitary authority in S Wales bordering Carmarthenshire, Neath Port Talbot and the sea. Main centres are the city of Swansea, Gorseinon, The Mumbles, Sketty, Cockett and Clydach. The area includes mountains in the N, the urban centre surrounding Swansea, and the Gower peninsula in the S. Swansea originally developed as a port serving the W coalfield of S Wales. The area gained an international reputation for tin-plating and copper and nickel production. Swansea is now a regional shopping and commercial centre, including a university and marina development. The Gower peninsula attracts many tourists with its fine beaches and cliff scenery; hang-gliding is popular at Rhossili Down, and there are associations with Dylan Thomas. The Mumbles is a popular resort, formerly connected to Swansea via a tramway. The chief river is the Tawe.

Swindon *Swin.* Population: 215,799
Unitary authority in SW England neighbouring Gloucestershire, Oxfordshire and Wiltshire. Main centres are Swindon, Stratton St. Margaret, Highworth and Wroughton. The area is located between the Cotswold Hills and Wiltshire Downs, on the fringes of the Thames Valley. Originally a railway town, Swindon has experienced rapid recent growth and is now a centre for car manufacture and central commercial operations. The town is a regional shopping centre with a redeveloped town centre and the Designer Outlet Village. The River Thames borders the area to the N and the River Cole to the E.

Telford & Wrekin *Tel. & W.* Population: 169,440
Unitary authority in W England bordered by Shropshire and Staffordshire. Main centres are Telford, Wellington, Madeley, Donnington, Oakengates, Hadley and Newport. The area was the cradle of the Industrial Revolution, with notable firsts including Darby's discovery of the iron smelting process at Coalbrookdale, the casting and construction of the first cold blast iron bridge at Ironbridge, and the construction of the first iron ship. The new town of Telford, named after the famous engineer, surveyor and road builder, Thomas Telford, is the major commercial centre. The River Severn runs S through the area.

Thurrock *Thur.* Population: 163,270
Unitary authority in SE England, N of the mouth of the River Thames. It is bounded by Essex and Greater London. The main centres are Grays, South Ockendon, Stanford-le-Hope, Corringham and Tilbury. The area is a mix of old and modern, rural and urban. In the N there are historic villages set in agricultural land, while in the S, there are the modern urban developments, and industrial activities surrounding oil refining and the container port of Tilbury. Grays is the commercial centre of Thurrock, with the major retail centre being Thurrock Lakeside. The area includes the N stretch of the Dartford Tunnel and Queen Elizabeth II Bridge, both of which cross the River Thames.

Torbay *Torbay* Population: 132,984
Unitary authority located on the SW coast of England neighbouring Devon. The major towns are Torquay, Paignton and Brixham. The area, situated on Tor Bay, is among Britain's main holiday resorts, and is widely regarded as the English Riviera. Tourism is the main industry, with Torbay receiving over 1.5 million visitors per year. Excellent leisure, recreation and conference facilities are added attractions.

Torfaen (Tor-faen). *Torfaen* Population: 91,609
Unitary authority in S Wales bounded by Blaenau Gwent, Caerphilly, Monmouthshire and Newport. The principal towns are Cwmbran, Pontypool and Blaenavon. Torfaen

contains rugged mountains with a 12-mile-long valley running N to S from Blaenavon to Cwmbran. The area is a manufacturing centre which includes electronics, engineering and automotive companies. The industrial past of the area has led to the growth of tourist attractions, with notable sites including The Valley Inheritance at Pontypool, and Big Pit National Mining Museum of Wales and 19c ironworks at Blaenavon. The river Afon Llwyd runs through the area.

Tyne & Wear *T. & W.* Population: 1,118,713
Maritime county of NE England bordered by Durham and Northumberland. It comprises the urban complex around the cities of Newcastle upon Tyne and Sunderland, South Shields, Gateshead, Washington and Wallsend. Named after its two important rivers, the area developed largely through the coal mining and ship-building industries. As these industries declined, the area has undergone urban and industrial regeneration. Newcastle upon Tyne is now a commercial, university and cultural centre, with a historic heart including a cathedral, 12c castle and the Tyne Bridge; the historic Quayside has recently been developed. Sunderland gained city status in 1992, and is now a centre for car manufacture and is also home to the National Glass Centre. Elsewhere, Wallsend has hi-tech and off-shore industries; South Tyneside has electronics industries, and tourism, via its Catherine Cookson links. Gateshead has an international athletics stadium, Europe's largest undercover shopping centre, the Metrocentre, and the modern symbol of renewal, the Angel of the North. The area is served by the Port of Tyne and Newcastle International Airport.
Districts: Gateshead; Newcastle upon Tyne; North Tyneside; South Tyneside; Sunderland.

Vale of Glamorgan (Bro Morgannwg). *V. of Glam.* Population: 127,685
Unitary authority on the S coast of Wales neighbouring Bridgend, Cardiff and Rhondda Cynon Taff. The chief towns are Barry, Penarth and Llantwit Major. Vale of Glamorgan is a lowland area between Cardiff and Bridgend, with some agricultural activities, and tourism at the resorts of Barry and Penarth. Cardiff International Airport is situated in the SE near Rhoose. Main river is the Ely, which passes through the area.

Warrington *Warr.* Population: 206,428
Unitary authority in NW England surrounding Warrington and bounded by Cheshire East, Cheshire West & Chester, Greater Manchester, Halton and Merseyside. The area developed as a main crossing point of the River Mersey and latterly the Manchester Ship Canal. During industrialisation it became an important strategic trading centre for the NW region. In 1968, Warrington was granted New Town status, leading to traditional industries such as chemicals, brewing and food processing being joined by hi-tech industries and research and development facilities. Warrington retains its importance as a regional shopping, leisure and commercial centre. The River Mersey flows through the area.

Warwickshire *Warks.* Population: 551,594
Midland county of England bounded by Gloucestershire, Leicestershire, Northamptonshire, Oxfordshire, Staffordshire, West Midlands and Worcestershire. Chief towns are Nuneaton, Rugby, Royal Leamington Spa, Bedworth, the county town of Warwick, Stratford-upon-Avon and Kenilworth. Warwickshire consists of mostly flat or undulating farmland, although the foothills of the Cotswold Hills spill over the SW border. Main manufacturing activites occur in an industrial belt extending NW from Rugby to the boundary with Staffordshire. They include motor and component industries, service sectors, electrical and general engineering. Tourism is centred on the historic town of Warwick with its medieval castle, and Stratford-upon-Avon with its Shakespeare associations. The principal river is the Avon.
Districts: North Warwickshire; Nuneaton & Bedworth; Rugby; Stratford-on-Avon; Warwick.

West Berkshire *W.Berks.* Population: 155,732
Unitary authority in S England bordered by Hampshire, Oxfordshire, Reading, Wiltshire and Wokingham. The chief centres are Newbury, Thatcham and Hungerford. West Berkshire is a mixture of old market towns, historic buildings and waterways, and includes the famous Newbury racecourse. Rivers include the Kennet and the Pang.

West Dunbartonshire *W.Dun.* Population: 89,730
Unitary authority in central Scotland bordered by Argyll & Bute, East Dunbartonshire, Glasgow, Inverclyde, Renfrewshire and Stirling. The chief towns are Clydebank, Dumbarton, Alexandria and Bonhill. The area is mountainous, containing the Kilpatrick Hills, and is bounded by Loch Lomond in the N and the Firth of Clyde in the S. The urban SE area of West Dunbartonshire forms part of the NW periphery of Greater Glasgow. There is a broad base of light manufacturing and service sector industries. Tourism and leisure are a feature, with the SE tip of Loch Lomond and the Trossachs National Park and the whole of Balloch Castle Country Park falling within the area. West Dunbartonshire includes the Erskine Bridge which spans the River Clyde, other rivers include the Leven.

West Lothian *W.Loth.* Population: 177,150
Unitary authority in central Scotland neighbouring Edinburgh, Falkirk, Midlothian, North Lanarkshire, Scottish Borders and South Lanarkshire. The chief towns are Livingston, Bathgate, Linlithgow, Broxburn, Whitburn and Armadale. The area undulates to the S of the Firth of Forth, and rises to moorland at the foot of the Pentland Hills in the S. The main urban areas are situated along commuter corridors between Glasgow, Edinburgh and Falkirk; elsewhere the area is mostly rural. Hi-tech and computing industries are in evidence.

West Midlands *W.Mid.* Population: 2,808,356
Former metropolitan county of central England bordered by Staffordshire, Warwickshire and Worcestershire. It comprises the urban complex around the cities of Birmingham and Coventry, and the towns of Wolverhampton, Dudley, Walsall, West Bromwich, Sutton Coldfield and Solihull. The West Midlands developed as a manufacturing and engineering centre which specialised in the metalworking and motor trades. The area around Dudley, Walsall and Wolverhampton became known as the Black Country, with heavy industry centred on the local deposits of coal, iron ore and limestone. Other local trades included glassware, saddlery and lock-making. Birmingham became Britain's second city by specialising in 1001 trades from confectionery to cars, and has developed into the major business, industrial, commercial and cultural centre for the area. As the traditional industries have declined, there has been a shift towards service, leisure and recreation sectors of the economy; several significant corporate service centres and venues, such as the National Exhibition Centre and the Indoor Arena, are in the West Midlands. The area is served by Birmingham International Airport. Rivers include the Tame and the Cole.
Districts: Birmingham; Coventry; Dudley; Sandwell; Solihull; Walsall; Wolverhampton.

West Sussex *W.Suss.* Population: 828,398
County of S England bounded by Brighton & Hove, East Sussex, Hampshire, Surrey and the English Channel. Main towns are Worthing, Crawley, Bognor Regis, Littlehampton, Horsham, Haywards Heath, East Grinstead, the cathedral city and county town of Chichester, Burgess Hill and Shoreham-by-Sea. N of a level coastal strip run the South Downs, a steep-sided chalk ridge which is thickly wooded in parts. The remaining inland area, The Weald, is largely well-wooded farmland, although there is industrial development around Crawley, Gatwick (London) Airport, Horsham, and Haywards Heath, as well as among the predominantly residential towns on the coast. Tourism is a major activity throughout the county. There are many castles and stately homes, such as Arundel Castle and Goodwood House, the popular seaside resorts of Bognor Regis and Worthing, race courses at Goodwood and Fontwell, Chichester Harbour, which is a centre for yachtsmen and wildfowl, historic Chichester itself, and numerous picturesque villages. The N of the county includes Gatwick (London) Airport. The rivers, none large, include the Adur and Arun, with its tributary the Rother; the Medway rises in the E of the county.
Districts: Adur; Arun; Chichester; Crawley; Horsham; Mid Sussex; Worthing.

West Yorkshire *W.Yorks.* Population: 2,264,329
Former metropolitan county of N England bordering Derbyshire, Greater Manchester, Lancashire, North Yorkshire and South Yorkshire. It comprises the area around the cities of Leeds, Bradford and Wakefield, and the towns of Huddersfield, Halifax, Dewsbury, Keighley, Batley, Morley, Castleford, Brighouse, Pudsey, Pontefract and Shipley. West Yorkshire developed as a centre for wool and textiles, manufacturing and engineering, creating an industrial urban landscape set against rural moorland. As the traditional industries have declined, the area has undergone regeneration and diversification, moving towards tertiary economic sectors. Leeds is the industrial, administrative, commercial and cultural centre of the area, containing regional government offices and many corporate service centres and head offices. Emerging economic activities across West Yorkshire have included printing, distribution, chemicals, food and drink production, hi-tech industries and financial services. Haworth with its Brontë associations, Holmfirth and the moorlands are the centres of tourism. The area includes Leeds Bradford International Airport. The chief rivers are the Aire and the Calder, while the Wharfe forms its N boundary below Addingham.
Districts: Bradford; Calderdale; Kirklees; Leeds; Wakefield.

Western Isles (Na h-Eileanan Siar. Also known as Outer Hebrides.) *W.Isles* Population: 27,250
String of islands off the W coast of Scotland and separated from Skye and the mainland by The Minch. They extend for some 130m/209km from Butt of Lewis in the N, to Barra Head in the S. Stornoway, situated on the Isle of Lewis, is the main town; elsewhere, there are mainly scattered coastal villages and settlements. The chief islands are Isle of Lewis, North Uist, Benbecula, South Uist and Barra. North Harris and South Harris form significant areas in the S part of the Isle of Lewis. The topography of the islands consists of undulating moorland, mountains and lochs. The main industries are fishing, grazing and, on the Isle of Lewis, tweed manufacture. There are airfields with scheduled passenger flights on the Isle of Lewis, Benbecula and Barra.

Wiltshire *Wilts.* Population: 483,143
Unitary authority of S England bounded by Bath & North East Somerset, Dorset, Gloucestershire, Hampshire, Oxfordshire, Somerset, South Gloucestershire, Swindon and West Berkshire. Main centres are the cathedral city of Salisbury, Trowbridge, Chippenham, Warminster, Devizes and Melksham. Wiltshire consists of extensive chalk uplands scattered with prehistoric remains, notably at Avebury and Stonehenge, and interspersed with wide, well-watered valleys. The N of the county is dominated by the Marlborough Downs which are much used for racehorse training, while in the S, the chalk plateau of Salisbury Plain is an important military training area. Between these two upland areas lies the fertile Vale of Pewsey where dairy production and bacon-curing are important agricultural activities. Other industries include electronics, computing, pharmaceuticals, plastics, telecommunications and service sector activities. Wiltshire attracts tourism with its prehistoric remains, stately houses and picturesque market towns and villages. Rivers include the so-called Bristol and Wiltshire Avons, Ebble, Kennet, Nadder, Wylye, and the upper reaches of the Thames.

Windsor & Maidenhead *W. & M.* Population: 147,400
Unitary authority in SE England to the W of Greater London, and bounded by Bracknell Forest, Buckinghamshire, Slough, Surrey and Wokingham. The towns of Maidenhead and Windsor are the main centres for industry, leisure and recreation. The area is particularly noted for its strong Royal connections as it includes Windsor Castle and the former Royal hunting estate of Windsor Great Park. Other popular tourist attractions include Ascot race course, Windsor Legoland and Eton College. The River Thames forms the N boundary.

Wokingham *W'ham* Population: 159,097
Unitary authority in SE England, to the W of Greater London. The area encompasses Wokingham and is bordered by Bracknell Forest, Buckinghamshire, Hampshire, Oxfordshire, Reading, West Berkshire and Windsor & Maidenhead. The area includes riverside villages in the N, with undulating ridges covered by woodlands and commons in the S. Wokingham is a growing centre for hi-tech and computer industries. The River Thames forms the N border, and the River Blackwater forms the border to the S.

Worcestershire *Worcs.* Population: 575,421
S midland county of England neighbouring Gloucestershire, Herefordshire, Shropshire, Warwickshire and West Midlands. Main centres are the cathedral city and the county town of Worcester, and the towns of Redditch, Kidderminster, Great Malvern, Bromsgrove, Droitwich Spa, Stourport-on-Severn and Evesham. The urban areas in the N of the county form part of the periphery and commuter belt of the West Midlands conurbation, and attract much of the industrial development. The central and S sections of the county are largely rural, containing the fertile Severn Valley and Vale of Evesham, with market gardening and orchard-growing being the main agricultural activities. Tourism is an important industry, much of it being centred on historic Worcester, with its cathedral, the triennial Three Choirs Festival, Worcester Sauce and china factories. Other popular attractions include boating on the River Severn and visiting the Vale of Evesham whilst the flowers are in full bloom. The main river is the Severn.
Districts: Bromsgrove; Malvern Hills; Redditch; Worcester; Wychavon; Wyre Forest.

Wrexham (Wrecsam). *Wrex.* Population: 136,714
Unitary authority in NE Wales bordering Denbighshire, Flintshire, Powys and the English counties of Cheshire and Shropshire. Main centres are Wrexham, Rhosllanerchrugog, Gwersyllt, Cefn-mawr and Coedpoeth. The area is mountainous in the SW, containing part of the Berwyn range; the Dee valley lies in the NE. The area was formerly dominated by the iron, coal and limestone industries. Food manufacture, brewing, plastics and hi-tech industries are now important to the local economy. Wrexham is the largest commercial and shopping centre in N Wales. The River Dee flows through the area.

York *York* Population: 204,439
Unitary authority in N England surrounding the historic cathedral city of York and bordered by East Riding of Yorkshire and North Yorkshire. York is a major archaeological, episcopal, industrial, commercial and cultural centre, situated at the confluence of the River Foss and the River Ouse. The city has a unique history dating from the original Roman military camp, which has led to it becoming one of the main museum and tourist centres in the country. The historic core, situated around the centrepiece of the medieval Minster, is well preserved. Other major attractions include the Jorvik Viking Centre, the medieval city walls and the National Railway Museum. Economic sectors include the confectionery industry, company head offices, Government departmental offices, and research and development establishments. The main river is the Ouse.

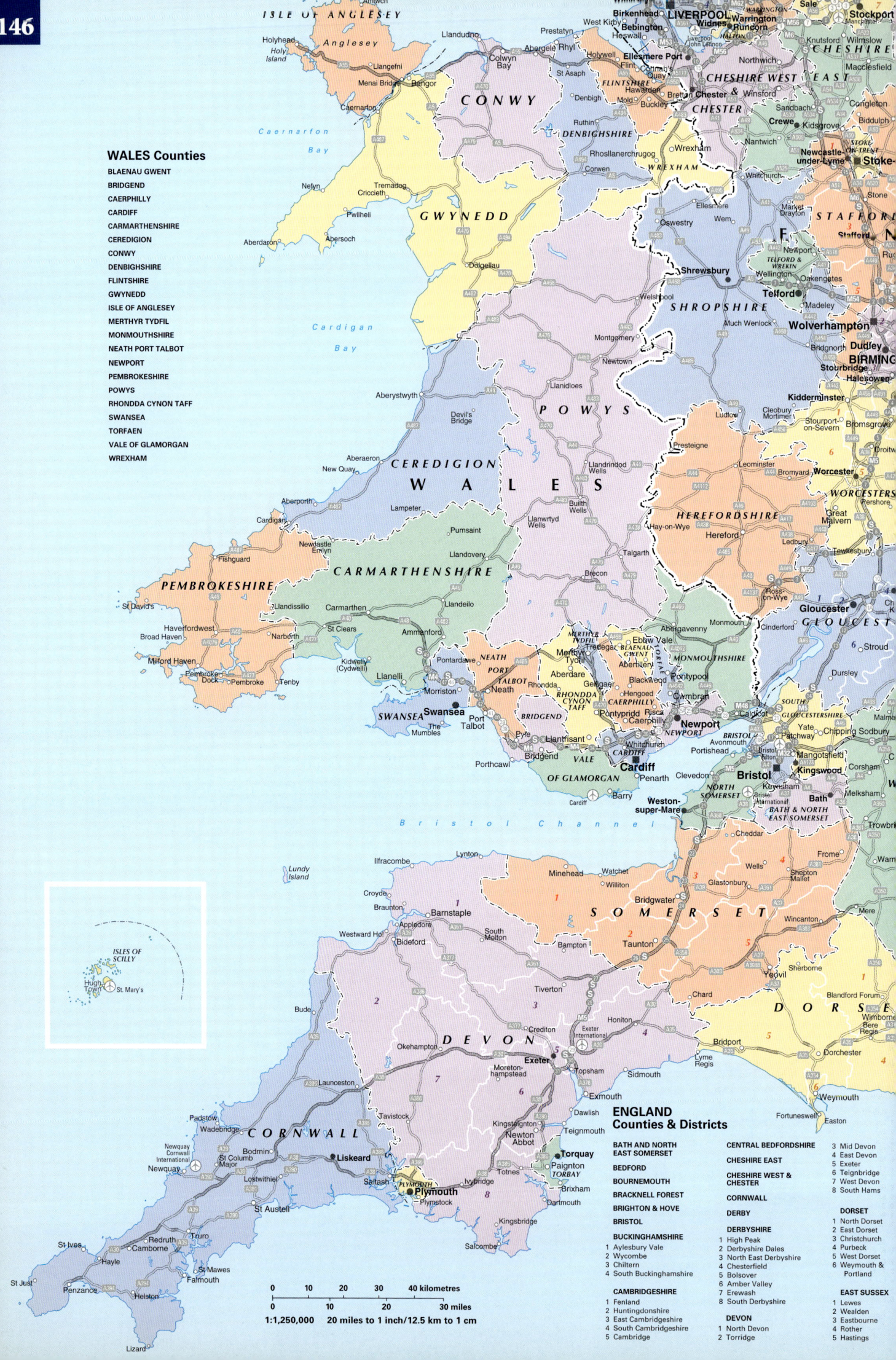

NORTHERN IRELAND
Districts

ANTRIM & NEWTOWNABBEY
ARDS & NORTH DOWN
ARMAGH CITY, BANBRIDGE & CRAIGAVON
BELFAST CITY
CAUSEWAY COAST & GLENS
DERRY CITY & STRABANE
FERMANAGH & OMAGH
LISBURN & CASTLEREAGH
MID & EAST ANTRIM
MID ULSTER
NEWRY, MOURNE & DOWN

150

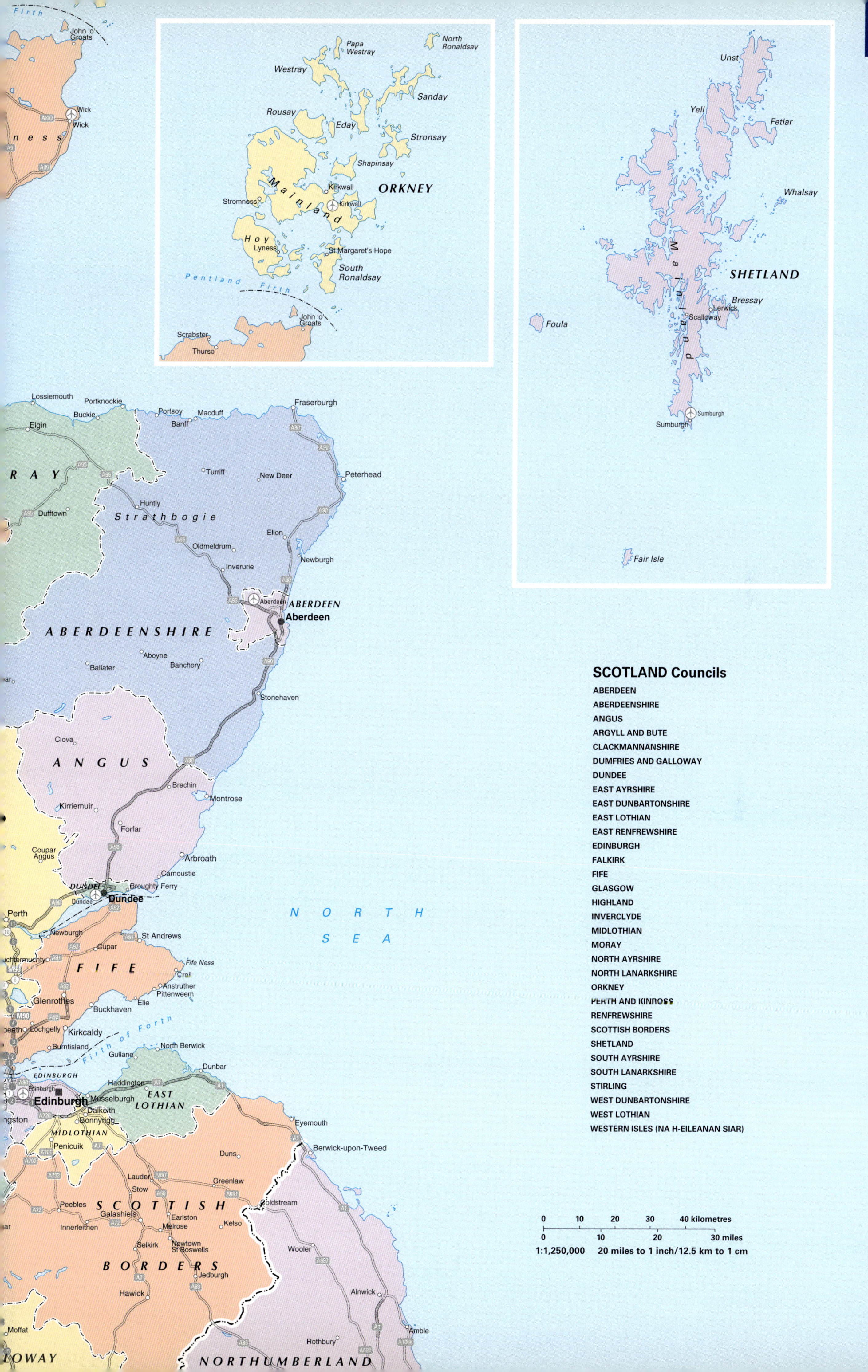

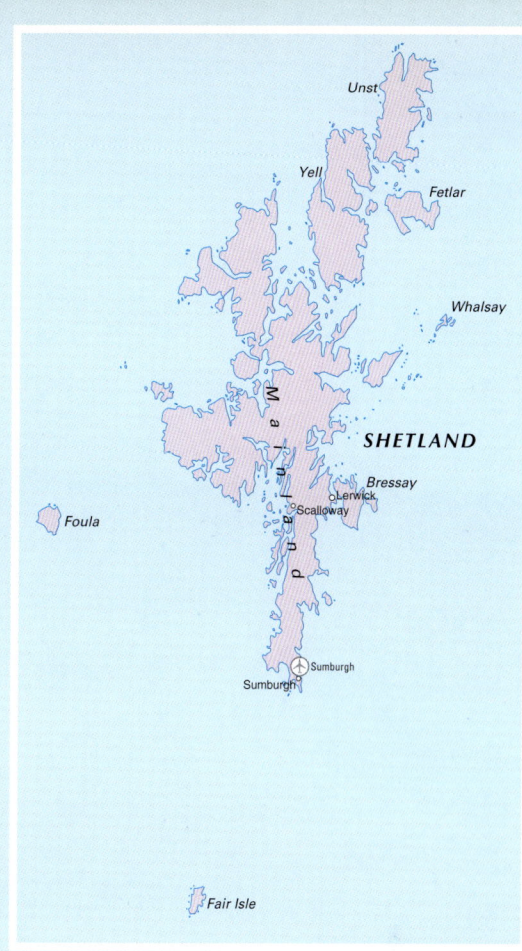

SCOTLAND Councils

ABERDEEN
ABERDEENSHIRE
ANGUS
ARGYLL AND BUTE
CLACKMANNANSHIRE
DUMFRIES AND GALLOWAY
DUNDEE
EAST AYRSHIRE
EAST DUNBARTONSHIRE
EAST LOTHIAN
EAST RENFREWSHIRE
EDINBURGH
FALKIRK
FIFE
GLASGOW
HIGHLAND
INVERCLYDE
MIDLOTHIAN
MORAY
NORTH AYRSHIRE
NORTH LANARKSHIRE
ORKNEY
PERTH AND KINROSS
RENFREWSHIRE
SCOTTISH BORDERS
SHETLAND
SOUTH AYRSHIRE
SOUTH LANARKSHIRE
STIRLING
WEST DUNBARTONSHIRE
WEST LOTHIAN
WESTERN ISLES (NA H-EILEANAN SIAR)

0 10 20 30 40 kilometres
0 10 20 30 miles
1:1,250,000 20 miles to 1 inch/12.5 km to 1 cm

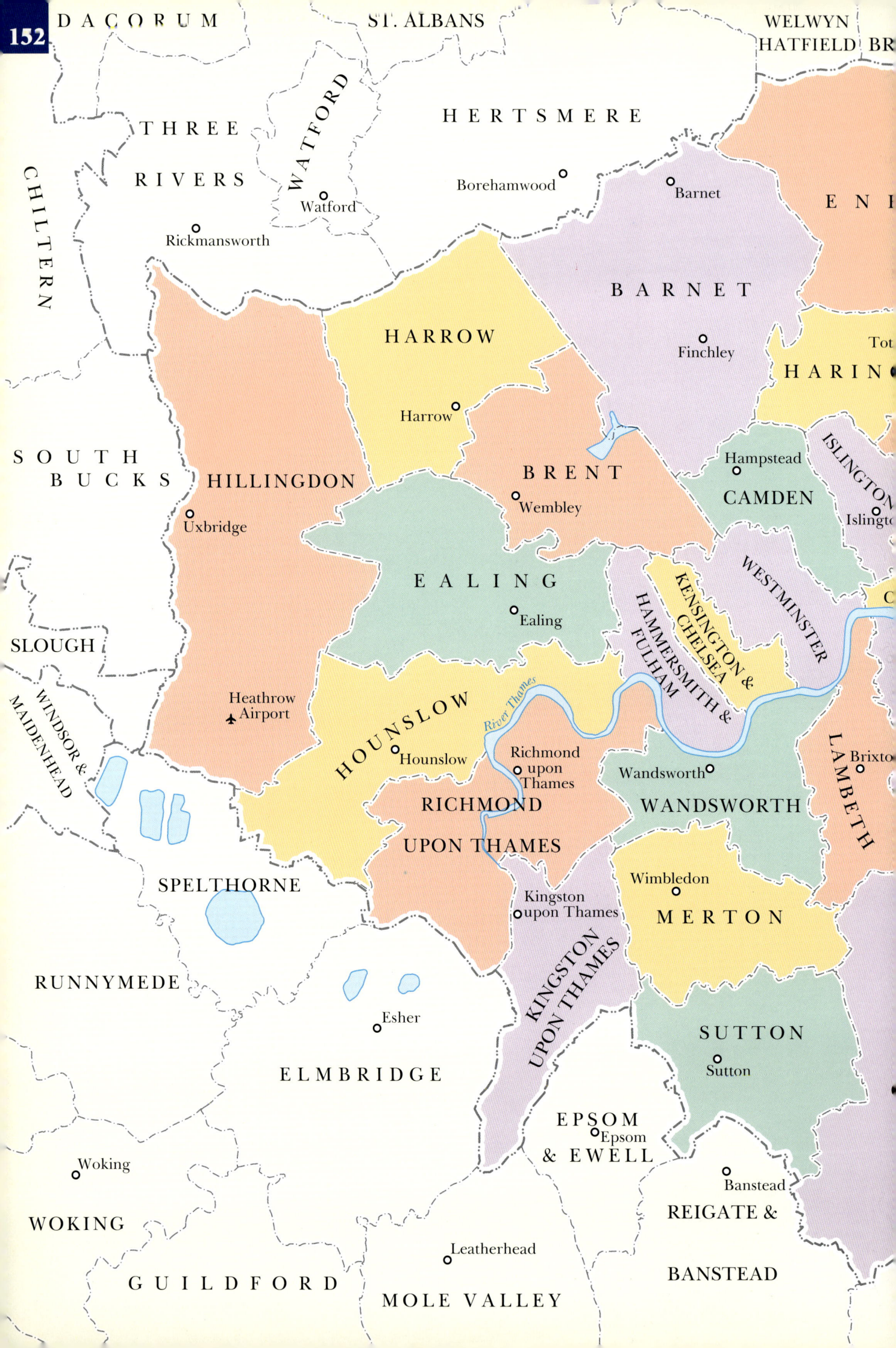

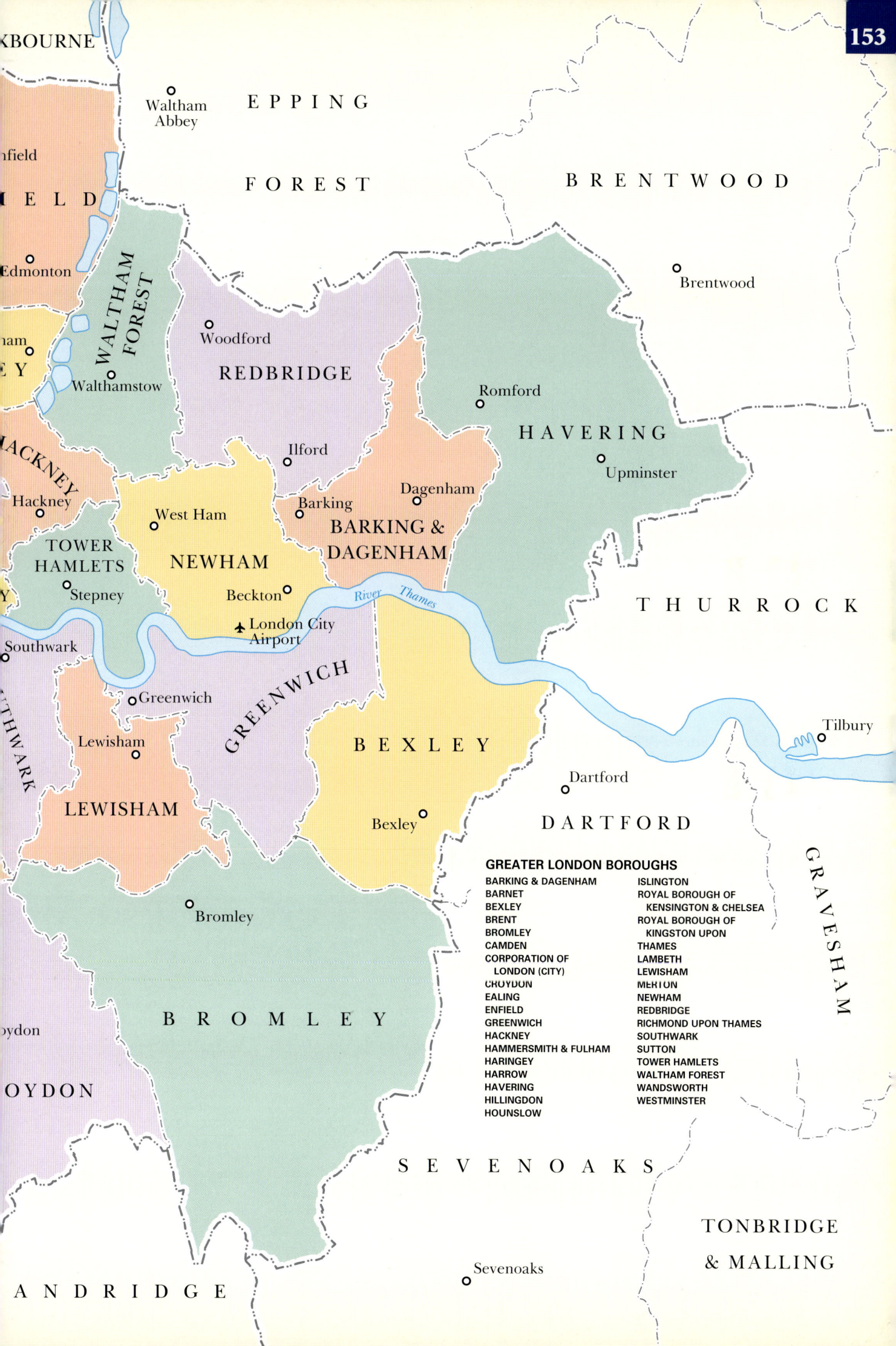

INDEX TO CENTRAL LONDON

General Abbreviations

All	Alley
Allot	Allotments
Amb	Ambulance
App	Approach
Apts	Apartments
Arc	Arcade
Av/Ave	Avenue
Bdy	Broadway
Bk	Bank
Bldgs	Buildings
Boul	Boulevard
Bowl	Bowling
Br/Bri	Bridge
Bus	Business
C of E	Church of England
Cath	Cathedral
Cem	Cemetery
Cen	Central, Centre
Cft	Croft
Cfts	Crofts
Ch	Church
Chyd	Churchyard
Cin	Cinema
Circ	Circus
Cl/Clo	Close
Co	County
Coll	College
Comm	Community
Conv	Convent
Cor	Corner
Coron	Coroners
Cors	Corners
Cotts	Cottages
Cov	Covered
Crem	Crematorium
Cres	Crescent
Ct	Court
Cts	Courts
Ctyd	Courtyard
Dep	Depot
Dev	Development
Dr	Drive
Dws	Dwellings
E	East
Ed	Education
Elec	Electricity
Embk	Embankment
Est	Estate
Ex	Exchange
Exhib	Exhibition
FB	Footbridge
FC	Football Club
Fld	Field
Flds	Fields
Fm	Farm
Gall	Gallery
Gar	Garage
Gdn	Garden
Gdns	Gardens
Govt	Government
Gra	Grange
Grd	Ground
Grds	Grounds
Grn	Green
Grns	Greens
Gro	Grove
Gros	Groves
Gt	Great
Ho	House
Hos	Houses
Hosp	Hospital
Hts	Heights
Ind	Industrial
Int	International
Junct	Junction
La	Lane
Las	Lanes
Lib	Library
Lo	Lodge
Lwr	Lower
Mag	Magistrates
Mans	Mansions
Mem	Memorial
Mkt	Market
Mkts	Markets
Ms	Mews
Mt	Mount
Mus	Museum
N	North
NT	National Trust
Nat	National
PH	Public House
PO	Post Office
Par	Parade
Pas	Passage
Pav	Pavilion
Pk	Park
Pl	Place
Pol	Police
Prec	Precinct
Prim	Primary
Prom	Promenade
Pt	Point
Quad	Quadrant
Rbt	Roundabout
RC	Roman Catholic
Rd	Road
Rds	Roads
Rec	Recreation
Res	Reservoir
Ri	Rise
S	South
Sch	School
Sec	Secondary
Shop	Shopping
Sq	Square
St.	Saint
St	Street
Sta	Station
Sts	Streets
Sub	Subway
Swim	Swimming
TA	Territorial Army
TH	Town Hall
Tenn	Tennis
Ter	Terrace
Thea	Theatre
Trd	Trading
Twr	Tower
Twrs	Towers
Uni	University
Vil	Villa, Villas
Vw	View
W	West
Wd	Wood
Wds	Woods
Wf	Wharf
Wk	Walk
Wks	Works
Yd	Yard

1 Canada Sq **E14** **127** M10
8 Walworth Rd - Strata **SE17** **130** A3

A

Abbeville Ms **SW4** **129** K10
Abbey Cl **SW8** **129** K7
Abbey Gdns **NW8** **124** C4
Abbey La **E15** **127** N4
Abbey Orchard St **SW1** **129** K2
Abbey Rd **NW6** **124** B2
Abbey Rd **NW8** **124** C4
Abbey Rd Est **NW8** **124** B3
Abbey St **SE1** **130** C2
Abbeyfield Rd **SE16** **130** G3
Abbot St **E8** **126** D1
Abbots Manor Est **SW1** **129** H3
Abbot's Pl **NW6** **124** B3
Abbotsbury Cl **E15** **127** N4
Abbotsbury Ms **SE15** **130** G9
Abbotshade Rd **SE16** **127** H10
Abbotswood Rd **SE22** **130** C10
Abbott Rd **E14** **127** N8
Abchurch La **EC4** **126** B9
Aberavon Rd **E3** **127** J5
Abercorn Cl **NW8** **124** C5
Abercorn Pl **NW8** **124** C5
Abercorn Way **SE1** **130** E4
Abercrombie St **SW11** **128** E8
Aberdare Gdns **NW6** **124** B2
Aberdeen Pl **NW8** **124** D6
Aberdeen Ter **SE3** **131** P8
Aberdour St **SE1** **130** C3
Aberfeldy St **E14** **127** N8
Abingdon Rd **W8** **128** A2
Abingdon St **SW1** **129** L2
Abingdon Vil **W8** **128** A2
Abinger Gro **SE8** **131** K5
Ablett St **SE16** **130** G4
Acacia Cl **SE8** **131** J3
Acacia Pl **NW8** **124** D4
Acacia Rd **NW8** **124** D4
Academy Gdns **W8** **128** A1
Acanthus Dr **SE1** **130** E4
Acanthus Rd **SW11** **128** G9
Acer Rd **E8** **126** D1
Acfold Rd **SW6** **128** B7
Achilles Cl **SE1** **130** E4
Achilles St **SE14** **131** J6
Ackmar Rd **SW6** **128** A7
Acland Cres **SE5** **130** B9
Acol Rd **NW6** **124** A2
Acorn Wk **SE16** **127** J10
Acre Dr **SE22** **130** E10
Acre La **SW2** **129** L10
Acton Ms **E8** **126** D3
Acton St **WC1** **125** M5
Ada Gdns **E14** **127** P8
Ada Pl **E2** **126** E3
Ada Rd **SE5** **130** C6
Ada St **E8** **126** F3
Adam & Eve Ms **W8** **128** A2
Adam St **WC2** **125** L9
Adams Row **W1** **124** G9
Adamson Rd **NW3** **124** D2
Adderley St **E14** **127** N8
Addington Rd **E3** **127** L5
Addington Sq **SE5** **130** A6
Adelaide Av **SE4** **131** K10
Adelaide Rd **NW3** **124** D2
Adelina Gro **E1** **126** G7
Adeline Pl **WC1** **125** K7
Adler St **E1** **126** E8
Admiral Pl **SE16** **127** J10
Admiral Sq **SW10** **128** C7
Admiral St **SE8** **131** L8
Admiral Wk **W9** **124** A7
Admirals Gate **SE10** **131** M7
Admirals Way **E14** **131** L1
Adolphus St **SE8** **131** K6
Adpar St **W2** **124** D6
Adrian Ms **SW10** **128** B5
Adys Rd **SE15** **130** D9
Afghan Rd **SW11** **128** E8
Agar Gro **NW1** **125** J2
Agar Gro Est **NW1** **125** K2
Agar Pl **NW1** **125** J2
Agar St **WC2** **125** L9
Agdon St **EC1** **125** P6
Agnes St **E14** **127** K8
Ailsa St **E14** **127** N7
Ainger Rd **NW3** **124** F2
Ainsdale Dr **SE1** **130** E4
Ainsley St **E2** **126** F5
Ainsty Est **SE16** **131** H1
Ainsworth Rd **E9** **126** G2
Ainsworth Way **NW8** **124** C3
Air St **W1** **125** J9
Airdrie Cl **N1** **125** M2
Akerman Rd **SW9** **129** P8
Albany Mans **SW11** **128** E6
Albany Rd **SE17** **130** B5
Albany Rd **SE5** **130** B5
Albany St **NW1** **125** H4
Albatross Way **SE16** **131** H1
Albatross Way **SE16** **130** G1
Albemarle St **W1** **125** H9
Albert Av **SW8** **129** M6
Albert Bigg Pt **E15** **127** N3
Albert Br **SW11** **128** E5
Albert Br **SW3** **128** E5
Albert Br Rd **SW11** **128** E6
Albert Embk **SE1** **129** L4
Albert Gdns **E1** **127** H8
Albert Gate **SW1** **128** F1
Albert Pl **W8** **128** B1
Albert Sq **SW8** **129** M6
Albert St **NW1** **125** H3
Albert Ter **NW1** **124** G3
Albert Way **SE15** **130** F6
Alberta Est **SE17** **129** P4
Alberta St **SE17** **129** P4
Albion Av **SW8** **129** K8
Albion Dr **E8** **126** D2
Albion Est **SE16** **130** G1
Albion Ms **N1** **125** N3
Albion Ms **W2** **124** E8
Albion Pl **EC1** **125** P7
Albion Riverside Building **SW11** **128** E6
Albion Sq **E8** **126** D2
Albion St **SE16** **130** G1
Albion St **W2** **124** E8
Albion Ter **E8** **126** D2
Albion Way **SE13** **131** N10
Albion Yd **E1** **126** F7
Albrighton Rd **SE22** **130** C9
Albury St **SE8** **131** L5
Albyn Rd **SE8** **131** L7
Aldbridge St **SE17** **130** C4
Aldebert Ter **SW8** **129** L6
Aldenham St **NW1** **125** K4
Alder Cl **SE15** **130** D5
Aldermanbury **EC2** **126** A8
Alderney Ms **SE1** **130** B2
Alderney Rd **E1** **127** H6
Alderney St **SW1** **129** H3
Aldersgate St **EC1** **126** A7
Alderton Rd **SE24** **130** A9
Aldford St **W1** **124** G10
Aldgate **EC3** **126** C8
Aldgate High St **EC3** **126** D8
Aldsworth Cl **W9** **124** B6
Aldwych **WC2** **125** M9
Alexander Pl **SW7** **128** E3
Alexander Sq **SW3** **128** E3
Alexander St **W2** **124** A8
Alexandra Av **SW11** **128** G7
Alexandra Cl **SE8** **131** K5
Alexandra Cotts **SE14** **131** K7
Alexandra Pl **NW8** **124** C3
Alexandra Rd **NW8** **124** C2
Alexandra St **SE14** **131** J6
Alexis St **SE16** **130** E3
Alfred Ms **W1** **125** K7
Alfred Pl **WC1** **125** K7
Alfred Rd **W2** **124** A7
Alfred St **E3** **127** K5
Alfreda St **SW11** **129** H7
Algernon Rd **NW6** **124** A3
Algernon Rd **SE13** **131** M10
Algiers Rd **SE13** **131** L10
Alice La **E3** **127** K3
Alice St **SE1** **130** C2
Alie St **E1** **126** D8
Aliwal Rd **SW11** **128** E10
All Saints St **N1** **125** M4
Allardyce St **SW4** **129** M10
Allen Edwards Dr **SW8** **129** L7
Allen Rd **E3** **127** K4
Allen St **W8** **128** A2
Allensbury Pl **NW1** **125** K2
Allingham St **N1** **126** A4
Allitsen Rd **NW8** **124** E4
Alloa Rd **SE8** **131** H4
Alloway Rd **E3** **127** J5
Allsop Pl **NW1** **124** F6
Alma Gro **SE1** **130** D3
Alma Sq **NW8** **124** C5
Alma St **E15** **127** P1
Alma St **NW5** **125** H1
Almeida St **N1** **125** P2
Almeric Rd **SW11** **128** F10
Almond Cl **SE15** **130** E8
Almond Rd **SE16** **130** F3
Almorah Rd **N1** **126** B2
Alpha Gro **E14** **131** L1
Alpha Pl **NW6** **124** A4
Alpha Pl **SW3** **128** E5
Alpha Rd **SE14** **131** K7
Alpha St **SE15** **130** E8
Alpine Gro **E9** **126** G2
Alpine Rd **SE16** **130** G3
Alsace Rd **SE17** **130** C4
Alscot Rd **SE1** **130** D3
Alscot Way **SE1** **130** D3
Altenburg Gdns **SW11** **128** F10
Althea St **SW6** **128** B8
Altius Wk **E20** **127** N1
Alton St **E14** **127** M7
Altura Twr **SW11** **128** D8
Alverton St **SE8** **131** K4
Alvey Est **SE17** **130** C3
Alvey St **SE17** **130** C4
Alwyne Pl **N1** **126** A1
Alwyne Rd **N1** **126** A2
Alwyne Sq **N1** **126** A1
Alwyne Vil **N1** **125** P2
Alzette Ho **E2** **127** H5
Ambergate St **SE17** **129** P4
Amberley Rd **W9** **124** A7
Ambrosden Av **SW1** **129** J2
Ambrose Ms **SW11** **128** E8
Ambrose St **SE16** **130** F3
Amelia St **SE17** **129** P4
Amersham Gro **SE14** **131** K6
Amersham Rd **SE14** **131** K6
Amersham Vale **SE14** **131** K6
Amiel St **E1** **126** G6
Amies St **SW11** **128** F9
Amott Rd **SE15** **130** E9
Amoy Pl **E14** **127** L8
Ampton St **WC1** **125** M5
Amsterdam Rd **E14** **131** N2
Amwell St **EC1** **125** N5
Anchor Retail Pk **E1** **126** G6
Anchor St **SE16** **130** F3
Andalus Rd **SW9** **129** L9
Anderson Rd **E9** **127** H1
Anderson St **SW3** **128** F4
Anderton Cl **SE5** **130** B9
Andover Pl **NW6** **124** B4
Andrew St **E14** **127** N8
Andrew's Rd **E8** **126** F3
Anerley St **SW11** **128** F8
Angel Cen **N1** **125** N4
Angel Ct **EC2** **126** B8
Angel La **E15** **127** P1
Angel Ms **N1** **125** N4
Angel St **EC1** **126** A8
Angell Pk Gdns **SW9** **129** N9
Angell Rd **SW9** **129** N9
Angler's La **NW5** **125** H1
Anglia Ho **E14** **127** J8
Anglo Rd **E3** **127** K4
Angrave Ct **E8** **126** D3
Angus St **SE14** **131** J6
Anhalt Rd **SW11** **128** E6
Ann La **SW10** **128** D6
Ann Moss Way **SE16** **130** G2
Anna Cl **E8** **126** D3
Annabel Cl **E14** **127** M8
Annie Besant Cl **E3** **127** K3
Annis Rd **E9** **127** J1
Ansdell Rd **SE15** **130** G8
Ansdell St **W8** **128** B2
Anselm Rd **SW6** **128** A5
Anstey Rd **SE15** **130** E9
Anthems Way **E20** **127** M1
Antill Rd **E3** **127** J5
Antill Ter **E1** **127** H8
Antrim Gro **NW3** **124** F1
Antrim Mans **NW3** **124** E1
Antrim Rd **NW3** **124** F1
Apollo Pl **SW10** **128** D6
Appleby Rd **E8** **126** E2
Appleby St **E2** **126** D4
Appold St **EC2** **126** C7
Approach Rd **E2** **126** G4
Aquila St **NW8** **124** D4
Aquinas St **SE1** **125** N10
Arabin Rd **SE4** **131** J10
Aragon Twr **SE8** **131** K3
Arbery Rd **E3** **127** J5
Arbour Sq **E1** **127** H8
Arbuthnot Rd **SE14** **131** H8
Arbutus St **E8** **126** C3
Arcadia St **E14** **127** L8
Arch St **SE1** **130** A2
Archangel St **SE16** **131** H1
Archer Ms **SW9** **129** L9
Archery Cl **W2** **124** E8
Archibald Ms **W1** **124** G9
Archibald St **E3** **127** L5
Arden Cres **E14** **131** L3
Arden Est **N1** **126** C4
Ardleigh Rd **N1** **126** B1
Argon Ms **SW6** **128** A6
Argyle Rd **E1** **127** H6
Argyle Sq **WC1** **125** L5
Argyle St **WC1** **125** L5
Argyle Way **SE16** **130** E4
Argyll Rd **W8** **128** A1
Argyll St **W1** **125** J8
Arica Ho **SE16** **130** F2
Arica Rd **SE4** **131** J10
Ariel Rd **NW6** **124** A1
Aristotle Rd **SW4** **129** K9
Arklow Rd **SE14** **131** K5
Arlesford Rd **SW9** **129** L9
Arlington Av **N1** **126** A4
Arlington Lo **SW2** **129** M10
Arlington Rd **NW1** **125** H3
Arlington Sq **N1** **126** A3
Arlington St **SW1** **125** J10
Arlington Way **EC1** **125** N5
Armadale Rd **SW6** **128** A6
Armagh Rd **E3** **127** K3
Armoury Rd **SE8** **131** M8
Armstrong Rd **SW7** **128** D2
Arne St **WC2** **125** L8
Arnhem Pl **E14** **131** L2
Arnold Circ **E2** **126** D5
Arnold Est **SE1** **130** D1
Arnold Rd **E3** **127** L5
Arnould Av **SE5** **130** B10
Arnside St **SE17** **130** A5
Arran Wk **N1** **126** A2
Arrow Rd **E3** **127** M5
Artesian Rd **W2** **124** A8
Artillery La **E1** **126** C7
Artillery Row **SW1** **129** K2
Arundel Pl **N1** **125** N1
Arundel Sq **N7** **125** N1
Arundel St **WC2** **125** M9
Ascalon St **SW8** **129** J6
Ash Gro **E8** **126** F3
Ashbridge St **NW8** **124** E6
Ashburn Gdns **SW7** **128** C3
Ashburn Pl **SW7** **128** C3
Ashburnham Gro **SE10** **131** M6
Ashburnham Pl **SE10** **131** M6
Ashburnham Retreat **SE10** **131** M6
Ashburnham Rd **SW10** **128** C6
Ashbury Rd **SW11** **128** F9
Ashby Gro **N1** **126** A2
Ashby Ms **SE4** **131** K8
Ashby Rd **SE4** **131** K8
Ashcombe St **SW6** **128** B8
Ashcroft Rd **E3** **127** J5
Ashdene **SE15** **130** F7
Ashdown Wk **E14** **131** L3
Asher Way **E1** **126** E10
Ashfield St **E1** **126** F7
Ashland Pl **W1** **124** G7
Ashleigh Ms **SE15** **130** D9
Ashley Cres **SW11** **128** G9
Ashley Gdns **SW1** **129** J2
Ashley Pl **SW1** **129** J2
Ashmead Rd **SE8** **131** L8
Ashmere Gro **SW2** **129** L10
Ashmill St **NW1** **124** E7
Ashmole Pl **SW8** **129** M5
Ashmole St **SW8** **129** M5
Ashmore Cl **SE15** **130** D6
Ashton St **E14** **127** N9
Ashwin St **E8** **126** D1
Ashworth Rd **W9** **124** B5
Aspen Way **E14** **127** L9
Aspinall Rd **SE4** **131** H9
Aspinden Rd **SE16** **130** F3
Assembly Pas **E1** **126** G7
Astbury Rd **SE15** **130** G7
Aste St **E14** **131** N1
Astell St **SW3** **128** E4
Astle St **SW11** **128** G8
Aston St **E14** **127** J8
Astoria Wk **SW9** **129** N9
Astwood Ms **SW7** **128** B3
Asylum Rd **SE15** **130** F6
Athelstane Gro **E3** **127** K4
Atherfold Rd **SW9** **129** L9
Atherstone Ms **SW7** **128** C3
Atherton St **SW11** **128** E8
Athlone St **NW5** **124** G1
Athol Sq **E14** **127** N8
Atlantic Rd **SW9** **129** N10
Atlas Ms **N7** **125** M1
Atley Rd **E3** **127** L3
Atterbury St **SW1** **129** K3
Aubrey Moore Pt **E15** **127** N4
Auburn Cl **SE14** **131** J6
Auckland Rd **SW11** **128** E10
Auden Pl **NW1** **124** G3
Audley Cl **SW11** **128** G9
Audrey St **E2** **126** E4
Augusta St **E14** **127** M8
Augustus St **NW1** **125** H4
Aulton Pl **SE11** **129** N4
Austen Ho **NW6** **124** A5
Austin Friars **EC2** **126** B8
Austin Rd **SW11** **128** G7
Austin St **E2** **126** D5
Austral St **SE11** **129** P3
Autumn St **E3** **127** L3
Avalon Rd **SW6** **128** B7
Ave Maria La **EC4** **125** P8
Aveline St **SE11** **129** N4
Avenue, The **SE10** **131** P6
Avenue Cl **NW8** **124** E3
Avenue Rd **NW3** **124** D2
Avenue Rd **NW8** **124** D2
Avery Row **W1** **125** H9
Avignon Rd **SE4** **131** H9
Avis Sq **E1** **127** H8
Avon Rd **SE4** **131** L9
Avondale Ri **SE15** **130** D9
Avondale Sq **SE1** **130** E4
Avonley Rd **SE14** **130** G6
Avonmouth St **SE1** **130** A2
Aybrook St **W1** **124** G7
Aylesbury Rd **SE17** **130** B4
Aylesbury St **EC1** **125** P6
Aylesford St **SW1** **129** K4
Aylesham Cen **SE15** **130** E7
Aylward St **E1** **126** G8
Aylwyn Est **SE1** **130** C2
Ayres St **SE1** **130** A1
Aytoun Pl **SW9** **129** M8
Aytoun Rd **SW9** **129** M8
Azenby Rd **SE15** **130** D8

B

Baches St **N1** **126** B5
Back Ch La **E1** **126** E9
Back Hill **EC1** **125** N6
Bacon Gro **SE1** **130** D2
Bacon St **E1** **126** D6
Bacon St **E2** **126** D6
Badsworth Rd **SE5** **130** A6
Bagley's La **SW6** **128** B7
Bagshot St **SE17** **130** C4
Baildon St **SE8** **131** K6
Bainbridge St **WC1** **125** K8
Baker St **NW1** **124** F6
Baker St **W1** **124** F7

Bak - Bro

Baker's Row EC1 125 N6
Bakery Cl SW9 129 M7
Balaclava Rd SE1 130 D3
Balcombe St NW1 124 F6
Balcorne St E9 126 G2
Balderton St W1 124 G8
Baldock St E3 127 M4
Baldwin Cres SE5 130 A7
Baldwin Ter N1 126 A4
Baldwin's Gdns EC1 125 N7
Bale Rd E1 127 J7
Balfe St N1 125 L4
Balfern St SW11 128 E7
Balfour St SE17 130 B3
Balladier Wk E14 127 M7
Ballance Rd E9 127 H1
Ballantine St SW18 128 C10
Ballast Quay SE10 131 P4
Ballater Rd SW2 129 L10
Balls Pond Rd N1 126 B1
Balmer Rd E3 127 K4
Balmes Rd N1 126 B3
Balmoral Gro N7 125 M1
Balmore Cl E14 127 N8
Balniel Gate SW1 129 K4
Baltic St E EC1 126 A6
Baltic St W EC1 126 A6
Baltimore Ho SW18 128 C10
Baltimore Wf E14 131 M1
Balvaird Pl SW1 129 K4
Banbury Rd E9 127 H2
Banbury St SW11 128 E8
Bancroft Rd E1 126 G5
Banfield Rd SE15 130 F9
Banister Ms NW6 124 B2
Bank End SE1 126 A10
Bank St E14 127 L10
Bankside Av SE13 131 M9
Bankton Rd SW2 129 N10
Banner St EC1 126 A6
Bannerman Ho SW8 129 L5
Banstead St SE15 130 G9
Bantry St SE5 130 B6
Barbican, The EC2 126 A7
Barchester St E14 127 M7
Barclay Cl SW6 128 A6
Barclay Rd SW6 128 A6
Bardsley La SE10 131 N5
Barfleur La SE8 131 K3
Barford St N1 125 N3
Barforth Rd SE15 130 F9
Baring St N1 126 B3
Bark Pl W2 124 B9
Barker Dr NW1 125 J2
Barker Ms SW4 129 H10
Barker St SW10 128 C5
Barkston Gdns SW5 128 B3
Barkworth Rd SE16 130 F4
Barlborough St SE14 130 G6
Barleycorn Way E14 127 K9
Barnaby Pl SW7 128 D3
Barnard Ms SW11 128 E10
Barnard Rd SW11 128 E10
Barnby St NW1 125 J4
Barnes St E14 127 J8
Barnes Ter SE8 131 K4
Barnet Gro E2 126 E5
Barnfield Pl E14 131 L3
Barnham St SE1 130 C1
Barnsbury Gro N7 125 M2
Barnsbury Pk N1 125 N2
Barnsbury Rd N1 125 N4
Barnsbury Sq N1 125 N2
Barnsbury St N1 125 N2
Barnsbury Ter N1 125 M2
Barnsdale Av E14 131 L3
Barnsley St E1 126 F6
Barnwood Cl W9 124 B6
Baron St N1 125 N4
Barons Pl SE1 129 N1
Barratt Ind Pk E3 127 N6
Barrett St W1 124 G8
Barriedale SE14 131 J7
Barrington Rd SW9 129 P9
Barrow Hill Rd NW8 124 E4
Barset Rd SE15 130 G9
Barter St WC1 125 L7
Bartholomew Cl EC1 126 A7
Bartholomew Cl SW18 128 C10
Bartholomew Rd NW5 125 J1
Bartholomew Sq EC1 126 A6
Bartholomew St SE1 130 B2
Bartholomew Vil NW5 125 J1
Bartlett Cl E14 127 L8
Basevi Way SE8 131 M5
Basil St SW3 128 F2
Basing Ct SE15 130 D7
Basingdon Way SE5 130 B10
Basinghall Av EC2 126 B7
Basinghall St EC2 126 A7
Basire St N1 126 A3
Basnett Rd SW11 128 G9
Bassett St NW5 124 G1
Bastwick St EC1 125 P6
Basuto Rd SW6 128 A7
Batavia Rd SE14 131 J6
Batchelor St N1 125 N4
Bateman's Row EC2 126 C6
Bath St EC1 126 B5
Bath Ter SE1 130 A2
Bathurst St W2 124 D9
Batten St SW11 128 E9
Battersea Br SW11 128 D6
Battersea Br SW3 128 D6
Battersea Br Rd SW11 128 E6
Battersea Ch Rd SW11 128 D7
Battersea High St SW11 128 D7
Battersea Pk SW11 128 F6
Battersea Pk Rd SW11 128 E8
Battersea Pk Rd SW8 129 H7
Batty St E1 126 E8
Bavent Rd SE5 130 A8
Bawtree Rd SE14 131 J6
Baxendale St E2 126 E5
Baxter Rd N1 126 B1
Bayford St E8 126 F2
Bayham Pl NW1 125 J3
Bayham St NW1 125 J3
Bayley St WC1 125 K7
Baylis Rd SE1 129 N1
Baynes St NW1 125 J2
Bayswater Rd W2 124 D9
Baythorne St E3 127 K7
Baytree Rd SW2 129 M10
Bazely St E14 127 N9
Beachy Rd E3 127 L2
Beacon Gate SE14 131 H9
Beaconsfield Rd SE17 130 B4
Beaconsfield St N1 125 L3
Beak St W1 125 J9
Beale Pl E3 127 K4
Beale Rd E3 127 K3
Beanacre Cl E9 127 K1
Bear Gdns SE1 126 A10
Bear La SE1 125 P10
Beaton Cl SE15 130 D7
Beatrice Pl W8 128 B2
Beatrice Rd SE1 130 E3
Beatson Wk SE16 127 H10
Beatty St NW1 125 J4
Beauchamp Pl SW3 128 E2
Beauchamp Rd SW11 128 E10
Beaufort Gdns SW3 128 E2
Beaufort St SW3 128 D5
Beaufoy Wk SE11 129 M3
Beaulieu Cl SE5 130 B9
Beaumont Gro E1 127 H6
Beaumont Pl W1 125 J6
Beaumont Sq E1 127 H6
Beaumont St W1 124 G7
Beaumont Wk NW3 124 F2
Beccles St E14 127 K9
Beck Cl SE13 131 M7
Beck Rd E8 126 F3
Beckett Ho SW9 129 L8
Beckway St SE17 130 B3
Bedale St SE1 126 B10
Bede Sq E3 127 K6
Bedford Av WC1 125 K7
Bedford Gdns W8 124 A10
Bedford Ho SW4 129 L10
Bedford Pl WC1 125 L7
Bedford Rd SW4 129 L9
Bedford Row WC1 125 M7
Bedford Sq WC1 125 K7
Bedford St WC2 125 L9
Bedford Way WC1 125 K6
Beech St EC2 126 A7
Beech Tree Cl N1 125 N2
Beechmore Rd SW11 128 F7
Beechwood Rd E8 126 D1
Beehive Cl E8 126 D2
Beehive Pl SW9 129 N9
Beeston Pl SW1 129 H2
Belfort Rd SE15 130 G8
Belgrave Ct E14 127 K9
Belgrave Gdns NW8 124 B3
Belgrave Ms N SW1 128 G2
Belgrave Ms S SW1 128 G2
Belgrave Ms W SW1 128 G2
Belgrave Pl SW1 128 G2
Belgrave Rd SW1 129 H3
Belgrave Sq SW1 128 G2
Belgrave St E1 127 H8
Belgrove St WC1 125 L5
Belinda Rd SW9 129 P9
Belitha Vil N1 125 M2
Bell La E1 126 D7
Bell St NW1 124 E7
Bell Wf La EC4 126 A9
Bell Yd WC2 125 N8
Bellefields Rd SW9 129 M9
Bellenden Rd SE15 130 D8
Bellevue Pl E1 126 G6
Bells All SW6 128 A8
Bellwood Rd SE15 131 H10
Belmont Cl SW4 129 J9
Belmont Gro SE13 131 P9
Belmont Hill SE13 131 P9
Belmont Pk SE13 131 P10
Belmont Pk Cl SE13 131 P10
Belmont Rd SW4 129 J9
Belmont St NW1 124 G2
Belmore St SW8 129 K7
Belsham St E9 126 G1
Belsize Av NW3 124 D1
Belsize Gro NW3 124 E1
Belsize La NW3 124 D1
Belsize Pk NW3 124 D1
Belsize Pk Gdns NW3 124 E1
Belsize Rd NW6 124 B3
Belsize Sq NW3 124 D1
Belsize Ter NW3 124 D1
Belton Way E3 127 L7
Beltran Rd SW6 128 B8
Belvedere, The SW10 128 C7
Belvedere Ms SE15 130 G9
Belvedere Pl SE1 129 P1
Belvedere Pl SW2 129 M10
Belvedere Rd SE1 129 M1
Bemerton Est N1 125 L2
Bemerton St N1 125 M3
Ben Jonson Rd E1 127 J7
Benbow St SE8 131 L5
Benedict Rd SW9 129 M9
Bengeworth Rd SE5 130 A9
Benham Cl SW11 128 D9
Benhill Rd SE5 130 B6
Benjamin Cl E8 126 E3
Benjamin St EC1 125 P7
Benledi Rd E14 127 P8
Benn St E9 127 J1
Bennett Gro SE13 131 M7
Bennett Rd SW9 129 N8
Bentham Rd E9 127 H1
Bentinck St W1 124 G8
Benworth St E3 127 K5
Berger Rd E9 127 H1
Berkeley Ho E3 127 K6
Berkeley Sq W1 125 H9
Berkeley St W1 125 H9
Berkeley Twr E14 127 K10
Berkley Rd NW1 124 F2
Berkshire Rd E9 127 K1
Bermondsey St SE1 130 C1
Bermondsey Wall E SE16 130 E1
Bermondsey Wall W SE16 130 E1
Bermuda Way E1 127 J7
Bernard St WC1 125 L6
Bernays Gro SW9 129 M10
Berners Ms W1 125 J7
Berners Pl W1 125 J8
Berners Rd N1 125 N4
Berners St W1 125 J7
Bernhardt Cres NW8 124 E6
Berry St EC1 125 P6
Berryfield Rd SE17 129 P4
Berthon St SE8 131 L6
Bertrand St SE13 131 M9
Berwick St W1 125 J8
Bessborough Gdns SW1 129 K4
Bessborough Pl SW1 129 K4
Bessborough St SW1 129 K4
Bessemer Rd SE5 130 A8
Besson St SE14 130 G7
Bestwood St SE8 131 H3
Bethnal Grn Rd E1 126 D6
Bethnal Grn Rd E2 126 D6
Bethwin Rd SE5 129 P6
Betterton St WC2 125 L8
Bevan St N1 126 A3
Bevenden St N1 126 B5
Beverley Ct SE4 131 K9
Beverly NW8 124 E5
Bevington St SE16 130 E1
Bevis Marks EC3 126 C8
Bewdley St N1 125 N2
Bewick Ms SE15 130 F6
Bewick St SW8 129 H8
Bézier Apts EC1 126 B6
Bianca Rd SE15 130 D5
Bibury Cl SE15 130 C5
Bickenhall St W1 124 F7
Bicknell Rd SE5 130 A9
Bicycle Ms SW4 129 K9
Bidborough St WC1 125 K5
Biddulph Rd W9 124 B5
Bidwell St SE15 130 F7
Biggerstaff Rd E15 127 N3
Bigland St E1 126 F8
Billing Pl SW10 128 B6
Billing Rd SW10 128 B6
Billing St SW10 128 B6
Billingford Cl SE4 131 H10
Billingsgate Mkt E14 127 M10
Billington Rd SE14 131 H6
Billiter St EC3 126 C8
Billson St E14 131 N3
Bina Gdns SW5 128 C3
Binfield Rd SW4 129 L7
Bingfield St N1 125 L3
Bingham St N1 126 B1
Binney St W1 124 G8
Birchfield St E14 127 L9
Birchin La EC3 126 B8
Birchington Rd NW6 124 A3
Bird in Bush Rd SE15 130 E6
Birdcage Wk SW1 129 J1
Birdhurst Rd SW18 128 C10
Birdsfield La E3 127 K3
Birkbeck St E2 126 F5
Birkenhead St WC1 125 L5
Birley St SW11 128 G8
Birrell Ho SW9 129 M8
Biscoe Way SE13 131 P9
Bishop St N1 126 A3
Bishops Br W2 124 C8
Bishops Br Rd W2 124 B8
Bishop's Rd SW11 128 E6
Bishops Sq E1 126 C7
Bishops Ter SE11 129 N3
Bishops Way E2 126 F4
Bishopsgate EC2 126 C8
Bisson Rd E15 127 N4
Black Friars La EC4 125 P9
Black Prince Rd SE1 129 M3
Black Prince Rd SE11 129 M3
Blackburn Rd NW6 124 B1
Blackburne's Ms W1 124 G9
Blackfriars Br EC4 125 P9
Blackfriars Br SE1 125 P9
Blackfriars Rd SE1 125 P10
Blackheath Av SE10 131 P6
Blackheath Hill SE10 131 N7
Blackheath Ri SE13 131 N8
Blackheath Rd SE10 131 M7
Blackhorse Rd SE8 131 J4
Blacklands Ter SW3 128 F3
Blackpool Rd SE15 130 F8
Blackstone Est E8 126 E2
Blackthorn St E3 127 L6
Blackthorne Av N7 125 N1
Blacktree Ms SW9 129 N9
Blackwall Trd Est E14 127 P7
Blackwall Tunnel E14 127 P10
Blackwall Tunnel Northern App E14 127 L4
Blackwall Tunnel Northern App E3 127 L4
Blackwall Way E14 127 N9
Blackwood St SE17 130 B4
Blair Cl N1 126 A1
Blair St E14 127 N8
Blake Gdns SW6 128 B7
Blaker Rd E15 127 N3
Blakes Rd SE15 130 C6
Blanchard Way E8 126 E1
Blanchedowne SE5 130 B10
Blandford Sq NW1 124 E6
Blandford St W1 124 G7
Blantyre St SW10 128 D6
Blashford NW3 124 F2
Blasker Wk E14 131 L4
Blenheim Gro SE15 130 E8
Blenheim Rd NW8 124 C4
Blenheim Ter NW8 124 C4
Blessington Cl SE13 131 P9
Blessington Rd SE13 131 P9
Bletchley Ct N1 126 B4
Bletchley St N1 126 A4
Bliss Cres SE13 131 M8
Blissett St SE10 131 N7
Blithfield St W8 128 B2
Blomfield Rd W9 124 C7
Blomfield St EC2 126 B7
Blomfield Vil W2 124 B7
Blondel St SW11 128 G8
Blondin St E3 127 L4
Bloomfield Pl W1 125 H9
Bloomfield Ter SW1 128 G4
Bloomsbury Pl WC1 125 L7
Bloomsbury Sq WC1 125 L7
Bloomsbury St WC1 125 K7
Bloomsbury Way WC1 125 L8
Blossom St E1 126 C7
Blount St E14 127 J8
Blucher Rd SE5 130 A6
Blue Anchor La SE16 130 E3
Blue Anchor Yd E1 126 E9
Blundell St N7 125 L2
Blythe St E2 126 F5
Boardwalk Pl E14 127 N10
Boathouse Wk SE15 130 D6
Bobbin Cl SW4 129 J9
Bocking St E8 126 F3
Bohemia Pl E8 126 F1
Bohn Rd E1 127 J7
Bolden St SE8 131 M8
Bolina Rd SE16 130 G4
Bolingbroke Gro SW11 128 E10
Bolingbroke Wk SW11 128 D6
Bolney St SW8 129 M6
Bolsover St W1 125 H6
Bolton Cres SE5 129 P5
Bolton Gdns SW5 128 B4
Bolton Gdns Ms SW10 128 B4
Bolton Rd NW8 124 B3
Bolton St W1 125 H10
Boltons, The SW10 128 C4
Boltons Pl SW5 128 C4
Bombay St SE16 130 F3
Bonar Rd SE15 130 E6
Bond Ct EC4 126 B8
Bondway SW8 129 L5
Bonfield Rd SE13 131 N10
Bonhill St EC2 126 B6
Bonita Ms SE4 131 H9
Bonner Rd E2 126 G4
Bonner St E2 126 G4
Bonnington Sq SW8 129 M5
Bonny St NW1 125 J2
Bonsor St SE5 130 C6
Boot St N1 126 C5
Borland Rd SE15 130 G10
Borough High St SE1 130 A1
Borough Rd SE1 129 P2
Borthwick St SE8 131 L4
Boscobel Pl SW1 128 G3
Boscobel St NW8 124 D6
Boston Pl NW1 124 F6
Boswell St WC1 125 L7
Boulcott St E1 127 H8
Boulevard, The SW6 128 C7
Boundary La SE17 130 A5
Boundary Rd NW8 124 B3
Boundary St E2 126 D6
Bourdon St W1 125 H9
Bourne Est EC1 125 N7
Bourne St SW1 128 G3
Bourne Ter W2 124 B7
Bournemouth Cl SE15 130 E8
Bournemouth Rd SE15 130 E8
Bousfield Rd SE14 131 H8
Boutflower Rd SW11 128 E10
Bouverie Pl W2 124 D8
Bouverie St EC4 125 N8
Bovingdon Rd SW6 128 B7
Bow Br Est E3 127 M5
Bow Common La E3 127 K6
Bow La EC4 126 A8
Bow Rd E3 127 K5
Bow St WC2 125 L8
Bowden St SE11 129 N4
Bowditch SE8 131 K4
Bowen St E14 127 M8
Bower St E1 127 H8
Bowerdean St SW6 128 B7
Bowerman Av SE14 131 J5
Bowhill Cl SW9 129 N6
Bowl Ct EC2 126 C6
Bowland Rd SW4 129 K10
Bowling Grn La EC1 125 N6
Bowling Grn Pl SE1 130 B1
Bowling Grn St SE11 129 N5
Bowood Rd SW11 128 G10
Bowsprit Pt E14 131 L2
Bowyer Pl SE5 130 B6
Bowyer St SE5 130 A6
Boyd St E1 126 E8
Boyfield St SE1 129 P1
Boyne Rd SE13 131 N9
Boyson Rd SE17 130 B5
Brabazon St E14 127 M8
Brabourn Gro SE15 130 G8
Brackley Av SE15 130 F9
Bracklyn St N1 126 B4
Bradbourne St SW6 128 A8
Bradenham Cl SE17 130 B5
Bradmead SW8 129 H6
Bradstock Rd E9 127 H1
Bradwell St E1 127 H5
Brady St E1 126 F6
Braes St N1 125 P2
Braganza St SE17 129 P4
Braintree St E2 126 G5
Braithwaite St E1 126 D6
Braithwaite Twr W2 124 D7
Bramah Grn SW9 129 N7
Bramah Rd SW9 129 N7
Bramcote Gro SE16 130 G4
Bramerton St SW3 128 E5
Bramford Rd SW18 128 C10
Bramham Gdns SW5 128 B4
Bramlands Cl SW11 128 E9
Bramshaw Rd E9 127 H1
Bramwell Ms N1 125 M3
Branch Pl N1 126 B3
Branch Rd E14 127 J9
Branch St SE15 130 C6
Brand St SE10 131 N6
Brandon Est SE17 129 P5
Brandon Rd N7 125 L2
Brandon St SE17 130 A3
Brangton Rd SE11 129 M4
Branscombe St SE13 131 M9
Brassey Sq SW11 128 G9
Braxfield Rd SE4 131 J10
Bray NW3 124 E2
Bray Pl SW3 128 F3
Brayards Rd SE15 130 F8
Brayburne Av SW4 129 J8
Bread St EC4 126 A9
Breakspears Rd SE4 131 K9
Bream St E3 127 L2
Bream's Bldgs EC4 125 N8
Breer St SW6 128 B9
Bremner Rd SW7 128 C1
Brendon St W1 124 E8
Brenthouse Rd E9 126 F2
Brenton St E14 127 J8
Bressenden Pl SW1 129 H2
Brewer St W1 125 J9
Brewery Rd N7 125 L2
Brewhouse La E1 126 F10
Brewhouse Wk SE16 127 J10
Brewster Ho E14 127 K9
Briant St SE14 131 H7
Brick La E1 126 D7
Brick La E2 126 D5
Brick St W1 125 H10
Bride St N7 125 M1
Bridewain St SE1 130 D2
Bridge App NW1 124 G2
Bridge La SW11 128 E7
Bridge Meadows SE14 131 H5
Bridge Pl SW1 129 H3
Bridge Rd E15 127 P2
Bridge St SW1 129 L1
Bridgefoot SE1 129 L4
Bridgeman Rd N1 125 M2
Bridgeman St NW8 124 E4
Bridgend Rd SW18 128 C10
Bridges Ct SW11 128 D9
Bridgeway St NW1 125 J4
Bridport Pl N1 126 B4
Brief St SE5 129 P7
Bright St E14 127 M8
Brightlingsea Pl E14 127 K9
Brighton Ter SW9 129 M10
Brill Pl NW1 125 K4
Brindley St SE14 131 K7
Brinklow Ho W2 124 B7
Brinkworth Way E9 127 K1
Brion Pl E14 127 N7
Brisbane St SE5 130 B6
Bristol Gdns W9 124 B6
Britannia Rd E14 131 L3
Britannia Rd SW6 128 B6
Britannia Row N1 125 P3
Britannia St WC1 125 M5
Britannia Wk N1 126 B4
British St E3 127 K5
Britten St SW3 128 E4
Britton St EC1 125 P6
Brixton Oval SW2 129 N10
Brixton Rd SW9 129 N8
Brixton Sta Rd SW9 129 N10
Broad La EC2 126 C7
Broad Sanctuary SW1 129 K1
Broad Wk NW1 125 H5
Broad Wk W1 124 G10
Broad Wk, The W8 124 B10
Broadfield La NW1 125 L2
Broadhinton Rd SW4 129 H9
Broadhurst Gdns NW6 124 B1
Broadley St NW8 124 E7
Broadley Ter NW1 124 E6
Broadwalk Ct W8 124 A10
Broadwall SE1 125 N10
Broadway E15 127 P2
Broadway SW1 129 K2
Broadway Mkt E8 126 F3
Broadwick St W1 125 J8
Brock Pl E3 127 M6
Brock St NW1 125 J6
Brockham St SE1 130 A2
Brockill Cres SE4 131 J10
Brocklehurst St SE14 131 H6
Brockley Footpath SE15 130 G10
Brockley Gdns SE4 131 K8
Brockley Rd SE4 131 K9
Brodlove La E1 127 H9
Broke Wk E8 126 E3
Brokesley St E3 127 K6
Bromar Rd SE5 130 C9
Bromell's Rd SW4 129 J10
Bromfelde Rd SW4 129 K9
Bromfelde Wk SW4 129 K8
Bromfield St N1 125 N4
Bromley Hall Rd E14 127 N7
Bromley High St E3 127 M5
Bromley St E1 127 H7
Brompton Pk Cres SW6 128 B5
Brompton Pl SW3 128 E2
Brompton Rd SW1 128 F2
Brompton Rd SW3 128 E2
Brompton Rd SW7 128 F2
Brompton Sq SW3 128 E2
Bromyard Ho SE15 130 F6
Bronte Ho NW6 124 A5
Bronti Cl SE17 130 A4
Bronze St SE8 131 L6
Brook Dr SE11 129 N3
Brook Gate W1 124 F9
Brook St W1 125 H8
Brook St W2 124 D9
Brookbank Rd SE13 131 L9
Brooke St EC1 125 N7
Brookfield Rd E9 127 J1
Brookmill Rd SE8 131 L7
Brook's Ms W1 125 H9
Brooksbank St E9 126 G1
Brooksby St N1 125 N2
Broome Way SE5 130 B6
Broomfield St E14 127 L7
Broomgrove Rd SW9 129 M8
Broomhouse La SW6 128 A8
Broomhouse Rd SW6 128 A8
Brougham Rd E8 126 E3
Brougham St SW11 128 F8
Broughton Dr SW9 129 N10
Broughton Rd SW6 128 B8
Broughton St SW8 128 G8
Brown Hart Gdns W1 124 G9
Brown St W1 124 F8
Brownfield St E14 127 M8
Browning St SE17 130 A4
Brownlow Ms WC1 125 M6
Brownlow Rd E8 126 D3
Brown's Bldgs EC3 126 C8

Bro - Cle

Broxwood Way NW8 124 E3
Bruce Rd E3 127 M5
Bruford Ct SE8 131 L5
Brune St E1 126 D7
Brunel Est W2 124 A7
Brunel Rd SE16 130 G1
Brunswick, The WC1 125 L6
Brunswick Ct SE1 130 C1
Brunswick Gdns W8 124 A10
Brunswick Pk SE5 130 B7
Brunswick Pl N1 126 B5
Brunswick Pl NW1 124 G6
Brunswick Quay SE16 131 H2
Brunswick Sq WC1 125 L6
Brunswick Vil SE5 130 C7
Brunton Pl E14 127 J8
Brushfield St E1 126 C7
Brussels Rd SW11 128 D10
Bruton La W1 125 H9
Bruton Pl W1 125 H9
Bruton St W1 125 H9
Bryan Rd SE16 131 K1
Bryanston Pl W1 124 F7
Bryanston Sq W1 124 F7
Bryanston St W1 124 F8
Bryant Ct E2 126 D4
Bryant St E15 127 P2
Brymay Cl E3 127 L4
Brynmaer Rd SW11 128 F7
Buchan Rd SE15 130 G9
Buck St NW1 125 H2
Buckfast St E2 126 E5
Buckhurst St E1 126 F6
Buckingham Gate SW1 129 J2
Buckingham Palace Rd SW1 129 H3
Buckingham Rd N1 126 C1
Buckland Cres NW3 124 D2
Buckland St N1 126 B4
Bucklersbury EC4 126 B8
Buckmaster Rd SW11 128 E10
Bucknall St WC2 125 K8
Bucknell Cl SW2 129 M10
Buckner Rd SW2 129 M10
Buckters Rents SE16 127 J10
Budge's Wk W2 124 C9
Bulinga St SW1 129 K3
Bullace Row SE5 130 B6
Bullards Pl E2 127 H5
Bullen St SW11 128 E8
Buller Cl SE15 130 E6
Bullivant St E14 127 N9
Bulmer Pl W11 124 A10
Bulstrode St W1 124 G8
Bunhill Row EC1 126 B6
Bunhouse Pl SW1 128 G4
Bunning Way N7 125 L2
Burbage Cl SE1 130 B2
Burcham St E14 127 M8
Burchell Rd SE15 130 F7
Burcher Gale Gro SE15 130 C6
Burder Cl N1 126 C1
Burdett Rd E14 127 K6
Burdett Rd E3 127 K6
Burford Rd E15 127 P2
Burge St SE1 130 B2
Burgess Bus Pk SE5 130 B6
Burgess St E14 127 L7
Burgh St N1 125 P4
Burgos Gro SE10 131 M7
Burgoyne Rd SW9 129 M9
Burlington Arc W1 125 J9
Burlington Gdns W1 125 J9
Burnaby St SW10 128 C6
Burne St NW1 124 E7
Burney St SE10 131 N6
Burnham NW3 124 E2
Burnham St E2 126 G5
Burnley Rd SW9 129 M8
Burns Rd SW11 128 F8
Burnsall St SW3 128 E4
Burnside Cl SE16 127 H10
Burnthwaite Rd SW6 128 A6
Burr Cl E1 126 E10
Burrell St SE1 125 P10
Burrells Wf Sq E14 131 M4
Burrow Rd SE22 130 C10
Burslem St E1 126 E8
Burton La SW9 129 N8
Burton Rd SW9 129 P8
Burton St WC1 125 K5
Burwell Wk E3 127 L6
Burwood Pl W2 124 E8
Bury Pl WC1 125 L7
Bury St EC3 126 C8
Bury St SW1 125 J10
Busby Pl NW5 125 K1
Bush Rd E8 126 F3
Bush Rd SE8 131 H3
Bushberry Rd E9 127 J1
Bushey Hill Rd SE5 130 C7
Bushwood Dr SE1 130 D3
Butcher Row E1 127 H9
Butcher Row E14 127 H9
Bute St SW7 128 D3
Butlers Wf SE1 126 D10
Butterfly Wk Shop Cen SE5 130 A7
Buttermere Wk E8 126 D1
Buttesland St N1 126 B5
Buxhall Cres E9 127 K1
Buxted Rd E8 126 D2
Buxted Rd SE22 130 C10
Buxton Ms SW4 129 K8
Buxton St E1 126 D6
Byam St SW6 128 C8
Byfield Cl SE16 131 J1
Bygrove St E14 127 M8
Byng Pl WC1 125 K6
Byng St E14 131 L1
Byron Cl E8 126 E3
Bythorn St SW9 129 M9
Byward St EC3 126 C9
Bywater Pl SE16 127 J10
Bywater St SW3 128 F4

C

Cabbell St NW1 124 E7
Cable St E1 126 E9
Cabot Pl E14 127 L10
Cabot Sq E14 127 L10
Cabul Rd SW11 128 E8
Cade Rd SE10 131 P7
Cadet Dr SE1 130 E3
Cadiz St SE17 130 A4
Cadogan Gdns SW3 128 F3
Cadogan Gate SW1 128 F3
Cadogan La SW1 128 G2
Cadogan Pl SW1 128 F2
Cadogan Sq SW1 128 F2
Cadogan St SW3 128 F3
Cadogan Ter E9 127 K1
Cahir St E14 131 M3
Calabria Rd N5 125 P1
Calais St SE5 129 P7
Caldecot Rd SE5 130 A8
Caldwell St SW9 129 M6
Cale St SW3 128 E4
Caledonia St N1 125 L4
Caledonian Rd N1 125 M4
Caledonian Wf E14 131 P3
Callendar Rd SW7 128 D2
Callow St SW3 128 D5
Calshot St N1 125 M4
Calthorpe St WC1 125 M6
Calvert Av E2 126 C5
Calverton SE5 130 C5
Calvin St E1 126 D6
Calypso Cres SE15 130 D6
Calypso Way SE16 131 K2
Cam Rd E15 127 P3
Camberwell Ch St SE5 130 B7
Camberwell Grn SE5 130 B7
Camberwell Gro SE5 130 B7
Camberwell New Rd SE5 129 N6
Camberwell Rd SE5 130 A5
Camberwell Sta Rd SE5 130 A7
Cambria Rd SE5 130 A9
Cambria St SW6 128 B6
Cambridge Av NW6 124 A4
Cambridge Circ WC2 125 K8
Cambridge Cres E2 126 F4
Cambridge Gdns NW6 124 A4
Cambridge Heath Rd E1 126 F4
Cambridge Heath Rd E2 126 F4
Cambridge Pl W8 128 B1
Cambridge Rd NW6 124 A5
Cambridge Rd SW11 128 F7
Cambridge Sq W2 124 E8
Cambridge St SW1 129 H4
Camden High St NW1 125 H3
Camden Ms NW1 125 K1
Camden Pk Rd NW1 125 K1
Camden Pas N1 125 P4
Camden Rd NW1 125 J3
Camden Sq NW1 125 K1
Camden St NW1 125 H2
Camden Wk N1 125 P3
Camdenhurst St E14 127 J8
Camera Pl SW10 128 D5
Camilla Rd SE16 130 F3
Camlet St E2 126 D6
Camley St NW1 125 K2
Camomile St EC3 126 C8
Campana Rd SW6 128 A7
Campbell Rd E3 127 L5
Campden Gro W8 128 A1
Campden Hill W8 128 A1
Campden Hill Gdns W8 124 A10
Campden Hill Rd W8 124 A10
Campden St W8 124 A10
Camplin St SE14 131 H6
Canada Est SE16 130 G2
Canada Sq E14 127 M10
Canada St SE16 131 H1
Canal App SE8 131 J5
Canal Boul NW1 125 K1
Canal Cl E1 127 J6
Canal Gro SE15 130 E5
Canal Path E2 126 D3
Canal Reach N1 125 L3
Canal St SE5 130 B5
Canal Wk N1 126 B3
Cancell Rd SW9 129 N7
Candahar Rd SW11 128 E8
Candle Gro SE15 130 F9
Candle St E1 127 J7
Candy St E3 127 K3
Canfield Gdns NW6 124 C2
Canning Cross SE5 130 C8
Canning Pas W8 128 C2
Canning Pl W8 128 C2
Cannon Dr E14 127 L9
Cannon St EC4 126 A8
Cannon St Rd E1 126 F8
Cannon Wf Bus Cen SE8 131 J3
Canon Beck Rd SE16 130 G1
Canon Row SW1 129 L1
Canon St N1 126 A3
Canonbury Cres N1 126 A2
Canonbury Gro N1 126 A2
Canonbury La N1 125 P2
Canonbury Pk N N1 126 A1
Canonbury Pk S N1 126 A1
Canonbury Pl N1 125 P1
Canonbury Rd N1 125 P1
Canonbury Sq N1 125 P2
Canonbury St N1 126 A2
Canonbury Vil N1 125 P2
Canrobert St E2 126 F5
Cantelowes Rd NW1 125 K1
Canterbury Cres SW9 129 N9
Canterbury Pl SE17 129 P3
Canterbury Rd NW6 124 A4
Canterbury Ter NW6 124 A4
Cantium Retail Pk SE1 130 E5
Canton St E14 127 L8
Canute Gdns SE16 131 H3
Capland St NW8 124 D6
Capper St WC1 125 J6
Capstan Rd SE8 131 K3
Capstan Sq E14 131 N1
Capstan Way SE16 127 J10
Caradoc Cl W2 124 A8
Carbis Rd E14 127 K8
Carburton St W1 125 H7
Carden Rd SE15 130 F9
Cardigan Rd E3 127 K4
Cardigan St SE11 129 N4
Cardinal Bourne St SE1 130 B2
Cardine Ms SE15 130 F6
Cardington St NW1 125 J5
Carew St SE5 130 A8
Carey Gdns SW8 129 J7
Carey St WC2 125 M8
Carfax Pl SW4 129 K10
Carlile Cl E3 127 K4
Carlisle La SE1 129 M2
Carlisle Ms NW8 124 D7
Carlisle Pl SW1 129 J2
Carlos Pl W1 124 G9
Carlton Gdns SW1 125 K10
Carlton Gro SE15 130 F7
Carlton Hill NW8 124 B4
Carlton Ho Ter SW1 125 K10
Carlton Twr Pl SW1 128 F2
Carlton Vale NW6 124 B4
Carlyle Sq SW3 128 D4
Carmelite St EC4 125 N9
Carmen St E14 127 M8
Carnaby St W1 125 J8
Carnegie St N1 125 M3
Carnoustie Dr N1 125 M2
Carnwath Rd SW6 128 A9
Carol St NW1 125 J3
Caroline Gdns SE15 130 F6
Caroline Pl SW11 128 G8
Caroline Pl W2 124 B9
Caroline St E1 127 H8
Caroline Ter SW1 128 G3
Carpenters Pl SW4 129 K10
Carpenters Rd E15 127 N2
Carr St E14 127 J7
Carrara Cl SW9 129 P10
Carrara Ms E8 126 E1
Carriage Dr E SW11 128 G6
Carriage Dr N SW11 128 G5
Carriage Dr S SW11 128 F7
Carriage Dr W SW11 128 F6
Carron Cl E14 127 M8
Carroun Rd SW8 129 M6
Carter La EC4 125 P8
Carter Pl SE17 130 A4
Carter St SE17 130 A5
Carteret St SW1 129 K1
Carteret Way SE8 131 J3
Cartier Circle E14 127 M10
Carting La WC2 125 L9
Cartwright Gdns WC1 125 L5
Cartwright St E1 126 D9
Casby Ho SE16 130 E2
Cascades Twr E14 127 K10
Casella Rd SE14 131 H6
Casey Cl NW8 124 E5
Caspian St SE5 130 B6
Cassidy Rd SW6 128 A6
Cassilis Rd E14 131 L1
Cassland Rd E9 126 G2
Casson St E1 126 E7
Castellain Rd W9 124 B6
Casterbridge NW6 124 B3
Castle La SW1 129 J2
Castle Pl NW1 125 H1
Castle Rd NW1 125 H1
Castlebrook Cl SE11 129 P3
Castlehaven Rd NW1 125 H2
Castlemain St E1 126 F7
Castlemaine Twr SW11 128 F7
Castlemead SE5 130 A6
Castor La E14 127 M9
Caterham Rd SE13 131 N9
Catesby St SE17 130 B3
Cathay St SE16 130 F1
Cathcart Rd SW10 128 C5
Cathcart St NW5 125 H1
Cathedral St SE1 126 B10
Catherine Gro SE10 131 M7
Catherine Pl SW1 129 J2
Catherine St WC2 125 M9
Catlin St SE16 130 E4
Cato Rd SW4 129 K9
Cato St W1 124 E7
Cator St SE15 130 D5
Catton St WC1 125 M7
Caulfield Rd SE15 130 F8
Causeway, The SW18 128 B10
Causton St SW1 129 K3
Cavell St E1 126 F7
Cavendish Av NW8 124 D4
Cavendish Cl NW8 124 D5
Cavendish Pl W1 125 H8
Cavendish Sq W1 125 H8
Cavendish St N1 126 B4
Caversham Rd NW5 125 J1
Caversham St SW3 128 F5
Caxton Gro E3 127 L5
Caxton St SW1 129 J2
Cedar Cl E3 127 K3
Cedar Way NW1 125 K2
Cedarne Rd SW6 128 B6
Cedars Cl SE13 131 P9
Cedars Ms SW4 129 H10
Cedars Rd SW4 129 H9
Celandine Cl E14 127 L7
Celandine Dr E8 126 D2
Celestial Gdns SE13 131 P10
Celtic St E14 127 M7
Centaur St SE1 129 M2
Central Av SW11 128 F6
Central St EC1 126 A5
Centre Pt WC1 125 K8
Centre St E2 126 F4
Centurion Cl N7 125 M2
Cephas Av E1 126 G6
Cephas St E1 126 G6
Cerise Rd SE15 130 E7
Chabot Dr SE15 130 F9
Chadbourn St E14 127 M7
Chadwell St EC1 125 N5
Chadwick Rd SE15 130 D8
Chadwick St SW1 129 K2
Chagford St NW1 124 F6
Chalbury Wk N1 125 M4
Chalcot Cres NW1 124 F3
Chalcot Gdns NW3 124 F1
Chalcot Rd NW1 124 G2
Chalcot Sq NW1 124 G2
Chalk Fm Rd NW1 124 G2
Chalsey Rd SE4 131 K10
Chalton St NW1 125 K4
Chamber St E1 126 D9
Chambers St SE16 130 E1
Chambord St E2 126 D5
Champion Gro SE5 130 B9
Champion Hill SE5 130 B9
Champion Hill Est SE5 130 C9
Champion Pk SE5 130 B8
Chance St E1 126 D6
Chance St E2 126 D6
Chancel St SE1 125 P10
Chancery La WC2 125 M7
Chandler Way SE15 130 C5
Chandlers Ms E14 131 L1
Chandos Pl WC2 125 L9
Chandos St W1 125 H7
Channelsea Rd E15 127 P3
Chant Sq E15 127 P2
Chant St E15 127 P2
Chantrey Rd SW9 129 M9
Chantry St N1 125 P3
Chapel Ho St E14 131 M4
Chapel Mkt N1 125 N4
Chapel Pl W1 125 H8
Chapel Side W2 124 B9
Chapel St NW1 124 E7
Chapel St SW1 128 G2
Chaplin Cl SE1 129 N1
Chapman Rd E9 127 K1
Chapman St E1 126 F9
Chapter Rd SE17 129 P4
Chapter St SW1 129 K3
Charing Cross Rd WC2 125 K8
Charlbert St NW8 124 E4
Charles Barry Cl SW4 129 J9
Charles Coveney Rd SE15 130 D7
Charles Dickens Ho E2 126 F5
Charles II St SW1 125 K10
Charles Sq N1 126 B5
Charles St W1 125 H10
Charleston St SE17 130 A3
Charlotte Despard Av SW11 128 G7
Charlotte Rd EC2 126 C6
Charlotte Row SW4 129 J9
Charlotte St W1 125 J7
Charlotte Ter N1 125 M3
Charlton Pl N1 125 P4
Charlwood Pl SW1 129 J3
Charlwood St SW1 129 J4
Charnwood Gdns E14 131 L3
Charrington St NW1 125 K4
Chart St N1 126 B5
Charterhouse, The EC1 125 P6
Charterhouse Sq EC1 125 P7
Charterhouse St EC1 125 P7
Chartham Ct SW9 129 N9
Chase, The SW4 129 H9
Chaseley St E14 127 J8
Chatfield Rd SW11 128 C9
Chatham Pl E9 126 G1
Chatham St SE17 130 B3
Chatsworth Ct W8 128 A3
Chaucer Dr SE1 130 D3
Cheapside EC2 126 A8
Cheapside Pas EC4 126 A8
Chelsea Br SW1 129 H5
Chelsea Br SW8 129 H5
Chelsea Br Rd SW1 128 G4
Chelsea Embk SW3 128 E5
Chelsea Harbour SW10 128 C7
Chelsea Harbour Dr SW10 128 C7
Chelsea Manor Gdns SW3 128 E5
Chelsea Manor St SW3 128 E4
Chelsea Pk Gdns SW3 128 D5
Chelsea Sq SW3 128 D4
Chelsea Wf SW10 128 D6
Chelsham Rd SW4 129 K9
Cheltenham Rd SE15 130 G10
Cheltenham Ter SW3 128 F4
Chelwood Wk SE4 131 J10
Chenies Pl NW1 125 K4
Chenies St WC1 125 K7
Cheniston Gdns W8 128 B2
Chepstow Cres W11 124 A9
Chepstow Pl W2 124 A8
Chepstow Rd W2 124 A8
Cherbury St N1 126 B4
Cherry Gdn St SE16 130 F1
Cherrywood Cl E3 127 J5
Cheryls Cl SW6 128 B7
Chesham Ms SW1 128 G2
Chesham Pl SW1 128 G2
Chesham St SW1 128 G2
Cheshire Cl SE4 131 K8
Cheshire St E2 126 D6
Chesil Ct E2 126 G4
Chesney St SW11 128 G7
Chester Cl SW1 128 G1
Chester Gate NW1 125 H5
Chester Ms SW1 129 H2
Chester Rd NW1 124 G5
Chester Row SW1 128 G3
Chester Sq SW1 129 H2
Chester St E2 126 E6
Chester St SW1 128 G2
Chester Way SE11 129 N3
Chesterfield Gdns W1 125 H10
Chesterfield Hill W1 125 H9
Chesterfield St W1 125 H10
Chesterfield Wk SE10 131 P7
Chesterfield Way SE15 130 G6
Chestnut Cl SE14 131 K7
Cheval Pl SW7 128 E2
Cheval St E14 131 L2
Cheyne Gdns SW3 128 E5
Cheyne Ms SW3 128 E5
Cheyne Pl SW3 128 F5
Cheyne Row SW3 128 E5
Cheyne Wk SW10 128 D6
Cheyne Wk SW3 128 E5
Chicheley St SE1 129 M1
Chichester Rd NW6 124 A4
Chichester Rd W2 124 B7
Chichester St SW1 129 J4
Chichester Way E14 131 P3
Chicksand St E1 126 D7
Chiddingstone St SW6 128 A8
Childeric Rd SE14 131 J6
Childers St SE8 131 J5
Child's Pl SW5 128 A3
Child's St SW5 128 A3
Chillington Dr SW11 128 C10
Chiltern Rd E3 127 L6
Chiltern St W1 124 G7
Chilton Gro SE8 131 H3
Chilton St E2 126 D6
Chilworth Ms W2 124 C8
Chilworth St W2 124 C8
Chip St SW4 129 K9
Chipka St E14 131 N1
Chipley St SE14 131 J5
Chippenham Gdns NW6 124 A5
Chippenham Ms W9 124 A6
Chippenham Rd W9 124 A6
Chipstead St SW6 128 A7
Chisenhale Rd E3 127 J4
Chiswell St EC1 126 A7
Chitty St W1 125 J7
Choumert Gro SE15 130 E8
Choumert Ms SE15 130 E8
Choumert Rd SE15 130 D9
Choumert Sq SE15 130 E8
Chris Pullen Way N7 125 L1
Chrisp St E14 127 M7
Christchurch St SW3 128 F5
Christian Ct SE16 127 K10
Christian St E1 126 E8
Christie Rd E9 127 J1
Christopher Cl SE16 131 H1
Christopher St EC2 126 B6
Chryssell Rd SW9 129 N6
Chubworthy St SE14 131 J5
Chudleigh St E1 127 H8
Chumleigh St SE5 130 C5
Church Cres E9 127 H2
Church Gro SE13 131 M10
Church Rd N1 126 A1
Church St NW8 124 D7
Church St W2 124 D7
Church St Est NW8 124 D6
Church Ter SW8 129 K8
Churchill Gdns SW1 129 J4
Churchill Gdns Rd SW1 129 H4
Churchill Pl E14 127 M10
Churchway NW1 125 K5
Churton Pl SW1 129 J3
Churton St SW1 129 J3
Cicely Rd SE15 130 E7
Cinnabar Wf E1 126 E10
Cinnamon Row SW11 128 C9
Cinnamon St E1 126 F10
Circus Rd NW8 124 D5
Circus St SE10 131 N6
Cirencester St W2 124 B7
Citius Wk E20 127 N1
City Gdn Row N1 125 P4
City Rd EC1 125 P4
Clabon Ms SW1 128 F2
Clack St SE16 130 G1
Claire Pl E14 131 L2
Clancarty Rd SW6 128 A8
Clandon St SE8 131 L8
Clanricarde Gdns W2 124 A9
Clapham Common SW4 129 J10
Clapham Common N Side SW4 129 H10
Clapham Common W Side SW4 128 G10
Clapham Cres SW4 129 K10
Clapham Est SW11 128 E10
Clapham High St SW4 129 K10
Clapham Junct Sta SW11 128 D10
Clapham Manor St SW4 129 J9
Clapham Pk Rd SW4 129 K10
Clapham Rd SW9 129 L9
Clapham Rd Est SW4 129 K9
Clare Ho E3 127 K3
Clare La N1 126 A2
Clare Rd SE14 131 K7
Clare St E2 126 F4
Claredale St E2 126 E4
Claremont Cl N1 125 N4
Claremont Sq N1 125 N4
Claremont St SE10 131 M5
Clarence Gdns NW1 125 H5
Clarence Ms SE16 127 H10
Clarence Rd SE8 131 M5
Clarence Wk SW4 129 L8
Clarence Way NW1 125 H2
Clarence Way Est NW1 125 H2
Clarendon Cl E9 126 G2
Clarendon Gdns W9 124 C6
Clarendon Pl W2 124 E9
Clarendon Ri SE13 131 N9
Clarendon St SW1 129 H4
Clareville Gro SW7 128 C3
Clareville St SW7 128 C3
Clarewood Wk SW9 129 P10
Clarges Ms W1 125 H10
Clarges St W1 125 H10
Claribel Rd SW9 129 P8
Clarissa St E8 126 D3
Clark St E1 126 F7
Clarkson Row NW1 125 J4
Clarkson St E2 126 F5
Clarnico La E20 127 L1
Claude Rd SE15 130 F8
Claude St E14 131 L3
Clavell St SE10 131 N5
Claverton St SW1 129 J4
Claylands Pl SW8 129 N6
Claylands Rd SW8 129 M5
Claypole Rd E15 127 N4
Clayton Cres N1 125 L3
Clayton Ms SE10 131 P7
Clayton Rd SE15 130 E7
Clayton St SE11 129 N5
Clearwell Dr W9 124 B6
Cleaver Sq SE11 129 N4
Cleaver St SE11 129 N4
Clemence St E14 127 K7
Clement Av SW4 129 K10
Clement's Inn WC2 125 M8
Clements La EC4 126 B9
Clements Rd SE16 130 E2
Clephane Rd N1 126 A1
Clerkenwell Cl EC1 125 N6
Clerkenwell Grn EC1 125 P6

Cle - Dou

Clerkenwell Rd EC1 125 N7
Clermont Rd E9 126 G3
Cleve Rd NW6 124 A2
Cleveland Gdns W2 124 C8
Cleveland Rd N1 126 B2
Cleveland Row SW1 125 J10
Cleveland Sq W2 124 C8
Cleveland St W1 125 H6
Cleveland Ter W2 124 C8
Cleveland Way E1 126 G6
Clichy Est E1 126 G7
Cliff Rd NW1 125 K1
Cliff Ter SE8 131 L8
Cliff Vil NW1 125 K1
Clifford Dr SW9 129 P10
Clifford St W1 125 J9
Cliffview Rd SE13 131 L9
Clifton Cres SE15 130 F6
Clifton Gdns W9 124 C6
Clifton Gro E8 126 E1
Clifton Hill NW8 124 B4
Clifton Pl W2 124 D9
Clifton Ri SE14 131 J6
Clifton Rd N1 126 A1
Clifton Rd W9 124 C6
Clifton St EC2 126 C7
Clifton Vil W9 124 B7
Clifton Way SE15 130 F6
Clink St SE1 126 A10
Clinton Rd E3 127 J5
Clipper Way SE13 131 N10
Clipstone Ms W1 125 J6
Clipstone St W1 125 J7
Clitheroe Rd SW9 129 L8
Cliveden Pl SW1 128 G3
Clock Vw Cres N7 125 L1
Cloth Fair EC1 125 P7
Cloudesley Pl N1 125 N3
Cloudesley Rd N1 125 N3
Cloudesley Sq N1 125 N3
Cloudesley St N1 125 N3
Clove Cres E14 127 P9
Clove Hitch Quay SW11 128 C9
Cloysters Grn E1 126 E10
Club Row E1 126 D6
Club Row E2 126 D6
Cluny Ms SW5 128 A3
Clutton St E14 127 M7
Clyde St SE8 131 K5
Clyston St SW8 129 J8
Coate St E2 126 E4
Cobb St E1 126 D7
Cobbett St SW8 129 M6
Cobblestone Sq E1 126 F10
Coborn Rd E3 127 K5
Coborn St E3 127 K5
Cobourg Rd SE5 130 D5
Cobourg St NW1 125 J5
Cochrane St NW8 124 D4
Cock La EC1 125 P7
Cockayne Way SE8 131 J4
Cockspur St SW1 125 K10
Code St E1 126 D6
Cody Rd E16 127 P6
Cody Rd Bus Cen E16 127 P6
Coffey St SE8 131 L6
Coin St SE1 125 N10
Coity Rd NW5 124 G1
Coke St E1 126 E8
Colbeck Ms SW7 128 B3
Cold Blow La SE14 131 H6
Cold Harbour E14 131 N1
Coldbath St SE13 131 M7
Coldharbour La SE5 129 N10
Coldharbour La SW9 129 N10
Cole St SE1 130 A1
Colebeck Ms N1 125 P1
Colebert Av E1 126 G6
Colebrooke Row N1 125 P4
Colegrove Rd SE15 130 D6
Coleherne Ct SW5 128 B4
Coleherne Ms SW10 128 B4
Coleherne Rd SW10 128 B4
Coleman Flds N1 126 A3
Coleman Rd SE5 130 C6
Coleman St EC2 126 B8
Coleridge Cl SW8 129 H8
Coleridge Gdns SW10 128 B6
Coleridge Sq SW10 128 C6
Colestown St SW11 128 E8
Coley St WC1 125 M6
College App SE10 131 N5
College Cres NW3 124 D1
College Cross N1 125 N2
College Pk Cl SE13 131 P10
College Pl NW1 125 J3
College Ter E3 127 K5
Collent St E9 126 G1
Collett Rd SE16 130 E2
Collier St N1 125 M4
Collingham Gdns SW5 128 B3
Collingham Pl SW5 128 B3
Collingham Rd SW5 128 B3
Collingwood St E1 126 F6
Colls Rd SE15 130 G7
Colmore Ms SE15 130 F7
Colnbrook St SE1 129 P2
Cologne Rd SW11 128 D10
Colombo St SE1 125 P10
Colonnade, The SE8 131 K3
Colonnade WC1 125 L6
Colonnade Wk SW1 129 H3
Coltman St E14 127 J7
Columbia Rd E2 126 D5
Columbine Way SE13 131 N8
Colville Est N1 126 B3
Colyer Cl N1 125 M4
Comber Gro SE5 130 A7
Combermere Rd SW9 129 M9
Comerford Rd SE4 131 J10
Comet Pl SE8 131 L6
Comet St SE8 131 L6
Comfort St SE15 130 C5
Commercial Rd E1 126 E8
Commercial Rd E14 126 G8
Commercial St E1 126 D6
Commercial Way SE15 130 D6
Commodore Ho SW18 128 C10
Commodore St E1 127 J6
Compayne Gdns NW6 124 B2
Compton Av N1 125 P1
Compton Cl E3 127 L7
Compton Rd N1 125 P1
Compton St EC1 125 P6
Compton Ter N1 125 P1
Comus Pl SE17 130 C3
Comyn Rd SW11 128 E10
Concanon Rd SW2 129 M10
Concert Hall App SE1 125 M10
Concorde Way SE16 131 H3
Condell Rd SW8 129 J7
Condray Pl SW11 128 E6
Conduit Ms W2 124 D8
Conduit Pl W2 124 D8
Conduit St W1 125 H9
Coney Way SW8 129 M5
Congreve St SE17 130 C3
Coniger Rd SW6 128 A8
Conington Rd SE13 131 M8
Coniston Ho SE5 130 A6
Conistone Way N7 125 L2
Connaught Pl W2 124 F9
Connaught Sq W2 124 F8
Connaught St W2 124 E8
Consort Rd SE15 130 F7
Constitution Hill SW1 129 H1
Content St SE17 130 B3
Conway St W1 125 J6
Conyer St E3 127 J4
Cook's Rd E15 127 M4
Cooks Rd SE17 129 P5
Coombs St N1 125 P4
Coopers Cl E1 126 G6
Coopers La NW1 125 K4
Coopers Rd SE1 130 D4
Cope Pl W8 128 A2
Cope St SE16 131 H3
Copeland Dr E14 131 L3
Copeland Rd SE15 130 E8
Copenhagen Pl E14 127 K8
Copenhagen St N1 125 L3
Copleston Pas SE15 130 D9
Copleston Rd SE15 130 D9
Copley St E1 127 H7
Copper Row SE1 126 D10
Copperas St SE8 131 M5
Copperfield Rd E3 127 J6
Copperfield St SE1 129 P1
Coppock Cl SW11 128 E8
Copthall Av EC2 126 B8
Copthall Ct EC2 126 B8
Coptic St WC1 125 L7
Coral St SE1 129 N1
Coram St WC1 125 L6
Corbden Cl SE15 130 D7
Corbiere Ho N1 126 C3
Corbridge Cres E2 126 F4
Cordelia Cl SE24 129 P10
Cordelia St E14 127 M8
Corfield St E2 126 F5
Coriander Av E14 127 P8
Cork St W1 125 J9
Corlett St NW1 124 E7
Cormont Rd SE5 129 P7
Cornelia St N7 125 M1
Cornhill EC3 126 B8
Cornwall Av E2 126 G5
Cornwall Gdns SW7 128 B2
Cornwall Ms S SW7 128 C2
Cornwall Rd SE1 125 N10
Cornwall Sq SE11 129 N4
Cornwood Dr E1 126 G8
Coronet St N1 126 C5
Corporation Row EC1 125 N6
Corrance Rd SW2 129 L10
Corry Dr SW9 129 P10
Corsham St N1 126 B5
Corsica St N5 125 P1
Corunna Rd SW8 129 J7
Corunna Ter SW8 129 J7
Cossall Wk SE15 130 F7
Cosser St SE1 129 N2
Costa St SE15 130 E8
Cosway St NW1 124 E7
Cotall St E14 127 L8
Cotleigh Rd NW6 124 A2
Cottage Grn SE5 130 B6
Cottage Gro SW9 129 L9
Cottage Pl SW3 128 E2
Cottage St E14 127 M9
Cottesmore Gdns W8 128 B2
Cottingham Rd SW8 129 M6
Cotton Row SW11 128 C9
Cotton St E14 127 N9
Coulgate St SE4 131 J9
Coulson St SW3 128 F4
Councillor St SE5 130 A6
County Gro SE5 130 A7
County St SE1 130 A2
Courland Gro SW8 129 K7
Courland Gro Hall SW8 129 K8
Courland St SW8 129 K7
Court Gdns N7 125 N1
Courtenay St SE11 129 N4
Courtfield Gdns SW5 128 B3
Courtfield Rd SW7 128 B3
Courthill Rd SE13 131 N10
Courtnell St W2 124 A8
Courtyard, The N1 125 M2
Covent Gdn WC2 125 L9
Covent Gdn Mkt WC2 125 L9
Coventry Rd E1 126 F6
Coventry Rd E2 126 F6
Coventry St W1 125 K9
Coverley Cl E1 126 E7
Cowcross St EC1 125 P7
Cowdenbeath Path N1 125 M3
Cowley Est SW9 129 N7
Cowley Rd SW9 129 N7
Cowper St EC2 126 B6
Cowthorpe Rd SW8 129 K7
Crabtree Cl E2 126 D4
Crampton St SE17 129 P3
Cranbourn St WC2 125 K9
Cranbrook Rd SE8 131 L7
Cranbury Rd SW6 128 B8
Crane Gro N7 125 N1
Crane Mead SE16 130 G4
Crane St SE10 131 P4
Crane St SE15 130 D7
Cranfield Rd SE4 131 K9
Cranford St E1 127 H9
Cranleigh Ms SW11 128 E8
Cranleigh St NW1 125 J4
Cranley Gdns SW7 128 C4
Cranley Ms SW7 128 C4
Cranley Pl SW7 128 D3
Cranmer Ct SW4 129 K9
Cranmer Rd SW9 129 N6
Cranston Est N1 126 B4
Cranswick Rd SE16 130 F4
Cranwell Cl E3 127 M6
Cranwood St EC1 126 B5
Cranworth Gdns SW9 129 N7
Craven Hill W2 124 C9
Craven Hill Gdns W2 124 C9
Craven Hill Ms W2 124 C9
Craven Pas WC2 125 L10
Craven Rd W2 124 C9
Craven St WC2 125 L10
Craven Ter W2 124 C9
Crawford Est SE5 130 A8
Crawford Pl W1 124 E8
Crawford Rd SE5 130 A7
Crawford St W1 124 F7
Crawshay Rd SW9 129 N7
Crawthew Gro SE22 130 D10
Creasy Est SE1 130 C2
Credon Rd SE16 130 F4
Creechurch La EC3 126 C8
Creek Rd SE10 131 L5
Creek Rd SE8 131 L5
Creekside SE8 131 M6
Cremer St E2 126 D4
Cremorne Rd SW10 128 C6
Crescent Gro SW4 129 J10
Crescent Pl SW3 128 E3
Crescent St N1 125 M2
Crescent Way SE4 131 L9
Cresford Rd SW6 128 B7
Cresset Rd E9 126 G1
Cresset St SW4 129 K9
Cressingham Rd SE13 131 N9
Cresswell Gdns SW5 128 C4
Cresswell Pl SW10 128 C4
Cressy Pl E1 126 G7
Crestfield St WC1 125 L5
Crewdson Rd SW9 129 N6
Crews St E14 131 L3
Crewys Rd SE15 130 F8
Cricketers Ct SE11 129 P3
Crimscott St SE1 130 C2
Crimsworth Rd SW8 129 K7
Crinan St N1 125 L4
Cringle St SW8 129 J6
Crispin St E1 126 D7
Croft St SE8 131 J3
Crofters Way NW1 125 K3
Crofton Rd SE5 130 C7
Crofts St E1 126 E9
Crogsland Rd NW1 124 G2
Cromer St WC1 125 L5
Crompton St W2 124 D6
Cromwell Cres SW5 128 A3
Cromwell Gdns SW7 128 D2
Cromwell Ms SW7 128 D3
Cromwell Pl SW7 128 D3
Cromwell Rd SW5 128 B3
Cromwell Rd SW7 128 B3
Cromwell Rd SW9 129 P7
Cromwell Twr EC2 126 A7
Crondace Rd SW6 128 A7
Crondall St N1 126 B4
Cronin St SE15 130 D6
Crooke Rd SE8 131 J4
Crooms Hill SE10 131 P6
Crooms Hill Gro SE10 131 N6
Cropley St N1 126 B4
Cropthorne Ct W9 124 C5
Crosby Row SE1 130 B1
Cross Av SE10 131 P5
Cross Rd SE5 130 C8
Cross St N1 125 P3
Crossfield Rd NW3 124 D2
Crossfield St SE8 131 L6
Crossford St SW9 129 M8
Crosslet Vale SE10 131 M7
Crossley St N7 125 N1
Crossmount Ho SE5 130 A6
Crossthwaite Av SE5 130 B10
Crosswall EC3 126 D9
Croston St E8 126 E3
Crowder St E1 126 F9
Crowhurst Cl SW9 129 N8
Crowland Ter N1 126 B2
Crown Cl E3 127 L3
Crown Cl NW6 124 B1
Crown Pas SW1 125 J10
Crown Pl EC2 126 C7
Crown St SE5 130 A6
Crowndale Rd NW1 125 J4
Crows Rd E15 127 P5
Crucifix La SE1 130 C1
Cruden St N1 125 P3
Cruikshank St WC1 125 N5
Crutched Friars EC3 126 C9
Crystal Palace Rd SE22 130 E10
Cuba St E14 131 L1
Cubitt St WC1 125 M5
Cubitt Ter SW4 129 J9
Cudworth St E1 126 F6
Cuff Pt E2 126 D5
Culford Gdns SW3 128 F3
Culford Gro N1 126 C1
Culford Rd N1 126 C2
Culloden Cl SE16 130 E4
Culloden St E14 127 N8
Culmore Rd SE15 130 F6
Culross St W1 124 G9
Culvert Pl SW11 128 G8
Culvert Rd SW11 128 F8
Cumberland Cl E8 126 D1
Cumberland Gate W1 124 F9
Cumberland Mkt NW1 125 H5
Cumberland St SW1 129 H4
Cumming St N1 125 M4
Cunard Wk SE16 131 J3
Cundy St SW1 128 G3
Cunningham Pl NW8 124 D6
Cupar Rd SW11 128 G7
Cureton St SW1 129 K3
Curlew St SE1 130 D1
Curness St SE13 131 N10
Cursitor St EC4 125 N8
Curtain Rd EC2 126 C6
Curtis St SE1 130 D3
Curtis Way SE1 130 D3
Curzon Gate W1 124 G10
Curzon St W1 124 G10
Custom Ho Reach SE16 131 K1
Custom Ho Wk EC3 126 C9
Cut, The SE1 129 N1
Cutcombe Rd SE5 130 A8
Cuthbert St W2 124 D6
Cuthill Wk SE5 130 B7
Cutler St E1 126 C8
Cyclops Ms E14 131 L3
Cynthia St N1 125 M4
Cyntra Pl E8 126 F2
Cyprus Pl E2 126 G4
Cyprus St E2 126 G4
Cyril Mans SW11 128 F7
Cyrus St EC1 125 P6
Czar St SE8 131 L5

D

Dabin Cres SE10 131 N7
Dacca St SE8 131 K5
Dace Rd E3 127 L2
Dacre St SW1 129 K2
Dagmar Rd SE5 130 C7
Dagmar Ter N1 125 P3
Dagnall St SW11 128 F8
Dairy Ms SW9 129 L9
Daisy La SW6 128 A9
Dalberg Rd SW2 129 N10
Dalby Rd SW18 128 C10
Dalby St NW5 125 H1
Dale Rd SE17 129 P5
Daleham Ms NW3 124 D1
Dalehead NW1 125 J4
Daley St E9 127 H1
Daley Thompson Way SW8 129 H8
Dalgleish St E14 127 J8
Daling Way E3 127 J3
Dallington St EC1 125 P6
Dalrymple Rd SE4 131 J10
Dalston La E8 126 D1
Dalston Sq E8 126 D1
Dalwood St SE5 130 C7
Dalyell Rd SW9 129 M9
Dame St N1 126 A4
Damien St E1 126 F8
Danbury St N1 125 P4
Danby St SE15 130 D9
Dance Sq EC1 126 A5
Danesdale Rd E9 127 J1
Danesfield SE5 130 C5
Daneville Rd SE5 130 B7
Daniel Gdns SE15 130 D6
Daniels Rd SE15 130 G9
Dante Rd SE11 129 P3
Danvers St SW3 128 D5
D'Arblay St W1 125 J8
Darien Rd SW11 128 D9
Darling Rd SE4 131 L9
Darling Row E1 126 F6
Darnley Ho E14 127 J8
Darnley Rd E9 126 F1
Darsley Dr SW8 129 K7
Dartford St SE17 130 A5
Dartmouth Gro SE10 131 N7
Dartmouth Hill SE10 131 N7
Dartmouth Row SE10 131 N8
Dartmouth St SW1 129 K1
Dartmouth Ter SE10 131 P7
Darwin St SE17 130 B3
Date St SE17 130 A4
Daubeney Twr SE8 131 K4
Davenant St E1 126 E7
Daventry St NW1 124 E7
Davey Cl N7 125 M1
Davey Rd E9 127 L2
Davey St SE15 130 D5
David Ms SE10 131 N6
David St E15 127 P1
Davidge St SE1 129 P1
Davidson Gdns SW8 129 L6
Davies St W1 125 H9
Dawes St SE17 130 B4
Dawson Pl W2 124 A9
Dawson St E2 126 D4
Dayton Gro SE15 130 G7
De Beauvoir Cres N1 126 C3
De Beauvoir Est N1 126 B3
De Beauvoir Rd N1 126 C3
De Beauvoir Sq N1 126 C2
De Courbetin St E20 127 N1
De Crespigny Pk SE5 130 B8
De Laune St SE17 129 P4
De Morgan Rd SW6 128 B9
De Vere Gdns W8 128 C1
Deacon Ms N1 126 B2
Deacon Way SE17 130 A3
Deal Porters Wk SE16 131 H1
Deal Porters Way SE16 130 G2
Deal St E1 126 E7
Deals Gateway SE13 131 L7
Dean Bradley St SW1 129 L2
Dean Farrar St SW1 129 K2
Dean Ryle St SW1 129 L3
Dean Stanley St SW1 129 L2
Dean St W1 125 K8
Dean Trench St SW1 129 L2
Deancross St E1 126 G8
Deanery St W1 124 G10
Deans Bldgs SE17 130 B3
Decima St SE1 130 C2
Dee St E14 127 N8
Deeley Rd SW8 129 K7
Deepdene Rd SE5 130 B10
Deerdale Rd SE24 130 A10
Delaford Rd SE16 130 F4
Delamere Ter W2 124 B7
Delancey St NW1 125 H3
Delaware Rd W9 124 B6
Delhi St N1 125 L3
Delius Gro E15 127 P4
Dell Cl E15 127 P3
Dellow St E1 126 F9
Delmare Cl SW9 129 M10
Deloraine St SE8 131 L7
Delverton Rd SE17 129 P4
Delvino Rd SW6 128 A7
Denbigh Pl SW1 129 J4
Denbigh St SW1 129 J3
Dene Cl SE4 131 J9
Denman Rd SE15 130 D7
Denmark Gro N1 125 N4
Denmark Hill SE5 130 B7
Denmark Hill Est SE5 130 B10
Denmark Rd SE5 130 A7
Denmark St WC2 125 K8
Denne Ter E8 126 D3
Dennetts Rd SE14 130 G7
Denning Cl NW8 124 C5
Dennington Pk Rd NW6 124 A1
Dennison Pt E15 127 N2
Denny St SE11 129 N4
Denyer St SW3 128 E3
Deptford Br SE8 131 L7
Deptford Bdy SE8 131 L7
Deptford Ch St SE8 131 L5
Deptford Ferry Rd E14 131 L3
Deptford Grn SE8 131 L5
Deptford High St SE8 131 L5
Deptford Strand SE8 131 K3
Deptford Trd Est SE8 131 J5
Deptford Wf SE8 131 K3
Derby Rd E9 127 H3
Derbyshire St E2 126 E5
Dericote St E8 126 E3
Dering St W1 125 H8
Derry St W8 128 B1
Derwent Gro SE22 130 D10
Desborough Cl W2 124 B7
Desmond St SE14 131 J5
Devas St E3 127 M6
Deverell St SE1 130 B2
Devon St SE15 130 F5
Devonia Rd N1 125 P4
Devonport St E1 126 G8
Devons Est E3 127 M5
Devons Rd E3 127 L7
Devonshire Cl W1 125 H7
Devonshire Dr SE10 131 M6
Devonshire Gro SE15 130 F5
Devonshire Ms S W1 125 H7
Devonshire Ms W W1 125 H7
Devonshire Pl W1 124 G6
Devonshire St W1 125 H7
Devonshire Ter W2 124 C8
Dewar St SE15 130 E9
Dewberry St E14 127 N7
Dewey Rd N1 125 N4
D'Eynsford Rd SE5 130 B7
Dial Wk, The W8 128 B1
Diamond St SE15 130 C6
Diamond Ter SE10 131 N7
Dibden St N1 125 P3
Dickens Est SE1 130 D1
Dickens Est SE16 130 D1
Dickens Ho NW6 124 A5
Dickens Sq SE1 130 A2
Dickens St SW8 129 H8
Digby Rd E9 127 H1
Digby St E2 126 G5
Dighton Ct SE5 130 A5
Dilke St SW3 128 F5
Dimson Cres E3 127 L6
Dingle Gdns E14 127 L9
Dingley Pl EC1 126 A5
Dingley Rd EC1 126 A5
Discovery Dock Apts E E14 131 M1
Discovery Dock Apts W E14 131 M1
Discovery Wk E1 126 F10
Diss St E2 126 D5
Distaff La EC4 126 A9
Distillery Twr SE8 131 L7
Distin St SE11 129 N3
Ditch All SE10 131 M7
Ditchburn St E14 127 N9
Dixon Rd SE14 131 J7
Dixon's All SE16 130 F1
Dobson Cl NW6 124 D2
Dock Hill Av SE16 131 H1
Dock St E1 126 E9
Dockers Tanner Rd E14 131 L2
Dockhead SE1 130 D1
Dockley Rd SE16 130 E2
Docwra's Bldgs N1 126 C1
Dod St E14 127 K8
Doddington Gro SE17 129 P5
Doddington Pl SE17 129 P5
Dodson St SE1 129 N1
Dog Kennel Hill SE22 130 C9
Dog Kennel Hill Est SE22 130 C9
Dolben St SE1 125 P10
Dolland St SE11 129 M4
Dolman St SW4 129 M10
Dolphin La E14 127 M9
Dolphin Sq SW1 129 J4
Dombey St WC1 125 M7
Domett Cl SE5 130 B10
Dominion Dr SE16 131 H1
Don Phelan Cl SE5 130 B7
Donegal St N1 125 M4
Dongola Rd E1 127 J7
Donne Pl SW3 128 E3
Dora St E14 127 K8
Dora Way SW9 129 N8
Doran Wk E15 127 N2
Doric Way NW1 125 K5
Dorking Cl SE8 131 K5
Dorman Way NW8 124 D3
Dorney NW3 124 E2
Dorothy Rd SW11 128 F9
Dorrington St EC1 125 N7
Dorset Est E2 126 D5
Dorset Ri EC4 125 P8
Dorset Rd SW8 129 M6
Dorset Sq NW1 124 F6
Dorset St W1 124 G7
Doughty Ms WC1 125 M6
Doughty St WC1 125 M6

Dou - Ful

Douglas Rd **N1** 126 A2
Douglas St **SW1** 129 K3
Douglas Way **SE8** 131 K6
Douro Pl **W8** 128 B2
Douro St **E3** 127 L4
Dove Ms **SW5** 128 C3
Dove Rd **N1** 126 B1
Dove Row **E2** 126 E3
Dovehouse St **SW3** 128 D4
Dover St **W1** 125 H9
Dovercourt Est **N1** 126 B1
Doves Yd **N1** 125 N3
Dowells St **SE10** 131 M5
Dowgate Hill **EC4** 126 B9
Dowlas St **SE5** 130 C6
Down St **W1** 125 H10
Downfield Cl **W9** 124 B6
Downham Rd **N1** 126 B2
Downing St **SW1** 129 L1
Downtown Rd **SE16** 131 J1
Dowson Cl **SE5** 130 B10
D'Oyley St **SW1** 128 G3
Draco St **SE17** 130 A5
Dragon Rd **SE15** 130 C5
Dragoon Rd **SE8** 131 K4
Drake Rd **SE4** 131 L9
Drakefell Rd **SE14** 131 H8
Drakefell Rd **SE4** 131 H8
Draper Ho **SE1** 129 P3
Drawdock Rd **SE10** 127 P10
Draycott Av **SW3** 128 E3
Draycott Pl **SW3** 128 F3
Draycott Ter **SW3** 128 F3
Drayson Ms **W8** 128 A1
Drayton Gdns **SW10** 128 C4
Dreadnought Wk **SE10** 131 M5
Dresden Cl **NW6** 124 B1
Driffield Rd **E3** 127 J4
Drovers Pl **SE15** 130 F6
Drovers Way **N7** 125 L1
Druid St **SE1** 130 C1
Drummond Cres **NW1** 125 K5
Drummond Gate **SW1** 129 K4
Drummond Rd **SE16** 130 F2
Drummond St **NW1** 125 J6
Drury La **WC2** 125 L8
Dryden Ct **SE11** 129 N3
Drysdale St **N1** 126 C5
Dublin Av **E8** 126 E3
Duchess of Bedford's Wk **W8** 128 A1
Duchess St **W1** 125 H7
Duchy St **SE1** 125 N10
Ducie St **SW4** 129 M10
Duckett St **E1** 127 H6
Dudley St **W2** 124 D7
Duff St **E14** 127 M8
Dufferin St **EC1** 126 A6
Dugard Way **SE11** 129 P3
Duke of Wellington Pl **SW1** 128 G1
Duke of York Sq **SW3** 128 F3
Duke of York St **SW1** 125 J10
Duke St **SW1** 125 J10
Duke St **W1** 124 G8
Dukes La **W8** 128 A1
Dukes Pl **EC3** 126 C8
Duke's Rd **WC1** 125 K5
Dunbridge St **E2** 126 E6
Duncan Rd **E8** 126 F3
Duncan St **N1** 125 P4
Duncan Ter **N1** 125 P4
Duncannon St **WC2** 125 L9
Dundalk Rd **SE4** 131 J9
Dundas Rd **SE15** 130 G8
Dundas Rd **SW9** 129 N7
Dundee St **E1** 126 F10
Dunelm St **E1** 127 H8
Dunloe St **E2** 126 D4
Dunston Rd **E8** 126 D3
Dunston Rd **SW11** 128 G8
Dunston St **E8** 126 D3
Dunton Rd **SE1** 130 D4
Durand Gdns **SW9** 129 M7
Durands Wk **SE16** 131 K1
Durant St **E2** 126 E4
Durham Row **E1** 127 J7
Durham St **SE11** 129 M4
Durham Ter **W2** 124 B8
Durward St **E1** 126 F7
Durweston St **W1** 124 F7
Dutton St **SE10** 131 N7
Dye Ho La **E3** 127 L3
Dylan Rd **SE24** 129 P10
Dylways **SE5** 130 B10
Dymock St **SW6** 128 B9
Dynham Rd **NW6** 124 A2
Dyott St **WC1** 125 K8

E

Eagle Ct **EC1** 125 P7
Eagle Pl **SW7** 128 C4
Eagle St **WC1** 125 M7
Eagle Wf Rd **N1** 126 A4
Eamont St **NW8** 124 E4
Eardley Cres **SW5** 128 A4
Earl St **EC2** 126 B7
Earlham St **WC2** 125 K8
Earls Ct Gdns **SW5** 128 B3
Earls Ct Rd **SW5** 128 A3
Earls Ct Rd **W8** 128 A3
Earls Ct Sq **SW5** 128 B4
Earls Wk **W8** 128 A2
Earlsferry Way **N1** 125 M2
Earlston Gro **E9** 126 F3
Earnshaw St **WC2** 125 K8
East Arbour St **E1** 127 H8
East Cross Route **E3** 127 K2
East Cross Route **E9** 127 K2
East Dulwich Rd **SE15** 130 D10
East Dulwich Rd **SE22** 130 D10
East Ferry Rd **E14** 131 M2
East India Dock Rd **E14** 127 L8
East La **SE16** 130 E1
East Mt St **E1** 126 F7
East Rd **N1** 126 B5
East Rd **SW3** 128 G4
East Smithfield **E1** 126 D9
East St **SE17** 130 A4
East Surrey Gro **SE15** 130 D6
East Tenter St **E1** 126 D8
Eastbourne Ms **W2** 124 C8
Eastbourne Ter **W2** 124 C8
Eastbury Ter **E1** 127 H6
Eastcastle St **W1** 125 J8
Eastcheap **EC3** 126 B9
Eastcote St **SW9** 129 M8
Eastdown Pk **SE13** 131 P10
Eastern Rd **SE4** 131 L10
Eastfield St **E14** 127 J7
Eastfields Av **SW18** 128 A10
Eastlake Rd **SE5** 129 P8
Eastney St **SE10** 131 P4
Eastside Ms **E3** 127 L4
Eastway **E9** 127 K1
Eaton Cl **SW1** 128 G3
Eaton Dr **SW9** 129 P10
Eaton Gate **SW1** 128 G3
Eaton Ho **E14** 127 K9
Eaton La **SW1** 129 H2
Eaton Ms N **SW1** 128 G2
Eaton Ms S **SW1** 129 H2
Eaton Ms W **SW1** 128 G3
Eaton Pl **SW1** 128 G2
Eaton Row **SW1** 129 H2
Eaton Sq **SW1** 128 G3
Eaton Ter **SW1** 128 G3
Ebbisham Dr **SW8** 129 M5
Ebenezer St **N1** 126 B5
Ebley Cl **SE15** 130 D5
Ebor St **E1** 126 D6
Ebury Br **SW1** 129 H4
Ebury Br Est **SW1** 129 H4
Ebury Br Rd **SW1** 128 G4
Ebury Ms **SW1** 129 H3
Ebury Sq **SW1** 128 G3
Ebury St **SW1** 128 G3
Eccles Rd **SW11** 128 F10
Ecclesbourne Rd **N1** 126 A2
Eccleston Br **SW1** 129 H3
Eccleston Ms **SW1** 128 G2
Eccleston Pl **SW1** 129 H3
Eccleston Sq **SW1** 129 H3
Eccleston Sq Ms **SW1** 129 H3
Eccleston St **SW1** 128 G2
Eckford St **N1** 125 N4
Eckstein Rd **SW11** 128 E10
Edbrooke Rd **W9** 124 A6
Eddystone Twr **SE8** 131 J3
Edenbridge Rd **E9** 127 H2
Edenvale St **SW6** 128 B8
Edgar Kail Way **SE22** 130 C10
Edgar Rd **E3** 127 M5
Edgar Wallace Cl **SE15** 130 C6
Edgeley Rd **SW4** 129 K9
Edgware Rd **W2** 124 E8
Edinburgh Gate **SW1** 128 F1
Edinburgh Ho **W9** 124 B5
Edis St **NW1** 124 G3
Edith Gro **SW10** 128 C5
Edith Row **SW6** 128 B7
Edith St **E2** 126 E4
Edith Ter **SW10** 128 C6
Edithna St **SW9** 129 L9
Edmeston Cl **E9** 127 J1
Edmund St **SE5** 130 B6
Edna St **SW11** 128 E7
Edric Rd **SE14** 131 H6
Edrich Ho **SW4** 129 L7
Edward Pl **SE8** 131 K5
Edward St **SE14** 131 J6
Edward St **SE8** 131 K5
Edwardes Sq **W8** 128 A2
Edwards Ms **N1** 125 N2
Edwards Ms **W1** 124 G8
Edwin St **E1** 126 G6
Effie Pl **SW6** 128 A6
Effie Rd **SW6** 128 A6
Effra Rd **SW2** 129 N10
Egbert St **NW1** 124 G3
Egeremont Rd **SE13** 131 M8
Egerton Cres **SW3** 128 E3
Egerton Dr **SE10** 131 M7
Egerton Gdns **SW3** 128 E2
Egerton Gdns Ms **SW3** 128 E2
Egerton Pl **SW3** 128 E2
Egerton Ter **SW3** 128 E2
Egmont St **SE14** 131 H6
Elam Cl **SE5** 129 P8
Elam St **SE5** 129 P8
Eland Rd **SW11** 128 F9
Elbe St **SW6** 128 C8
Elcho St **SW11** 128 E6
Elcot Av **SE15** 130 F6
Elder St **E1** 126 D7
Eldon Rd **W8** 128 B2
Eldon St **EC2** 126 B7
Eleanor Cl **SE16** 131 H1
Eleanor Rd **E8** 126 F1
Eleanor Rd **SW9** 129 N7
Eleanor St **E3** 127 L5
Electra Bus Pk **E16** 127 P7
Electric Av **SW9** 129 N10
Electric La **SW9** 129 N10
Elephant & Castle **SE1** 129 P3
Elephant La **SE16** 130 G1
Elephant Rd **SE17** 130 A3
Elf Row **E1** 126 G9
Elgar St **SE16** 131 J2
Elgin Av **W9** 124 B5
Elia Ms **N1** 125 P4
Elia St **N1** 125 P4
Elias Pl **SW8** 129 N5
Elim Est **SE1** 130 B2
Eliot Hill **SE13** 131 N8
Eliot Ms **NW8** 124 C4
Eliot Pk **SE13** 131 N9
Eliot Vale **SE3** 131 P8
Elizabeth Av **N1** 126 A2
Elizabeth Br **SW1** 129 H3
Elizabeth Est **SE17** 130 B5
Elizabeth Ms **NW3** 124 E1
Elizabeth St **SW1** 128 G3
Elland Rd **SE15** 130 G10
Ellen St **E1** 126 E8
Ellerdale St **SE13** 131 M10
Ellery St **SE15** 130 F8
Ellesmere Rd **E3** 127 J4
Ellesmere St **E14** 127 M8
Ellingfort Rd **E8** 126 F2
Ellington St **N7** 125 N1
Elliott Rd **SW9** 129 P6
Elliott Sq **NW3** 124 E2
Elliotts Row **SE11** 129 P3
Ellis St **SW1** 128 F3
Ellsworth St **E2** 126 F5
Elm Friars Wk **NW1** 125 K2
Elm Gro **SE15** 130 D8
Elm Pk Gdns **SW10** 128 D4
Elm Pk La **SW3** 128 D4
Elm Pk Rd **SW3** 128 D5
Elm Pl **SW7** 128 D4
Elm Quay Ct **SW8** 129 K5
Elm St **WC1** 125 M6
Elm Tree Cl **NW8** 124 D5
Elm Tree Rd **NW8** 124 D5
Elmfield Way **W9** 124 A7
Elmhurst St **SW4** 129 K9
Elmington Est **SE5** 130 B6
Elmington Rd **SE5** 130 B7
Elmira St **SE13** 131 M9
Elmore St **N1** 126 A2
Elms Ms **W2** 124 D9
Elmslie Pt **E3** 127 K7
Elmstone Rd **SW6** 128 A7
Elmwood Ct **SW11** 129 H7
Elrington Rd **E8** 126 E1
Elsa St **E1** 127 J7
Elsdale St **E9** 126 G1
Elsie Rd **SE22** 130 D10
Elsley Rd **SW11** 128 F9
Elspeth Rd **SW11** 128 F10
Elsted St **SE17** 130 B3
Elswick Rd **SE13** 131 M8
Elswick St **SW6** 128 C8
Elsworthy Ri **NW3** 124 E2
Elsworthy Rd **NW3** 124 E3
Elsworthy Ter **NW3** 124 E2
Elthiron Rd **SW6** 128 A7
Elton Ho **E3** 127 K3
Eltringham St **SW18** 128 C10
Elvaston Ms **SW7** 128 C2
Elvaston Pl **SW7** 128 C2
Elverson Ms **SE8** 131 M8
Elverson Rd **SE8** 131 M8
Elverton St **SW1** 129 K3
Elwin St **E2** 126 E5
Elysian Ms **N7** 125 M1
Elystan Pl **SW3** 128 E4
Elystan St **SW3** 128 E3
Emba St **SE16** 130 E1
Embankment Gdns **SW3** 128 F5
Embankment Pier **WC2** 125 M10
Embankment Pl **WC2** 125 L10
Emberton **SE5** 130 C5
Embleton Rd **SE13** 131 M9
Emden St **SW6** 128 B7
Emerald St **WC1** 125 M7
Emerson St **SE1** 126 A10
Emery Hill St **SW1** 129 J2
Emma St **E2** 126 F4
Emmott Cl **E1** 127 J6
Emperor's Gate **SW7** 128 B2
Empire Wf Rd **E14** 131 P3
Empress Ms **SE5** 130 A8
Empress Pl **SW6** 128 A4
Empress St **SE17** 130 A5
Empson St **E3** 127 M6
Emu Rd **SW8** 129 H8
Endell St **WC2** 125 L8
Endsleigh Gdns **WC1** 125 K6
Endsleigh Pl **WC1** 125 K6
Endsleigh St **WC1** 125 K6
Endwell Rd **SE4** 131 J8
Enfield Rd **N1** 126 C2
Enford St **W1** 124 F7
Engate St **SE13** 131 N10
Englands La **NW3** 124 F1
Englefield Rd **N1** 126 B1
English St **E3** 127 K6
Enid St **SE16** 130 D2
Ennerdale Ho **E3** 127 K6
Ennismore Gdns **SW7** 128 E1
Ennismore Gdns Ms **SW7** 128 E2
Ennismore Ms **SW7** 128 E1
Ennismore St **SW7** 128 E2
Ensign Ho **SW18** 128 C9
Ensign St **E1** 126 E9
Enterprise Way **SW18** 128 A10
Enterprize Way **SE8** 131 K3
Epirus Ms **SW6** 128 A6
Epping Cl **E14** 131 L3
Epworth St **EC2** 126 B6
Erasmus St **SW1** 129 K3
Eresby Pl **NW6** 124 A2
Eric St **E3** 127 K6
Erlanger Rd **SE14** 131 H7
Ermine Rd **SE13** 131 M9
Ernest St **E1** 127 H6
Errol St **EC1** 126 A6
Erskine Rd **NW3** 124 F2
Esmeralda Rd **SE1** 130 E3
Essendine Rd **W9** 124 A6
Essex Rd **N1** 125 P3
Essex Vil **W8** 128 A1
Essian St **E1** 127 J7
Este Rd **SW11** 128 E9
Esterbrooke St **SW1** 129 K3
Ethelburga St **SW11** 128 E7
Ethelred Est **SE11** 129 N3
Ethnard Rd **SE15** 130 F5
Eton Av **NW3** 124 D2
Eton Coll Rd **NW3** 124 F1
Eton Rd **NW3** 124 F2
Eton Vil **NW3** 124 F1
Etta St **SE8** 131 J5
Ettrick St **E14** 127 N8
Eugenia Rd **SE16** 130 G3
Europa Trade Pk **E16** 127 P6
Eustace Rd **SW6** 128 A6
Euston Gro **NW1** 125 K5
Euston Rd **N1** 125 K6
Euston Rd **NW1** 125 H6
Euston Sq **NW1** 125 K5
Euston Sta **NW1** 125 J5
Euston St **NW1** 125 J5
Euston Twr **NW1** 125 J6
Evan Cook Cl **SE15** 130 G7
Evandale Rd **SW9** 129 N8
Evelina Rd **SE15** 130 G9
Eveline Lowe Est **SE16** 130 E2
Evelyn Gdns **SW7** 128 D4
Evelyn St **SE8** 131 J4
Evelyn Wk **N1** 126 B4
Everest Pl **E14** 127 N7
Evergreen Sq **E8** 126 D2
Everilda St **N1** 125 M3
Eversholt St **NW1** 125 J4
Eversleigh Rd **SW11** 128 F9
Everthorpe Rd **SE15** 130 D9
Evesham Wk **SW9** 129 N8
Evesham Way **SW11** 128 G9
Ewe Cl **N7** 125 L1
Ewer St **SE1** 126 A10
Ewhurst Cl **E1** 126 G7
Excelsior Gdns **SE13** 131 N8
Exchange Sq **EC2** 126 C7
Exeter St **WC2** 125 L9
Exeter Way **SE14** 131 K6
Exhibition Rd **SW7** 128 D2
Exmouth Mkt **EC1** 125 N6
Exmouth Pl **E8** 126 F2
Exon St **SE17** 130 C3
Exton St **SE1** 125 N10
Ezra St **E2** 126 D5

F

Fairbairn Grn **SW9** 129 P7
Fairbairn Grn **SW9** 129 N7
Faircharm Trd Est **SE8** 131 L6
Fairclough St **E1** 126 E8
Fairfax Pl **NW6** 124 C2
Fairfax Rd **NW6** 124 C2
Fairfield Rd **E3** 127 L4
Fairfoot Rd **E3** 127 L6
Fairhazel Gdns **NW6** 124 B1
Fairmont Av **E14** 127 P10
Fakruddin St **E1** 126 E6
Falcon Ct **EC4** 125 N8
Falcon Gro **SW11** 128 E9
Falcon La **SW11** 128 E9
Falcon Rd **SW11** 128 E8
Falcon Ter **SW11** 128 E9
Falcon Way **E14** 131 M3
Falcon Wf **SW11** 128 D8
Falkirk Ho **W9** 124 B5
Falkirk St **N1** 126 C4
Falmouth Rd **SE1** 130 A2
Fann St **EC1** 126 A6
Fann St **EC2** 126 A6
Fanshaw St **N1** 126 C5
Farm La **SW6** 128 A5
Farm St **W1** 125 H9
Farmers Rd **SE5** 129 P6
Farncombe St **SE16** 130 E1
Farrance St **E14** 127 K8
Farrell Ho **E1** 126 G8
Farrier St **NW1** 125 H2
Farrier Wk **SW10** 128 C5
Farringdon La **EC1** 125 N6
Farringdon Rd **EC1** 125 N6
Farringdon St **EC4** 125 P7
Farrins Rents **SE16** 127 J10
Farrow La **SE14** 130 G6
Farthingale Wk **E15** 127 P2
Fashion St **E1** 126 D7
Fassett Rd **E8** 126 E1
Fassett Sq **E8** 126 E1
Faulkner St **SE14** 130 G7
Favart Rd **SW6** 128 A7
Fawcett Cl **SW11** 128 D8
Fawcett St **SW10** 128 C5
Fawe St **E14** 127 M7
Feathers Pl **SE10** 131 P5
Featherstone St **EC1** 126 B6
Featley Rd **SW9** 129 P9
Fellows Ct **E2** 126 D4
Fellows Rd **NW3** 124 D2
Felmersham Cl **SW4** 129 K10
Felstead St **E9** 127 K1
Felton St **N1** 126 B3
Fenchurch Av **EC3** 126 C8
Fenchurch St **EC3** 126 C9
Fendall St **SE1** 130 C2
Fenham Rd **SE15** 130 E6
Fentiman Rd **SW8** 129 L5
Fenton Cl **SW9** 129 M8
Fenwick Gro **SE15** 130 E9
Fenwick Pl **SW9** 129 L9
Fenwick Rd **SE15** 130 E9
Ferdinand Dr **SE15** 130 C6
Ferdinand St **NW1** 124 G1
Ferguson Cl **E14** 131 L3
Fern St **E3** 127 L6
Ferndale Rd **SW4** 129 L10
Ferndale Rd **SW9** 129 M9
Ferndene Rd **SE24** 130 A10
Fernshaw Rd **SW10** 128 C5
Ferrey Ms **SW9** 129 N8
Ferrier Ind Est **SW18** 128 B10
Ferrier St **SW18** 128 B10
Ferris Rd **SE22** 130 E10
Ferry St **E14** 131 N4
Ferryman's Quay **SW6** 128 C8
Fetter La **EC4** 125 N8
Ffinch St **SE8** 131 L6
Field St **WC1** 125 M5
Fieldgate St **E1** 126 E7
Fielding Ho **NW6** 124 A5
Fielding St **SE17** 130 A5
Fields Est **E8** 126 E2
Fife Ter **N1** 125 M4
Finborough Rd **SW10** 128 B4
Finch Ms **SE15** 130 D6
Finchley Pl **NW8** 124 D4
Finchley Rd **NW3** 124 C1
Finchley Rd **NW8** 124 D3
Findhorn St **E14** 127 N8
Finland Rd **SE4** 131 J9
Finland St **SE16** 131 J2
Finnis St **E2** 126 F5
Finsbury Circ **EC2** 126 B7
Finsbury Est **EC1** 125 N5
Finsbury Mkt **EC2** 126 C6
Finsbury Pavement **EC2** 126 B7
Finsbury Sq **EC2** 126 B7
Finsbury St **EC2** 126 B7
Finsen Rd **SE5** 130 A9
Fir Trees Cl **SE16** 127 J10
Firbank Rd **SE15** 130 F8
First St **SW3** 128 E3
Fish St Hill **EC3** 126 B9
Fisher St **WC1** 125 M7
Fishermans Dr **SE16** 131 H1
Fisherman's Wk **E14** 127 L10
Fisherton St **NW8** 124 D6
Fitzalan St **SE11** 129 M3
Fitzgerald Ho **E14** 127 M8
Fitzhardinge St **W1** 124 G8
Fitzmaurice Pl **W1** 125 H9
Fitzpatrick Rd **SW9** 129 P7
Fitzroy Rd **NW1** 124 G3
Fitzroy Sq **W1** 125 J6
Fitzroy St **W1** 125 J6
Fitzwilliam Rd **SW4** 129 J9
Fiveways Rd **SW9** 129 N8
Flamborough St **E14** 127 J8
Flanders Way **E9** 127 H1
Flaxman Rd **SE5** 129 P8
Flaxman Ter **WC1** 125 K5
Fleet St **EC4** 125 N8
Fleming Rd **SE17** 129 P5
Fleur de Lis St **E1** 126 C6
Flint St **SE17** 130 B3
Flinton St **SE17** 130 C4
Flodden Rd **SE5** 130 A7
Flood St **SW3** 128 E4
Flood Wk **SW3** 128 E5
Flora Cl **E14** 127 M8
Floral St **WC2** 125 L9
Florence Rd **SE14** 131 K7
Florence St **N1** 125 P2
Florence Ter **SE14** 131 K7
Florida St **E2** 126 E5
Flower Wk, The **SW7** 128 C1
Foley St **W1** 125 J7
Folgate St **E1** 126 C7
Follett St **E14** 127 N8
Folly Wall **E14** 131 N1
Fontarabia Rd **SW11** 128 G10
Ford Rd **E3** 127 J3
Ford Sq **E1** 126 F7
Ford St **E3** 127 J3
Fordham St **E1** 126 E8
Fore St **EC2** 126 A7
Foreign St **SE5** 129 P8
Foreshore **SE8** 131 K3
Forest Gro **E8** 126 D1
Forest Rd **E8** 126 D1
Forester Rd **SE15** 130 F10
Forfar Rd **SW11** 128 G7
Formosa St **W9** 124 B7
Forset St **W1** 124 E8
Forsyth Gdns **SE17** 129 P5
Fort Rd **SE1** 130 D3
Forthbridge Rd **SW11** 128 G10
Fortius Wk **E20** 127 N1
Fortrose Cl **E14** 127 P8
Fortune Pl **SE1** 130 D4
Fortune St **EC1** 126 A6
Forum Magnum Sq **SE1** 129 M1
Fossil Rd **SE13** 131 L9
Foster La **EC2** 126 A8
Foubert's Pl **W1** 125 J8
Foulis Ter **SW7** 128 D4
Foundry Cl **SE16** 127 J10
Fount St **SW8** 129 K6
Fountain Ms **NW3** 124 F1
Fountain Pl **SW9** 129 N7
Fountain Sq **SW1** 129 H3
Four Seasons Cl **E3** 127 L4
Fournier St **E1** 126 D7
Fowler Cl **SW11** 128 D9
Fownes St **SW11** 128 E9
Fox Cl **E1** 126 G6
Foxberry Rd **SE4** 131 J9
Foxcote **SE5** 130 C4
Foxley Rd **SW9** 129 N6
Foxmore St **SW11** 128 F7
Foxwell St **SE4** 131 J9
Frampton Pk Est **E9** 126 G2
Frampton Pk Rd **E9** 126 G1
Frampton St **NW8** 124 D6
Francis Chichester Way **SW11** 128 G7
Francis St **SW1** 129 J3
Frank Ms **SE16** 130 F3
Frankham St **SE8** 131 L6
Frankland Cl **SE16** 130 F2
Franklin Cl **SE13** 131 M7
Franklin Pl **SE13** 131 M7
Franklin St **E3** 127 M5
Franklin's Row **SW3** 128 F4
Frazier St **SE1** 129 N1
Frean St **SE16** 130 E2
Frederick Cl **W2** 124 E9
Frederick Cres **SW9** 129 P6
Frederick St **WC1** 125 M5
Freedom St **SW11** 128 F8
Freemantle St **SE17** 130 C4
Freight La **N1** 125 K2
Freke Rd **SW11** 128 G9
Fremont St **E9** 126 G3
Frendsbury Rd **SE4** 131 J10
Frensham St **SE15** 130 E5
Frere St **SW11** 128 E8
Freshfield Av **E8** 126 D2
Friars Mead **E14** 131 N2
Friary Est **SE15** 130 E5
Friary Rd **SE15** 130 E6
Friday St **EC4** 126 A8
Friend St **EC1** 125 P5
Friendly St **SE8** 131 L7
Friendly St Ms **SE8** 131 L8
Frimley Way **E1** 127 H6
Friston St **SW6** 128 B8
Frith St **W1** 125 K8
Frogley Rd **SE22** 130 D10
Frognal Ct **NW3** 124 C1
Frome St **N1** 126 A4
Frostic Wk **E1** 126 D7
Froude St **SW8** 129 H8
Fulford St **SE16** 130 F1
Fulham Bdy **SW6** 128 A6
Fulham Rd **SW10** 128 B6
Fulham Rd **SW3** 128 C5

Ful - Her

Fulmead St **SW6** **128** B7
Fulwood Pl **WC1** **125** M7
Furley Rd **SE15** **130** E6
Furlong Rd **N7** **125** N1
Furness Rd **SW6** **128** B8
Furnival St **EC4** **125** N8
Furze St **E3** **127** L7
Fyfield Rd **SW9** **129** N9
Fynes St **SW1** **129** K3

G

Gables Cl **SE5** **130** C7
Gabrielle Ct **NW3** **124** D1
Gainsford St **SE1** **130** D1
Gairloch Rd **SE5** **130** C8
Gaisford St **NW5** **125** J1
Gaitskell Ct **SW11** **128** D8
Galbraith St **E14** **131** N2
Gale St **E3** **127** L7
Gales Gdns **E2** **126** F5
Galleywall Rd **SE16** **130** F3
Galsworthy Av **E14** **127** J7
Galway St **EC1** **126** A5
Gambetta St **SW8** **129** H8
Ganley Rd **SW11** **128** D9
Garden Rd **NW8** **124** C5
Garden Row **SE1** **129** P2
Garden St **E1** **127** H7
Gardens, The **SE22** **130** E10
Garfield Rd **SW11** **128** G9
Garford St **E14** **127** L9
Garlick Hill **EC4** **126** A9
Garnet St **E1** **126** G9
Garnies Cl **SE15** **130** D6
Garrick Cl **SW18** **128** C10
Garrick St **WC2** **125** L9
Garrison Rd **E3** **127** L3
Garsington Ms **SE4** **131** K9
Gartons Way **SW11** **128** C9
Garway Rd **W2** **124** B8
Gascoigne Pl **E2** **126** D5
Gascony Av **NW6** **124** A2
Gascoyne Rd **E9** **127** H2
Gaselee St **E14** **127** N9
Gaskell St **SW4** **129** L8
Gaskin St **N1** **125** P3
Gataker St **SE16** **130** F2
Gate Ms **SW7** **128** E1
Gateforth St **NW8** **124** E6
Gateley Rd **SW9** **129** M9
Gateway **SE17** **130** A5
Gateways, The **SW3** **128** F3
Gatliff Rd **SW1** **128** G4
Gatonby St **SE15** **130** D7
Gauden Cl **SW4** **129** K9
Gauden Rd **SW4** **129** K8
Gautrey Rd **SE15** **130** G8
Gawber St **E2** **126** G5
Gay Rd **E15** **127** P4
Gaydon Ho **W2** **124** B7
Gayfere St **SW1** **129** L2
Gayhurst Rd **E8** **126** E2
Gayton Ho **E3** **127** L6
Gaywood Est **SE1** **129** P2
Gedling Pl **SE1** **130** D1
Gee St **EC1** **126** A6
Geffrye St **E2** **126** D4
Geldart Rd **SE15** **130** F6
Gellatly Rd **SE14** **130** G8
General Wolfe Rd **SE10** **131** P7
Geneva Dr **SW9** **129** N10
Geoff Cade Way **E3** **127** K7
Geoffrey Cl **SE5** **130** A8
Geoffrey Rd **SE4** **131** K9
George Beard Rd **SE8** **131** K3
George Mathers Rd **SE11** **129** P3
George Row **SE16** **130** E1
George St **W1** **124** G8
George Yd **W1** **124** G9
Georgiana St **NW1** **125** J3
Gerald Rd **SW1** **128** G3
Geraldine St **SE11** **129** P2
Gerards Cl **SE16** **130** G4
Gernon Rd **E3** **127** J4
Gerrard Rd **N1** **125** P4
Gerrard St **W1** **125** K9
Gerridge St **SE1** **129** N1
Gertrude St **SW10** **128** C5
Gervase St **SE15** **130** F6
Gibbins Rd **E15** **127** N2
Gibbon Rd **SE15** **130** G8
Gibraltar Wk **E2** **126** D5
Gibson Rd **SE11** **129** M3
Gibson Sq **N1** **125** N3
Gideon Rd **SW11** **128** G9
Giffin St **SE8** **131** L6
Gifford St **N1** **125** L2
Gilbert Rd **SE11** **129** N3
Gilbert St **W1** **124** G8
Gilbeys Yd **NW1** **124** G2
Gill St **E14** **127** K8
Gillender St **E14** **127** N6
Gillender St **E3** **127** N6
Gillfoot **NW1** **125** J4
Gilling Ct **NW3** **124** E1
Gillingham St **SW1** **129** H3
Gilmore Rd **SE13** **131** P10
Gilstead Rd **SW6** **128** B8
Gilston Rd **SW10** **128** C4
Giltspur St **EC1** **125** P8
Giraud St **E14** **127** M8
Glade Wk **E20** **127** M1
Gladstone St **SE1** **129** P2
Gladys Rd **NW6** **124** A2
Glaisher St **SE8** **131** L5
Glamis Pl **E1** **126** G9
Glamis Rd **E1** **126** G9
Glasgow Ho **W9** **124** B4
Glasgow Ter **SW1** **129** J4
Glasshill St **SE1** **129** P1
Glasshouse Flds **E1** **127** H9
Glasshouse St **W1** **125** J9
Glasshouse Wk **SE11** **129** L4
Glaucus St **E3** **127** M7
Glebe Pl **SW3** **128** E5
Gledhow Gdns **SW5** **128** C3
Glenaffric Av **E14** **131** P3
Glendall St **SW9** **129** M10
Glendower Pl **SW7** **128** D3
Glenfinlas Way **SE5** **129** P6
Glengall Gro **E14** **131** M2
Glengall Rd **SE15** **130** D5
Glengall Ter **SE15** **130** D5
Glengarnock Av **E14** **131** N3
Glenilla Rd **NW3** **124** E1
Glenloch Rd **NW3** **124** E1
Glenmore Rd **NW3** **124** E1
Glenrosa St **SW6** **128** C8
Glensdale Rd **SE4** **131** K9
Glenton Ms **SE15** **130** G8
Glentworth St **NW1** **124** F6
Glenville Gro **SE8** **131** K6
Glenworth Av **E14** **131** P3
Globe Pond Rd **SE16** **127** J10
Globe Rd **E1** **126** G5
Globe Rd **E2** **126** G5
Globe Rope Wk **E14** **131** N3
Globe St **SE1** **130** A2
Gloucester Av **NW1** **124** G2
Gloucester Circ **SE10** **131** N6
Gloucester Cres **NW1** **125** H3
Gloucester Gate **NW1** **125** H4
Gloucester Ho **NW6** **124** A4
Gloucester Ms **W2** **124** C8
Gloucester Pl **NW1** **124** F6
Gloucester Pl **W1** **124** F7
Gloucester Rd **SW7** **128** C2
Gloucester Sq **W2** **124** D8
Gloucester St **SW1** **129** J4
Gloucester Ter **W2** **124** D9
Gloucester Wk **W8** **128** A1
Gloucester Way **EC1** **125** N5
Glycena Rd **SW11** **128** F9
Godalming Rd **E14** **127** M7
Godfrey St **E15** **127** N4
Godfrey St **SW3** **128** E4
Goding St **SE11** **129** L4
Godliman St **EC4** **126** A8
Godman Rd **SE15** **130** F8
Godson St **N1** **125** N4
Goffers Rd **SE3** **131** P7
Golden Jubilee Br **SE1** **125** L10
Golden Jubilee Br **WC2** **125** L10
Golden La **EC1** **126** A6
Golden Sq **W1** **125** J9
Goldhurst Ter **NW6** **124** B2
Golding St **E1** **126** E8
Goldington Cres **NW1** **125** K4
Goldington St **NW1** **125** K4
Goldman Cl **E2** **126** E6
Goldney Rd **W9** **124** A6
Goldsboro Rd **SW8** **129** K7
Goldsmith Rd **SE15** **130** E7
Goldsmith's Row **E2** **126** E4
Goldsmith's Sq **E2** **126** E4
Goldsworthy Gdns **SE16** **130** G3
Goldwin Cl **SE14** **130** G7
Gomm Rd **SE16** **130** G2
Gonson St **SE8** **131** M5
Goodge St **W1** **125** J7
Goodhart Pl **E14** **127** J9
Goodinge Cl **N7** **125** L1
Goodman's Stile **E1** **126** E8
Goodman's Yd **E1** **126** D9
Goods Way **NW1** **125** L4
Goodway Gdns **E14** **127** P8
Goodwin Cl **SE16** **130** E2
Goodwood Rd **SE14** **131** J6
Gopsall St **N1** **126** B3
Gordon Gro **SE5** **129** P8
Gordon Pl **W8** **128** A1
Gordon Rd **SE15** **130** F8
Gordon Sq **WC1** **125** K6
Gordon St **WC1** **125** K6
Gore Rd **E9** **126** G3
Gore St **SW7** **128** C2
Gorefield Pl **NW6** **124** A4
Goring St **EC3** **126** C8
Gorsuch St **E2** **126** D5
Gosfield St **W1** **125** J7
Gosling Way **SW9** **129** N7
Gosset St **E2** **126** D5
Gosterwood St **SE8** **131** J5
Goswell Rd **EC1** **125** P5
Gough Sq **EC4** **125** N8
Gough St **WC1** **125** M6
Goulden Ho App **SW11** **128** E8
Goulston St **E1** **126** D8
Gower Ms **WC1** **125** K7
Gower Pl **WC1** **125** J6
Gower St **WC1** **125** J6
Gower's Wk **E1** **126** E8
Gowlett Rd **SE15** **130** E9
Gowrie Rd **SW11** **128** G9
Grace St **E3** **127** M5
Gracechurch St **EC3** **126** B9
Grace's All **E1** **126** E9
Graces Ms **SE5** **130** C8
Graces Rd **SE5** **130** C8
Grafton Cres **NW1** **125** H1
Grafton Ho **E3** **127** L5
Grafton Pl **NW1** **125** K5
Grafton Sq **SW4** **129** J9
Grafton St **W1** **125** H9
Grafton Way **W1** **125** J6
Grafton Way **WC1** **125** J6
Graham Rd **E8** **126** D1
Graham St **N1** **125** P4
Graham Ter **SW1** **128** G3
Granary Rd **E1** **126** F6
Granary St **NW1** **125** K3
Granby St **E2** **126** D6
Granby Ter **NW1** **125** J4
Grand Junct Wf **N1** **126** A4
Grand Union Cres **E8** **126** E2
Grand Union Wk **NW1** **125** H2
Granfield St **SW11** **128** D7
Grange, The **SE1** **130** D2
Grange Gro **N1** **125** P1
Grange Pl **NW6** **124** A2
Grange Rd **SE1** **130** C2
Grange St **N1** **126** B3
Grange Wk **SE1** **130** C2
Grange Yd **SE1** **130** D2
Gransden Av **E8** **126** F2
Grant Rd **SW11** **128** D10
Grantbridge St **N1** **125** P4
Grantham Rd **SW9** **129** L8
Grantley St **E1** **127** H5
Grantully Rd **W9** **124** B5
Granville Ct **N1** **126** B3
Granville Gro **SE13** **131** N9
Granville Pk **SE13** **131** N9
Granville Pl **W1** **124** G8
Granville Rd **NW6** **124** A4
Granville Sq **SE15** **130** C6
Granville Sq **WC1** **125** M5
Grayling Sq **E2** **126** E5
Gray's Inn **WC1** **125** M7
Gray's Inn Rd **WC1** **125** M5
Grayshott Rd **SW11** **128** G8
Great Castle St **W1** **125** J8
Great Cen St **NW1** **124** F7
Great Chapel St **W1** **125** K8
Great Chart St **SW11** **128** D10
Great Coll St **SW1** **129** L2
Great Cumberland Pl **W1** **124** F8
Great Dover St **SE1** **130** A1
Great Eastern Enterprise Cen
E14 **131** M1
Great Eastern Rd **E15** **127** P2
Great Eastern St **EC2** **126** C5
Great George St **SW1** **129** K1
Great Guildford St **SE1** **126** A10
Great James St **WC1** **125** M7
Great Marlborough St **W1** **125** J8
Great Maze Pond **SE1** **130** B1
Great Northern Hotel **N1** **125** L4
Great Ormond St **WC1** **125** L7
Great Percy St **WC1** **125** M5
Great Peter St **SW1** **129** K2
Great Portland St **W1** **125** H6
Great Pulteney St **W1** **125** J9
Great Queen St **WC2** **125** L8
Great Russell St **WC1** **125** L7
Great St. Helens **EC3** **126** C8
Great Scotland Yd **SW1** **125** L10
Great Smith St **SW1** **129** K2
Great Suffolk St **SE1** **125** P10
Great Sutton St **EC1** **125** P6
Great Titchfield St **W1** **125** J8
Great Twr St **EC3** **126** C9
Great Winchester St **EC2** **126** B8
Great Windmill St **W1** **125** K9
Greatfield Cl **SE4** **131** L10
Greatorex St **E1** **126** E7
Greek St **W1** **125** K8
Green Bk **E1** **126** F10
Green Dale **SE5** **130** B10
Green Hundred Rd **SE15** **130** E5
Green St **W1** **124** G9
Greenbanks Cl **SE13** **131** M9
Greenberry St **NW8** **124** E4
Greencoat Pl **SW1** **129** J3
Greencroft Gdns **NW6** **124** B2
Greenfield Rd **E1** **126** E7
Greenham Cl **SE1** **129** N1
Greenland Quay **SE16** **131** H3
Greenland Rd **NW1** **125** J3
Greenman St **N1** **126** A2
Greenwell St **W1** **125** H6
Greenwich Ch St **SE10** **131** N5
Greenwich Foot Tunnel
E14 **131** N4
Greenwich Foot Tunnel
SE10 **131** N4
Greenwich High Rd **SE10** **131** M7
Greenwich Pk **SE10** **131** P6
Greenwich Pk St **SE10** **131** P4
Greenwich Quay **SE8** **131** M5
Greenwich S St **SE10** **131** M7
Greenwich Vw Pl **E14** **131** M2
Greenwood Ct **SW1** **129** J4
Greenwood Rd **E8** **126** E1
Greet St **SE1** **125** N10
Gregory Pl **W8** **128** B1
Grenade St **E14** **127** K9
Grenard Cl **SE15** **130** E6
Grendon St **NW8** **124** E6
Grenville Pl **SW7** **128** C2
Grenville St **WC1** **125** L6
Gresham Rd **SW9** **129** N9
Gresham St **EC2** **126** A8
Gresse St **W1** **125** K7
Greville Pl **NW6** **124** B4
Greville Rd **NW6** **124** B3
Greville St **EC1** **125** N7
Grey Eagle St **E1** **126** D7
Greycoat Pl **SW1** **129** K2
Greycoat St **SW1** **129** K2
Grimwade Cl **SE15** **130** G9
Grinling Pl **SE8** **131** L5
Grinstead Rd **SE8** **131** J4
Grittleton Rd **W9** **124** A6
Groom Pl **SW1** **128** G2
Groombridge Rd **E9** **127** H2
Grosvenor Br **SW1** **129** H5
Grosvenor Br **SW8** **129** H5
Grosvenor Cres **SW1** **128** G1
Grosvenor Cres Ms **SW1** **128** G1
Grosvenor Est **SW1** **129** K3
Grosvenor Gdns **SW1** **129** H2
Grosvenor Gate **W1** **124** F9
Grosvenor Hill **W1** **125** H9
Grosvenor Pk **SE5** **130** A5
Grosvenor Pl **SW1** **128** G1
Grosvenor Rd **SW1** **129** H5
Grosvenor Sq **W1** **124** G9
Grosvenor St **W1** **125** H9
Grosvenor Ter **SE5** **129** P6
Grosvenor Wf Rd **E14** **131** P3
Grove Cotts **SW3** **128** E5
Grove Cres Rd **E15** **127** P1
Grove End Rd **NW8** **124** D5
Grove Hill Rd **SE5** **130** C9
Grove La **SE5** **130** B7
Grove Pk **SE5** **130** C8
Grove Pas **E2** **126** F4
Grove Rd **E3** **127** H3
Grove St **SE8** **131** K3
Grove Vale **SE22** **130** C10
Grove Vil **E14** **127** M9
Grovelands Cl **SE5** **130** C8
Groveway **SW9** **129** M7
Grummant Rd **SE15** **130** D7
Grundy St **E14** **127** M8
Guerin Sq **E3** **127** K5
Guildford Gro **SE10** **131** M7
Guildford Rd **SW8** **129** L7
Guildhouse St **SW1** **129** J3
Guilford Pl **WC1** **125** M6
Guilford St **WC1** **125** L6
Guinness Cl **E9** **127** J2
Guinness Trust Bldgs
SE11 **129** P4
Guinness Trust Bldgs
SW9 **129** P10
Gulliver St **SE16** **131** K2
Gun St **E1** **126** D7
Gunmakers La **E3** **127** J3
Gunter Gro **SW10** **128** C5
Gunthorpe St **E1** **126** D7
Gunwhale Cl **SE16** **127** H10
Gurney Rd **SW6** **128** C9
Gutter La **EC2** **126** A8
Guy St **SE1** **130** B1
Gwen Morris Ho **SE5** **130** A6
Gwyn Cl **SW6** **128** C6
Gwynne Rd **SW11** **128** D8
Gylcote Cl **SE5** **130** B10

H

Haberdasher St **N1** **126** B5
Hackford Rd **SW9** **129** M7
Hackford Wk **SW9** **129** M7
Hackney Rd **E2** **126** D5
Haddo St **SE10** **131** M5
Haddonfield **SE8** **131** H3
Hadleigh St **E2** **126** G6
Hadley St **NW1** **125** H1
Hadrian Est **E2** **126** E4
Hafer Rd **SW11** **128** F10
Haggerston Rd **E8** **126** D2
Haines Cl **N1** **126** C2
Hainford Cl **SE4** **131** H10
Hainton Cl **E1** **126** F8
Halcomb St **N1** **126** C3
Hale St **E14** **127** M9
Halesworth Rd **SE13** **131** M9
Half Moon Cres **N1** **125** M4
Half Moon St **W1** **125** H10
Halford Rd **SW6** **128** A5
Halkin Arc **SW1** **128** F2
Halkin Pl **SW1** **128** G2
Halkin St **SW1** **128** G1
Hall Pl **W2** **124** D6
Hall Rd **NW8** **124** C5
Hall St **EC1** **125** P5
Hall Twr **W2** **124** D7
Hallam St **W1** **125** H7
Halley Gdns **SE13** **131** P10
Halley St **E14** **127** J7
Hallfield Est **W2** **124** C8
Halliford St **N1** **126** A2
Halo **E15** **127** N3
Halsey St **SW3** **128** F3
Halsmere Rd **SE5** **129** P7
Halton Cross St **N1** **125** P3
Halton Rd **N1** **125** P2
Hamble St **SW6** **128** B9
Hamilton Cl **NW8** **124** D5
Hamilton Gdns **NW8** **124** C5
Hamilton Pl **W1** **124** G10
Hamilton Ter **NW8** **124** B4
Hamlet, The **SE5** **130** B9
Hamlets Way **E3** **127** K6
Hammond St **NW5** **125** J1
Hampson Way **SW8** **129** M7
Hampstead Rd **NW1** **125** J4
Hampton Cl **NW6** **124** A5
Hampton St **SE1** **129** P3
Hampton St **SE17** **129** P3
Hanbury St **E1** **126** D7
Hancock Rd **E3** **127** N5
Hand Ct **WC1** **125** M7
Handel St **WC1** **125** L6
Handforth Rd **SW9** **129** N6
Handley Rd **E9** **126** G2
Handyside St **N1** **125** L3
Hankey Pl **SE1** **130** B1
Hannaford Wk **E3** **127** M6
Hannibal Rd **E1** **126** G7
Hannington Rd **SW4** **129** H9
Hanover Gdns **SE11** **129** N5
Hanover Gate **NW1** **124** E5
Hanover Pk **SE15** **130** E7
Hanover Sq **W1** **125** H8
Hanover St **W1** **125** H8
Hanover Ter **NW1** **124** E5
Hans Cres **SW1** **128** F2
Hans Pl **SW1** **128** F2
Hans Rd **SW3** **128** F2
Hanson St **W1** **125** J7
Hanway St **W1** **125** K8
Hanworth Ho **SE5** **129** P6
Harben Rd **NW6** **124** C2
Harbet Rd **W2** **124** D7
Harbinger Rd **E14** **131** M3
Harbledown Rd **SW6** **128** A7
Harbour Av **SW10** **128** C7
Harbour Ex Sq **E14** **131** M1
Harbour Rd **SE5** **130** A9
Harbut Rd **SW11** **128** D10
Harcourt Rd **SE4** **131** J10
Harcourt St **W1** **124** E7
Harcourt Ter **SW10** **128** B4
Harders Rd **SE15** **130** F8
Hardinge St **E1** **126** G8
Hardwick St **EC1** **125** N5
Hare & Billet Rd **SE3** **131** P7
Hare Row **E2** **126** F4
Hare Wk **N1** **126** C4
Harecourt Rd **N1** **126** A1
Haredale Rd **SE24** **130** A10
Harefield Ms **SE4** **131** K9
Harefield Rd **SE4** **131** K9
Harewood Av **NW1** **124** E6
Harfield Gdns **SE5** **130** C9
Harford St **E1** **127** J6
Hargwyne St **SW9** **129** M9
Harlescott Rd **SE15** **131** H10
Harley Gdns **SW10** **128** C4
Harley Gro **E3** **127** K5
Harley Pl **W1** **125** H7
Harley Rd **NW3** **124** D2
Harley St **W1** **125** H6
Harleyford Rd **SE11** **129** M5
Harleyford St **SE11** **129** N5
Harmony Pl **SE1** **130** D4
Harmood St **NW1** **125** H2
Harmsworth St **SE17** **129** P4
Harold Est **SE1** **130** C2
Harper Rd **SE1** **130** A2
Harpley Sq **E1** **126** G5
Harpur St **WC1** **125** M7
Harrap St **E14** **127** N9
Harriet Cl **E8** **126** E3
Harriet Wk **SW1** **128** F1
Harrington Gdns **SW7** **128** B3
Harrington Rd **SW7** **128** D3
Harrington Sq **NW1** **125** J4
Harrington St **NW1** **125** J5
Harris St **SE5** **130** B6
Harrison St **WC1** **125** L5
Harrow La **E14** **127** N9
Harrow Pl **E1** **126** C8
Harroway Rd **SW11** **128** D8
Harrowby St **W1** **124** E8
Harrowgate Rd **E9** **127** J1
Harston Wk **E3** **127** M6
Hartfield Ter **E3** **127** L4
Hartington Rd **SW8** **129** L7
Hartlake Rd **E9** **127** H1
Hartland Rd **NW1** **125** H2
Hartley St **E2** **126** G5
Harton St **SE8** **131** L7
Harts La **SE14** **131** J6
Harvey Rd **SE5** **130** B7
Harvey St **N1** **126** B3
Harwood Rd **SW6** **128** A6
Harwood Ter **SW6** **128** B7
Haselrigge Rd **SW4** **129** K10
Hasker St **SW3** **128** E3
Haslam Cl **N1** **125** N2
Haslam St **SE15** **130** D6
Hassett Rd **E9** **127** H1
Hastings Cl **SE15** **130** E6
Hastings St **WC1** **125** L5
Hatcham Pk Rd **SE14** **131** H7
Hatcham Rd **SE15** **130** G5
Hatfields **SE1** **125** P10
Hatherley Gro **W2** **124** B8
Hathorne Cl **SE15** **130** F8
Hatton Gdn **EC1** **125** N7
Hatton Pl **EC1** **125** N6
Hatton Wall **EC1** **125** N7
Havannah St **E14** **131** L1
Havelock St **N1** **125** L3
Havelock Ter **SW8** **129** H6
Haven Way **SE11** **130** D2
Haverfield Rd **E3** **127** J5
Haverstock St **N1** **125** P4
Havil St **SE5** **130** C6
Hawes St **N1** **125** P2
Hawgood St **E3** **127** L7
Hawkstone Est **SE16** **130** G3
Hawkstone Rd **SE16** **130** G3
Hawley Cres **NW1** **125** H2
Hawley Rd **NW1** **125** H2
Hawley St **NW1** **125** H2
Hawthorn Av **E3** **127** K3
Hawthorne Cl **N1** **126** C1
Hawtrey Rd **NW3** **124** E2
Hay Currie St **E14** **127** M8
Hay Hill **W1** **125** H9
Hay St **E2** **126** E3
Haydon Way **SW11** **128** D10
Hayes Gro **SE22** **130** D10
Hayles St **SE11** **129** P3
Haymarket **SW1** **125** K9
Haymerle Rd **SE15** **130** E5
Hay's Galleria **SE1** **126** C10
Hay's Ms **W1** **125** H9
Hazel Cl **SE15** **130** E8
Hazelmere Rd **NW6** **124** A3
Hazlebury Rd **SW6** **128** B8
Hazlewood Ms **SW9** **129** L9
Head St **E1** **127** H8
Headfort Pl **SW1** **128** G1
Headlam St **E1** **126** F6
Heald St **SE14** **131** K7
Healey St **NW1** **125** H1
Hearn St **EC2** **126** C6
Hearnshaw St **E14** **127** J7
Heath La **SE3** **131** P8
Heath Rd **SW8** **129** H8
Heathcote St **WC1** **125** M6
Heather Cl **SW8** **129** H9
Heathfield Ct **SE14** **130** G6
Heathwall St **SW11** **128** F9
Heaton Rd **SE15** **130** E9
Heddon St **W1** **125** J9
Hedgers Gro **E9** **127** J1
Heiron St **SE17** **129** P5
Helmet Row **EC1** **126** A6
Helmsley Pl **E8** **126** F2
Helsinki Sq **SE16** **131** J2
Hemans St **SW8** **129** K6
Hemberton Rd **SW9** **129** L9
Hemingford Rd **N1** **125** M3
Hemming St **E1** **126** E6
Hemp Wk **SE17** **130** B3
Hemstal Rd **NW6** **124** A2
Hemsworth St **N1** **126** C4
Heneage St **E1** **126** D7
Henley Dr **SE1** **130** D3
Henley St **SW11** **128** G8
Henning St **SW11** **128** E7
Henrietta Cl **SE8** **131** L5
Henrietta Pl **W1** **125** H8
Henrietta St **WC2** **125** L9
Henriques St **E1** **126** E8
Henry Dent Cl **SE5** **130** B9
Henshall St **N1** **126** B1
Henshaw St **SE17** **130** B3
Henstridge Pl **NW8** **124** E4
Henty Cl **SW11** **128** E6
Hepscott Rd **E9** **127** L2
Herbal Hill **EC1** **125** N6
Herbert St **NW5** **124** G1
Herbrand St **WC1** **125** L6
Hercules Rd **SE1** **129** M2
Hereford Ho **NW6** **124** A4
Hereford Rd **E3** **127** K4
Hereford Rd **W2** **124** A8
Hereford Sq **SW7** **128** C3

Her - Law

Hereford St E2 126 E6
Heritage Cl SW9 129 P9
Hermit St EC1 125 P5
Hermitage St W2 124 D7
Hermitage Wall E1 126 E10
Herne Hill Rd SE24 130 A9
Heron Pl SE16 127 J10
Heron Quay E14 127 L10
Heron Rd SE24 130 A10
Herrick St SW1 129 K3
Hertford Rd N1 126 C3
Hertford St W1 125 H10
Hertsmere Rd E14 127 L9
Hesper Ms SW5 128 B4
Hesperus Cres E14 131 M3
Hessel St E1 126 F8
Hester Rd SW11 128 E6
Heston St SE14 131 K7
Hetherington Rd SW4 129 L10
Hewison St E3 127 K4
Hewlett Rd E3 127 J4
Heyford Av SW8 129 L6
Heygate St SE17 130 A3
Hibbert St SW11 128 C9
Hickmore Wk SW4 129 J9
Hicks Cl SW11 128 E9
Hicks St SE8 131 J4
Hide Pl SW1 129 K3
Hide Twr SW1 129 K3
High Br SE10 131 P4
High Br Wf SE10 131 P4
High Holborn WC1 125 L8
High St E15 127 N4
High Timber St EC4 126 A9
Highbury Cor N5 125 N1
Highbury Gro N5 125 P1
Highbury Pl N5 125 P1
Highbury Sta Rd N1 125 N1
Highshore Rd SE15 130 D8
Highway, The E1 126 F9
Highway, The E14 126 F9
Hilary Cl SW6 128 B6
Hildyard Rd SW6 128 A5
Hilgrove Rd NW6 124 C2
Hill Rd NW8 124 C4
Hill St W1 125 H10
Hillbeck Cl SE15 130 G6
Hillcrest SE24 130 B10
Hillgate Pl W8 124 A10
Hillgate St W8 124 A10
Hillingdon St SE17 129 P6
Hillman St E8 126 F1
Hillmead Dr SW9 129 P10
Hillside Cl NW8 124 B4
Hilltop Rd NW6 124 A2
Hilly Flds Cres SE4 131 L9
Hillyard St SW9 129 N7
Hinckley Rd SE15 130 E10
Hind Gro E14 127 L8
Hinde St W1 124 G8
Hinton Rd SE24 129 P9
Hitchcock La E15 127 N1
Hitchin Sq E3 127 J4
Hobart Pl SW1 129 H2
Hobday St E14 127 M7
Hobury St SW10 128 C5
Hodnet Gro SE16 131 H3
Hogarth Rd SW5 128 B3
Holbeck Row SE15 130 E6
Holbein Ms SW1 128 G4
Holbein Pl SW1 128 G3
Holborn EC1 125 N7
Holborn Viaduct EC1 125 N7
Holcroft Rd E9 126 G2
Holden St SW11 128 G8
Holford St WC1 125 N5
Holgate Av SW11 128 D9
Holland Gro SW9 129 N6
Holland St SE1 125 P10
Holland St W8 128 A1
Hollen St W1 125 J8
Holles St W1 125 H8
Holly Gro SE15 130 D8
Holly St E8 126 D1
Hollybush Gdns E2 126 F5
Hollydale Rd SE15 130 G7
Hollydene SE15 130 F7
Hollymount Cl SE10 131 N7
Hollywood Rd SW10 128 C5
Holman Rd SW11 128 D8
Holmead Rd SW6 128 B6
Holmefield Ct NW3 124 E1
Holmes Cl SE22 130 E10
Holmes Ter SE1 129 N1
Holms St E2 126 E4
Holton St E1 127 H6
Holwood Pl SW4 129 K10
Holywell La EC2 126 C6
Holywell Row EC2 126 C6
Home Rd SW11 128 E8
Homefield St N1 126 C4
Homer Dr E14 131 L3
Homer Rd E9 127 J1
Homer Row W1 124 E7
Homer St W1 124 E7
Hooper St E1 126 E8
Hope St SW11 128 D9
Hopewell St SE5 130 B6
Hopton St SE1 125 P10
Hopton's Gdns SE1 125 P10
Hopwood Rd SE17 130 B5
Horatio St E2 126 D4
Horbury Cres W11 124 A9
Horle Wk SE5 129 P8
Hornby Cl NW3 124 D2
Hornshay St SE15 130 G5
Hornton Pl W8 128 A1
Hornton St W8 124 A10
Horse Guards Av SW1 125 L10
Horse Guards Rd SW1 125 K10
Horse Ride SW1 125 J10
Horseferry Pl SE10 131 N5
Horseferry Rd E14 127 J9
Horseferry Rd SW1 129 K3
Horselydown La SE1 130 D1
Horseshoe Cl E14 131 N4
Horsley St SE17 130 B5
Hortensia Rd SW10 128 C6
Horton Rd E8 126 F1
Hosier La EC1 125 P7
Hoskins St SE10 131 P4
Hotel Verta SW11 128 D8
Hotspur St SE11 129 N4
Houndsditch EC3 126 C8
Howard Building SW8 129 H5
Howbury Rd SE15 130 G9
Howden St SE15 130 E9
Howick Pl SW1 129 J2
Howie St SW11 128 E6
Howitt Rd NW3 124 E1
Howland Est SE16 130 G2
Howland St W1 125 J7
Howland Way SE16 131 J1
Howley Pl W2 124 C7
Hows St E2 126 D4
Howson Rd SE4 131 J10
Hoxton Sq N1 126 C5
Hoxton St N1 126 C3
Hubert Gro SW9 129 L9
Huddart St E3 127 K7
Huddleston Cl E2 126 G4
Hugh St SW1 129 H3
Hugon Rd SW6 128 B9
Huguenot Pl E1 126 D7
Hull Cl SE16 131 H1
Humphrey St SE1 130 D4
Hungerford Br SE1 125 L10
Hungerford Br WC2 125 L10
Hunsdon Rd SE14 131 H5
Hunter St WC1 125 L6
Huntingdon St N1 125 M2
Huntley St WC1 125 J6
Hunton St E1 126 E6
Hunts La E15 127 N4
Huntsman St SE17 130 B3
Hurlingham Bus Pk SW6 128 A9
Hurlingham Sq SW6 128 A9
Huson Cl NW3 124 E2
Hutchings St E14 131 L1
Hyde Pk SW7 124 E10
Hyde Pk W1 124 E10
Hyde Pk W2 124 E10
Hyde Pk Cor W1 128 G1
Hyde Pk Cres W2 124 E8
Hyde Pk Gdns W2 124 D9
Hyde Pk Gate SW7 128 C1
Hyde Pk Pl W2 124 E9
Hyde Pk Sq W2 124 E8
Hyde Pk St W2 124 E8
Hyde Rd N1 126 B3
Hyde Vale SE10 131 N6
Hyndman St SE15 130 F5

I

Iceland Rd E3 127 L3
Ida St E14 127 N8
Idonia St SE8 131 K6
Ifield Rd SW10 128 B5
Ilchester Gdns W2 124 B9
Ilderton Rd SE15 130 G6
Ilderton Rd SE16 130 F4
Iliffe St SE17 129 P4
Ilminster Gdns SW11 128 E10
Imber St N1 126 B3
Imperial Coll London SW7 128 D2
Imperial Coll Rd SW7 128 D2
Imperial Cres SW6 128 C8
Imperial Rd SW6 128 B7
Imperial Sq SW6 128 B7
Imperial St E3 127 N5
Imperial Wf SW6 128 C8
Indescon Ct E14 131 M1
Ingate Pl SW8 129 H7
Ingelow Rd SW8 129 H8
Ingleborough St SW9 129 N8
Inglesham Wk E9 127 K1
Ingleton St SW9 129 N8
Inglewood Cl E14 131 L3
Inglis St SE5 129 P7
Ingrave St SW11 128 D9
Inkerman Rd NW5 125 H1
Inner Circle NW1 124 G5
Innes St SE15 130 C6
International Way E15 127 N1
Inverness Pl W2 124 B9
Inverness St NW1 125 H3
Inverness Ter W2 124 B9
Inverton Rd SE15 131 H10
Inville Rd SE17 130 B4
Inwen Ct SE8 131 J4
Inworth St SW11 128 E8
Ireland Yd EC4 125 P8
Irene Rd SW6 128 A7
Iron Wks E3 127 L3
Ironmonger Row EC1 126 A5
Irving Gro SW9 129 M8
Irving St WC2 125 K9
Isabella St SE1 125 P10
Isambard Ms E14 131 N2
Island Ho E3 127 N5
Island Rd SE16 131 H3
Island Row E14 127 K8
Islington Grn N1 125 P3
Islington High St N1 125 P4
Islington Pk St N1 125 N2
Ivanhoe Rd SE5 130 D9
Iveagh Cl E9 127 H3
Iveley Rd SW4 129 J8
Iverna Ct W8 128 A2
Iverna Gdns W8 128 A2
Ives St SW3 128 E3
Ivimey St E2 126 E5
Ivor Pl NW1 124 F6
Ivor St NW1 125 J2
Ivy Rd SE4 131 K10
Ivy St N1 126 C4
Ivydale Rd SE15 131 H9
Ixworth Pl SW3 128 E4

J

Jacaranda Gro E8 126 D2
Jackman St E8 126 F3
Jackson Cl E9 126 G2
Jacob St SE1 130 D1
Jago Wk SE5 130 B6
Jamaica Rd SE1 130 D1
Jamaica Rd SE16 130 D1
Jamaica St E1 126 G8
James St W1 124 G8
Jameson Ct E2 126 G4
Jameson St W8 124 A10
Jamestown Rd NW1 125 H3
Jamestown Way E14 127 P9
Jamuna Cl E14 127 J7
Janet St E14 131 L2
Janeway St SE16 130 E1
Jardine Rd E1 127 H9
Jarrow Rd SE16 130 G3
Jay Ms SW7 128 C1
Jebb St E3 127 L4
Jedburgh St SW11 128 G10
Jefferson Plaza E3 127 N6
Jeffreys Rd SW4 129 L8
Jeffreys St NW1 125 H2
Jeffreys Wk SW4 129 L8
Jeger Av E2 126 D3
Jerdan Pl SW6 128 A6
Jeremiah St E14 127 M8
Jermyn St SW1 125 K9
Jerningham Rd SE14 131 J8
Jerome Cres NW8 124 E6
Jerrard St SE13 131 M9
Jewry St EC3 126 D8
Jew's Row SW18 128 B10
Jim Veal Dr N7 125 L1
Joan St SE1 125 P10
Jocelyn St SE15 130 E7
Jockey's Flds WC1 125 M7
Jodane St SE8 131 K3
Jodrell Rd E3 127 K3
John Adam St WC2 125 L9
John Aird Ct W2 124 C7
John Carpenter St EC4 125 P9
John Felton Rd SE16 130 E1
John Fisher St E1 126 E9
John Islip St SW1 129 L3
John Maurice Cl SE17 130 B3
John Penn St SE13 131 M7
John Princes St W1 125 H8
John Roll Way SE16 130 E2
John Ruskin St SE5 129 P6
John Silkin La SE8 131 H4
John Spencer Sq N1 125 P1
John St WC1 125 M6
John Williams Cl SE14 131 H5
John's Ms WC1 125 M6
Johnson Cl E8 126 E3
Johnson's Pl SW1 129 J4
Jonathan St SE11 129 M4
Joseph Hardcastle Cl SE14 131 H6
Joseph Ms N7 125 N1
Joseph St E3 127 K6
Joubert St SW11 128 F8
Jowett St SE15 130 D6
Jubilee Pl SW3 128 E4
Jubilee St E1 126 G8
Judd St WC1 125 L5
Juer St SW11 128 E6
Julian Pl E14 131 M4
Junction App SE13 131 N9
Junction App SW11 128 E9
Juniper Cres NW1 124 G2
Juniper Dr SW18 128 C10
Juniper St E1 126 G9
Juno Way SE14 131 H5
Jupiter Way N7 125 M1
Jupp Rd E15 127 P2
Jupp Rd W E15 127 N3
Juxon St SE11 129 M3

K

Kambala Rd SW11 128 D8
Kassala Rd SW11 128 F7
Kathleen Rd SW11 128 F9
Kay Rd SW9 129 L8
Kay St E2 126 E4
Kean St WC2 125 M8
Keel Cl SE16 127 H10
Keeley St WC2 125 M8
Keetons Rd SE16 130 F2
Keildon Rd SW11 128 F10
Kellett Rd SW2 129 N10
Kelly Av SE15 130 D6
Kelly St NW1 125 H1
Kelman Cl SW4 129 K8
Kelmore Gro SE22 130 E10
Kelsey St E2 126 E6
Kelso Pl W8 128 B2
Kelson Ho E14 131 N2
Kelvedon Ho SW8 129 L7
Kemble St WC2 125 M8
Kemerton Rd SE5 130 A9
Kempsford Gdns SW5 128 A4
Kempsford Rd SE11 129 N3
Kempson Rd SW6 128 A7
Kempthorne Rd SE8 131 J3
Kenbury St SE5 130 A8
Kenchester Cl SW8 129 L6
Kendal Cl SW9 129 P6
Kendal St W2 124 E8
Kender St SE14 130 G6
Kendoa Rd SW4 129 K10
Kendrick Pl SW7 128 D3
Kenilworth Rd E3 127 J4
Kennard Rd E15 127 P2
Kennard St SW11 128 G7
Kennet St E1 126 E10
Kenning Ter N1 126 C3
Kennings Way SE11 129 N4
Kennington La SE11 129 M4
Kennington Oval SE11 129 M5
Kennington Pk SW9 129 N6
Kennington Pk Gdns SE11 129 P5
Kennington Pk Pl SE11 129 N5
Kennington Pk Rd SE11 129 N5
Kennington Rd SE1 129 N2
Kennington Rd SE11 129 N3
Kensington Ch Ct W8 128 B1
Kensington Ch St W8 124 A10
Kensington Ch Wk W8 128 B1
Kensington Ct W8 128 B1
Kensington Ct Pl W8 128 B2
Kensington Gdns W2 124 C10
Kensington Gdns Sq W2 124 B8
Kensington Gate W8 128 C2
Kensington Gore SW7 128 D1
Kensington High St W8 128 A2
Kensington Mall W8 124 A10
Kensington Palace W8 128 B1
Kensington Palace Gdns W8 124 B10
Kensington Pl W8 124 A10
Kensington Rd SW7 128 D1
Kensington Rd W8 128 B1
Kensington Sq W8 128 B1
Kent Pk Ind Est SE15 130 F5
Kent Pas NW1 124 F5
Kent St E2 126 D4
Kent Ter NW1 124 E5
Kentish Town Rd NW1 125 H2
Kentish Town Rd NW5 125 H2
Kenton Rd E9 127 H1
Kenton St WC1 125 L6
Kentwell Cl SE4 131 J10
Kenway Rd SW5 128 B3
Kenwyn Rd SW4 129 K10
Kepler Rd SW4 129 L10
Keppel St WC1 125 K7
Kerbey St E14 127 M8
Kerfield Cres SE5 130 B7
Kerfield Pl SE5 130 B7
Kerridge Ct N1 126 C1
Kerrison Rd E15 127 P3
Kerrison Rd SW11 128 E9
Kerry Path SE14 131 K5
Kerry Rd SE14 131 K5
Kersley Ms SW11 128 F8
Kersley St SW11 128 F8
Keston Rd SE15 130 E9
Kestrel Ho EC1 125 P5
Kevan Ho SE5 130 A6
Key Cl E1 126 F6
Keyworth St SE1 129 P2
Khyber Rd SW11 128 E8
Kibworth St SW8 129 M6
Kilburn Pk Rd NW6 124 A5
Kilburn Pl NW6 124 A3
Kilburn Priory NW6 124 B3
Kilburn Sq NW6 124 A3
Kildare Gdns W2 124 A8
Kildare Ter W2 124 A8
Kilkie St SW6 128 C8
Killick St N1 125 M4
Killick Way E1 127 H7
Killowen Rd E9 127 H1
Killyon Rd SW8 129 J8
Killyon Ter SW8 129 J8
Kilner St E14 127 L7
Kimberley Av SE15 130 F8
Kimberley Rd SW9 129 L8
Kimpton Rd SE5 130 B7
Kinburn St SE16 131 H1
Kincaid Rd SE15 130 F6
King & Queen St SE17 130 A4
King Arthur Cl SE15 130 G6
King Charles St SW1 129 K1
King David La E1 126 G9
King Edward St EC1 126 A8
King Edward Wk SE1 129 N2
King Edwards Rd E9 126 F3
King Frederik IX Twr SE16 131 K2
King George St SE10 131 N6
King Henry's Rd NW3 124 E2
King Henry's Wk N1 126 C1
King James St SE1 129 P1
King John St E1 127 H7
King St EC2 126 A8
King St SW1 125 J10
King St WC2 125 L9
King William St EC4 126 B9
King William Wk SE10 131 N5
Kingdom St W2 124 C7
Kingdon Rd NW6 124 A1
Kingfield St E14 131 N3
Kingfisher Ho SW18 128 C9
Kingfisher Ms SE13 131 M10
Kingfisher Sq SE8 131 K5
Kinglake St SE17 130 C4
Kingly St W1 125 J8
Kings Coll Rd NW3 124 E2
King's Cross Rd WC1 125 M5
King's Cross Sta N1 125 L4
Kings Gro SE15 130 F6
King's Ms WC1 125 M7
King's Rd SW10 128 B7
King's Rd SW3 128 F4
King's Rd SW6 128 B7
Kingsbury Rd N1 126 C1
Kingsbury Ter N1 126 C1
Kingsgate Pl NW6 124 A2
Kingsgate Rd NW6 124 A2
Kingshold Rd E9 126 G2
Kingsland Grn E8 126 C1
Kingsland Rd E2 126 C4
Kingsland Rd E8 126 C4
Kingsland Shop Cen E8 126 D1
Kingsley St SW11 128 F9
Kingsmill Ter NW8 124 D4
Kingstown St NW1 124 G3
Kingsway WC2 125 M8
Kingswood Cl SW8 129 L6
Kinnerton St SW1 128 G1
Kinsale Rd SE15 130 E9
Kipling Est SE1 130 B1
Kipling St SE1 130 B1
Kirby Est SE16 130 F2
Kirby Gro SE1 130 C1
Kirkland Wk E8 126 D1
Kirkwall Pl E2 126 G5
Kirkwood Rd SE15 130 F8
Kirtling St SW8 129 J6
Kirwyn Way SE5 129 P6
Kitcat Ter E3 127 L5
Kitson Rd SE5 130 A6
Kitto Rd SE14 131 H8
Knapp Rd E3 127 L6
Knaresborough Pl SW5 128 B3
Knatchbull Rd SE5 130 A7
Kneller Rd SE4 131 J10
Knighten St E1 126 F10
Knighthead Pt E14 131 L2
Knights Twr SE8 131 L4
Knightsbridge SW1 128 F1
Knightsbridge SW7 128 E1
Knivet Rd SW6 128 A5
Knottisford St E2 126 G5
Knowle Cl SW9 129 N9
Knowles Wk SW4 129 J9
Knowsley Rd SW11 128 F8
Knox St W1 124 F7
Knoyle St SE14 131 J5
Kylemore Rd NW6 124 A2
Kynance Ms SW7 128 B2
Kynance Pl SW7 128 C2

L

Laburnum Ct E2 126 D3
Laburnum St E2 126 D3
Lacey Ms E3 127 L4
Lackington St EC2 126 B7
Lacon Rd SE22 130 E10
Ladycroft Rd SE13 131 M9
Lafone St SE1 130 D1
Lagado Ms SE16 127 H10
Laird Ho SE5 130 A6
Lamb La E8 126 F2
Lamb St E1 126 D7
Lambert St N1 125 N2
Lambeth Br SE1 129 L3
Lambeth Br SW1 129 L3
Lambeth High St SE1 129 M3
Lambeth Hill EC4 126 A9
Lambeth Palace Rd SE1 129 M2
Lambeth Pier SE1 129 L2
Lambeth Rd SE1 129 M3
Lambeth Rd SE11 129 M3
Lambeth Wk SE1 129 M3
Lambeth Wk SE11 129 M3
Lambolle Pl NW3 124 E1
Lambolle Rd NW3 124 E1
Lambourn Rd SW4 129 H9
Lambourne Gro SE16 131 H3
Lamb's Conduit St WC1 125 M6
Lambs Pas EC1 126 B7
Lamerton St SE8 131 L5
Lammas Rd E9 127 H2
Lamont Rd SW10 128 C5
Lanark Pl W9 124 C6
Lanark Rd W9 124 B4
Lanark Sq E14 131 M2
Lanbury Rd SE15 131 H10
Lancaster Dr NW3 124 E1
Lancaster Gate W2 124 C9
Lancaster Gro NW3 124 D1
Lancaster Ms W2 124 C9
Lancaster Pl WC2 125 M9
Lancaster St SE1 129 P1
Lancaster Ter W2 124 D9
Lancaster Wk W2 124 C10
Lancelot Pl SW7 128 F1
Lanchester Way SE14 130 G7
Lancresse Ct N1 126 C3
Landmann Ho SE16 130 F3
Landmann Way SE14 131 H5
Landmark E Twr E14 131 L1
Landmark W Twr E14 131 L1
Landon Pl SW1 128 F2
Landons Cl E14 127 N10
Landor Rd SW9 129 L9
Lanfranc Rd E3 127 J4
Lang St E1 126 G6
Langbourne Pl E14 131 M4
Langdale Cl SE17 130 A5
Langdale Rd SE10 131 N6
Langford Ct NW8 124 C4
Langford Grn SE5 130 C9
Langford Pl NW8 124 C4
Langford Rd SW6 128 B8
Langham Pl W1 125 H7
Langham St W1 125 H7
Langley La SW8 129 L5
Langley St WC2 125 L8
Langton Rd SW9 129 P6
Langton St SW10 128 C5
Langtry Rd NW8 124 B3
Lanhill Rd W9 124 A6
Lanrick Rd E14 127 P8
Lansbury Est E14 127 M8
Lansbury Gdns E14 127 P8
Lanscombe Wk SW8 129 L7
Lansdowne Dr E8 126 E1
Lansdowne Gdns SW8 129 L7
Lansdowne Ter WC1 125 L6
Lansdowne Way SW8 129 K7
Lant St SE1 130 A1
Lanvanor Rd SE15 130 G8
Lapis Ms E15 127 N3
Larcom St SE17 130 A3
Lark Row E2 126 G3
Larkhall La SW4 129 K8
Larkhall Ri SW4 129 J9
Lassell St SE10 131 P4
Latchmere Rd SW11 128 F8
Latchmere St SW11 128 F8
Latham Ho E1 127 H8
Latona Rd SE15 130 E5
Lauderdale Rd W9 124 B5
Lauderdale Twr EC2 126 A7
Launceston Pl W8 128 C2
Launch St E14 131 N2
Laurel Ms SE5 130 A9
Laurel St E8 126 D1
Laurie Gro SE14 131 J7
Lauriston Rd E9 127 H3
Lausanne Rd SE15 130 G7
Lavender Gdns SW11 128 F10
Lavender Gro E8 126 D2
Lavender Hill SW11 128 E10
Lavender Rd SE16 127 J10
Lavender Rd SW11 128 D9
Lavender Sq SW9 129 M7
Lavender Sweep SW11 128 F10
Lavender Wk SW11 128 F10
Laverton Pl SW5 128 B3
Lavington Cl E9 127 K1
Lavington St SE1 125 P10
Law St SE1 130 B2
Lawford Rd N1 126 C2
Lawford Rd NW5 125 J1
Lawless St E14 127 M9
Lawn Ho Cl E14 131 N1

Law - Mil

Lawn La SW8 129 L5
Lawrence Cl E3 127 L4
Lawrence St SW3 128 E5
Lawson Est SE1 130 B2
Lawton Rd E3 127 J5
Laxley Cl SE5 129 P6
Layard Rd SE16 130 F3
Layard Sq SE16 130 F3
Laycock St N1 125 N1
Laystall St EC1 125 N6
Lea Valley Wk E14 127 M7
Lea Valley Wk E15 127 N5
Lea Valley Wk E3 127 N6
Leabank Sq E9 127 L1
Leadenhall St EC3 126 C8
Leake St SE1 129 M1
Leamouth Rd E14 127 P8
Leander Ct SE8 131 L7
Leather La EC1 125 N7
Leather Rd SE16 131 H3
Leathermarket Ct SE1 130 C1
Leathermarket St SE1 130 C1
Leathwaite Rd SW11 128 F10
Leathwell Rd SE8 131 M8
Lecky St SW7 128 D4
Ledbury Est SE15 130 F6
Ledbury St SE15 130 E6
Lee Br SE13 131 N9
Lee High Rd SE12 131 P9
Lee High Rd SE13 131 P9
Lee St E8 126 D3
Leeke St WC1 125 M5
Leerdam Dr E14 131 N2
Lees Pl W1 124 G9
Leeson Rd SE24 129 N10
Leeway SE8 131 K4
Lefevre Wk E3 127 L4
Leggatt Rd E15 127 N4
Legion Cl N1 125 N1
Legion Ter E3 127 K3
Leicester Sq WC2 125 K9
Leigh St WC1 125 L6
Leinster Gdns W2 124 C8
Leinster Ms W2 124 C9
Leinster Pl W2 124 C8
Leinster Sq W2 124 A8
Leinster Ter W2 124 C9
Leirum St N1 125 M4
Leman St E1 126 D8
Lendal Ter SW4 129 K9
Lennox Gdns SW1 128 F2
Lennox Gdns Ms SW1 128 F2
Lensbury Av SW6 128 C8
Lenthall Rd E8 126 D2
Leo St SE15 130 F6
Leonard St EC2 126 B6
Leontine Cl SE15 130 E6
Leopold St E3 127 K7
Leroy St SE1 130 C3
Lethbridge Cl SE13 131 N7
Lett Rd E15 127 P2
Lett Rd SW9 129 M7
Lettsom St SE5 130 C8
Levehurst Way SW4 129 L8
Leven Rd E14 127 N7
Lever St EC1 125 P5
Lewey Ho E3 127 K6
Lewis Gro SE13 131 N9
Lewis St NW1 125 H1
Lewisham Cen SE13 131 N9
Lewisham High St SE13 131 N9
Lewisham Hill SE13 131 N8
Lewisham Rd SE13 131 M7
Lewisham Way SE14 131 K7
Lewisham Way SE4 131 K7
Lexham Gdns W8 128 B2
Lexham Gdns Ms W8 128 B2
Lexham Ms W8 128 A3
Lexington St W1 125 J8
Leybourne Rd NW1 125 H2
Leylang Rd SE14 131 H6
Liardet St SE14 131 J5
Liberia Rd N5 125 P1
Liberty St SW9 129 M7
Libra Rd E3 127 K3
Library St SE1 129 P1
Lichfield Rd E3 127 J5
Lidcote Gdns SW9 129 N8
Liddell Rd NW6 124 A1
Lidlington Pl NW1 125 J4
Lighter Cl SE16 131 J3
Lighterman Ms E1 127 H8
Lightermans Rd E14 131 L1
Lilac Pl SE11 129 M3
Lilestone St NW8 124 E6
Lilford Rd SE5 129 P8
Lillie Yd SW6 128 A5
Lillieshall Rd SW4 129 H9
Lily Pl EC1 125 N7
Limburg Rd SW11 128 F10
Lime Cl E1 126 E10
Lime St EC3 126 C9
Limeburner La EC4 125 P8
Limeharbour E14 131 M2
Limehouse Causeway E14 127 K9
Limehouse Link E14 127 J9
Limerston St SW10 128 C5
Limes Gro SE13 131 N10
Limes Wk SE15 130 F10
Limesford Rd SE15 131 H10
Linacre Cl SE15 130 F9
Linberry Wk SE8 131 K3
Lincoln St SW3 128 F3
Lincoln's Inn WC2 125 M8
Lincoln's Inn Flds WC2 125 M8
Lind St SE8 131 M8
Linden Gdns W2 124 A9
Linden Gro SE15 130 F9
Lindfield St E14 127 L8
Lindley St E1 126 G7
Lindore Rd SW11 128 F10
Lindrop St SW6 128 C8
Lindsay Sq SW1 129 K4
Lindsell St SE10 131 N7
Lindsey Ms N1 126 A2
Lindsey St EC1 125 P7
Linford St SW8 129 J7
Lingards Rd SE13 131 N10
Lingham St SW9 129 L8
Linhope St NW1 124 F6
Link St E9 126 G1
Linnell Rd SE5 130 C8
Linom Rd SW4 129 L10
Linsey St SE16 130 E3
Linstead St NW6 124 A2
Linton St N1 126 A3
Linver Rd SW6 128 A8
Linwood Cl SE5 130 D8
Lisford St SE15 130 D7
Lisle St WC2 125 K9
Lisson Grn Est NW8 124 E6
Lisson Gro NW1 124 E6
Lisson Gro NW8 124 D5
Lisson St NW1 124 E7
Liston Rd SW4 129 J9
Litchfield St WC2 125 K9
Lithos Rd NW3 124 B1
Little Boltons, The SW10 128 B4
Little Boltons, The SW5 128 B4
Little Britain EC1 125 P7
Little Chester St SW1 129 H2
Little Dorrit Ct SE1 130 A1
Little Newport St WC2 125 K9
Little Portland St W1 125 H8
Little Russell St WC1 125 L7
Little St. James's St SW1 125 J10
Littlebury Rd SW4 129 K9
Livermere Rd E8 126 D3
Liverpool Gro SE17 130 B4
Liverpool Rd N1 125 N4
Liverpool St EC2 126 C7
Livingstone Wk SW11 128 D9
Lizard St EC1 126 A5
Llewellyn St SE16 130 E1
Lloyd Baker St WC1 125 M5
Lloyd Sq WC1 125 N5
Lloyd St WC1 125 N5
Lloyd Vil SE4 131 L8
Lloyd's Av EC3 126 C8
Loampit Hill SE13 131 L8
Loampit Vale SE13 131 M9
Lochnagar St E14 127 N7
Lockesfield Pl E14 131 M4
Lockhart Cl N7 125 M1
Lockhart St E3 127 K6
Lockington Rd SW8 129 H7
Lockmead Rd SE13 131 N9
Locksley Est E14 127 K8
Locksley St E14 127 K7
Lockwood Sq SE16 130 F2
Loddiges Rd E9 126 G2
Loder St SE15 130 G7
Lodge Rd NW8 124 D5
Lodore St E14 127 N8
Loftie St SE16 130 E1
Lofting Rd N1 125 M2
Logan Ms W8 128 A3
Logan Pl W8 128 A3
Lollard St SE11 129 M3
Loman St SE1 129 P1
Lomas Dr E8 126 D2
Lomas St E1 126 E7
Lombard Rd SW11 128 D8
Lombard St EC3 126 B8
Lomond Gro SE5 130 B6
Loncroft Rd SE5 130 C5
London Br EC4 126 B10
London Br SE1 126 B10
London Br St SE1 126 B10
London Cen Mkts EC1 125 P7
London Flds E8 126 F2
London Flds E Side E8 126 F2
London Flds W Side E8 126 E2
London La E8 126 F2
London Rd SE1 129 P2
London Silver Vaults WC2 125 M7
London St W2 124 D8
London Trocadero, The
W1 125 K9
London Wall EC2 126 A7
Long Acre WC2 125 L9
Long La EC1 125 P7
Long La SE1 130 B1
Long Rd SW4 129 H10
Long St E2 126 D5
Long Yd WC1 125 M6
Longbeach Rd SW11 128 F9
Longfield Est SE1 130 D3
Longford St NW1 125 H6
Longhedge St SW11 128 G8
Longhope Cl SE15 130 D5
Longley St SE1 130 E3
Longmoore St SW1 129 J3
Longnor Rd E1 127 H5
Longridge Rd SW5 128 A3
Long's Ct WC2 125 K9
Longshore SE8 131 K3
Lonsdale Sq N1 125 N2
Lord Amory Way E14 131 N1
Lord Hills Rd W2 124 B7
Lord N St SW1 129 L2
Lorden Wk E2 126 E5
Lorenzo St WC1 125 M5
Loring Rd SE14 131 J7
Lorn Ct SW9 129 N8
Lorn Rd SW9 129 M8
Lorrimore Rd SE17 129 P5
Lorrimore Sq SE17 129 P5
Lothbury EC2 126 B8
Lothian Rd SW9 129 P7
Lots Rd SW10 128 C6
Loudoun Rd NW8 124 C3
Lough Rd N7 125 M1
Loughborough Pk SW9 129 P10
Loughborough Rd SW9 129 N8
Loughborough St SE11 129 M4
Louisa St E1 127 H6
Louvaine Rd SW11 128 D10
Love La EC2 126 A8
Love Wk SE5 130 B8
Lovegrove St SE1 130 E4
Lovegrove Wk E14 127 N10
Lovelinch Cl SE15 130 G5
Lovell Ho E8 126 E3
Lover's Wk W1 124 G10
Lowden Rd SE24 129 P10
Lowell St E14 127 J8
Lower Belgrave St SW1 129 H2
Lower Grosvenor Pl SW1 129 H2
Lower Marsh SE1 129 N1
Lower Merton Ri NW3 124 E2
Lower Rd SE16 130 G2
Lower Rd SE8 130 G2
Lower Sloane St SW1 128 G3
Lower Thames St EC3 126 B9
Lowfield Rd NW6 124 A2
Lowndes Cl SW1 128 G2
Lowndes Pl SW1 128 G2
Lowndes Sq SW1 128 F1
Lowndes St SW1 128 F2
Lowth Rd SE5 130 A8
Lowther Gdns SW7 128 D1
Lubbock St SE14 130 G6
Lucan Pl SW3 128 E3
Lucas St SE8 131 L7
Lucey Rd SE16 130 E2
Lucey Way SE16 130 E2
Ludgate Hill EC4 125 P8
Ludgate Sq EC4 125 P8
Ludwick Ms SE14 131 J6
Lugard Rd SE15 130 F8
Luke Ho E1 126 F8
Luke St EC2 126 C6
Lukin St E1 126 G8
Lulworth Rd SE15 130 F8
Lupin Pt SE1 130 D2
Lupus St SW1 129 H5
Lurline Gdns SW11 128 G7
Luscombe Way SW8 129 L6
Luton Pl SE10 131 N6
Luton St NW8 124 D6
Luxborough St W1 124 G6
Luxford St SE16 131 H3
Luxmore St SE4 131 K7
Luxor St SE5 129 P8
Lyal Rd E3 127 J4
Lyall Ms SW1 128 G2
Lyall St SW1 128 G2
Lydon Rd SW4 129 J9
Lyme St NW1 125 J2
Lymington Rd NW6 124 B1
Lympstone Gdns SE15 130 E6
Lynbrook Gro SE15 130 C6
Lyncott Cres SW4 129 H10
Lyndhurst Gro SE15 130 C8
Lyndhurst Sq SE15 130 D7
Lyndhurst Way SE15 130 D7
Lynton Rd SE1 130 D3
Lyons Pl NW8 124 D6
Lytham St SE17 130 B4
Lyttelton Cl NW3 124 E2

M

Mabledon Pl WC1 125 K5
Mabley St E9 127 J1
Macaulay Ct SW4 129 H9
Macaulay Rd SW4 129 H9
Macaulay Sq SW4 129 J10
Macauley Ms SE13 131 N8
Macclesfield Br NW1 124 E4
Macclesfield St W1 125 K9
Macduff Rd SW11 128 G7
Mace St E2 127 H4
Macfarland Gro SE15 130 C6
Machell Rd SE15 130 G9
Mackay Rd SW4 129 H9
Mackennal St NW8 124 E4
Mackenzie Rd N7 125 M1
Mackenzie Wk E14 127 L10
Macklin St WC2 125 L8
Macks Rd SE16 130 E3
Mackworth St NW1 125 J5
Macleod St SE17 130 A4
Maconochies Rd E14 131 M4
Macquarie Way E14 131 M3
Maddams St E3 127 M6
Maddock Way SE17 129 P5
Maddox St W1 125 H9
Madeleine Ter SE5 130 D7
Madinah Rd E8 126 E1
Madras Pl N7 125 N1
Madrigal La SE5 129 P6
Madron St SE17 130 C4
Magee St SE11 129 N5
Maguire St SE1 130 D1
Mahogany Cl SE16 127 J10
Maida Av W2 124 C7
Maida Vale W9 124 B4
Maiden La NW1 125 K2
Maiden La WC2 125 L9
Maidenstone Hill SE10 131 N7
Maitland Cl SE10 131 M6
Maitland Pk Est NW3 124 F1
Maitland Pk Rd NW3 124 F1
Maitland Pk Vil NW3 124 F1
Major Cl SW9 129 P9
Makins St SW3 128 E3
Malabar St E14 131 L1
Malcolm Pl E2 126 G6
Malcolm Rd E1 126 G6
Malden Cres NW1 124 G1
Maldon Cl N1 126 A3
Maldon Cl SE5 130 C9
Malet Pl WC1 125 K6
Malet St WC1 125 K6
Malfort Rd SE5 130 C9
Mall, The SW1 129 J1
Mallard Cl NW6 124 A4
Mallord St SW3 128 D5
Mallory Cl SE4 131 J10
Mallory St NW8 124 E6
Malmesbury Rd E3 127 K5
Malpas Rd E8 126 F1
Malpas Rd SE4 131 K8
Malt St SE1 130 E5
Maltby St SE1 130 D1
Malting Ho E14 127 K9
Maltings Cl E3 127 N5
Maltings Pl SW6 128 B7
Malvern Ct SW7 128 D3
Malvern Rd E8 126 E2
Malvern Rd NW6 124 A5
Malvern Ter N1 125 N3
Manaton Cl SE15 130 F9
Manchester Gro E14 131 N4
Manchester Rd E14 131 N4
Manchester Sq W1 124 G8
Manchester St W1 124 G7
Manciple St SE1 130 B2
Mandela St NW1 125 J3
Mandela St SW9 129 N6
Mandela Way SE1 130 C3
Mandeville Pl W1 124 G8
Manette St W1 125 K8
Manger Rd N7 125 L1
Manilla St E14 131 L1
Manley St NW1 124 G3
Manor Av SE4 131 K8
Manor Est SE16 130 F3
Manor Gro SE15 130 G5
Manor Ms SE4 131 K8
Manor Pk SE13 131 P10
Manor Pl SE17 129 P4
Manresa Rd SW3 128 E4
Mansell St E1 126 D8
Mansfield St W1 125 H7
Mansford St E2 126 E4
Mansion Ho EC4 126 B8
Manson Ms SW7 128 C3
Manson Pl SW7 128 D3
Mantle Rd SE4 131 J9
Mantua St SW11 128 D9
Mantus Rd E1 126 G6
Mape St E2 126 F6
Maple St E2 126 E4
Maple St W1 125 J7
Mapledene Rd E8 126 D2
Maplin St E3 127 K5
Marble Arch W1 124 F9
Marble Quay E1 126 E10
Marcella Rd SW9 129 N8
Marchant St SE14 131 J5
Marchmont St WC1 125 L6
Marchwood Cl SE5 130 C6
Marcia Rd SE1 130 C3
Marcon Pl E8 126 F1
Marcus Garvey Way SE24 129 N10
Marden Sq SE16 130 F2
Mare St E8 126 F3
Margaret St W1 125 H8
Margaretta Ter SW3 128 E5
Margery St WC1 125 N5
Maria Ter E1 127 H7
Marian Pl E2 126 F4
Marigold St SE16 130 F1
Marinefield Rd SW6 128 B8
Mariners Ms E14 131 P3
Marischal Rd SE13 131 P9
Maritime Quay E14 131 L4
Maritime St E3 127 K6
Marjorie Gro SW11 128 F10
Mark La EC3 126 C9
Market Est N7 125 L1
Market Ms W1 125 H10
Market Pl W1 125 J8
Market Rd N7 125 L1
Markham Sq SW3 128 F4
Markham St SW3 128 E4
Marl Rd SW18 128 B10
Marlborough Av E8 126 E3
Marlborough Ct W8 128 A3
Marlborough Gro SE1 130 E4
Marlborough Hill NW8 124 C3
Marlborough Pl NW8 124 C4
Marlborough Rd SW1 125 J10
Marlborough St SW3 128 E3
Marley St SE16 131 H3
Marloes Rd W8 128 B2
Marlow Way SE16 131 H1
Marlowes, The NW8 124 D3
Marmion Rd SW11 128 G10
Marmont Rd SE15 130 E7
Marney Rd SW11 128 G10
Maroon St E14 127 J7
Marquess Est N1 126 A1
Marquess Rd N1 126 B1
Marquis Rd NW1 125 K1
Marsala Rd SE13 131 M10
Marsden Rd SE15 130 D9
Marsden St NW5 124 G1
Marsh Wall E14 127 L10
Marshall St W1 125 J8
Marshalsea Rd SE1 130 A1
Marsham St SW1 129 K2
Marshfield St E14 131 N2
Marshgate La E15 127 N4
Marsland Cl SE17 129 P4
Marston Cl NW6 124 C2
Martello St E8 126 F2
Martello Ter E8 126 F2
Martha Ct E2 126 F4
Martha St E1 126 F8
Martineau St E1 126 G9
Mary Ann Gdns SE8 131 L5
Mary Datchelor Cl SE5 130 B7
Mary Grn NW8 124 B3
Mary St N1 126 A3
Mary Ter NW1 125 H3
Marylands Rd W9 124 A6
Marylebone High St W1 124 G7
Marylebone La W1 124 G8
Marylebone Ms W1 125 H7
Marylebone Rd NW1 124 E7
Marylebone St W1 124 G7
Marylee Way SE11 129 M3
Maskelyne Cl SW11 128 E7
Mason St SE17 130 B3
Masons Pl EC1 125 P5
Massingham St E1 127 H6
Mast Ho Ter E14 131 L3
Mast Leisure Pk SE16 131 H2
Masterman Ho SE5 130 A6
Masters Dr SE16 130 F4
Masters St E1 127 H7
Mastmaker Rd E14 131 L1
Matham Gro SE22 130 D10
Matilda Gdns E3 127 L4
Matilda St N1 125 M3
Matlock Cl SE24 130 A10
Matlock St E14 127 J8
Matrimony Pl SW8 129 J8
Matthew Parker St SW1 129 K1
Matthews St SW11 128 F8
Maude Rd SE5 130 C7
Maunsel St SW1 129 K3
Maverton Rd E3 127 L3
Mawbey Est SE1 130 D4
Mawbey Pl SE1 130 D4
Mawbey St SW8 129 L6
Maxted Rd SE15 130 D9
Maxwell Rd SW6 128 B6
Maya Cl SE15 130 F8
Maydew Ho SE16 130 G3
Mayfair Pl W1 125 H10
Mayfield Rd E8 126 D2
Mayflower Rd SW9 129 L9
Mayflower St SE16 130 G1
Maygood St N1 125 M4
Maysoule Rd SW11 128 D10
Mazenod Av NW6 124 A2
McAuley Cl SE1 129 N2
McCullum Rd E3 127 K3
McDermott Cl SW11 128 E9
McDermott Rd SE15 130 E9
McDowall Rd SE5 130 A7
McEwen Way E15 127 P3
McKerrell Rd SE15 130 E7
McMillan St SE8 131 L5
McNeil Rd SE5 130 C8
Mead Pl E9 126 G1
Meadcroft Rd SE11 129 P5
Meadow Ms SW8 129 M5
Meadow Pl SW8 129 L6
Meadow Rd SW8 129 M5
Meadow Row SE1 130 A2
Meadowbank NW3 124 F2
Meakin Est SE1 130 C2
Meard St W1 125 K8
Meath Cres E2 127 H5
Meath St SW11 129 H7
Mecklenburgh Pl WC1 125 M6
Mecklenburgh Sq WC1 125 M6
Medburn St NW1 125 K4
Medlar St SE5 130 A7
Medley Rd NW6 124 A1
Medway Rd E3 127 J4
Medway St SW1 129 K2
Medwin St SW4 129 M10
Meeting Ho La SE15 130 F7
Mehetabel Rd E9 126 G1
Melbourne Gro SE22 130 C10
Melbourne Ms SW9 129 N7
Melbourne Pl WC2 125 M9
Melcombe Pl NW1 124 F7
Melcombe St NW1 124 F6
Melina Pl NW8 124 D5
Melior St SE1 130 B1
Mellish St E14 131 L2
Melon Rd SE15 130 E7
Melton St NW1 125 K5
Mendip Rd SW11 128 C9
Mentmore Ter E8 126 F2
Mepham St SE1 125 M10
Mercator Rd SE13 131 P10
Mercer St WC2 125 L8
Merceron St E1 126 F6
Merchant St E3 127 K5
Mercia Gro SE13 131 N10
Mercury Way SE14 131 H5
Mercy Ter SE13 131 M10
Meredith Ms SE4 131 K10
Meretone Cl SE4 131 J10
Meridian Gate E14 131 N1
Meridian Pl E14 131 M1
Meridian Sq E15 127 P2
Mermaid Ct SE1 130 B1
Mermaid Ct SE16 127 K10
Merriam Av E9 127 K1
Merrick Sq SE1 130 A2
Merrington Rd SW6 128 A5
Merrow St SE17 130 A5
Merton Ri NW3 124 E2
Mervan Rd SW2 129 N10
Messina Av NW6 124 A2
Meteor St SW11 128 G10
Methley St SE11 129 N4
Mews St E1 126 E10
Meymott St SE1 125 P10
Meynell Cres E9 127 H2
Meynell Gdns E9 127 H2
Meynell Rd E9 127 H2
Meyrick Rd SW11 128 D9
Micawber St N1 126 A5
Michael Rd SW6 128 B7
Micklethwaite Rd SW6 128 A5
Middle Fld NW8 124 D3
Middle Temple La EC4 125 N8
Middlesex St E1 126 C7
Middleton Dr SE16 131 H1
Middleton Rd E8 126 D2
Middleton St E2 126 F5
Middleton Way SE13 131 P10
Midland Rd NW1 125 K4
Midnight Av SE5 129 P6
Midship Pt E14 131 L1
Milborne Gro SW10 128 C4
Milborne St E9 126 G1
Milcote St SE1 129 P1
Mildmay Av N1 126 B1
Mildmay St N1 126 B1
Mile End Pl E1 127 H6
Mile End Rd E1 126 G7
Mile End Rd E3 126 G7
Miles St SW8 129 L5
Milford La WC2 125 M9
Milk Yd E1 126 G9
Mill Row N1 126 C3
Mill St SE1 130 D1
Mill St W1 125 J9
Millard Rd SE8 131 K4
Millbank SW1 129 L2
Millbank Twr SW1 129 L3
Millbrook Rd SW9 129 P9
Millender Wk SE16 130 G3
Millennium Br EC4 126 A9
Millennium Br SE1 126 A9
Millennium Dr E14 131 P3
Millennium Harbour E14 131 K1
Millennium Pl E2 126 F4
Miller St NW1 125 J4
Millgrove St SW11 128 G8
Millharbour E14 131 M2
Milligan St E14 127 K9
Millman Ms WC1 125 M6
Millman St WC1 125 M6
Millmark Gro SE14 131 J8
Millstone Cl E15 127 P1
Millstream Rd SE1 130 D1

Mil - Par

Millwall Dock Rd **E14** 131 L2
Milman's St **SW10** 128 D5
Milner Pl **N1** 125 N3
Milner Sq **N1** 125 P2
Milner St **SW3** 128 F3
Milton Cl **SE1** 130 D3
Milton Ct Rd **SE14** 131 J5
Milton St **EC2** 126 B7
Milverton St **SE11** 129 N4
Mina Rd **SE17** 130 C4
Mincing La **EC3** 126 C9
Minera Ms **SW1** 128 G3
Minerva Cl **SW9** 129 N6
Minerva St **E2** 126 F4
Minet Rd **SW9** 129 P8
Ming St **E14** 127 L9
Minories **EC3** 126 D8
Minson Rd **E9** 127 H3
Mintern St **N1** 126 B4
Mission Pl **SE15** 130 E7
Mitchell St **EC1** 126 A6
Mitchison Rd **N1** 126 B1
Mitre Rd **SE1** 129 N1
Mitre St **EC3** 126 C8
Moat Pl **SW9** 129 M9
Modling Ho **E2** 126 G4
Molesford Rd **SW6** 128 A7
Molesworth St **SE13** 131 N9
Molyneux St **W1** 124 E7
Mona Rd **SE15** 130 G8
Monck St **SW1** 129 K2
Monclar Rd **SE5** 130 B10
Moncrieff St **SE15** 130 E8
Monier Rd **E3** 127 L2
Monkton St **SE11** 129 N3
Monmouth Rd **W2** 124 A8
Monmouth St **WC2** 125 L9
Monnow Rd **SE1** 130 E3
Monson Rd **SE14** 131 H6
Montagu Ms N **W1** 124 F7
Montagu Pl **W1** 124 F7
Montagu Sq **W1** 124 F7
Montagu St **W1** 124 F8
Montague Av **SE4** 131 K10
Montague Pl **WC1** 125 K7
Montague St **EC1** 126 A8
Montague St **WC1** 125 L7
Monteagle Way **SE15** 130 F9
Montefiore St **SW8** 129 H8
Montevetro **SW11** 128 D7
Montfichet Rd **E15** 127 N2
Montford Pl **SE11** 129 N4
Montgomery St **E14** 127 M10
Montpelier Pl **E1** 126 G8
Montpelier Pl **SW7** 128 E2
Montpelier Rd **SE15** 130 F7
Montpelier Sq **SW7** 128 E1
Montpelier St **SW7** 128 E1
Montpelier Wk **SW7** 128 E2
Montreal Ho **SE16** 131 H1
Montrose Ct **SW7** 128 D1
Montrose Pl **SW1** 128 G1
Monument St **EC3** 126 B9
Monza St **E1** 126 G9
Moodkee St **SE16** 130 G2
Moody Rd **SE15** 130 D6
Moody St **E1** 127 H5
Moon St **N1** 125 P3
Moor La **EC2** 126 B7
Moore Pk Rd **SW6** 128 B6
Moore St **SW3** 128 F3
Moorfields **EC2** 126 B7
Moorgate **EC2** 126 B8
Moorhouse Rd **W2** 124 A8
Moorland Rd **SW9** 129 P10
Moorlands Est **SW9** 129 N10
Mora St **EC1** 126 A5
Morant St **E14** 127 L9
Morat St **SW9** 129 M7
Moravian St **E2** 126 G4
Mordaunt St **SW9** 129 M9
Morden Hill **SE13** 131 N8
Morden La **SE13** 131 N7
Morden St **SE13** 131 M7
Morecambe Cl **E1** 127 H7
Morecambe St **SE17** 130 A3
Moreland St **EC1** 125 P5
Moresby Wk **SW8** 129 H8
Moreton Pl **SW1** 129 J4
Moreton St **SW1** 129 K4
Moreton Ter **SW1** 129 J4
Morgan St **E3** 127 J5
Morgans La **SE1** 126 C10
Morie St **SW18** 128 B10
Morley Rd **SE13** 131 N10
Morley St **SE1** 129 N1
Morna Rd **SE5** 130 A8
Morning La **E9** 126 G1
Mornington Cres **NW1** 125 J4
Mornington Gro **E3** 127 L5
Mornington Ms **SE5** 130 A7
Mornington Rd **SE8** 131 K6
Mornington St **NW1** 125 H4
Mornington Ter **NW1** 125 H3
Morocco St **SE1** 130 C1
Morpeth Gro **E9** 127 H3
Morpeth Rd **E9** 126 G3
Morpeth St **E2** 127 H5
Morpeth Ter **SW1** 129 J2
Morris Rd **E14** 127 M7
Morris St **E1** 126 F8
Morrison St **SW11** 128 G9
Morshead Rd **W9** 124 A5
Mortham St **E15** 127 P3
Mortimer Cres **NW6** 124 B3
Mortimer Est **NW6** 124 B3
Mortimer Pl **NW6** 124 B3
Mortimer Rd **N1** 126 C2
Mortimer St **W1** 125 H8
Morton Rd **N1** 126 A2
Morville St **E3** 127 L4
Moscow Rd **W2** 124 A9
Mossbury Rd **SW11** 128 E9
Mossford St **E3** 127 K6
Mossop St **SW3** 128 E3
Mostyn Gro **E3** 127 K4
Mostyn Rd **SW9** 129 N7
Motcomb St **SW1** 128 G2
Moulins Rd **E9** 126 G3
Mount Pleasant **WC1** 125 N6
Mount Row **W1** 125 H9
Mount St **W1** 124 G9
Mountague Pl **E14** 127 N9
Mounts Pond Rd **SE3** 131 P8
Mowlem St **E2** 126 F4
Mowll St **SW9** 129 N6
Moxon St **W1** 124 G7
Mozart Ter **SW1** 128 G3
Mulberry Rd **E8** 126 D2
Mulberry Wk **SW3** 128 D5
Mulvaney Way **SE1** 130 B1
Mundy St **N1** 126 C5
Munkenbeck Building **W2** 124 D7
Munro Ter **SW10** 128 D5
Munster Sq **NW1** 125 H5
Munton Rd **SE17** 130 A3
Murdock St **SE15** 130 F5
Muriel St **N1** 125 M4
Murillo Rd **SE13** 131 P10
Murphy St **SE1** 129 N1
Murray Gro **N1** 126 A4
Murray Ms **NW1** 125 K2
Murray St **NW1** 125 K2
Mursell Est **SW8** 129 M7
Musbury St **E1** 126 G8
Muschamp Rd **SE15** 130 D9
Museum St **WC1** 125 L8
Musgrave Cres **SW6** 128 A7
Musgrove Rd **SE14** 131 H7
Mutrix Rd **NW6** 124 A3
Myatt Rd **SW9** 129 P7
Myddelton Sq **EC1** 125 N5
Myddelton St **EC1** 125 N5
Myers La **SE14** 131 H5
Mylne St **EC1** 125 N5
Myrdle St **E1** 126 E7
Myron Pl **SE13** 131 N9
Myrtle Wk **N1** 126 C4
Mysore Rd **SW11** 128 F9

N

Nairn St **E14** 127 N7
Nankin St **E14** 127 L8
Nansen Rd **SW11** 128 G10
Nantes Cl **SW18** 128 C10
Napier Av **E14** 131 L4
Napier Gro **N1** 126 A4
Napier Ter **N1** 125 P2
Narborough St **SW6** 128 B8
Narrow St **E14** 127 J9
Naseby Cl **NW6** 124 C2
Nash Rd **SE4** 131 H10
Nassau St **W1** 125 J7
Naval Row **E14** 127 N9
Navarino Gro **E8** 126 E1
Navarino Mans **E8** 126 E1
Navarino Rd **E8** 126 E1
Navarre Rd **SW9** 129 P7
Navarre St **E2** 126 D6
Navy St **SW4** 129 K9
Naylor Rd **SE15** 130 F6
Nazareth Gdns **SE15** 130 F8
Nazrul St **E2** 126 D5
Neal St **WC2** 125 L8
Nealden St **SW9** 129 M9
Neate St **SE5** 130 D5
Nebraska St **SE1** 130 B1
Neckinger **SE16** 130 D2
Neckinger Est **SE16** 130 D2
Neckinger St **SE1** 130 D1
Nectarine Way **SE13** 131 M8
Needleman St **SE16** 131 H1
Nelldale Rd **SE16** 130 G3
Nelson Cl **NW6** 124 A5
Nelson Gdns **E2** 126 E5
Nelson Pl **N1** 125 P4
Nelson Rd **SE10** 131 N5
Nelson Sq **SE1** 129 P1
Nelson St **E1** 126 F8
Nelson Ter **N1** 125 P4
Nelson Wk **E3** 127 M6
Nelson's Row **SW4** 129 K10
Nepaul Rd **SW11** 128 E8
Neptune St **SE16** 130 G2
Nesham St **E1** 126 E10
Netherford Rd **SW4** 129 J8
Netherhall Gdns **NW3** 124 C1
Netherton Gro **SW10** 128 C5
Nettleton Rd **SE14** 131 H7
Nevada St **SE10** 131 N5
Nevern Pl **SW5** 128 A3
Nevern Rd **SW5** 128 A3
Nevern Sq **SW5** 128 A3
Neville Cl **SE15** 130 E6
Neville St **SW7** 128 D4
Neville Ter **SW7** 128 D4
New Bond St **W1** 125 H9
New Br St **EC4** 125 P8
New Broad St **EC2** 126 C7
New Burlington St **W1** 125 J9
New Butt La **SE8** 131 L6
New Caledonian Wf **SE16** 131 K2
New Cavendish St **W1** 124 G7
New Change **EC4** 126 A8
New Change Pas **EC4** 126 A8
New Ch Rd **SE5** 130 A6
New Clocktower Pl **N7** 125 L1
New Compton St **WC2** 125 K8
New Covent Gdn Flower Mkt **SW8** 129 K5
New Covent Gdn Mkt **SW8** 129 K6
New Cross **SE14** 131 J6
New Cross Rd **SE14** 130 G6
New Fetter La **EC4** 125 N8
New Globe Wk **SE1** 126 A10
New Inn Yd **EC2** 126 C6
New Kent Rd **SE1** 130 A2
New King St **SE8** 131 L5
New Mt St **E15** 127 P2
New N Rd **N1** 126 B4
New N St **WC1** 125 M7
New Oxford St **WC1** 125 K8
New Pl Sq **SE16** 130 F2
New Providence Wf **E14** 127 P10
New Quebec St **W1** 124 F8
New Ride **SW7** 128 E1
New River Wk **N1** 126 A1
New Rd **E1** 126 F7
New Row **WC2** 125 L9
New Sq **WC2** 125 M8
New St **EC2** 126 C7
New Union Cl **E14** 131 N2
New Union St **EC2** 126 B7
New Wf Rd **N1** 125 L4
Newark St **E1** 126 F7
Newburgh St **W1** 125 J8
Newburn St **SE11** 129 M4
Newby Pl **E14** 127 N9
Newby St **SW8** 129 H9
Newcastle Pl **W2** 124 D7
Newcomen Rd **SW11** 128 D9
Newcomen St **SE1** 130 B1
Newcourt St **NW8** 124 E4
Newell St **E14** 127 K8
Newent Cl **SE15** 130 C6
Newgate St **EC1** 125 P8
Newington Butts **SE1** 129 P3
Newington Butts **SE11** 129 P3
Newington Causeway **SE1** 129 P2
Newington Grn Rd **N1** 126 B1
Newlands Quay **E1** 126 G9
Newman St **W1** 125 J7
Newport Av **E14** 127 P9
Newport Pl **WC2** 125 K9
Newport St **SE11** 129 M3
Newton Pl **E14** 131 L3
Newton Rd **W2** 124 A8
Newton St **WC2** 125 L8
Nicholas Rd **E1** 126 G6
Nicholl St **E2** 126 E3
Nicholson St **SE1** 125 P10
Nido Twr **E1** 126 D7
Nigel Rd **SE15** 130 E9
Nightingale Rd **N1** 126 A1
Nile St **N1** 126 B5
Nile Ter **SE15** 130 D4
Nine Elms La **SW8** 129 J6
Noble St **EC2** 126 A8
Noel Rd **N1** 125 P4
Noel St **W1** 125 J8
Norbiton Rd **E14** 127 K8
Norfolk Cres **W2** 124 E8
Norfolk Pl **W2** 124 D8
Norfolk Rd **NW8** 124 D3
Norfolk Sq **W2** 124 D8
Norman Gro **E3** 127 J4
Norman Rd **SE10** 131 M6
Normandy Rd **SW9** 129 N7
North Audley St **W1** 124 G8
North Bk **NW8** 124 E5
North Carriage Dr **W2** 124 E9
North Colonnade, The **E14** 127 L10
North Cres **E16** 127 P6
North Cres **WC1** 125 K7
North Gower St **NW1** 125 J5
North Ms **WC1** 125 M6
North Pas **SW18** 128 A10
North Ride **W2** 124 E9
North Rd **N7** 125 L1
North Row **W1** 124 F9
North Several **SE3** 131 P8
North St **SW4** 129 J9
North Tenter St **E1** 126 D8
North Ter **SW3** 128 E2
North Vil **NW1** 125 K1
North Wf Rd **W2** 124 D7
Northampton Pk **N1** 126 A1
Northampton Rd **EC1** 125 N6
Northampton Sq **EC1** 125 P5
Northampton St **N1** 126 A2
Northbourne Rd **SW4** 129 K10
Northburgh St **EC1** 125 P6
Northchurch Rd **N1** 126 B2
Northchurch Ter **N1** 126 C2
Northcote Rd **SW11** 128 E10
Northdown St **N1** 125 M4
Northey St **E14** 127 J9
Northfields **SW18** 128 A10
Northiam St **E9** 126 F3
Northington St **WC1** 125 M6
Northlands St **SE5** 130 A8
Northpoint Sq **NW1** 125 K1
Northport St **N1** 126 B3
Northumberland All **EC3** 126 C8
Northumberland Av **WC2** 125 L10
Northumberland Pl **W2** 124 A8
Northumberland St **WC2** 125 L10
Northumbria St **E14** 127 L8
Northway Rd **SE5** 130 A9
Northwick Ter **NW8** 124 D6
Norton Folgate **E1** 126 C7
Norway Gate **SE16** 131 J2
Norway St **SE10** 131 M5
Norwich St **EC4** 125 N8
Notley St **SE5** 130 B6
Notre Dame Est **SW4** 129 J10
Notting Hill Gate **W11** 124 A10
Nottingham Pl **W1** 124 G6
Nottingham St **W1** 124 G7
Novello St **SW6** 128 A7
Nuding Cl **SE13** 131 L9
Nugent Ter **NW8** 124 C4
Nunhead Cres **SE15** 130 F9
Nunhead Est **SE15** 130 F10
Nunhead Grn **SE15** 130 F9
Nunhead Gro **SE15** 130 F9
Nunhead La **SE15** 130 F9
Nursery Cl **SE4** 131 K8
Nursery La **E2** 126 D3
Nursery Rd **SW9** 129 M10
Nutbrook St **SE15** 130 E9
Nutcroft Rd **SE15** 130 F6
Nutford Pl **W1** 124 F8
Nutley Ter **NW3** 124 C1
Nutmeg La **E14** 127 P8
Nutt St **SE15** 130 D6
Nuttall St **N1** 126 C4
Nynehead St **SE14** 131 J6

O

O2 Cen **NW3** 124 C1
Oak La **E14** 127 K9
Oak Sq **SW9** 129 M8
Oak Tree Rd **NW8** 124 E5
Oakbank Gro **SE24** 130 A10
Oakbury Rd **SW6** 128 B8
Oakcroft Rd **SE13** 131 P8
Oakdale Rd **SE15** 130 G9
Oakden St **SE11** 129 N3
Oakey La **SE1** 129 N2
Oakfield St **SW10** 128 C5
Oakhurst Gro **SE22** 130 E10
Oakington Rd **W9** 124 A6
Oakley Gdns **SW3** 128 E5
Oakley Pl **SE1** 130 D4
Oakley Rd **N1** 126 B2
Oakley Sq **NW1** 125 J4
Oakley St **SW3** 128 E5
Oat La **EC2** 126 A8
Oban St **E14** 127 P8
Oberstein Rd **SW11** 128 D10
Observatory Gdns **W8** 128 A1
Occupation Rd **SE17** 130 A4
Ocean Est **E1** 127 H6
Ocean St **E1** 127 H7
Ocean Wf **E14** 131 K1
Ockendon Rd **N1** 126 B1
Octavia St **SW11** 128 E7
Octavius St **SE8** 131 L6
Odessa St **SE16** 131 K1
Odger St **SW11** 128 F8
Offenbach Ho **E2** 127 H4
Offerton Rd **SW4** 129 J9
Offley Rd **SW9** 129 N6
Offord Rd **N1** 125 M2
Offord St **N1** 125 M2
Oglander Rd **SE15** 130 D10
Ogle St **W1** 125 J7
Old Bailey **EC4** 125 P8
Old Bellgate Pl **E14** 131 L2
Old Bethnal Grn Rd **E2** 126 E5
Old Bond St **W1** 125 J9
Old Broad St **EC2** 126 B8
Old Brompton Rd **SW5** 128 A4
Old Brompton Rd **SW7** 128 A4
Old Burlington St **W1** 125 J9
Old Castle St **E1** 126 D7
Old Cavendish St **W1** 125 H8
Old Ch Rd **E1** 127 H8
Old Ch St **SW3** 128 D4
Old Compton St **W1** 125 K9
Old Ct Pl **W8** 128 B1
Old Ford Rd **E2** 126 G4
Old Ford Rd **E3** 127 J4
Old Gloucester St **WC1** 125 L7
Old Jamaica Rd **SE16** 130 E2
Old James St **SE15** 130 F9
Old Jewry **EC2** 126 B8
Old Kent Rd **SE1** 130 C3
Old Kent Rd **SE15** 130 C3
Old Marylebone Rd **NW1** 124 E7
Old Montague St **E1** 126 E7
Old Nichol St **E2** 126 D6
Old Palace Yd **SW1** 129 L2
Old Paradise St **SE11** 129 M3
Old Pk La **W1** 124 G10
Old Pearson St **SE10** 131 M6
Old Pye St **SW1** 129 K2
Old Quebec St **W1** 124 F8
Old Queen St **SW1** 129 K1
Old Royal Free Sq **N1** 125 N3
Old S Lambeth Rd **SW8** 129 L6
Old Spitalfields Mkt **E1** 126 D7
Old Sq **WC2** 125 M8
Old St **EC1** 126 A6
Old Town **SW4** 129 J9
Old Watermen's Wk **EC3** 126 B9
Old Woolwich Rd **SE10** 131 P5
Oldbury Pl **W1** 124 G6
Oldfield Gro **SE16** 131 H3
O'Leary Sq **E1** 126 G7
Olga St **E3** 127 J4
Oliver Ms **SE15** 130 E8
Oliver-Goldsmith Est **SE15** 130 E7
Ollerton Grn **E3** 127 K3
Olliffe St **E14** 131 N2
Olmar St **SE1** 130 E5
Olney Rd **SE17** 129 P5
Olympic Pk **E15** 127 M1
Olympic Pk Av **E20** 127 M1
O'Meara St **SE1** 126 A10
Omega St **SE14** 131 L7
Omega Wks **E3** 127 L2
Ommaney Rd **SE14** 131 H7
Ondine Rd **SE15** 130 D10
One Hyde Pk **SW1** 128 F1
One New Change **EC4** 126 A8
Onega Gate **SE16** 131 J2
Ongar Rd **SW6** 128 A5
Onslow Cres **SW7** 128 D3
Onslow Gdns **SW7** 128 D4
Onslow Sq **SW7** 128 D3
Ontario St **SE1** 129 P2
Ontario Twr **E14** 127 P9
Ontario Way **E14** 127 L9
Opal St **SE11** 129 P3
Ophir Ter **SE15** 130 E7
Oppenheim Rd **SE13** 131 N8
Oppidans Rd **NW3** 124 F2
Orange St **WC2** 125 K9
Oransay Rd **N1** 126 A1
Orb St **SE17** 130 B3
Orbel St **SW11** 128 E7
Orchard, The **SE3** 131 P8
Orchard Cl **N1** 126 A2
Orchard Dr **SE3** 131 P8
Orchard St **W1** 124 G8
Orchardson St **NW8** 124 D6
Orde Hall St **WC1** 125 M6
Ordell Rd **E3** 127 K4
Ordnance Cres **SE10** 131 P1
Ordnance Hill **NW8** 124 D3
Oregano Dr **E14** 127 P8
Oriel Rd **E9** 127 H1
Oriens Ms **E20** 127 N1
Orion Bus Cen **SE14** 131 H4
Orkney St **SW11** 128 G8
Orlando Rd **SW4** 129 J9
Orleston Ms **N7** 125 N1
Orleston Rd **N7** 125 N1
Orme Ct **W2** 124 B9
Orme La **W2** 124 B9
Ormonde Gate **SW3** 128 F4
Ormonde Ter **NW8** 124 F3
Ormsby St **E2** 126 D4
Ormside St **SE15** 130 G5
Orpheus St **SE5** 130 B7
Orsett St **SE11** 129 M4
Orsett Ter **W2** 124 C8
Orsman Rd **N1** 126 C3
Orville Rd **SW11** 128 D8
Osborn Cl **E8** 126 E3
Osborn St **E1** 126 D7
Osborne Rd **E9** 127 K1
Oscar St **SE8** 131 L7
Oseney Cres **NW5** 125 J1
Osier St **E1** 126 G6
Osiers Rd **SW18** 128 A10
Oslo Sq **SE16** 131 J2
Osnaburgh St **NW1** 125 H6
Osnaburgh St **NW1** 125 H5
Osric Path **N1** 126 C4
Ossington St **W2** 124 A9
Ossory Rd **SE1** 130 E4
Ossulston St **NW1** 125 K5
Oswell Ho **E1** 126 F10
Oswin St **SE11** 129 P3
Oswyth Rd **SE5** 130 C8
Otis St **E3** 127 N5
Otter Cl **E15** 127 N3
Otterburn Ho **SE5** 130 A6
Otto St **SE17** 129 P5
Outer Circle **NW1** 125 H4
Outram Pl **N1** 125 L3
Oval, The **E2** 126 F4
Oval Pl **SW8** 129 M6
Oval Rd **NW1** 125 H3
Oval Way **SE11** 129 M4
Overcliff Rd **SE13** 131 L9
Oversley Ho **W2** 124 A7
Overton Rd **SW9** 129 N8
Ovex Cl **E14** 131 N1
Ovington Gdns **SW3** 128 E2
Ovington Ms **SW3** 128 E2
Ovington Sq **SW3** 128 E2
Ovington St **SW3** 128 E2
Oxendon St **SW1** 125 K9
Oxenford St **SE15** 130 D9
Oxenholme **NW1** 125 J4
Oxestalls Rd **SE8** 131 J4
Oxford Rd **E15** 127 P1
Oxford Rd **NW6** 124 A4
Oxford Sq **W2** 124 E8
Oxford St **W1** 125 H8
Oxley Cl **SE1** 130 D4
Oxley Sq **E3** 127 M6
Oxo Twr Wf **SE1** 125 N9
Oxonian St **SE22** 130 D10
Oyster Wf **SW11** 128 D8

P

Packington Sq **N1** 126 A3
Packington St **N1** 125 P3
Padbury **SE17** 130 C4
Padbury Ct **E2** 126 D5
Paddington Grn **W2** 124 D7
Paddington Sta **W2** 124 C8
Paddington St **W1** 124 G7
Paddington Underground Sta **W2** 124 C8
Padfield Rd **SE5** 130 A9
Pagden St **SW8** 129 H7
Page St **SW1** 129 K3
Pages Wk **SE1** 130 C3
Pagnell St **SE14** 131 K6
Pagoda Gdns **SE3** 131 P8
Pakenham St **WC1** 125 M5
Palace Av **W8** 124 B10
Palace Cl **E9** 127 K1
Palace Ct **W2** 124 B9
Palace Gdns Ms **W8** 124 A10
Palace Gdns Ter **W8** 124 A10
Palace Gate **W8** 128 C1
Palace Grn **W8** 128 B1
Palace St **SW1** 129 J2
Palfrey Pl **SW8** 129 M6
Palgrave Gdns **NW1** 124 E6
Pall Mall **SW1** 125 J10
Pall Mall E **SW1** 125 K10
Palmer St **SW1** 129 K2
Palmers Rd **E2** 127 H4
Palmerston Rd **NW6** 124 A2
Pancras Rd **N1** 125 K4
Pancras Rd **NW1** 125 K4
Pancras Way **E3** 127 L4
Pandora Rd **NW6** 124 A1
Parade, The **SW11** 128 F6
Paradise Rd **SW4** 129 L8
Paradise St **SE16** 130 F1
Paradise Wk **SW3** 128 F5
Paragon Rd **E9** 126 G1
Pardoner St **SE1** 130 B2
Parfett St **E1** 126 E7
Paris Gdn **SE1** 125 P10
Park Cl **E9** 126 G3
Park Cl **SW1** 128 F1
Park Cres **W1** 125 H6
Park La **W1** 124 G9
Park Pl **E14** 127 L10
Park Pl **N1** 126 B3
Park Pl **SW1** 125 J10
Park Pl Vil **W2** 124 C7
Park Rd **NW1** 124 E5
Park Rd **NW8** 124 E5
Park Row **SE10** 131 P5
Park Sq E **NW1** 125 H6
Park Sq Ms **NW1** 125 H6
Park Sq W **NW1** 125 H6
Park St **SE1** 126 A10
Park St **W1** 124 G9
Park Vw Est **E2** 126 G4
Park Vw Ms **SW9** 129 M8
Park Village E **NW1** 125 H4
Park Village W **NW1** 125 H4
Park Vista **SE10** 131 P5
Park Wk **SW10** 128 C5
Parker St **WC2** 125 L8
Parkfield Rd **SE14** 131 K7
Parkgate Rd **SW11** 128 E6
Parkham St **SW11** 128 E7
Parkholme Rd **E8** 126 D1
Parkhouse St **SE5** 130 B6
Parkside Av **SE10** 131 N7

Par - Ros

Parkside Rd **SW11** 128 G7
Parkway **NW1** 125 H3
Parliament Sq **SW1** 129 L1
Parliament St **SW1** 129 L1
Parliament Vw Apts **SE1** 129 M3
Parma Cres **SW11** 128 F10
Parmiter St **E2** 126 F4
Parnell Rd **E3** 127 K3
Parr St **N1** 126 B4
Parry St **SW8** 129 L5
Parsonage St **E14** 131 N3
Parsons Grn **SW6** 128 A7
Parsons Grn La **SW6** 128 A7
Parson's Ho **W2** 124 D6
Parthenia Rd **SW6** 128 A7
Parvin St **SW8** 129 K7
Pascal St **SW8** 129 K6
Passmore St **SW1** 128 G3
Pastor St **SE11** 129 P3
Patcham Ter **SW8** 129 H7
Pater St **W8** 128 A2
Patience Rd **SW11** 128 E8
Patmore Est **SW8** 129 J7
Patmore St **SW8** 129 J7
Patmos Rd **SW9** 129 P6
Paton Cl **E3** 127 L5
Patriot Sq **E2** 126 F4
Patshull Rd **NW5** 125 J1
Patterdale Rd **SE15** 130 G6
Pattina Wk **SE16** 127 K10
Paul Julius Cl **E14** 127 P9
Paul St **E15** 127 P3
Paul St **EC2** 126 B6
Paulet Rd **SE5** 129 P8
Pauline Ho **E1** 126 E7
Paul's Wk **EC4** 126 A9
Paultons Sq **SW3** 128 D5
Paultons St **SW3** 128 D5
Paveley Dr **SW11** 128 E6
Paveley St **NW8** 124 E6
Pavement, The **SW4** 129 J10
Pavilion Rd **SW1** 128 F1
Paxton Ter **SW1** 129 H5
Payne Rd **E3** 127 M4
Payne St **SE8** 131 K5
Peabody Est **SE1** 125 N10
Peabody Est **SW1** 129 J3
Peabody Sq **SE1** 129 P1
Peabody Trust **SE1** 126 A10
Pear Tree Cl **E2** 126 D3
Pear Tree Ct **EC1** 125 N6
Pear Tree St **EC1** 125 P6
Peardon St **SW8** 129 H9
Pearman St **SE1** 129 N1
Pearscroft Ct **SW6** 128 B7
Pearscroft Rd **SW6** 128 B7
Pearson Ms **SW4** 129 K9
Pearson St **E2** 126 C4
Peckford Pl **SW9** 129 N8
Peckham Gro **SE15** 130 C6
Peckham High St **SE15** 130 E7
Peckham Hill St **SE15** 130 E6
Peckham Pk Rd **SE15** 130 E6
Peckham Rd **SE15** 130 C7
Peckham Rd **SE5** 130 C7
Peckham Rye **SE15** 130 E9
Peckham Rye **SE22** 130 E10
Pedlars Wk **N7** 125 L1
Pedley St **E1** 126 D6
Peel Gro **E2** 126 G4
Peel Prec **NW6** 124 A4
Peel St **W8** 124 A10
Peerless St **EC1** 126 B5
Pekin St **E14** 127 L8
Pelham Cl **SE5** 130 C8
Pelham Cres **SW7** 128 E3
Pelham Pl **SW7** 128 E3
Pelham St **SW7** 128 E3
Pelican Est **SE15** 130 D7
Pelling St **E14** 127 L8
Pelter St **E2** 126 D5
Pembridge Cres **W11** 124 A9
Pembridge Gdns **W2** 124 A9
Pembridge Ms **W11** 124 A9
Pembridge Pl **W2** 124 A9
Pembridge Rd **W11** 124 A9
Pembridge Sq **W2** 124 A9
Pembridge Vil **W11** 124 A9
Pembridge Vil **W2** 124 A9
Pembroke Av **N1** 125 L3
Pembroke Cl **SW1** 128 G1
Pembroke Gdns Cl **W8** 128 A2
Pembroke Pl **W8** 128 A2
Pembroke Rd **W8** 128 A3
Pembroke Sq **W8** 128 A2
Pembroke St **N1** 125 L2
Pembroke Vil **W8** 128 A3
Pembroke Wk **W8** 128 A3
Pembry Cl **SW9** 129 N7
Penang St **E1** 126 F10
Penarth St **SE15** 130 G5
Pencraig Way **SE15** 130 F5
Pendrell Rd **SE4** 131 J8
Penfold Pl **NW1** 124 E7
Penfold St **NW1** 124 D6
Penfold St **NW8** 124 D6
Penford St **SE5** 129 P8
Peninsula Hts **SE1** 129 L4
Penn St **N1** 126 B3
Pennack Rd **SE15** 130 D5
Pennant Ms **W8** 128 B3
Pennethorne Rd **SE15** 130 F6
Pennington St **E1** 126 E9
Penny Brookes St **E15** 127 P1
Pennyfields **E14** 127 L9
Penpoll Rd **E8** 126 F1
Penrose Gro **SE17** 130 A4
Penrose Ho **SE17** 130 A4
Penrose St **SE17** 130 A4
Penryn St **NW1** 125 K4
Pensbury Pl **SW8** 129 J8
Pensbury St **SW8** 129 J8
Penshurst Rd **E9** 127 H2
Pentland Rd **NW6** 124 A5
Penton Pl **SE17** 129 P4
Penton Ri **WC1** 125 M5
Penton St **N1** 125 N4
Pentonville Rd **N1** 125 M4
Pentridge St **SE15** 130 D6
Penywern Rd **SW5** 128 A4
Pepper St **E14** 131 M2
Pepys Rd **SE14** 131 H7
Pepys St **EC3** 126 C9
Percival St **EC1** 125 P6
Percy Circ **WC1** 125 M5
Percy St **W1** 125 K7
Peregrine Ho **EC1** 125 P5
Perkin's Rents **SW1** 129 K2
Perrymead St **SW6** 128 A7
Perseverance Pl **SW9** 129 N6
Peter St **W1** 125 K9
Peterborough Ms **SW6** 128 A8
Peterborough Rd **SW6** 128 A8
Peterborough Vil **SW6** 128 B7
Petergate **SW11** 128 C10
Petersham La **SW7** 128 C2
Petersham Ms **SW7** 128 C2
Petersham Pl **SW7** 128 C2
Peto Pl **NW1** 125 H6
Petticoat La **E1** 126 C7
Petticoat Sq **E1** 126 C8
Petty France **SW1** 129 J2
Petworth St **SW11** 128 E7
Petyward **SW3** 128 E3
Peyton Pl **SE10** 131 N6
Phelp St **SE17** 130 B5
Phene St **SW3** 128 E5
Philbeach Gdns **SW5** 128 A4
Philip Wk **SE15** 130 E9
Phillimore Gdns **W8** 128 A1
Phillimore Pl **W8** 128 A1
Phillimore Wk **W8** 128 A2
Phillipp St **N1** 126 C3
Philpot St **E1** 126 F8
Phipp St **EC2** 126 C6
Phoenix Pl **WC1** 125 M6
Phoenix Rd **NW1** 125 K5
Piccadilly **W1** 125 H10
Piccadilly Circ **W1** 125 K9
Pickfords Wf **N1** 126 A4
Picton St **SE5** 130 B6
Pier St **E14** 131 N3
Pigott St **E14** 127 L8
Pigsty All **SE10** 131 N7
Pilgrimage St **SE1** 130 B1
Pilkington Rd **SE15** 130 F8
Pilot Cl **SE8** 131 K5
Pilton Pl **SE17** 130 A4
Pimlico Rd **SW1** 128 G4
Pinchin St **E1** 126 E9
Pincott Pl **SE4** 131 H10
Pindar St **EC2** 126 C7
Pine St **EC1** 125 N6
Pinefield Cl **E14** 127 L9
Pioneer St **SE15** 130 E7
Piper Cl **N7** 125 M1
Pitchford St **E15** 127 P2
Pitfield Est **N1** 126 B5
Pitfield St **N1** 126 C5
Pitman St **SE5** 130 A6
Pitsea St **E1** 127 H8
Pitt St **W8** 128 A1
Pitt's Head Ms **W1** 124 G10
Pixley St **E14** 127 K8
Plantation Wf **SW11** 128 C9
Plato Rd **SW2** 129 L10
Platt St **NW1** 125 K4
Plaza Shop Cen, The **W1** 125 J8
Pleasant Pl **N1** 125 P2
Pleasant Row **NW1** 125 H3
Plender St **NW1** 125 J3
Plevna St **E14** 131 N2
Plough Rd **SW11** 128 D9
Plough Ter **SW11** 128 D10
Plough Way **SE16** 131 H3
Plough Yd **EC2** 126 C6
Plover Way **SE16** 131 J2
Plumbers Row **E1** 126 E7
Plymouth Wf **E14** 131 P3
Plympton St **NW8** 124 E6
Pocock St **SE1** 129 P1
Podmore Rd **SW18** 128 C10
Point Hill **SE10** 131 N6
Point Pleasant **SW18** 128 A10
Pointers Cl **E14** 131 M4
Poland Ho **E15** 127 P3
Poland St **W1** 125 J8
Polesworth Ho **W2** 124 A7
Pollard Row **E2** 126 E5
Pollard St **E2** 126 E5
Polygon Rd **NW1** 125 K4
Pomeroy St **SE14** 130 G7
Ponder St **N7** 125 M2
Ponler St **E1** 126 F8
Ponsford St **E9** 126 G1
Ponsonby Pl **SW1** 129 K4
Ponsonby Ter **SW1** 129 K4
Pont St **SW1** 128 F2
Pont St Ms **SW1** 128 F2
Ponton Rd **SW8** 129 K5
Poole Rd **E9** 127 H1
Poole St **N1** 126 B3
Poolmans St **SE16** 131 H1
Pope St **SE1** 130 C1
Popes Rd **SW9** 129 N9
Popham Rd **N1** 126 A3
Popham St **N1** 125 P3
Poplar Bus Pk **E14** 127 N9
Poplar High St **E14** 127 L9
Poplar Pl **W2** 124 B9
Poplar Rd **SE24** 130 A10
Poplar Wk **SE24** 130 A10
Porchester Cl **SE5** 130 A10
Porchester Gdns **W2** 124 B9
Porchester Ms **W2** 124 B8
Porchester Pl **W2** 124 E8
Porchester Rd **W2** 124 B7
Porchester Sq **W2** 124 B8
Porchester Sq Ms **W2** 124 B8
Porchester Ter **W2** 124 C9
Porchester Ter N **W2** 124 B8
Porden Rd **SW2** 129 M10
Porlock St **SE1** 130 B1
Portelet Rd **E1** 127 H5
Porteus Rd **W2** 124 C7
Portia Way **E3** 127 K6
Portland Gro **SW8** 129 M7
Portland Pl **W1** 125 H7
Portland St **SE17** 130 B4
Portman Cl **W1** 124 F8
Portman Ms S **W1** 124 G8
Portman Pl **E2** 126 G5
Portman Sq **W1** 124 F8
Portman St **W1** 124 G8
Portpool La **EC1** 125 N7
Portree St **E14** 127 P8
Portslade Rd **SW8** 129 J8
Portsoken St **E1** 126 D9
Portugal St **WC2** 125 M8
Post Office Way **SW8** 129 K6
Potier St **SE1** 130 B2
Pott St **E2** 126 F5
Potters Rd **SW6** 128 C8
Pottery St **SE16** 130 F1
Poultry **EC2** 126 B8
Pountney Rd **SW11** 128 G9
Powis Pl **WC1** 125 L6
Powis Rd **E3** 127 M5
Pownall Rd **E8** 126 D3
Poyntz Rd **SW11** 128 F8
Poyser St **E2** 126 F4
Praed St **W2** 124 D8
Prairie St **SW8** 128 G8
Pratt St **NW1** 125 J3
Pratt Wk **SE11** 129 M3
Prebend St **N1** 126 A3
Prescot St **E1** 126 D9
Prescott Pl **SW4** 129 K9
Prestage Way **E14** 127 N9
Prestons Rd **E14** 131 N1
Price's Ct **SW11** 128 D9
Prices Ms **N1** 125 M3
Price's St **SE1** 125 P10
Prideaux Pl **WC1** 125 M5
Prideaux Rd **SW9** 129 L9
Prima Rd **SW9** 129 N6
Primrose Cl **E3** 127 L4
Primrose Gdns **NW3** 124 E1
Primrose Hill Ct **NW3** 124 F2
Primrose Hill Rd **NW3** 124 E2
Primrose Sq **E9** 126 G2
Primrose St **EC2** 126 C7
Prince Albert Rd **NW1** 124 E4
Prince Albert Rd **NW8** 124 E4
Prince Consort Rd **SW7** 128 C2
Prince Edward Rd **E9** 127 K1
Prince of Wales Dr **SW11** 128 F7
Prince of Wales Dr **SW8** 129 H6
Prince of Wales Gate **SW7** 128 E1
Prince of Wales Rd **NW5** 124 G1
Prince St **SE8** 131 K5
Princelet St **E1** 126 D7
Princes Ct **SE16** 131 K2
Princes Ct Bus Cen **E1** 126 F9
Princes Gdns **SW7** 128 D2
Princes Gate **SW7** 128 E1
Princes Gate Ms **SW7** 128 D2
Princes Ri **SE13** 131 N8
Princes Riverside Rd **SE16** 127 H10
Princes Sq **W2** 124 B9
Princes St **EC2** 126 B8
Princes St **W1** 125 H8
Princess Rd **NW1** 124 G3
Princess Rd **NW6** 124 A4
Princess St **SE1** 129 P2
Princethorpe Ho **W2** 124 A7
Princeton St **WC1** 125 M7
Printers Ms **E3** 127 J3
Printers Rd **SW9** 129 M7
Prior Bolton St **N1** 125 P1
Prior St **SE10** 131 N6
Prioress St **SE1** 130 B2
Priory Ct **SW8** 129 K7
Priory Grn Est **N1** 125 M4
Priory Gro **SW8** 129 L7
Priory Ms **SW8** 129 K7
Priory Rd **NW6** 124 B3
Priory St **E3** 127 M5
Priory Ter **NW6** 124 B3
Priory Wk **SW10** 128 C4
Pritchard's Rd **E2** 126 E3
Priter Rd **SE16** 130 E2
Procter St **WC1** 125 M7
Prospect Cotts **SW18** 128 A10
Prospect Pl **E1** 126 G10
Prospect Quay **SW18** 128 A10
Providence Ct **W1** 124 G9
Province Dr **SE16** 131 H1
Provost Est **N1** 126 B4
Provost Rd **NW3** 124 F2
Provost St **N1** 126 B5
Prusom St **E1** 126 F10
Pudding La **EC3** 126 B9
Pudding Mill La **E15** 127 M3
Pulross Rd **SW9** 129 M9
Pulteney Cl **E3** 127 K3
Pulteney Ter **N1** 125 M3
Pulton Pl **SW6** 128 A6
Pump La **SE14** 130 G6
Pundersons Gdns **E2** 126 F5
Purbrook St **SE1** 130 C2
Purcell St **N1** 126 C4
Purchese St **NW1** 125 K4
Purdy St **E3** 127 M6
Purelake Ms **SE13** 131 P9
Puteaux Ho **E2** 127 H4
Pytchley Rd **SE22** 130 C9

Q

Quaker St **E1** 126 D6
Quality Ct **WC2** 125 N8
Quarrendon St **SW6** 128 A8
Quarterdeck, The **E14** 131 L1
Quebec Way **SE16** 131 H1
Queen Anne Rd **E9** 127 H1
Queen Anne St **W1** 125 H8
Queen Anne's Gate **SW1** 129 K1
Queen Elizabeth St **SE1** 130 D1
Queen of Denmark Ct **SE16** 131 K2
Queen Sq **WC1** 125 L6
Queen St **EC4** 126 A9
Queen St **W1** 125 H10
Queen Victoria St **EC4** 125 P9
Queenhithe **EC4** 126 A9
Queen's Cres **NW5** 124 G1
Queens Gdns **W2** 124 C9
Queen's Gate **SW7** 128 D3
Queen's Gate Gdns **SW7** 128 C2
Queen's Gate Ms **SW7** 128 C1
Queen's Gate Pl **SW7** 128 C2
Queen's Gate Pl Ms **SW7** 128 C2
Queen's Gate Ter **SW7** 128 C2
Queen's Gro **NW8** 124 D3
Queen's Gro Ms **NW8** 124 D3
Queen's Head St **N1** 125 P3
Queens Ms **W2** 124 B9
Queens Rd **SE14** 130 F7
Queens Rd **SE15** 130 F7
Queen's Row **SE17** 130 B5
Queen's Ter **NW8** 124 D4
Queen's Wk, The **SE1** 125 M10
Queen's Wk **SW1** 125 J10
Queensberry Pl **SW7** 128 D3
Queensborough Ter **W2** 124 B9
Queensbridge Rd **E2** 126 D3
Queensbridge Rd **E8** 126 D2
Queensbury St **N1** 126 A2
Queensgate Pl **NW6** 124 A2
Queensmead **NW8** 124 D3
Queenstown Rd **SW8** 129 H5
Queensway **W2** 124 B8
Querrin St **SW6** 128 C8
Quex Rd **NW6** 124 A3
Quick St **N1** 125 P4
Quilter St **E2** 126 E5
Quince Rd **SE13** 131 M8
Quixley St **E14** 127 P9
Quorn Rd **SE22** 130 C10

R

Racton Rd **SW6** 128 A5
Radcot St **SE11** 129 N4
Radlett Pl **NW8** 124 E3
Radley Ms **W8** 128 A2
Radnor Pl **W2** 124 E8
Radnor Rd **SE15** 130 E6
Radnor St **EC1** 126 A5
Radnor Wk **SW3** 128 E4
Radstock St **SW11** 128 E6
Raeburn St **SW2** 129 L10
Raglan St **NW5** 125 H1
Railton Rd **SE24** 129 N10
Railway App **SE1** 126 B10
Railway Av **SE16** 130 G1
Railway St **N1** 125 L4
Rainbow Av **E14** 131 M4
Rainbow Quay **SE16** 131 J2
Rainbow St **SE5** 130 C6
Raine St **E1** 126 F10
Rainhill Way **E3** 127 L5
Rainsborough Av **SE8** 131 J3
Raleana Rd **E14** 127 N10
Raleigh St **N1** 125 P3
Ramillies Pl **W1** 125 J8
Rampayne St **SW1** 129 K4
Ramsey St **E2** 126 E6
Ramsey Wk **N1** 126 B1
Randall Cl **SW11** 128 E7
Randall Pl **SE10** 131 N6
Randall Rd **SE11** 129 M4
Randell's Rd **N1** 125 L3
Randolph Av **W9** 124 C6
Randolph Cres **W9** 124 C6
Randolph Gdns **NW6** 124 B4
Randolph Ms **W9** 124 C6
Randolph Rd **W9** 124 C6
Randolph St **NW1** 125 J2
Ranelagh Gdns **SW3** 128 G4
Ranelagh Gro **SW1** 128 G4
Rangers Sq **SE10** 131 P7
Ranwell St **E3** 127 K3
Raphael St **SW7** 128 F1
Ratcliffe Cross St **E1** 127 H8
Ratcliffe La **E14** 127 J8
Rathbone Pl **W1** 125 K8
Rathbone St **W1** 125 J7
Rattray Rd **SW2** 129 N10
Raul Rd **SE15** 130 E8
Raven Row **E1** 126 F7
Ravens Wk **E20** 127 M1
Ravensbourne Pl **SE13** 131 M8
Ravenscroft St **E2** 126 D4
Ravensdon St **SE11** 129 N4
Ravenstone **SE17** 130 C4
Rawlings St **SW3** 128 F3
Rawstorne St **EC1** 125 P5
Ray St **EC1** 125 N6
Raymouth Rd **SE16** 130 F3
Reading La **E8** 126 F1
Reardon Path **E1** 126 F10
Reardon St **E1** 126 F10
Reaston St **SE14** 130 G6
Record St **SE15** 130 G5
Rector St **N1** 126 A3
Rectory Gro **SW4** 129 J9
Rectory Sq **E1** 127 H7
Reculver Rd **SE16** 131 H4
Red Lion Row **SE17** 130 A5
Red Lion Sq **WC1** 125 M7
Red Lion St **WC1** 125 M7
Red Path **E9** 127 K1
Red Post Hill **SE24** 130 B10
Redan Pl **W2** 124 B8
Redbridge Gdns **SE5** 130 C6
Redburn St **SW3** 128 F5
Redcar St **SE5** 130 A6
Redcastle Cl **E1** 126 G9
Redchurch St **E2** 126 D6
Redcliffe Gdns **SW10** 128 B4
Redcliffe Gdns **SW5** 128 B4
Redcliffe Ms **SW10** 128 B4
Redcliffe Pl **SW10** 128 C5
Redcliffe Rd **SW10** 128 C4
Redcliffe Sq **SW10** 128 B4
Redcliffe St **SW10** 128 B5
Redcross Way **SE1** 130 A1
Reddins Rd **SE15** 130 E5
Redesdale St **SW3** 128 F5
Redfield La **SW5** 128 A3
Redhill St **NW1** 125 H4
Redmans Rd **E1** 126 G7
Redriff Est **SE16** 131 J2
Redriff Rd **SE16** 131 H3
Redruth Rd **E9** 127 H3
Redwood Cl **E3** 127 L4
Redwood Cl **SE16** 127 J10
Reece Ms **SW7** 128 D3
Reed Pl **SW4** 129 K10
Reedham St **SE15** 130 E8
Reedworth St **SE11** 129 N3
Rees St **N1** 126 A3
Reeves Ms **W1** 124 G9
Reeves Rd **E3** 127 M6
Reform St **SW11** 128 F8
Regan Way **N1** 126 C4
Regency St **SW1** 129 K3
Regeneration Rd **SE16** 131 H3
Regent Sq **E3** 127 M5
Regent Sq **WC1** 125 L5
Regent St **SW1** 125 K9
Regent St **W1** 125 H8
Regents Br Gdns **SW8** 129 L6
Regent's Pk, The **NW1** 124 F4
Regents Pk Rd **NW1** 124 F3
Regents Row **E8** 126 E3
Reginald Rd **SE8** 131 L6
Reginald Sq **SE8** 131 L6
Regis Pl **SW2** 129 M10
Relf Rd **SE15** 130 E9
Rembrandt Cl **E14** 131 P2
Remington St **N1** 125 P4
Renforth St **SE16** 130 G1
Renfrew Rd **SE11** 129 P3
Rennell St **SE13** 131 N9
Rennie Est **SE16** 130 F3
Rennie St **SE1** 125 P10
Repton St **E14** 127 J8
Reservoir Rd **SE4** 131 J8
Retreat Pl **E9** 126 G1
Revelon Rd **SE4** 131 J10
Reverdy Rd **SE1** 130 E3
Rheidol Ter **N1** 125 P4
Rhodesia Rd **SW9** 129 L8
Rhodeswell Rd **E14** 127 J7
Rhondda Gro **E3** 127 J5
Rhyl St **NW5** 124 G1
Ricardo St **E14** 127 M8
Rich St **E14** 127 K9
Richborne Ter **SW8** 129 M6
Richmond Av **N1** 125 M3
Richmond Cres **N1** 125 M3
Richmond Gro **N1** 125 P2
Richmond Rd **E8** 126 D2
Richmond Ter **SW1** 129 L1
Rick Roberts Way **E15** 127 N3
Rickett St **SW6** 128 A5
Ridgdale St **E3** 127 M4
Ridgeway Rd **SW9** 129 P9
Riding Ho St **W1** 125 H7
Rifle Ct **SE11** 129 N5
Rifle St **E14** 127 M7
Rigden St **E14** 127 M8
Rigge Pl **SW4** 129 K10
Riley Rd **SE1** 130 D2
Riley St **SW10** 128 D5
Ring, The **W2** 124 D9
Ripplevale Gro **N1** 125 M2
Risdon St **SE16** 130 G1
Risinghill St **N1** 125 M4
Rita Rd **SW8** 129 M6
Ritchie St **N1** 125 N4
Ritson Rd **E8** 126 E1
Rivaz Pl **E9** 126 G1
River Pl **N1** 126 A2
River St **EC1** 125 N5
Riverside Ct **SW8** 129 K5
Riverside Rd **E15** 127 N4
Riverside Twr **SW6** 128 C8
Rivington St **EC2** 126 C5
Roach Rd **E3** 127 L2
Roan St **SE10** 131 N5
Robert Adam St **W1** 124 G8
Robert Dashwood Way **SE17** 130 A3
Robert Lowe Cl **SE14** 131 H6
Robert Sq **SE13** 131 N10
Robert St **NW1** 125 H5
Roberta St **E2** 126 E5
Roberts Cl **SE16** 131 H1
Robertson St **SW8** 129 H9
Robin Ct **SE16** 130 E3
Robin Hood La **E14** 127 N9
Robinson Rd **E2** 126 G4
Robsart St **SW9** 129 M8
Rochdale Way **SE8** 131 L6
Rochelle Cl **SW11** 128 D10
Rochester Ms **NW1** 125 J2
Rochester Pl **NW1** 125 J1
Rochester Rd **NW1** 125 J1
Rochester Row **SW1** 129 J3
Rochester Sq **NW1** 125 J2
Rochester St **SW1** 129 K2
Rochester Ter **NW1** 125 J1
Rockingham Est **SE1** 130 A2
Rockingham St **SE1** 130 A2
Rodmarton St **W1** 124 F7
Rodney Pl **SE17** 130 A3
Rodney Rd **SE17** 130 A3
Rodney St **N1** 125 M4
Roffey St **E14** 131 N1
Roger Dowley Ct **E2** 126 G4
Roger St **WC1** 125 M6
Rokeby Rd **SE4** 131 K8
Roland Gdns **SW7** 128 C4
Roland Way **SE17** 130 B4
Rollins St **SE15** 130 G5
Rolls Rd **SE1** 130 D4
Rolt St **SE8** 131 J5
Roman Rd **E2** 126 G5
Roman Rd **E3** 127 J4
Roman Way **N7** 125 M1
Romford St **E1** 126 E7
Romilly St **W1** 125 K9
Romney Rd **SE10** 131 N5
Romney St **SW1** 129 K2
Rood La **EC3** 126 C9
Rookery Rd **SW4** 129 J10
Rope St **SE16** 131 J3
Ropemaker Pl **EC2** 126 B7
Ropemaker Rd **SE16** 131 J2
Ropemaker St **EC2** 126 B7
Ropery St **E3** 127 K6
Ropley St **E2** 126 E4
Rosary Gdns **SW7** 128 C3
Rose All **SE1** 126 A10

Ros - Sta

Rose Sq SW3 128 D4
Rosebank Gdns E3 127 K4
Roseberry Pl E8 126 D1
Roseberry St SE16 130 F3
Rosebery Av EC1 125 N6
Rosebury Rd SW6 128 B8
Rosefield Gdns E14 127 L9
Rosemary Dr E14 127 P8
Rosemary Rd SE15 130 D6
Rosemont Rd NW3 124 C1
Rosemoor St SW3 128 F3
Rosenau Cres SW11 128 E7
Rosenau Rd SW11 128 E7
Roserton St E14 131 N1
Rosetta Cl SW8 129 L6
Rosher Cl E15 127 P2
Rosoman St EC1 125 N5
Rossendale Way NW1 125 J2
Rossetti Rd SE16 130 F4
Rossiter Gro SW9 129 N9
Rossmore Rd NW1 124 E6
Rothbury Rd E9 127 K2
Rotherfield St N1 126 A2
Rotherhithe New Rd SE16 130 E4
Rotherhithe Old Rd SE16 131 H3
Rotherhithe St SE16 130 G1
Rotherhithe Tunnel E1 126 G10
Rotherhithe Tunnel App E14 127 J9
Rotherhithe Tunnel App SE16 130 G1
Rothery Ter SW9 129 P6
Rothsay St SE1 130 C2
Rothwell St NW1 124 F3
Rotten Row SW1 128 F1
Rotten Row SW7 128 E1
Rotterdam Dr E14 131 N2
Rouel Rd SE16 130 E2
Roundel Cl SE4 131 K10
Roundhouse La E15 127 N1
Rounton Rd E3 127 L6
Roupell St SE1 125 N10
Rousden St NW1 125 J2
Rowcross St SE1 130 D4
Rowditch La SW11 128 G8
Rowena Cres SW11 128 E8
Rowington Cl W2 124 B7
Rowley Way NW8 124 B3
Rowse Cl E15 127 N3
Roxby Pl SW6 128 A5
Royal Av SW3 128 F4
Royal Cl SE8 131 K5
Royal Coll St NW1 125 J2
Royal Ct SE16 131 K2
Royal Ex EC3 126 B8
Royal Hill SE10 131 N6
Royal Hosp Rd SW3 128 F5
Royal Ms, The SW1 129 H2
Royal Mint Ct EC3 126 D9
Royal Mint St E1 126 D9
Royal Naval Pl SE14 131 K6
Royal Oak Rd E8 126 F1
Royal Oak Yd SE1 130 C1
Royal Opera Arc SW1 125 K10
Royal Pl SE10 131 N6
Royal Rd SE17 129 P5
Royal St SE1 129 M2
Royal Victor Pl E3 127 H4
Royston St E2 126 G4
Rozel Ct N1 126 C3
Rozel Rd SW4 129 J8
Ruby St SE15 130 F5
Rudolph Rd NW6 124 A4
Rufford St N1 125 L3
Rugby St WC1 125 M6
Rugg St E14 127 L9
Rum Cl E1 126 G9
Rumbold Rd SW6 128 B6
Rumsey Rd SW9 129 M9
Rupert Gdns SW9 129 P8
Rupert St W1 125 K9
Rush Hill Rd SW11 128 G9
Rushcroft Rd SW2 129 N10
Rushton St N1 126 B4
Rushworth St SE1 129 P1
Ruskin Pk Ho SE5 130 B9
Russell Gro SW9 129 N6
Russell Sq WC1 125 L7
Russell St WC2 125 M8
Russia Dock Rd SE16 127 J10
Russia La E2 126 G4
Russia Wk SE16 131 J1
Rust Sq SE5 130 B6
Ruston St E3 127 K3
Rutherford St SW1 129 K3
Rutland Gdns SW7 128 E1
Rutland Gate SW7 128 E1
Rutland Rd E9 126 G3
Rutland St SW7 128 E2
Rutts Ter SE14 131 H7
Ryder Dr SE16 130 F4
Ryder St SW1 125 J10
Rye Hill Pk SE15 130 G10
Rye La SE15 130 E7
Rye Pas SE15 130 E9
Rye Rd SE15 131 H10
Ryecroft St SW6 128 B7
Ryland Rd NW5 125 H1
Rysbrack St SW3 128 F2

S

Sabine Rd SW11 128 F9
Sable St N1 125 P2
Sackville St W1 125 J9
Saffron Av E14 127 P9
Saffron Hill EC1 125 N6
Sail St SE11 129 M3
St. Agnes Pl SE11 129 N5
St. Albans Gro W8 128 B2
St. Alban's Pl N1 125 P3
St. Alfege Pas SE10 131 N5
St. Alphonsus Rd SW4 129 J10
St. Andrew St EC4 125 N7
St. Andrew's Hill EC4 125 P9
St. Andrews Pl NW1 125 H6
St. Andrews Way E3 127 M6
St. Ann's St SW1 129 K2
St. Ann's Ter NW8 124 D4
St. Anthonys Cl E1 126 E10
St. Asaph Rd SE4 131 H9
St. Augustines Rd NW1 125 K2
St. Austell Rd SE13 131 N8
St. Barnabas St SW1 128 G4
St. Barnabas Vil SW8 129 L7
St. Botolph St EC3 126 D8
St. Bride St EC4 125 P8
St. Chad's Pl WC1 125 L5
St. Chad's St WC1 125 L5
St. Clements St N7 125 N1
St. Cross St EC1 125 N7
St. Davids Sq E14 131 M4
St. Donatts Rd SE14 131 K7
St. Edmunds Ter NW8 124 E3
St. Elmos Rd SE16 131 J1
St. Francis Rd SE22 130 C10
St. George St W1 125 H8
St. George Wf SW8 129 L4
St. Georges Circ SE1 129 P2
St. George's Dr SW1 129 H3
St. Georges Flds W2 124 E8
St. Georges Rd SE1 129 N2
St. Georges Sq SE8 131 K3
St. George's Sq SW1 129 K4
St. George's Sq Ms SW1 129 K4
St. Georges Way SE15 130 C5
St. Giles High St WC2 125 K8
St. Giles Rd SE5 130 C6
St. Gilles Ho E2 127 H4
St. Helena Rd SE16 131 H3
St. James Ms E14 131 N2
St. James's SE14 131 J7
St. James's Av E2 126 G4
St. James's Ct SW1 129 J2
St. James's Cres SW9 129 N9
St. James's Palace SW1 125 J10
St. James's Pk SW1 129 J1
St. James's Pl SW1 125 J10
St. James's Rd SE1 130 E4
St. James's Rd SE16 130 E2
St. James's Sq SW1 125 J10
St. James's St SW1 125 J10
St. James's Ter Ms NW8 124 F3
St. James's Wk EC1 125 P6
St. John St EC1 125 P6
St. John's Cres SW9 129 N9
St. John's Est N1 126 B4
St. John's Hill SW11 128 D10
St. John's Hill Gro SW11 128 D10
St. John's La EC1 125 P6
St. John's Rd SW11 128 E10
St. John's Vale SE8 131 L8
St. John's Wd High St NW8 124 D4
St. John's Wd Pk NW8 124 D3
St. John's Wd Rd NW8 124 D6
St. John's Wd Ter NW8 124 D4
St. Joseph's Vale SE3 131 P8
St. Jude's Rd E2 126 F4
St. Katharine's Way E1 126 D10
St. Lawrence St E14 127 N10
St. Lawrence Way SW9 129 N7
St. Leonards Ct N1 126 B5
St. Leonards Rd E14 127 M7
St. Leonards Sq NW5 124 G1
St. Leonards St E3 127 M5
St. Leonard's Ter SW3 128 F4
St. Loo Av SW3 128 E5
St. Luke's Av SW4 129 K10
St. Luke's Cl EC1 126 A6
St. Luke's Est EC1 126 B5
St. Luke's St SW3 128 E4
St. Margarets La W8 128 B2
St. Margarets Rd SE4 131 K10
St. Margaret's St SW1 129 L1
St. Mark St E1 126 D8
St. Marks Cres NW1 124 G3
St. Mark's Gro SW10 128 B5
St. Marks Sq NW1 124 G3
St. Martins Cl NW1 125 J3
St. Martin's Ctyd WC2 125 L9
St. Martin's La WC2 125 L9
St. Martin's Pl WC2 125 L9
St. Martin's Rd SW9 129 M8
St. Martin's-le-Grand EC1 126 A8
St. Mary at Hill EC3 126 C9
St. Mary Axe EC3 126 C8
St. Marychurch St SE16 130 G1
St. Mary's Gdns SE11 129 N3
St. Mary's Gate W8 128 B2
St. Mary's Gro N1 125 P1
St. Marys Mans W2 124 C7
St. Marys Path N1 125 P3
St. Mary's Pl W8 128 B2
St. Mary's Rd SE15 130 G7
St. Marys Sq W2 124 D7
St. Marys Ter W2 124 D7
St. Mary's Wk SE11 129 N3
St. Matthew's Rd SW2 129 M10
St. Matthew's Row E2 126 E5
St. Michael's Rd SW9 129 M8
St. Michaels St W2 124 D8
St. Norbert Grn SE4 131 J10
St. Norbert Rd SE4 131 J10
St. Olav's Sq SE16 130 G1
St. Oswald's Pl SE11 129 M4
St. Pancras Way NW1 125 J2
St. Paul St N1 126 A3
St. Paul's Av SE16 127 H10
St. Paul's Chyd EC4 125 P8
St. Paul's Cres NW1 125 K2
St. Paul's Ms NW1 125 K2
St. Paul's Pl N1 126 B1
St. Paul's Rd N1 125 P1
St. Paul's Shrubbery N1 126 B1
St. Pauls Way E14 127 K7
St. Pauls Way E3 127 K7
St. Peter's Cl E2 126 E4
St. Peters St N1 125 P3
St. Peter's Way N1 126 C2
St. Petersburgh Ms W2 124 B9
St. Petersburgh Pl W2 124 B9
St. Philip Sq SW8 129 H8
St. Philip St SW8 129 H8
St. Philip's Rd E8 126 E1
St. Rule St SW8 129 J8
St. Saviour's Est SE1 130 D1
St. Silas Pl NW5 124 G1
St. Silas St Est NW5 124 G1
St. Stephens Cres W2 124 A8
St. Stephens Gdns W2 124 A8
St. Stephens Gro SE13 131 N9
St. Stephens Rd E3 127 K4
St. Stephens Ter SW8 129 M6
St. Stephen's Wk SW7 128 C3
St. Swithin's La EC4 126 B9
St. Thomas St SE1 126 B10
St. Thomas's Pl E9 126 G2
St. Thomas's Sq E9 126 F2
Salamanca St SE1 129 L3
Sale Pl W2 124 E7
Salem Rd W2 124 B9
Salisbury Ct EC4 125 P8
Salisbury Pl SW9 129 P6
Salisbury Pl W1 124 F7
Salisbury St NW8 124 E6
Salisbury Ter SE15 130 G9
Salmon La E14 127 J8
Salter Rd SE16 127 H10
Salter St E14 127 L9
Saltoun Rd SW2 129 N10
Saltwell St E14 127 L9
Samford St NW8 124 E6
Sampson St E1 126 E10
Samuel Cl SE14 131 H5
Samuel Lewis Trust Dws SW6 128 A6
Samuel St SE15 130 D6
Sancroft St SE11 129 M4
Sandall Rd NW5 125 J1
Sandbourne Rd SE4 131 J8
Sandgate St SE15 130 F5
Sandilands Rd SW6 128 B7
Sandison St SE15 130 D9
Sandland St WC1 125 M7
Sandmere Rd SW4 129 L10
Sandpiper Cl SE16 131 K1
Sandrock Rd SE13 131 L9
Sand's End La SW6 128 B7
Sandwell Cres NW6 124 A1
Sandwich St WC1 125 L5
Sandy's Row E1 126 C7
Sanford St SE14 131 J5
Sangora Rd SW11 128 D10
Sans Wk EC1 125 N6
Sansom St SE5 130 B6
Santley St SW4 129 M10
Saperton Wk SE11 129 M3
Sapphire Rd SE8 131 J3
Saracen St E14 127 L8
Sartor Rd SE15 131 H10
Satchwell Rd E2 126 E5
Saunders Ness Rd E14 131 N4
Saunders St SE11 129 N3
Savile Row W1 125 J9
Savona Est SW8 129 J6
Savona St SW8 129 J6
Savoy Ms SW9 129 L9
Savoy Pl WC2 125 L9
Savoy St WC2 125 M9
Sawmill Yd E3 127 J3
Sawyer St SE1 130 A1
Saxon Rd E3 127 K4
Saxton Cl SE13 131 P9
Sayes Ct St SE8 131 K5
Scala St W1 125 J7
Scandrett St E1 126 F10
Scarsdale Vil W8 128 A2
Scawen Rd SE8 131 J4
Scawfell St E2 126 D4
Sceaux Gdns SE5 130 C7
Sceptre Rd E2 126 G5
Schoolhouse La E1 127 H9
Schooner Cl E14 131 P2
Sclater St E1 126 D6
Scoresby St SE1 125 P10
Scott Ellis Gdns NW8 124 D5
Scott Lidgett Cres SE16 130 E1
Scott St E1 126 F6
Scriven St E8 126 D3
Scrutton St EC2 126 C6
Scylla Rd SE15 130 F9
Seacon Twr E14 131 K1
Seaford St WC1 125 M5
Seager Pl SE8 131 L7
Seagrave Rd SW6 128 A5
Searles Cl SW11 128 E6
Searles Rd SE1 130 B3
Sears St SE5 130 B6
Sebastian St EC1 125 P5
Sebbon St N1 125 P2
Sedding St SW1 128 G3
Sedgmoor Pl SE5 130 C6
Seething La EC3 126 C9
Sekforde St EC1 125 P6
Selby St E1 126 E6
Selden Rd SE15 130 G8
Selsdon Way E14 131 M2
Selsey St E14 127 L7
Selwood Pl SW7 128 D4
Selworthy Ho SW11 128 D7
Selwyn Rd E3 127 K4
Semley Pl SW1 128 G3
Senate St SE15 130 G8
Sendall Ct SW11 128 D9
Senior St W2 124 B7
Senrab St E1 127 H8
Serenaders Rd SW9 129 N8
Serle St WC2 125 M8
Serpentine Rd W2 124 F10
Settles St E1 126 E7
Settrington Rd SW6 128 B8
Seven Sea Gdns E3 127 M7
Severnake Cl E14 131 L3
Severus Rd SW11 128 E10
Seville Ms N1 126 C2
Seville St SW1 128 F1
Sevington St W9 124 B6
Seward St EC1 126 A5
Sewardstone Rd E2 126 G4
Sextant Av E14 131 P3
Seymour Gdns SE4 131 J9
Seymour Ms W1 124 G8
Seymour Pl W1 124 E7
Seymour St W1 124 F8
Seymour St W2 124 F8
Seymour Wk SW10 128 C5
Seyssel St E14 131 N3
Shacklewell St E2 126 D6
Shad Thames SE1 126 D10
Shaftesbury Av W1 125 K9
Shaftesbury Av WC2 125 K9
Shaftesbury St N1 126 A4
Shafton Rd E9 127 H3
Shalcomb St SW10 128 C5
Shamrock St SW4 129 K9
Shand St SE1 130 C1
Shandy St E1 127 H7
Shannon Gro SW9 129 M10
Shard London Br SE1 126 B10
Shardeloes Rd SE14 131 K9
Shardeloes Rd SE4 131 K9
Sharon Gdns E9 126 G3
Sharpleshall St NW1 124 F2
Sharratt St SE15 130 G5
Sharsted St SE17 129 P4
Shaw Ct SW11 128 D9
Shaw Cres E14 127 J7
Shaw Rd SE22 130 C10
Shawfield St SW3 128 E4
Shearling Way N7 125 L1
Sheep La E8 126 F3
Sheepcote La SW11 128 F8
Sheffield Ter W8 124 A10
Sheldon Pl E2 126 E4
Sheldon Sq W2 124 C7
Shell Rd SE13 131 M9
Shelley Cl SE15 130 F8
Shellwood Rd SW11 128 F8
Shelmerdine Cl E3 127 L7
Shelton St WC2 125 L8
Shenfield St N1 126 C4
Shenley Rd SE5 130 C7
Shepherdess Wk N1 126 A4
Sheppard Dr SE16 130 F4
Shepperton Rd N1 126 A3
Sherborne St N1 126 B3
Sheringham Rd N7 125 M1
Sherriff Rd NW6 124 A1
Sherwin Rd SE14 131 H7
Sherwood Gdns E14 131 L3
Sherwood Gdns SE16 130 E4
Shetland Rd E3 127 K4
Ship St SE8 131 L7
Shipton St E2 126 D5
Shipwright Rd SE16 131 J1
Shirbutt St E14 127 M9
Shirland Rd W9 124 A5
Shirley Gro SW11 128 G9
Shoe La EC4 125 N8
Shooters Hill Rd SE10 131 P7
Shore Pl E9 126 G2
Shore Rd E9 126 G2
Shore Way SW9 129 N8
Shoreditch High St E1 126 C6
Shorncliffe Rd SE1 130 D4
Short Wall E15 127 N5
Shorter St E1 126 D9
Shorts Gdns WC2 125 L8
Shottendane Rd SW6 128 A7
Shouldham St W1 124 E7
Shrewsbury Rd W2 124 A8
Shroton St NW1 124 E7
Shrubland Rd E8 126 E3
Shuttleworth Rd SW11 128 E8
Sibella Rd SW4 129 K8
Sidmouth St WC1 125 L5
Sidney Rd SW9 129 M8
Sidney Sq E1 126 G8
Sidney St E1 126 F7
Sidworth St E8 126 F2
Silex St SE1 129 P1
Silk Mills Sq E9 127 K1
Silk St EC2 126 A7
Silk Weaver Way E2 126 F4
Silver Rd SE13 131 M9
Silver Wk SE16 127 K10
Silverthorne Rd SW8 129 H8
Silvocea Way E14 127 P8
Silwood St SE16 130 G3
Simms Rd SE1 130 E3
Simpson St SW11 128 E8
Simpsons Rd E14 127 M9
Sirinham Pt SW8 129 M5
Sisters Av SW11 128 F10
Sisulu Pl SW9 129 N9
Sivill Ho E2 126 D5
Six Bridges Trd Est SE1 130 E4
Sketchley Gdns SE16 131 H4
Skinner St EC1 125 N5
Skipworth Rd E9 126 G3
Skylines Village E14 131 N1
Slaidburn St SW10 128 C5
Slaithwaite Rd SE13 131 N10
Sleaford Ho E3 127 L6
Sleaford St SW8 129 J6
Slippers Pl SE16 130 F2
Sloane Av SW3 128 E3
Sloane Ct W SW3 128 G4
Sloane Gdns SW1 128 G3
Sloane Sq SW1 128 F3
Sloane St SW1 128 F1
Sloane Ter SW1 128 F3
Smart St E2 127 H5
Smead Way SE13 131 M9
Smeaton St E1 126 F10
Smedley St SW4 129 K8
Smedley St SW8 129 K8
Smeed Rd E3 127 L2
Smiles Pl SE13 131 N8
Smith Cl SE16 127 H10
Smith Sq SW1 129 L2
Smith St SW3 128 F4
Smith Ter SW3 128 F4
Smithy St E1 126 G7
Smokehouse Yd EC1 125 P7
Smugglers Way SW18 128 B10
Smyrks Rd SE17 130 C4
Smyrna Rd NW6 124 A2
Smythe St E14 127 M9
Snow Hill EC1 125 P7
Snowbury Rd SW6 128 B8
Snowman Ho NW6 124 B3
Snowsfields SE1 130 B1
Soames St SE15 130 D9
Soho Sq W1 125 K8
Solebay St E1 127 J6
Solomon's Pas SE15 130 F10
Solon New Rd SW4 129 L10
Solon Rd SW2 129 L10
Solway Rd SE22 130 E10
Somerfield St SE16 131 H4
Somerford St E1 126 F6
Somerford Way SE16 131 J1
Somerleyton Pas SW9 129 P10
Somerleyton Rd SW9 129 N10
Somers Cres W2 124 E8
Somerset Est SW11 128 D7
Somerset Gdns SE13 131 M8
Somerton Rd SE15 130 F10
Sondes St SE17 130 B5
Sopwith Way SW8 129 H6
Sorrel La E14 127 P8
Sotheran Cl E8 126 E3
Sotheron Rd SW6 128 B6
Soudan Rd SW11 128 F7
South Audley St W1 124 G9
South Bk Twr SE1 125 N10
South Bolton Gdns SW5 128 B4
South Carriage Dr SW1 128 E1
South Carriage Dr SW7 128 E1
South Colonnade, The E14 127 L10
South Cres E16 127 P6
South Cres WC1 125 K7
South Eaton Pl SW1 128 G3
South End Row W8 128 B2
South Island Pl SW9 129 M6
South Kensington Underground Sta SW7 128 D3
South Lambeth Pl SW8 129 L5
South Lambeth Rd SW8 129 L5
South Molton La W1 125 H8
South Molton St W1 125 H8
South Par SW3 128 D4
South Pk SW6 128 A8
South Pk Ms SW6 128 B9
South Pl EC2 126 B7
South Quay Plaza E14 131 M1
South Sea St SE16 131 K2
South St W1 124 G10
South Tenter St E1 126 D9
South Ter SW7 128 E3
South Vil NW1 125 K1
South Wf Rd W2 124 D8
Southall Pl SE1 130 B1
Southampton Pl WC1 125 L7
Southampton Row WC1 125 L7
Southampton St WC2 125 L9
Southampton Way SE5 130 B6
Southbank Bus Cen SW8 129 K5
Southborough Rd E9 126 G3
Southern Gro E3 127 K5
Southern St N1 125 M4
Southerngate Way SE14 131 J6
Southey Rd SW9 129 N7
Southgate Gro N1 126 B2
Southgate Rd N1 126 B3
Southmoor Way E9 127 K1
Southolm St SW11 129 H7
Southville SW8 129 K7
Southwark Br EC4 126 A10
Southwark Br SE1 126 A10
Southwark Br Rd SE1 129 P2
Southwark Pk SE16 130 G3
Southwark Pk Est SE16 130 F3
Southwark Pk Rd SE16 130 D3
Southwark St SE1 125 P10
Southwater Cl E14 127 K8
Southwell Gdns SW7 128 C3
Southwell Rd SE5 130 A9
Southwick Pl W2 124 E8
Southwick St W2 124 E8
Sovereign Cl E1 126 F9
Spa Grn Est EC1 125 N5
Spa Rd SE16 130 D2
Spanby Rd E3 127 L6
Spanish Pl W1 124 G8
Sparkford Ho SW11 128 D7
Sparta St SE10 131 M7
Spear Ms SW5 128 A3
Speldhurst Rd E9 127 H2
Spelman St E1 126 E7
Spencer Rd SW18 128 D10
Spencer St EC1 125 P5
Spenser St SW1 129 J2
Spert St E14 127 J9
Spey St E14 127 N7
Spicer Cl SW9 129 P8
Spindrift Av E14 131 M3
Spital Sq E1 126 C7
Spital St E1 126 E6
Sporle Ct SW11 128 D9
Sprimont Pl SW3 128 F4
Spring St W2 124 D8
Spring Way SE5 130 A7
Springall St SE15 130 F6
Springfield La NW6 124 B3
Springfield Rd NW8 124 C3
Springfield Wk NW6 124 B3
Springhill Cl SE5 130 B9
Springwood Cl E3 127 L4
Sprules Rd SE4 131 J8
Spur Rd SE1 129 N1
Spur Rd SW1 129 J1
Spurgeon St SE1 130 B2
Spurling Rd SE22 130 D10
Squirrels, The SE13 131 P9
Squirries St E2 126 E5
Stable St N1 125 L3
Stable Yd Rd SW1 125 J10
Stables Way SE11 129 N4
Stacey St WC2 125 K8
Stadium St SW10 128 C6
Stafford Cl NW6 124 A5
Stafford Ct W8 128 A2
Stafford Pl SW1 129 J2
Stafford Rd E3 127 K4
Stafford Rd NW6 124 A5
Stafford St W1 125 J10
Stafford Ter W8 128 A2
Staffordshire St SE15 130 E7
Staining La EC2 126 A8
Stainsby Rd E14 127 L8
Stalham St SE16 130 F2
Stamford Rd N1 126 C2
Stamford St SE1 125 N10
Stamp Pl E2 126 D5
Stanbury Rd SE15 130 F7

Sta - Vic

Stane Gro **SW9** 129 L8
Stanfield Rd **E3** 127 J4
Stanford Rd **W8** 128 B2
Stanhope Gdns **SW7** 128 C3
Stanhope Gate **W1** 124 G10
Stanhope Ms E **SW7** 128 C3
Stanhope Ms W **SW7** 128 C3
Stanhope Pl **W2** 124 F8
Stanhope St **NW1** 125 J5
Stanhope Ter **W2** 124 D9
Stanley Cl **SW8** 129 M5
Stanley Gro **SW8** 128 G8
Stanley St **SE8** 131 K6
Stanmer St **SW11** 128 E7
Stannard Ms **E8** 126 E1
Stannard Rd **E8** 126 E1
Stannary St **SE11** 129 N5
Stansfield Rd **SW9** 129 M9
Stanswood Gdns **SE5** 130 C6
Stanway St **N1** 126 C4
Stanworth St **SE1** 130 D1
Staple Inn Bldgs **WC1** 125 N7
Staple St **SE1** 130 B1
Staples Cl **SE16** 127 J10
Star St **W2** 124 E7
Starboard Way **E14** 131 L2
Starcross St **NW1** 125 J5
Station App **SE1** 129 N1
Station Ct **SW6** 128 C7
Station La **E20** 127 N1
Station Pas **SE15** 130 G7
Station Rd **SE13** 131 N9
Station St **E15** 127 P2
Station Ter **SE5** 130 A7
Staunton St **SE8** 131 K5
Stave Yd Rd **SE16** 127 J10
Stayner's Rd **E1** 127 H6
Stead St **SE17** 130 B3
Stean St **E8** 126 D3
Stebondale St **E14** 131 N4
Steeles Rd **NW3** 124 F1
Steers Way **SE16** 131 J1
Stephan Cl **E8** 126 E3
Stephen St **W1** 125 K7
Stephendale Rd **SW6** 128 B8
Stephenson Way **NW1** 125 J6
Stepney Causeway **E1** 127 H8
Stepney Grn **E1** 126 G7
Stepney High St **E1** 127 H7
Stepney Way **E1** 126 F7
Sterling Gdns **SE14** 131 J5
Sternhall La **SE15** 130 E9
Sterry St **SE1** 130 B1
Stevens Av **E9** 126 G1
Stevenson Cres **SE16** 130 E4
Stewart St **E14** 131 N1
Stewart's Gro **SW3** 128 D3
Stewart's Rd **SW8** 129 J6
Stillington St **SW1** 129 J3
Stirling Rd **SW9** 129 L8
Stockholm Ho **E1** 126 E9
Stockholm Rd **SE16** 130 G4
Stockholm Way **E1** 126 E10
Stockwell Av **SW9** 129 M9
Stockwell Gdns **SW9** 129 M8
Stockwell Gdns Est **SW9** 129 L8
Stockwell Grn **SW9** 129 M8
Stockwell La **SW9** 129 M8
Stockwell Pk Cres **SW9** 129 M8
Stockwell Pk Est **SW9** 129 M9
Stockwell Pk Rd **SW9** 129 M7
Stockwell Pk Wk **SW9** 129 M9
Stockwell Rd **SW9** 129 M8
Stockwell St **SE10** 131 N5
Stockwell Ter **SW9** 129 M7
Stokenchurch St **SW6** 128 B7
Stone Bldgs **WC2** 125 M7
Stonecutter St **EC4** 125 P8
Stonefield St **N1** 125 N3
Stones End St **SE1** 130 A1
Stoney St **SE1** 126 B10
Stonhouse St **SW4** 129 K9
Stopes St **SE15** 130 D6
Stopford Rd **SE17** 129 P4
Store St **WC1** 125 K7
Storers Quay **E14** 131 P3
Storey's Gate **SW1** 129 K1
Stories Ms **SE5** 130 C8
Stories Rd **SE5** 130 C9
Storks Rd **SE16** 130 E2
Stormont Rd **SW11** 128 G9
Stour Rd **E3** 127 L2
Stourcliffe St **W1** 124 F8
Stowage **SE8** 131 L5
Strafford St **E14** 131 L1
Strahan Rd **E3** 127 J5
Straightsmouth **SE10** 131 N6
Straker's Rd **SE15** 130 F10
Strand **WC2** 125 L9
Stranraer Way **N1** 125 L2
Strasburg Rd **SW11** 129 H7
Stratford Cen, The **E15** 127 P2
Stratford Pl **W1** 125 H8
Stratford Rd **W8** 128 A2
Stratford Vil **NW1** 125 J2
Strath Ter **SW11** 128 E10
Strathblaine Rd **SW11** 128 D10
Strathearn Pl **W2** 124 E9
Strathnairn St **SE1** 130 E3
Strathray Gdns **NW3** 124 E1
Stratton St **W1** 125 H10
Strattondale St **E14** 131 N2
Streatham St **WC1** 125 K8
Streimer Rd **E15** 127 N4
Strickland St **SE8** 131 L8
Stroudley Wk **E3** 127 M5
Strutton Grd **SW1** 129 K2
Stuart Rd **NW6** 124 A5
Stuart Rd **SE15** 130 G10
Stuart Twr **W9** 124 C6
Stubbs Dr **SE16** 130 F4
Studd St **N1** 125 P3
Studdridge St **SW6** 128 A8
Studholme St **SE15** 130 F6
Studley Est **SW4** 129 L7
Studley Rd **SW4** 129 L7
Stukeley St **WC1** 125 L8
Stukeley St **WC2** 125 L8
Sturdy Rd **SE15** 130 F8
Sturgeon Rd **SE17** 130 A4
Sturry St **E14** 127 M8
Sturt St **N1** 126 A4
Stutfield St **E1** 126 E8
Styles Gdns **SW9** 129 P9
Sudeley St **N1** 125 P4
Sudlow Rd **SW18** 128 A10
Sugar Ho La **E15** 127 N4
Sugar Quay Wk **EC3** 126 C9
Sugden Rd **SW11** 128 G9
Sulivan Ct **SW6** 128 A9
Sulivan Rd **SW6** 128 A9
Sullivan Cl **SW11** 128 E9
Sullivan Rd **SE11** 129 N3
Sultan St **SE5** 130 A6
Summercourt Rd **E1** 126 G8
Sumner Pl **SW7** 128 D3
Sumner Rd **SE15** 130 D6
Sumner St **SE1** 126 A10
Sumpter Cl **NW3** 124 C1
Sun St **EC2** 126 B7
Sunbury La **SW11** 128 D7
Sunderland Ter **W2** 124 B8
Sunlight Sq **E2** 126 F5
Sunninghill Rd **SE13** 131 M8
Sunray Av **SE24** 130 B10
Sunset Rd **SE5** 130 A10
Surma Cl **E1** 126 F6
Surrendale Pl **W9** 124 A6
Surrey Canal Rd **SE14** 130 G5
Surrey Canal Rd **SE15** 130 G5
Surrey La **SW11** 128 E7
Surrey La Est **SW11** 128 E7
Surrey Quays Rd **SE16** 130 G2
Surrey Quays Shop Cen **SE16** 131 H2
Surrey Row **SE1** 129 P1
Surrey Sq **SE17** 130 C4
Surrey St **WC2** 125 M9
Surrey Ter **SE17** 130 C4
Surrey Water Rd **SE16** 127 H10
Susannah St **E14** 127 M8
Sussex Gdns **W2** 124 D8
Sussex Pl **NW1** 124 F5
Sussex Pl **W2** 124 D8
Sussex Sq **W2** 124 D9
Sussex St **SW1** 129 H4
Sutherland Av **W9** 124 C5
Sutherland Pl **W2** 124 A8
Sutherland Row **SW1** 129 H4
Sutherland Sq **SE17** 130 A4
Sutherland St **SW1** 129 H4
Sutherland Wk **SE17** 130 A4
Sutterton St **N7** 125 M1
Sutton Est, The **N1** 125 P2
Sutton Est **SW3** 128 E4
Sutton Row **W1** 125 K8
Sutton St **E1** 126 G8
Swain St **NW8** 124 E6
Swallow Cl **SE14** 130 G7
Swan Mead **SE1** 130 C2
Swan Rd **SE16** 130 G1
Swan St **SE1** 130 A2
Swan Wk **SW3** 128 F5
Swanfield St **E2** 126 D5
Swaton Rd **E3** 127 L6
Sweden Gate **SE16** 131 J2
Swedenborg Gdns **E1** 126 E9
Sweeney Cres **SE1** 130 D1
Swinford Gdns **SW9** 129 P9
Swinton Pl **WC1** 125 M5
Swinton St **WC1** 125 M5
Swiss Ter **NW6** 124 D2
Sybil Phoenix Cl **SE8** 131 H4
Sycamore Av **E3** 127 K3
Sycamore Ms **SW4** 129 J9
Sydney Cl **SW3** 128 D3
Sydney Ms **SW3** 128 D3
Sydney Pl **SW7** 128 D3
Sydney St **SW3** 128 E3
Sylvan Gro **SE15** 130 F6
Sylvester Rd **E8** 126 F1
Symons Cl **SE15** 130 G8
Symons St **SW3** 128 F3

T

Tabard Gdns Est **SE1** 130 B1
Tabard St **SE1** 130 B1
Tabernacle St **EC2** 126 B6
Tachbrook Est **SW1** 129 K4
Tachbrook St **SW1** 129 J3
Tack Ms **SE4** 131 L9
Tadema Rd **SW10** 128 C6
Taeping St **E14** 131 M3
Tait St **E1** 126 F8
Talacre Rd **NW5** 124 G1
Talbot Rd **SE22** 130 C10
Talbot Rd **W2** 124 A8
Talbot Sq **W2** 124 D8
Talfourd Pl **SE15** 130 D7
Talfourd Rd **SE15** 130 D7
Tallis St **EC4** 125 N9
Talma Rd **SW2** 129 N10
Talwin St **E3** 127 M5
Tamworth St **SW6** 128 A5
Tanner St **SE1** 130 C1
Tanners Hill **SE8** 131 K7
Taplow **NW3** 124 D2
Taplow St **N1** 126 A4
Tapp St **E1** 126 F6
Tapper Wk **N1** 125 L3
Tappesfield Rd **SE15** 130 G9
Tariff Cres **SE8** 131 K3
Tarling St **E1** 126 F8
Tarling St Est **E1** 126 G8
Tarragon Cl **SE14** 131 J6
Tarrant Pl **W1** 124 F7
Tarver Rd **SE17** 129 P4
Tarves Way **SE10** 131 M6
Tasman Rd **SW9** 129 L9
Tatham Pl **NW8** 124 D4
Tatum St **SE17** 130 B3
Taunton Pl **NW1** 124 F6
Tavern La **SW9** 129 N8
Tavistock Pl **WC1** 125 L6
Tavistock Sq **WC1** 125 K6
Tavistock St **WC2** 125 L9
Taviton St **WC1** 125 K6
Tavy Cl **SE11** 129 N4
Tawny Way **SE16** 131 H3
Taybridge Rd **SW11** 128 G9
Tayburn Cl **E14** 127 N8
Taylor Cl **SE8** 131 K5
Tayport Cl **N1** 125 L2
Teak Cl **SE16** 127 J10
Teale St **E2** 126 E4
Tedworth Sq **SW3** 128 F4
Teesdale Cl **E2** 126 E4
Teesdale St **E2** 126 F4
Teignmouth Cl **SW4** 129 K10
Telegraph Pl **E14** 131 M3
Telford Ter **SW1** 129 J5
Tell Gro **SE22** 130 D10
Templar St **SE5** 129 P8
Temple, The **EC4** 125 N9
Temple Av **EC4** 125 N9
Temple Pl **WC2** 125 M9
Temple St **E2** 126 F4
Temple W Ms **SE11** 129 P2
Templecombe Rd **E9** 126 G3
Templeton Pl **SW5** 128 A3
Tench St **E1** 126 F10
Tenison Way **SE1** 125 M10
Tennis St **SE1** 130 B1
Tennyson St **SW8** 129 H8
Tent St **E1** 126 F6
Tenterden St **W1** 125 H8
Teredo St **SE16** 131 H2
Terminus Pl **SW1** 129 H2
Terrace, The **NW6** 124 A3
Terrace Rd **E9** 126 G2
Tessa Sanderson Pl **SW8** 129 H9
Tetcott Rd **SW10** 128 C6
Teversham La **SW8** 129 L7
Teviot St **E14** 127 N7
Thackeray Rd **SW8** 129 H8
Thackeray St **W8** 128 B1
Thalia Cl **SE10** 131 P5
Thame Rd **SE16** 131 H1
Thames Av **SW10** 128 C7
Thames St **SE10** 131 M5
Thanet St **WC1** 125 L5
Thayer St **W1** 124 G7
Theatre St **SW11** 128 F9
Theberton St **N1** 125 N3
Theed St **SE1** 125 N10
Theobald's Rd **WC1** 125 M7
Thermopylae Gate **E14** 131 M3
Thessaly Rd **SW8** 129 J6
Thirleby Rd **SW1** 129 J2
Thirsk Rd **SW11** 128 G9
Thistle Gro **SW10** 128 C4
Thomas Baines Rd **SW11** 128 D9
Thomas Doyle St **SE1** 129 P2
Thomas More St **E1** 126 E9
Thomas Rd **E14** 127 K8
Thompsons Av **SE5** 130 A6
Thorburn Sq **SE1** 130 E3
Thoresby St **N1** 126 A5
Thorncroft St **SW8** 129 L6
Thorndike Cl **SW10** 128 C6
Thorndike Rd **N1** 126 A1
Thorndike St **SW1** 129 J3
Thorne Rd **SW8** 129 L6
Thorney Cres **SW11** 128 D6
Thorney St **SW1** 129 L3
Thorngate Rd **W9** 124 A6
Thornham St **SE10** 131 M5
Thornhaugh St **WC1** 125 K6
Thornhill Cres **N1** 125 M2
Thornhill Rd **N1** 125 N2
Thornhill Sq **N1** 125 M2
Thornton Pl **W1** 124 F7
Thornton St **SW9** 129 N8
Thornville St **SE8** 131 L7
Thorparch Rd **SW8** 129 K7
Thoydon Rd **E3** 127 J4
Thrale St **SE1** 126 A10
Thrawl St **E1** 126 D7
Threadneedle St **EC2** 126 B8
Three Colt St **E14** 127 K9
Three Colts La **E2** 126 F6
Three Kings Yd **W1** 125 H9
Three Mill La **E3** 127 N5
Three Quays Wk **EC3** 126 C9
Throgmorton Av **EC2** 126 B8
Throgmorton St **EC2** 126 B8
Thurland Rd **SE16** 130 E2
Thurloe Cl **SW7** 128 E2
Thurloe Pl **SW7** 128 D3
Thurloe Sq **SW7** 128 E3
Thurloe St **SW7** 128 D3
Thurlow St **SE17** 130 B4
Thurston Rd **SE13** 131 M8
Thurtle Rd **E2** 126 D3
Tibbatts Rd **E3** 127 M6
Tidemill Way **SE8** 131 L6
Tideway Ind Est **SW8** 129 J5
Tidey St **E3** 127 L7
Tidworth Rd **E3** 127 L6
Tileyard Rd **N7** 125 L2
Tilia Wk **SW9** 129 P10
Tiller Rd **E14** 131 L2
Tillings Cl **SE5** 130 A7
Tilney Gdns **N1** 126 B1
Tilney St **W1** 124 G10
Timber Mill Way **SW4** 129 K9
Timber Pond Rd **SE16** 131 H1
Tindal St **SW9** 129 P7
Tinsley Rd **E1** 126 G7
Tintagel Cres **SE22** 130 D10
Tintern St **SW4** 129 L10
Tinworth St **SE11** 129 M4
Tipthorpe Rd **SW11** 128 G9
Tisdall Pl **SE17** 130 B3
Titchfield Rd **NW8** 124 F3
Tite St **SW3** 128 F4
Tiverton St **SE1** 130 A2
Tivoli Ct **SE16** 131 K1
Tobacco Dock **E1** 126 F9
Tobin Cl **NW3** 124 E2
Toby La **E1** 127 J6
Tollet St **E1** 127 H6
Tollgate Gdns **NW6** 124 B4
Tolpuddle St **N1** 125 N4
Tomlins Gro **E3** 127 L5
Tomlinson Cl **E2** 126 D5
Tonbridge St **WC1** 125 L5
Tooley St **SE1** 126 C10
Topmast Pt **E14** 131 L1
Tor Gdns **W8** 128 A1
Toronto Ho **SE16** 131 H1
Torrens St **EC1** 125 N4
Torridge Gdns **SE15** 130 G10
Torrington Pl **E1** 126 E10
Torrington Pl **WC1** 125 K7
Torrington Sq **WC1** 125 K6
Tothill St **SW1** 129 K1
Tottan Ter **E1** 127 H8
Tottenham Ct Rd **W1** 125 J6
Tottenham Rd **N1** 126 C1
Tottenham St **W1** 125 J7
Totteridge Ho **SW11** 128 D8
Toulmin St **SE1** 130 A1
Toulon St **SE5** 130 A6
Tours Pas **SW11** 128 D10
Towcester Rd **E3** 127 M6
Tower 42 **EC2** 126 C8
Tower Br **E1** 126 D10
Tower Br **SE1** 126 D10
Tower Br App **E1** 126 D10
Tower Br Rd **SE1** 130 C2
Tower Br Wf **E1** 126 E10
Tower Mill Rd **SE15** 130 C6
Tower St **WC2** 125 K8
Town Hall Rd **SW11** 128 F9
Townmead Rd **SW6** 128 C8
Townsend St **SE17** 130 B3
Townshend Est **NW8** 124 E4
Townshend Rd **NW8** 124 E3
Toynbee St **E1** 126 D7
Tradescant Rd **SW8** 129 L6
Trafalgar Av **SE15** 130 D4
Trafalgar Gdns **E1** 127 H7
Trafalgar Gro **SE10** 131 P5
Trafalgar Ms **E9** 127 K1
Trafalgar Rd **SE10** 131 P5
Trafalgar Sq **SW1** 125 K10
Trafalgar Sq **WC2** 125 K10
Trafalgar St **SE17** 130 B4
Trafalgar Way **E14** 127 N10
Trahorn Cl **E1** 126 F6
Transept St **NW1** 124 E7
Transom Sq **E14** 131 M3
Tranton Rd **SE16** 130 E2
Treadway St **E2** 126 F4
Treaty St **N1** 125 M3
Trebovir Rd **SW5** 128 A4
Treby St **E3** 127 K6
Tredegar Rd **E3** 127 K4
Tredegar Sq **E3** 127 K5
Tredegar Ter **E3** 127 K5
Trederwen Rd **E8** 126 E3
Tregarvon Rd **SW11** 128 G10
Trego Rd **E9** 127 L2
Tregothnan Rd **SW9** 129 L9
Tregunter Rd **SW10** 128 C5
Trelawney Est **E9** 126 G1
Tremadoc Rd **SW4** 129 K10
Tremaine Cl **SE4** 131 L8
Trematon Wk **N1** 125 L4
Trenchard St **SE10** 131 P4
Trenchold St **SW8** 129 L5
Tresco Rd **SE15** 130 F10
Tresham Cres **NW8** 124 E6
Tressillian Cres **SE4** 131 L9
Tressillian Rd **SE4** 131 K10
Trevithick Ho **SE16** 130 F3
Trevithick St **SE8** 131 L4
Trevithick Way **E3** 127 L5
Trevor Pl **SW7** 128 E1
Trevor Sq **SW7** 128 F1
Trevor St **SW7** 128 E1
Triangle Pl **SW4** 129 K10
Triangle Rd **E8** 126 F3
Trident St **SE16** 131 H3
Trigon Rd **SW8** 129 M6
Trim St **SE14** 131 K5
Trinity Ch Sq **SE1** 130 A2
Trinity Cl **E8** 126 D1
Trinity Cl **SE13** 131 P10
Trinity Gdns **SW9** 129 M10
Trinity Gro **SE10** 131 N7
Trinity Sq **EC3** 126 C9
Trinity St **SE1** 130 A1
Trinity Wk **NW3** 124 C1
Triton Sq **NW1** 125 J6
Trothy Rd **SE1** 130 E3
Trott St **SW11** 128 E7
Troutbeck Rd **SE14** 131 J7
Trowbridge Rd **E9** 127 K1
Troy Town **SE15** 130 E9
Truman Wk **E3** 127 M6
Trundleys Rd **SE8** 131 H4
Trundleys Ter **SE8** 131 H3
Truro St **NW5** 124 G1
Tryon Cres **E9** 126 G3
Tryon St **SW3** 128 F4
Tudor Gro **E9** 126 G2
Tudor Rd **E9** 126 F3
Tudor St **EC4** 125 N9
Tufton St **SW1** 129 K2
Tuilerie St **E2** 126 E4
Tunnel Av **SE10** 131 P1
Tunstall Rd **SW9** 129 M10
Turenne Cl **SW18** 128 C10
Turin St **E2** 126 E5
Turks Row **SW3** 128 F4
Turner Cl **SW9** 129 P7
Turner St **E1** 126 F7
Turners Rd **E3** 127 K7
Turnmill St **EC1** 125 N6
Turnpike Ho **EC1** 125 P5
Turnpin La **SE10** 131 N5
Turret Gro **SW4** 129 J9
Tuscan Ho **E2** 126 G5
Tustin Est **SE15** 130 G5
Twelvetrees Cres **E3** 127 N6
Twine Ct **E1** 126 G9
Twyford St **N1** 125 M3
Tyburn Way **W1** 124 F9
Tyers Est **SE1** 130 C1
Tyers St **SE11** 129 M4
Tyers Ter **SE11** 129 M4
Tyler Cl **E2** 126 D4
Tyndale Ct **E14** 131 M4
Tyneham Rd **SW11** 128 G8
Tynemouth St **SW6** 128 C8
Type St **E2** 127 H4
Tyrawley Rd **SW6** 128 B7
Tyrrell Rd **SE22** 130 E10
Tyrwhitt Rd **SE4** 131 L9
Tyssen Pas **E8** 126 D1
Tyssen St **E8** 126 D1

U

Uamvar St **E14** 127 M7
Ufford St **SE1** 129 N1
Ufton Gro **N1** 126 B2
Ufton Rd **N1** 126 B2
Undercliff Rd **SE13** 131 L9
Underwood Rd **E1** 126 E6
Underwood Row **N1** 126 A5
Underwood St **N1** 126 A5
Undine Rd **E14** 131 M3
Union Gro **SW8** 129 K8
Union Jack Club **SE1** 129 N1
Union Rd **SW4** 129 K8
Union Rd **SW8** 129 K8
Union Sq **N1** 126 A3
Union St **SE1** 125 P10
University St **WC1** 125 J6
Unwin Cl **SE15** 130 E5
Upcerne Rd **SW10** 128 C6
Upper Bk St **E14** 127 M10
Upper Belgrave St **SW1** 128 G2
Upper Berkeley St **W1** 124 F8
Upper Brockley Rd **SE4** 131 K8
Upper Brook St **W1** 124 G9
Upper Cheyne Row **SW3** 128 E5
Upper Grosvenor St **W1** 124 G9
Upper Grd **SE1** 125 N10
Upper Marsh **SE1** 129 M2
Upper Montagu St **W1** 124 F7
Upper N St **E14** 127 L7
Upper Phillimore Gdns **W8** 128 A1
Upper St **N1** 125 N4
Upper Tachbrook St **SW1** 129 J3
Upper Thames St **EC4** 126 A9
Upper Wimpole St **W1** 124 G7
Upper Woburn Pl **WC1** 125 K5
Upstall St **SE5** 129 P7
Urlwin St **SE5** 130 A5
Ursula Gould Way **E14** 127 L7
Ursula St **SW11** 128 E7
Usborne Ms **SW8** 129 M6
Usher Rd **E3** 127 K4
Usk Rd **SW11** 128 C10
Usk St **E2** 127 H5
Uverdale Rd **SW10** 128 C6
Uxbridge St **W8** 124 A10

V

Vale, The **SW3** 128 D5
Vale Ct **W9** 124 C5
Vale Royal **N7** 125 L2
Valentine Pl **SE1** 129 P1
Valentine Rd **E9** 127 H1
Valentine Row **SE1** 129 P1
Valette St **E9** 126 F1
Vallance Rd **E1** 126 E6
Vallance Rd **E2** 126 E5
Valmar Rd **SE5** 130 A7
Van Gogh Wk **SW9** 129 M7
Vandon St **SW1** 129 J2
Vanguard St **SE8** 131 L7
Vansittart St **SE14** 131 J6
Vanston Pl **SW6** 128 A6
Varcoe Rd **SE16** 130 F4
Varden St **E1** 126 F8
Vardens Rd **SW11** 128 D10
Varndell St **NW1** 125 J5
Vassall Rd **SW9** 129 N6
Vauban Est **SE16** 130 D2
Vauban St **SE16** 130 D2
Vaughan Rd **SE5** 130 A8
Vaughan St **SE16** 131 K1
Vaughan Way **E1** 126 E9
Vauxhall Br **SE1** 129 L4
Vauxhall Br **SW1** 129 L4
Vauxhall Br Rd **SW1** 129 J3
Vauxhall Gdns Est **SE11** 129 M4
Vauxhall Gro **SW8** 129 L5
Vauxhall St **SE11** 129 M4
Vauxhall Wk **SE11** 129 M4
Vawdrey Cl **E1** 126 G6
Veda Rd **SE13** 131 L10
Velletri Ho **E2** 127 H4
Venables St **NW8** 124 D6
Venetian Rd **SE5** 130 A8
Venn St **SW4** 129 J10
Ventnor Rd **SE14** 131 H6
Venue St **E14** 127 N7
Vere St **W1** 125 H8
Verney Rd **SE16** 130 E5
Verney Way **SE16** 130 F4
Vernon Pl **WC1** 125 L7
Vernon Ri **WC1** 125 M5
Vernon Rd **E3** 127 K4
Vesta Rd **SE4** 131 J8
Vestry Ms **SE5** 130 C7
Vestry Rd **SE5** 130 C7
Vestry St **N1** 126 B5
Viaduct St **E2** 126 F5
Vian St **SE13** 131 M9
Vicarage Cres **SW11** 128 D7
Vicarage Gdns **W8** 124 A10
Vicarage Gate **W8** 128 B1
Vicarage Gro **SE5** 130 B7
Vicars Hill **SE13** 131 M10
Viceroy Rd **SW8** 129 L7
Victoria Embk **EC4** 125 M9
Victoria Embk **SW1** 129 L1
Victoria Embk **WC2** 129 L1
Victoria Gdns **W11** 124 A10
Victoria Gate Gdns **SE10** 131 M6
Victoria Gro **W8** 128 C2
Victoria Ms **NW6** 124 A3
Victoria Par **SE10** 131 M5
Victoria Pk **E9** 127 J2
Victoria Pk Rd **E9** 126 G3
Victoria Pk Sq **E2** 126 G5
Victoria Pl **SW1** 129 H3
Victoria Ri **SW4** 129 H9

Vic - Zul

Victoria Rd **W8** 128 C2
Victoria Sta **SW1** 129 H3
Victoria St **SW1** 129 J2
Victoria Wf **E14** 127 J9
Victory Par **E20** 127 M1
Victory Pl **SE17** 130 A3
Victory Way **SE16** 131 J1
Vigo St **W1** 125 J9
Viking Ct **SW6** 128 A5
Villa Rd **SW9** 129 N9
Villa St **SE17** 130 B4
Villiers St **WC2** 125 L9
Vince St **EC1** 126 B5
Vincent Cl **SE16** 131 J1
Vincent Sq **SW1** 129 J3
Vincent St **SW1** 129 K3
Vincent Ter **N1** 125 P4
Vincentia Quay **SW11** 128 C8
Vine St Br **EC1** 125 N6
Viney Rd **SE13** 131 M9
Vineyard Wk **EC1** 125 N6
Vining St **SW9** 129 N10
Violet Hill **NW8** 124 C4
Violet Rd **E3** 127 M6
Virgil St **SE1** 129 M2
Virginia Rd **E2** 126 D5
Virginia St **E1** 126 E9
Vivian Rd **E3** 127 J4
Voltaire Rd **SW4** 129 K9
Voss St **E2** 126 E5
Voysey Sq **E3** 127 M7
Vulcan Rd **SE4** 131 K8
Vulcan Ter **SE4** 131 K8
Vulcan Way **N7** 125 M1
Vyner St **E2** 126 F3

W

Wadding St **SE17** 130 B3
Waddington St **E15** 127 P1
Wades Pl **E14** 127 M9
Wadeson St **E2** 126 F4
Wadham Gdns **NW3** 124 E3
Wadhurst Rd **SW8** 129 J7
Wager St **E3** 127 K6
Waghorn St **SE15** 130 E9
Wagner St **SE15** 130 G6
Waite St **SE15** 130 D5
Wakefield St **WC1** 125 L5
Wakeham St **N1** 126 B1
Wakeling St **E14** 127 J8
Wakley St **EC1** 125 P5
Walberswick St **SW8** 129 L6
Walbrook **EC4** 126 B9
Walcot Sq **SE11** 129 N3
Walden St **E1** 126 F8
Walerand Rd **SE13** 131 N8
Wales Cl **SE15** 130 F6
Waley St **E1** 127 H7
Walham Gro **SW6** 128 A6
Wall St **N1** 126 B1
Wallace Rd **N1** 126 A1
Wallbutton Rd **SE4** 131 J8
Waller Rd **SE14** 131 H7
Wallgrave Rd **SW5** 128 B3
Wallis Cl **SW11** 128 D9
Wallis Rd **E9** 127 K1
Wallwood St **E14** 127 K7
Walnut Tree Wk **SE11** 129 N3
Walpole St **SW3** 128 F4
Walsham Rd **SE14** 131 H8
Walter St **E2** 127 H5
Walter Ter **E1** 127 H8
Walton Cl **SW8** 129 L6
Walton Pl **SW3** 128 F2
Walton St **SW3** 128 E3
Walworth Pl **SE17** 130 A4
Walworth Rd **SE1** 130 A3
Walworth Rd **SE17** 130 A3
Wandon Rd **SW6** 128 B6
Wandsworth Br **SW18** 128 B9
Wandsworth Br **SW6** 128 B9
Wandsworth Br Rd **SW6** 128 B7
Wandsworth Rd **SW8** 129 K6
Wanless Rd **SE24** 130 A9
Wanley Rd **SE5** 130 B10
Wansbeck Rd **E3** 127 K2
Wansbeck Rd **E9** 127 K2
Wansey St **SE17** 130 A3
Wapping High St **E1** 126 E10
Wapping La **E1** 126 F9
Wapping Wall **E1** 126 G10
Warburton Rd **E8** 126 F3
Ward Rd **E15** 127 P3
Wardalls Gro **SE14** 130 G6
Warden Rd **NW5** 124 G1
Wardour St **W1** 125 K9
Warham St **SE5** 129 P6
Warley St **E2** 127 H5
Warlock Rd **W9** 124 A6
Warndon St **SE16** 131 H3
Warneford St **E9** 126 F3
Warner Pl **E2** 126 E4
Warner Rd **SE5** 130 A7
Warner St **EC1** 125 N6
Warren St **W1** 125 J6
Warriner Gdns **SW11** 128 F7
Warrington Cres **W9** 124 C6
Warton Rd **E15** 127 N3
Warwick Av **W2** 124 C6
Warwick Av **W9** 124 C6
Warwick Building **SW8** 129 H5
Warwick Ct **SE15** 130 E8
Warwick Cres **W2** 124 C7
Warwick Est **W2** 124 B7
Warwick Ho St **SW1** 125 K10
Warwick La **EC4** 125 P8
Warwick Pl **W9** 124 C7
Warwick Pl N **SW1** 129 J3
Warwick Row **SW1** 129 H2
Warwick Sq **SW1** 129 J4
Warwick Sq Ms **SW1** 129 J3
Warwick St **W1** 125 J9
Warwick Way **SW1** 129 J3
Warwickshire Path **SE8** 131 K6
Washington Cl **E3** 127 M5
Wat Tyler Rd **SE10** 131 N8
Wat Tyler Rd **SE3** 131 N8
Water Gdns, The **W2** 124 E8
Water Gdns Sq **SE16** 131 H1
Water La **SE14** 130 G6
Water Ms **SE15** 130 G10
Waterford Rd **SW6** 128 B7
Waterfront Apts **W9** 124 B6
Watergate St **SE8** 131 L5
Waterloo Br **SE1** 125 M9
Waterloo Br **WC2** 125 M9
Waterloo Est **E2** 126 G4
Waterloo Gdns **E2** 126 G4
Waterloo Pl **SW1** 125 K10
Waterloo Rd **SE1** 129 N1
Waterloo Sta **SE1** 129 N1
Waterloo Ter **N1** 125 P2
Waterman Way **E1** 126 F10
Watermans Wk **SE16** 131 J2
Watermeadow La **SW6** 128 C8
Waterside Cl **E3** 127 K3
Waterside Pt **SW11** 128 E6
Waterside Twr **SW6** 128 C7
Waterson St **E2** 126 C5
Waterview Ho **E14** 127 J7
Waterway Av **SE13** 131 M9
Watkinson Rd **N7** 125 M1
Watling St **EC4** 126 A8
Watney St **E1** 126 F8
Watson's St **SE8** 131 L6
Watts Gro **E3** 127 M7
Watts St **E1** 126 F10
Watts St **SE15** 130 D7
Waveney Av **SE15** 130 F10
Waverley Pl **NW8** 124 D4
Waverley Wk **W2** 124 A7
Waverton Ho **E3** 127 K3
Waverton St **W1** 124 G10
Wayford St **SW11** 128 E8
Wayland Ho **SW9** 129 N8
Wayman Ct **E8** 126 F1
Wear Pl **E2** 126 F5
Weardale Rd **SE13** 131 P10
Wearside Rd **SE13** 131 M10
Weatherley Cl **E3** 127 K7
Weaver St **E1** 126 E6
Weavers Ter **SW6** 128 A5
Weavers Way **NW1** 125 K3
Webb St **SE1** 130 C2
Webber Row **SE1** 129 N1
Webber St **SE1** 129 N1
Webster Rd **SE16** 130 E2
Weighhouse St **W1** 124 G8
Weir's Pas **NW1** 125 K5
Welbeck St **W1** 125 H8
Welbeck Way **W1** 125 H8
Welby St **SE5** 129 P7
Well St **E9** 126 F2
Welland St **SE10** 131 N5
Wellclose Sq **E1** 126 E9
Wellesley Ct **W9** 124 C5
Wellesley St **E1** 127 H7
Wellesley Ter **N1** 126 A5
Wellington Ms **SE22** 130 E10
Wellington Pl **NW8** 124 E4
Wellington Rd **NW8** 124 D4
Wellington Row **E2** 126 D5
Wellington Sq **SW3** 128 F4
Wellington St **WC2** 125 L9
Wellington Ter **E1** 126 F10
Wellington Way **E3** 127 L5
Wells Ri **NW8** 124 F3
Wells St **W1** 125 J8
Wells Way **SE5** 130 B5
Wells Way **SW7** 128 D2
Welmar Ms **SW4** 129 K10
Welsford St **SE1** 130 E4
Wendle Ct **SW8** 129 L5
Wendon St **E3** 127 K3
Wendover **SE17** 130 C4
Wenlock Rd **N1** 126 A4
Wenlock St **N1** 126 A4
Wennington Rd **E3** 127 H4
Wentworth Cres **SE15** 130 E6
Wentworth St **E1** 126 D8
Werrington St **NW1** 125 J4
Wesley Cl **SE17** 129 P3
Wessex St **E2** 126 G5
West Arbour St **E1** 127 H8
West Carriage Dr **W2** 124 D9
West Eaton Pl **SW1** 128 G3
West End La **NW6** 124 A2
West Gdns **E1** 126 F9
West Gro **SE10** 131 N7
West Halkin St **SW1** 128 G2
West Hampstead Ms **NW6** 124 B1
West India Av **E14** 127 L10
West India Dock Rd **E14** 127 K8
West La **SE16** 130 F1
West One Shop Cen **W1** 125 H8
West Pk Wk **E20** 127 N1
West Rd **SW3** 128 F5
West Smithfield **EC1** 125 P7
West Sq **SE11** 129 P2
West St **E2** 126 F4
West Tenter St **E1** 126 D8
Westbourne Br **W2** 124 C7
Westbourne Cres **W2** 124 D9
Westbourne Gdns **W2** 124 B8
Westbourne Gro **W11** 124 A8
Westbourne Gro **W2** 124 A8
Westbourne Gro Ter **W2** 124 B8
Westbourne Pk Rd **W2** 124 A7
Westbourne Pk Vil **W2** 124 A7
Westbourne Rd **N7** 125 N1
Westbourne St **W2** 124 D9
Westbourne Ter **W2** 124 D8
Westbourne Ter Ms **W2** 124 C8
Westbourne Ter Rd **W2** 124 C7
Westbridge Rd **SW11** 128 D7
Westbury St **SW8** 129 J8
Westcott Rd **SE17** 129 P5
Western Rd **SW9** 129 N9
Westferry Circ **E14** 127 K10
Westferry Rd **E14** 127 L10
Westfield Av **E20** 127 M1
Westfield Cl **SW10** 128 C6
Westfield Stratford City **E15** 127 N1
Westfield Way **E1** 127 J5
Westgate St **E8** 126 F3
Westgate Ter **SW10** 128 B4
Westgrove La **SE10** 131 N7
Westminster Br **SE1** 129 L1
Westminster Br **SW1** 129 L1
Westminster Br Rd **SE1** 129 N2
Westminster Gdns **SW1** 129 K3
Westmoreland Pl **SW1** 129 H4
Westmoreland Rd **SE17** 130 A5
Westmoreland St **W1** 124 G7
Westmoreland Ter **SW1** 129 H4
Westmoreland Wk **SE17** 130 B5
Weston Ri **WC1** 125 M4
Weston St **SE1** 130 C1
Westport St **E1** 127 H8
Westway **W2** 124 A7
Westway **W9** 124 A7
Wetherby Gdns **SW5** 128 C3
Wetherby Pl **SW7** 128 C3
Wetherell Rd **E9** 127 H3
Weybridge Pt **SW11** 128 G8
Weymouth Ms **W1** 125 H7
Weymouth St **W1** 125 H7
Weymouth Ter **E2** 126 D4
Wharf Pl **E2** 126 E3
Wharf Rd **N1** 125 L4
Wharf Rd **N1** 126 A4
Wharf St **SE8** 131 L4
Wharfdale Rd **N1** 125 L4
Wharfedale St **SW10** 128 B4
Wharfside Pt S **E14** 127 N9
Wharton St **WC1** 125 M5
Wheat Sheaf Cl **E14** 131 M3
Wheatsheaf La **SW8** 129 L6
Wheelwright St **N7** 125 M2
Wheler St **E1** 126 D6
Whidborne St **WC1** 125 L5
Whiskin St **EC1** 125 P5
Whistlers Av **SW11** 128 D6
Whiston Rd **E2** 126 D4
Whitbread Rd **SE4** 131 J10
Whitburn Rd **SE13** 131 M10
Whitcher Cl **SE14** 131 J5
Whitcomb St **WC2** 125 K9
White Ch La **E1** 126 E8
White Hart St **SE11** 129 N4
White Horse La **E1** 127 H6
White Horse Rd **E1** 127 J8
White Horse St **W1** 125 H10
White Lion Hill **EC4** 125 P9
White Lion St **N1** 125 N4
White Post La **E9** 127 K2
White Post La **SE13** 131 L9
White Post St **SE15** 130 G6
White Twr Way **E1** 127 J7
Whiteadder Way **E14** 131 M3
Whitear Wk **E15** 127 P1
Whitechapel High St **E1** 126 D8
Whitechapel Rd **E1** 126 E8
Whitecross St **EC1** 126 A6
Whitefriars St **EC4** 125 N8
Whitehall **SW1** 125 L10
Whitehall Ct **SW1** 125 L10
Whitehall Pl **SW1** 125 L10
Whitehead's Gro **SW3** 128 E4
Whiteleys Shop Cen **W2** 124 B8
Whites Grds **SE1** 130 C1
White's Row **E1** 126 D7
Whitethorn St **E3** 127 L6
Whitfield St **W1** 125 K7
Whitgift St **SE11** 129 M3
Whitmore Est **N1** 126 C3
Whitmore Rd **N1** 126 C3
Whittaker St **SW1** 128 G3
Whitton Wk **E3** 127 L4
Whorlton Rd **SE15** 130 F9
Wick La **E3** 127 L4
Wick Rd **E9** 127 H1
Wickersley Rd **SW11** 128 G8
Wickford St **E1** 126 G6
Wickham Cl **E1** 126 G7
Wickham Gdns **SE4** 131 K9
Wickham Ms **SE4** 131 K8
Wickham Rd **SE4** 131 K10
Wickham St **SE11** 129 M4
Wicklow St **WC1** 125 M5
Wickwood St **SE5** 129 P8
Widdin St **E15** 127 P2
Widley Rd **W9** 124 A5
Wigmore Pl **W1** 125 H8
Wigmore St **W1** 124 G8
Wilberforce Ms **SW4** 129 K10
Wilbraham Pl **SW1** 128 F3
Wilcox Cl **SW8** 129 L6
Wilcox Rd **SW8** 129 L6
Wild Ct **WC2** 125 M8
Wild Goose Dr **SE14** 130 G7
Wild St **WC2** 125 L8
Wilde Cl **E8** 126 E3
Wilderness Ms **SW4** 129 H10
Wild's Rents **SE1** 130 C2
Wilfred St **SW1** 129 J2
Wilkes St **E1** 126 D7
Wilkin St **NW5** 125 H1
Wilkinson St **SW8** 129 M6
Willard St **SW8** 129 H9
Willes Rd **NW5** 125 H1
William Bonney Est **SW4** 129 K10
William Cl **SE13** 131 N8
William IV St **WC2** 125 L9
William Morris Way **SW6** 128 C9
William Rd **NW1** 125 H5
William St **SW1** 128 F1
Williams Bldgs **E2** 126 G6
Willington Rd **SW9** 129 L9
Willis St **E14** 127 M8
Willoughby Pas **E14** 127 L10
Willow Br Rd **N1** 126 A1
Willow Pl **SW1** 129 J3
Willow St **EC2** 126 C6
Willow Wk **SE1** 130 C3
Willowbrook Rd **SE15** 130 D5
Wilman Gro **E8** 126 E2
Wilmcote Ho **W2** 124 A7
Wilmer Gdns **N1** 126 C3
Wilmer Lea Cl **E15** 127 N2
Wilmington Sq **WC1** 125 N5
Wilmington St **WC1** 125 N5
Wilmot Cl **SE15** 130 E6
Wilmot Pl **NW1** 125 J2
Wilmot St **E2** 126 F6
Wilshaw St **SE14** 131 L7
Wilson Gro **SE16** 130 F1
Wilson Rd **SE5** 130 C7
Wilson St **EC2** 126 B7
Wilton Cres **SW1** 128 G1
Wilton Ms **SW1** 128 G2
Wilton Pl **SW1** 128 G1
Wilton Rd **SW1** 129 H2
Wilton Row **SW1** 128 G1
Wilton Sq **N1** 126 B3
Wilton St **SW1** 129 H2
Wilton Ter **SW1** 128 G2
Wilton Vil **N1** 126 B3
Wilton Way **E8** 126 E1
Wiltshire Rd **SW9** 129 N9
Wiltshire Row **N1** 126 B3
Wimbolt St **E2** 126 E5
Wimbourne St **N1** 126 B4
Wimpole Ms **W1** 125 H7
Wimpole St **W1** 125 H8
Winans Wk **SW9** 129 N8
Winchester Cl **SE17** 129 P3
Winchester Rd **NW3** 124 D2
Winchester Sq **SE1** 126 B10
Winchester St **SW1** 129 H4
Winchester Wk **SE1** 126 B10
Wincott St **SE11** 129 N3
Winders Rd **SW11** 128 E8
Windlass Pl **SE8** 131 J3
Windmill Cl **SE13** 131 N8
Windmill La **E15** 127 P1
Windmill Row **SE11** 129 N4
Windmill St **W1** 125 K7
Windmill Wk **SE1** 125 N10
Windrose Cl **SE16** 131 H1
Windsock Cl **SE16** 131 K3
Windsor Gdns **W9** 124 A7
Windsor St **N1** 125 P3
Windsor Ter **N1** 126 A5
Windsor Wk **SE5** 130 B8
Wine Cl **E1** 126 G9
Winford Ho **E3** 127 K2
Winforton St **SE10** 131 N7
Wingfield St **SE15** 130 E9
Wingmore Rd **SE24** 130 A9
Winifred Gro **SW11** 128 F10
Winkley St **E2** 126 F4
Winsland St **W2** 124 D8
Winsley St **W1** 125 J8
Winslow **SE17** 130 C4
Winstanley Est **SW11** 128 D9
Winstanley Rd **SW11** 128 D9
Winterslow Rd **SW9** 129 P7
Winterton Ho **E1** 126 G8
Winthrop St **E1** 126 F7
Wise Rd **E15** 127 P3
Wisteria Rd **SE13** 131 P10
Witan St **E2** 126 F5
Witcombe Pt **SE15** 130 E7
Wivenhoe Cl **SE15** 130 F9
Wixs La **SW4** 129 H10
Woburn Pl **WC1** 125 K6
Woburn Sq **WC1** 125 K6
Woburn Wk **WC1** 125 K5
Wodeham Gdns **E1** 126 E7
Wodehouse Av **SE5** 130 D7
Wolfe Cres **SE16** 131 H1
Wolftencroft Cl **SW11** 128 D9
Wolseley St **SE1** 130 D1
Wolsey Ms **NW5** 125 J1
Wood Cl **E2** 126 E6
Wood St **EC2** 126 A8
Wood Wf **SE10** 131 M5
Wood Wf **SE10** 131 M5
Woodbridge St **EC1** 125 P6
Woodchester Sq **W2** 124 B7
Woodchurch Rd **NW6** 124 A2
Woodfall St **SW3** 128 F4
Woodfarrs **SE5** 130 B10
Woodhouse Cl **SE22** 130 E10
Woodland Cres **SE16** 131 H1
Woodpecker Rd **SE14** 131 J5
Woods Ms **W1** 124 F9
Woods Rd **SE15** 130 F7
Woodseer St **E1** 126 D7
Woodstock Ter **E14** 127 M9
Wooler St **SE17** 130 B4
Woolmore St **E14** 127 N9
Woolneigh St **SW6** 128 B9
Woolstaplers Way **SE16** 130 E3
Wooster Gdns **E14** 127 P8
Wootton St **SE1** 125 N10
Worfield St **SW11** 128 E6
Worgan St **SE11** 129 M4
Worgan St **SE16** 131 H2
Worlingham Rd **SE22** 130 D10
Wormwood St **EC2** 126 C8
Woronzow Rd **NW8** 124 D3
Worship St **EC2** 126 B6
Wotton Rd **SE8** 131 K5
Wren Rd **SE5** 130 B7
Wren St **WC1** 125 M6
Wrexham Rd **E3** 127 L4
Wrigglesworth St **SE14** 131 H6
Wrights Cl **SE13** 131 P10
Wrights La **W8** 128 B1
Wrights Rd **E3** 127 K4
Wroxton Rd **SE15** 130 F8
Wyatt Cl **SE16** 131 K1
Wycliffe Rd **SW11** 128 G8
Wye St **SW11** 128 D8
Wyke Rd **E3** 127 L2
Wyllen Cl **E1** 126 G6
Wymering Rd **W9** 124 A5
Wynan Rd **E14** 131 M4
Wyndham Est **SE5** 130 A6
Wyndham Pl **W1** 124 F7
Wyndham Rd **SE5** 129 P6
Wyndham St **W1** 124 F7
Wynford Rd **N1** 125 M4
Wynne Rd **SW9** 129 N8
Wynnstay Gdns **W8** 128 A2
Wynter St **SW11** 128 C10
Wynyard Ter **SE11** 129 M4
Wyvil Rd **SW8** 129 L5
Wyvis St **E14** 127 M7

Y

Yabsley St **E14** 127 N10
Yalding Rd **SE16** 130 E2
Yardley St **WC1** 125 N5
Yeate St **N1** 126 B2
Yelverton Rd **SW11** 128 D8
Yeo St **E3** 127 M7
Yeoman St **SE8** 131 J3
Yeoman's Row **SW3** 128 E2
York Br **NW1** 124 G6
York Gate **NW1** 124 G6
York Gro **SE15** 130 G7
York Ho Pl **W8** 128 B1
York Pl **SW11** 128 D9
York Rd **SE1** 129 M1
York Rd **SW11** 128 C10
York Rd **SW18** 128 C10
York Sq **E14** 127 J8
York St **W1** 124 F7
York Ter E **NW1** 124 G6
York Ter W **NW1** 124 G6
York Way **N1** 125 L3
York Way **N7** 125 K1
York Way Ct **N1** 125 L3
Yorkshire Rd **E14** 127 J8
Yorkton St **E2** 126 E4
Young St **W8** 128 B1
Yvon Ho **SW11** 128 G7

Z

Zampa Rd **SE16** 130 G4
Zealand Rd **E3** 127 J4
Zenoria St **SE22** 130 D10
Zetland St **E14** 127 M7
Zulu Ms **SW11** 128 E8

INDEX TO GREAT BRITAIN

Administrative area abbreviations

Aber.	Aberdeenshire
Arg. & B.	Argyll & Bute
B'burn.	Blackburn with Darwen
B'pool	Blackpool
B. & H.	Brighton & Hove
B. & N.E.Som.	Bath & North East Somerset
B.Gwent	Blaenau Gwent
Bed.	Bedford
Bourne.	Bournemouth
Brack.F.	Bracknell Forest
Bucks.	Buckinghamshire
Caerp.	Caerphilly
Cambs.	Cambridgeshire
Carmar.	Carmarthenshire
Cen.Beds.	Central Bedfordshire
Cere.	Ceredigion
Chan.I.	Channel Islands
Ches.E.	Cheshire East
Ches.W. & C.	Cheshire West & Chester
Cornw.	Cornwall
Cumb.	Cumbria
D. & G.	Dumfries & Galloway
Darl.	Darlington
Denb.	Denbighshire
Derbys.	Derbyshire
Dur.	Durham
E.Ayr.	East Ayrshire
E.Dun.	East Dunbartonshire
E.Loth.	East Lothian
E.Renf.	East Renfrewshire
E.Riding	East Riding of Yorkshire
E.Suss.	East Sussex
Edin.	Edinburgh
Falk.	Falkirk
Flints.	Flintshire
Glas.	Glasgow
Glos.	Gloucestershire
Gt.Lon.	Greater London
Gt.Man.	Greater Manchester
Gwyn.	Gwynedd
Hants.	Hampshire
Hart.	Hartlepool
Here.	Herefordshire
Herts.	Hertfordshire
High.	Highland
Hull	Kingston upon Hull
I.o.A.	Isle of Anglesey
I.o.M.	Isle of Man
I.o.S.	Isles of Scilly
I.o.W.	Isle of Wight
Inclyde	Inverclyde
Lancs.	Lancashire
Leic.	Leicester
Leics.	Leicestershire
Lincs.	Lincolnshire
M.K.	Milton Keynes
M.Tyd.	Merthyr Tydfil
Med.	Medway
Mersey.	Merseyside
Middbro.	Middlesbrough
Midloth.	Midlothian
Mon.	Monmouthshire
N.Ayr.	North Ayrshire
N.E.Lincs.	North East Lincolnshire
N.Lan.	North Lanarkshire
N.Lincs.	North Lincolnshire
N.P.T.	Neath Port Talbot
N.Som.	North Somerset
N.Yorks.	North Yorkshire
Norf.	Norfolk
Northants.	Northamptonshire
Northumb.	Northumberland
Nott.	Nottingham
Notts.	Nottinghamshire
Ork.	Orkney
Oxon.	Oxfordshire
P. & K.	Perth & Kinross
Pembs.	Pembrokeshire
Peter.	Peterborough
Plym.	Plymouth
Ports.	Portsmouth
R. & C.	Redcar & Cleveland
R.C.T.	Rhondda Cynon Taff
Read.	Reading
Renf.	Renfrewshire
Rut.	Rutland
S'end	Southend
S'ham.	Southampton
S.Ayr.	South Ayrshire
S.Glos.	South Gloucestershire
S.Lan.	South Lanarkshire
S.Yorks.	South Yorkshire
Sc.Bord.	Scottish Borders
Shet.	Shetland
Shrop.	Shropshire
Slo.	Slough
Som.	Somerset
Staffs.	Staffordshire
Stir.	Stirling
Stock.	Stockton-on-Tees
Stoke	Stoke-on-Trent
Suff.	Suffolk
Surr.	Surrey
Swan.	Swansea
Swin.	Swindon
T. & W.	Tyne & Wear
Tel. & W.	Telford & Wrekin
Thur.	Thurrock
V. of Glam.	Vale of Glamorgan
W'ham	Wokingham
W. & M.	Windsor & Maidenhead
W.Berks.	West Berkshire
W.Dun.	West Dunbartonshire
W.Isles	Western Isles (Na h-Eileanan Siar)
W.Loth.	West Lothian
W.Mid.	West Midlands
W.Suss.	West Sussex
W.Yorks.	West Yorkshire
Warks.	Warwickshire
Warr.	Warrington
Wilts.	Wiltshire
Worcs.	Worcestershire
Wrex.	Wrexham

Notes

This index reads in the sequence: Place Name / Postal District / Map Page Number / Grid Reference.

Example: Bishop's Cleeve **GL52** **29** J6

Where there is more than one place with the same name, the index reads in the sequence: Place Name / Administrative Area / Postal District / Map Page Number / Grid Reference.

Example: Prestbury, *Ches.* **SK10** **49** H5
Prestbury, *Glos.* **GL52** **29** J6

Entries in the index shown in **BOLD CAPITALS** indicate the principal post town within a postcode area.
Entries in the index shown in **bold** indicate other post towns.

Example: **GLOUCESTER GL** **29 H7**
Example: **Cheltenham GL50** **29 J6**

A

Ab Kettleby **LE14** **42** A3
Ab Lench **WR10** **30** B3
Abbas Combe **BA8** **9** G2
Abberley **WR6** **29** G2
Abberley Common **WR6** **29** G2
Abberton *Essex* **CO5** **34** E7
Abberton *Worcs.* **WR10** **29** J3
Abberwick **NE66** **71** G2
Abbess Roding **CM5** **33** J7
Abbey Dore **HR2** **28** C5
Abbey Hulton **ST2** **40** B1
Abbey St. Bathans **TD11** **77** F4
Abbey Town **CA7** **60** C1
Abbey Village **PR6** **56** B7
Abbey Wood **SE2** **23** H4
Abbeycwmhir **LD1** **27** K1
Abbeydale **S7** **51** F4
Abbeystead **LA2** **55** J4
Abbotrule **TD9** **70** B2
Abbots Bickington **EX22** **6** B4
Abbots Bromley **WS15** **40** C3
Abbots Langley WD5 **22 D1**
Abbots Leigh **BS8** **19** J4
Abbots Morton **WR7** **30** B3
Abbots Ripton **PE28** **33** F1
Abbot's Salford **WR11** **30** B3
Abbots Worthy **SO21** **11** F1
Abbotsbury **DT3** **8** E6
Abbotsfield Farm **WA9** **48** E3
Abbotsham **EX39** **6** C3
Abbotskerswell **TQ12** **5** J4
Abbotsley **PE19** **33** F3
Abbotstone **SO24** **11** G1
Abbotts Ann **SP11** **21** G7
Abbott's Barton **SO23** **11** F1
Abbottswood **SO51** **10** E2
Abdon **SY7** **38** E7
Abdy **S62** **51** G3
Abenhall **GL17** **29** F7
Aber **SA40** **17** H1
Aber Village **LD3** **28** A6
Aberaeron SA46 **26 D2**
Aberaman **CF44** **18** D1
Aberangell **SY20** **37** H5
Aber-Arad **SA38** **17** G1
Aberarder **PH20** **88** B6
Aberarder House **IV2** **88** D2
Aberargie **PH2** **82** C6
Aberarth **SA46** **26** D2
Aberavon **SA12** **18** A3
Aber-banc **SA44** **17** G1
Aberbargoed **CF81** **18** E1
Aberbeeg **NP13** **19** F1
Aberbowlan **SA19** **17** K2
Aberbran **LD3** **27** J6
Abercanaid **CF48** **18** D1
Abercarn **NP11** **19** F2
Abercastle **SA62** **16** B2
Abercegir **SY20** **37** H5
Aberchalder **PH35** **87** K4
Aberchirder **AB54** **98** E5
Abercorn **EH30** **75** J3
Abercraf **SA9** **27** H7
Abercregan **SA13** **18** B2
Abercrombie **KY10** **83** G7
Abercrychan **SA20** **27** G5
Abercwmboi **CF44** **18** D1
Abercych **SA37** **17** F1
Abercynafon **LD3** **27** K7
Abercynon **CF45** **18** D2
Abercywarch **SY20** **37** H4
Aberdalgie **PH2** **82** B5
Aberdare CF44 **18 C1**
Aberdaron **LL53** **36** A3
Aberdaugleddau (Milford Haven) SA73 **16 B5**
ABERDEEN AB **91 H4**
Aberdeen Airport **AB21** **91** G3
Aberdesach **LL54** **46** C7
Aberdour **KY3** **75** K2
Aberdovey (Aberdyfi) LL35 **37 F6**
Aberduhonw **LD2** **27** K3
Aberdulais **SA10** **18** A2
Aberdyfi (Aberdovey) LL35 **37 F6**
Aberedw **LD2** **27** K4
Abereiddy **SA62** **16** A2
Abererch **LL53** **36** C2
Aberfan **CF48** **18** D1
Aberfeldy PH15 **81 K3**
Aberffraw **LL63** **46** B6
Aberffrwd **SY23** **27** F1
Aberford **LS25** **57** K6
Aberfoyle **FK8** **81** G7
Abergavenny (Y Fenni) NP7 **28 B7**
Abergele LL22 **47 H5**
Aber-Giâr **SA40** **17** J1
Abergorlech **SA32** **17** J2
Abergwaun (Fishguard) SA65 **16 C2**
Abergwesyn **LD5** **27** H3
Abergwili **SA31** **17** H3
Abergwydol **SY20** **37** G5
Abergwynant **LL40** **37** F4
Abergwynfi **SA13** **18** B2
Abergwyngregyn **LL33** **46** E5
Abergynolwyn **LL36** **37** F5
Aberhafesp **SY16** **37** K6
Aberhonddu (Brecon) LD3 **27 K6**
Aberhosan **SY20** **37** H6
Aberkenfig **CF32** **18** B3
Aberlady **EH32** **76** C2
Aberlemno **DD8** **83** G2
Aberllefenni **SY20** **37** G5
Aber-Ilia **CF44** **27** J7
Aberllynfi (Three Cocks) **LD3** **28** A5
Aberlour (Charlestown of Aberlour) AB38 **97 K7**
Abermad **SY23** **26** E1
Abermaw (Barmouth) LL42 **37 F4**
Abermeurig **SA48** **26** E3
Aber-miwl (Abermule) **SY15** **38** A6
Abermule (Aber-miwl) **SY15** **38** A6
Abernaint **SY22** **38** A3
Abernant *Carmar.* **SA33** **17** G3
Aber-nant *R.C.T.* **CF44** **18** D1
Abernethy **PH2** **82** C6
Abernyte **PH14** **82** D4
Aberpennar (Mountain Ash) CF45 **18 D2**
Aberporth **SA43** **26** B3
Aberriw (Berriew) **SY21** **38** A5
Aberscross **KW10** **96** E2
Abersky **IV2** **88** C2
Abersoch **LL53** **36** C3
Abersychan **NP4** **19** F1
ABERTAWE (SWANSEA) SA **17 K6**
Aberteifi (Cardigan) SA43 **16 E1**
Aberthin **CF71** **18** D4
Abertillery NP13 **19 F1**
Abertridwr *Caerp.* **CF83** **18** E3
Abertridwr *Powys* **SY10** **37** K4
Abertysswg **NP22** **18** E1
Aberuthven **PH3** **82** A6
Aberyscir **LD3** **27** J6
Aberystwyth SY23 **36 E7**
Abhainnsuidhe **HS3** **100** C7
Abingdon OX14 **21 H2**
Abinger Common **RH5** **22** E7
Abinger Hammer **RH5** **22** D7
Abington **ML12** **68** E1
Abington Pigotts **SG8** **33** G4
Abingworth **RH20** **12** E5
Ablington *Glos.* **GL7** **20** E1
Ablington *Wilts.* **SP4** **20** E7
Abney **S32** **50** D5
Above Church **ST10** **50** C7
Aboyne AB34 **90 D5**
Abram **WN2** **49** F2
Abriachan **IV3** **88** C1
Abridge **RM4** **23** H2
Abronhill **G67** **75** F3
Abson **BS30** **20** A4
Abthorpe **NN12** **31** H4
Abune-the-Hill **KW17** **106** B5
Aby **LN13** **53** H5
Acaster Malbis **YO23** **58** B5
Acaster Selby **YO23** **58** B5
Accrington BB5 **56 C7**
Accurrach **PA33** **80** C5
Acha **PA78** **78** C2
Achacha **PA37** **80** A3
Achadacaie **PA29** **73** G4
Achadh Mòr **HS2** **101** F5
Achadh-chaorrunn **PA29** **73** F5
Achadunan **PA26** **80** D6
Achagavel **PH33** **79** J2
Achaglass **PA29** **73** F6
Achahoish **PA31** **73** F3
Achalader **PH10** **82** C3
Achallader **PA36** **80** E3
Achamore **PA60** **72** D3
Achandunie **IV17** **96** D4
Achany **IV27** **96** C1
Achaphubuil **PH33** **87** G7
Acharacle PH36 **79 H1**
Achargary **KW11** **104** C3
Acharn *Arg. & B.* **PA35** **80** C4
Acharn *P. & K.* **PH15** **81** J3
Acharonich **PA73** **79** F4
Acharosson **PA21** **73** H3
Achateny **PH36** **86** B7
Achath **AB32** **91** F3
Achavanich **KW5** **105** G4
Achddu **SA16** **17** H5
Achduart **IV26** **95** G1
Achentoul **KW11** **104** D5
Achfary **IV27** **102** E5
Achgarve **IV22** **94** E2
Achiemore *High.* **IV27** **103** F2
Achiemore *High.* **KW13** **104** D3
Achies **KW12** **105** G3
A'Chill **PH44** **85** H4
Achiltibuie **IV26** **95** G1
Achina **KW14** **104** C2
Achindown **IV12** **97** F7
Achinduich **IV27** **96** C2
Achingills **KW12** **105** G2
Achintee **IV54** **95** F7
Achintee House **PH33** **87** H7
Achintraid **IV54** **86** E1
Achlean **PH21** **89** F5
Achleanan **PA34** **79** G2
Achleek **PH33** **79** J2
Achlian **PA33** **80** C5
Achlyness **IV27** **102** E3
Achmelvich **IV27** **102** C6
Achmony **IV63** **88** C1
Achmore *High.* **IV53** **86** E1
Achmore *High.* **IV23** **95** G2
Achmore *Stir.* **FK21** **81** G4
Achnaba **PA31** **73** H2
Achnabat **IV2** **88** C1
Achnabourin **KW14** **104** C3
Achnacairn **PA37** **80** A4
Achnacarnin **IV27** **102** C5
Achnacarry **PH34** **87** H6
Achnaclerach **IV23** **96** B5
Achnacloich *Arg. & B.* **PA37** **80** A4
Achnacloich *High.* **IV46** **86** B4
Achnaclyth **KW6** **105** F5
Achnacraig **PA73** **79** F3
Achnacroish **PA34** **79** K3
Achnadrish **PA75** **79** F2
Achnafalnich **PA33** **80** D5
Achnafauld **PH8** **81** K4
Achnagairn **IV5** **96** C7
Achnagarron **IV18** **96** D4
Achnaha *High.* **PA34** **79** H3
Achnaha *High.* **PH36** **79** F1
Achnahanat **IV24** **96** C2
Achnahannet **PH26** **89** G2
Achnairn **IV27** **103** H7
Achnalea **PH33** **79** K1
Achnamara **PA31** **73** F2
Achnanellan **PH37** **79** J1
Achnasaul **PH34** **87** H6
Achnasheen IV22 **95 H6**
Achnashelloch **PA31** **73** G1
Achnastank **AB55** **89** K1
Achorn **KW6** **105** G5
Achosnich *High.* **IV25** **96** E2
Achosnich *High.* **PH36** **79** F1
Achreamie **KW14** **105** F2
Achriabhach **PH33** **80** C1
Achriesgill **IV27** **102** E3
Achtoty **KW14** **103** J2
Achurch **PE8** **42** D7
Achuvoldrach **IV27** **103** H3
Achvaich **IV25** **96** E2
Achvarasdal **KW14** **104** E2
Achvlair **PA38** **80** A2
Achvraie **IV26** **95** G1
Ackenthwaite **LA7** **55** J1
Ackergill **KW1** **105** J3
Acklam *Middbro.* **TS5** **63** F5
Acklam *N.Yorks.* **YO17** **58** D3
Ackleton **WV6** **39** G6
Acklington **NE65** **71** H3
Ackton **WF7** **57** K7
Ackworth Moor Top **WF7** **51** G1
Acle **NR13** **45** J4
Acock's Green **B27** **40** D7
Acol **CT7** **25** K5
Acomb *Northumb.* **NE46** **70** E7
Acomb *York* **YO24** **58** B4
Aconbury **HR2** **28** E5
Acre **BB5** **56** C7
Acrefair **LL14** **38** B1
Acrise Place **CT18** **15** G3
Acton *Ches.E.* **CW5** **49** F7
Acton *Dorset* **BH19** **9** J7
Acton *Gt.Lon.* **W3** **22** E3
Acton *Shrop.* **SY9** **38** C7
Acton *Staffs.* **ST5** **40** A1
Acton *Suff.* **CO10** **34** C4
Acton *Worcs.* **DY13** **29** H2
Acton *Wrex.* **LL12** **48** C7
Acton Beauchamp **WR6** **29** F3
Acton Bridge **CW8** **48** E5
Acton Burnell **SY5** **38** E5
Acton Green **WR6** **29** F3
Acton Pigott **SY5** **38** E5
Acton Round **WV16** **39** F6
Acton Scott **SY6** **38** D7
Acton Trussell **ST17** **40** B4
Acton Turville **GL9** **20** B3
Adamhill **KA1** **74** C7
Adbaston **ST20** **39** G3
Adber **DT9** **8** E2
Adderbury **OX17** **31** F5
Adderley **TF9** **39** F2
Adderstone **NE70** **77** K7
Addiewell **EH55** **75** H4
Addingham **LS29** **57** F5
Addington *Bucks.* **MK18** **31** J6
Addington *Gt.Lon.* **CR0** **23** G5
Addington *Kent* **ME19** **23** K6
Addiscombe **CR0** **23** G5
Addlestone KT15 **22 D5**
Addlethorpe **PE24** **53** J6
Adel **LS16** **57** H6
Adeney **TF10** **39** G4
Adeyfield **HP2** **22** D1
Adfa **SY16** **37** K5
Adforton **SY7** **28** D1
Adisham **CT3** **15** H2
Adlestrop **GL56** **30** D6
Adlingfleet **DN14** **58** E7
Adlington *Ches.E.* **SK10** **49** J4
Adlington *Lancs.* **PR7** **48** E1
Admaston *Staffs.* **WS15** **40** C3
Admaston *Tel. & W.* **TF5** **39** F4
Admington **CV36** **30** D4
Adsborough **TA2** **8** B2
Adscombe **TA5** **7** K2
Adstock **MK18** **31** J5
Adstone **NN12** **31** G3
Adversane **RH14** **12** D4
Advie **PH26** **89** J1
Adwalton **BD11** **57** H7
Adwell **OX9** **21** K2

Adw - Ard

Adwick le Street DN6 51 H2
Adwick upon Dearne S64 51 G2
Adziel AB43 99 H5
Ae Village DG1 68 E5
Affetside BL8 49 G1
Affleck AB21 91 G2
Affpuddle DT2 9 H5
Afon Wen LL53 36 D2
Afon-wen CH7 47 K5
Afton PO40 10 E6
Afton Bridgend KA18 68 B2
Agglethorpe DL8 57 F1
Aigburth L17 48 C4
Aiginis HS2 101 G4
Aike YO25 59 G5
Aikerness KW17 106 D2
Aikers KW17 106 D8
Aiketgate CA4 61 F2
Aikshaw CA7 60 C2
Aikton CA7 60 D1
Aikwood Tower TD7 69 K1
Ailby LN13 53 H5
Ailey HR3 28 C4
Ailsworth PE5 42 E6
Aimes Green EN9 23 H1
Aimster KW14 105 G2
Ainderby Quernhow YO7 57 J1
Ainderby Steeple DL7 62 E7
Aingers Green CO7 35 F6
Ainsdale PR8 48 C1
Ainsdale-on-Sea PR8 48 C1
Ainstable CA4 61 G2
Ainsworth BL2 49 G1
Ainthorpe YO21 63 J6
Aintree L10 48 C3
Aird W.Isles HS7 92 C6
Aird W.Isles HS2 101 H4
Aird a' Mhachair HS8 92 C7
Aird a' Mhulaidh HS3 100 D6
Aird Asaig HS3 100 D7
Aird Dhail HS2 101 G1
Aird Leimhe HS3 93 G3
Aird Mhige HS3 93 G2
Aird Mhighe HS3 93 F3
Aird of Sleat IV45 86 B4
Aird Thunga HS2 101 G4
Aird Uig HS2 100 C4
Airdrie Fife KY10 83 G7
Airdrie N.Lan. ML6 75 F4
Aire View BD20 56 E5
Airidh a' Bhruaich HS2 100 E6
Airieland DG7 65 H5
Airies DG9 66 D7
Airigh-drishaig IV54 86 D1
Airmyn DN14 58 D7
Airntully PH1 82 B4
Airor PH41 86 D4
Airth FK2 75 G2
Airton BD23 56 E4
Airyhassen DG8 64 D6
Aisby Lincs. DN21 52 B3
Aisby Lincs. NG32 42 D2
Aisgernis (Askernish) HS8 84 C2
Aisgill CA17 61 J7
Aish Devon TQ10 5 G4
Aish Devon TQ9 5 J5
Aisholt TA5 7 K2
Aiskew DL8 57 H1
Aislaby N.Yorks. YO21 63 K6
Aislaby N.Yorks. YO18 58 D1
Aislaby Stock. TS16 63 F5
Aisthorpe LN1 52 C4
Aith Ork. KW16 106 B6
Aith Ork. KW17 106 F5
Aith Shet. ZE2 107 M7
Aith Shet. ZE2 107 Q3
Aithsetter ZE2 107 N9
Aitnoch PH26 89 G1
Akeld NE71 70 E1
Akeley MK18 31 J5
Akenham IP1 35 F4
Albaston PL18 4 E3
Albecq GY5 3 H5
Alberbury SY5 38 C4
Albert Town SA61 16 C4
Albourne BN6 13 F5
Albourne Green BN6 13 F5
Albrighton Shrop. WV7 40 A5
Albrighton Shrop. SY4 38 D4
Alburgh IP20 45 G7
Albury Herts. SG11 33 H6
Albury Oxon. OX9 21 K1
Albury Surr. GU5 22 D7
Albury End SG11 33 H6
Albury Heath GU5 22 D7
Albyfield CA8 61 G1
Alcaig IV7 96 C6
Alcaston SY6 38 D7
Alcester B49 30 B3
Alciston BN26 13 J6
Alcombe TA24 7 H1
Alconbury PE28 32 E1
Alconbury Hill PE28 32 E1
Alconbury Weston PE28 32 E1
Aldborough N.Yorks. YO51 57 K3
Aldborough Norf. NR11 45 F2
Aldbourne SN8 21 F4
Aldbrough HU11 59 J6
Aldbrough St. John DL11 62 D5
Aldbury HP23 32 C7
Aldclune PH16 82 A1
Aldeburgh IP15 35 J3
Aldeby NR34 45 J6
Aldenham WD25 22 E2
Alderbury SP5 10 C2
Alderford NR9 45 F4
Alderholt SP6 10 C3
Alderley GL12 20 A2
Alderley Edge SK9 49 H5
Aldermaston RG7 21 J5
Aldermaston Wharf RG7 21 K5
Alderminster CV37 30 D4
Alderney GY9 3 K4
Alderney Airport GY9 3 J4
Alder's End HR1 29 F5
Aldersey Green CH3 48 D7
Aldershot GU11 22 B6
Alderton Glos. GL20 29 J5
Alderton Northants. NN12 31 J4
Alderton Suff. IP12 35 H4
Alderton Wilts. SN14 20 B3
Alderwasley DE56 51 F7
Aldfield HG4 57 H3
Aldford CH3 48 D7
Aldham Essex CO6 34 D6
Aldham Suff. IP7 34 E4
Aldie Aber. AB42 99 J6
Aldie High. IV19 96 E3
Aldingbourne PO20 12 C6
Aldingham LA12 55 F2
Aldington Kent TN25 15 F4
Aldington Worcs. WR11 30 B4
Aldivalloch AB54 90 B2
Aldochlay G83 74 B1
Aldons KA26 67 F5
Aldous's Corner IP19 45 H7
Aldreth CB6 33 H1
Aldridge WS9 40 C5
Aldringham IP16 35 J2
Aldsworth Glos. GL54 20 E1
Aldsworth W.Suss. PO10 11 J4
Aldunie AB54 90 B2
Aldville PH8 82 A4
Aldwark Derbys. DE4 50 E7
Aldwark N.Yorks. YO61 57 K3
Aldwick PO21 12 C7
Aldwincle NN14 42 D7
Aldworth RG8 21 J4
Alexandria G83 74 B3
Aley TA5 7 K2
Aley Green LU1 32 D7
Alfardisworthy EX22 6 A4
Alfington EX11 7 K6
Alfold GU6 12 D3
Alfold Crossways GU6 12 D3
Alford Aber. AB33 90 D3
Alford Lincs. LN13 53 H5
Alford Som. BA7 9 F1
Alfreton DE55 51 G7
Alfrick WR6 29 G3
Alfrick Pound WR6 29 G3
Alfriston BN26 13 J6
Algarkirk PE20 43 F2
Alhampton BA4 9 F1
Alkborough DN15 58 E7
Alkerton OX15 30 E4
Alkham CT15 15 H3
Alkington SY13 38 E2
Alkmonton DE6 40 D2
All Cannings SN10 20 D5
All Saints South Elmham IP19 45 H7
All Stretton SY6 38 D6
Allaleigh TQ9 5 J5
Allanaquoich AB35 89 J5
Allancreich AB34 90 D5
Allanfearn IV2 96 E7
Allangillfoot DG13 69 H4
Allanton D. & G. DG2 68 E5
Allanton E.Ayr. KA17 74 E7
Allanton N.Lan. ML7 75 G5
Allanton S.Lan. ML3 75 F5
Allanton Sc.Bord. TD11 77 G5
Allardice DD10 91 G7
Allathasdal HS9 84 B4
Allbrook SO50 11 F2
Allendale Town NE47 61 K1
Allenheads NE47 61 K2
Allen's Green CM21 33 H7
Allensford DH8 62 B1
Allensmore HR2 28 D5
Allenton DE24 41 F2
Aller TA10 8 D2
Allerby CA7 60 B3
Allercombe EX5 7 J6
Allerford Devon EX20 6 C7
Allerford Som. TA24 7 H1
Allerston YO18 58 E1
Allerthorpe YO42 58 D5
Allerton Mersey. L18 48 D4
Allerton W.Yorks. BD15 57 G6
Allerton Bywater WF10 57 K7
Allerton Mauleverer HG5 57 K4
Allesley CV5 40 E7
Allestree DE22 41 F2
Allet Common TR4 2 E4
Allexton LE15 42 B5
Allgreave SK11 49 J6
Allhallows ME3 24 E4
Allhallows-on-Sea ME3 24 E4
Alligin Shuas IV22 94 E6
Allimore Green ST18 40 A4
Allington Dorset DT6 8 D5
Allington Lincs. NG32 42 B1
Allington Wilts. SP4 10 D1
Allington Wilts. SN10 20 D5
Allington Wilts. SN14 20 B4
Allithwaite LA11 55 G2
Allnabad IV27 103 G4
Alloa FK10 75 G1
Allonby CA15 60 B2
Allostock WA16 49 G5
Alloway KA7 67 H2
Allowenshay TA17 8 C3
Allscot WV15 39 G6
Allscott TF6 39 F4
Allt na h-Airbhe IV26 95 H2
Alltachonaich PA34 79 J2
Alltbeithe IV4 87 G2
Alltforgan SY10 37 J3
Alltmawr LD2 27 K4
Alltnacaillich IV27 103 G4
Allt-na-subh IV40 87 F1
Alltsigh IV63 88 B3
Alltwalis SA32 17 H2
Alltwen SA8 18 A1
Alltyblaca SA40 17 J1
Allwood Green IP22 34 E1
Almeley HR3 28 C3
Almeley Wootton HR3 28 C3
Almer DT11 9 J5
Almington TF9 39 G2
Almiston Cross EX39 6 B3
Almondbank PH1 82 B5
Almondbury HD4 50 D1
Almondsbury BS32 19 K3
Alne YO61 57 K3
Alness IV17 96 D5
Alnham NE66 70 E2
Alnmouth NE66 71 H2
Alnwick NE66 71 G2
Alperton HA0 22 E3
Alphamstone CO8 34 C5
Alpheton CO10 34 C3
Alphington EX2 7 H6
Alport DE45 50 E6
Alpraham CW6 48 E7
Alresford CO7 34 E6
Alrewas DE13 40 D4
Alrick PH11 82 C1
Alsager ST7 49 G7
Alsagers Bank ST7 40 A1
Alsop en le Dale DE6 50 D7
Alston Cumb. CA9 61 J2
Alston Devon EX13 8 C4
Alston Sutton BS26 19 H6
Alstone Glos. GL20 29 J5
Alstone Som. TA9 19 G7
Alstone Staffs. ST18 40 A4
Alstonefield DE6 50 D7
Alswear EX36 7 F3
Alt OL8 49 J2
Altandhu IV26 102 B7
Altanduin KW11 104 D6
Altarnun PL15 4 C2
Altass IV27 96 C1
Altens AB12 91 H4
Alterwall KW1 105 H2
Altham BB5 56 C6
Althorne CM3 25 F2
Althorpe DN17 52 B2
Alticry DG8 64 C5
Altnafeadh PH49 80 D2
Altnaharra IV27 103 H5
Altofts WF6 57 J7
Alton Derbys. S42 51 F6
Alton Hants. GU34 11 J1
Alton Staffs. ST10 40 C1
Alton Barnes SN8 20 E5
Alton Pancras DT2 9 G4
Alton Priors SN8 20 E5
Altonside IV30 97 K6
Altrincham WA14 49 G4
Altura PH34 87 J5
Alva FK12 75 G1
Alvanley WA6 48 D5
Alvaston DE24 41 F2
Alvechurch B48 30 B1
Alvecote B79 40 E5
Alvediston SP5 9 J2
Alveley WV15 39 G7
Alverdiscott EX31 6 D3
Alverstoke PO12 11 H5
Alverstone PO36 11 G6
Alverthorpe WF2 57 J7
Alverton NG13 42 A1
Alves IV30 97 J5
Alvescot OX18 21 F1
Alveston S.Glos. BS35 19 K3
Alveston Warks. CV37 30 D3
Alvie PH21 89 F4
Alvingham LN11 53 G3
Alvington GL15 19 K1
Alwalton PE2 42 E6
Alweston DT9 9 F3
Alwington EX39 6 C3
Alwinton NE65 70 E3
Alwoodley LS17 57 J5
Alwoodley Gates LS17 57 J5
Alyth PH11 82 D3
Amalebra TR20 2 B5
Ambaston DE72 41 G2
Amber Hill PE20 43 F1
Ambergate DE56 51 F7
Amberley Glos. GL5 20 B1
Amberley W.Suss. BN18 12 D5
Amble NE65 71 H3
Amblecote DY8 40 A7
Ambleside LA22 60 E6
Ambleston SA62 16 D3
Ambrismore PA20 73 J5
Ambrosden OX25 31 H7
Amcotts DN17 52 B1
Amersham HP6 22 C2
Amerton ST18 40 B3
Amesbury SP4 20 E7
Ameysford BH22 10 B4
Amington B77 40 E5
Amisfield Town DG1 69 F5
Amlwch LL68 46 C3
Amlwch Port LL68 46 C3
Ammanford (Rhydaman) SA18 17 K4
Amotherby YO17 58 D2
Ampfield SO51 10 E2
Ampleforth YO62 58 B2
Ampleforth College YO62 58 B2
Ampney Crucis GL7 20 D1
Ampney St. Mary GL7 20 D1
Ampney St. Peter GL7 20 D1
Amport SP11 21 G7
Ampthill MK45 32 D5
Ampton IP31 34 C1
Amroth SA67 16 E5
Amulree PH8 81 K4
An T-Òb (Leverburgh) HS5 93 F3
Anaboard PH26 89 H1
Anaheilt PH36 79 K1
Ancaster NG32 42 C1
Anchor SY7 38 A7
Anchor Corner NR17 44 E6
Ancroft TD15 77 H6
Ancrum TD8 70 B1
Ancton PO22 12 C6
Anderby PE24 53 J5
Anderby Creek PE24 53 J5
Andersea TA7 8 C1
Andersfield TA5 8 B1
Anderson DT11 9 H5
Anderton CW9 49 F5
Andover SP10 21 G7
Andover Down SP11 21 G7
Andoversford GL54 30 B7
Andreas IM7 54 D4
Anelog LL53 36 A3
Anfield L4 48 C3
Angarrack TR27 2 C5
Angarrick TR3 2 E5
Angelbank SY8 28 E1
Angerton CA7 60 D1
Angle SA71 16 B5
Angler's Retreat SY20 37 G6
Anglesey (Ynys Môn) LL 46 B4
Angmering BN16 12 D6
Angmering-on-Sea BN16 12 D6
Angram N.Yorks. YO23 58 B5
Angram N.Yorks. DL11 61 K7
Anick NE46 70 E7
Anie FK17 81 G6
Ankerville IV19 97 F4
Anlaby HU10 59 G7
Anmer PE31 44 B3
Anmore PO7 11 H3
Anna Valley SP11 21 G7
Annan DG12 69 G7
Annaside LA19 54 D1
Annat Arg. & B. PA35 80 B5
Annat High. IV22 94 E6
Annbank KA6 67 J1
Annesley NG15 51 H7
Annesley Woodhouse NG17 51 G7
Annfield Plain DH9 62 C1
Anniesland G13 74 D4
Annscroft SY5 38 D5
Ansdell FY8 55 G7
Ansford BA7 9 F1
Ansley CV10 40 E6
Anslow DE13 40 E3
Anslow Gate DE13 40 D3
Ansteadbrook GU27 12 C3
Anstey Herts. SG9 33 H5
Anstey Leics. LE7 41 H5
Anstruther KY10 83 G7
Ansty W.Suss. RH17 13 F4
Ansty Warks. CV7 41 F7
Ansty Wilts. SP3 9 J2
Ansty Coombe SP3 9 J2
Ansty Cross DT2 9 G4
Anthill Common PO7 11 H3
Anthorn CA7 60 C1
Antingham NR28 45 G2
Anton's Gowt PE22 43 F1
Antony PL11 4 D5
Antrobus CW9 49 F5
Anvil Corner EX22 6 B5
Anvil Green CT4 15 G3
Anwick NG34 52 E7
Anwoth DG7 65 F5
Aoradh PA44 72 A4
Apethorpe PE8 42 D6
Apeton ST20 40 A4
Apley LN8 52 E5
Apperknowle S18 51 F5
Apperley GL19 29 H6
Apperley Bridge BD10 57 G6
Appersett DL8 61 K7
Appin PA38 80 A3
Appin House PA38 80 A3
Appleby DN15 52 C1
Appleby Magna DE12 41 F4
Appleby Parva DE12 41 F5
Appleby-in-Westmorland CA16 61 H4
Applecross IV54 94 D7
Appledore Devon EX39 6 C2
Appledore Devon EX16 7 J4
Appledore Kent TN26 14 E5
Appledore Heath TN26 14 E4
Appleford OX14 21 J2
Appleshaw SP11 21 G7
Applethwaite CA12 60 D4
Appleton Halton WA8 48 E4
Appleton Oxon. OX13 21 H1
Appleton Roebuck YO23 58 B5
Appleton Thorn WA4 49 F4
Appleton Wiske DL6 62 E6
Appleton-le-Moors YO62 58 D1
Appleton-le-Street YO17 58 D2
Appletreehall TD9 70 A2
Appletreewick BD23 57 F3
Appley TA21 7 J3
Appley Bridge WN6 48 E2
Apse Heath PO36 11 G6
Apsey Green IP13 35 G2
Apsley HP3 22 D1
Apsley End SG5 32 E5
Apuldram PO20 12 B6
Arberth (Narberth) SA67 16 E4
Arbirlot DD11 83 G3
Arborfield RG2 22 A5
Arborfield Cross RG2 22 A5
Arborfield Garrison RG2 22 A5
Arbourthorne S2 51 F4
Arbroath DD11 83 H3
Arbuthnott AB30 91 F7
Archdeacon Newton DL2 62 D5
Archiestown AB38 97 K7
Arclid CW11 49 G6
Ard a' Chapuill PA22 73 J2
Ardachearanbeg PA22 73 J2
Ardacheranmor PA22 73 J2
Ardachoil PA65 79 J4
Ardachu IV28 96 D1
Ardailly PA41 72 E5
Ardalanish PA67 78 E6
Ardallie AB42 91 J1
Ardanaiseig PA35 80 B5
Ardaneaskan IV54 86 E1
Ardanstur PA34 79 K6
Ardantiobairt PA34 79 H2
Ardantrive PA34 79 K5
Ardarroch IV54 94 E7
Ardbeg Arg. & B. PA20 73 J4
Ardbeg Arg. & B. PA42 72 C6
Ardbeg Arg. & B. PA23 73 K2
Ardblair IV4 88 C1
Ardbrecknish PA33 80 B5
Ardcharnich IV23 95 H3
Ardchiavaig PA67 78 E6
Ardchonnel PA37 80 A4
Ardchonnell PA33 80 A6
Ardchrishnish PA70 79 F5
Ardchronie IV24 96 D3
Ardchuilk IV4 87 J1
Ardchullarie More FK18 81 G6
Ardchyle FK21 81 G5
Arddlin SY22 38 B4
Ardechvie PH34 87 H5
Ardeley SG2 33 G6
Ardelve IV40 86 E2
Arden G83 74 B2
Ardencaple House PA34 79 J6
Ardens Grafton B49 30 C3
Ardentallan PA34 79 K5
Ardentinny PA23 73 K2
Ardeonaig FK21 81 H4
Ardersier IV2 96 E6
Ardery PH36 79 J1
Ardessie IV23 95 G3
Ardfad PA34 79 J6
Ardfern PA31 79 K7
Ardfin PA60 72 C4
Ardgartan G83 80 D7
Ardgay IV24 96 D2
Ardgenavan PA26 80 C6
Ardgour (Corran) PH33 80 B1
Ardgowan PA16 74 A3
Ardgowse AB33 90 E3
Ardgye IV30 97 J5
Ardhallow PA23 73 K3
Ardheslaig IV54 94 D6
Ardiecow AB45 98 D4
Ardinamar PA34 79 J6
Ardindrean IV23 95 H3
Ardingly RH17 13 G4
Ardington OX12 21 H3
Ardington Wick OX12 21 H3
Ardintoul IV40 86 E2
Ardkinglas House PA26 80 C6
Ardlair AB52 90 D2
Ardlamont PA21 73 H4
Ardleigh CO7 34 E6
Ardleigh Green RM2 23 J3
Ardleigh Heath CO7 34 E5
Ardleish G83 80 E6
Ardler PH12 82 D3
Ardley OX27 31 G6
Ardley End CM22 33 J7
Ardlui G83 80 E6
Ardlussa PA60 72 E2
Ardmaddy PA35 80 B4
Ardmair IV26 95 H2
Ardmaleish PA20 73 J4
Ardmay G83 80 D7
Ardmenish PA60 72 D3
Ardmhòr HS9 84 C4
Ardminish PA41 72 E6
Ardmolich PH36 86 D7
Ardmore Arg. & B. PA42 72 C5
Ardmore Arg. & B. PA34 79 J5
Ardmore Arg. & B. G82 74 B3
Ardmore High. IV19 96 E3
Ardnackaig PA31 73 F1
Ardnacross PA72 79 G3
Ardnadam PA23 73 K2
Ardnadrochit PA64 79 J4
Ardnagoine IV26 95 F1
Ardnagowan PA25 80 C7
Ardnahein PA24 73 K1
Ardnahoe PA46 72 C3
Ardnarff IV53 86 E1
Ardnastang PH36 79 K1
Ardnave PA44 72 A3
Ardno PA26 80 C7
Ardo AB41 91 G1
Ardoch D. & G. DG3 68 D3
Ardoch Moray IV36 97 J6
Ardoch P. & K. PH1 82 B4
Ardochrig G75 74 E6
Ardoyne AB52 90 E2
Ardpatrick PA29 73 F4
Ardpeaton G84 74 A2
Ardradnaig PH15 81 J3
Ardrishaig PA30 73 G2
Ardroe IV27 102 C6
Ardross IV17 96 D4
Ardrossan KA22 74 A6
Ardscalpsie PA20 73 J5

Ard - Bab

Ardshave **IV25** 96 E2
Ardshealach **PH36** 79 H1
Ardshellach **PA34** 79 J6
Ardsley **S71** 51 F2
Ardslignish **PH36** 79 G1
Ardtalla **PA42** 72 C5
Ardtalnaig **PH15** 81 J4
Ardtaraig **PA23** 73 J2
Ardteatle **PA33** 80 C5
Ardtoe **PH36** 86 C7
Ardtornish **PA34** 79 J3
Ardtrostan **PH6** 81 H5
Ardtur **PA38** 80 A3
Arduaine **PA34** 79 K6
Ardullie **IV15** 96 C5
Ardura **PA65** 79 H4
Ardvar **IV27** 102 D5
Ardvasar **IV45** 86 C4
Ardveenish **HS9** 84 C4
Ardveich **FK19** 81 H5
Ardverikie **PH20** 88 C6
Ardvorlich *Arg. & B.* **G83** 80 E6
Ardvorlich *P. & K.* **FK19** 81 H5
Ardwall **DG7** 65 F5
Ardwell *D. & G.* **DG9** 64 B6
Ardwell *Moray* **AB54** 90 B1
Ardwell *S.Ayr.* **KA26** 67 F4
Ardwick **M12** 49 H3
Areley Kings **DY13** 29 H1
Arford **GU35** 12 B3
Argaty **FK16** 81 J7
Argoed **NP12** 18 E2
Argoed Mill **LD1** 27 J2
Argos Hill **TN20** 13 J4
Argrennan House **DG7** 65 H5
Arichamish **PA31** 80 A7
Arichastlich **PA33** 80 D4
Arichonan **PA31** 73 F1
Aridhglas **PA66** 78 E5
Arienskill **PH38** 86 D6
Arileod **PA78** 78 C2
Arinacrinachd **IV54** 94 D6
Arinafad Beg **PA31** 73 F2
Arinagour **PA78** 78 D2
Arinambane **HS8** 84 C2
Arisaig PH39 86 **C6**
Arivegaig **PH36** 79 H1
Arkendale **HG5** 57 J3
Arkesden **CB11** 33 H5
Arkholme **LA6** 55 J2
Arkle Town **DL11** 62 B6
Arkleby **CA7** 60 C3
Arkleside **DL8** 57 F1
Arkleton **DG13** 69 J4
Arkley **EN5** 23 F2
Arksey **DN5** 51 H2
Arkwright Town **S44** 51 G5
Arlary **KY13** 82 C7
Arle **GL51** 29 J6
Arlecdon **CA26** 60 B5
Arlesey SG15 32 **E5**
Arleston **TF1** 39 F4
Arley **CW9** 49 F4
Arlingham **GL2** 29 G7
Arlington *Devon* **EX31** 6 E1
Arlington *E.Suss.* **BN26** 13 J6
Arlington *Glos.* **GL7** 20 E1
Arlington Beccott **EX31** 6 E1
Armadale *High.* **KW14** 104 C2
Armadale *High.* **IV45** 86 C4
Armadale *W.Loth.* **EH48** 75 H4
Armathwaite **CA4** 61 G2
Arminghall **NR14** 45 G5
Armitage **WS15** 40 C4
Armitage Bridge **HD4** 50 D1
Armley **LS12** 57 H6
Armscote **CV37** 30 D4
Armshead **ST9** 40 B1
Armston **PE8** 42 D7
Armthorpe **DN3** 51 J2
Arnabost **PA78** 78 D1
Arnaby **LA18** 54 E1
Arncliffe **BD23** 56 E2
Arncliffe Cote **BD23** 56 E2
Arncroach **KY10** 83 G7
Arne **BH20** 9 J6
Arnesby **LE8** 41 J6
Arngask **PH2** 82 C6
Arngibbon **FK8** 74 E1
Arngomery **FK8** 74 E1
Arnhall **DD9** 83 H1
Arnicle **PA29** 73 F7
Arnipol **PH38** 86 D6
Arnisdale **IV40** 86 E3
Arnish **IV40** 94 B7
Arniston Engine **EH23** 76 B4
Arnol **HS2** 101 F3
Arnold *E.Riding* **HU11** 59 H5
Arnold *Notts.* **NG5** 41 H1
Arnprior **FK8** 74 E1
Arnside **LA5** 55 H2
Arowry **SY13** 38 D2
Arrad Foot **LA12** 55 G1
Arradoul **AB56** 98 C4
Arram **HU17** 59 G5
Arran KA27 73 **H7**
Arras **YO43** 59 F5
Arrat **DD9** 83 H2
Arrathorne **DL8** 62 D7
Arreton **PO30** 11 G6
Arrington **SG8** 33 G3
Arrivain **FK20** 80 D4
Arrochar G83 80 **E7**
Arrow **B49** 30 B3
Arscaig **IV27** 103 H7
Arscott **SY5** 38 D5
Arthington **LS21** 57 H5
Arthingworth **LE16** 42 A7
Arthog LL39 37 **F4**
Arthrath **AB41** 91 H1
Arthurstone **PH12** 82 D3
Artrochie **AB41** 91 J1
Aruadh **PA49** 72 A4
Arundel BN18 12 **D6**
Aryhoulan **PH33** 80 B1
Asby **CA14** 60 B4
Ascog **PA20** 73 K4
Ascot SL5 22 **C5**
Ascott **CV36** 30 E5
Ascott d'Oyley **OX7** 30 E7
Ascott Earl **OX7** 30 D7
Ascott-under-Wychwood **OX7** 30 E7
Ascreavie **DD8** 82 E2
Asenby **YO7** 57 K2
Asfordby **LE14** 42 A4
Asfordby Hill **LE14** 42 A4
Asgarby *Lincs.* **PE23** 53 G6
Asgarby *Lincs.* **NG34** 42 E1
Ash *Dorset* **DT11** 9 H3
Ash *Kent* **CT3** 15 H2
Ash *Kent* **TN15** 24 C5
Ash *Som.* **TA12** 8 D2
Ash *Surr.* **GU12** 22 B6
Ash Barton **EX20** 6 D5
Ash Bullayne **EX17** 7 F5
Ash Green *Surr.* **GU12** 22 C7
Ash Green *Warks.* **CV7** 41 F7
Ash Magna **SY13** 38 E2
Ash Mill **EX36** 7 F3
Ash Parva **SY13** 38 E2
Ash Priors **TA4** 7 K3
Ash Street **IP7** 34 E4
Ash Thomas **EX16** 7 J4
Ash Vale **GU12** 22 B6
Ashampstead **RG8** 21 J4
Ashbocking **IP6** 35 F3
Ashbourne DE6 40 **D1**
Ashbrittle **TA21** 7 J3
Ashburnham Place **TN33** 13 K5
Ashburton **TQ13** 5 H4
Ashbury *Devon* **EX20** 6 D6
Ashbury *Oxon.* **SN6** 21 F3
Ashby **DN16** 52 B2
Ashby by Partney **PE23** 53 H6
Ashby cum Fenby **DN37** 53 F2
Ashby de la Launde **LN4** 52 D7
Ashby de la Zouch LE65 41 **F4**
Ashby Dell **NR32** 45 J6
Ashby Folville **LE14** 42 A4
Ashby Hill **DN37** 53 F2
Ashby Magna **LE17** 41 H6
Ashby Parva **LE17** 41 H7
Ashby Puerorum **LN9** 53 G5
Ashby St. Ledgers **CV23** 31 G2
Ashby St. Mary **NR14** 45 H5
Ashchurch **GL20** 29 J5
Ashcombe *Devon* **EX7** 5 K3
Ashcombe *N.Som.* **BS22** 19 G5
Ashcott **TA7** 8 D1
Ashdon **CB10** 33 J4
Ashe **RG25** 21 J7
Asheldham **CM0** 25 F1
Ashen **CO10** 34 B4
Ashenden **TN30** 14 D4
Ashendon **HP18** 31 J7
Ashens **PA29** 73 G3
Ashfield *Arg. & B.* **PA31** 73 F2
Ashfield *Here.* **HR9** 28 E6
Ashfield *Stir.* **FK15** 81 J7
Ashfield *Suff.* **IP14** 35 G2
Ashfield Green *Suff.* **IP21** 35 G1
Ashfield Green *Suff.* **CB8** 34 B3
Ashfold Crossways **RH13** 13 F4
Ashford *Devon* **TQ7** 5 G6
Ashford *Devon* **EX31** 6 D2
Ashford *Hants.* **SP6** 10 C3
Ashford *Kent* TN23 15 **F3**
Ashford *Surr.* TW15 22 **D4**
Ashford Bowdler **SY8** 28 E1
Ashford Carbonel **SY8** 28 E1
Ashford Hill **RG19** 21 J5
Ashford in the Water **DE45** 50 D5
Ashgill **ML9** 75 F5
Ashiestiel **TD1** 76 C7
Ashill *Devon* **EX15** 7 J4
Ashill *Norf.* **IP25** 44 C5
Ashill *Som.* **TA19** 8 C3
Ashingdon **SS4** 24 E2
Ashington *Northumb.* NE63 71 **H5**
Ashington *Som.* **BA22** 8 E2
Ashington *W.Suss.* **RH20** 12 E5
Ashkirk **TD7** 69 K1
Ashlett **SO45** 11 F4
Ashleworth **GL19** 29 H6
Ashleworth Quay **GL19** 29 H6
Ashley *Cambs.* **CB8** 33 K2
Ashley *Ches.E.* **WA15** 49 G4
Ashley *Devon* **EX18** 6 E4
Ashley *Glos.* **GL8** 20 C2
Ashley *Hants.* **SO20** 10 E1
Ashley *Hants.* **BH25** 10 D5
Ashley *Kent* **CT15** 15 J3
Ashley *Northants.* **LE16** 42 A6
Ashley *Staffs.* **TF9** 39 G2
Ashley *Wilts.* **SN13** 20 B5
Ashley Down **BS7** 19 J4
Ashley Green **HP5** 22 C1
Ashley Heath *Dorset* **BH24** 10 C4
Ashley Heath *Staffs.* **TF9** 39 G2
Ashmanhaugh **NR12** 45 H3
Ashmansworth **RG20** 21 H6
Ashmansworthy **EX39** 6 B4
Ashmore *Dorset* **SP5** 9 J3
Ashmore *P. & K.* **PH10** 82 C2
Ashmore Green **RG18** 21 J5
Ashorne **CV35** 30 E3
Ashover **S45** 51 F6
Ashover Hay **S45** 51 F6
Ashow **CV8** 30 E1
Ashperton **HR8** 29 F4
Ashprington **TQ9** 5 J5
Ashreigney **EX18** 6 E4
Ashtead KT21 22 **E6**
Ashton *Ches.W. & C.* **CH3** 48 E6
Ashton *Cornw.* **TR13** 2 D6
Ashton *Cornw.* **PL17** 4 D4
Ashton *Hants.* **SO32** 11 G3
Ashton *Here.* **HR6** 28 E2
Ashton *Inclyde* **PA19** 74 A3
Ashton *Northants.* **PE8** 42 D7
Ashton *Northants.* **NN7** 31 J4
Ashton *Peter.* **PE9** 42 E5
Ashton Common **BA14** 20 B6
Ashton Keynes **SN6** 20 D2
Ashton under Hill **WR11** 29 J5
Ashton upon Mersey **M33** 49 G3
Ashton-in-Makerfield **WN4** 48 E3
Ashton-under-Lyne OL7 49 **J3**
Ashurst *Hants.* **SO40** 10 E3
Ashurst *Kent* **TN3** 13 J3
Ashurst *W.Suss.* **BN44** 12 E5
Ashurst Bridge **SO40** 10 E3
Ashurstwood **RH18** 13 H3
Ashwater **EX21** 6 B6
Ashwell *Herts.* **SG7** 33 F5
Ashwell *Rut.* **LE15** 42 B4
Ashwell End **SG7** 33 F4
Ashwellthorpe **NR16** 45 F6
Ashwick **BA3** 19 K7
Ashwicken **PE32** 44 B4
Ashybank **TD9** 70 A2
Askam in Furness LA16 55 **F2**
Askern **DN6** 51 H1
Askernish (Aisgernis) **HS8** 84 C2
Askerswell **DT2** 8 E5
Askett **HP27** 22 B1
Askham *Cumb.* **CA10** 61 G4
Askham *Notts.* **NG22** 51 K5
Askham Bryan **YO23** 58 B5
Askham Richard **YO23** 58 B5
Asknish **PA31** 73 H1
Askrigg **DL8** 62 A7
Askwith **LS21** 57 G5
Aslackby **NG34** 42 D2
Aslacton **NR15** 45 F6
Aslockton **NG13** 42 A1
Asloun **AB33** 90 D3
Aspall **IP14** 35 F2
Aspatria **CA7** 60 C2
Aspenden **SG9** 33 G6
Asperton **PE20** 43 F2
Aspley Guise **MK17** 32 C5
Aspley Heath **MK17** 32 C5
Aspull **WN2** 49 F2
Asselby **DN14** 58 D7
Asserby **LN13** 53 H5
Assington **CO10** 34 D5
Assington Green **CO10** 34 B3
Astbury **CW12** 49 H6
Astcote **NN12** 31 H3
Asterby **LN11** 53 F5
Asterley **SY5** 38 C5
Asterton **SY7** 38 D6
Asthall **OX18** 30 D7
Asthall Leigh **OX29** 30 E7
Astle **SK10** 49 H5
Astley *Gt.Man.* **M29** 49 G2
Astley *Shrop.* **SY4** 38 E4
Astley *Warks.* **CV10** 41 F7
Astley *Worcs.* **DY13** 29 G2
Astley Abbotts **WV16** 39 G6
Astley Bridge **BL1** 49 G1
Astley Cross **DY13** 29 H2
Astley Green **M29** 49 G3
Astley Lodge **SY4** 38 E4
Aston *Ches.E.* **CW5** 39 F1
Aston *Ches.W. & C.* **WA7** 48 E5
Aston *Derbys.* **S33** 50 D4
Aston *Derbys.* **DE6** 40 D2
Aston *Flints.* **CH5** 48 C6
Aston *Here.* **SY8** 28 D1
Aston *Here.* **HR6** 28 D2
Aston *Herts.* **SG2** 33 F6
Aston *Oxon.* **OX18** 21 G1
Aston *S.Yorks.* **S26** 51 G4
Aston *Shrop.* **SY4** 38 E3
Aston *Shrop.* **WV5** 40 A6
Aston *Staffs.* **TF9** 39 G1
Aston *Tel. & W.* **TF6** 39 F5
Aston *W'ham* **RG9** 22 A3
Aston *W.Mid.* **B6** 40 C7
Aston Abbotts **HP22** 32 B6
Aston Botterell **WV16** 39 F7
Aston Cantlow **B95** 30 C3
Aston Clinton **HP22** 32 B7
Aston Crews **HR9** 29 F6
Aston Cross **GL20** 29 J5
Aston End **SG2** 33 F6
Aston Eyre **WV16** 39 F6
Aston Fields **B60** 29 J2
Aston Flamville **LE10** 41 G6
Aston Heath **WA7** 48 E5
Aston Ingham **HR9** 29 F6
Aston juxta Mondrum **CW5** 49 F7
Aston le Walls **NN11** 31 F3
Aston Magna **GL56** 30 C5
Aston Munslow **SY7** 38 E7
Aston on Carrant **GL20** 29 J5
Aston on Clun **SY7** 38 C7
Aston Pigott **SY5** 38 C5
Aston Rogers **SY5** 38 C5
Aston Rowant **OX49** 22 A2
Aston Sandford **HP17** 22 A1
Aston Somerville **WR12** 30 B5
Aston Subedge **GL55** 30 C4
Aston Tirrold **OX11** 21 J3
Aston Upthorpe **OX11** 21 J3
Aston-by-Stone **ST15** 40 B2
Aston-on-Trent **DE72** 41 G3
Astwick **SG5** 33 F5
Astwood **MK16** 32 C4
Astwood Bank **B96** 30 B2
Aswarby **NG34** 42 D1
Aswardby **PE23** 53 G5
Aswick Grange **PE12** 43 G4
Atch Lench **WR11** 30 B3
Atcham **SY5** 38 E5
Ath Linne **HS2** 100 D6
Athelhampton **DT2** 9 G5
Athelington **IP21** 35 G1
Athelney **TA7** 8 C2
Athelstaneford **EH39** 76 D3
Atherington *Devon* **EX37** 6 D3
Atherington *W.Suss.* **BN17** 12 D6
Athersley North **S71** 51 F2
Atherstone CV9 41 **F6**
Atherstone on Stour **CV37** 30 D3
Atherton **M46** 49 F2
Atlow **DE6** 40 E1
Attadale **IV54** 87 F1
Attenborough **NG9** 41 H2
Atterby **LN8** 52 C3
Attercliffe **S9** 51 F4
Atterley **TF13** 39 F6
Atterton **CV13** 41 F6
Attleborough *Norf.* NR17 44 **E6**
Attleborough *Warks.* **CV11** 41 F6
Attlebridge **NR9** 45 F4
Attleton Green **CB8** 34 B3
Atwick **YO25** 59 H4
Atworth **SN12** 20 B5
Auberrow **HR4** 28 D4
Aubourn **LN5** 52 C6
Auch **PA36** 80 E4
Auchairne **KA26** 67 F5
Auchallater **AB35** 89 J6
Auchameanach **PA29** 73 G5
Auchamore **KA27** 73 G6
Aucharnie **AB54** 98 E6
Aucharrigill **IV27** 96 B1
Auchattie **AB31** 90 E5
Auchavan **PH11** 82 C1
Auchbraad **PA30** 73 G2
Auchbreck **AB37** 89 K2
Auchenback **G78** 74 D5
Auchenblae **AB30** 91 F7
Auchenbothie **PA13** 74 B3
Auchenbrack **DG3** 68 C4
Auchenbreck **PA22** 73 J2
Auchencairn **DG7** 65 H5
Auchencrow **TD14** 77 G4
Auchendinny **EH26** 76 A4
Auchendolly **DG7** 65 H4
Auchenfoyle **PA13** 74 B3
Auchengillan **G63** 74 D2
Auchengray **ML11** 75 H5
Auchenhalrig **IV32** 98 B4
Auchenheath **ML11** 75 G6
Auchenhessnane **DG3** 68 D4
Auchenlochan **PA21** 73 H3
Auchenmalg **DG8** 64 C5
Auchenrivock **DG14** 69 J5
Auchentiber **KA13** 74 B6
Auchenvennel **G84** 74 A2
Auchessan **FK20** 81 F5
Auchgourish **PH24** 89 G3
Auchinafaud **PA29** 73 F5
Auchincruive **KA6** 67 H1
Auchindarrach **PA31** 73 G2
Auchindarroch **PA38** 80 B2
Auchindrain **PA32** 80 B7
Auchindrean **IV23** 95 H3
Auchininna **AB53** 98 E6
Auchinleck **KA18** 67 K1
Auchinloch **G66** 74 E3
Auchinner **PH6** 81 H6
Auchinroath **AB38** 97 K6
Auchintoul *Aber.* **AB33** 90 D3
Auchintoul *Aber.* **AB54** 98 E5
Auchintoul *High.* **IV27** 96 C2
Auchiries **AB42** 91 J1
Auchleven **AB52** 90 E2
Auchlochan **ML11** 75 G7
Auchlunachan **IV23** 95 H3
Auchlunies **AB12** 91 G5
Auchlunkart **AB55** 98 B6
Auchlyne **FK21** 81 G5
Auchmacoy **AB41** 91 H1
Auchmair **AB54** 90 B2
Auchmantle **DG8** 64 B4
Auchmithie **DD11** 83 H3
Auchmuirbridge **KY6** 82 D7
Auchmull **DD9** 90 D7
Auchnabony **DG6** 65 H6
Auchnabreac **PA32** 80 B7
Auchnacloich **PH8** 81 K4
Auchnacraig **PA64** 79 J4
Auchnacree **DD8** 83 F1
Auchnafree **PH8** 81 K4
Auchnagallin **PH26** 89 H1
Auchnagatt **AB41** 99 H6
Auchnaha **PA21** 73 H2
Auchnangoul **PA32** 80 B7
Aucholzie **AB35** 90 B5
Auchorrie **AB51** 90 E4
Auchrannie **PH11** 82 D2
Auchraw **FK19** 81 G5
Auchreoch **FK20** 80 E5
Auchronie **DD9** 90 C6
Auchterarder PH3 82 **A6**
Auchtercairn **IV21** 94 E4
Auchterderran **KY5** 76 A1
Auchterhouse **DD3** 82 E4
Auchtermuchty **KY14** 82 D6
Auchterneed **IV14** 96 B6
Auchtertool **KY2** 76 A1
Auchtertyre *Angus* **PH12** 82 D3
Auchtertyre *High.* **IV40** 86 E2
Auchtertyre *Moray* **IV30** 97 J6
Auchtertyre *Stir.* **FK20** 80 E5
Auchtubh **FK19** 81 G5
Auckengill **KW1** 105 J2
Auckley **DN9** 51 J2
Audenshaw **M34** 49 J3
Audlem **CW3** 39 F1
Audley **ST7** 49 G7
Audley End *Essex* **CB11** 33 J5
Audley End *Essex* **CO9** 34 C5
Audley End *Suff.* **IP29** 34 C3
Audmore **ST20** 40 A3
Auds **AB45** 98 E4
Aughton *E.Riding* **YO42** 58 D6
Aughton *Lancs.* **L39** 48 C2
Aughton *Lancs.* **LA2** 55 J3
Aughton *S.Yorks.* **S26** 51 G4
Aughton *Wilts.* **SN8** 21 F6
Aughton Park **L39** 48 D2
Auldearn **IV12** 97 G6
Aulden **HR6** 28 D3
Auldgirth **DG2** 68 E5
Auldhame **EH39** 76 D2
Auldhouse **G75** 74 E5
Aulich **PH17** 81 H2
Ault a'chruinn **IV40** 87 F2
Ault Hucknall **S44** 51 G6
Aultanrynie **IV27** 103 F5
Aultbea **IV22** 94 E3
Aultgrishan **IV21** 94 D3
Aultguish Inn **IV23** 95 K4
Aultibea **KW7** 105 F6
Aultiphurst **KW14** 104 D2
Aultmore **AB55** 98 C5
Ault-na-goire **IV2** 88 C2
Aultnamain Inn **IV19** 96 D3
Aultnapaddock **AB55** 98 B6
Aulton **AB52** 90 E2
Aultvaich **IV4** 96 C7
Aultvoulin **PH41** 86 D4
Aunby **PE9** 42 D4
Aundorach **PH25** 89 G3
Aunk **EX15** 7 J5
Aunsby **NG34** 42 D2
Auquhorthies **AB51** 91 G2
Aust **BS35** 19 J3
Austerfield **DN10** 51 J3
Austrey **CV9** 40 E5
Austwick **LA2** 56 C3
Authorpe **LN11** 53 H4
Authorpe Row **PE24** 53 J5
Avebury **SN8** 20 E4
Avebury Trusloe **SN8** 20 D5
Aveley **RM15** 23 J3
Avening **GL8** 20 B2
Averham **NG23** 51 K7
Avery Hill **SE9** 23 H4
Aveton Gifford **TQ7** 5 G6
Avielochan **PH22** 89 G3
Aviemore PH22 89 **F3**
Avington *Hants.* **SO21** 11 G1
Avington *W.Berks.* **RG17** 21 G5
Avoch IV9 96 **D6**
Avon **BH23** 10 C5
Avon Dassett **CV47** 31 F3
Avonbridge **FK1** 75 H3
Avoncliff **BA15** 20 B6
Avonmouth **BS11** 19 J4
Avonwick **TQ10** 5 H5
Awbridge **SO51** 10 E2
Awhirk **DG9** 64 A5
Awkley **BS35** 19 J3
Awliscombe **EX14** 7 K5
Awre **GL14** 20 A1
Awsworth **NG16** 41 G1
Axbridge BS26 19 **H6**
Axford *Hants.* **RG25** 21 K7
Axford *Wilts.* **SN8** 21 F4
Axminster EX13 8 **B5**
Axmouth **EX12** 8 B5
Axton **CH8** 47 K4
Axtown **PL20** 5 F4
Aycliffe **DL5** 62 D4
Aydon **NE45** 71 F7
Aylburton **GL15** 19 K1
Ayle **CA9** 61 J2
Aylesbeare **EX5** 7 J6
Aylesbury HP20 32 **B7**
Aylesby **DN37** 53 F2
Aylesford ME20 14 **C2**
Aylesham **CT3** 15 H2
Aylestone **LE2** 41 H5
Aylmerton **NR11** 45 F2
Aylsham **NR11** 45 F3
Aylton **HR8** 29 F5
Aymestrey **HR6** 28 D2
Aynho **OX17** 31 G5
Ayot Green **AL6** 33 F7
Ayot St. Lawrence **AL6** 32 E7
Ayot St. Peter **AL6** 33 F7
Ayr KA7 67 **H1**
Aysgarth **DL8** 57 F1
Ayshford **EX16** 7 J4
Ayside **LA11** 55 G1
Ayston **LE15** 42 B5
Aythorpe Roding **CM6** 33 J7
Ayton *P. & K.* **PH2** 82 C6
Ayton *Sc.Bord.* **TD14** 77 H4
Aywick **ZE2** 107 P4
Azerley **HG4** 57 H2

B

Babbacombe **TQ1** 5 K4
Babbinswood **SY11** 38 C2
Babb's Green **SG12** 33 G7
Babcary **TA11** 8 E2

Bab - Bar

Babel SA20 27 H5
Babell CH8 47 K5
Babeny TQ13 5 G3
Bablock Hythe OX29 21 H1
Babraham CB22 33 J3
Babworth DN22 51 J4
Baby's Hill AB37 89 K1
Bac HS2 101 G3
Bachau LL71 46 C4
Back of Keppoch PH39 86 C6
Back Street CB8 34 B3
Backaland KW17 106 E4
Backaskaill KW17 106 D2
Backbarrow LA12 55 G1
Backburn AB54 90 D1
Backe SA33 17 F4
Backfolds AB42 99 J5
Backford CH2 48 D5
Backhill AB53 91 F1
Backhill of Clackriach AB42 99 H6
Backhill of Trustach AB31 90 E5
Backies *High.* KW10 97 F1
Backies *Moray* AB56 98 D5
Backlass KW1 105 H3
Backside AB54 90 C1
Backwell BS48 19 H5
Backworth NE23 71 H6
Bacon End CM6 33 J7
Baconend Green CM6 33 J7
Baconsthorpe NR25 45 F2
Bacton *Here.* HR2 28 C5
Bacton *Norf.* NR12 45 H2
Bacton *Suff.* IP14 34 E2
Bacton Green IP14 34 E2
Bacup OL13 56 D7
Badachro IV21 94 D4
Badanloch Lodge KW11 104 C5
Badavanich IV22 95 H6
Badbea KW7 105 F7
Badbury SN4 20 E3
Badbury Wick SN4 20 E3
Badby NN11 31 G3
Badcall *High.* IV27 102 D4
Badcall *High.* IV27 102 E3
Badcaul IV23 95 G2
Baddeley Green ST2 49 J7
Badden PA31 73 G2
Baddesley Clinton B93 30 C1
Baddesley Ensor CV9 40 E6
Baddidarach IV27 102 C6
Badenscoth AB51 91 F1
Badenyon AB36 90 B3
Badgall PL15 4 C2
Badger WV6 39 G6
Badgerbank SK11 49 H5
Badgers Mount TN14 23 H5
Badgeworth GL51 29 J7
Badgworth BS26 19 G6
Badicaul IV40 86 D2
Badingham IP13 35 H2
Badintagairt IV27 103 G7
Badlesmere ME13 15 F2
Badley IP6 34 E3
Badlipster KW1 105 H4
Badluarach IV23 95 F2
Badminton GL9 20 B3
Badnaban IV27 102 C6
Badnabay IV27 102 E4
Badnafrave AB37 89 K3
Badnagie KW6 105 G5
Badnambiast PH18 88 E7
Badninish IV25 96 E2
Badrallach IV23 95 G2
Badsey WR11 30 B4
Badshot Lea GU9 22 B7
Badsworth WF9 51 G1
Badwell Ash IP31 34 D2
Badworthy TQ10 5 G4
Badyo PH16 82 A1
Bae Cinmel (Kinmel Bay) LL18 47 H4
Bae Colwyn (Colwyn Bay) LL29 47 G5
Bae Penrhyn (Penrhyn Bay) LL30 47 G4
Bag Enderby PE23 53 G5
Bagber DT10 9 G3
Bagby YO7 57 K1
Bagendon GL7 20 D1
Bagginswood DY14 39 F7
Baggrave Hall LE7 41 J5
Baggrow CA7 60 C2
Bàgh a'Chaisteil (Castlebay) HS9 84 B5
Bàgh Mòr HS6 92 D6
Baghasdal HS8 84 C3
Bagillt CH6 48 B5
Baginton CV8 30 E1
Baglan SA12 18 A2
Bagley *Shrop.* SY12 38 D3
Bagley *Som.* BS28 19 H7
Bagmore RG25 21 K7
Bagnall ST9 49 J7
Bagnor RG20 21 H5
Bagpath GL8 20 B2
Bagshot *Surr.* GU19 22 C5
Bagshot *Wilts.* RG17 21 G5
Bagstone GL12 19 K3
Bagthorpe *Norf.* PE31 44 B2
Bagthorpe *Notts.* NG16 51 G7
Baguley M23 49 H4
Bagworth LE67 41 G5
Bagwyllydiart HR2 28 D6
Baildon BD17 57 G6
Baile Ailein (Balallan) HS2 100 E5
Baile a'Mhanaich (Balivanich) HS7 92 C6
Baile an Truiseil HS2 101 F2
Baile Boidheach PA31 73 F3
Baile Gharbhaidh HS8 92 C7
Baile Glas HS6 92 D6
Baile Lion (Balelone) HS6 92 C4
Baile Mhartainn HS6 92 C4
Baile Mhic Phail HS6 92 D4
Baile Mòr *Arg. & B.* PA76 78 D5
Baile Mòr (Balemore) *W.Isles* HS6 92 C5
Baile nan Cailleach HS7 92 C6
Baile Raghaill HS6 92 C5
Bailebeag IV2 88 C3
Baileguish PH21 89 F5
Baile-na-Cille HS6 92 D3
Bailetonach PH36 86 C7
Bailiesward AB54 90 C1
Bailiff Bridge HD6 57 G7
Baillieston G69 74 E4
Bainbridge DL8 62 A7
Bainsford FK2 75 G2
Bainshole AB54 90 E1
Bainton *E.Riding* YO25 59 F4
Bainton *Oxon.* OX27 31 G6
Bainton *Peter.* PE9 42 D5
Bairnkine TD8 70 B2
Bakebare AB55 90 B1
Baker Street RM16 24 C3
Baker's End SG12 33 G7
Bakewell DE45 50 E6
Bala (Y Bala) LL23 37 J2
Balachuirn IV40 94 B7
Balado KY13 82 B7
Balafark G63 74 E1
Balaldie IV20 97 F4
Balallan (Baile Ailein) HS2 100 E5
Balavil PH21 88 E4
Balbeg *High.* IV63 88 B1
Balbeg *High.* IV63 88 B2
Balbeggie PH2 82 C5
Balbirnie KY7 82 D7
Balbithan AB51 91 F3
Balblair *High.* IV7 96 E5
Balblair *High.* IV24 96 C2
Balblair *High.* IV19 96 E3
Balby DN4 51 H2
Balcharn IV27 96 C1
Balcherry IV19 97 F3
Balchers AB45 99 F5
Balchladich IV27 102 C5
Balchraggan *High.* IV5 96 C7
Balchraggan *High.* IV3 88 C1
Balchrick IV27 102 D3
Balcombe RH17 13 G3
Balcurvie KY8 82 E7
Baldernock G62 74 D3
Baldersby YO7 57 J2
Baldersby St. James YO7 57 J2
Balderstone *Gt.Man.* OL16 49 J1
Balderstone *Lancs.* BB2 56 B6
Balderton *Ches.W. & C.* CH4 48 C6
Balderton *Notts.* NG24 52 B7
Baldhu TR3 2 E4
Baldinnie KY15 83 F6
Baldock SG7 33 F5
Baldon Row OX44 21 J1
Baldovan DD3 82 E4
Baldovie *Angus* DD8 82 E2
Baldovie *Dundee* DD5 83 F4
Baldrine IM4 54 D5
Baldslow TN37 14 C6
Baldwin IM4 54 C5
Baldwinholme CA5 60 E1
Baldwin's Gate ST5 39 G1
Baldwins Hill RH19 13 G3
Bale NR21 44 E2
Balelone (Baile Lion) HS6 92 C4
Balemartine PA77 78 A3
Balemore (Baile Mòr) HS6 92 C5
Balendoch PH12 82 D3
Balephuil PA77 78 A3
Balerno EH14 75 K4
Balernock G84 74 A2
Baleromindubh PA61 72 B1
Balerominmore PA61 72 B1
Baleshare (Bhaleshear) HS6 92 C5
Balevulin PA69 79 F5
Balfield DD9 83 G1
Balfour *Aber.* AB34 90 D5
Balfour *Ork.* KW17 106 D6
Balfron G63 74 D2
Balfron Station G63 74 D2
Balgonar KY12 75 J1
Balgove AB51 91 G1
Balgowan *D. & G.* DG9 64 B6
Balgowan *High.* PH20 88 D5
Balgown IV51 93 J5
Balgreen AB45 99 F5
Balgreggan DG9 64 A5
Balgy IV54 94 E6
Balhaldie FK15 81 K7
Balhalgardy AB51 91 F2
Balham SW12 23 F4
Balhary PH11 82 D3
Balhelvie KY14 82 E5
Balhousie KY8 83 F7
Baliasta ZE2 107 Q2
Baligill KW14 104 D2
Baligrundle PA34 79 K3
Balindore PA35 80 A4
Balintore *Angus* DD8 82 D2
Balintore *High.* IV20 97 F4
Balintraid IV18 96 E4
Balintyre PH15 81 H3
Balivanich (Baile a'Mhanaich) HS7 92 C6
Balkeerie DD8 82 E3
Balkholme DN14 58 D7
Balkissock KA26 67 F5
Ball SY10 38 C3
Ball Haye Green ST13 49 J7
Ball Hill RG20 21 H5
Balla HS8 84 C3
Ballabeg IM9 54 B6
Ballacannell IM4 54 D5
Ballacarnane Beg IM6 54 B5
Ballachulish PH49 80 B2
Balladoole IM9 54 B7
Ballafesson IM9 54 B6
Ballagyr IM5 54 B5
Ballajora IM7 54 D4
Ballakilpheric IM9 54 B6
Ballamodha IM9 54 B6
Ballantrae KA26 66 E5
Ballards Gore SS4 25 F2
Ballasalla *I.o.M.* IM9 54 B6
Ballasalla *I.o.M.* IM7 54 C4
Ballater AB35 90 B5
Ballaterach AB34 90 C5
Ballaugh IM7 54 C4
Ballaveare IM4 54 C6
Ballchraggan IV18 96 E4
Ballechin PH9 82 A2
Balleich FK8 81 G7
Ballencrieff EH32 76 C3
Ballidon DE6 50 E7
Balliekine KA27 73 G6
Balliemeanoch PA27 80 C7
Balliemore *Arg. & B.* PA27 73 K1
Balliemore *Arg. & B.* PA34 79 K5
Ballig IM4 54 B5
Ballimeanoch PA33 80 B6
Ballimore *Arg. & B.* PA21 73 H2
Ballimore *Stir.* FK19 81 G6
Ballinaby PA44 72 A4
Ballindean PH14 82 D5
Ballingdon CO10 34 C4
Ballinger Common HP16 22 C1
Ballingham HR2 28 E5
Ballingry KY5 75 K1
Ballinlick PH8 82 A3
Ballinluig *P. & K.* PH9 82 A2
Ballinluig *P. & K.* PH10 82 B2
Ballintuim PH10 82 C2
Balloch *Angus* DD8 82 E2
Balloch *High.* IV2 96 E7
Balloch *N.Lan.* G68 75 F3
Balloch *W.Dun.* G83 74 B2
Ballochan AB31 90 D5
Ballochandrain PA22 73 H2
Ballochford AB54 90 B1
Ballochgair PA28 66 B1
Ballochmartin KA28 73 K5
Ballochmorrie KA26 67 G5
Ballochmyle KA5 67 K1
Ballochroy PA29 73 F5
Ballogie AB34 90 D5
Balls Cross GU28 12 C4
Balls Green *Essex* CO7 34 E6
Ball's Green *Glos.* GL6 20 B2
Balls Hill B71 40 B6
Ballyaurgan PA31 73 F3
Ballygown PA73 79 F3
Ballygrant PA45 72 B4
Ballyhaugh PA78 78 C2
Ballymeanoch PA31 73 G1
Ballymichael KA27 73 H7
Balmacara IV40 86 E2
Balmaclellan DG7 65 G3
Balmacneil PH8 82 A2
Balmadies DD8 83 G3
Balmae DG6 65 G6
Balmaha G63 74 C1
Balmalcolm KY15 82 E7
Balmaqueen IV51 93 K4
Balmeanach *Arg. & B.* PA65 79 H3
Balmeanach *Arg. & B.* PA68 79 F4
Balmedie AB23 91 H3
Balmer Heath SY12 38 D2
Balmerino DD6 82 E5
Balmerlawn SO42 10 E4
Balminnoch DG8 64 C4
Balmore *E.Dun.* G64 74 E3
Balmore *High.* IV12 97 F7
Balmore *High.* IV4 87 K1
Balmore *High.* IV55 93 H7
Balmore *P. & K.* PH16 81 J2
Balmullo KY16 83 F5
Balmungie IV10 96 E6
Balmyle PH10 82 B2
Balnaboth DD8 82 E1
Balnabruaich IV19 96 E4
Balnacra IV54 95 F7
Balnafoich IV2 88 D1
Balnagall IV20 97 F3
Balnagown Castle IV18 96 E4
Balnaguard PH9 82 A2
Balnaguisich IV18 96 D4
Balnahard *Arg. & B.* PA61 72 C1
Balnahard *Arg. & B.* PA68 79 F4
Balnain IV63 88 B1
Balnakeil IV27 103 F2
Balnaknock IV51 93 K5
Balnamoon DD9 83 G1
Balnapaling IV19 96 E5
Balnespick PH21 89 F4
Balquhidder FK19 81 G5
Balsall CV7 30 D1
Balsall Common CV7 30 D1
Balsall Heath B12 40 C7
Balscote OX15 30 E4
Balsham CB21 33 J3
Baltasound ZE2 107 Q2
Balterley CW2 49 G7
Balterley Heath CW2 49 G7
Baltersan DG8 64 E4
Balthangie AB53 99 G5
Balthayock PH2 82 C5
Baltonsborough BA6 8 E1
Baluachraig PA31 73 G1
Balulive PA45 72 C4
Balure *Arg. & B.* PA35 80 B4
Balure *Arg. & B.* PA37 79 K4
Balvaird IV6 96 C6
Balvarran PH10 82 B1
Balvicar PA34 79 J6
Balvraid *High.* IV40 86 E3
Balvraid *High.* IV13 89 F1
Bamber Bridge PR5 55 J7
Bamber's Green CM22 33 J6
Bamburgh NE69 77 K7
Bamff PH11 82 D2
Bamford *Derbys.* S33 50 E4
Bamford *Gt.Man.* OL11 49 H1
Bampton *Cumb.* CA10 61 G5
Bampton *Devon* EX16 7 H3
Bampton *Oxon.* OX18 21 G1
Bampton Grange CA10 61 G5
Banavie PH33 87 H7
Banbury OX16 31 F4
Bancffosfelen SA15 17 H4
Banchor IV12 97 G7
Banchory AB31 91 F5
Banchory-Devenick AB12 91 H4
Bancycapel SA32 17 H4
Bancyfelin SA33 17 G4
Bancyffordd SA44 17 H2
Bandon KY7 82 D7
Bandrake Head LA12 55 G1
Banff AB45 98 E4
Bangor LL57 46 D5
Bangor-is-y-coed LL13 38 C1
Bangor's Green L39 48 C2
Banham NR16 44 E7
Bank SO43 10 D4
Bank End LA20 54 E1
Bank Newton BD23 56 E4
Bank Street WR15 29 F2
Bank Top *Lancs.* WN8 48 E2
Bank Top *W.Yorks.* HX3 57 G7
Bankend DG1 69 F7
Bankfoot PH1 82 B4
Bankglen KA18 67 K2
Bankhead *Aber.* AB33 90 E3
Bankhead *Aber.* AB51 90 E4
Bankhead *Aberdeen* AB21 91 G3
Bankhead *D. & G.* DG6 65 H6
Bankland TA7 8 C2
Banknock FK4 75 F3
Banks *Cumb.* CA8 70 A7
Banks *Lancs.* PR9 55 G7
Bankshill DG11 69 G5
Banningham NR11 45 G3
Bannister Green CM6 33 K6
Bannockburn FK7 75 G1
Banstead SM7 23 F6
Bantam Grove LS27 57 H7
Bantham TQ7 5 G6
Banton G65 75 F3
Banwell BS29 19 G6
Banwen Pyrddin SA10 18 B1
Banyard's Green IP13 35 H1
Bapchild ME9 25 F5
Baptiston G63 74 D2
Bapton BA12 9 J1
Bar End SO23 11 F2
Bar Hill CB23 33 G2
Barabhas (Barvas) HS2 101 F3
Barachander PA35 80 B5
Barassie KA10 74 B7
Barbaraville IV18 96 E4
Barber Booth S33 50 D4
Barber Green LA11 55 G1
Barber's Moor PR26 48 D1
Barbon LA6 56 B1
Barbridge CW5 49 F7
Barbrook EX35 7 F1
Barby CV23 31 G1
Barcaldine PA37 80 A3
Barcaple DG7 65 G5
Barcheston CV36 30 D5
Barclose CA6 69 K7
Barcombe BN8 13 H5
Barcombe Cross BN8 13 H5
Barden DL8 62 C7
Barden Park TN9 23 J7
Bardennoch DG7 67 K4
Bardfield End Green CM6 33 K5
Bardfield Saling CM7 33 K6
Bardister ZE2 107 M5
Bardney LN3 52 E6
Bardon *Leics.* LE67 41 G4
Bardon *Moray* IV30 97 K6
Bardon Mill NE47 70 C7
Bardowie G62 74 D3
Bardsea LA12 55 F2
Bardsey LS17 57 J5
Bardsey Island (Ynys Enlli) LL53 36 A3
Bardwell IP31 34 D1
Bare LA4 55 H3
Barewood HR6 28 C3
Barfad PA29 73 G4
Barford *Norf.* NR9 45 F5
Barford *Warks.* CV35 30 D2
Barford St. John OX15 31 F5
Barford St. Martin SP3 10 B1
Barford St. Michael OX15 31 F5
Barfrestone CT15 15 H2
Bargaly DG8 64 E4
Bargany Mains KA26 67 G3
Bargeddie G69 74 E4
Bargoed CF81 18 E2
Bargrennan DG8 64 D3
Barham *Cambs.* PE28 32 E1
Barham *Kent* CT4 15 H2
Barham *Suff.* IP6 35 F3
Barharrow DG7 65 G5
Barholm PE9 42 D4
Barholm Mains DG8 64 E5
Barkby LE7 41 J5
Barkby Thorpe LE7 41 J5
Barkers Green SY4 38 E3
Barkestone-le-Vale NG13 42 A2
Barkham RG41 22 A5
Barking *Gt.Lon.* IG11 23 H3
Barking *Suff.* IP6 34 E3
Barking Tye IP6 34 E3
Barkisland HX4 57 F7
Barkston NG32 42 C1
Barkston Ash LS24 57 K6
Barkway SG8 33 G5
Barlae DG8 64 C4
Barland LD8 28 B2
Barlaston ST12 40 A2
Barlavington GU28 12 C5
Barlborough S43 51 G5
Barlby YO8 58 C6
Barlestone CV13 41 G5
Barley *Herts.* SG8 33 G5
Barley *Lancs.* BB12 56 D5
Barley Green IP21 35 G1
Barleycroft End SG9 33 H6
Barleyhill NE44 62 B1
Barleythorpe LE15 42 B5
Barling SS3 25 F3
Barlings LN3 52 D5
Barlow *Derbys.* S18 51 F5
Barlow *N.Yorks.* YO8 58 C7
Barlow *T. & W.* NE21 71 G7
Barmby Moor YO42 58 D5
Barmby on the Marsh DN14 58 C7
Barmer PE31 44 C2
Barmolloch PA31 73 G1
Barmoor Lane End TD15 77 J7
Barmouth (Abermaw) LL42 37 F4
Barmpton DL1 62 E5
Barmston YO25 59 H4
Barnaby Green NR34 35 J1
Barnacabber PA23 73 K2
Barnacarry PA27 73 H1
Barnack PE9 42 D5
Barnacle CV7 41 F7
Barnamuc PA38 80 B3
Barnard Castle DL12 62 B5
Barnard Gate OX29 31 F7
Barnardiston CB9 34 B4
Barnard's Green WR14 29 G4
Barnbarroch *D. & G.* DG8 64 D5
Barnbarroch *D. & G.* DG5 65 J5
Barnburgh DN5 51 G2
Barnby NR34 45 J7
Barnby Dun DN3 51 J2
Barnby in the Willows NG24 52 B7
Barnby Moor DN22 51 J4
Barndennoch DG2 68 D5
Barne Barton PL5 4 E5
Barnehurst DA7 23 J4
Barnes SW13 23 F4
Barnes Street TN11 23 K7
Barnet EN5 23 F2
Barnet Gate EN5 23 F2
Barnetby le Wold DN38 52 D2
Barney NR21 44 D2
Barnham *Suff.* IP24 34 C1
Barnham *W.Suss.* PO22 12 C6
Barnham Broom NR9 44 E5
Barnhead DD10 83 H2
Barnhill *Ches.W. & C.* CH3 48 D7
Barnhill *Dundee* DD5 83 F4
Barnhill *Moray* IV30 97 J6
Barnhills DG9 66 D6
Barningham *Dur.* DL11 62 B5
Barningham *Suff.* IP31 34 D1
Barningham Green NR11 45 F2
Barnoldby le Beck DN37 53 F2
Barnoldswick BB18 56 D5
Barns Green RH13 12 E4
Barnsdale Bar WF8 51 H1
Barnsley *Glos.* GL7 20 D1
Barnsley *S.Yorks.* S70 51 F2
Barnsole CT3 15 H2
Barnstaple EX31 6 D2
Barnston *Essex* CM6 33 K7
Barnston *Mersey.* CH61 48 B4
Barnstone NG13 42 A2
Barnt Green B45 30 B1
Barnton *Ches.W. & C.* CW8 49 F5
Barnton *Edin.* EH4 75 K3
Barnwell All Saints PE8 42 D7
Barnwell St. Andrew PE8 42 D7
Barnwood GL4 29 H7
Barons' Cross HR6 28 D3
Barr *Arg. & B.* PA44 72 B4
Barr *High.* PA34 79 H2
Barr *S.Ayr.* KA26 67 G4
Barr *Som.* TA4 7 K3
Barr Hall CO9 34 B5
Barra (Barraigh) HS9 84 B4
Barra (Tràigh Mhòr) Airport HS9 84 B4
Barrachan DG8 64 D6
Barrackan PA31 79 J7
Barraer DG8 64 D4
Barraglom HS2 100 D4
Barrahormid PA31 73 F2
Barraigh (Barra) HS9 84 B4
Barran PA33 80 C5
Barrapoll PA77 78 A3
Barrasford NE48 70 E6
Barravullin PA31 79 K7
Barregarrow IM6 54 C5

Bar - Bet

Barrets Green CW6 48 E7
Barrhead G78 74 C5
Barrhill KA26 67 G5
Barrington Cambs. CB22 33 G4
Barrington Som. TA19 8 C3
Barripper TR14 2 D5
Barrisdale PH35 86 E4
Barrmill KA15 74 B5
Barrnacarry PA34 79 K5
Barrock KW14 105 H1
Barrow Glos. GL51 29 H6
Barrow Lancs. BB7 56 C6
Barrow Rut. LE15 42 B4
Barrow Shrop. TF12 39 F5
Barrow Som. BA9 9 G1
Barrow Som. BA4 19 J7
Barrow Suff. IP29 34 B2
Barrow Gurney BS48 19 J5
Barrow Hann DN19 59 G7
Barrow Haven DN19 59 G7
Barrow Hill S43 51 G5
Barrow Nook L39 48 D2
Barrow Street BA12 9 H1
Barrow upon Humber DN19 59 G7
Barrow upon Soar LE12 41 H4
Barrow upon Trent DE73 41 F3
Barroway Drove PE38 43 J5
Barrowby NG32 42 B2
Barrowcliff YO12 59 G1
Barrowden LE15 42 C5
Barrowford BB9 56 D6
Barrow-in-Furness LA14 55 F3
Barrows Green LA8 55 J1
Barry Angus DD7 83 G4
Barry V. of Glam. CF62 18 E5
Barsby LE7 41 J4
Barsham NR34 45 H7
Barskimming KA5 67 J1
Barsloisnoch PA31 73 G1
Barston B92 30 D1
Bartestree HR1 28 E4
Barthol Chapel AB51 91 G1
Bartholomew Green CM77 34 B6
Barthomley CW2 49 G7
Bartley SO40 10 E3
Bartley Green B32 40 C7
Bartlow CB21 33 J4
Barton Cambs. CB23 33 H3
Barton Ches.W. & C. SY14 48 D7
Barton Cumb. CA10 61 F4
Barton Glos. GL54 30 B6
Barton Lancs. PR3 55 J6
Barton Lancs. L39 48 C2
Barton N.Yorks. DL10 62 D6
Barton Oxon. OX3 21 J1
Barton Torbay TQ2 5 K4
Barton Warks. B50 30 C3
Barton Bendish PE33 44 B5
Barton End GL6 20 B2
Barton Green DE13 40 D4
Barton Hartshorn MK18 31 H5
Barton Hill YO60 58 D3
Barton in Fabis NG11 41 H2
Barton in the Beans CV13 41 F5
Barton Mills IP28 34 B1
Barton on Sea BH25 10 D5
Barton St. David TA11 8 E1
Barton Seagrave NN15 32 B1
Barton Stacey SO21 21 H7
Barton Town EX31 6 E1
Barton Turf NR12 45 H3
Bartongate OX7 31 F6
Barton-le-Clay MK45 32 D5
Barton-le-Street YO17 58 D2
Barton-le-Willows YO60 58 D3
Barton-on-the-Heath GL56 30 D5
Barton-under-Needwood DE13 40 D4
Barton-upon-Humber DN18 59 G7
Barvas (Barabhas) HS2 101 F3
Barway CB7 33 J1
Barwell LE9 41 G6
Barwhinnock DG6 65 G5
Barwick Herts. SG11 33 G7
Barwick Som. BA22 8 E3
Barwick in Elmet LS15 57 J6
Barwinnock DG8 64 D6
Baschurch SY4 38 D3
Bascote CV47 31 F2
Base Green IP14 34 E2
Basford Green ST13 49 J7
Bashall Eaves BB7 56 B5
Bashall Town BB7 56 C5
Bashley BH25 10 D5
Basildon Essex SS14 24 D3
Basildon W.Berks. RG8 21 K4
Basingstoke RG21 21 K6
Baslow DE45 50 E5
Bason Bridge TA9 19 G7
Bassaleg NP10 19 F3
Bassenthwaite CA12 60 D3
Basset's Cross EX20 6 D5
Bassett SO16 11 F3
Bassingbourn SG8 33 G4
Bassingfield NG12 41 J2
Bassingham LN5 52 C7
Bassingthorpe NG33 42 C3
Basta ZE2 107 P3
Baston PE6 42 E4
Bastonford WR2 29 H3
Bastwick NR29 45 J4
Batavaime FK21 81 F4
Batch BS24 19 G6
Batchley B97 30 B2
Batchworth WD3 22 D2
Batchworth Heath WD3 22 D2
Batcombe Dorset DT2 9 F4
Batcombe Som. BA4 9 F1
Bate Heath CW9 49 F5
BATH BA 20 A5
Bathampton BA2 20 A5
Bathealton TA4 7 J3
Batheaston BA1 20 A5
Bathford BA1 20 A5
Bathgate EH48 75 H4
Bathley NG23 51 K7
Bathpool Cornw. PL15 4 C3
Bathpool Som. TA2 8 B2
Bathway BA3 19 J6
Batley WF17 57 H7
Batsford GL56 30 C5
Batson TQ8 5 H7
Battersby TS9 63 G6
Battersea SW11 23 F4
Battisborough Cross PL8 5 G6
Battisford IP14 34 E3
Battisford Tye IP14 34 E3
Battle E.Suss. TN33 14 C6
Battle Powys LD3 27 K5
Battledown GL52 29 J6
Battlefield SY1 38 E4
Battlesbridge SS11 24 D2
Battlesden MK17 32 C6
Battlesea Green IP21 35 G1
Battleton TA22 7 H3
Battlies Green IP30 34 D2
Battramsley SO41 10 E5
Batt's Corner GU10 22 B7
Bauds of Cullen AB56 98 C4
Baugh PA77 78 B3
Baughton WR8 29 H4
Baughurst RG26 21 J5
Baulds AB31 90 E5
Baulking SN7 21 G2
Baumber LN9 53 F5
Baunton GL7 20 D1
Baveney Wood DY14 29 F1
Baverstock SP3 10 B1
Bawburgh NR9 45 F5
Bawdeswell NR20 44 E3
Bawdrip TA7 8 C1
Bawdsey IP12 35 H4
Bawdsey Manor IP12 35 H5
Bawsey PE32 44 A4
Bawtry DN10 51 J3
Baxenden BB5 56 C7
Baxterley CV9 40 E6
Baxter's Green CB8 34 B3
Bay SP8 9 H2
Baybridge DH8 62 A2
Baycliff LA12 55 F2
Baydon SN8 21 F4
Bayford Herts. SG13 23 G1
Bayford Som. BA9 9 G2
Bayfordbury SG13 33 G7
Bayham Abbey TN3 13 K3
Bayles CA9 61 J2
Baylham IP6 35 F3
Baynards Green OX27 31 G6
Baysham HR9 28 E6
Bayston Hill SY3 38 D5
Bayswater W2 23 F3
Baythorn End CO9 34 B4
Bayton DY14 29 F1
Bayworth OX13 21 J1
Beach High. PA34 79 J2
Beach S.Glos. BS30 20 A4
Beachampton MK19 31 J5
Beachamwell PE37 44 B5
Beacharr PA29 72 E6
Beachley NP16 19 J2
Beacon Devon EX14 7 K5
Beacon Devon EX14 8 B4
Beacon Hill Dorset BH16 9 J5
Beacon Hill Essex CM8 34 C7
Beacon Hill Surr. GU26 12 B3
Beacon's Bottom HP14 22 A2
Beaconsfield HP9 22 C2
Beacravik HS3 93 G2
Beadlam YO62 58 C1
Beadlow SG17 32 E5
Beadnell NE67 71 H1
Beaford EX19 6 D4
Beal N.Yorks. DN14 58 B7
Beal Northumb. TD15 77 J6
Bealach PA38 80 A2
Bealsmill PL17 4 D3
Beambridge CW5 49 F7
Beamhurst ST14 40 C2
Beaminster DT8 8 D4
Beamish DH9 62 D1
Beamsley BD23 57 F4
Bean DA2 23 J4
Beanacre SN12 20 C5
Beanley NE66 71 F2
Beaquoy KW17 106 C5
Beardon EX20 6 D7
Beardwood BB2 56 B7
Beare EX5 7 H5
Beare Green RH5 22 E7
Bearley CV37 30 C2
Bearnie AB41 91 H1
Bearnock IV63 88 B1
Bearnus PA73 78 E3
Bearpark DH7 62 D2
Bearsbridge NE47 61 J1
Bearsden G61 74 D3
Bearsted ME14 14 C2
Bearstone TF9 39 G2
Bearwood Poole BH11 10 B5
Bearwood W.Mid. B66 40 C7
Beattock DG10 69 F3
Beauchamp Roding CM5 33 J7
Beauchief S8 51 F4
Beaudesert B95 30 C2
Beaufort NP23 28 A7
Beaulieu SO42 10 E4
Beauly IV4 96 C7
Beaumaris (Biwmares) LL58 46 E5
Beaumont Chan.I. JE3 3 J7
Beaumont Cumb. CA5 60 E1
Beaumont Essex CO16 35 F6
Beaumont Hill DL1 62 D5
Beaumont Leys LE4 41 H5
Beausale CV35 30 D1
Beauvale NG16 41 G1
Beauworth SO24 11 G2
Beaver Green TN23 14 E3
Beaworthy EX21 6 C6
Beazley End CM7 34 B6
Bebington CH63 48 C4
Bebside NE24 71 H5
Beccles NR34 45 J6
Beccles Heliport NR34 45 J7
Becconsall PR4 55 H7
Beck Foot LA8 61 H7
Beck Hole YO22 63 K6
Beck Row IP28 33 K1
Beck Side Cumb. LA17 55 F1
Beck Side Cumb. LA11 55 G1
Beckbury TF11 39 G5
Beckenham BR3 23 G5
Beckering LN8 52 E4
Beckermet CA21 60 B6
Beckermonds BD23 56 D1
Beckett End IP26 44 B6
Beckfoot Cumb. CA19 60 C6
Beckfoot Cumb. CA7 60 B2
Beckford GL20 29 J5
Beckhampton SN8 20 D5
Beckingham Lincs. LN5 52 B7
Beckingham Notts. DN10 51 K4
Beckington BA11 20 B6
Beckley E.Suss. TN31 14 D5
Beckley Oxon. OX3 31 G7
Beck's Green NR34 45 H7
Beckside LA6 56 B1
Beckton E6 23 H3
Beckwithshaw HG3 57 H4
Becontree RM8 23 H3
Bedale DL8 57 H1
Bedburn DL13 62 B3
Bedchester SP7 9 H3
Beddau CF38 18 D3
Beddgelert LL55 36 E1
Beddingham BN8 13 H6
Beddington SM6 23 F5
Beddington Corner CR4 23 F5
Bedfield IP13 35 G2
Bedfield Little Green IP13 35 G2
Bedford MK40 32 D4
Bedgebury Cross TN17 14 C4
Bedgrove HP21 32 B7
Bedham RH20 12 D4
Bedhampton PO9 11 J4
Bedingfield IP23 35 F2
Bedingfield Green IP23 35 F2
Bedingfield Street IP23 35 F2
Bedingham Green NR35 45 G6
Bedlam Lancs. BB5 56 C7
Bedlam N.Yorks. HG3 57 H3
Bedlar's Green CM22 33 J6
Bedlington NE22 71 H5
Bedlinog CF46 18 D1
Bedminster BS3 19 J4
Bedmond WD5 22 D1
Bednall ST17 40 B4
Bedol CH6 48 B5
Bedrule TD9 70 B2
Bedstone SY7 28 C1
Bedwas CF83 18 E3
Bedwell SG1 33 F6
Bedwellty NP12 18 E1
Bedworth CV12 41 F7
Bedworth Woodlands CV12 41 F7
Beeby LE7 41 J5
Beech Hants. GU34 11 H1
Beech Staffs. ST4 40 A2
Beech Hill RG7 21 K5
Beechingstoke SN9 20 D6
Beechwood WA7 48 E4
Beedon RG20 21 H4
Beeford YO25 59 H4
Beeley DE4 50 E6
Beelsby DN37 53 F2
Beenham RG7 21 J5
Beeny PL35 4 B1
Beer EX12 8 B6
Beer Hackett DT9 9 F3
Beercrocombe TA3 8 C2
Beesands TQ7 5 J6
Beesby Lincs. LN13 53 H4
Beesby N.E.Lincs. DN36 53 F3
Beeson TQ7 5 J6
Beeston Cen.Beds. SG19 32 E4
Beeston Ches.W. & C. CW6 48 E7
Beeston Norf. PE32 44 D4
Beeston Notts. NG9 41 H2
Beeston W.Yorks. LS11 57 H6
Beeston Regis NR26 45 F1
Beeston St. Lawrence NR12 45 H3
Beeswing DG2 65 J4
Beetham Cumb. LA7 55 H2
Beetham Som. TA20 8 B3
Beetley NR20 44 D4
Beffcote ST20 40 A4
Began CF3 19 F3
Begbroke OX5 31 F7
Begdale PE14 43 H5
Begelly SA68 16 E5
Beggar's Bush LD8 28 B2
Beggearn Huish TA23 7 J2
Beggshill AB54 90 D1
Beguildy (Bugeildy) LD7 28 A1
Beighton Norf. NR13 45 H5
Beighton S.Yorks. S20 51 G4
Beili-glas NP7 19 G1
Beinn na Faoghla (Benbecula) HS7 92 D6
Beith KA15 74 B5
Bekesbourne CT4 15 G2
Belaugh NR12 45 G4
Belbroughton DY9 29 J1
Belchalwell DT11 9 G4
Belchalwell Street DT11 9 G4
Belchamp Otten CO10 34 C4
Belchamp St. Paul CO10 34 B4
Belchamp Walter CO10 34 C4
Belchford LN9 53 F5
Belfield OL16 49 J1
Belford NE70 77 K7
Belgrave LE4 41 H5
Belhaven EH42 76 E3
Belhelvie AB23 91 H3
Belhinnie AB54 90 C2
Bell Bar AL9 23 F1
Bell Busk BD23 56 E4
Bell End DY9 29 J1
Bell Heath DY9 29 J1
Bell Hill GU32 11 J2
Bell o' th' Hill SY13 38 E1
Bellabeg AB36 90 B3
Belladrum IV4 96 C7
Bellanoch PA31 73 G1
Bellaty PH11 82 D2
Belle Isle LS10 57 J7
Belle Vue CA2 60 E1
Belleau LN13 53 H5
Belleheiglash AB37 89 J1
Bellerby DL8 62 C7
Bellever PL20 5 G3
Bellfields GU1 22 C6
Belliehill DD9 83 G1
Bellingdon HP5 22 C1
Bellingham Gt.Lon. SE6 23 G4
Bellingham Northumb. NE48 70 D5
Belloch PA29 72 E7
Bellochantuy PA28 72 E7
Bell's Cross IP6 35 F3
Bells Yew Green TN3 13 K3
Bellsbank KA6 67 J3
Bellshill N.Lan. ML4 75 F4
Bellshill Northumb. NE70 77 K7
Bellside ML1 75 G5
Bellsmyre G82 74 C3
Bellsquarry EH54 75 J4
Belluton BS39 19 K5
Belmaduthy IV8 96 D6
Belmesthorpe PE9 42 D4
Belmont B'burn. BL7 49 F1
Belmont Gt.Lon. SM2 23 F5
Belmont Gt.Lon. HA7 22 E2
Belmont Shet. ZE2 107 P2
Belnie PE11 43 F2
Belowda PL26 3 G2
Belper DE56 41 F1
Belper Lane End DE56 41 F1
Belsay NE20 71 G6
Belsford TQ9 5 H5
Belsize WD3 22 D1
Belstead IP8 35 F4
Belston KA6 67 H1
Belstone EX20 6 E6
Belstone Corner EX20 6 E6
Belsyde EH49 75 H3
Belthorn BB1 56 C7
Beltinge CT6 25 H5
Beltingham NE47 70 C7
Beltoft DN9 52 B2
Belton Leics. LE12 41 G3
Belton Lincs. NG32 42 C2
Belton N.Lincs. DN9 51 K2
Belton Norf. NR31 45 J5
Belton Rut. LE15 42 B5
Beltring TN12 23 K7
Belvedere DA17 23 H4
Belvoir NG32 42 B2
Bembridge PO35 11 H6
Bemersyde TD6 76 D7
Bemerton SP2 10 C1
Bempton YO15 59 H2
Ben Alder Cottage PH17 81 F1
Ben Alder Lodge PH19 88 C7
Ben Rhydding LS29 57 G5
Benacre NR34 45 K7
Benbecula (Beinn na Faoghla) HS7 92 D6
Benbecula (Balivanich) Airport HS7 92 C6
Benbuie DG3 68 C4
Benderloch PA37 80 A4
Bendish SG4 32 E6
Benenden TN17 14 D4
Benfield DG8 64 D4
Benfieldside DH8 62 B1
Bengate NR28 45 H3
Bengeo SG14 33 G7
Bengeworth WR11 30 B4
Benhall GL51 29 J6
Benhall Green IP17 35 H2
Benhall Street IP17 35 H2
Benholm DD10 83 K1
Beningbrough YO30 58 B4
Benington Herts. SG2 33 F6
Benington Lincs. PE22 43 G1
Benington Sea End PE22 43 H1
Benllech LL74 46 D4
Benmore Arg. & B. PA23 73 K2
Benmore Stir. FK20 81 F5
Bennacott PL15 4 C1
Bennan Cottage DG7 65 G3
Bennett End HP14 22 A2
Bennetts End HP3 22 D1
Benniworth LN8 53 F4
Benover ME18 14 C3
Benson OX10 21 K2
Benston ZE2 107 N7
Benthall Northumb. NE67 71 H1
Benthall Shrop. TF12 39 F5
Bentham GL51 29 J7
Benthoul AB14 91 G4
Bentlawnt SY5 38 C5
Bentley E.Riding HU17 59 G6
Bentley Essex CM15 23 J2
Bentley Hants. GU10 22 A7
Bentley S.Yorks. DN5 51 H2
Bentley Suff. IP9 35 F5
Bentley W.Mid. WS2 40 B6
Bentley W.Yorks. LS6 57 H6
Bentley Warks. CV9 40 E6
Bentley Heath Herts. EN5 23 F2
Bentley Heath W.Mid. B93 30 C1
Bentley Rise DN5 51 H2
Benton EX32 6 E2
Benton Square NE12 71 J6
Bentpath DG13 69 J4
Bentworth GU34 21 K7
Benvie DD2 82 E4
Benville Lane DT2 8 E4
Benwell NE15 71 H7
Benwick PE15 43 G6
Beoley B98 30 B2
Beoraidbeg PH40 86 C5
Bepton GU29 12 B5
Berden CM23 33 H6
Bere Alston PL20 4 E4
Bere Ferrers PL20 4 E4
Bere Regis BH20 9 H5
Berea SA62 16 A2
Berepper TR12 2 D6
Bergh Apton NR15 45 H5
Berinsfield OX10 21 J2
Berkeley GL13 19 K2
Berkhamsted HP4 22 C1
Berkley BA11 20 B7
Berkswell CV7 30 D1
Bermondsey SE16 23 G4
Bernera IV40 86 E2
Berneray (Eilean Bhearnaraigh) HS6 92 E3
Berners Roding CM5 24 C1
Bernice PA23 73 K1
Bernisdale IV51 93 K6
Berrick Prior OX10 21 K2
Berrick Salome OX10 21 K2
Berriedale KW7 105 G6
Berriew (Aberriw) SY21 38 A5
Berrington Northumb. TD15 77 J6
Berrington Shrop. SY5 38 E5
Berrington Worcs. WR15 28 E2
Berrington Green WR15 28 E2
Berriowbridge PL15 4 C3
Berrow Som. TA8 19 F6
Berrow Worcs. WR13 29 G5
Berrow Green WR6 29 G3
Berry Cross EX38 6 C4
Berry Down Cross EX34 6 D1
Berry Hill Glos. GL16 28 E7
Berry Hill Pembs. SA42 16 D1
Berry Pomeroy TQ9 5 J4
Berryhillock AB56 98 D4
Berrynarbor EX34 6 D1
Berry's Green TN16 23 H6
Bersham LL14 38 C1
Berstane KW15 106 D6
Berthlwyd SA4 17 J6
Berwick BN26 13 J6
Berwick Bassett SN4 20 E4
Berwick Hill NE20 71 G6
Berwick St. James SP3 10 B1
Berwick St. John SP7 9 J2
Berwick St. Leonard SP3 9 J1
Berwick-upon-Tweed TD15 77 H5
Bescar L40 48 C1
Bescot WS2 40 C6
Besford Shrop. SY4 38 E3
Besford Worcs. WR8 29 J4
Bessacarr DN4 51 J2
Bessels Leigh OX13 21 H1
Besses o' th' Barn M45 49 H2
Bessingby YO16 59 H3
Bessingham NR11 45 F2
Best Beech Hill TN5 13 K3
Besthorpe Norf. NR17 44 E6
Besthorpe Notts. NG23 52 B6
Bestwood Village NG5 41 H1
Beswick E.Riding YO25 59 G5
Beswick Gt.Man. M11 49 H3
Betchworth RH3 23 F6
Bethania Cere. SY23 26 E2
Bethania Gwyn. LL41 37 G1
Bethel Gwyn. LL55 46 D6
Bethel Gwyn. LL23 37 J2
Bethel I.o.A. LL62 46 B5
Bethersden TN26 14 E4
Bethesda Gwyn. LL57 46 E6
Bethesda Pembs. SA67 16 D4
Bethlehem SA19 17 K3
Bethnal Green E2 23 G3
Betley CW3 39 G1
Betley Common CW3 39 G1
Betsham DA13 24 C4
Betteshanger CT14 15 J2
Bettiscombe DT6 8 C5
Bettisfield SY13 38 D2
Betton Shrop. SY5 38 C5
Betton Shrop. TF9 39 F2
Betton Strange SY5 38 E5
Bettws Bridgend CF32 18 C3
Bettws Newport NP20 19 F2
Bettws Bledrws SA48 26 E3
Bettws Cedewain SY16 38 A6

Bet - Bli

Bettws Gwerfil Goch LL21 37 K1
Bettws Newydd NP15 19 G1
Bettws-y-crwyn SY7 38 B7
Bettyhill KW14 104 C2
Betws SA18 17 K4
Betws Disserth LD1 28 A3
Betws Garmon LL54 46 D7
Betws Ifan SA38 17 G1
Betws-y-coed LL24 47 F7
Betws-yn-Rhos LL22 47 H5
Beulah Cere. SA38 17 F1
Beulah Powys LD5 27 J3
Bevendean BN2 13 G6
Bevercotes NG22 51 J5
Beverley HU17 59 G6
Beverstone GL8 20 B2
Bevington GL13 19 K2
Bewaldeth CA13 60 D3
Bewcastle CA6 70 A6
Bewdley DY12 29 G1
Bewerley HG3 57 G3
Bewholme YO25 59 H4
Bewley Common SN15 20 C5
Bexhill TN40 14 C7
Bexley DA5 23 H4
Bexleyheath DA6 23 H4
Bexwell PE38 44 A5
Beyton IP30 34 D2
Beyton Green IP30 34 D2
Bhalamus HS2 100 E7
Bhaleshear (Baleshare) HS6 92 C5
Bhaltos HS2 100 C4
Bhatarsaigh (Vatersay) HS9 84 B5
Biallaid PH20 88 E5
Bibury GL7 20 E1
Bicester OX26 31 G6
Bickenhall TA3 8 B3
Bickenhill B92 40 D7
Bicker PE20 43 F2
Bickershaw WN2 49 F2
Bickerstaffe L39 48 D2
Bickerton Ches.W. & C. SY14 48 E7
Bickerton Devon TQ7 5 J7
Bickerton N.Yorks. LS22 57 K4
Bickerton Northumb. NE65 70 E3
Bickford ST19 40 A4
Bickham TA24 7 H1
Bickham Bridge TQ9 5 H5
Bickham House EX6 7 H7
Bickington Devon EX31 6 D2
Bickington Devon TQ12 5 H3
Bickleigh Devon PL6 5 F4
Bickleigh Devon EX16 7 H5
Bickleton EX31 6 D2
Bickley BR1 23 H5
Bickley Moss SY13 38 E1
Bickley Town SY14 38 E1
Bicknacre CM3 24 D1
Bicknoller TA4 7 K2
Bicknor ME9 14 D2
Bickton SP6 10 C3
Bicton Here. HR6 28 D2
Bicton Shrop. SY3 38 D4
Bicton Shrop. SY7 38 B7
Bicton Heath SY3 38 D4
Bidborough TN4 23 J7
Biddenden TN27 14 D4
Biddenden Green TN27 14 D3
Biddenham MK40 32 D4
Biddestone SN14 20 B4
Biddick NE38 62 E1
Biddisham BS26 19 G6
Biddlesden NN13 31 H4
Biddlestone NE65 70 E3
Biddulph ST8 49 H7
Biddulph Moor ST8 49 J7
Bideford EX39 6 C3
Bidford-on-Avon B50 30 C3
Bidham Dock ME9 25 F5
Bidlake EX20 6 C7
Bidston CH43 48 B3
Bidwell LU5 32 D6
Bielby YO42 58 D5
Bieldside AB15 91 G4
Bierley I.o.W. PO38 11 G7
Bierley W.Yorks. BD4 57 G6
Bierton HP22 32 B7
Big Sand IV21 94 D4
Bigbury TQ7 5 G6
Bigbury-on-Sea TQ7 5 G6
Bigby DN38 52 D2
Bigert Mire LA20 60 C7
Biggar Cumb. LA14 54 E3
Biggar S.Lan. ML12 75 J7
Biggin Derbys. DE6 40 E1
Biggin Derbys. SK17 50 D7
Biggin N.Yorks. LS25 58 B6
Biggin Hill TN16 23 H6
Biggings ZE2 107 K6
Biggleswade SG18 32 E4
Bigholms DG13 69 J5
Bighouse KW14 104 D2
Bighton SO24 11 H1
Biglands CA7 60 D1
Bignor RH20 12 C5
Bigrigg CA24 60 B5
Bigton ZE2 107 M10
Bilberry PL26 4 A5
Bilborough NG8 41 H1
Bilbrook Som. TA24 7 J1
Bilbrook Staffs. WV8 40 A5
Bilbrough YO23 58 B5
Bilbster KW1 105 H3
Bilby DN22 51 J4
Bildershaw DL14 62 C4
Bildeston IP7 34 D4
Billericay CM12 24 C2
Billesdon LE7 42 A5
Billesley B49 30 C3
Billholm DG13 69 H4
Billingborough NG34 42 E2
Billinge WN5 48 E3
Billingford Norf. IP21 35 F1
Billingford Norf. NR20 44 E3
Billingham TS23 63 F4
Billinghay LN4 52 E7
Billingley S72 51 G2
Billingshurst RH14 12 D4
Billingsley WV16 39 G7
Billington Cen.Beds. LU7 32 C6
Billington Lancs. BB7 56 C6
Billington Staffs. ST18 40 A3
Billister ZE2 107 N6
Billockby NR29 45 J4
Billy Row DL15 62 C3
Bilsborrow PR3 55 J6
Bilsby LN13 53 H5
Bilsby Field LN13 53 H5
Bilsdean TD13 77 F3
Bilsham BN18 12 C6
Bilsington TN25 15 F4
Bilson Green GL14 29 F7
Bilsthorpe NG22 51 J6
Bilsthorpe Moor NG22 51 J7
Bilston Midloth. EH25 76 A4
Bilston W.Mid. WV14 40 B6
Bilstone CV13 41 F5
Bilting TN25 15 F3
Bilton E.Riding HU11 59 H6
Bilton N.Yorks. HG1 57 J4
Bilton Northumb. NE66 71 H2
Bilton Warks. CV22 31 F1
Bilton-in-Ainsty YO26 57 K5
Bimbister KW17 106 C6
Binbrook LN8 53 F3
Bincombe DT3 9 F6
Bindal IV20 97 G3
Bindon TA21 7 K3
Binegar BA3 19 K7
Bines Green RH13 12 E5
Binfield RG42 22 B4
Binfield Heath RG9 22 A4
Bingfield NE19 70 E6
Bingham NG13 42 A2
Bingham's Melcombe DT2 9 G4
Bingley BD16 57 G6
Bings Heath SY4 38 E4
Binham NR21 44 D2
Binley Hants. SP11 21 H6
Binley W.Mid. CV3 30 E1
Binniehill FK1 75 G3
Binsoe HG4 57 H2
Binstead PO33 11 G5
Binsted Hants. GU34 22 A7
Binsted W.Suss. BN18 12 C6
Binton CV37 30 C3
Bintree NR20 44 E3
Binweston SY5 38 C5
Birch Essex CO2 34 D6
Birch Gt.Man. M24 49 H2
Birch Cross ST14 40 D2
Birch Green Essex CO2 34 D7
Birch Green Herts. SG14 33 F7
Birch Grove RH17 13 H4
Birch Heath CW6 48 E6
Birch Vale SK22 50 C4
Birch Wood TA20 8 B3
Bircham Newton PE31 44 B2
Bircham Tofts PE31 44 B2
Birchanger CM23 33 J6
Bircher HR6 28 D2
Bircher Common HR6 28 D2
Birchfield IV24 96 B2
Birchgrove Cardiff CF14 18 E3
Birchgrove Swan. SA7 18 A2
Birchington CT7 25 K5
Birchmoor B78 40 E5
Birchover DE4 50 E6
Birchwood Lincs. LN6 52 C6
Birchwood Warr. WA3 49 F3
Bircotes DN11 51 J3
Bird Street IP7 34 E3
Birdbrook CO9 34 B4
Birdbush SP7 9 J2
Birdfield PA32 73 H1
Birdforth YO7 57 K2
Birdham PO20 12 B6
Birdingbury CV23 31 F2
Birdlip GL4 29 J7
Birdoswald CA8 70 B7
Birds Green CM5 23 J1
Birdsall YO17 58 E3
Birdsgreen WV15 39 G7
Birdsmoor Gate DT6 8 C4
Birdston G66 74 E3
Birdwell S70 51 F2
Birdwood GL19 29 G7
Birgham TD12 77 F7
Birichen IV25 96 E2
Birkby Cumb. CA15 60 B3
Birkby N.Yorks. DL7 62 E6
Birkdale Mersey. PR8 48 C1
Birkdale N.Yorks. DL11 61 K6
Birkenhead CH41 48 C4
Birkenhills AB53 99 F6
Birkenshaw BD11 57 H7
Birkhall AB35 90 B5
Birkhill Angus DD2 82 E4
Birkhill Sc.Bord. TD4 76 D6
Birkhill Sc.Bord. TD7 69 H2
Birkholme NG33 42 C3
Birkin WF11 58 B7
Birks LS27 57 H7
Birkwood ML11 75 G7
Birley HR4 28 D3
Birley Carr S6 51 F3
Birling Kent ME19 24 C5
Birling Northumb. NE65 71 H3
Birling Gap BN20 13 J7
Birlingham WR10 29 J4
BIRMINGHAM B 40 C7
Birmingham International Airport B26 40 D7
Birnam PH8 82 B3
Birsay KW17 106 B5
Birse AB34 90 D5
Birsemore AB34 90 D5
Birstall LE4 41 H5
Birstall Smithies WF17 57 H7
Birstwith HG3 57 H4
Birthorpe NG34 42 E2
Birtle OL11 49 H1
Birtley Here. SY7 28 C2
Birtley Northumb. NE48 70 D6
Birtley T. & W. DH3 62 D1
Birts Street WR13 29 G5
Birtsmorton WR13 29 H5
Bisbrooke LE15 42 B6
Biscathorpe LN11 53 F4
Bish Mill EX36 7 F3
Bisham SL7 22 B3
Bishampton WR10 29 J3
Bishop Auckland DL14 62 D4
Bishop Burton HU17 59 F5
Bishop Middleham DL17 62 E3
Bishop Monkton HG3 57 J3
Bishop Norton LN8 52 C3
Bishop Sutton BS39 19 J6
Bishop Thornton HG3 57 H3
Bishop Wilton YO42 58 D4
Bishopbridge LN8 52 D3
Bishopbriggs G64 74 E3
Bishopmill IV30 97 K5
Bishops Cannings SN10 20 D5
Bishop's Castle SY9 38 C7
Bishop's Caundle DT9 9 F3
Bishop's Cleeve GL52 29 J6
Bishop's Frome WR6 29 F4
Bishops Gate TW20 22 C4
Bishop's Green Essex CM6 33 K7
Bishop's Green Hants. RG19 21 J5
Bishop's Hull TA1 8 B2
Bishop's Itchington CV47 30 E3
Bishops Lydeard TA4 7 K3
Bishop's Norton GL2 29 H6
Bishops Nympton EX36 7 F3
Bishop's Offley ST21 39 G3
Bishop's Stortford CM23 33 H6
Bishop's Sutton SO24 11 H1
Bishop's Tachbrook CV33 30 E2
Bishop's Tawton EX32 6 D2
Bishop's Waltham SO32 11 G3
Bishop's Wood ST19 40 A5
Bishopsbourne CT4 15 G2
Bishopsteignton TQ14 5 K3
Bishopstoke SO50 11 F3
Bishopston Bristol BS6 19 J4
Bishopston Swan. SA3 17 J7
Bishopstone Bucks. HP17 32 B7
Bishopstone E.Suss. BN25 13 H6
Bishopstone Here. HR4 28 D4
Bishopstone Swin. SN6 21 F3
Bishopstone Wilts. SP5 10 B2
Bishopstrow BA12 20 B7
Bishopswood TA20 8 B3
Bishopsworth BS13 19 J5
Bishopthorpe YO23 58 B5
Bishopton Darl. TS21 62 E4
Bishopton N.Yorks. HG4 57 H2
Bishopton Renf. PA7 74 C3
Bishopton Warks. CV37 30 C3
Bishton NP18 19 G3
Bisley Glos. GL6 20 C1
Bisley Surr. GU24 22 C6
Bispham FY2 55 G6
Bispham Green L40 48 D1
Bissoe TR4 2 E4
Bisterne BH24 10 C4
Bisterne Close BH24 10 D4
Bitchet Green TN15 23 J6
Bitchfield NG33 42 C3
Bittadon EX31 6 D1
Bittaford PL21 5 G5
Bittering NR19 44 D4
Bitterley SY8 28 E1
Bitterne SO18 11 F3
Bitteswell LE17 41 H7
Bitton BS30 19 K5
Biwmares (Beaumaris) LL58 46 E5
Bix RG9 22 A3
Bixter ZE2 107 M7
Blaby LE8 41 H6
Black Bourton OX18 21 F1
Black Bridge SA73 16 C5
Black Callerton NE5 71 G7
Black Carr NR17 44 E6
Black Clauchrie KA26 67 G5
Black Corries Lodge PH49 80 D2
Black Crofts PA37 80 A4
Black Cross TR8 3 G2
Black Dog EX17 7 G5
Black Heddon NE20 71 F6
Black Hill CV37 30 D3
Black Marsh SY5 38 C6
Black Moor LS17 57 H5
Black Mount PA36 80 D3
Black Notley CM77 34 B6
Black Pill SA3 17 K6
Black Street NR33 45 K7
Black Torrington EX21 6 C5
Blackaburn NE48 70 C6
Blackacre DG11 69 F4
Blackadder TD11 77 G5
Blackawton TQ9 5 J5
Blackborough Devon EX15 7 J5
Blackborough Norf. PE32 44 A4
Blackborough End PE32 44 A4
Blackboys TN22 13 J4
Blackbraes Aber. AB21 91 G3
Blackbraes Falk. FK1 75 H3
Blackbrook Derbys. DE56 41 F1
Blackbrook Leics. LE12 41 G4
Blackbrook Mersey. WA11 48 E3
Blackbrook Staffs. ST5 39 G2
Blackburn Aber. AB21 91 G3
BLACKBURN B'burn. BB 56 B7
Blackburn W.Loth. EH47 75 H4
Blackbushe GU17 22 A6
Blackcastle IV2 97 F6
Blackchambers AB32 91 F3
Blackcraig D. & G. DG8 64 E4
Blackcraig D. & G. DG7 68 C5
Blackden Heath CW4 49 G5
Blackdog AB23 91 H3
Blackdown Devon PL19 5 F3
Blackdown Dorset DT8 8 C4
Blackdown Warks. CV32 30 E2
Blacker Hill S74 51 F2
Blackfen DA15 23 H4
Blackfield SO45 11 F4
Blackford Aber. AB51 91 F1
Blackford Cumb. CA6 69 J7
Blackford P. & K. PH4 81 K7
Blackford Som. BA22 9 F2
Blackford Som. BS28 19 H7
Blackford Bridge BL9 49 H2
Blackfordby DE11 41 F4
Blackgang PO38 11 F7
Blackhall Edin. EH4 76 A3
Blackhall Renf. PA1 74 C4
Blackhall Colliery TS27 63 F3
Blackhall Mill NE17 62 C1
Blackhall Rocks TS27 63 F3
Blackham TN3 13 J3
Blackhaugh TD1 76 C7
Blackheath Essex CO2 34 E6
Blackheath Gt.Lon. SE3 23 G4
Blackheath Suff. IP19 35 J1
Blackheath Surr. GU4 22 D7
Blackheath W.Mid. B65 40 B7
Blackhill Aber. AB42 99 J5
Blackhill Aber. AB42 99 J6
Blackhillock AB55 98 C6
Blackhills IV30 97 K6
Blackland SN11 20 D5
Blacklands TA24 7 G2
Blackleach PR4 55 H6
Blackley M9 49 H2
Blacklunans PH10 82 C1
Blackmill CF35 18 C3
Blackmoor Hants. GU33 11 J1
Blackmoor Som. TA21 7 K4
Blackmoor Gate EX31 6 E1
Blackmoorfoot HD7 50 C1
Blackmore CM4 24 C1
Blackmore End Essex CM7 34 B5
Blackmore End Herts. AL4 32 E7
Blackness Aber. AB31 90 E5
Blackness Falk. EH49 75 J3
Blackness High. KW3 105 H5
Blacknest GU34 22 A7
Blackney DT6 8 D5
Blacko BB9 56 D5
Blackpole WR3 29 H3
BLACKPOOL B'pool FY 55 G6
Blackpool Devon TQ6 5 J6
Blackpool Bridge SA67 16 D4
Blackpool Gate CA6 70 A6
Blackridge EH48 75 G4
Blackrock Arg. & B. PA44 72 B4
Blackrock Mon. NP7 28 B7
Blackrod BL6 49 F1
Blackshaw DG1 69 F7
Blackshaw Head HX7 56 E7
Blacksmith's Green IP14 35 F2
Blacksnape BB3 56 C7
Blackstone BN5 13 F5
Blackthorn OX25 31 H7
Blackthorpe IP30 34 D2
Blacktoft DN14 58 E7
Blacktop AB15 91 G4
Blacktown CF3 19 F3
Blackwater Cornw. TR4 2 E4
Blackwater Hants. GU17 22 B6
Blackwater I.o.W. PO30 11 G6
Blackwater Norf. NR9 44 E3
Blackwater Som. TA20 8 B3
Blackwaterfoot KA27 66 D1
Blackwell Darl. DL3 62 D5
Blackwell Derbys. SK17 50 D5
Blackwell Derbys. DE55 51 G7
Blackwell W.Suss. RH19 13 G3
Blackwell Warks. CV36 30 D4
Blackwell Worcs. B60 29 J1
Blackwells End GL19 29 G6
Blackwood (Coed-duon) Caerp. NP12 18 E2
Blackwood D. & G. DG2 68 E5
Blackwood S.Lan. ML11 75 F6
Blackwood Hill ST9 49 J7
Blacon CH1 48 C6
Bladbean CT4 15 G3
Bladnoch DG8 64 E5
Bladon OX20 31 F7
Blaen Clydach CF40 18 C2
Blaenannerch SA43 17 F1
Blaenau Dolwyddelan LL25 46 E7
Blaenau Ffestiniog LL41 37 F1
Blaenavon NP4 19 F1
Blaenawey NP7 28 B7
Blaencelyn SA44 26 C3
Blaencwm CF42 18 C1
Blaendyryn LD3 27 J5
Blaenffos SA37 16 E2
Blaengarw CF32 18 C2
Blaengeuffordd SY23 37 F7
Blaengweche SA18 17 K4
Blaengwrach SA11 18 B1
Blaengwynfi SA13 18 B2
Blaenllechau CF43 18 D2
Blaenos SA20 27 G5
Blaenpennal SY23 27 F2
Blaenplwyf SY23 26 E1
Blaenporth SA43 17 F1
Blaenrhondda CF42 18 C1
Blaenwaun SA34 17 F3
Blaen-y-coed SA33 17 G3
Blagdon N.Som. BS40 19 H6
Blagdon Torbay TQ3 5 J4
Blagdon Hill TA3 8 B3
Blaguegate WN8 48 D2
Blaich PH33 87 G7
Blaina NP13 18 E1
Blair KA24 74 B6
Blair Atholl PH18 81 K1
Blair Drummond FK9 75 F1
Blairannaich G83 80 E7
Blairbuie PA23 73 K3
Blairgowrie PH10 82 C3
Blairhall KY12 75 J2
Blairhoyle FK8 81 H7
Blairhullichan FK8 81 F7
Blairingone FK14 75 H1
Blairkip KA5 74 D7
Blairlogie FK9 75 G1
Blairmore Arg. & B. PA23 73 K2
Blairmore High. IV28 96 E1
Blairmore High. IV27 102 D3
Blairnairn G84 74 A2
Blairnamarrow AB37 89 K3
Blairpark KA24 74 A5
Blairquhan KA19 67 H3
Blairquhosh G63 74 D2
Blair's Ferry PA21 73 H4
Blairshinnoch AB45 98 E4
Blairuskinmore FK8 81 F7
Blairvadach G84 74 A2
Blairydryne AB31 91 F5
Blairythan Cottage AB41 91 H2
Blaisdon GL17 29 G7
Blake End CM77 34 B6
Blakebrook DY11 29 H1
Blakedown DY10 29 H1
Blakelaw Sc.Bord. TD5 77 F7
Blakelaw T. & W. NE5 71 H7
Blakeley WV5 40 A6
Blakelow CW5 49 F7
Blakemere HR2 28 C4
Blakeney Glos. GL15 19 K1
Blakeney Norf. NR25 44 E1
Blakenhall Ches.E. CW5 39 G1
Blakenhall W.Mid. WV2 40 B6
Blakeshall DY11 40 A7
Blakesley NN12 31 H3
Blanchland DH8 62 A1
Bland Hill HG3 57 H4
Blandford Camp DT11 9 J4
Blandford Forum DT11 9 H4
Blandford St. Mary DT11 9 H4
Blanefield G63 74 D3
Blanerne TD11 77 G5
Blankney LN4 52 D6
Blantyre G72 74 E5
Blar a' Chaorainn PH33 80 C1
Blargie PH20 88 D5
Blarglas G83 74 B2
Blarmachfoldach PH33 80 B1
Blarnalearoch IV23 95 H2
Blashford BH24 10 C4
Blaston LE16 42 B6
Blathaisbhal HS6 92 D4
Blatherwycke PE8 42 C6
Blawith LA12 55 F1
Blaxhall IP12 35 H3
Blaxton DN9 51 J2
Blaydon NE21 71 G7
Bleadney BA5 19 H7
Bleadon BS24 19 G6
Bleak Hey Nook OL3 50 C2
Blean CT2 25 H5
Bleasby Lincs. LN8 52 E4
Bleasby Notts. NG14 42 A1
Bleasby Moor LN8 52 E4
Bleatarn CA16 61 J5
Bleathwood Common SY8 28 E2
Blebocraigs KY15 83 F6
Bleddfa LD7 28 B2
Bledington OX7 30 D6
Bledlow HP27 22 A1
Bledlow Ridge HP14 22 A2
Blencarn CA10 61 H3
Blencogo CA7 60 C2
Blencow CA11 61 F3
Blendworth PO8 11 J3
Blennerhasset CA7 60 C2
Blervie Castle IV36 97 H6
Bletchingdon OX5 31 G7
Bletchingley RH1 23 G6
Bletchley M.K. MK3 32 B5
Bletchley Shrop. TF9 39 F2
Bletherston SA63 16 D3
Bletsoe MK44 32 D3
Blewbury OX11 21 J3
Blickling NR11 45 F3
Blidworth NG21 51 H7
Blidworth Bottoms NG21 51 H7
Blindburn Aber. AB41 91 H1
Blindburn Northumb. NE65 70 D2
Blindcrake CA13 60 C3
Blindley Heath RH7 23 G7
Blisland PL30 4 A3

Bli - Bra

Bliss Gate **DY14** **29** G1
Blissford **SP6** **10** C3
Blisworth **NN7** **31** J3
Blithbury **WS15** **40** C3
Blitterlees **CA7** **60** C1
Blo' Norton **IP22** **34** E1
Blockley **GL56** **30** C5
Blofield **NR13** **45** H5
Blofield Heath **NR13** **45** H4
Blore **DE6** **40** D1
Blossomfield **B91** **30** C1
Blount's Green **ST14** **40** C2
Blowick **PR9** **48** C1
Bloxham **OX15** **31** F5
Bloxholm **LN4** **52** D7
Bloxwich **WS3** **40** B5
Bloxworth **BH20** **9** H5
Blubberhouses **LS21** **57** G4
Blue Anchor *Cornw.* **TR9** **3** G3
Blue Anchor *Som.* **TA24** **7** J1
Blue Bell Hill **ME5** **24** D5
Bluewater **DA9** **23** J4
Blundellsands **L23** **48** C3
Blundeston **NR32** **45** K6
Blunham **MK44** **32** E3
Blunsdon St. Andrew **SN26** **20** E3
Bluntington **DY10** **29** H1
Bluntisham **PE28** **33** G1
Blunts **PL12** **4** D4
Blurton **ST3** **40** A1
Blyborough **DN21** **52** C3
Blyford **IP19** **35** J1
Blymhill **TF11** **40** A4
Blymhill Common **TF11** **39** G4
Blymhill Lawn **TF11** **40** A4
Blyth *Northumb.* NE24 **71** **J5**
Blyth *Notts.* **S81** **51** J4
Blyth Bridge **EH46** **75** K6
Blyth End **B46** **40** E6
Blythburgh **IP19** **35** J1
Blythe Bridge **ST11** **40** B1
Blythe Marsh **ST11** **40** B1
Blyton **DN21** **52** B3
Boarhills **KY16** **83** G6
Boarhunt **PO17** **11** H4
Boars Hill **OX1** **21** H1
Boarsgreave **BB4** **56** D7
Boarshead **TN6** **13** J3
Boarstall **HP18** **31** H7
Boarzell **TN19** **14** C5
Boasley Cross **EX20** **6** D6
Boat o' Brig **IV32** **98** B5
Boat of Garten PH24 **89** **G3**
Boath **IV17** **96** C4
Bobbing **ME9** **24** E5
Bobbington **DY7** **40** A6
Bobbingworth **CM5** **23** J1
Bocaddon **PL13** **4** B5
Bochastle **FK17** **81** H7
Bockhampton **RG17** **21** G4
Bocking **CM7** **34** B6
Bocking Churchstreet **CM7** **34** B6
Bockleton **WR15** **28** E2
Boconnoc **PL22** **4** B4
Boddam *Aber.* **AB42** **99** K6
Boddam *Shet.* **ZE2** **107** M11
Bodden **BA4** **19** K7
Boddington **GL51** **29** H6
Bodedern **LL65** **46** B4
Bodelwyddan **LL18** **47** J5
Bodenham *Here.* **HR1** **28** E3
Bodenham *Wilts.* **SP5** **10** C2
Bodenham Moor **HR1** **28** E3
Bodesbeck **DG10** **69** G3
Bodewryd **LL66** **46** B3
Bodfari **LL16** **47** J5
Bodffordd **LL77** **46** C5
Bodfuan **LL53** **36** C2
Bodham **NR25** **45** F1
Bodiam **TN32** **14** C5
Bodicote **OX15** **31** F5
Bodieve **PL27** **3** G1
Bodinnick **PL23** **4** B5
Bodior **LL65** **46** A5
Bodle Street Green **BN27** **13** K5
Bodmin PL31 **4** **A4**
Bodney **IP26** **44** C6
Bodorgan LL62 **46** **B6**
Bodrane **PL14** **4** C4
Bodsham Green **TN25** **15** G3
Bodwen **PL26** **4** A4
Bodymoor Heath **B76** **40** D6
Bogallan **IV1** **96** D6
Bogbain **IV2** **96** E7
Bogbrae **AB42** **91** J1
Bogbuie **IV7** **96** C6
Bogend **KA1** **74** B7
Bogfern **AB33** **90** D4
Bogfields **AB33** **90** D4
Bogfold **AB43** **99** G5
Boghead *Aber.* **AB45** **98** E5
Boghead *E.Ayr.* **KA18** **68** B1
Boghead *S.Lan.* **ML11** **75** F6
Boghole Farm **IV12** **97** G6
Bogmoor **IV32** **98** B4
Bogniebrae **AB54** **98** D6
Bognor Regis PO21 **12** **C7**
Bograxie **AB51** **91** F3
Bogroy **PH23** **89** G2
Bogside **FK10** **75** H1
Bogston **AB36** **90** B4
Bogton **AB53** **98** E5
Bogue **DG7** **68** B5
Bohemia **SP5** **10** D3
Bohenie **PH31** **87** J6
Bohetherick **PL12** **4** E4
Bohortha **TR2** **3** F5
Bohuntine **PH31** **87** J6
Boirseam **HS3** **93** F3
Bojewyan **TR19** **2** A5
Bokiddick **PL30** **4** A4
Bolam *Dur.* **DL2** **62** C4
Bolam *Northumb.* **NE61** **71** F5
Bolberry **TQ7** **5** G7
Bold Heath **WA8** **48** E4
Bolderwood **SO43** **10** D4
Boldon **NE36** **71** J7
Boldon Colliery NE35 **71** **J7**
Boldre **SO41** **10** E5
Boldron **DL12** **62** B5
Bole **DN22** **51** K4
Bolehill **DE4** **50** E7
Boleigh **TR19** **2** B6
Bolenowe **TR14** **2** D5
Boleside **TD1** **76** C7
Bolfracks **PH15** **81** K3
Bolgoed **SA4** **17** K5
Bolham *Devon* **EX16** **7** H4
Bolham *Notts.* **DN22** **51** K4
Bolham Water **EX15** **7** K4
Bolingey **TR6** **2** E3
Bollington **SK10** **49** J5
Bolney **RH17** **13** F4
Bolnhurst **MK44** **32** D3
Bolshan **DD11** **83** H2
Bolsover **S44** **51** G5
Bolsterstone **S36** **50** E3
Bolstone **HR2** **28** E5
Boltby **YO7** **57** K1
Bolter End **HP14** **22** A2
Bolton *Cumb.* **CA16** **61** H4
Bolton *E.Loth.* **EH41** **76** D3
Bolton *E.Riding* **YO42** **58** D4
BOLTON *Gt.Man.* BL **49** **G2**
Bolton *Northumb.* **NE66** **71** G2
Bolton Abbey **BD23** **57** F4
Bolton Bridge **BD23** **57** F4
Bolton by Bowland **BB7** **56** C5
Bolton Houses **PR4** **55** H6
Bolton Low Houses **CA7** **60** D2
Bolton Percy **YO23** **58** B5
Bolton upon Dearne **S63** **51** G2
Bolton Wood Lane **CA7** **60** D2
Boltonfellend **CA6** **69** K7
Boltongate **CA7** **60** D2
Bolton-le-Sands **LA5** **55** H3
Bolton-on-Swale **DL10** **62** D7
Bolventor **PL15** **4** B3
Bombie **DG6** **65** H6
Bomere Heath **SY4** **38** D4
Bonar Bridge **IV24** **96** D2
Bonawe **PA37** **80** B4
Bonby **DN20** **52** D1
Boncath SA37 **17** **F2**
Bonchester Bridge **TD9** **70** A2
Bonchurch **PO38** **11** G7
Bondleigh **EX20** **6** E5
Bonds **PR3** **55** H5
Bonehill **B78** **40** D5
Bo'ness EH51 **75** **H2**
Bonhill **G83** **74** B3
Boningale **WV7** **40** A5
Bonjedward **TD8** **70** B1
Bonkle **ML2** **75** G5
Bonning Gate **LA8** **61** F7
Bonnington *Edin.* **EH27** **75** K4
Bonnington *Kent* **TN25** **15** F4
Bonnybank **KY8** **82** E7
Bonnybridge FK4 **75** **G2**
Bonnykelly **AB53** **99** G5
Bonnyrigg EH19 **76** **B4**
Bonnyton *Aber.* **AB52** **90** E1
Bonnyton *Angus* **DD11** **83** G4
Bonnyton *Angus* **DD10** **83** H2
Bonnyton *Angus* **DD3** **82** E4
Bonsall **DE4** **50** E7
Bont **NP7** **28** C7
Bont Dolgadfan **SY19** **37** H5
Bont Newydd **LL40** **37** G3
Bontddu **LL40** **37** F4
Bont-goch (Elerch) **SY24** **37** F7
Bonthorpe **LN13** **53** H5
Bont-newydd *Conwy* **LL17** **47** J5
Bontnewydd *Gwyn.* **LL55** **46** C6
Bontuchel **LL15** **47** J7
Bonvilston **CF5** **18** D4
Bon-y-maen **SA1** **17** K6
Boode **EX33** **6** D2
Boohay **TQ6** **5** K5
Booker **HP12** **22** B2
Booley **SY4** **38** E3
Boor **IV22** **94** E3
Boorley Green **SO32** **11** G3
Boosbeck **TS12** **63** H5
Boose's Green **CO6** **34** C5
Boot **CA19** **60** C6
Boot Street **IP6** **35** G4
Booth **HX2** **57** F7
Booth Bank **HD7** **50** C1
Booth Green **SK10** **49** J4
Booth Wood **HX6** **50** C1
Boothby Graffoe **LN5** **52** C7
Boothby Pagnell **NG33** **42** C2
Boothstown **M28** **49** G2
Boothville **NN3** **31** J2
Bootle *Cumb.* **LA19** **54** E1
Bootle *Mersey.* L20 **48** **C3**
Booton **NR10** **45** F3
Boots Green **WA16** **49** G5
Booze **DL11** **62** B6
Boquhan **G63** **74** D2
Boraston **WR15** **29** F1
Bordeaux **GY3** **3** J5
Borden *Kent* **ME9** **24** E5
Borden *W.Suss.* **GU30** **12** B4
Bordley **BD23** **56** E3
Bordon GU35 **11** **J1**
Boreham *Essex* **CM3** **24** D1
Boreham *Wilts.* **BA12** **20** B7
Boreham Street **BN27** **13** K5
Borehamwood WD6 **22** **E2**
Boreland *D. & G.* **DG11** **69** G4
Boreland *D. & G.* **DG8** **64** D4
Boreland *Stir.* **FK21** **81** G4
Boreley **WR9** **29** H2
Boreraig **IV55** **93** G6
Borgh *W.Isles* **HS6** **92** E3
Borgh *W.Isles* **HS9** **84** B4
Borgh (Borve) *W.Isles* **HS2** **101** G2
Borghastan **HS2** **100** D3
Borgie **KW14** **103** J3
Borgue *D. & G.* **DG6** **65** G6
Borgue *High.* **KW7** **105** G6
Borley **CO10** **34** C4
Borley Green *Essex* **CO10** **34** C4
Borley Green *Suff.* **IP30** **34** D2
Bornais **HS8** **84** C2
Borness **DG6** **65** G6
Bornisketaig **IV51** **93** J4
Borough Green **TN15** **23** K6
Boroughbridge **YO51** **57** J3
Borras Head **LL13** **48** C7
Borrowash **DE72** **41** G2
Borrowby *N.Yorks.* **TS13** **63** J5
Borrowby *N.Yorks.* **YO7** **57** K1
Borrowdale **CA12** **60** D5
Borrowfield **AB39** **91** G5
Borstal **ME1** **24** D5
Borth SY24 **37** **F7**
Borthwick **EH23** **76** B5
Borthwickbrae **TD9** **69** K2
Borthwickshiels **TD9** **69** K2
Borth-y-Gest **LL49** **36** E2
Borve *High.* **IV51** **93** K7
Borve (Borgh) *W.Isles* **HS2** **101** G2
Borwick **LA6** **55** J2
Borwick Rails **LA18** **54** E2
Bosavern **TR19** **2** A5
Bosbury **HR8** **29** F4
Boscarne **PL30** **4** A4
Boscastle PL35 **4** **A1**
Boscombe *Bourne.* **BH5** **10** C5
Boscombe *Wilts.* **SP4** **10** D1
Bosham **PO18** **12** B6
Bosham Hoe **PO18** **12** B6
Bosherston **SA71** **16** C6
Bosley **SK11** **49** J6
Bossall **YO60** **58** D3
Bossiney **PL34** **4** A2
Bossingham **CT4** **15** G3
Bossington *Hants.* **SO20** **10** E1
Bossington *Som.* **TA24** **7** G1
Bostadh **HS2** **100** D4
Bostock Green **CW10** **49** F6
Boston PE21 **43** **G1**
Boston Spa **LS23** **57** K5
Boswarthan **TR20** **2** B5
Boswinger **PL26** **3** G4
Botallack **TR19** **2** A5
Botany Bay **EN2** **23** G2
Botcheston **LE9** **41** G5
Botesdale **IP22** **34** E1
Bothal **NE61** **71** H5
Bothamsall **DN22** **51** J5
Bothel **CA7** **60** C3
Bothenhampton **DT6** **8** D5
Bothwell **G71** **75** F5
Botley *Bucks.* **HP5** **22** C1
Botley *Hants.* **SO30** **11** G3
Botley *Oxon.* **OX2** **21** H1
Botloe's Green **GL18** **29** G6
Botolph Claydon **MK18** **31** J6
Botolphs **BN44** **12** E6
Botolph's Bridge **CT21** **15** G4
Bottacks **IV14** **96** B5
Bottesford *Leics.* **NG13** **42** B2
Bottesford *N.Lincs.* **DN16** **52** B2
Bottisham **CB25** **33** J2
Bottlesford **SN9** **20** E6
Bottom Boat **WF3** **57** J7
Bottom of Hutton **PR4** **55** H7
Bottom o'th'Moor **BL6** **49** F1
Bottomcraig **DD6** **82** E5
Bottoms **OL14** **56** E7
Botton Head **LA2** **56** B3
Botusfleming **PL12** **4** E4
Botwnnog **LL53** **36** B2
Bough Beech **TN8** **23** H7
Boughrood **LD3** **28** A5
Boughspring **NP16** **19** J2
Boughton *Norf.* **PE33** **44** B5
Boughton *Northants.* **NN2** **31** J2
Boughton *Notts.* **NG22** **51** J6
Boughton Aluph **TN25** **15** F3
Boughton Green **ME17** **14** C2
Boughton Lees **TN25** **15** F3
Boughton Malherbe **ME17** **14** D3
Boughton Street **ME13** **15** F2
Boulby **TS13** **63** J5
Bouldnor **PO41** **10** E6
Bouldon **SY7** **38** E7
Boulge **IP13** **35** G3
Boulmer **NE66** **71** H2
Boulston **SA62** **16** C4
Boultenstone Hotel **AB36** **90** C3
Boultham **LN6** **52** C6
Boundary *Derbys.* **DE11** **41** F4
Boundary *Staffs.* **ST10** **40** B1
Bourn **CB23** **33** G3
Bourne PE10 **42** **D3**
Bourne End *Bucks.* SL8 **22** **B3**
Bourne End *Cen.Beds.* **MK43** **32** C4
Bourne End *Herts.* **HP1** **22** D1
Bournebridge **RM4** **23** J2
BOURNEMOUTH BH **10** **B5**
Bournemouth Airport **BH23** **10** C5
Bournheath **B61** **29** J1
Bournmoor **DH4** **62** E1
Bournville **B30** **40** C7
Bourton *Bucks.* **MK18** **31** J5
Bourton *Dorset* **SP8** **9** G1
Bourton *N.Som.* **BS22** **19** G5
Bourton *Oxon.* **SN6** **21** F3
Bourton *Shrop.* **TF13** **38** E6
Bourton *Wilts.* **SN10** **20** D5
Bourton on Dunsmore **CV23** **31** F1
Bourton-on-the-Hill **GL56** **30** C5
Bourton-on-the-Water **GL54** **30** C6
Bousd **PA78** **78** D1
Boustead Hill **CA5** **60** D1
Bouth **LA12** **55** G1
Bouthwaite **HG3** **57** G2
Bovain **FK21** **81** G4
Boveney **SL4** **22** C4
Boveridge **BH21** **10** B3
Boverton **CF61** **18** C5
Bovey Tracey **TQ13** **5** J3
Bovingdon **HP3** **22** D1
Bovinger **CM5** **23** J1
Bovington Camp **BH20** **9** H6
Bow *Cumb.* **CA5** **60** E1
Bow *Devon* **EX17** **7** F5
Bow *Devon* **TQ9** **5** J5
Bow *Ork.* **KW16** **106** C8
Bow Brickhill **MK17** **32** C5
Bow of Fife **KY15** **82** E6
Bow Street *Cere.* SY24 **37** **F7**
Bow Street *Norf.* **NR17** **44** E6
Bowbank **DL12** **62** A4
Bowburn **DH6** **62** E3
Bowcombe **PO30** **11** F6
Bowd **EX10** **7** K6
Bowden *Devon* **TQ6** **5** J6
Bowden *Sc.Bord.* **TD6** **76** D7
Bowden Hill **SN15** **20** C5
Bowdon **WA14** **49** G4
Bower **NE48** **70** C5
Bower Hinton **TA12** **8** D3
Bower House Tye **CO6** **34** D4
Bowerchalke **SP5** **10** B2
Bowerhill **SN12** **20** C5
Bowermadden **KW1** **105** H2
Bowers **ST21** **40** A2
Bowers Gifford **SS13** **24** D3
Bowershall **KY12** **75** J1
Bowertower **KW1** **105** H2
Bowes **DL12** **62** A5
Bowgreave **PR3** **55** H5
Bowhousebog **ML7** **75** G5
Bowithick **PL15** **4** B2
Bowker's Green **L39** **48** D2
Bowland Bridge **LA11** **55** H1
Bowley **HR1** **28** E3
Bowley Town **HR1** **28** E3
Bowlhead Green **GU8** **12** C3
Bowling *W.Dun.* **G60** **74** C3
Bowling *W.Yorks.* **BD4** **57** G6
Bowling Bank **LL13** **38** C1
Bowlish **BA4** **19** K7
Bowmanstead **LA21** **60** E7
Bowmore **PA43** **72** B4
Bowness-on-Solway **CA7** **69** H7
Bowness-on-Windermere **LA23** **60** F7
Bowscale **CA11** **60** E3
Bowsden **TD15** **77** H6
Bowside Lodge **KW14** **104** D2
Bowston **LA8** **61** F7
Bowthorpe **NR5** **45** F5
Bowtrees **FK2** **75** H2
Box *Glos.* **GL6** **20** B1
Box *Wilts.* **SN13** **20** B5
Box End **MK43** **32** D4
Boxbush *Glos.* **GL14** **29** G7
Boxbush *Glos.* **GL17** **29** F6
Boxford *Suff.* **CO10** **34** D4
Boxford *W.Berks.* **RG20** **21** H4
Boxgrove **PO18** **12** C6
Boxley **ME14** **14** C2
Boxmoor **HP1** **22** D1
Box's Shop **EX23** **6** A5
Boxted *Essex* **CO4** **34** D5
Boxted *Suff.* **IP29** **34** C3
Boxted Cross **CO4** **34** D5
Boxwell **GL8** **20** D2
Boxworth **CB23** **33** G2
Boxworth End **CB24** **33** G2
Boyden Gate **CT3** **25** J5
Boydston **KA1** **74** C7
Boylestone **DE6** **40** D2
Boyndie **AB45** **98** E4
Boynton **YO16** **59** H3
Boys Hill **DT9** **9** F4
Boysack **DD11** **83** H3
Boyton *Cornw.* **PL15** **6** B6
Boyton *Suff.* **IP12** **35** H4
Boyton *Wilts.* **BA12** **9** J1
Boyton Cross **CM1** **24** C1
Boyton End **CO10** **34** B4
Bozeat **NN29** **32** C3
Braaid **IM4** **54** C6
Braal Castle **KW12** **105** G2
Brabling Green **IP13** **35** G2
Brabourne **TN25** **15** F3
Brabourne Lees **TN25** **15** F3
Brabster **KW1** **105** J2
Bracadale **IV56** **85** J1
Braceborough **PE9** **42** D4
Bracebridge Heath **LN4** **52** C6
Braceby **NG34** **42** D2
Bracewell **BD23** **56** D5
Brachla **IV3** **88** C1
Bracken Hill **WF14** **57** G7
Brackenber **CA16** **61** J5
Brackenbottom **BD24** **56** D2
Brackenfield **DE55** **51** F7
Brackens **AB53** **99** F5
Bracklach **AB54** **90** B2
Bracklamore **AB43** **99** G5
Bracklesham **PO20** **12** B7
Brackletter **PH34** **87** H6
Brackley *Arg. & B.* **PA31** **73** G2
Brackley *High.* **IV2** **97** F6
Brackley *Northants.* NN13 **31** **G5**
Brackley Gate **DE7** **41** F1
Brackley Hatch **NN13** **31** H4
Bracknell RG12 **22** **B5**
Braco **FK15** **81** K7
Bracobrae **AB55** **98** D5
Bracon Ash **NR14** **45** F6
Bracora **PH40** **86** D5
Bracorina **PH40** **86** D5
Bradbourne **DE6** **50** E7
Bradbury **TS21** **62** E4
Bradda **IM9** **54** A6
Bradden **NN12** **31** H4
Braddock **PL22** **4** B4
Bradenham *Bucks.* **HP14** **22** B2
Bradenham *Norf.* **IP25** **44** D5
Bradenstoke **SN15** **20** D4
Bradfield *Devon* **EX15** **7** J5
Bradfield *Essex* **CO11** **35** F5
Bradfield *Norf.* **NR28** **45** G2
Bradfield *W.Berks.* **RG7** **21** K4
Bradfield Combust **IP30** **34** C3
Bradfield Green **CW1** **49** F7
Bradfield Heath **CO11** **35** F6
Bradfield St. Clare **IP30** **34** D3
Bradfield St. George **IP30** **34** D2
Bradford *Cornw.* **PL30** **4** B3
Bradford *Derbys.* **DE45** **50** E6
Bradford *Devon* **EX22** **6** C5
Bradford *Northumb.* **NE70** **77** K7
Bradford *Northumb.* **NE20** **71** F6
BRADFORD *W.Yorks.* BD **57** **G6**
Bradford Abbas **DT9** **8** E3
Bradford Leigh **BA15** **20** B5
Bradford Peverell **DT2** **9** F5
Bradford-on-Avon BA15 **20** **B5**
Bradford-on-Tone **TA4** **7** K3
Bradiford **EX31** **6** D2
Brading **PO36** **11** H6
Bradley *Ches.W. & C.* **WA6** **48** E5
Bradley *Derbys.* **DE6** **40** E1
Bradley *Hants.* **SO24** **21** K7
Bradley *N.E.Lincs.* **DN37** **53** F2
Bradley (Low Bradley) *N.Yorks.* **BD20** **57** F5
Bradley *Staffs.* **ST18** **40** A4
Bradley *W.Mid.* **WV14** **40** B6
Bradley Fold **BL2** **49** G2
Bradley Green *Warks.* **CV9** **40** E5
Bradley Green *Worcs.* **B96** **29** J2
Bradley in the Moors **ST10** **40** C1
Bradley Mills **HD5** **50** D1
Bradley Stoke **BS32** **19** K3
Bradmore *Notts.* **NG11** **41** H2
Bradmore *W.Mid.* **WV3** **40** A6
Bradney **TA7** **8** C1
Bradninch **EX5** **7** J5
Bradnop **ST13** **50** C7
Bradnor Green **HR5** **28** B3
Bradpole **DT6** **8** D5
Bradshaw *Gt.Man.* **BL2** **49** G1
Bradshaw *W.Yorks.* **HX2** **57** F6
Bradstone **PL19** **6** B7
Bradwall Green **CW11** **49** G6
Bradwell *Derbys.* **S33** **50** D4
Bradwell *Devon* **EX34** **6** C1
Bradwell *Essex* **CM77** **34** C6
Bradwell *M.K.* **MK13** **32** B5
Bradwell *Norf.* **NR31** **45** K5
Bradwell Grove **OX18** **21** F1
Bradwell Waterside **CM0** **25** F1
Bradwell-on-Sea **CM0** **25** G1
Bradworthy **EX22** **6** B4
Brae *D. & G.* **DG2** **65** J3
Brae *High.* **IV24** **96** B1
Brae *Shet.* **ZE2** **107** M6
Brae of Achnahaird **IV26** **102** C7
Braeantra **IV17** **96** C4
Braedownie **DD8** **89** K7
Braefoot **AB53** **99** F6
Braegrum **PH1** **82** B5
Braehead *D. & G.* **DG8** **64** E5
Braehead *Glas.* **G51** **74** D4
Braehead *Moray* **AB55** **98** B6
Braehead *Ork.* **KW17** **106** E7
Braehead *Ork.* **KW17** **106** D3
Braehead *S.Lan.* **ML11** **75** G7
Braehead *S.Lan.* **ML11** **75** H5
Braehead of Lunan **DD10** **83** H2
Braehoulland **ZE2** **107** L5
Braeleny **FK17** **81** H6
Braemar **AB35** **89** J5
Braemore *High.* **IV27** **96** C1
Braemore *High.* **KW6** **105** F5
Braemore *High.* **IV23** **95** H4
Braenaloin **AB35** **89** K5
Braes of Enzie **AB56** **98** B5
Braes of Foss **PH16** **81** J2
Braes of Ullapool **IV26** **95** H2
Braeswick **KW17** **106** F4
Braeval **FK8** **81** G7
Braewick **ZE2** **107** M7
Brafferton *Darl.* **DL1** **62** D4
Brafferton *N.Yorks.* **YO61** **57** K2
Brafield-on-the-Green **NN7** **32** B3
Bragar **HS2** **100** E3
Bragbury End **SG2** **33** F6
Bragenham **LU7** **32** C6
Bragleenbeg **PA34** **80** A5
Braichmelyn **LL57** **46** E6

Bra - Bro

Braides LA2 55 H4
Braidley DL8 57 F1
Braidwood ML8 75 G6
Braigo PA44 72 A4
Brailsford DE6 40 E1
Brain's Green GL15 19 K1
Braintree CM7 34 B6
Braiseworth IP23 35 F1
Braishfield SO51 10 E2
Braithwaite Cumb. CA12 60 D4
Braithwaite S.Yorks. DN7 51 J1
Braithwaite W.Yorks. BD22 57 F5
Braithwell S66 51 H3
Bramber BN44 12 E5
Brambletye RH18 13 H3
Brambridge SO50 11 F2
Bramcote Notts. NG9 41 H2
Bramcote Warks. CV11 41 G7
Bramdean SO24 11 H2
Bramerton NR14 45 G5
Bramfield Herts. SG14 33 F7
Bramfield Suff. IP19 35 H1
Bramford IP8 35 F4
Bramhall SK7 49 H4
Bramham LS23 57 K5
Bramhope LS16 57 H5
Bramley Hants. RG26 21 K6
Bramley S.Yorks. S66 51 G3
Bramley Surr. GU5 22 D7
Bramley Corner RG26 21 K6
Bramley Head HG3 57 G4
Bramley Vale S44 51 G6
Bramling CT3 15 H2
Brampford Speke EX5 7 H6
Brampton Cambs. PE28 33 F1
Brampton Cumb. CA8 70 A7
Brampton Cumb. CA16 61 H4
Brampton Derbys. S40 51 F5
Brampton Lincs. LN1 52 B5
Brampton Norf. NR10 45 G3
Brampton S.Yorks. S73 51 G2
Brampton Suff. NR34 45 J7
Brampton Abbotts HR9 29 F6
Brampton Ash LE16 42 A7
Brampton Bryan SY7 28 C1
Brampton en le Morthen S66 51 G4
Brampton Street NR34 45 J7
Bramshall ST14 40 C2
Bramshaw SO43 10 D3
Bramshill RG27 22 A5
Bramshott GU30 12 B3
Bramwell TA10 8 D2
Bran End CM6 33 K6
Branault PH36 79 G1
Brancaster PE31 44 B1
Brancaster Staithe PE31 44 B1
Brancepeth DH7 62 D3
Branchill IV36 97 H6
Brand Green GL19 29 G6
Brandelhow CA12 60 D4
Branderburgh IV31 97 K4
Brandesburton YO25 59 H5
Brandeston IP13 35 G2
Brandis Corner EX22 6 C5
Brandiston NR10 45 F3
Brandon Dur. DH7 62 D3
Brandon Lincs. NG32 42 C1
Brandon Northumb. NE66 71 F2
Brandon Suff. IP27 44 B7
Brandon Warks. CV8 31 F1
Brandon Bank PE38 44 A7
Brandon Creek PE38 44 A6
Brandon Parva NR9 44 E5
Brandsby YO61 58 B2
Brandy Wharf DN21 52 D3
Brane TR20 2 B6
Branksome BH12 10 B5
Branksome Park BH13 10 B5
Bransbury SO21 21 H7
Bransby LN1 52 B5
Branscombe EX12 7 K7
Bransford WR6 29 G3
Bransford Bridge WR6 29 H3
Bransgore BH23 10 C5
Bransholme HU7 59 H6
Branson's Cross B98 30 B1
Branston Leics. NG32 42 B3
Branston Lincs. LN4 52 D6
Branston Staffs. DE14 40 E3
Branston Booths LN4 52 D6
Brant Broughton LN5 52 C7
Brantham CO11 35 F5
Branthwaite Cumb. CA7 60 D3
Branthwaite Cumb. CA14 60 B4
Brantingham HU15 59 F7
Branton Northumb. NE66 71 F2
Branton S.Yorks. DN3 51 J2
Brantwood LA21 60 E7
Branxholm Bridgend TD9 69 K2
Branxholme TD9 69 K2
Branxton TD12 77 G7
Brassey Green CW6 48 E6
Brassington DE4 50 E7
Brasted TN16 23 H6
Brasted Chart TN16 23 H6
Brathens AB31 90 E5
Bratoft PE24 53 H6
Brattleby LN1 52 C4
Bratton Som. TA24 7 H1
Bratton Tel. & W. TF5 39 F4
Bratton Wilts. BA13 20 C6
Bratton Clovelly EX20 6 C6
Bratton Fleming EX31 6 E2
Bratton Seymour BA9 9 F2
Braughing SG11 33 G6
Brauncewell NG34 52 D7
Braunston Northants. NN11 31 G2
Braunston Rut. LE15 42 B5
Braunstone LE3 41 H5
Braunton EX33 6 C2
Brawby YO17 58 D2
Brawdy SA62 16 B3
Brawith TS9 63 G6
Brawl KW14 104 D2
Brawlbin KW12 105 F3
Bray SL6 22 C4
Bray Shop PL17 4 D3
Bray Wick SL6 22 B4
Braybrooke LE16 42 A7
Braydon Side SN15 20 D3
Brayford EX32 6 E2
Brayshaw BD23 56 C4
Braythorn LS21 57 H5
Brayton YO8 58 C6
Braywoodside SL6 22 B4
Brazacott PL15 4 C1
Brea TR15 2 D4
Breach Kent CT4 15 G3
Breach Kent ME9 24 E5
Breachwood Green SG4 32 E6
Breacleit HS2 100 D4
Breaden Heath SY13 38 D2
Breadsall DE21 41 F2
Breadstone GL13 20 A1
Breage TR13 2 D6
Breakon ZE2 107 P2
Bream GL15 19 K1
Breamore SP6 10 C3
Brean TA8 19 F6
Breanais HS2 100 B5
Brearton HG3 57 J3
Breascleit HS2 100 E4
Breaston DE72 41 G2
Brechfa SA32 17 J2
Brechin DD9 83 H1
Brecklate PA28 66 A2
Breckles NR17 44 D6
Brecon (Aberhonddu) LD3 27 K6
Breconside DG3 68 D3
Bredbury SK6 49 J3
Brede TN31 14 D6
Bredenbury HR7 29 F3
Bredfield IP13 35 G3
Bredgar ME9 24 E5
Bredhurst ME7 24 D5
Bredon GL20 29 J5
Bredon's Hardwick GL20 29 J5
Bredon's Norton GL20 29 J5
Bredwardine HR3 28 C4
Breedon on the Hill DE73 41 G3
Breibhig HS2 101 G4
Breich EH55 75 H4
Breightmet BL2 49 G2
Breighton YO8 58 D6
Breinton HR4 28 D5
Breinton Common HR4 28 D5
Bremhill SN11 20 C4
Bremhill Wick SN11 20 C4
Brenachoille PA32 80 B7
Brenchley TN12 23 K7
Brendon Devon EX35 7 F1
Brendon Devon EX22 6 B4
Brendon Devon EX22 6 B5
Brenkley NE13 71 H6
Brent Eleigh CO10 34 D4
Brent Knoll TA9 19 G6
Brent Pelham SG9 33 H5
Brentford TW8 22 E4
Brentingby LE14 42 A4
Brentwood CM14 23 J2
Brenzett TN29 15 F5
Brenzett Green TN29 15 F5
Breoch DG7 65 H5
Brereton WS15 40 C4
Brereton Green CW11 49 G6
Brereton Heath CW12 49 H6
Breretonhill WS15 40 C4
Bressay ZE2 107 P8
Bressingham IP22 44 E7
Bressingham Common IP22 44 E7
Bretby DE15 40 E3
Bretford CV23 31 F1
Bretforton WR11 30 B4
Bretherdale Head CA10 61 G6
Bretherton PR26 55 H7
Brettabister ZE2 107 N7
Brettenham Norf. IP24 44 D7
Brettenham Suff. IP7 34 D3
Bretton Derbys. S32 50 D5
Bretton Flints. CH4 48 C6
Brevig HS9 84 B5
Brewood ST19 40 A5
Briach IV36 97 H6
Briantspuddle DT2 9 H5
Brick End CM6 33 J6
Brickendon SG13 23 G1
Bricket Wood AL2 22 E1
Brickkiln Green CM7 34 B5
Bricklehampton WR10 29 J4
Bride IM7 54 D3
Bridekirk CA13 60 C3
Bridell SA43 16 E1
Bridestones CW12 49 J6
Bridestowe EX20 6 D7
Brideswell AB54 90 D1
Bridford EX6 7 G7
Bridge Cornw. TR16 2 D4
Bridge Kent CT4 15 G2
Bridge End Cumb. LA20 55 F1
Bridge End Devon TQ7 5 G6
Bridge End Essex CM7 33 K5
Bridge End Lincs. NG34 42 E2
Bridge End Shet. ZE2 107 M9
Bridge Hewick HG4 57 J2
Bridge o'Ess AB34 90 D5
Bridge of Alford AB33 90 D3
Bridge of Allan FK9 75 F1
Bridge of Avon AB37 89 J1
Bridge of Balgie PH15 81 G3
Bridge of Bogendreip AB31 90 E5
Bridge of Brewlands PH11 82 C1
Bridge of Brown AB37 89 J2
Bridge of Cally PH10 82 C2
Bridge of Canny AB31 90 E5
Bridge of Craigisla PH11 82 D2
Bridge of Dee Aber. AB35 89 J5
Bridge of Dee Aber. AB31 90 E5
Bridge of Dee D. & G. DG7 65 H4
Bridge of Don AB23 91 H4
Bridge of Dun DD10 83 H2
Bridge of Dye AB31 90 E6
Bridge of Earn PH2 82 C6
Bridge of Ericht PH17 81 G2
Bridge of Feugh AB31 90 E5
Bridge of Forss KW14 105 F2
Bridge of Gairn AB35 90 B5
Bridge of Gaur PH17 81 G2
Bridge of Muchalls AB39 91 G5
Bridge of Muick AB35 90 B5
Bridge of Orchy PA36 80 D4
Bridge of Tynet AB56 98 B4
Bridge of Walls ZE2 107 L7
Bridge of Weir PA11 74 B4
Bridge Reeve EX18 6 E4
Bridge Sollers HR4 28 D4
Bridge Street CO10 34 C4
Bridge Trafford CH2 48 D5
Bridgefoot Angus DD3 82 E4
Bridgefoot Cambs. SG8 33 H4
Bridgefoot Cumb. CA14 60 B4
Bridgehampton BA22 8 E2
Bridgehaugh AB55 90 B1
Bridgehill DH8 62 B1
Bridgemary PO13 11 G4
Bridgemere CW5 39 G1
Bridgend Aber. AB54 90 D1
Bridgend Aber. AB53 99 F6
Bridgend Angus DD9 83 G1
Bridgend Arg. & B. PA44 72 B4
Bridgend Arg. & B. PA31 73 G1
Bridgend (Pen-y-bont ar Ogwr) Bridgend CF31 18 C4
Bridgend Cornw. PL22 4 B5
Bridgend Cumb. CA11 60 F5
Bridgend Fife KY15 82 E6
Bridgend Moray AB54 90 B1
Bridgend P. & K. PH2 82 C5
Bridgend W.Loth. EH49 75 J3
Bridgend of Lintrathen DD8 82 D2
Bridgerule EX22 6 A5
Bridges SY5 38 C6
Bridgeton Aber. AB33 90 D3
Bridgeton Glas. G40 74 E4
Bridgetown Cornw. PL15 6 B7
Bridgetown Som. TA22 7 H2
Bridgeyate BS30 19 K4
Bridgham NR16 44 D7
Bridgnorth WV16 39 G6
Bridgtown WS11 40 B5
Bridgwater TA6 8 B1
Bridlington YO16 59 H3
Bridport DT6 8 D5
Bridstow HR9 28 E6
Brierfield BB9 56 D6
Brierley Glos. GL17 29 F7
Brierley Here. HR6 28 D3
Brierley S.Yorks. S72 51 G1
Brierley Hill DY5 40 B7
Brierton TS22 63 F4
Briestfield WF12 50 E1
Brig o'Turk FK17 81 G7
Brigg DN20 52 D2
Briggate NR28 45 H3
Briggswath YO21 63 K6
Brigham Cumb. CA13 60 B3
Brigham E.Riding YO25 59 G4
Brighouse HD6 57 G7
Brighstone PO30 11 F6
Brightgate DE4 50 E7
Brighthampton OX29 21 G1
Brightholmlee S35 50 E3
Brightling TN32 13 K4
Brightlingsea CO7 34 E7
BRIGHTON B. & H. BN 13 G6
Brighton Cornw. TR2 3 G3
Brightons FK2 75 H3
Brightwalton RG20 21 H4
Brightwalton Green RG20 21 H4
Brightwell IP10 35 G4
Brightwell Baldwin OX49 21 K2
Brightwell Upperton OX49 21 K2
Brightwell-cum-Sotwell OX10 21 J2
Brignall DL12 62 B5
Brigsley DN37 53 F2
Brigsteer LA8 55 H1
Brigstock NN14 42 C7
Brill Bucks. HP18 31 H7
Brill Cornw. TR11 2 E6
Brilley HR3 28 B4
Brilley Mountain HR3 28 B3
Brimaston SA62 16 C3
Brimfield SY8 28 E2
Brimington S43 51 G5
Brimington Common S43 51 G5
Brimley TQ13 5 J3
Brimpsfield GL4 29 J7
Brimpton RG7 21 J5
Brims KW16 106 B9
Brimscombe GL5 20 B1
Brimstage CH63 48 C4
Brinacory PH41 86 D5
Brindham BA6 19 J7
Brindister Shet. ZE2 107 N9
Brindister Shet. ZE2 107 L7
Brindle PR6 55 J7
Brindley Ford ST8 49 H7
Brineton TF11 40 A4
Bringhurst LE16 42 B6
Brington PE28 32 D1
Brinian KW17 106 D5
Briningham NR24 44 E2
Brinkhill LN11 53 G5
Brinkley Cambs. CB8 33 K3
Brinkley Notts. NG25 51 K7
Brinklow CV23 31 F1
Brinkworth SN15 20 D3
Brinmore IV2 88 D2
Brinscall PR6 56 B7
Brinsea BS49 19 H5
Brinsley NG16 41 G1
Brinsop HR4 28 D4
Brinsworth S60 51 G3
Brinton NR24 44 E2
Brisco CA4 60 F1
Brisley NR20 44 D3
Brislington BS4 19 K4
Brissenden Green TN26 14 E4
BRISTOL BS 19 J4
Bristol Filton Airport BS10 19 J3
Bristol International Airport BS48 19 J5
Briston NR24 44 E2
Britannia OL13 56 D7
Britford SP5 10 C2
Brithdir Caerp. NP24 18 E1
Brithdir Gwyn. LL40 37 G4
Brithem Bottom EX15 7 J4
Briton Ferry (Llansawel) SA11 18 A2
Britwell SL2 22 C3
Britwell Salome OX49 21 K2
Brixham TQ5 5 K5
Brixton Devon PL8 5 F5
Brixton Gt.Lon. SW2 23 G4
Brixton Deverill BA12 9 H1
Brixworth NN6 31 J1
Brize Norton OX18 21 F1
Broad Alley WR9 29 H2
Broad Blunsdon SN26 20 E2
Broad Campden GL55 30 C5
Broad Carr HX4 50 C1
Broad Chalke SP5 10 B2
Broad Ford TN12 14 C4
Broad Green Cambs. CB8 33 K3
Broad Green Cen.Beds. MK43 32 C4
Broad Green Essex CO6 34 C6
Broad Green Essex SG8 33 H5
Broad Green Mersey. L14 48 D3
Broad Green Suff. IP6 34 E3
Broad Green Worcs. WR6 29 G3
Broad Haven SA62 16 B4
Broad Hill CB7 33 J1
Broad Hinton SN4 20 E4
Broad Laying RG20 21 H5
Broad Marston CV37 30 C4
Broad Oak Carmar. SA32 17 J3
Broad Oak Cumb. CA18 60 C7
Broad Oak E.Suss. TN31 14 D6
Broad Oak E.Suss. TN21 13 K4
Broad Oak Here. HR2 28 D6
Broad Road IP21 35 G1
Broad Street E.Suss. TN36 14 D6
Broad Street Kent ME17 14 D2
Broad Street Wilts. SN9 20 E6
Broad Street Green CM9 24 E1
Broad Town SN4 20 D4
Broadbottom SK14 49 J3
Broadbridge PO18 12 B6
Broadbridge Heath RH12 12 E3
Broadclyst EX5 7 H6
Broadfield Lancs. BB5 56 C7
Broadfield Lancs. PR25 55 J7
Broadford IV49 86 C2
Broadford Bridge RH14 12 D4
Broadgate LA18 54 E1
Broadhaugh TD9 69 K3
Broadhaven KW1 105 J3
Broadheath Gt.Man. WA14 49 G4
Broadheath Worcs. WR15 29 F2
Broadhembury EX14 7 K5
Broadhempston TQ9 5 J4
Broadholme LN1 52 B5
Broadland Row TN31 14 D6
Broadlay SA17 17 G5
Broadley Lancs. OL12 49 H1
Broadley Moray AB56 98 B4
Broadley Common EN9 23 H1
Broadmayne DT2 9 G6
Broadmeadows TD7 76 C7
Broadmere RG25 21 K7
Broadmoor SA68 16 D5
Broadnymett EX17 7 F5
Broadoak Dorset DT6 8 D5
Broadoak Glos. GL14 29 F7
Broadoak Kent CT2 25 H5
Broadoak End SG14 33 G7
Broadrashes AB55 98 C5
Broad's Green CM3 33 K7
Broadsea AB43 99 H4
Broadstairs CT10 25 K5
Broadstone Poole BH18 10 B5
Broadstone Shrop. SY7 38 E7
Broadstreet Common NP18 19 G3
Broadwas WR6 29 G3
Broadwater Herts. SG2 33 F6
Broadwater W.Suss. BN14 12 E6
Broadwater Down TN2 13 J3
Broadwaters DY10 29 H1
Broadway Carmar. SA17 17 G5
Broadway Carmar. SA33 17 F4
Broadway Pembs. SA62 16 B4
Broadway Som. TA19 8 C3
Broadway Suff. IP19 35 H1
Broadway Worcs. WR12 30 C5
Broadwell Glos. GL56 30 D6
Broadwell Oxon. GL7 21 F1
Broadwell Warks. CV23 31 F2
Broadwell House NE47 62 A1
Broadwey DT3 9 F6
Broadwindsor DT8 8 D4
Broadwood Kelly EX19 6 E5
Broadwoodwidger PL16 6 C7
Brobury HR3 28 C4
Brocastle CF35 18 C4
Brochel IV40 94 B7
Brochloch DG7 67 K4
Brock PA77 78 B3
Brockamin WR6 29 G3
Brockbridge SO32 11 H3
Brockdish IP21 35 G1
Brockenhurst SO42 10 D4
Brockford Green IP14 35 F2
Brockford Street IP14 35 F2
Brockhall NN7 31 H2
Brockham RH3 22 E7
Brockhampton Glos. GL54 30 B6
Brockhampton Glos. GL51 29 J6
Brockhampton Here. HR1 28 E5
Brockhampton Here. WR6 29 F3
Brockhampton Green DT2 9 G4
Brockholes HD9 50 D1
Brockhurst Hants. PO12 11 G4
Brockhurst W.Suss. RH19 13 H3
Brocklebank CA7 60 E2
Brocklesby DN41 52 E1
Brockley N.Som. BS48 19 H5
Brockley Suff. IP29 34 C3
Brockley Green CO10 34 B4
Brock's Green RG20 21 H5
Brockton Shrop. TF13 38 E6
Brockton Shrop. TF11 39 G5
Brockton Shrop. SY5 38 C5
Brockton Shrop. SY7 38 C7
Brockton Tel. & W. TF10 39 G4
Brockweir NP16 19 J1
Brockwood Park SO24 11 H2
Brockworth GL3 29 H7
Brocton ST17 40 B4
Brodick KA27 73 J7
Brodsworth DN5 51 H2
Brogborough MK43 32 C5
Brogden BB18 56 D5
Brogyntyn SY10 38 B2
Broken Cross Ches.E. SK11 49 H5
Broken Cross Ches.W. & C. CW9 49 F5
Brokenborough SN16 20 C3
Brokes DL11 62 C7
Bromborough CH62 48 C4
Brome IP23 35 F1
Brome Street IP23 35 F1
Bromeswell IP12 35 H3
Bromfield Cumb. CA7 60 C2
Bromfield Shrop. SY8 28 D1
Bromham Bed. MK43 32 D3
Bromham Wilts. SN15 20 C5
BROMLEY Gt.Lon. BR 23 H5
Bromley S.Yorks. S35 51 F3
Bromley Cross BL7 49 G1
Bromley Green TN26 14 E4
Brompton Med. ME7 24 D5
Brompton N.Yorks. DL6 62 E7
Brompton N.Yorks. YO13 59 F1
Brompton Shrop. SY5 38 E5
Brompton on Swale DL10 62 D7
Brompton Ralph TA4 7 J2
Brompton Regis TA22 7 H2
Bromsash HR9 29 F6
Bromsberrow HR8 29 G5
Bromsberrow Heath HR8 29 G5
Bromsgrove B61 29 J1
Bromstead Heath TF10 40 A4
Bromyard HR7 29 F3
Bromyard Downs HR7 29 F3
Bronaber LL41 37 G2
Brondesbury NW6 23 F3
Brongest SA38 17 G1
Bronington SY13 38 D2
Bronllys LD3 28 A5
Bronnant SY23 27 F2
Bronwydd Arms SA33 17 H3
Bronydd HR3 28 B4
Bron-y-gaer SA33 17 G4
Bronygarth SY10 38 B2
Brook Carmar. SA33 17 F5
Brook Hants. SO43 10 D3
Brook Hants. SO20 10 E2
Brook I.o.W. PO30 10 E6
Brook Kent TN25 15 F3
Brook Surr. GU8 12 C3
Brook Surr. GU5 22 D7
Brook Bottom OL5 49 J2
Brook End Bed. MK44 32 D2
Brook End Herts. SG9 33 G6
Brook End M.K. MK16 32 C4
Brook End Worcs. WR5 29 H4
Brook Hill SO43 10 D3
Brook Street Essex CM14 23 J2
Brook Street Kent TN26 14 E4
Brook Street Suff. CO10 34 C4
Brook Street W.Suss. RH17 13 G4
Brooke Norf. NR15 45 G6
Brooke Rut. LE15 42 B5
Brookend Glos. GL15 19 J2
Brookend Glos. GL13 19 K1
Brookfield SK14 50 C3
Brookhampton OX44 21 K2

Bro - Bus

Brookhouse *Ches.E.* **SK10** 49 J5
Brookhouse *Denb.* **LL16** 47 J6
Brookhouse *Lancs.* **LA2** 55 J3
Brookhouse *S.Yorks.* **S25** 51 H4
Brookhouse Green **CW11** 49 H6
Brookhouses **ST10** 40 B1
Brookland **TN29** 14 E5
Brooklands *D. & G.* **DG2** 65 J3
Brooklands *Shrop.* **SY13** 38 E1
Brookmans Park **AL9** 23 F1
Brooks **SY21** 38 A6
Brooks Green **RH13** 12 E4
Brooksby **LE14** 41 J4
Brookthorpe **GL4** 29 H7
Brookwood **GU24** 22 C6
Broom *Cen.Beds.* **SG18** 32 E4
Broom *Fife* **KY8** 82 E7
Broom *Warks.* **B50** 30 B3
Broom Green **NR20** 44 D3
Broom Hill *Dorset* **BH21** 10 B4
Broom Hill *Worcs.* **DY9** 29 J1
Broom of Dalreach **PH3** 82 B6
Broomcroft **SY5** 38 E5
Broome *Norf.* **NR35** 45 H6
Broome *Shrop.* **SY7** 38 D7
Broome *Worcs.* **DY9** 29 J1
Broome Wood **NE66** 71 G2
Broomedge **WA13** 49 G4
Broomer's Corner **RH13** 12 E4
Broomfield *Aber.* **AB41** 91 H1
Broomfield *Essex* **CM1** 34 B7
Broomfield *Kent* **CT6** 25 H5
Broomfield *Kent* **ME17** 14 D2
Broomfield *Som.* **TA5** 8 B1
Broomfleet **HU15** 58 E7
Broomhall Green **CW5** 39 F1
Broomhaugh **NE44** 71 F7
Broomhead **AB43** 99 H4
Broomhill *Bristol* **BS16** 19 K4
Broomhill *Northumb.* **NE65** 71 H3
Broomielaw **DL12** 62 B5
Broomley **NE43** 71 F7
Broompark **DH7** 62 D2
Broom's Green **GL18** 29 G5
Brora KW9 97 G1
Broseley TF12 39 F5
Brotherlee **DL13** 62 A3
Brothertoft **PE20** 43 F1
Brotherton **WF11** 57 K7
Brotton **TS12** 63 H5
Broubster **KW14** 105 F2
Brough *Cumb.* **CA17** 61 J5
Brough *Derbys.* **S33** 50 D4
Brough *E.Riding* HU15 59 F7
Brough *High.* **KW14** 105 H1
Brough *Notts.* **NG23** 52 B7
Brough *Ork.* **KW17** 106 C6
Brough *Shet.* **ZE2** 107 P8
Brough *Shet.* **ZE2** 107 N5
Brough *Shet.* **ZE2** 107 P6
Brough *Shet.* **ZE2** 107 P5
Brough Lodge **ZE2** 107 P3
Brough Sowerby **CA17** 61 J5
Broughall **SY13** 38 E1
Brougham **CA10** 61 G4
Broughton *Bucks.* **HP20** 32 B7
Broughton *Cambs.* **PE28** 33 F1
Broughton *Flints.* **CH4** 48 C6
Broughton *Hants.* **SO20** 10 E1
Broughton *Lancs.* **PR3** 55 J6
Broughton *M.K.* **MK16** 32 B4
Broughton *N.Lincs.* **DN20** 52 C2
Broughton *N.Yorks.* **YO17** 58 D2
Broughton *N.Yorks.* **BD23** 56 E4
Broughton *Northants.* **NN14** 32 B1
Broughton *Ork.* **KW17** 106 D3
Broughton *Oxon.* **OX15** 31 F5
Broughton *Sc.Bord.* **ML12** 75 K7
Broughton *V. of Glam.* **CF71** 18 C4
Broughton Astley **LE9** 41 H6
Broughton Beck **LA12** 55 F1
Broughton Gifford **SN12** 20 B5
Broughton Green **WR9** 29 J2
Broughton Hackett **WR7** 29 J3
Broughton in Furness LA20 55 F1
Broughton Mills **LA20** 60 D7
Broughton Moor **CA15** 60 B3
Broughton Poggs **GL7** 21 F1
Broughtown **KW17** 106 F3
Broughty Ferry **DD5** 83 F4
Browland **ZE2** 107 L7
Brown Candover **SO24** 11 G1
Brown Edge *Lancs.* **PR8** 48 C1
Brown Edge *Staffs.* **ST6** 49 J7
Brown Heath **CH3** 48 D6
Brown Lees **ST8** 49 H7
Brown Street **IP14** 34 E2
Brownber **CA17** 61 J6
Browndown **PO13** 11 G5
Brownheath **SY12** 38 D3
Brownhill **AB41** 99 G6
Brownhills *Fife* **KY16** 83 G6
Brownhills *W.Mid.* **WS8** 40 C5
Brownieside **NE67** 71 G1
Brownlow **CW12** 49 H6
Brownlow Heath **CW12** 49 H6
Brown's Bank **CW3** 39 F1
Brownsea Island **BH15** 10 B6
Brownshill **GL6** 20 B1
Brownshill Green **CV5** 41 F7
Brownsover **CV21** 31 G1
Brownston **PL21** 5 G5
Browston Green **NR31** 45 J5
Broxa **YO13** 63 J3
Broxbourne EN10 23 G1
Broxburn *E.Loth.* **EH42** 76 E3
Broxburn *W.Loth.* EH52 75 J3
Broxholme **LN1** 52 C5
Broxted **CM6** 33 J6
Broxton **CH3** 48 D7
Broxwood **HR6** 28 C3
Broyle Side **BN8** 13 H5
Bru (Brue) **HS2** 101 F3
Bruachmary **IV12** 97 F7
Bruan **KW2** 105 J5
Brue (Bru) **HS2** 101 F3
Bruera **CH3** 48 D6
Bruern **OX7** 30 D6
Bruernish **HS9** 84 C4
Bruichladdich **PA49** 72 A4
Bruisyard **IP17** 35 H2
Bruisyard Street **IP17** 35 H2
Brumby **DN16** 52 C1
Brund **SK17** 50 D6
Brundall **NR13** 45 H5
Brundish *Norf.* **NR14** 45 H6
Brundish *Suff.* **IP13** 35 G2
Brundish Street **IP13** 35 G1
Brunstock **CA6** 60 F1
Brunswick Village **NE13** 71 H6
Bruntingthorpe **LE17** 41 J6
Bruntland **AB54** 90 C2
Brunton *Fife* **KY15** 82 E5
Brunton *Northumb.* **NE66** 71 H1
Brunton *Wilts.* **SN8** 21 F6
Brushfield **SK17** 50 D5
Brushford *Devon* **EX18** 6 E5
Brushford *Som.* **TA22** 7 H3
Bruton BA10 9 F1
Bryanston **DT11** 9 H4
Bryant's Bottom **HP16** 22 B2
Brydekirk **DG12** 69 G6
Brymbo **LL11** 48 B7
Brympton **BA22** 8 E3
Bryn *Caerp.* **NP12** 18 E2
Bryn *Carmar.* **SA14** 17 J5
Bryn *Ches.W. & C.* **CW8** 49 F5
Bryn *Gt.Man.* **WN4** 48 E2
Bryn *N.P.T.* **SA13** 18 B2
Bryn *Shrop.* **SY9** 38 B7
Bryn Bwbach **LL47** 37 F2
Bryn Gates **WN2** 48 E2
Bryn Pen-y-lan **LL14** 38 C1
Brynamman **SA18** 27 G7
Brynberian **SA41** 16 E2
Brynbuga (Usk) NP15 19 G1
Bryncae **CF72** 18 C3
Bryncethin **CF32** 18 C3
Bryncir **LL51** 36 D1
Bryncoch *Bridgend* **CF32** 18 C3
Bryn-côch *N.P.T.* **SA10** 18 A2
Bryncroes **LL53** 36 B2
Bryncrug **LL36** 37 F5
Bryneglwys **LL21** 38 A1
Brynford **CH8** 47 K5
Bryngwran **LL65** 46 B5
Bryngwyn *Mon.* **NP15** 19 G1
Bryngwyn *Powys* **HR5** 28 A4
Bryn-henllan **SA42** 16 D2
Brynhoffnant **SA44** 26 C3
Bryning **PR4** 55 H6
Brynithel **NP13** 19 F1
Brynmawr *B.Gwent* **NP23** 28 A7
Bryn-mawr *Gwyn.* **LL53** 36 B2
Brynmelyn **LD1** 28 A1
Brynmenyn **CF32** 18 C3
Brynna **CF72** 18 C3
Brynnau Gwynion **CF72** 18 C3
Brynog **SA48** 26 E3
Bryn-penarth **SY21** 38 A5
Brynrefail *Gwyn.* **LL55** 46 D6
Brynrefail *I.o.A.* **LL70** 46 C4
Brynsadler **CF72** 18 D3
Brynsaithmarchog **LL21** 47 J7
Brynsiencyn **LL61** 46 C6
Bryn-teg *I.o.A.* **LL78** 46 C4
Brynteg *Wrex.* LL11 48 C7
Bryn-y-cochin **SY12** 38 C2
Brynygwenin **NP7** 28 C7
Bryn-y-maen **LL28** 47 G5
Buaile nam Bodach **HS9** 84 C4
Bualadubh **HS8** 92 C7
Bualintur **IV47** 85 K2
Bualnaluib **IV22** 94 E2
Bubbenhall **CV8** 30 E1
Bubnell **DE45** 50 E5
Bubwith **YO8** 58 D6
Buccleuch **TD9** 69 J2
Buchan **DG7** 65 H4
Buchanan Castle **G63** 74 C2
Buchanhaven **AB42** 99 K6
Buchanty **PH1** 82 A5
Buchlyvie **FK8** 74 D1
Buckabank **CA5** 60 E2
Buckby Wharf **NN11** 31 H2
Buckden *Cambs.* **PE19** 32 E2
Buckden *N.Yorks.* **BD23** 56 E2
Buckenham **NR13** 45 H5
Buckerell **EX14** 7 K5
Buckfast **TQ11** 5 H4
Buckfastleigh TQ11 5 H4
Buckhaven **KY8** 76 B1
Buckholm **TD1** 76 C7
Buckholt **NP25** 28 E7
Buckhorn Weston **SP8** 9 G2
Buckhurst Hill IG9 23 H2
Buckie AB56 98 C4
Buckies **KW14** 105 G2
Buckingham MK18 31 H5
Buckland *Bucks.* **HP22** 32 B7
Buckland *Devon* **TQ7** 5 G6
Buckland *Glos.* **WR12** 30 B5
Buckland *Hants.* **SO41** 10 E5
Buckland *Here.* **HR6** 28 E3
Buckland *Herts.* **SG9** 33 G5
Buckland *Kent* **CT16** 15 J3
Buckland *Oxon.* **SN7** 21 G2
Buckland *Surr.* **RH3** 23 F6
Buckland Brewer **EX39** 6 C3
Buckland Common **HP23** 22 C1
Buckland Dinham **BA11** 20 A6
Buckland Filleigh **EX21** 6 C5
Buckland in the Moor **TQ13** 5 H3
Buckland Monachorum **PL20** 4 E4
Buckland Newton **DT2** 9 F4
Buckland Ripers **DT3** 9 F6
Buckland St. Mary **TA20** 8 B3
Buckland-tout-Saints **TQ7** 5 H6
Bucklebury **RG7** 21 J4
Bucklerheads **DD5** 83 F4
Bucklers Hard **SO42** 11 F5
Bucklesham **IP10** 35 G4
Buckley (Bwcle) CH7 48 B6
Buckley Green **B95** 30 C2
Bucklow Hill **WA16** 49 G4
Buckman Corner **RH14** 12 E4
Buckminster **NG33** 42 B3
Bucknall *Lincs.* **LN10** 52 E6
Bucknall *Stoke* **ST2** 40 B1
Bucknell *Oxon.* **OX27** 31 G6
Bucknell *Shrop.* SY7 28 C1
Buckridge **DY14** 29 G1
Buck's Cross **EX39** 6 B3
Bucks Green **RH12** 12 D3
Bucks Hill **WD4** 22 D1
Bucks Horn Oak **GU10** 22 B7
Buck's Mills **EX39** 6 B3
Bucksburn **AB21** 91 G4
Buckspool **SA71** 16 C6
Buckton *E.Riding* **YO15** 59 H2
Buckton *Here.* **SY7** 28 C1
Buckton *Northumb.* **NE70** 77 J7
Buckton Vale **SK15** 49 J2
Buckworth **PE28** 32 E1
Budbrooke **CV35** 30 D2
Budby **NG22** 51 J5
Buddon **DD7** 83 G4
Bude EX23 6 A5
Budge's Shop **PL12** 4 D5
Budlake **EX5** 7 H5
Budle **NE69** 77 K7
Budleigh Salterton EX9 7 J7
Budock Water **TR11** 2 E5
Budworth Heath **CW9** 49 F5
Buerton **CW3** 39 F1
Bugbrooke **NN7** 31 H3
Bugeildy (Beguildy) **LD7** 28 A1
Buglawton **CW12** 49 H6
Bugle **PL26** 4 A5
Bugthorpe **YO41** 58 D4
Building End **SG8** 33 H5
Buildwas **TF8** 39 F5
Builth Road **LD2** 27 K3
Builth Wells (Llanfair-ym-Muallt) LD2 27 K3
Bulby **PE10** 42 D3
Bulcote **NG14** 41 J1
Buldoo **KW14** 104 E2
Bulford **SP4** 20 E7
Bulford Camp **SP4** 20 E7
Bulkeley **SY14** 48 E7
Bulkington *Warks.* **CV12** 41 F7
Bulkington *Wilts.* **SN10** 20 C6
Bulkworthy **EX22** 6 B4
Bull Bay (Porth Llechog) **LL68** 46 C3
Bull Green **TN26** 14 E4
Bullbridge **DE56** 51 F7
Bullbrook **RG12** 22 B5
Bullen's Green **AL4** 23 F1
Bulley **GL2** 29 G7
Bullington **LN3** 52 D5
Bullpot Farm **LA6** 56 B1
Bulls Cross **EN2** 23 G2
Bull's Green *Herts.* **SG3** 33 F7
Bull's Green *Norf.* **NR34** 45 J6
Bullwood **PA23** 73 K3
Bulmer *Essex* **CO10** 34 C4
Bulmer *N.Yorks.* **YO60** 58 C3
Bulmer Tye **CO10** 34 C5
Bulphan **RM14** 24 C3
Bulstone **EX12** 7 K7
Bulverhythe **TN38** 14 C7
Bulwark **AB42** 99 H6
Bulwell **NG6** 41 H1
Bulwick **NN17** 42 C6
Bumble's Green **EN9** 23 H1
Bun Abhainn Eadarra **HS3** 100 D7
Bun Loyne **IV63** 87 J4
Bunarkaig **PH34** 87 H6
Bunbury **CW6** 48 E7
Bunbury Heath **CW6** 48 E7
Bunchrew **IV3** 96 D7
Bundalloch **IV40** 86 E2
Buness **ZE2** 107 Q2
Bunessan **PA67** 78 E5
Bungay NR35 45 H7
Bunker's Hill **LN4** 53 F7
Bunlarie **PA28** 73 F7
Bunloit **IV63** 88 C2
Bunmhullin **HS8** 84 C3
Bunnahabhain **PA46** 72 C3
Bunny **NG11** 41 H3
Buntait **IV63** 87 K1
Buntingford SG9 33 G6
Bunwell **NR16** 45 F6
Bunwell Street **NR16** 45 F6
Burbage *Derbys.* **SK17** 50 C5
Burbage *Leics.* **LE10** 41 G6
Burbage *Wilts.* **SN8** 21 F5
Burchett's Green **SL6** 22 B3
Burcombe **SP2** 10 B1
Burcot *Oxon.* **OX14** 21 J2
Burcot *Worcs.* **B60** 29 J1
Burcott **LU7** 32 B6
Burdale **YO17** 58 E3
Burdocks **RH14** 12 D4
Burdon **SR3** 62 E1
Burdrop **OX15** 30 E5
Bures CO8 34 D5
Bures Green **CO8** 34 D5
Burfa **LD8** 28 B2
Burford *Oxon.* OX18 30 D7
Burford *Shrop.* **WR15** 28 E2
Burg **PA74** 78 E3
Burgate **IP22** 34 E1
Burgates **GU33** 11 J2
Burge End **SG5** 32 E5
Burgess Hill RH15 13 G5
Burgh **IP13** 35 G3
Burgh by Sands **CA5** 60 E1
Burgh Castle **NR31** 45 J5
Burgh Heath **KT20** 23 F6
Burgh le Marsh **PE24** 53 H6
Burgh next Aylsham **NR11** 45 G3
Burgh on Bain **LN8** 53 F4
Burgh St. Margaret (Fleggburgh) **NR29** 45 J4
Burgh St. Peter **NR34** 45 J6
Burghclere **RG20** 21 H5
Burghead **IV30** 97 J5
Burghfield **RG30** 21 K5
Burghfield Common **RG7** 21 K5
Burghfield Hill **RG7** 21 K5
Burghill **HR4** 28 D4
Burghwallis **DN6** 51 H1
Burham **ME1** 24 D5
Buriton **GU31** 11 J2
Burland **CW5** 49 F7
Burlawn **PL27** 3 G2
Burleigh **SL5** 22 C5
Burlescombe **EX16** 7 J4
Burleston **DT2** 9 G5
Burley *Hants.* **BH24** 10 D4
Burley *Rut.* **LE15** 42 B4
Burley *W.Yorks.* **LS6** 57 H6
Burley Gate **HR1** 28 E4
Burley in Wharfedale **LS29** 57 G5
Burley Street **BH24** 10 D4
Burley Woodhead **LS29** 57 G5
Burleydam **SY13** 39 F1
Burlingjobb **LD8** 28 B3
Burlow **TN21** 13 J5
Burlton **SY4** 38 D3
Burmarsh **TN29** 15 G4
Burmington **CV36** 30 D5
Burn **YO8** 58 B7
Burn Farm **DD9** 90 E7
Burn Naze **FY5** 55 G5
Burn of Cambus **FK16** 81 J7
Burnage **M19** 49 H3
Burnaston **DE65** 40 E2
Burnby **YO42** 58 E5
Burncross **S35** 51 F3
Burndell **BN18** 12 C6
Burnden **BL3** 49 G2
Burnedge **OL2** 49 J1
Burnend **AB41** 99 G6
Burneside **LA9** 61 G7
Burness **KW17** 106 F3
Burneston **DL8** 57 J1
Burnett **BS31** 19 K5
Burnfoot *High.* **KW11** 104 D6
Burnfoot *P. & K.* **FK14** 82 A7
Burnfoot *Sc.Bord.* **TD9** 70 A2
Burnfoot *Sc.Bord.* **TD9** 69 K2
Burnham *Bucks.* **SL1** 22 C3
Burnham *N.Lincs.* **DN18** 52 D1
Burnham Deepdale **PE31** 44 C1
Burnham Green **AL6** 33 F7
Burnham Market **PE31** 44 C1
Burnham Norton **PE31** 44 C1
Burnham Overy Staithe **PE31** 44 C1
Burnham Overy Town **PE31** 44 C1
Burnham Thorpe **PE31** 44 C1
Burnham-on-Crouch CM0 25 F2
Burnham-on-Sea TA8 19 G7
Burnhaven **AB42** 99 K6
Burnhead *D. & G.* **DG7** 67 K5
Burnhead *D. & G.* **DG3** 68 D4
Burnhervie **AB51** 91 F3
Burnhill Green **WV6** 39 G5
Burnhope **DH7** 62 C2
Burnhouse **KA15** 74 B5
Burniston **YO13** 63 K3
Burnley BB11 56 D6
Burnmouth **TD14** 77 H4
Burnopfield **NE16** 62 C1
Burn's Green **SG2** 33 G6
Burnsall **BD23** 57 F3
Burnside *Aber.* **AB32** 91 F3
Burnside *Angus* **DD8** 83 G2
Burnside *E.Ayr.* **KA18** 67 K2
Burnside *Fife* **KY13** 82 C7
Burnside *Shet.* **ZE2** 107 L5
Burnside *W.Loth.* **EH52** 75 J3
Burnside of Duntrune **DD4** 83 F4
Burnstones **CA8** 61 H1
Burnswark **DG11** 69 G6
Burnt Hill **RG18** 21 J4
Burnt Houses **DL13** 62 C4
Burnt Oak **HA8** 23 F2
Burnt Yates **HG3** 57 H3
Burntcliff Top **SK11** 49 J6
Burntisland KY3 76 A2
Burnton *E.Ayr.* **KA18** 67 J2
Burnton *E.Ayr.* **KA6** 67 J3
Burntwood WS7 40 C5
Burntwood Green **WS7** 40 C5
Burnworthy **TA3** 7 K4
Burpham *Surr.* **GU1** 22 D6
Burpham *W.Suss.* **BN18** 12 D6
Burra **ZE2** 107 M9
Burradon *Northumb.* **NE65** 70 E3
Burradon *T. & W.* **NE23** 71 H6
Burrafirth **ZE2** 107 Q1
Burraland **ZE2** 107 M5
Burras **TR13** 2 D5
Burraton *Cornw.* **PL12** 4 E4
Burraton *Cornw.* **PL12** 4 E5
Burravoe *Shet.* **ZE2** 107 P5
Burravoe *Shet.* **ZE2** 107 M6
Burray **KW17** 106 D8
Burrells **CA16** 61 H5
Burrelton **PH13** 82 D4
Burridge *Devon* **EX13** 8 C4
Burridge *Hants.* **SO31** 11 G3
Burrill **DL8** 57 H1
Burringham **DN17** 52 B2
Burrington *Devon* **EX37** 6 E4
Burrington *Here.* **SY8** 28 D1
Burrington *N.Som.* **BS40** 19 H6
Burrough Green **CB8** 33 K3
Burrough on the Hill **LE14** 42 A4
Burrow *Som.* **TA24** 7 H1
Burrow *Som.* **TA12** 8 D3
Burrow Bridge **TA7** 8 C1
Burrowhill **GU24** 22 C5
Burrows Cross **GU5** 22 D7
Burry **SA3** 17 H6
Burry Green **SA3** 17 H6
Burry Port SA16 17 H5
Burscough **L40** 48 D1
Burscough Bridge **L40** 48 D1
Bursea **YO43** 58 E6
Burshill **YO25** 59 G5
Bursledon **SO31** 11 F4
Burslem **ST6** 40 A1
Burstall **IP8** 35 F4
Burstock **DT8** 8 D4
Burston *Norf.* **IP22** 45 F7
Burston *Staffs.* **ST18** 40 B2
Burstow **RH6** 23 G7
Burstwick **HU12** 59 J7
Burtersett **DL8** 56 D1
Burthorpe **IP29** 34 B2
Burthwaite **CA4** 60 F2
Burtle **TA7** 19 H7
Burtle Hill **TA7** 19 G7
Burton *Ches.W. & C.* **CH64** 48 C5
Burton *Ches.W. & C.* **CW6** 48 E6
Burton *Dorset* **BH23** 10 C5
Burton *Lincs.* **LN1** 52 C5
Burton *Northumb.* **NE69** 77 K7
Burton *Pembs.* **SA73** 16 C5
Burton *Som.* **TA5** 7 K1
Burton *Wilts.* **SN14** 20 B4
Burton *Wilts.* **BA12** 9 H1
Burton Agnes **YO25** 59 H3
Burton Bradstock **DT6** 8 D6
Burton Coggles **NG33** 42 C3
Burton End **CM24** 33 J6
Burton Ferry **SA73** 16 C5
Burton Fleming **YO25** 59 G2
Burton Green *Warks.* **CV8** 30 D1
Burton Green *Wrex.* **LL12** 48 C7
Burton Hastings **CV11** 41 G7
Burton in Lonsdale **LA6** 56 B2
Burton Joyce **NG14** 41 J1
Burton Latimer **NN15** 32 C1
Burton Lazars **LE14** 42 A4
Burton Leonard **HG3** 57 J3
Burton on the Wolds **LE12** 41 H3
Burton Overy **LE8** 41 J6
Burton Pedwardine **NG34** 42 E1
Burton Pidsea **HU12** 59 J6
Burton Salmon **LS25** 57 K7
Burton Stather **DN15** 52 B1
Burton upon Stather **DN15** 52 B1
Burton upon Trent DE14 40 E3
Burton-in-Kendal **LA6** 55 J2
Burton's Green **CM77** 34 C6
Burtonwood **WA5** 48 E3
Burwardsley **CH3** 48 E7
Burwarton **WV16** 39 F7
Burwash **TN19** 13 K4
Burwash Common **TN19** 13 K4
Burwash Weald **TN19** 13 K4
Burwell *Cambs.* **CB25** 33 J2
Burwell *Lincs.* **LN11** 53 G5
Burwen **LL68** 46 C3
Burwick *Ork.* **KW17** 106 D9
Burwick *Shet.* **ZE1** 107 M8
Bury *Cambs.* **PE26** 43 F7
Bury *Gt.Man.* BL9 49 H1
Bury *Som.* **TA22** 7 H3
Bury *W.Suss.* **RH20** 12 D5
Bury End **WR12** 30 B5
Bury Green **SG11** 33 H6
Bury St. Edmunds IP33 34 C2
Buryas Bridge **TR19** 2 B6
Burythorpe **YO17** 58 D3
Busbridge **GU7** 22 C7
Busby *E.Renf.* **G76** 74 D5
Busby *P. & K.* **PH1** 82 B5
Buscot **SN7** 21 F2
Bush **EX23** 6 A5
Bush Bank **HR4** 28 D3
Bush Crathie **AB35** 89 K5
Bush Green **IP21** 45 G7
Bushbury **WV10** 40 B5
Bushby **LE7** 41 J5
Bushey WD23 22 E2
Bushey Heath **WD23** 22 E2
Bushley **GL20** 29 H5
Bushley Green **GL20** 29 H5
Bushton **SN4** 20 D4
Bushy Common **NR19** 44 D4
Busk **CA10** 61 H2
Buslingthorpe **LN8** 52 D4
Bussage **GL6** 20 B1
Busta **ZE2** 107 M6

But - Car

Butcher's Common **NR12** 45 H3
Butcher's Cross **TN20** 13 J4
Butcher's Pasture **CM6** 33 K6
Butcombe **BS40** 19 J5
Bute PA20 73 J4
Bute Town **NP22** 18 E1
Buthill **IV30** 97 J5
Butleigh **BA6** 8 E1
Butleigh Wootton **BA6** 8 E1
Butler's Cross **HP17** 22 B1
Butler's Hill **NG15** 41 H1
Butlers Marston **CV35** 30 E4
Butlersbank **SY4** 38 E3
Butley **IP12** 35 H3
Butley Abbey **IP12** 35 H4
Butley Low Corner **IP12** 35 H4
Butley Mills **IP12** 35 H3
Butley Town **SK10** 49 J5
Butt Green **CW5** 49 F7
Butt Lane **ST7** 49 H7
Butterburn **CA8** 70 B6
Buttercrambe **YO41** 58 D4
Butterknowle **DL13** 62 C4
Butterleigh **EX15** 7 H5
Butterley **DE5** 51 G7
Buttermere *Cumb.* **CA13** 60 C5
Buttermere *Wilts.* **SN8** 21 G5
Butters Green **ST7** 49 H7
Buttershaw **BD6** 57 G6
Butterstone **PH8** 82 B3
Butterton *Staffs.* **ST13** 50 C7
Butterton *Staffs.* **ST5** 40 A1
Butterwick *Dur.* **TS21** 62 E4
Butterwick *Lincs.* **PE22** 43 G1
Butterwick *N.Yorks.* **YO17** 59 F2
Butterwick *N.Yorks.* **YO17** 58 D2
Buttington **SY21** 38 B5
Buttonbridge **DY12** 29 G1
Buttonoak **DY12** 29 G1
Buttons' Green **IP30** 34 D3
Butts **EX6** 7 G7
Butt's Green *Essex* **CM2** 24 D1
Butt's Green *Hants.* **SO51** 10 D2
Buttsash **SO45** 11 F4
Buxhall **IP14** 34 E3
Buxted **TN22** 13 H4
Buxton *Derbys.* SK17 50 C5
Buxton *Norf.* **NR10** 45 G3
Buxton Heath **NR10** 45 F3
Buxworth **SK23** 50 C4
Bwcle (Buckley) CH7 48 B6
Bwlch **LD3** 28 A6
Bwlch-clawdd **SA44** 17 G2
Bwlch-derwin **LL51** 36 D1
Bwlchgwyn **LL11** 48 B7
Bwlch-llan **SA48** 26 E3
Bwlchnewydd **SA33** 17 G3
Bwlchtocyn **LL53** 36 C3
Bwlch-y-cibau **SY22** 38 A4
Bwlch-y-ddar **SY10** 38 A3
Bwlchyfadfa **SA44** 17 H1
Bwlch-y-ffridd **SY16** 37 K6
Bwlch-y-groes **SA35** 17 F2
Bwlchyllyn **LL54** 46 D7
Bwlchymynydd **SA4** 17 J6
Bwlch-y-sarnau **LD6** 27 K1
Byers Green **DL16** 62 D3
Byfield **NN11** 31 G3
Byfleet **KT14** 22 D5
Byford **HR4** 28 C4
Bygrave **SG7** 33 F5
Byker **NE6** 71 H7
Byland Abbey **YO61** 58 B2
Bylane End **PL14** 4 C5
Bylchau **LL16** 47 H6
Byley **CW10** 49 G6
Bynea **SA14** 17 J6
Byrness **NE19** 70 C3
Bystock **EX8** 7 J7
Bythorn **PE28** 32 D1
Byton **LD8** 28 C2
Bywell **NE43** 71 F7
Byworth **GU28** 12 C4

C

Cabharstadh **HS2** 101 F5
Cabourne **LN7** 52 E2
Cabourne Parva **LN7** 52 E2
Cabrach *Arg. & B.* **PA60** 72 C4
Cabrach *Moray* **AB54** 90 B2
Cabus **PR3** 55 H5
Cackle Street *E.Suss.* **TN31** 14 D6
Cackle Street *E.Suss.* **TN22** 13 H4
Cacrabank **TD7** 69 J2
Cadboll **IV20** 97 F4
Cadbury **EX5** 7 H5
Cadbury Barton **EX18** 6 E4
Cadbury Heath **BS30** 19 K4
Cadder **G64** 74 E3
Cadderlie **PA35** 80 B4
Caddington **LU1** 32 D7
Caddleton **PA34** 79 J6
Caddonfoot **TD1** 76 C7
Cade Street **TN21** 13 K4
Cadeby *Leics.* **CV13** 41 G5
Cadeby *S.Yorks.* **DN5** 51 H2
Cadeleigh **EX16** 7 H5
Cader **LL16** 47 J6
Cadgwith **TR12** 2 E7
Cadham **KY7** 82 D7
Cadishead **M44** 49 G3
Cadle **SA5** 17 K6
Cadley *Lancs.* **PR2** 55 J6
Cadley *Wilts.* **SN8** 21 F5
Cadmore End **HP14** 22 A2
Cadnam **SO40** 10 D3
Cadney **DN20** 52 D2
Cadole **CH7** 48 B6
Cadover Bridge **PL7** 5 F4
Cadoxton **CF63** 18 E5
Cadoxton-Juxta-Neath **SA10** 18 A2
Cadwell **SG5** 32 E5
Cadwst **LL21** 37 K2
Cadzow **ML3** 75 F5
Cae Ddafydd **LL48** 37 F1
Caeathro **LL55** 46 D6
Caehopkin **SA9** 27 H7
Caen **KW8** 105 F7
Caenby **LN8** 52 D4
Caenby Corner **LN8** 52 C4
Caer Llan **NP25** 19 H1
Caerau *Bridgend* **CF34** 18 B2
Caerau *Cardiff* **CF5** 18 E4
Caerdeon **LL42** 37 F4
CAERDYDD (CARDIFF) CF 18 E4
Caerfarchell **SA62** 16 A3
Caerfyrddin (Carmarthen) SA31 17 H3
Caergeiliog **LL65** 46 B5
Caergwrle **LL12** 48 C7
Caergybi (Holyhead) LL65 46 A4
Caerhun **LL32** 47 F5
Caer-Lan **SA9** 27 H7
Caerleon **NP18** 19 G2
Caernarfon LL55 46 C6
Caerphilly CF83 18 E3
Caersws SY17 37 K6
Caerwedros **SA44** 26 C3
Caerwent **NP26** 19 H2
Caerwys **CH7** 47 K5
Caethle Farm **LL36** 37 F6
Caggan **PH22** 89 F3
Caggle Street **NP7** 28 C7
Caim *High.* **PH36** 79 G1
Caim *I.o.A.* **LL58** 46 E4
Caio **SA19** 17 K2
Cairinis (Carinish) **HS6** 92 D5
Cairisiadar **HS2** 100 C4
Cairminis **HS5** 93 F3
Cairnargat **AB54** 90 C1
Cairnbaan **PA31** 73 G1
Cairnbeathie **AB31** 90 D4
Cairnbrogie **AB51** 91 G2
Cairnbulg **AB43** 99 J4
Cairncross *Angus* **DD9** 90 D7
Cairncross *Sc.Bord.* **TD14** 77 G4
Cairncurran **PA13** 74 B4
Cairndoon **DG8** 64 D7
Cairndow PA26 80 C6
Cairness **AB43** 99 J4
Cairney Lodge **KY15** 82 E6
Cairneyhill **KY12** 75 J2
Cairnhill *Aber.* **AB41** 91 H2
Cairnhill *Aber.* **AB52** 90 E1
Cairnie *Aber.* **AB32** 91 G4
Cairnie *Aber.* **AB54** 98 C6
Cairnorrie **AB41** 99 G6
Cairnryan **DG9** 64 A4
Cairnsmore **DG8** 64 E4
Caister-on-Sea **NR30** 45 K4
Caistor **LN7** 52 E2
Caistor St. Edmund **NR14** 45 G5
Caistron **NE65** 70 E3
Cake Street **NR17** 44 E6
Cakebole **DY10** 29 H1
Calanais (Callanish) **HS2** 100 E4
Calbost **HS2** 101 G6
Calbourne **PO30** 11 F6
Calceby **LN13** 53 G5
Calcoed **CH8** 47 K5
Calcot **RG31** 21 K4
Calcott *Kent* **CT3** 25 H5
Calcott *Shrop.* **SY3** 38 D4
Calcotts Green **GL2** 29 G7
Calcutt **SN6** 20 E2
Caldarvan **G83** 74 C2
Caldback **ZE2** 107 Q2
Caldbeck **CA7** 60 E3
Caldbergh **DL8** 57 F1
Caldecote *Cambs.* **CB23** 33 G3
Caldecote *Cambs.* **PE7** 42 E7
Caldecote *Herts.* **SG7** 33 F5
Caldecote *Northants.* **NN12** 31 H3
Caldecote *Warks.* **CV10** 41 F6
Caldecott *Northants.* **NN9** 32 C2
Caldecott *Oxon.* **OX14** 21 H2
Caldecott *Rut.* **LE16** 42 B6
Calder Bridge **CA20** 60 B6
Calder Grove **WF4** 51 F1
Calder Mains **KW12** 105 F3
Calder Vale **PR3** 55 J5
Calderbank **ML6** 75 F4
Calderbrook **OL15** 49 J1
Caldercruix **ML6** 75 G4
Calderglen **G72** 74 E5
Caldermill **ML10** 74 E6
Caldey Island **SA70** 16 E6
Caldhame **DD8** 83 F3
Caldicot NP26 19 H3
Caldwell *Derbys.* **DE12** 40 E4
Caldwell *E.Renf.* **G78** 74 C5
Caldwell *N.Yorks.* **DL11** 62 C5
Caldy **CH48** 48 B4
Calebreck **CA7** 60 E3
Caledrhydiau **SA48** 26 D3
Calford Green **CB9** 33 K4
Calfsound **KW17** 106 E4
Calgary **PA75** 78 E2
Califer **IV36** 97 H6
California *Falk.* **FK1** 75 H3
California *Norf.* **NR29** 45 K4
California *Suff.* **IP4** 35 G4
Calke **DE73** 41 F3
Callakille **IV54** 94 C6
Callaly **NE66** 71 F3
Callander FK17 81 H7
Callanish (Calanais) **HS2** 100 E4
Callaughton **TF13** 39 F6
Callerton Lane End **NE5** 71 G7
Calliburn **PA28** 66 B1
Calligarry **IV45** 86 C4
Callington PL17 4 D4
Callingwood **DE13** 40 D3
Callisterhall **DG11** 69 H5
Callow **HR2** 28 D5
Callow End **WR2** 29 H4
Callow Hill *Wilts.* **SN15** 20 D3
Callow Hill *Worcs.* **DY14** 29 G1
Callow Hill *Worcs.* **B97** 30 B2
Callows Grave **WR15** 28 E2
Calmore **SO40** 10 E3
Calmsden **GL7** 20 D1
Calne SN11 20 C4
Calow **S44** 51 G5
Calrossie **IV19** 96 E4
Calshot **SO45** 11 F4
Calstock PL18 4 E4
Calstone Wellington **SN11** 20 D5
Calthorpe **NR11** 45 F2
Calthwaite **CA11** 61 F2
Calton *N.Yorks.* **BD23** 56 E4
Calton *Staffs.* **ST10** 50 D7
Calveley **CW6** 48 E7
Calver **S32** 50 E5
Calver Hill **HR4** 28 C4
Calverhall **SY13** 39 F2
Calverleigh **EX16** 7 H4
Calverley **LS28** 57 H6
Calvert **MK18** 31 H6
Calverton *M.K.* **MK19** 31 J5
Calverton *Notts.* **NG14** 41 J1
Calvine **PH18** 81 J1
Calvo **CA7** 60 C1
Cam **GL11** 20 A2
Camasnacroise **PH33** 79 K2
Camastianavaig **IV51** 86 B1
Camasunary **IV49** 86 B3
Camault Muir **IV4** 96 C7
Camb **ZE2** 107 P3
Camber **TN31** 14 E6
Camberley GU15 22 B5
Camberwell **SE15** 23 G4
Camblesforth **YO8** 58 C7
Cambo **NE61** 71 F5
Cambois **NE24** 71 J5
Camborne TR14 2 D5
CAMBRIDGE *Cambs.* CB 33 H3
Cambridge *Glos.* **GL2** 20 A1
Cambridge City Airport **CB5** 33 H3
Cambus **FK10** 75 G1
Cambus o'May **AB35** 90 C5
Cambusbarron **FK7** 75 F1
Cambuskenneth **FK9** 75 G1
Cambuslang **G72** 74 E4
Cambusnethan **ML2** 75 G5
Camden Town **NW1** 23 F3
Camel Hill **BA22** 8 E2
Cameley **BS39** 19 K6
Camelford PL32 4 B2
Camelon **FK1** 75 G2
Camelsdale **GU27** 12 B3
Camer **DA13** 24 C5
Cameron House **G83** 74 B2
Camerory **PH26** 89 H1
Camer's Green **WR13** 29 G5
Camerton *B. & N.E.Som.* **BA2** 19 K6
Camerton *Cumb.* **CA14** 60 B3
Camerton *E.Riding* **HU12** 59 J7
Camghouran **PH17** 81 G2
Camis Eskan **G84** 74 B2
Cammachmore **AB39** 91 H5
Cammeringham **LN1** 52 C4
Camore **IV25** 96 E2
Camp Hill *Pembs.* **SA67** 16 E4
Camp Hill *Warks.* **CV10** 41 F6
Campbeltown PA28 66 B1
Campbeltown Airport **PA28** 66 A1
Camperdown **NE12** 71 H6
Campmuir **PH13** 82 D4
Camps **EH27** 75 K4
Camps End **CB21** 33 K4
Camps Heath **NR32** 45 K6
Campsall **DN6** 51 H1
Campsea Ashe **IP13** 35 H3
Campton **SG17** 32 E5
Camptown **TD8** 70 B2
Camquhart **PA22** 73 H2
Camrose **SA62** 16 C3
Camserney **PH15** 81 K3
Camstraddan House **G83** 74 B1
Camus Croise **IV43** 86 C3
Camus-luinie **IV40** 87 F2
Camusnagaul *High.* **PH33** 87 G7
Camusnagaul *High.* **IV23** 95 G3
Camusrory **PH41** 86 E5
Camusteel **IV54** 94 D7
Camusterrach **IV54** 94 D7
Camusurich **FK21** 81 H4
Camusvrachan **PH15** 81 H3
Canada **SO51** 10 D3
Canaston Bridge **SA67** 16 D4
Candacraig **AB35** 90 B5
Candlesby **PE23** 53 H6
Candy Mill **ML12** 75 J6
Cane End **RG4** 21 K4
Canewdon **SS4** 24 E2
Canfield End **CM6** 33 J6
Canford Bottom **BH21** 10 B4
Canford Cliffs **BH13** 10 B6
Canford Magna **BH21** 10 B5
Canham's Green **IP14** 34 E2
Canisbay **KW1** 105 J1
Canley **CV4** 30 E1
Cann **SP7** 9 H2
Cann Common **SP7** 9 H2
Canna PH44 85 H4
Cannard's Grave **BA4** 19 K7
Cannich **IV4** 87 K1
Canning Town **E16** 23 H3
Cannington **TA5** 8 B1
Cannock WS11 40 B5
Cannock Wood **WS15** 40 C4
Cannop **GL16** 29 F7
Canon Bridge **HR2** 28 D4
Canon Frome **HR8** 29 F4
Canon Pyon **HR4** 28 D4
Canonbie DG14 69 J6
Canons Ashby **NN11** 31 G3
Canon's Town **TR27** 2 C5
Canterbury *Aber.* **AB45** 98 D5
CANTERBURY *Kent* CT 15 G2
Cantley *Norf.* **NR13** 45 H5
Cantley *S.Yorks.* **DN3** 51 J2
Cantlop **SY5** 38 E5
Canton **CF11** 18 E4
Cantray **IV2** 96 E7
Cantraydoune **IV12** 96 E7
Cantraywood **IV2** 96 E7
Cantsfield **LA6** 56 B2
Canvey Island SS8 24 D3
Canwell Hall **B75** 40 D5
Canwick **LN4** 52 C6
Canworthy Water **PL15** 4 C1
Caol **PH33** 87 H7
Caolas *Arg. & B.* **PA77** 78 B3
Caolas *W.Isles* **HS9** 84 B5
Caolas Scalpaigh (Kyles Scalpay) **HS3** 93 H2
Caolasnacon **PH50** 80 C1
Capel *Kent* **TN12** 23 K7
Capel *Surr.* **RH5** 22 E7
Capel Bangor **SY23** 37 F7
Capel Betws Lleucu **SY25** 27 F3
Capel Carmel **LL53** 36 A3
Capel Celyn **LL23** 37 H1
Capel Coch **LL77** 46 C4
Capel Curig **LL24** 47 F7
Capel Cynon **SA44** 17 G1
Capel Dewi *Carmar.* **SA32** 17 H3
Capel Dewi *Cere.* **SY23** 37 F7
Capel Dewi *Cere.* **SA44** 17 H1
Capel Garmon **LL26** 47 G7
Capel Gwyn *Carmar.* **SA32** 17 H3
Capel Gwyn *I.o.A.* **LL65** 46 B5
Capel Gwynfe **SA19** 27 G6
Capel Hendre **SA18** 17 J4
Capel Isaac **SA19** 17 J3
Capel Iwan **SA38** 17 F2
Capel le Ferne **CT18** 15 H4
Capel Llanilltern **CF5** 18 D3
Capel Mawr **LL62** 46 C5
Capel Parc **LL71** 46 C4
Capel St. Andrew **IP12** 35 H4
Capel St. Mary **IP9** 34 E5
Capel St. Silin **SA48** 26 E3
Capel Seion **SY23** 27 F1
Capel Tygwydd **SA38** 17 F1
Capeluchaf **LL54** 36 D1
Capelulo **LL34** 47 F5
Capel-y-ffin **NP7** 28 B5
Capel-y-graig **LL57** 46 D6
Capenhurst **CH1** 48 C5
Capernwray **LA6** 55 J2
Capheaton **NE19** 71 F5
Caplaw **G78** 74 C5
Capon's Green **IP13** 35 G2
Cappercleuch **TD7** 69 H1
Capplegill **DG10** 69 G3
Capstone **ME7** 24 D5
Capton *Devon* **TQ6** 5 J5
Capton *Som.* **TA4** 7 J2
Caputh **PH1** 82 B3
Car Colston **NG13** 42 A1
Caradon Town **PL14** 4 C3
Carbellow **KA18** 68 B1
Carbeth **G63** 74 D3
Carbis Bay **TR26** 2 C5
Carbost *High.* **IV51** 93 K7
Carbost *High.* **IV47** 85 J1
Carbrain **G67** 75 F3
Carbrooke **IP25** 44 D5
Carburton **S80** 51 J5
Carcary **DD9** 83 H2
Carco **DG4** 68 C2
Carcroft **DN6** 51 H1
Cardenden **KY5** 76 A1
Cardeston **SY5** 38 C4
Cardew **CA5** 60 E2
CARDIFF (CAERDYDD) CF 18 E4
Cardiff Airport **CF62** 18 D5
Cardigan (Aberteifi) SA43 16 E1
Cardinal's Green **CB21** 33 K4
Cardington *Bed.* **MK44** 32 D4
Cardington *Shrop.* **SY6** 38 E6
Cardinham **PL30** 4 B4
Cardno **AB43** 99 H4
Cardonald **G52** 74 D4
Cardoness **DG7** 65 F5
Cardow **AB38** 97 J7
Cardrona **EH45** 76 A7
Cardross **G82** 74 B3
Cardurnock **CA7** 60 C1
Careby **PE9** 42 D4
Careston **DD9** 83 G2
Carew **SA70** 16 D5
Carew Cheriton **SA70** 16 D5
Carew Newton **SA68** 16 D5
Carey **HR2** 28 E5
Carfin **ML1** 75 F5
Carfrae **EH41** 76 D4
Carfraemill **TD2** 76 D5
Cargate Green **NR13** 45 H4
Cargen **DG2** 65 K3
Cargenbridge **DG2** 65 K3
Cargill **PH2** 82 C4
Cargo **CA6** 60 E1
Cargreen **PL12** 4 E4
Carham **TD12** 77 F7
Carhampton **TA24** 7 J1
Carharrack **TR16** 2 E4
Carie *P. & K.* **PH17** 81 H2
Carie *P. & K.* **PH15** 81 H4
Carines **TR8** 2 E3
Carinish (Cairinis) **HS6** 92 D5
Carisbrooke **PO30** 11 F6
Cark **LA11** 55 G2
Carkeel **PL12** 4 E4
Carlabhagh (Carloway) **HS2** 100 E3
Carland Cross **TR8** 3 F3
Carlatton **CA8** 61 G1
Carlby **PE9** 42 D4
Carlecotes **S36** 50 D2
Carleen **TR13** 2 D6
Carleton *Cumb.* **CA1** 60 F1
Carleton *Cumb.* **CA11** 61 G4
Carleton *Lancs.* **FY6** 55 G5
Carleton *N.Yorks.* **BD23** 56 E5
Carleton *W.Yorks.* **WF8** 57 K7
Carleton Fishery **KA26** 67 F5
Carleton Forehoe **NR9** 44 E5
Carleton Rode **NR16** 45 F6
Carleton St. Peter **NR14** 45 H5
Carlin How **TS13** 63 J5
CARLISLE CA 60 F1
Carloggas **TR8** 3 F2
Carlops **EH26** 75 K5
Carloway (Carlabhagh) **HS2** 100 E3
Carlton *Bed.* **MK43** 32 C3
Carlton *Cambs.* **CB8** 33 K3
Carlton *Leics.* **CV13** 41 F5
Carlton *N.Yorks.* **DN14** 58 C7
Carlton *N.Yorks.* **DL8** 57 F1
Carlton *N.Yorks.* **YO62** 58 C1
Carlton *Notts.* **NG4** 41 J1
Carlton *S.Yorks.* **S71** 51 F1
Carlton *Stock.* **TS21** 62 E4
Carlton *Suff.* **IP17** 35 H2
Carlton *W.Yorks.* **WF3** 57 J7
Carlton Colville **NR33** 45 K6
Carlton Curlieu **LE8** 41 J6
Carlton Green **CB8** 33 K3
Carlton Husthwaite **YO7** 57 K2
Carlton in Lindrick **S81** 51 H4
Carlton Miniott **YO7** 57 J1
Carlton Scroop **NG32** 42 C1
Carlton-in-Cleveland **TS9** 63 G6
Carlton-le-Moorland **LN5** 52 C7
Carlton-on-Trent **NG23** 52 B6
Carluke ML8 75 G5
Carlyon Bay **PL25** 4 A5
Carmacoup **ML11** 68 C1
Carmarthen (Caerfyrddin) SA31 17 H3
Carmel *Carmar.* **SA14** 17 J4
Carmel *Flints.* **CH8** 47 K5
Carmel *Gwyn.* **LL54** 46 C7
Carmel *I.o.A.* **LL71** 46 B4
Carmichael **ML12** 75 H7
Carmont **AB39** 91 G6
Carmunnock **G76** 74 D5
Carmyle **G32** 74 E4
Carmyllie **DD11** 83 G3
Carn **PA48** 72 A5
Carn Brea Village **TR15** 2 D4
Carn Dearg **IV21** 94 D4
Carnaby **YO16** 59 H3
Carnach *High.* **IV40** 87 G2
Carnach *High.* **IV23** 95 G2
Carnach *W.Isles* **HS6** 92 D5
Carnan **HS8** 92 C7
Carnassarie **PA31** 79 K7
Carnbee **KY10** 83 G7
Carnbo **KY13** 82 B7
Carndu **IV40** 86 E2
Carnduncan **PA44** 72 A4
Carnforth LA5 55 H2
Carnhedryn **SA62** 16 B3
Carnhell Green **TR14** 2 D5
Carnichal **AB42** 99 H5
Carnkie *Cornw.* **TR16** 2 D5
Carnkie *Cornw.* **TR13** 2 E5
Carnmore **PA42** 72 B6
Carno **SY17** 37 J6
Carnoch *High.* **IV4** 87 K1
Carnoch *High.* **IV6** 95 J6
Carnoch *High.* **IV12** 97 F7
Carnock **KY12** 75 J2
Carnon Downs **TR3** 3 F4
Carnousie **AB53** 98 E5
Carnoustie DD7 83 G4
Carntyne **G32** 74 E4
Carnwath **ML11** 75 H6
Carnyorth **TR19** 2 A5
Carol Green **CV7** 30 D1
Carperby **DL8** 57 F1
Carr **S66** 51 H3
Carr Hill **DN10** 51 J3
Carr Houses **L38** 48 C2
Carr Shield **NE47** 61 K2
Carr Vale **S44** 51 G5
Carradale **PA28** 73 F7
Carradale East **PA28** 73 G7
Carragrich **HS3** 93 G2
Carrbridge PH23 89 G2
Carrefour Selous **JE3** 3 J7
Carreg-lefn **LL68** 46 B4
Carreg-wen **SA37** 17 F1
Carrhouse **DN9** 51 K2
Carrick **PA31** 73 H2
Carrick Castle **PA24** 73 K1
Carriden **EH51** 75 J2
Carrine **PA28** 66 A3
Carrington *Gt.Man.* **M31** 49 G3

Car - Che

Carrington *Lincs.* **PE22** 53 G7
Carrington *Midloth.* **EH23** 76 B4
Carroch **DG7** 68 B4
Carrog *Conwy* **LL24** 37 G1
Carrog *Denb.* **LL21** 38 A1
Carroglen **PH6** 81 J5
Carrol **KW9** 97 F1
Carron *Arg. & B.* **PA31** 73 H1
Carron *Falk.* **FK2** 75 G2
Carron *Moray* **AB38** 97 K7
Carron Bridge **FK6** 75 F2
Carronbridge **DG3** 68 D4
Carronshore **FK2** 75 G2
Carrot **DD8** 83 F3
Carrow Hill **NP26** 19 H2
Carrutherstown **DG1** 69 G6
Carruthmuir **PA10** 74 B4
Carrville **DH1** 62 E2
Carry **PA21** 73 H4
Carsaig **PA70** 79 G5
Carscreugh **DG8** 64 C4
Carse **PA29** 73 F4
Carse of Ardersier **IV2** 97 F6
Carsegowan **DG8** 64 E5
Carseriggan **DG8** 64 D4
Carsethorn **DG2** 65 K5
Carsgoe **KW12** 105 G2
Carshalton SM5 23 **F5**
Carshalton Beeches **SM5** 23 F5
Carsie **PH10** 82 C3
Carsington **DE4** 50 E7
Carsluith **DG8** 64 E5
Carsphairn **DG7** 67 K4
Carstairs **ML11** 75 H6
Carstairs Junction **ML11** 75 H6
Carswell Marsh **SN7** 21 G2
Carter's Clay **SO51** 10 E2
Carterton OX18 21 **F1**
Carterway Heads **DH8** 62 B1
Carthew **PL26** 4 A5
Carthorpe **DL8** 57 J1
Cartington **NE65** 71 F3
Cartland **ML11** 75 G6
Cartmel **LA11** 55 G2
Cartmel Fell **LA11** 55 H1
Cartworth **HD9** 50 D2
Carway **SA17** 17 H5
Cascob **LD8** 28 B2
Cas-gwent (Chepstow) NP16 19 **J2**
Cashel Farm **G63** 74 C1
Cashes Green **GL5** 20 B1
Cashlie **PH15** 81 F3
Cashmoor **DT11** 9 J3
Caskieberran **KY6** 82 D7
CASNEWYDD (NEWPORT) NP 19 **G3**
Cassencarie **DG8** 64 E5
Cassington **OX29** 31 F7
Cassop **DH6** 62 E3
Castell **LL32** 47 F6
Castell Gorfod **SA33** 17 F3
Castell Howell **SA44** 17 H1
Castell Newydd Emlyn (Newcastle Emlyn) SA38 17 **G1**
Castellau **CF38** 18 D3
Castell-nedd (Neath) SA11 18 **A2**
Castell-y-bwch **NP44** 19 F2
Casterton **LA6** 56 B2
Castle Acre **PE32** 44 C4
Castle Ashby **NN7** 32 B3
Castle Bolton **DL8** 62 B7
Castle Bromwich **B36** 40 D7
Castle Bytham **NG33** 42 C4
Castle Caereinion **SY21** 38 A5
Castle Camps **CB21** 33 K4
Castle Carrock **CA8** 61 G1
Castle Cary BA7 9 **F1**
Castle Combe **SN14** 20 B4
Castle Donington **DE74** 41 G3
Castle Douglas DG7 65 **H4**
Castle Eaton **SN6** 20 E2
Castle Eden **TS27** 63 F3
Castle End **CV8** 30 D1
Castle Frome **HR8** 29 F4
Castle Gate **TR20** 2 B5
Castle Goring **BN13** 12 E6
Castle Green **GU24** 22 C5
Castle Gresley **DE11** 40 E4
Castle Heaton **TD12** 77 H6
Castle Hedingham **CO9** 34 B5
Castle Hill *Kent* **TN12** 23 K7
Castle Hill *Suff.* **IP1** 35 F4
Castle Kennedy **DG9** 64 B5
Castle Leod **IV14** 96 B6
Castle Levan **PA19** 74 A3
Castle Madoc **LD3** 27 K5
Castle Morris **SA62** 16 C2
Castle O'er **DG13** 69 H4
Castle Rising **PE31** 44 A3
Castle Stuart **IV2** 96 E7
Castlebay (Bàgh a'Chaisteil) **HS9** 84 B5
Castlebythe **SA62** 16 D3
Castlecary **G68** 75 F3
Castlecraig *High.* **IV19** 97 F5
Castlecraig *Sc.Bord.* **EH46** 75 K6
Castlefairn **DG3** 68 C5
Castleford WF10 57 **K7**
Castlemartin **SA71** 16 C6
Castlemilk *D. & G.* **DG11** 69 G6
Castlemilk *Glas.* **G45** 74 E5
Castlemorton **WR13** 29 G5
Castlerigg **CA12** 60 D4
Castleside **DH8** 62 B2
Castlesteads **CA8** 70 A7
Castlethorpe **MK19** 31 J4
Castleton *Aber.* **AB45** 99 F5
Castleton *Angus* **DD8** 82 E3
Castleton *Arg. & B.* **PA31** 73 G2
Castleton *Derbys.* **S33** 50 D4
Castleton *Gt.Man.* **OL11** 49 H1
Castleton *N.Yorks.* **YO21** 63 H6
Castleton *Newport* **CF3** 19 F3
Castleton *Sc.Bord.* **TD9** 70 A4
Castletown *Dorset* **DT5** 9 F7
Castletown *High.* **KW14** 105 G2
Castletown *High.* **IV2** 96 E7
Castletown *I.o.M.* **IM9** 54 B7
Castletown *T. & W.* **SR5** 62 E1
Castleweary **TD9** 69 K3
Castlewigg **DG8** 64 E6
Castley **LS21** 57 H5
Caston **NR17** 44 D6
Castor **PE5** 42 E6
Castramont **DG7** 65 F4
Caswell **SA3** 17 J7
Cat & Fiddle Inn **SK11** 50 C5
Catacol **KA27** 73 H6
Catbrain **BS10** 19 J3
Catbrook **NP16** 19 J1
Catchall **TR19** 2 B6
Catcleugh **NE19** 70 C3
Catcliffe **S60** 51 G4
Catcott **TA7** 8 C1
Caterham CR3 23 **G6**
Catfield **NR29** 45 H3
Catfirth **ZE2** 107 N7
Catford **SE6** 23 G4
Catforth **PR4** 55 H6
Cathays **CF24** 18 E4
Cathcart **G44** 74 D4
Cathedine **LD3** 28 A6
Catherine-de-Barnes **B91** 40 D7
Catherington **PO8** 11 H3
Catherston Leweston **DT6** 8 C5
Catherton **DY14** 29 F1
Cathkin **G73** 74 E5
Catisfield **PO15** 11 G4
Catlodge **PH20** 88 D5
Catlowdy **CA6** 69 K6
Catmere End **CB11** 33 H5
Catmore **RG20** 21 H3
Caton *Devon* **TQ13** 5 H3
Caton *Lancs.* **LA2** 55 J3
Caton Green **LA2** 55 J3
Cator Court **TQ13** 5 G3
Catrine **KA5** 67 K1
Catsfield **TN33** 14 C6
Catsfield Stream **TN33** 14 C6
Catshaw **S36** 50 E2
Catshill **B61** 29 J1
Cattadale **PA44** 72 B4
Cattal **YO26** 57 K4
Cattawade **CO11** 34 E5
Catterall **PR3** 55 J5
Catterick **DL10** 62 D7
Catterick Bridge **DL10** 62 D7
Catterick Garrison DL9 62 **C7**
Catterlen **CA11** 61 F3
Catterline **AB39** 91 G7
Catterton **LS24** 58 B5
Catteshall **GU7** 22 C7
Catthorpe **LE17** 31 G1
Cattishall **IP31** 34 C2
Cattistock **DT2** 8 E4
Catton *N.Yorks.* **YO7** 57 J2
Catton *Norf.* **NR6** 45 G4
Catton *Northumb.* **NE47** 61 K1
Catton Hall **DE12** 40 E4
Catwick **HU17** 59 H5
Catworth **PE28** 32 D1
Caudle Green **GL53** 29 J7
Caulcott *Cen.Beds.* **MK43** 32 D4
Caulcott *Oxon.* **OX25** 31 G6
Cauldcots **DD11** 83 H3
Cauldhame *Stir.* **FK8** 74 E1
Cauldhame *Stir.* **FK15** 81 K7
Cauldon **ST10** 40 C1
Caulkerbush **DG2** 65 K5
Caulside **DG14** 69 K5
Caundle Marsh **DT9** 9 F3
Caunsall **DY11** 40 A7
Caunton **NG23** 51 K6
Causeway End *D. & G.* **DG8** 64 E4
Causeway End *Essex* **CM6** 33 K7
Causeway End *Lancs.* **L40** 48 D1
Causewayhead *Cumb.* **CA7** 60 C1
Causewayhead *Stir.* **FK9** 75 G1
Causey **DH9** 62 D1
Causey Park **NE61** 71 G4
Causeyend **AB23** 91 H3
Cautley **LA10** 61 H7
Cavendish **CO10** 34 C4
Cavendish Bridge **DE72** 41 G3
Cavenham **IP28** 34 B2
Cavens **DG2** 65 K5
Cavers **TD9** 70 A2
Caversfield **OX27** 31 G6
Caversham **RG4** 22 A4
Caverswall **ST11** 40 B1
Cawdor **IV12** 97 F6
Cawkeld **YO25** 59 F4
Cawkwell **LN11** 53 F4
Cawood **YO8** 58 B6
Cawsand **PL10** 4 E6
Cawston *Norf.* **NR10** 45 F3
Cawston *Warks.* **CV22** 31 F1
Cawthorn **YO18** 58 D1
Cawthorne **S75** 50 E2
Cawthorpe **PE10** 42 D3
Cawton **YO62** 58 C2
Caxton **CB23** 33 G3
Caxton Gibbet **CB23** 33 F2
Caynham **SY8** 28 E1
Caythorpe *Lincs.* **NG32** 42 C1
Caythorpe *Notts.* **NG14** 41 J1
Cayton **YO11** 59 G1
Ceallan **HS6** 92 D6
Ceann a' Bhàigh *W.Isles* **HS6** 92 C5
Ceann a' Bhàigh *W.Isles* **HS3** 93 F3
Ceann Loch Shiphoirt **HS2** 100 E6
Ceann Lochroag (Kinlochroag) **HS2** 100 D5
Ceannaridh **HS6** 92 D6
Cearsiadar **HS2** 101 F6
Ceathramh Meadhanach (Middlequarter) **HS6** 92 D4
Cedig **SY10** 37 J3
Cefn Berain **LL16** 47 H6
Cefn Bychan (Newbridge) **NP11** 19 F2
Cefn Canol **SY10** 38 B2
Cefn Cantref **LD3** 27 K6
Cefn Coch **LL15** 47 K7
Cefn Cribwr **CF32** 18 B3
Cefn Cross **CF32** 18 B3
Cefn Einion **SY9** 38 B7
Cefn Hengoed **CF82** 18 E2
Cefn Llwyd **SY23** 37 F7
Cefn Rhigos **CF44** 18 C1
Cefn-brith **LL21** 47 H7
Cefn-caer-Ferch **LL53** 36 D1
Cefn-coch **SY10** 38 A3
Cefn-coed-y-cymmer **CF48** 18 D1
Cefn-ddwysarn **LL23** 37 J2
Cefndeuddwr **LL40** 37 G3
Cefneithin **SA14** 17 J4
Cefn-gorwydd **LD4** 27 J4
Cefn-gwyn **SY16** 38 A7
Cefn-mawr **LL14** 38 B1
Cefnpennar **CF45** 18 D1
Cefn-y-bedd **LL12** 48 C7
Cefn-y-pant **SA34** 16 E3
Cegidfa (Guilsfield) **SY21** 38 B4
Ceidio **LL71** 46 C4
Ceidio Fawr **LL53** 36 B2
Ceinewydd (New Quay) SA45 26 **C2**
Ceint **LL77** 46 C5
Cellan **SA48** 17 K1
Cellardyke **KY10** 83 G7
Cellarhead **ST9** 40 B1
Cemaes LL67 46 **B3**
Cemmaes **SY20** 37 H5
Cemmaes Road (Glantwymyn) **SY20** 37 H5
Cenarth **SA38** 17 F1
Cennin **LL51** 36 D1
Ceos (Keose) **HS2** 101 F5
Ceres **KY15** 83 F6
Ceri (Kerry) **SY16** 38 A6
Cerist **SY17** 37 J7
Cerne Abbas **DT2** 9 F4
Cerney Wick **GL7** 20 D2
Cerrigceinwen **LL62** 46 C5
Cerrigydrudion **LL21** 37 J1
Cessford **TD5** 70 C1
Ceunant **LL55** 46 D6
Chaceley **GL19** 29 H5
Chacewater **TR4** 2 E4
Chackmore **MK18** 31 H5
Chacombe **OX17** 31 F4
Chad Valley **B15** 40 C7
Chadderton **OL9** 49 J2
Chadderton Fold **OL9** 49 H2
Chaddesden **DE21** 41 F2
Chaddesley Corbett **DY10** 29 H1
Chaddleworth **RG20** 21 H4
Chadlington **OX7** 30 E6
Chadshunt **CV35** 30 E3
Chadstone **NN7** 32 B3
Chadwell *Leics.* **LE14** 42 A3
Chadwell *Shrop.* **TF10** 39 G4
Chadwell St. Mary **RM16** 24 C4
Chadwick End **B93** 30 D1
Chaffcombe **TA20** 8 C3
Chafford Hundred **RM16** 24 C4
Chagford **TQ13** 7 F7
Chailey **BN8** 13 G5
Chainhurst **TN12** 14 C3
Chalbury **BH21** 10 B4
Chalbury Common **BH21** 10 B4
Chaldon **CR3** 23 G6
Chaldon Herring (East Chaldon) **DT2** 9 G6
Chale **PO38** 11 F7
Chale Green **PO38** 11 F6
Chalfont Common **SL9** 22 D2
Chalfont St. Giles HP8 22 **C2**
Chalfont St. Peter **SL9** 22 D2
Chalford *Glos.* **GL6** 20 B1
Chalford *Wilts.* **BA13** 20 B7
Chalgrove **OX44** 21 K2
Chalk **DA12** 24 C4
Chalk End **CM1** 33 K7
Challacombe **EX31** 6 E1
Challister **ZE2** 107 P6
Challoch **DG8** 64 D4
Challock **TN25** 15 F2
Chalmington **DT2** 8 E4
Chalton *Cen.Beds.* **LU4** 32 D6
Chalton *Hants.* **PO8** 11 J3
Chalvey **SL1** 22 C4
Chalvington **BN27** 13 J6
Champany **EH49** 75 J3
Chancery **SY23** 26 E1
Chandler's Cross **WD3** 22 D2
Chandler's Ford **SO53** 11 F2
Channel Islands **GYJE** 3 G7
Channel's End **MK44** 32 E3
Channerwick **ZE2** 107 N10
Chantry *Som.* **BA11** 20 A7
Chantry *Suff.* **IP2** 35 F4
Chapel **KY2** 76 A1
Chapel Allerton *Som.* **BS26** 19 H6
Chapel Allerton *W.Yorks.* **LS7** 57 J6
Chapel Amble **PL27** 3 G1
Chapel Brampton **NN6** 31 J2
Chapel Chorlton **ST5** 40 A2
Chapel Cleeve **TA24** 7 J1
Chapel Cross **TN21** 13 K4
Chapel End **MK45** 32 D4
Chapel Green *Warks.* **CV7** 40 E7
Chapel Green *Warks.* **CV47** 31 F2
Chapel Haddlesey **YO8** 58 B7
Chapel Hill *Aber.* **AB42** 91 J1
Chapel Hill *Lincs.* **LN4** 53 F7
Chapel Hill *Mon.* **NP16** 19 J1
Chapel Hill *N.Yorks.* **LS22** 57 J5
Chapel Knapp **SN13** 20 B5
Chapel Lawn **SY7** 28 C1
Chapel Leigh **TA4** 7 K3
Chapel Milton **SK23** 50 C4
Chapel of Garioch **AB51** 91 F2
Chapel Rossan **DG9** 64 B6
Chapel Row *Essex* **CM3** 24 D1
Chapel Row *W.Berks.* **RG7** 21 J5
Chapel St. Leonards **PE24** 53 J5
Chapel Stile **LA22** 60 E6
Chapel Town **TR8** 3 F3
Chapelbank **PH3** 82 B6
Chapeldonan **KA26** 67 F3
Chapelend Way **CO9** 34 B5
Chapel-en-le-Frith **SK23** 50 C4
Chapelgate **PE12** 43 H3
Chapelhall **ML6** 75 F4
Chapelhill *High.* **IV20** 97 F4
Chapelhill *P. & K.* **PH2** 82 D5
Chapelhill *P. & K.* **PH1** 82 B4
Chapelknowe **DG14** 69 J6
Chapel-le-Dale **LA6** 56 C2
Chapelthorpe **WF2** 51 F1
Chapelton *Aber.* **AB39** 91 G6
Chapelton *Angus* **DD11** 83 H3
Chapelton *Devon* **EX37** 6 D3
Chapelton *S.Lan.* **ML10** 74 E6
Chapeltown *B'burn.* **BL7** 49 G1
Chapeltown *Cumb.* **CA6** 69 K6
Chapeltown *Moray* **AB37** 89 K2
Chapeltown *S.Yorks.* **S35** 51 F3
Chapmans Well **PL15** 6 B6
Chapmanslade **BA13** 20 B7
Chapmore End **SG12** 33 G7
Chappel **CO6** 34 C6
Charaton **PL14** 4 D4
Chard TA20 8 **C4**
Chard Junction **TA20** 8 C4
Chardleigh Green **TA20** 8 C3
Chardstock **EX13** 8 C4
Charfield **GL12** 20 A2
Charing **TN27** 14 E3
Charing Cross **SP6** 10 C3
Charing Heath **TN27** 14 E3
Charingworth **GL55** 30 C5
Charlbury **OX7** 30 E7
Charlcombe **BA1** 20 A5
Charlecote **CV35** 30 D3
Charles **EX32** 6 E2
Charles Tye **IP14** 34 E3
Charlesfield **TD6** 70 A1
Charleshill **GU10** 22 B7
Charleston **DD8** 82 E3
Charlestown *Aber.* **AB43** 99 J4
Charlestown *Aberdeen* **AB12** 91 H4
Charlestown *Cornw.* **PL25** 4 A5
Charlestown *Derbys.* **SK13** 50 C3
Charlestown *Dorset* **DT3** 9 F7
Charlestown *Fife* **KY11** 75 J2
Charlestown *Gt.Man.* **M7** 49 H2
Charlestown *High.* **IV21** 94 E4
Charlestown *High.* **IV1** 96 D7
Charlestown *W.Yorks.* **BD17** 57 G6
Charlestown *W.Yorks.* **HX7** 56 E7
Charlestown of Aberlour (Aberlour) AB38 97 **K7**
Charlesworth **SK13** 50 C3
Charleton **KY9** 83 F7
Charlinch **TA5** 8 B1
Charlotteville **GU1** 22 D7
Charlton *Gt.Lon.* **SE7** 23 H4
Charlton *Hants.* **SP10** 21 G7
Charlton *Herts.* **SG5** 32 E6
Charlton *Northants.* **OX17** 31 G5
Charlton *Northumb.* **NE48** 70 D5
Charlton *Oxon.* **OX12** 21 H3
Charlton *Som.* **BA3** 19 K6
Charlton *Som.* **BA4** 19 K7
Charlton *Som.* **TA3** 8 B2
Charlton *Tel. & W.* **TF6** 38 E4
Charlton *W.Suss.* **PO18** 12 B5
Charlton *Wilts.* **SP7** 9 J2
Charlton *Wilts.* **SN16** 20 C3
Charlton *Wilts.* **SN9** 20 E6
Charlton *Worcs.* **WR10** 30 B4
Charlton Abbots **GL54** 30 B6
Charlton Adam **TA11** 8 E2
Charlton Down **DT2** 9 F5
Charlton Horethorne **DT9** 9 F2
Charlton Kings **GL52** 29 J6
Charlton Mackrell **TA11** 8 E2
Charlton Marshall **DT11** 9 H4
Charlton Musgrove **BA9** 9 G2
Charlton on the Hill **DT11** 9 H4
Charlton-All-Saints **SP5** 10 C2
Charlton-on-Otmoor **OX5** 31 G7
Charltons **TS12** 63 H5
Charlwood **RH6** 23 F7
Charminster **DT2** 9 F5
Charmouth **DT6** 8 C5
Charndon **OX27** 31 H6
Charney Bassett **OX12** 21 G2
Charnock Richard **PR7** 48 E1
Charsfield **IP13** 35 G3
Chart Corner **ME17** 14 C3
Chart Sutton **ME17** 14 D3
Charter Alley **RG26** 21 J6
Charterhouse **BS40** 19 H6
Charterville Allotments **OX29** 21 G1
Chartham **CT4** 15 G2
Chartham Hatch **CT4** 15 G2
Chartridge **HP5** 22 C1
Charvil **RG10** 22 A4
Charwelton **NN11** 31 G3
Chase End Street **HR8** 29 G5
Chase Terrace **WS7** 40 C5
Chasetown **WS7** 40 C5
Chastleton **GL56** 30 D6
Chasty **EX22** 6 B5
Chatburn **BB7** 56 C5
Chatcull **ST21** 39 G2
Chatham ME4 24 **D5**
Chatham Green **CM3** 34 B7
Chathill NE67 71 **G1**
Chattenden **ME3** 24 D4
Chatteris PE16 43 **G7**
Chattisham **IP8** 34 E4
Chatto **TD5** 70 C2
Chatton **NE66** 71 F1
Chaul End **LU1** 32 D6
Chavey Down **SL5** 22 B5
Chawleigh **EX18** 7 F4
Chawley **OX2** 21 H1
Chawston **MK44** 32 E3
Chawton **GU34** 11 J1
Chazey Heath **RG4** 21 K4
Cheadle *Gt.Man.* SK8 49 **H4**
Cheadle *Staffs.* **ST10** 40 C1
Cheadle Heath **SK3** 49 H4
Cheadle Hulme **SK8** 49 H4
Cheam **SM3** 23 F5
Cheapside **SL5** 22 C5
Chearsley **HP18** 31 J7
Chebsey **ST21** 40 A3
Checkendon **RG8** 21 K3
Checkley *Ches.E.* **CW5** 39 G1
Checkley *Here.* **HR1** 28 E5
Checkley *Staffs.* **ST10** 40 C2
Checkley Green **CW5** 39 G1
Chedburgh **IP29** 34 B3
Cheddar BS27 19 **H6**
Cheddington **LU7** 32 C7
Cheddleton **ST13** 49 J7
Cheddon Fitzpaine **TA2** 8 B2
Chedglow **SN16** 20 C2
Chedgrave **NR14** 45 H6
Chedington **DT8** 8 D4
Chediston **IP19** 35 H1
Chediston Green **IP19** 35 H1
Chedworth **GL54** 30 B7
Chedzoy **TA7** 8 C1
Cheesden **OL12** 49 H1
Cheeseman's Green **TN24** 15 F4
Cheetham Hill **M8** 49 H2
Cheglinch **EX34** 6 D1
Cheldon **EX18** 7 F4
Chelford **SK11** 49 H5
Chellaston **DE73** 41 F2
Chells **SG2** 33 F6
Chelmarsh **WV16** 39 G7
Chelmondiston **IP9** 35 G5
Chelmorton **SK17** 50 D6
CHELMSFORD CM 24 **D1**
Chelmsley Wood **B37** 40 D7
Chelsea **SW3** 23 F4
Chelsfield **BR6** 23 H5
Chelsham **CR6** 23 G6
Chelston Heath **TA21** 7 K3
Chelsworth **IP7** 34 D4
Cheltenham GL50 29 **J6**
Chelveston **NN9** 32 C2
Chelvey **BS48** 19 H5
Chelwood **BS39** 19 K5
Chelwood Common **RH17** 13 H4
Chelwood Gate **RH17** 13 H4
Chelworth **SN16** 20 C2
Cheney Longville **SY7** 38 D7
Chenies **WD3** 22 D2
Chepstow (Cas-gwent) NP16 19 **J2**
Chorhill **SN11** 20 D4
Cherington *Glos.* **GL8** 20 C2
Cherington *Warks.* **CV36** 30 D5
Cheriton *Devon* **EX35** 7 F1
Cheriton *Hants.* **SO24** 11 G2
Cheriton *Kent* **CT19** 15 G4
Cheriton *Pembs.* **SA71** 16 C6
Cheriton *Swan.* **SA3** 17 H6
Cheriton Bishop **EX6** 7 F6
Cheriton Cross **EX6** 7 F6
Cheriton Fitzpaine **EX17** 7 G5
Cherrington **TF6** 39 F3
Cherry Burton **HU17** 59 F5
Cherry Green **CM6** 33 J6
Cherry Hinton **CB1** 33 H3
Cherry Willingham **LN3** 52 D5
Chertsey KT16 22 **D5**
Cheselbourne **DT2** 9 G5
Chesham HP5 22 **C1**
Chesham Bois **HP6** 22 C1
Cheshunt **EN8** 23 G1
Cheslyn Hay **WS6** 40 B5
Chessington KT9 22 **E5**
Chestall **WS15** 40 C4
CHESTER CH 48 **D6**
Chester Moor **DH2** 62 D2
Chesterblade **BA4** 19 K7

Che - Clo

Chesterfield *Derbys.* S40 51 F5
Chesterfield *Staffs.* WS14 40 D5
Chester-le-Street DH3 62 D1
Chesters *Sc.Bord.* TD9 70 B2
Chesters *Sc.Bord.* TD8 70 B1
Chesterton *Cambs.* CB4 33 H2
Chesterton *Cambs.* PE7 42 E6
Chesterton *Oxon.* OX26 31 G6
Chesterton *Shrop.* WV15 39 G6
Chesterton *Staffs.* ST5 40 A1
Chesterton *Warks.* CV33 30 E3
Chesterton Green CV33 30 E3
Chestfield CT5 25 H5
Cheston TQ10 5 G5
Cheswardine TF9 39 G3
Cheswick TD15 77 J6
Cheswick Buildings TD15 77 J6
Cheswick Green B90 30 C1
Chetnole DT9 9 F4
Chettiscombe EX16 7 H4
Chettisham CB6 43 J7
Chettle DT11 9 J3
Chetton WV16 39 F6
Chetwode MK18 31 H6
Chetwynd Aston TF10 39 G4
Chetwynd Park TF10 39 G3
Cheveley CB8 33 K2
Chevening TN14 23 H6
Cheverell's Green AL3 32 D7
Chevington IP29 34 B3
Chevington Drift NE61 71 H4
Chevithorne EX16 7 H4
Chew Magna BS40 19 J5
Chew Moor BL6 49 F2
Chew Stoke BS40 19 J5
Chewton Keynsham BS31 19 K5
Chewton Mendip BA3 19 J6
Chichacott EX20 6 E6
Chicheley MK16 32 C4
Chichester PO19 12 B6
Chickerell DT3 9 F6
Chickering IP21 35 G1
Chicklade SP3 9 J1
Chickney CM6 33 J6
Chicksands SG17 32 E5
Chidden PO7 11 H3
Chidden Holt PO7 11 H3
Chiddingfold GU8 12 C3
Chiddingly BN8 13 J5
Chiddingstone TN8 23 H7
Chiddingstone Causeway TN11 23 J7
Chiddingstone Hoath TN8 23 H7
Chideock DT6 8 D5
Chidham PO18 11 J4
Chidswell WF12 57 H7
Chieveley RG20 21 H4
Chignall St. James CM1 24 C1
Chignall Smealy CM1 33 K7
Chigwell IG7 23 H2
Chigwell Row IG7 23 H2
Chilbolton SO20 21 G7
Chilcomb SO21 11 G2
Chilcombe DT6 8 E5
Chilcompton BA3 19 K6
Chilcote DE12 40 E4
Child Okeford DT11 9 H3
Childer Thornton CH66 48 C5
Childerditch CM13 24 C3
Childrey OX12 21 G3
Child's Ercall TF9 39 F3
Childs Hill NW3 23 F3
Childswickham WR12 30 B5
Childwall L16 48 D4
Childwick Green AL3 32 E7
Chilfrome DT2 8 E5
Chilgrove PO18 12 B5
Chilham CT4 15 F2
Chilhampton SP2 10 B1
Chilla EX21 6 C5
Chillaton PL16 6 C7
Chillenden CT3 15 H2
Chillerton PO30 11 F6
Chillesford IP12 35 H3
Chilley TQ9 5 H5
Chillingham NE66 71 F1
Chillington *Devon* TQ7 5 H6
Chillington *Som.* TA19 8 C3
Chilmark SP3 9 J1
Chilson *Oxon.* OX7 30 E7
Chilson *Som.* TA20 8 C4
Chilsworthy *Cornw.* PL18 4 E3
Chilsworthy *Devon* EX22 6 B5
Chilthorne Domer BA22 8 E3
Chilton *Bucks.* HP18 31 H7
Chilton *Devon* EX17 7 G5
Chilton *Dur.* DL17 62 D4
Chilton *Oxon.* OX11 21 H3
Chilton *Suff.* CO10 34 C4
Chilton Candover SO24 11 G1
Chilton Cantelo BA22 8 E2
Chilton Foliat RG17 21 G4
Chilton Polden TA7 8 C1
Chilton Street CO10 34 B4
Chilton Trinity TA5 8 B1
Chilvers Coton CV10 41 F6
Chilwell NG9 41 H2
Chilworth *Hants.* SO16 11 F3
Chilworth *Surr.* GU4 22 D7
Chimney OX18 21 G1
Chimney Street CO10 34 B4
Chineham RG24 21 K6
Chingford E4 23 G2
Chinley SK23 50 C4
Chinley Head SK23 50 C4
Chinnor OX39 22 A1
Chipchase Castle NE48 70 D6
Chipley TA21 7 K3
Chipnall TF9 39 G2
Chippenham *Cambs.* CB7 33 K2
Chippenham *Wilts.* SN15 20 C4
Chipperfield WD4 22 D1
Chipping *Herts.* SG9 33 G5
Chipping *Lancs.* PR3 56 B5
Chipping Campden GL55 30 C5
Chipping Hill CM8 34 C7
Chipping Norton OX7 30 E6
Chipping Ongar CM5 23 J1
Chipping Sodbury BS37 20 A3
Chipping Warden OX17 31 F4
Chipstable TA4 7 J3
Chipstead *Kent* TN13 23 H6
Chipstead *Surr.* CR5 23 F6
Chirbury SY15 38 B6
Chirk (Y Waun) LL14 38 B2
Chirk Green LL14 38 B2
Chirmorrie KA26 64 C3
Chirnside TD11 77 G5
Chirnsidebridge TD11 77 G5
Chirton *T. & W.* NE29 71 J7
Chirton *Wilts.* SN10 20 D6
Chisbury SN8 21 F5
Chiscan PA28 66 A2
Chiselborough TA14 8 D3
Chiseldon SN4 20 E4
Chiserley HX7 57 F7
Chislehampton OX44 21 J2
Chislehurst BR7 23 H4
Chislet CT3 25 J5
Chiswell Green AL2 22 E1
Chiswick W4 23 F4
Chiswick End SG8 33 G4
Chisworth SK13 49 J3
Chithurst GU31 12 B4
Chittering CB25 33 H1
Chitterne BA12 20 C7
Chittlehamholt EX37 6 E3
Chittlehampton EX37 6 E3
Chittoe SN15 20 C5
Chivelstone TQ7 5 H7
Chivenor EX31 6 D2
Chobham GU24 22 C5
Choicelee TD11 77 F5
Cholderton SP4 21 F7
Cholesbury HP23 22 C1
Chollerford NE46 70 E6
Chollerton NE46 70 E6
Cholsey OX10 21 J3
Cholstrey HR6 28 D3
Cholwell *B. & N.E.Som.* BS39 19 K6
Cholwell *Devon* PL19 4 E3
Chop Gate TS9 63 G7
Choppington NE62 71 H5
Chopwell NE17 62 C1
Chorley *Ches.E.* CW5 48 E7
Chorley *Lancs.* PR7 48 E1
Chorley *Shrop.* WV16 39 F7
Chorley *Staffs.* WS13 40 C4
Chorleywood WD3 22 D2
Chorlton CW2 49 G7
Chorlton Lane SY14 38 D1
Chorlton-cum-Hardy M21 49 H3
Chowley CH3 48 D7
Chrishall SG8 33 H5
Chrishall Grange SG8 33 H4
Chrisswell PA16 74 A3
Christchurch *Cambs.* PE14 43 H6
Christchurch *Dorset* BH23 10 C5
Christchurch *Glos.* GL16 28 E7
Christchurch *Newport* NP18 19 G3
Christian Malford SN15 20 C4
Christleton CH3 48 D6
Christmas Common OX49 22 A2
Christon BS26 19 G6
Christon Bank NE66 71 H1
Christow EX6 7 G7
Christskirk AB52 90 E2
Chryston G69 74 E3
Chudleigh TQ13 5 J3
Chudleigh Knighton TQ13 5 J3
Chulmleigh EX18 6 E4
Chunal SK13 50 C3
Church BB5 56 C7
Church Aston TF10 39 G4
Church Brampton NN6 31 J2
Church Brough CA17 61 J5
Church Broughton DE65 40 E2
Church Charwelton NN11 31 G3
Church Common IP17 35 H3
Church Crookham GU52 22 B6
Church Eaton ST20 40 A4
Church End *Cambs.* PE13 43 G5
Church End *Cambs.* CB24 33 G1
Church End *Cambs.* PE28 43 F7
Church End *Cen.Beds.* SG15 32 E5
Church End *Cen.Beds.* MK17 32 C5
Church End *Cen.Beds.* LU6 32 C6
Church End *Cen.Beds.* SG19 32 E3
Church End *Cen.Beds.* MK43 32 C5
Church End *E.Riding* YO25 59 G4
Church End *Essex* CB10 33 J4
Church End *Essex* CM7 34 B6
Church End *Glos.* GL20 29 H5
Church End *Hants.* RG27 21 K6
Church End *Herts.* AL3 32 E7
Church End *Herts.* SG11 33 H6
Church End *Lincs.* PE11 43 F2
Church End *Lincs.* LN11 53 H3
Church End *Warks.* CV10 40 E6
Church End *Wilts.* SN15 20 D4
Church Enstone OX7 30 E6
Church Fenton LS24 58 B6
Church Green EX24 7 K6
Church Gresley DE11 40 E4
Church Hanborough OX29 31 F7
Church Hill *Ches.W. & C.* CW7 49 F6
Church Hill *Derbys.* S42 51 G6
Church Houses YO62 63 H7
Church Knowle BH20 9 J6
Church Laneham DN22 52 B5
Church Langley CM17 23 H1
Church Langton LE16 42 A6
Church Lawford CV23 31 F1
Church Lawton ST7 49 H7
Church Leigh ST10 40 C2
Church Lench WR11 30 B3
Church Mayfield DE6 40 D1
Church Minshull CW5 49 F6
Church Norton PO20 12 B7
Church Preen SY6 38 E6
Church Pulverbatch SY5 38 D5
Church Stoke SY15 38 B6
Church Stowe NN7 31 H3
Church Street *Essex* CO10 34 B4
Church Street *Kent* ME3 24 D4
Church Stretton SY6 38 D6
Church Town *Leics.* LE67 41 F4
Church Town *Surr.* RH9 23 G6
Church Village CF38 18 D3
Church Warsop NG20 51 H6
Church Westcote OX7 30 D6
Church Wilne DE72 41 G2
Churcham GL2 29 G7
Churchdown GL3 29 H7
Churchend *Essex* SS3 25 G2
Churchend *Essex* CM6 33 K6
Churchend *S.Glos.* GL12 20 A2
Churchfield B71 40 C6
Churchgate EN7 23 G1
Churchgate Street CM17 33 H7
Churchill *Devon* EX31 6 D1
Churchill *Devon* EX13 8 B4
Churchill *N.Som.* BS25 19 H6
Churchill *Oxon.* OX7 30 D6
Churchill *Worcs.* WR7 29 J3
Churchill *Worcs.* DY10 29 H1
Churchingford TA3 8 B3
Churchover CV23 41 H7
Churchstanton TA3 7 K4
Churchstow TQ7 5 H6
Churchtown *Devon* EX31 6 E1
Churchtown *I.o.M.* IM7 54 D4
Churchtown *Lancs.* PR3 55 H5
Churchtown *Mersey.* PR9 48 C1
Churnsike Lodge NE48 70 B6
Churston Ferrers TQ5 5 K5
Churt GU10 12 B3
Churton CH3 48 D7
Churwell LS27 57 H7
Chute Cadley SP11 21 G6
Chute Standen SP11 21 G6
Chwilog LL53 36 D2
Chwitffordd (Whitford) CH8 47 K5
Chyandour TR18 2 B5
Chysauster TR20 2 B5
Cilan Uchaf LL53 36 B3
Cilcain CH7 47 K6
Cilcennin SA48 26 E2
Cilcewydd SY21 38 B5
Cilfrew SA10 18 A1
Cilfynydd CF37 18 D2
Cilgerran SA43 16 E1
Cilgwyn *Carmar.* SA20 27 G5
Cilgwyn *Pembs.* SA42 16 D2
Ciliau Aeron SA48 26 E3
Cilldonnain (Kildonan) HS8 84 C2
Cille Bhrighde HS8 84 C3
Cille Pheadair HS8 84 C3
Cilmaengwyn SA8 18 A1
Cilmery LD2 27 K3
Cilrhedyn SA35 17 F2
Cilrhedyn Bridge SA65 16 D2
Cilsan SA19 17 J3
Ciltalgarth LL23 37 H1
Cilwendeg SA37 17 F2
Cilybebyll SA8 18 A1
Cilycwm SA20 27 G4
Cimla SA11 18 A2
Cinderford GL14 29 F7
Cippenham SL1 22 C3
Cippyn SA43 16 E1
Cirbhig HS2 100 D3
Circebost (Kirkibost) HS2 100 D4
Cirencester GL7 20 D1
City *Gt.Lon.* EC3M 23 G3
City *V. of Glam.* CF71 18 C4
City Airport E16 23 H3
City Dulas LL70 46 C4
Clabhach PA78 78 C2
Clachaig PA23 73 K2
Clachan *Arg. & B.* PA26 80 C6
Clachan *Arg. & B.* PA34 79 K3
Clachan *Arg. & B.* PA29 73 F5
Clachan *High.* IV40 86 B1
Clachan *W.Isles* HS8 92 C7
Clachan Mòr PA77 78 A3
Clachan of Campsie G66 74 E3
Clachan of Glendaruel PA22 73 H2
Clachan Strachur (Strachur) PA27 80 B7
Clachan-a-Luib HS6 92 D5
Clachandhu PA68 79 F4
Clachaneasy DG8 64 D3
Clachanmore DG9 64 A6
Clachan-Seil PA34 79 J6
Clachanturn AB35 89 K5
Clachbreck PA31 73 F3
Clachnabrain DD8 82 E1
Clachnaharry IV3 96 D7
Clachtoll IV27 102 C6
Clackmannan FK10 75 H1
Clackmarras IV30 97 K6
Clacton-on-Sea CO15 35 F7
Cladach a Bhale Shear HS6 92 D5
Cladach a' Chaolais HS6 92 C5
Cladach Chircebost HS6 92 C5
Cladach Chnoc a Lin HS6 92 C5
Cladich PA33 80 B5
Cladswell B49 30 B3
Claggan *High.* PH33 87 H7
Claggan *High.* PA34 79 J3
Claigan IV55 93 H6
Claines WR3 29 H3
Clandown BA3 19 K6
Clanfield *Hants.* PO8 11 J3
Clanfield *Oxon.* OX18 21 F1
Clannaborough Barton EX17 7 F5
Clanville SP11 21 G7
Claonaig PA29 73 G5
Claonairigh PA32 80 B7
Claonel IV27 96 C1
Clapgate SG11 33 H6
Clapham *Bed.* MK41 32 D3
Clapham *Devon* EX2 7 G7
Clapham *Gt.Lon.* SW4 23 F4
Clapham *N.Yorks.* LA2 56 C3
Clapham *W.Suss.* BN13 12 D6
Clapham Green MK41 32 D3
Clapham Hill CT5 25 H5
Clappers TD15 77 H5
Clappersgate LA22 60 E6
Clapton *Som.* TA18 8 D4
Clapton *Som.* BA3 19 K6
Clapton-in-Gordano BS20 19 H4
Clapton-on-the-Hill GL54 30 C7
Clapworthy EX36 6 E3
Clara Vale NE40 71 G7
Clarach SY23 37 F7
Clarbeston SA63 16 D3
Clarbeston Road SA63 16 D3
Clarborough DN22 51 K4
Clardon KW14 105 G2
Clare CO10 34 B4
Clarebrand DG7 65 H4
Clarencefield DG1 69 F7
Clarilaw TD9 70 A2
Clark's Green RH5 12 E3
Clarkston G76 74 D5
Clashban IV24 96 D2
Clashcoig IV24 96 D2
Clashdorran IV4 96 C7
Clashgour PA36 80 D3
Clashindarroch AB54 90 C1
Clashmore *High.* IV25 96 E3
Clashmore *High.* IV27 102 C5
Clashnessie IV27 102 C5
Clashnoir AB37 89 K2
Clatford SN8 20 E5
Clathy PH7 82 A5
Clatt AB54 90 D2
Clatter SY17 37 J6
Clattercote OX17 31 F4
Clatterford PO30 11 F6
Clatterford End CM5 23 J1
Clatterin Brig AB30 90 E7
Clatteringshaws DG7 65 F3
Clatworthy TA4 7 J2
Claughton *Lancs.* LA2 55 J3
Claughton *Lancs.* PR3 55 J5
Clavelshay TA6 8 B1
Claverdon CV35 30 C2
Claverham BS49 19 H5
Clavering CB11 33 H5
Claverley WV5 39 G6
Claverton BA2 20 A5
Claverton Down BA2 20 A5
Clawdd-côch CF71 18 D4
Clawdd-newydd LL15 47 J7
Clawfin KA6 67 K3
Clawthorpe LA6 55 J2
Clawton EX22 6 B6
Claxby LN8 52 E3
Claxby Pluckacre LN9 53 G6
Claxby St. Andrew LN13 53 H5
Claxton *N.Yorks.* YO60 58 C3
Claxton *Norf.* NR14 45 H5
Claxton Grange TS22 63 F4
Clay Common NR34 45 J7
Clay Coton NN6 31 G1
Clay Cross S45 51 F6
Clay End SG2 33 G6
Clay Hill BS16 19 K4
Clay of Allan IV20 97 F4
Claybrooke Magna LE17 41 G7
Claybrooke Parva LE17 41 G7
Claydene TN8 23 H7
Claydon *Oxon.* OX17 31 F3
Claydon *Suff.* IP6 35 F4
Claygate *Kent* TN12 14 C3
Claygate *Surr.* KT10 22 E5
Claygate Cross TN15 23 K6
Clayhanger *Devon* EX16 7 J3
Clayhanger *W.Mid.* WS8 40 C5
Clayhidon EX15 7 K4
Clayhill *E.Suss.* TN31 14 D5
Clayhill *Hants.* SO43 10 E4
Clayhithe CB25 33 J2
Clayock KW12 105 G3
Claypit Hill CB23 33 G3
Claypits GL10 20 A1
Claypole NG23 42 B1
Claythorpe LN13 53 H5
Clayton *S.Yorks.* DN5 51 G2
Clayton *Staffs.* ST5 40 A1
Clayton *W.Suss.* BN6 13 F5
Clayton *W.Yorks.* BD14 57 G6
Clayton Green PR6 55 J7
Clayton West HD8 50 E1
Clayton-le-Moors BB5 56 C6
Clayton-le-Woods PR25 55 J7
Clayworth DN22 51 K4
Cleadale PH42 85 K6
Cleadon SR6 71 J7
Clearbrook PL20 5 F4
Clearwell GL16 19 J1
Cleasby DL2 62 D5
Cleat *Ork.* KW17 106 D9
Cleat *W.Isles* HS9 84 B4
Cleatlam DL2 62 C5
Cleatop BD24 56 D3
Cleator CA23 60 B5
Cleator Moor CA25 60 B5
Cleckheaton BD19 57 G7
Clee St. Margaret SY7 38 E7
Cleedownton SY8 38 E7
Cleehill SY8 28 E1
Cleestanton SY8 28 E1
Cleethorpes DN35 53 G2
Cleeton St. Mary DY14 29 F1
Cleeve *N.Som.* BS49 19 H5
Cleeve *Oxon.* RG8 21 K3
Cleeve Hill GL52 29 J6
Cleeve Prior WR11 30 B4
Cleghorn ML11 75 G6
Clehonger HR2 28 D5
Cleigh PA34 79 K5
Cleish KY13 75 J1
Cleland ML1 75 F5
Clement's End LU6 32 D7
Clench Common SN8 20 E5
Clenchwarton PE34 43 J3
Clennell NE65 70 E3
Clent DY9 29 J1
Cleobury Mortimer DY14 29 F1
Cleobury North WV16 39 F7
Clephanton IV2 97 F6
Clerklands TD6 70 A1
Clermiston EH4 75 K3
Clestrain KW17 106 C7
Cleuch Head TD9 70 A2
Cleughbrae DG1 69 F6
Clevancy SN11 20 D4
Clevedon BS21 19 H4
Cleveland Tontine Inn DL6 63 F7
Cleveley OX7 30 E6
Cleveleys FY5 55 G5
Clevelode WR13 29 H4
Cleverton SN15 20 C3
Clewer BS28 19 H6
Clewer Green SL4 22 C4
Clewer Village SL4 22 C4
Cley next the Sea NR25 44 E1
Cliburn CA10 61 G4
Cliddesden RG25 21 K7
Cliff *Carmar.* SA17 17 G5
Cliff *High.* PH36 79 H1
Cliff End TN35 14 D6
Cliff Grange TF9 39 F2
Cliffe *Lancs.* BB6 56 C6
Cliffe *Med.* ME3 24 D4
Cliffe *N.Yorks.* YO8 58 C6
Cliffe Woods ME3 24 D4
Clifford *Here.* HR3 28 B4
Clifford *W.Yorks.* LS23 57 K5
Clifford Chambers CV37 30 C3
Clifford's Mesne GL18 29 F6
Cliffs End CT12 25 K5
Clifton *Bristol* BS8 19 J4
Clifton *Cen.Beds.* SG17 32 E5
Clifton *Cumb.* CA10 61 G4
Clifton *Derbys.* DE6 40 D1
Clifton *Devon* EX31 6 D1
Clifton *Lancs.* PR4 55 H6
Clifton *N.Yorks.* LS21 57 G5
Clifton *Northumb.* NE61 71 H5
Clifton *Nott.* NG11 41 H2
Clifton *Oxon.* OX15 31 F5
Clifton *S.Yorks.* S66 51 H3
Clifton *Stir.* FK20 80 E4
Clifton *W.Yorks.* HD6 57 G7
Clifton *Worcs.* WR8 29 H4
Clifton *York* YO30 58 B4
Clifton Campville B79 40 E4
Clifton Hampden OX14 21 J2
Clifton Maybank BA22 8 E3
Clifton Reynes MK46 32 C3
Clifton upon Dunsmore CV23 31 G1
Clifton upon Teme WR6 29 G2
Cliftonville CT9 25 K4
Climping BN17 12 D6
Climpy ML11 75 H5
Clink BA11 20 A7
Clint HG3 57 H4
Clint Green NR19 44 E4
Clinterty AB21 91 G3
Clintmains TD6 76 E7
Clippesby NR29 45 J4
Clippings Green NR20 44 E4
Clipsham LE15 42 C4
Clipston *Northants.* LE16 42 A7
Clipston *Notts.* NG12 41 J2
Clipstone NG21 51 H6
Clitheroe BB7 56 C5
Cliuthar (Cluer) HS3 93 G2
Clive SY4 38 E3
Clivocast ZE2 107 Q2
Clixby DN38 52 E2
Clocaenog LL15 47 J7
Clochan AB56 98 C4
Clochtow AB41 91 J1
Clock Face WA9 48 E3
Clockhill AB42 99 G6
Cloddach IV30 97 J6
Cloddiau SY21 38 A5
Clodock HR2 28 C6

Clo - Cor

Cloford **BA11** 20 A7
Cloichran **FK21** 81 H4
Clola **AB42** 99 J6
Clonrae **DG3** 68 D4
Clophill **MK45** 32 D5
Clopton **NN14** 42 D7
Clopton Corner **IP13** 35 G3
Clopton Green *Suff.* **CB8** 34 B3
Clopton Green *Suff.* **IP13** 35 G3
Close Clark **IM9** 54 B6
Closeburn **DG3** 68 D4
Closworth **BA22** 8 E3
Clothall **SG7** 33 F5
Clothan **ZE2** 107 N4
Clotton **CW6** 48 E6
Clough *Cumb.* **LA10** 61 J7
Clough *Gt.Man.* **OL15** 49 J1
Clough *Gt.Man.* **OL2** 49 J2
Clough *W.Yorks.* **HD7** 50 C1
Clough Foot **OL14** 56 E7
Clough Head **HX6** 57 F7
Cloughfold **BB4** 56 D7
Cloughton **YO13** 63 K3
Cloughton Newlands **YO13** 63 K3
Clounlaid **PA34** 79 J2
Clousta **ZE2** 107 M7
Clouston **KW16** 106 C6
Clova *Aber.* **AB54** 90 C2
Clova *Angus* **DD8** 90 B7
Clove Lodge **DL12** 62 A5
Clovelly **EX39** 6 B3
Clovelly Cross **EX39** 6 B3
Clovenfords **TD1** 76 C7
Clovenstone **AB51** 91 F3
Cloverhill **AB23** 91 H3
Cloves **IV36** 97 J5
Clovullin **PH33** 80 B1
Clow Bridge **BB11** 56 D7
Clowne **S43** 51 G5
Clows Top **DY14** 29 G1
Cloyntie **KA19** 67 H3
Cluanach **PA43** 72 B5
Clubworthy **PL15** 4 C1
Cluddley **TF6** 39 F4
Cluer (Cluthar) **HS3** 93 G2
Clun **SY7** 38 B7
Clunas **IV12** 97 F7
Clunbury **SY7** 38 C7
Clune *High.* **IV13** 88 E2
Clune *Moray* **AB56** 98 D4
Clunes **PH34** 87 J6
Clungunford **SY7** 28 C1
Clunie *Aber.* **AB53** 98 E5
Clunie *P. & K.* **PH10** 82 C3
Clunton **SY7** 38 C7
Cluny **KY2** 76 A1
Clutton *B. & N.E.Som.* **BS39** 19 K6
Clutton *Ches.W. & C.* **CH3** 48 D7
Clwt-y-bont **LL55** 46 D6
Clwydyfagwyr **CF48** 18 D1
Clydach *Mon.* **NP7** 28 B7
Clydach *Swan.* **SA6** 17 K5
Clydach Terrace **NP23** 28 A7
Clydach Vale **CF40** 18 C2
Clydebank G81 74 C3
Clydey **SA35** 17 F2
Clyffe Pypard **SN4** 20 D4
Clynder **G84** 74 A2
Clynderwen SA66 16 E4
Clyne **SA11** 18 B1
Clynelish **KW9** 97 F1
Clynfyw **SA37** 17 F2
Clynnog-fawr **LL54** 36 D1
Clyro **HR3** 28 B4
Clyst Honiton **EX5** 7 H6
Clyst Hydon **EX15** 7 J5
Clyst St. George **EX3** 7 H7
Clyst St. Lawrence **EX15** 7 J5
Clyst St. Mary **EX5** 7 H6
Clyst William **EX15** 7 J5
Cnewr **LD3** 27 H6
Cnoc **HS2** 101 G4
Cnoc an Torrain (Knockintorran) **HS6** 92 C5
Cnwch Coch **SY23** 27 F1
Coachford **AB54** 98 C6
Coad's Green **PL15** 4 C3
Coal Aston **S18** 51 F5
Coalbrookdale **TF8** 39 F5
Coalbrookvale **NP13** 18 E1
Coalburn **ML11** 75 G7
Coalburns **NE40** 71 G7
Coalcleugh **NE47** 61 K2
Coaley **GL11** 20 A1
Coalmoor **TF6** 39 F5
Coalpit Heath **BS36** 19 K3
Coalpit Hill **ST7** 49 H7
Coalport **TF8** 39 G5
Coalsnaughton **FK13** 75 H1
Coaltown of Balgonie **KY7** 76 B1
Coaltown of Wemyss **KY1** 76 B1
Coalville LE67 41 G4
Coalway **GL16** 28 E7
Coanwood **NE49** 61 H1
Coast **IV22** 95 F2
Coat **TA12** 8 D2
Coatbridge ML5 75 F4
Coate *Swin.* **SN3** 20 E3
Coate *Wilts.* **SN10** 20 D5
Coates *Cambs.* **PE7** 43 G6
Coates *Glos.* **GL7** 20 C1
Coates *Lincs.* **LN1** 52 C4
Coates *Notts.* **DN22** 52 B4
Coates *W.Suss.* **RH20** 12 C5
Coatham **TS10** 63 G4
Coatham Mundeville **DL3** 62 D4
Cobairdy **AB54** 98 D6
Cobbaton **EX37** 6 E3
Cobbler's Plain **NP16** 19 H1
Cobby Syke **HG3** 57 G4
Cobden **EX5** 7 J6
Coberley **GL53** 29 J7
Cobhall Common **HR2** 28 D5
Cobham *Kent* **DA12** 24 C5
Cobham *Surr.* KT11 22 E5
Cobleland **FK8** 74 D1
Cobley Hill **B60** 30 B1
Cobnash **HR6** 28 D2
Coburty **AB43** 99 H4
Cochno **G81** 74 C3
Cock Alley **S44** 51 G6
Cock Bank **LL13** 38 C1
Cock Bevington **WR11** 30 B3
Cock Bridge **AB36** 89 K4
Cock Clarks **CM3** 24 E1
Cock Green **CM6** 34 B7
Cock Marling **TN31** 14 D6
Cockayne **YO62** 63 H7
Cockayne Hatley **SG19** 33 F4
Cockburnspath TD13 77 F3
Cockenzie & Port Seton **EH32** 76 C3
Cocker Bar **PR26** 55 J7
Cockerham **LA2** 55 H4
Cockermouth CA13 60 C3
Cockernhoe **LU2** 32 E6
Cockerton **DL3** 62 D5
Cockett **SA2** 17 K6
Cockfield *Dur.* **DL13** 62 C4
Cockfield *Suff.* **IP30** 34 D3
Cockfosters **EN4** 23 F2
Cocking **GU29** 12 B5
Cockington **TQ2** 5 J4
Cocklake **BS28** 19 H7
Cocklaw **NE46** 70 E6
Cockle Park **NE61** 71 H4
Cockleford **GL53** 29 J7
Cockley Beck **LA20** 60 D6
Cockley Cley **PE37** 44 B5
Cockpen **EH19** 76 B4
Cockpole Green **RG10** 22 A3
Cockshutt **SY12** 38 D3
Cockthorpe **NR23** 44 D1
Cockwood *Devon* **EX6** 7 H7
Cockwood *Som.* **TA5** 19 F7
Cockyard **SK23** 50 C4
Codda **PL15** 4 B3
Coddenham **IP6** 35 F3
Coddington *Ches.W. & C.* **CH3** 48 D7
Coddington *Here.* **HR8** 29 G4
Coddington *Notts.* **NG24** 52 B7
Codford St. Mary **BA12** 9 J1
Codford St. Peter **BA12** 9 J1
Codicote **SG4** 33 F7
Codmore Hill **RH20** 12 D5
Codnor **DE5** 41 G1
Codnor Park **NG16** 51 G7
Codrington **BS37** 20 A4
Codsall **WV8** 40 A5
Codsall Wood **WV8** 40 A5
Coed Morgan **NP7** 28 C7
Coed Ystumgwern **LL44** 36 E3
Coedcae **NP4** 19 F1
Coed-duon (Blackwood) NP12 18 E2
Coedely **CF39** 18 D3
Coedkernew **NP10** 19 F3
Coedpoeth **LL11** 48 B7
Coedway **SY5** 38 C4
Coed-y-bryn **SA44** 17 G1
Coed-y-caerau **NP18** 19 G2
Coed-y-paen **NP4** 19 G2
Coed-y-parc **LL57** 46 E6
Coed-yr-ynys **NP8** 28 A6
Coelbren **SA10** 27 H7
Coffinswell **TQ12** 5 J4
Cofton **EX6** 7 H7
Cofton Hackett **B45** 30 B1
Cogan **CF64** 18 E4
Cogenhoe **NN7** 32 B2
Cogges **OX28** 21 G1
Coggeshall **CO6** 34 C6
Coggeshall Hamlet **CO6** 34 C6
Coggins Mill **TN20** 13 J4
Cóig Peighinnean **HS2** 101 H1
Coilantogle **FK17** 81 G7
Coileitir **PH49** 80 C3
Coilessan **G83** 80 D7
Coillaig **PA35** 80 B5
Coille Mhorgil **PH35** 87 G4
Coille-righ **IV40** 87 F2
Coillore **IV56** 85 J1
Coity **CF35** 18 C3
Col **HS2** 101 G4
Col Uarach **HS2** 101 G4
Colaboll **IV27** 103 H7
Colan **TR8** 3 F2
Colaton Raleigh **EX10** 7 J7
Colbost **IV55** 93 H7
Colburn **DL9** 62 D7
Colbury **SO40** 10 E3
Colby *Cumb.* **CA16** 61 H4
Colby *I.o.M.* **IM9** 54 B6
Colby *Norf.* **NR11** 45 G2
COLCHESTER CO 34 D6
Colchester Green **IP30** 34 D3
Colcot **CF62** 18 E5
Cold Ash **RG18** 21 J5
Cold Ashby **NN6** 31 H1
Cold Ashton **SN14** 20 A4
Cold Aston **GL54** 30 C7
Cold Blow **SA67** 16 E4
Cold Brayfield **MK46** 32 C3
Cold Chapel **ML12** 68 E1
Cold Cotes **LA2** 56 C2
Cold Hanworth **LN8** 52 D4
Cold Harbour **RG8** 21 K3
Cold Hatton **TF6** 39 F3
Cold Hatton Heath **TF6** 39 F3
Cold Hesledon **SR7** 63 F2
Cold Higham **NN12** 31 H3
Cold Inn **SA68** 16 E5
Cold Kirby **YO7** 58 B1
Cold Newton **LE7** 42 A5
Cold Northcott **PL15** 4 C2
Cold Norton **CM3** 24 E1
Cold Overton **LE15** 42 B5
Cold Row **FY6** 55 G5
Coldbackie **IV27** 103 J2
Coldblow **DA5** 23 J4
Coldean **BN1** 13 G6
Coldeast **TQ12** 5 J3
Coldeaton **DE6** 50 D7
Coldeaton Common **SO21** 11 F2
Coldfair Green **IP17** 35 J2
Coldham **PE14** 43 H5
Coldharbour *Glos.* **GL15** 19 J1
Coldharbour *Surr.* **RH5** 22 E7
Coldingham **TD14** 77 H4
Coldrain **KY13** 82 B7
Coldred **CT15** 15 H3
Coldrey **GU34** 22 A7
Coldridge **EX17** 6 E5
Coldrife **NE61** 71 F4
Coldstream TD12 77 G7
Coldvreath **PL26** 3 G3
Coldwaltham **RH20** 12 D5
Coldwells **AB42** 99 K6
Cole **BA10** 9 F1
Cole End **B46** 40 D7
Cole Green **SG14** 33 F7
Cole Henley **RG28** 21 H6
Colebatch **SY9** 38 C7
Colebrook **EX15** 7 J5
Colebrooke **EX17** 7 F5
Coleburn **IV30** 97 K6
Coleby *Lincs.* **LN5** 52 C6
Coleby *N.Lincs.* **DN15** 52 B1
Coleford *Devon* **EX17** 7 F5
Coleford *Glos.* GL16 28 E7
Coleford *Som.* **BA3** 19 K7
Colegate End **IP21** 45 F7
Colehill **BH21** 10 B4
Coleman Green **AL4** 32 E7
Coleman's Hatch **TN7** 13 H3
Colemere **SY12** 38 D2
Colemore **GU34** 11 J1
Colemore Green **WV16** 39 G6
Colenden **PH2** 82 C5
Coleorton **LE67** 41 G4
Colerne **SN14** 20 B4
Cole's Common **IP21** 45 G7
Cole's Cross **TQ9** 5 H6
Cole's Green **IP13** 35 G2
Colesbourne **GL53** 30 B7
Colesden **MK44** 32 E3
Coleshill *Bucks.* **HP7** 22 C2
Coleshill *Oxon.* **SN6** 21 F2
Coleshill *Warks.* **B46** 40 E7
Colestocks **EX14** 7 J5
Coley *B. & N.E.Som.* **BS40** 19 J6
Coley *Staffs.* **ST18** 40 C3
Colfin **DG9** 64 A5
Colgate **RH12** 13 F3
Colgrain **G82** 74 B3
Colindale **NW9** 23 F3
Colinsburgh **KY9** 83 F7
Colinton **EH13** 76 A4
Colintraive PA22 73 J3
Colkirk **NR21** 44 D3
Coll PA78 78 C2
Collace **PH2** 82 D4
Collafirth **ZE2** 107 N6
Collamoor Head **PL32** 4 B1
Collaton St. Mary **TQ3** 5 J4
Collessie **KY15** 82 D6
Colleton Mills **EX37** 6 E4
Collett's Green **WR2** 29 H3
Collier Row **RM5** 23 J2
Collier Street **TN12** 14 C3
Collier's End **SG11** 33 G6
Collier's Wood **SW19** 23 F4
Colliery Row **DH4** 62 E2
Collieston **AB41** 91 J2
Collin **DG1** 69 F6
Collingbourne Ducis **SN8** 21 F6
Collingbourne Kingston **SN8** 21 F6
Collingham *Notts.* **NG23** 52 B6
Collingham *W.Yorks.* **LS22** 57 J5
Collington **HR7** 29 F2
Collingtree **NN4** 31 J3
Collins End **RG8** 21 K4
Collins Green *Warr.* **WA5** 48 E3
Collins Green *Worcs.* **WR6** 29 G3
Colliston **DD11** 83 H3
Colliton **EX14** 7 J5
Collmuir **AB31** 90 D4
Collycroft **CV12** 41 F7
Collyhurst **M8** 49 H2
Collynie **AB41** 91 G1
Collyweston **PE9** 42 C5
Colmonell **KA26** 67 F5
Colmworth **MK44** 32 E3
Coln Rogers **GL54** 20 D1
Coln St. Aldwyns **GL7** 20 E1
Coln St. Dennis **GL54** 30 B7
Colnabaichin **AB36** 89 K4
Colnbrook **SL3** 22 D4
Colne *Cambs.* **PE28** 33 G1
Colne *Lancs.* BB8 56 D6
Colne Engaine **CO6** 34 C5
Colney **NR4** 45 F5
Colney Heath **AL4** 23 F1
Colney Street **AL2** 22 E1
Colonsay PA61 72 B1
Colonsay House **PA61** 72 B1
Colpy **AB52** 90 E1
Colquhar **EH44** 76 B6
Colsterdale **HG4** 57 G1
Colsterworth **NG33** 42 C3
Colston Bassett **NG12** 42 A2
Coltfield **IV36** 97 J5
Colthouse **LA22** 60 E7
Coltishall **NR12** 45 G4
Coltness **ML2** 75 G5
Colton *Cumb.* **LA12** 55 G1
Colton *N.Yorks.* **LS24** 58 B5
Colton *Norf.* **NR9** 45 F5
Colton *Staffs.* **WS15** 40 C3
Colton *W.Yorks.* **LS15** 57 J6
Colva **HR5** 28 B3
Colvend **DG5** 65 J5
Colvister **ZE2** 107 P3
Colwall **WR13** 29 G4
Colwall Green **WR13** 29 G4
Colwall Stone **WR13** 29 G4
Colwell **NE46** 70 E6
Colwich **ST18** 40 C3
Colwick **NG4** 41 J2
Colwinston **CF71** 18 C4
Colworth **PO20** 12 C6
Colwyn Bay (Bae Colwyn) LL29 47 G5
Colyford **EX24** 8 B5
Colyton EX24 8 B5
Combe *Here.* **LD8** 28 C2
Combe *Oxon.* **OX29** 31 F7
Combe *Som.* **TA10** 8 D2
Combe *W.Berks.* **RG17** 21 G5
Combe Common **GU8** 12 C3
Combe Cross **TQ13** 5 H3
Combe Down **BA2** 20 A5
Combe Florey **TA4** 7 K2
Combe Hay **BA2** 20 A6
Combe Martin **EX34** 6 D1
Combe Pafford **TQ1** 5 K4
Combe Raleigh **EX14** 7 K5
Combe St. Nicholas **TA20** 8 C3
Combeinteignhead **TQ12** 5 J3
Comberbach **CW9** 49 F5
Comberford **B79** 40 D5
Comberton *Cambs.* **CB23** 33 G3
Comberton *Here.* **SY8** 28 D2
Combpyne **EX13** 8 B5
Combridge **ST14** 40 C2
Combrook **CV35** 30 E3
Combs *Derbys.* **SK23** 50 C5
Combs *Suff.* **IP14** 34 E3
Combs Ford **IP14** 34 E3
Combwich **TA5** 19 F7
Comer **FK8** 80 E7
Comers **AB51** 90 E4
Comhampton **DY13** 29 H2
Comins Coch **SY23** 37 F7
Commercial End **CB25** 33 J2
Commins Coch **SY20** 37 H5
Common Edge **FY4** 55 G6
Common Moor **PL14** 4 C3
Common Platt **SN5** 20 E3
Common Side **S18** 51 F5
Common Square **LN4** 52 D5
Commondale **YO21** 63 H5
Commonside **DE6** 40 E1
Compstall **SK6** 49 J3
Compton *Devon* **TQ3** 5 J4
Compton *Hants.* **SO21** 11 F2
Compton *Plym.* **PL3** 4 E5
Compton *Staffs.* **DY7** 40 A7
Compton *Surr.* **GU3** 22 C7
Compton *W.Berks.* **RG20** 21 J4
Compton *W.Suss.* **PO18** 11 J3
Compton *W.Yorks.* **LS22** 57 K5
Compton *Wilts.* **SN9** 20 E6
Compton Abbas **SP7** 9 H3
Compton Abdale **GL54** 30 B7
Compton Bassett **SN11** 20 D4
Compton Beauchamp **SN6** 21 F3
Compton Bishop **BS26** 19 G6
Compton Chamberlayne **SP3** 10 B2
Compton Dando **BS39** 19 K5
Compton Dundon **TA11** 8 D1
Compton Martin **BS40** 19 J6
Compton Pauncefoot **BA22** 9 F2
Compton Valence **DT2** 8 E5
Compton Verney **CV35** 30 E3
Compton Wynyates **CV35** 30 E4
Comra **PH20** 88 C5
Comrie *Fife* **KY12** 75 J2
Comrie *P. & K.* **PH6** 81 J5
Conchra *Arg. & B.* **PA22** 73 J2
Conchra *High.* **IV40** 86 E2
Concraigie **PH10** 82 C3
Conder Green **LA2** 55 H4
Conderton **GL20** 29 J5
Condicote **GL54** 30 C6
Condorrat **G67** 75 F3
Condover **SY5** 38 D5
Coney Weston **IP31** 34 D1
Coneyhurst **RH14** 12 E4
Coneysthorpe **YO60** 58 D2
Coneythorpe **HG5** 57 J4
Conford **GU30** 12 B3
Congash **PH26** 89 H2
Congdon's Shop **PL15** 4 C3
Congerstone **CV13** 41 F5
Congham **PE32** 44 B3
Congleton CW12 49 H6
Congresbury **BS49** 19 H5
Congreve **ST19** 40 B4
Conicavel **IV36** 97 G6
Coningsby **LN4** 53 F7
Conington *Cambs.* **PE7** 42 E7
Conington *Cambs.* **CB23** 33 G2
Conisbrough **DN12** 51 H3
Conisby **PA49** 72 A4
Conisholme **LN11** 53 H3
Coniston *Cumb.* LA21 60 E7
Coniston *E.Riding* **HU11** 59 H6
Coniston Cold **BD23** 56 E4
Conistone **BD23** 56 E3
Conland **AB54** 98 E6
Connah's Quay **CH5** 48 B6
Connel **PA37** 80 A4
Connel Park **KA18** 68 B2
Connor Downs **TR27** 2 C5
Conock **SN10** 20 D6
Conon Bridge **IV7** 96 C6
Cononish **FK20** 80 E5
Cononley **BD20** 56 E5
Cononsyth **DD11** 83 G3
Consall **ST9** 40 B1
Consett DH8 62 C1
Constable Burton **DL8** 62 C7
Constantine **TR11** 2 E6
Constantine Bay **PL28** 3 F1
Contin **IV14** 96 B6
Contlaw **AB13** 91 G4
Contullich **IV17** 96 D4
Conwy LL32 47 F5
Conyer **ME9** 25 F5
Conyer's Green **IP31** 34 C2
Cooden **TN39** 14 C7
Coodham **KA1** 74 C7
Cooil **IM4** 54 C6
Cookbury **EX22** 6 C5
Cookbury Wick **EX22** 6 B5
Cookham **SL6** 22 B3
Cookham Dean **SL6** 22 B3
Cookham Rise **SL6** 22 B3
Cookhill **B49** 30 B3
Cookley *Suff.* **IP19** 35 H1
Cookley *Worcs.* **DY10** 40 A7
Cookley Green *Oxon.* **RG9** 21 K2
Cookley Green *Suff.* **IP19** 35 H1
Cookney **AB39** 91 G5
Cook's Green **IP7** 34 D3
Cooksbridge **BN7** 13 G5
Cooksey Green **B61** 29 J2
Cookshill **ST11** 40 B1
Cooksmill Green **CM1** 24 C1
Cookston **AB41** 91 H1
Coolham **RH13** 12 E4
Cooling **ME3** 24 D4
Cooling Street **ME3** 24 D4
Coombe *Cornw.* **PL26** 3 G3
Coombe *Cornw.* **EX23** 6 A4
Coombe *Cornw.* **TR14** 2 D4
Coombe *Cornw.* **TR3** 3 F4
Coombe *Devon* **TQ9** 5 H6
Coombe *Devon* **EX6** 7 G7
Coombe *Devon* **EX10** 7 K6
Coombe *Som.* **TA2** 8 B2
Coombe *Som.* **TA18** 8 D4
Coombe *Wilts.* **SN9** 20 E6
Coombe Bissett **SP5** 10 C2
Coombe End **TA4** 7 J3
Coombe Hill **GL19** 29 H6
Coombe Keynes **BH20** 9 H6
Coombes **BN15** 12 E6
Coombes Moor **LD8** 28 C2
Cooper's Corner *E.Suss.* **TN19** 14 C5
Cooper's Corner *Kent* **TN14** 23 H7
Cooper's Green **AL4** 22 E1
Coopersale **CM16** 23 H1
Coopersale Street **CM16** 23 H1
Cootham **RH20** 12 D5
Cop Street **CT3** 15 H2
Copdock **IP8** 35 F4
Copford Green **CO6** 34 D6
Copgrove **HG3** 57 J3
Copister **ZE2** 107 N5
Cople **MK44** 32 E4
Copley *Dur.* **DL13** 62 B4
Copley *W.Yorks.* **HX4** 57 F7
Coplow Dale **SK17** 50 D5
Copmanthorpe **YO23** 58 B5
Copmere End **ST21** 40 A3
Copp **PR3** 55 H6
Coppathorne **EX23** 6 A5
Coppenhall **ST18** 40 B4
Coppenhall Moss **CW1** 49 G7
Copperhouse **TR27** 2 C5
Coppicegate **DY12** 39 G7
Coppingford **PE28** 42 E7
Coppleridge **SP7** 9 H2
Copplestone **EX17** 7 F5
Coppull **PR7** 48 E1
Coppull Moor **PR7** 48 E1
Copsale **RH13** 12 E4
Copse Hill **SW20** 23 F5
Copster Green **BB1** 56 B6
Copston Magna **LE10** 41 G7
Copt Heath **B93** 30 C1
Copt Hewick **HG4** 57 J2
Copt Oak **LE67** 41 G4
Copthall Green **EN9** 23 H1
Copthorne **RH10** 13 G3
Copy Lake **EX18** 6 E4
Copythorne **SO40** 10 E3
Coralhill **AB43** 99 J4
Corbets Tey **RM14** 23 J3
Corbiegoe **KW1** 105 J4
Corbridge NE45 70 E7
Corby NN17 42 B7
Corby Glen **NG33** 42 D3
Cordach **AB34** 90 E5
Cordorcan **DG8** 64 D3
Coreley **SY8** 29 F1
Corfcott Green **EX22** 6 B6
Corfe **TA3** 8 B3
Corfe Castle **BH20** 9 J6
Corfe Mullen **BH21** 9 J5
Corfton **SY7** 38 D7
Corgarff **AB36** 89 K4

Cor - Cro

Corhampton SO32 11 H2
Corley CV7 40 E7
Corley Ash CV7 40 E7
Corley Moor CV7 40 E7
Cornabus PA42 72 B6
Cornard Tye CO10 34 D4
Corndon TQ13 6 E7
Corney LA19 60 C7
Cornforth DL17 62 E3
Cornhill AB45 98 D5
Cornhill on Tweed TD12 77 G7
Cornholme OL14 56 E7
Cornish Hall End CM7 33 K5
Cornquoy KW17 106 E7
Cornriggs DL13 61 K2
Cornsay DH7 62 C2
Cornsay Colliery DH7 62 C2
Corntown High. IV7 96 C6
Corntown V. of Glam. CF35 18 C4
Cornwell OX7 30 D6
Cornwood PL21 5 G5
Cornworthy TQ9 5 J5
Corpach PH33 87 H7
Corpusty NR11 45 F3
Corrachree AB34 90 C4
Corran Arg. & B. PA24 74 A1
Corran High. IV40 86 E4
Corran (Ardgour) High. PH33 80 B1
Corranbuie PA29 73 G4
Corranmore PA31 79 J7
Corrany IM7 54 D5
Corribeg PH33 87 F7
Corrie KA27 73 J6
Corrie Common DG11 69 H5
Corriechrevie PA29 73 F5
Corriecravie KA27 66 D1
Corriedoo DG7 68 B5
Corriekinloch IV27 103 F6
Corrielorne PA34 79 K6
Corrievorrie IV13 88 E2
Corrimony IV63 87 K1
Corringham Lincs. DN21 52 B3
Corringham Thur. SS17 24 D3
Corris SY20 37 G5
Corris Uchaf SY20 37 G5
Corrlarach PH33 87 F7
Corrour Shooting Lodge PH30 81 F1
Corrow PA24 80 C7
Corry IV49 86 C2
Corrychurrachan PH33 80 B1
Corrylach PA28 73 F7
Corrymuckloch PH8 81 K4
Corsback KW1 105 H2
Corscombe DT2 8 E4
Corse Aber. AB54 98 E6
Corse Glos. GL19 29 G6
Corse Lawn GL19 29 H5
Corse of Kinnoir AB54 98 D6
Corsebank DG4 68 D2
Corsegight AB53 99 G5
Corsehill DG11 69 G5
Corsewall DG9 64 A4
Corsham SN13 20 B4
Corsindae AB51 90 E4
Corsley BA12 20 B7
Corsley Heath BA12 20 B7
Corsock DG7 65 H3
Corston B. & N.E.Som. BA2 19 K5
Corston Wilts. SN16 20 C3
Corstorphine EH12 75 K3
Cortachy DD8 82 E2
Corton Suff. NR32 45 K6
Corton Wilts. BA12 20 C7
Corton Denham DT9 9 F2
Corwar House KA26 67 G5
Corwen LL21 37 K1
Coryton Devon EX20 6 C7
Coryton Thur. SS17 24 D3
Cosby LE9 41 H6
Coscote OX11 21 J3
Coseley WV14 40 B6
Cosford Shrop. WV7 39 G5
Cosford Warks. CV21 31 G1
Cosgrove MK19 31 J4
Cosham PO6 11 H4
Cosheston SA72 16 D5
Coshieville PH15 81 J3
Coskills DN38 52 D2
Cosmeston CF64 18 E5
Cossall NG16 41 G1
Cossington Leics. LE7 41 J4
Cossington Som. TA7 19 G7
Costa KW17 106 C5
Costessey NR8 45 F4
Costock LE12 41 H3
Coston Leics. LE14 42 B3
Coston Norf. NR9 44 E5
Cote Oxon. OX18 21 G1
Cote Som. TA9 19 G7
Cotebrook CW6 48 E6
Cotehill CA4 61 F1
Cotes Cumb. LA8 55 H1
Cotes Leics. LE12 41 H3
Cotes Staffs. ST15 40 A2
Cotesbach LE17 41 H7
Cotgrave NG12 41 J2
Cothall AB21 91 G3
Cotham NG23 42 A1
Cothelstone TA4 7 K2
Cotheridge WR6 29 G3
Cotherstone DL12 62 B5
Cothill OX13 21 H2
Cotleigh EX14 8 B4
Cotmanhay DE7 41 G1
Coton Cambs. CB23 33 H3
Coton Northants. NN6 31 H1
Coton Staffs. ST18 40 B2
Coton Staffs. B79 40 D5
Coton Staffs. ST20 40 A3
Coton Clanford ST18 40 A3
Coton Hill ST18 40 B2
Coton in the Clay DE6 40 D3
Coton in the Elms DE12 40 E4
Cotonwood Shrop. SY13 38 E2
Cotonwood Staffs. ST20 40 A3
Cott TQ9 5 H4
Cottam Lancs. PR4 55 H6
Cottam Notts. DN22 52 B4
Cottartown PH26 89 H1
Cottenham CB24 33 H2
Cotterdale DL8 61 K7
Cottered SG9 33 G6
Cotteridge B30 30 B1
Cotterstock PE8 42 D6
Cottesbrooke NN6 31 J1
Cottesmore LE15 42 C4
Cottingham E.Riding HU16 59 G6
Cottingham Northants. LE16 42 B6
Cottingley BD16 57 G6
Cottisford NN13 31 G5
Cotton Staffs. ST10 40 C1
Cotton Suff. IP14 34 E2
Cotton End MK45 32 D4
Cottonworth SP11 10 E1
Cottown Aber. AB54 90 D2
Cottown Aber. AB53 99 G6
Cottown Aber. AB51 91 F3
Cotts PL20 4 E4
Cottwood EX18 6 E4
Cotwall TF6 39 F4
Cotwalton ST15 40 B2
Couch's Mill PL22 4 B5
Coughton Here. HR9 28 E6
Coughton Warks. B49 30 B2
Cougie IV4 87 J2
Coulaghailtro PA29 73 F4
Coulags IV54 95 F7
Coulby Newham TS8 63 G5
Coulderton CA22 60 A6
Coull AB34 90 D4
Coulport G84 74 A2
Coulsdon CR5 23 F6
Coulston BA13 20 C6
Coulter ML12 75 J7
Coultershaw Bridge GU28 12 C5
Coultings TA5 19 F7
Coulton YO62 58 C2
Coultra KY15 82 E5
Cound SY5 38 E5
Coundlane SY5 38 E5
Coundon Dur. DL14 62 D4
Coundon W.Mid. CV6 41 F7
Countersett DL8 56 E1
Countess Wear EX2 7 H6
Countesthorpe LE8 41 H6
Countisbury EX35 7 F1
Coup Green PR5 55 J7
Coupar Angus PH13 82 D3
Coupland Cumb. CA16 61 J5
Coupland Northumb. NE71 77 H7
Cour PA28 73 G6
Court Colman CF32 18 B3
Court Henry SA32 17 J3
Court House Green CV6 41 F7
Court-at-Street CT21 15 F4
Courteenhall NN7 31 J3
Courtsend SS3 25 G2
Courtway TA5 8 B1
Cousland EH22 76 B4
Cousley Wood TN5 13 K3
Coustonn PA22 73 J3
Cove Arg. & B. G84 74 A2
Cove Devon EX16 7 H4
Cove Hants. GU14 22 B6
Cove High. IV22 94 E2
Cove Sc.Bord. TD13 77 F3
Cove Bay AB12 91 H4
Cove Bottom NR34 45 J7
Covehithe NR34 45 K7
Coven WV9 40 B5
Coveney CB6 43 H7
Covenham St. Bartholomew LN11 53 G3
Covenham St. Mary LN11 53 G3
COVENTRY CV 30 E1
Coventry Airport CV3 30 E1
Coverack TR12 2 E7
Coverham DL8 57 G1
Covesea IV30 97 J4
Covingham SN3 20 E3
Covington Cambs. PE28 32 D1
Covington S.Lan. ML12 75 H7
Cowan Bridge LA6 56 B2
Cowbeech BN27 13 K5
Cowbit PE12 43 F4
Cowbridge Som. TA24 7 H1
Cowbridge V. of Glam. CF71 18 C4
Cowden TN8 23 H7
Cowden Pound TN8 23 H7
Cowdenbeath KY4 75 K1
Cowdenburn EH46 76 A5
Cowers Lane DE56 41 F1
Cowes PO31 11 F5
Cowesby YO7 57 K1
Cowesfield Green SP5 10 D2
Cowey Green CO7 34 E6
Cowfold RH13 13 F4
Cowgill LA10 56 C1
Cowie Aber. AB39 91 G6
Cowie Stir. FK7 75 G2
Cowlam Manor YO25 59 F3
Cowley Devon EX4 7 H6
Cowley Glos. GL53 29 J7
Cowley Gt.Lon. UB8 22 D3
Cowley Oxon. OX4 21 J1
Cowling Lancs. PR7 48 E1
Cowling N.Yorks. BD22 56 E5
Cowling N.Yorks. DL8 57 H1
Cowlinge CB8 34 B3
Cowmes HD8 50 D1
Cowpe BB4 56 D7
Cowpen NE24 71 H5
Cowpen Bewley TS23 63 F4
Cowplain PO8 11 H3
Cowsden WR7 29 J3
Cowshill DL13 61 K2
Cowthorpe LS22 57 K4
Cox Common IP19 45 H7
Coxbank CW3 39 F1
Coxbench DE21 41 F1
Coxbridge BA6 8 E1
Coxford PE31 44 C3
Coxheath ME17 14 C2
Coxhoe DH6 62 E3
Coxley BA5 19 J7
Coxley Wick BA5 19 J7
Coxpark PL18 4 E3
Coxtie Green CM14 23 J2
Coxwold YO61 58 B2
Coychurch CF35 18 C4
Coylet PA23 73 K2
Coylton KA6 67 J2
Coylumbridge PH22 89 G3
Coynach AB34 90 C4
Coynachie AB54 90 C1
Coytrahen CF32 18 B3
Crabbet Park RH10 13 G3
Crabgate NR11 44 E3
Crabtree Plym. PL3 5 F5
Crabtree S.Yorks. S4 51 F4
Crabtree W.Suss. RH13 13 F4
Crabtree Green LL13 38 C1
Crackaig PA60 72 D4
Crackenthorpe CA16 61 H4
Crackington EX23 4 B1
Crackington Haven EX23 4 B1
Crackley ST5 40 A1
Crackleybank TF11 39 G4
Crackpot DL11 62 A7
Crackthorn Corner IP22 34 E1
Cracoe BD23 56 E3
Craddock EX15 7 J4
Cradhlastadh HS2 100 C4
Cradley Here. WR13 29 G4
Cradley W.Mid. B63 40 B7
Cradley Heath B64 40 B7
Crafthole PL11 4 D5
Crafton LU7 32 B7
Cragg HX7 57 F7
Cragg Hill LS18 57 H6
Craggan Moray AB37 89 J1
Craggan P. & K. PH5 81 K6
Cragganruar PH15 81 H3
Craggie High. IV2 88 E1
Craggie High. KW8 104 D7
Craghead DH9 62 D1
Craibstone Aberdeen AB21 91 G3
Craibstone Moray AB56 98 C5
Craichie DD8 83 G3
Craig Arg. & B. PA35 80 B4
Craig Arg. & B. PA70 79 G4
Craig D. & G. DG7 65 G4
Craig High. IV22 94 D5
Craig High. IV54 95 G7
Craig S.Ayr. KA19 67 H3
Craig Berthlŵyd CF46 18 D2
Craigans PA31 73 G1
Craigbeg PH31 88 B6
Craig-cefn-parc SA6 17 K5
Craigcleuch DG13 69 J5
Craigculter AB43 99 H5
Craigdallie PH14 82 D5
Craigdam AB41 91 G1
Craigdarroch D. & G. DG3 68 C4
Craigdarroch E.Ayr. KA18 68 B3
Craigdhu D. & G. DG8 64 D6
Craigdhu High. IV4 96 B7
Craigearn AB51 91 F3
Craigellachie AB38 97 K7
Craigellie AB43 99 J4
Craigencallie DG7 65 F3
Craigend Moray IV36 97 J6
Craigend P. & K. PH2 82 C5
Craigendive PA23 73 J2
Craigendoran G84 74 B2
Craigengillan KA6 67 J3
Craigenputtock DG2 68 C5
Craigens PA44 72 A4
Craigglas PA31 73 G1
Craighall KY15 83 F6
Craighat G63 74 C2
Craighead Fife KY10 83 H6
Craighead High. IV11 96 E5
Craighlaw DG8 64 D4
Craighouse PA60 72 D4
Craigie Aber. AB23 91 H3
Craigie Dundee DD4 83 F4
Craigie P. & K. PH10 82 C3
Craigie S.Ayr. KA1 74 C7
Craigie Brae AB41 91 G1
Craigieburn DG10 69 G3
Craigieholm PH13 82 C4
Craigielaw EH32 76 C3
Craiglockhart EH14 76 A3
Craiglug AB31 91 G5
Craigmaud AB43 99 G5
Craigmillar EH16 76 A3
Craigmore PA20 73 K4
Craigmyle House AB31 90 E4
Craignafeoch PA21 73 H3
Craignant SY10 38 B2
Craignavie FK21 81 G4
Craigneil KA26 67 F5
Craigneuk ML1 75 F5
Craignure PA65 79 J4
Craigo DD10 83 H1
Craigoch KA19 67 G3
Craigow KY13 82 B7
Craigrothie KY15 82 E6
Craigroy IV36 97 J6
Craigroy Farm AB37 89 J1
Craigruie FK19 81 F5
Craigsanquhar KY15 82 E6
Craigton Aberdeen AB14 91 G4
Craigton Angus DD5 83 G4
Craigton Angus DD8 82 E2
Craigton High. IV1 96 D7
Craigton Stir. G63 74 E2
Craigtown KW13 104 D3
Craig-y-nos SA9 27 H7
Craik Aber. AB54 90 C2
Craik Sc.Bord. TD9 69 J3
Crail KY10 83 H7
Crailing TD8 70 B1
Crailinghall TD8 70 B1
Crakehill YO7 57 K2
Crakemarsh ST14 40 C2
Crambe YO60 58 D3
Cramlington NE23 71 H6
Cramond EH4 75 K3
Cranage CW4 49 G6
Cranberry ST21 40 A2
Cranborne BH21 10 B3
Cranbourne SL4 22 C4
Cranbrook Gt.Lon. IG1 23 H3
Cranbrook Kent TN17 14 C4
Cranbrook Common TN17 14 C4
Crane Moor S35 51 F2
Cranfield MK43 32 C4
Cranford Devon EX39 6 B3
Cranford Gt.Lon. TW5 22 E4
Cranford St. Andrew NN14 32 C1
Cranford St. John NN14 32 C1
Cranham Glos. GL4 29 H7
Cranham Gt.Lon. RM14 23 J3
Crank WA11 48 E3
Cranleigh GU6 12 D3
Cranmer Green IP31 34 E1
Cranmore I.o.W. PO41 10 E6
Cranmore Som. BA4 19 K7
Cranna AB54 98 E5
Crannoch AB55 98 C5
Cranoe LE16 42 A6
Cransford IP13 35 H2
Cranshaws TD11 76 E4
Cranstal IM7 54 D3
Crantock TR8 2 E2
Cranwell NG34 42 D1
Cranwich IP26 44 B6
Cranworth IP25 44 D5
Craobh Haven PA31 79 J7
Crapstone PL20 5 F4
Crarae PA32 73 H1
Crask Inn IV27 103 H6
Crask of Aigas IV4 96 B7
Craskins AB34 90 D4
Craster NE66 71 H2
Craswall HR2 28 B5
Crateford ST19 40 B5
Cratfield IP19 35 H1
Crathes AB31 91 F5
Crathie Aber. AB35 89 K5
Crathie High. PH20 88 C5
Crathorne TS15 63 F6
Craven Arms SY7 38 D7
Craw KA27 73 G6
Crawcrook NE40 71 G7
Crawford Lancs. WN8 48 D2
Crawford S.Lan. ML12 68 E1
Crawfordjohn ML12 68 D1
Crawfordton DG3 68 C4
Crawick DG4 68 C2
Crawley Devon EX14 8 B4
Crawley Hants. SO21 11 F1
Crawley Oxon. OX29 30 E7
Crawley W.Suss. RH11 13 F3
Crawley Down RH10 13 G3
Crawleyside DL13 62 A2
Crawshawbooth BB4 56 D7
Crawton AB39 91 G7
Crawyn IM7 54 C4
Cray N.Yorks. BD23 56 E2
Cray P. & K. PH10 82 C1
Cray Powys LD3 27 H6
Crayford DA1 23 J4
Crayke YO61 58 B2
Crays Hill CM11 24 D2
Cray's Pond RG8 21 K3
Crazies Hill RG10 22 A3
Creacombe EX16 7 G4
Creag Ghoraidh (Creagorry) HS7 92 C7
Creagan PA38 80 A3
Creagbheitheachain PH33 80 A1
Creagorry (Creag Ghoraidh) HS7 92 C7
Creamore Bank SY4 38 E2
Creaton NN6 31 J1
Creca DG12 69 H6
Credenhill HR4 28 D4
Crediton EX17 7 G5
Creebridge DG8 64 E4
Creech Heathfield TA3 8 B2
Creech St. Michael TA3 8 B2
Creed TR2 3 G4
Creedy Park EX17 7 G5
Creekmouth IG11 23 H3
Creeting St. Mary IP6 34 E3
Creeton NG33 42 D3
Creetown DG8 64 E5
Creggans PA27 80 B7
Cregneash IM9 54 A7
Cregrina LD1 28 A3
Creich KY15 82 E5
Creigau NP16 19 H2
Creigiau CF15 18 D3
Crelevan IV4 87 K1
Crelly TR13 2 D5
Cremyll PL10 4 E5
Crendell SP6 10 B3
Cressage SY5 38 E5
Cressbrook SK17 50 D5
Cresselly SA68 16 D5
Cressing CM77 34 B6
Cresswell Northumb. NE61 71 H4
Cresswell Staffs. ST11 40 B2
Cresswell Quay SA68 16 D5
Creswell S80 51 H5
Cretingham IP13 35 G3
Cretshengan PA29 73 F4
CREWE Ches.E. CW 49 G7
Crewe Ches.W. & C. CH3 48 D7
Crewe Green CW1 49 G7
Crewgreen SY5 38 C4
Crewkerne TA18 8 D4
Crew's Hole BS5 19 K4
Crewton DE24 41 F2
Crianlarich FK20 80 E5
Cribbs Causeway BS34 19 J3
Cribyn SA48 26 E3
Criccieth LL52 36 D2
Crich DE4 51 F7
Crich Carr DE4 51 F7
Crich Common DE4 51 F7
Crichie AB42 99 H6
Crichton EH37 76 B4
Crick Mon. NP26 19 H2
Crick Northants. NN6 31 G1
Crickadarn LD2 27 K4
Cricket Hill GU46 22 B5
Cricket St. Thomas TA20 8 C4
Crickham BS28 19 H7
Crickheath SY10 38 B3
Crickhowell NP8 28 B7
Cricklade SN6 20 E2
Cricklewood NW2 23 F3
Crick's Green HR7 29 F3
Criddlestyle SP6 10 C3
Cridling Stubbs WF11 58 B7
Crieff PH7 81 K5
Criftins (Dudleston Heath) SY12 38 C2
Criggan PL26 4 A4
Criggion SY5 38 B4
Crigglestone WF4 51 F1
Crimble OL11 49 H1
Crimchard TA20 8 C4
Crimdon Park TS27 63 F3
Crimond AB43 99 J5
Crimonmogate AB43 99 J5
Crimplesham PE33 44 A5
Crinan PA31 73 F1
Crinan Ferry PA31 73 F1
Cringleford NR4 45 F5
Cringletie EH45 76 A6
Crinow SA67 16 E4
Cripplestyle SP6 10 B3
Cripp's Corner TN32 14 C5
Crix CM3 34 B7
Crizeley HR2 28 D5
Croalchapel DG3 68 E4
Croasdale CA23 60 B5
Crock Street TA19 8 C3
Crockenhill BR8 23 J5
Crocker End RG9 22 A3
Crockerhill PO17 11 G4
Crockernwell EX6 7 F6
Crockerton BA12 20 B7
Crockerton Green BA12 20 B7
Crocketford (Ninemile Bar) DG2 65 J3
Crockey Hill YO19 58 C5
Crockham Hill TN8 23 H6
Crockhurst Street TN11 23 K7
Crockleford Heath CO7 34 E6
Croes Hywel NP7 28 C7
Croes y pant NP4 19 G1
Croesau Bach SY10 38 B3
Croeserw SA13 18 B2
Croesgoch SA62 16 B2
Croes-lan SA44 17 G1
Croesor LL48 37 F1
Croespenmaen NP11 18 E2
Croesyceiliog Carmar. SA32 17 H4
Croesyceiliog Torfaen NP44 19 G2
Croes-y-mwyalch NP44 19 G2
Croesywaun LL55 46 D7
Croford TA4 7 K3
Croft Here. HR6 28 D2
Croft Leics. LE9 41 H6
Croft Lincs. PE24 53 J6
Croft Warr. WA3 49 F3
Croftamie G63 74 C2
Crofthead CA6 69 J6
Croftmore PH18 81 K1
Crofton W.Yorks. WF4 51 F1
Crofton Wilts. SN8 21 F5
Croft-on-Tees DL2 62 D6
Crofts DG7 65 H3
Crofts of Benachielt KW5 105 G5
Crofts of Buinach IV30 97 J6
Crofts of Haddo AB41 91 G1
Crofty SA4 17 J6
Crogen LL21 37 K2
Croggan PA63 79 J5
Croglin CA4 61 G2
Croick High. IV24 96 B2
Croick High. KW13 104 D3

Cro - Dar

Croig **PA75** 79 F2
Crois Dughaill **HS8** 84 C3
Croit e Caley **IM9** 54 B7
Cromarty IV11 96 E5
Crombie **KY12** 75 J2
Crombie Mill **DD7** 83 G4
Cromblet **AB51** 91 F1
Cromdale **PH26** 89 H2
Cromer *Herts.* **SG2** 33 F6
Cromer *Norf.* NR27 45 G1
Cromford **DE4** 50 E7
Cromhall **GL12** 19 K2
Cromhall Common **GL12** 19 K3
Cromore **HS2** 101 G5
Crompton Fold **OL2** 49 J1
Cromwell **NG23** 51 K6
Cronberry **KA18** 68 B1
Crondall **GU10** 22 A7
Cronk-y-Voddy **IM6** 54 C5
Cronton **WA8** 48 D4
Crook *Cumb.* **LA8** 61 F7
Crook *Dur.* DL15 62 C3
Crook of Devon **KY13** 82 B7
Crooked Soley **RG17** 21 G4
Crookedholm **KA3** 74 C7
Crookham *Northumb.* **TD12** 77 H7
Crookham *W.Berks.* **RG19** 21 J5
Crookham Eastfield **TD12** 77 H7
Crookham Village **GU51** 22 A6
Crooklands **LA7** 55 J1
Cropredy **OX17** 31 F4
Cropston **LE7** 41 H4
Cropthorne **WR10** 29 J4
Cropton **YO18** 58 D1
Cropwell Bishop **NG12** 41 J2
Cropwell Butler **NG12** 41 J2
Cros (Cross) **HS2** 101 H1
Crosbie **KA23** 74 A6
Crosbost **HS2** 101 F5
Crosby *Cumb.* **CA15** 60 B3
Crosby *I.o.M.* **IM4** 54 C6
Crosby *Mersey.* **L23** 48 C3
Crosby *N.Lincs.* **DN15** 52 B1
Crosby Court **DL6** 62 E7
Crosby Garrett **CA17** 61 J6
Crosby Ravensworth **CA10** 61 H5
Crosby Villa **CA15** 60 B3
Crosby-on-Eden **CA6** 60 F1
Croscombe **BA5** 19 J7
Crosemere **SY12** 38 D3
Crosland Hill **HD4** 50 D1
Cross *Som.* **BS26** 19 H6
Cross (Cros) *W.Isles* **HS2** 101 H1
Cross Ash **NP7** 28 D7
Cross Bank **DY12** 29 G1
Cross End *Bed.* **MK44** 32 D3
Cross End *Essex* **CO9** 34 C5
Cross Foxes Inn **LL40** 37 G4
Cross Gates **LS15** 57 J6
Cross Green *Devon* **PL15** 6 B7
Cross Green *Staffs.* **WV10** 40 B5
Cross Green *Suff.* **IP7** 34 D3
Cross Green *Suff.* **IP30** 34 C3
Cross Green *Suff.* **IP29** 34 C3
Cross Hands *Carmar.* **SA14** 17 J4
Cross Hands *Pembs.* **SA67** 16 D4
Cross Hill **DE5** 41 G1
Cross Hills **BD20** 57 F5
Cross Houses **SY5** 38 E5
Cross in Hand **TN21** 13 J4
Cross Inn *Cere.* **SA44** 26 C3
Cross Inn *Cere.* **SY23** 26 E2
Cross Inn *R.C.T.* **CF72** 18 D3
Cross Keys **SN13** 20 B4
Cross Lane **PO30** 11 G6
Cross Lane Head **WV16** 39 G6
Cross Lanes *Cornw.* **TR4** 2 E4
Cross Lanes *Cornw.* **TR12** 2 D6
Cross Lanes *N.Yorks.* **YO61** 58 B3
Cross Lanes *Wrex.* **LL13** 38 C1
Cross of Jackston **AB51** 91 F1
Cross o'th'hands **DE56** 40 E1
Cross Street **IP21** 35 F1
Crossaig **PA29** 73 G5
Crossapol **PA78** 78 C2
Crossapoll **PA77** 78 A3
Cross-at-Hand **TN12** 14 C3
Crossbush **BN18** 12 D6
Crosscanonby **CA15** 60 B3
Crossdale Street **NR27** 45 G2
Crossens **PR9** 48 C1
Crossflatts **BD16** 57 G5
Crossford *D. & G.* **DG3** 68 D5
Crossford *Fife* **KY12** 75 J2
Crossford *S.Lan.* **ML8** 75 G6
Crossgate *Lincs.* **PE11** 43 F3
Crossgate *Staffs.* **ST11** 40 B2
Crossgatehall **EH22** 76 B4
Crossgates *Fife* **KY4** 75 K2
Crossgates *N.Yorks.* **YO12** 59 G1
Crossgates *P. & K.* **PH2** 82 B5
Crossgates *Powys* **LD1** 27 K2
Crossgill **LA2** 55 J3
Crosshands **KA5** 74 C7
Crosshill *Fife* **KY5** 75 K1
Crosshill *S.Ayr.* **KA19** 67 H3
Crosshouse **KA2** 74 B7
Crosskeys **NP11** 19 F2
Crosskirk **KW14** 105 F1
Crosslanes **SY10** 38 C4
Crosslee *Renf.* **PA6** 74 C4
Crosslee *Sc.Bord.* **TD7** 69 J2
Crossmichael **DG7** 65 H4
Crossmoor **PR4** 55 H6
Crossroads *Aber.* **AB31** 91 F5
Crossroads *E.Ayr.* **KA1** 74 C7
Crossway *Mon.* **NP25** 28 D7
Crossway *Powys* **LD1** 27 K3
Crossway Green *Mon.* **NP16** 19 J2
Crossway Green *Worcs.* **DY13** 29 H2
Crossways *Dorset* **DT2** 9 G6
Crossways *Glos.* **GL16** 28 E7
Crosswell **SA41** 16 E2
Crosswood **SY23** 27 F1
Crosthwaite **LA8** 60 F7
Croston **PR26** 48 D1
Crostwick **NR12** 45 G4
Crostwight **NR28** 45 H3
Crothair **HS2** 100 D4
Crouch **TN15** 23 K6
Crouch End **N8** 23 F3
Crouch Hill **DT9** 9 G3
Croucheston **SP5** 10 B2
Croughton **NN13** 31 G5
Crovie **AB45** 99 G4
Crow **BH24** 10 C4
Crow Edge **S36** 50 D2
Crow Green **CM15** 23 J2
Crow Hill **HR9** 29 F6
Crowan **TR14** 2 D5
Crowborough TN6 13 J3
Crowborough Warren **TN6** 13 J3
Crowcombe **TA4** 7 K2
Crowdecote **SK17** 50 D6
Crowden **SK13** 50 C3
Crowdhill **SO50** 11 F2
Crowell **OX39** 22 A2
Crowfield *Northants.* **NN13** 31 H4
Crowfield *Suff.* **IP6** 35 F3
Crowhurst *E.Suss.* **TN33** 14 C6
Crowhurst *Surr.* **RH7** 23 G7
Crowhurst Lane End **RH7** 23 G7
Crowland *Lincs.* **PE6** 43 F4
Crowland *Suff.* **IP31** 34 E1
Crowlas **TR20** 2 C5
Crowle *N.Lincs.* **DN17** 51 K1
Crowle *Worcs.* **WR7** 29 J3
Crowle Green **WR7** 29 J3
Crowmarsh Gifford **OX10** 21 K3
Crown Corner **IP13** 35 G1
Crownhill **PL6** 4 E5
Crownthorpe **NR18** 44 E5
Crowntown **TR13** 2 D5
Crow's Nest **PL14** 4 C4
Crows-an-wra **TR19** 2 A6
Crowsnest **SY5** 38 C5
Crowthorne RG45 22 B5
Crowton **CW8** 48 E5
Croxall **WS13** 40 E4
Croxby **LN7** 52 E3
Croxdale **DH6** 62 D3
Croxden **ST14** 40 C2
Croxley Green **WD3** 22 D2
Croxton *Cambs.* **PE19** 33 F3
Croxton *N.Lincs.* **DN39** 52 D1
Croxton *Norf.* **IP24** 44 C7
Croxton *Staffs.* **ST21** 39 G2
Croxton Green **SY14** 48 E7
Croxton Kerrial **NG32** 42 B3
Croxtonbank **ST21** 39 G2
Croy *High.* **IV2** 96 E7
Croy *N.Lan.* **G65** 75 F3
Croyde **EX33** 6 C2
Croyde Bay **EX33** 6 C2
Croydon *Cambs.* **SG8** 33 G4
CROYDON *Gt.Lon.* CR 23 G5
Cruach **PA43** 72 B5
Cruchie **AB54** 98 D6
Cruckmeole **SY5** 38 D5
Cruckton **SY5** 38 D4
Cruden Bay **AB42** 91 J1
Crudgington **TF6** 39 F4
Crudwell **SN16** 20 C2
Crug **LD1** 28 A1
Crugmeer **PL28** 3 G1
Crugybar **SA19** 17 K2
Crùlabhig **HS2** 100 D4
Crumlin **NP11** 19 F2
Crumpsall **M8** 49 H2
Crumpsbrook **DY14** 29 F1
Crundale *Kent* **CT4** 15 F3
Crundale *Pembs.* **SA62** 16 C4
Crunwere Farm **SA67** 16 E4
Crutherland Farm **G75** 74 E5
Cruwys Morchard **EX16** 7 G4
Crux Easton **RG20** 21 H6
Crwbin **SA17** 17 H4
Cryers Hill **HP15** 22 B2
Crymlyn **LL33** 46 E5
Crymych SA41 16 E2
Crynant **SA10** 18 A1
Crystal Palace **SE26** 23 G4
Cuaig **IV54** 94 D6
Cubbington **CV32** 30 E2
Cubert **TR8** 2 E3
Cubley **S36** 50 E2
Cublington *Bucks.* **LU7** 32 B6
Cublington *Here.* **HR2** 28 D5
Cuckfield **RH17** 13 G4
Cucklington **BA9** 9 G2
Cuckney **NG20** 51 H5
Cuckold's Green **NR34** 45 J7
Cuckoo Bridge **PE11** 43 F3
Cuckoo's Corner **GU34** 22 A7
Cuckoo's Nest **CH4** 48 C6
Cuddesdon **OX44** 21 J1
Cuddington *Bucks.* **HP18** 31 J7
Cuddington *Ches.W. & C.* **CW8** 48 E5
Cuddington Heath **SY14** 38 D1
Cuddy Hill **PR4** 55 H6
Cudham **TN14** 23 H6
Cudlipptown **PL19** 5 F3
Cudworth *S.Yorks.* **S72** 51 F2
Cudworth *Som.* **TA19** 8 C3
Cuerdley Cross **WA5** 48 E4
Cuffley **EN6** 23 G1
Cuidhaseadair **HS2** 101 H2
Cuidhir **HS9** 84 B4
Cuidhtinis (Quidinish) **HS3** 93 F3
Cuidrach **IV51** 93 J6
Cuilmuich **PA24** 73 K1
Cuil-uaine **PA37** 80 A4
Culag **G83** 74 B1
Culbo **IV7** 96 D5
Culbokie **IV7** 96 D6
Culbone **TA24** 7 G1
Culburnie **IV4** 96 B7
Culcabock **IV2** 96 D7
Culcharan **PA37** 80 A4
Culcharry **IV12** 97 F6
Culcheth **WA3** 49 F3
Culdrain **AB54** 90 D1
Culduie **IV54** 94 D7
Culford **IP28** 34 C2
Culfordheath **IP31** 34 C1
Culgaith **CA10** 61 H4
Culgower **KW8** 104 E7
Culham **OX14** 21 J2
Culindrach **PA29** 73 H5
Culkein **IV27** 102 C5
Culkerton **GL8** 20 C2
Cullachie **PH24** 89 G2
Cullen **AB56** 98 D4
Cullercoats **NE30** 71 J6
Cullicudden **IV7** 96 D5
Culligran **IV4** 95 K7
Cullingworth **BD13** 57 F6
Cullipool **PA34** 79 J6
Cullivoe **ZE2** 107 P2
Culloch **PH6** 81 J6
Culloden **IV2** 96 E7
Cullompton EX15 7 J5
Culmaily **KW10** 97 F2
Culmalzie **DG8** 64 D5
Culmington **SY8** 38 D7
Culmstock **EX15** 7 K4
Culnacraig **IV26** 95 G1
Culnadalloch **PA37** 80 A4
Culnaknock **IV51** 94 B5
Culnamean **IV47** 85 K2
Culpho **IP6** 35 G4
Culquhirk **DG8** 64 E5
Culrain **IV24** 96 C2
Culross **KY12** 75 H2
Culroy **KA19** 67 H2
Culsh **AB35** 90 B5
Culshabbin **DG8** 64 D5
Culswick **ZE2** 107 L8
Culter Allers Farm **ML12** 75 J7
Cultercullen **AB41** 91 H2
Cults *Aber.* **AB54** 90 D1
Cults *Aberdeen* **AB15** 91 G4
Cults *D. & G.* **DG8** 64 E6
Cultybraggan Camp **PH6** 81 J6
Culverhouse Cross **CF5** 18 E4
Culverstone Green **DA13** 24 C5
Culverthorpe **NG32** 42 D1
Culvie **AB54** 98 D5
Culworth **OX17** 31 G4
Cumberhead **ML11** 75 F7
Cumberlow Green **SG9** 33 G5
Cumbernauld **G67** 75 F3
Cumberworth **LN13** 53 J5
Cuminestown **AB53** 99 G6
Cumloden **DG8** 64 E4
Cummersdale **CA2** 60 E1
Cummertrees **DG12** 69 G7
Cummingstown **IV30** 97 J5
Cumnock KA18 67 K1
Cumnor **OX2** 21 H1
Cumrew **CA8** 61 G1
Cumrue **DG11** 69 F5
Cumstoun **DG6** 65 G5
Cumwhinton **CA4** 61 F1
Cumwhitton **CA8** 61 G1
Cundall **YO61** 57 K2
Cunninghamhead **KA3** 74 B6
Cunningsburgh **ZE2** 107 N10
Cunnister **ZE2** 107 P3
Cunnoquhie **KY15** 82 E6
Cupar KY15 82 E6
Cupar Muir **KY15** 82 E6
Curbar **S32** 50 E5
Curborough **WS13** 40 D4
Curbridge *Hants.* **SO30** 11 G3
Curbridge *Oxon.* **OX29** 21 G1
Curdridge **SO32** 11 G3
Curdworth **B76** 40 D6
Curland **TA3** 8 B3
Curlew Green **IP17** 35 H2
Curling Tye Green **CM9** 24 E1
Curload **TA3** 8 C2
Curridge **RG18** 21 H4
Currie EH14 75 K4
Curry Mallet **TA3** 8 C2
Curry Rivel **TA10** 8 C2
Curteis' Corner **TN27** 14 D4
Curtisden Green **TN17** 14 C3
Curtisknowle **TQ9** 5 H5
Cury **TR12** 2 D6
Cusgarne **TR4** 2 E4
Cushnie **AB45** 99 F4
Cushuish **TA2** 7 K2
Cusop **HR3** 28 B4
Cusworth **DN5** 51 H2
Cutcloy **DG8** 64 E7
Cutcombe **TA24** 7 H2
Cutgate **OL11** 49 H1
Cuthill **IV25** 96 E3
Cutiau **LL42** 37 F4
Cutlers Green **CM6** 33 J5
Cutnall Green **WR9** 29 H2
Cutsdean **GL54** 30 B5
Cutsyke **WF10** 57 K7
Cutthorpe **S42** 51 F5
Cutts **ZE1** 107 N9
Cuttyhill **AB42** 99 J5
Cuxham **OX49** 21 K2
Cuxton **ME2** 24 D5
Cuxwold **LN7** 52 E2
Cwm *B.Gwent* **NP23** 18 E1
Cwm *Denb.* **LL18** 47 J5
Cwm Ffrwd-oer **NP4** 19 F1
Cwm Gwaun **SA65** 16 D2
Cwm Head **SY6** 38 D7
Cwm Irfon **LD5** 27 H4
Cwm Penmachno **LL24** 37 G1
Cwm Plysgog **SA43** 16 E1
Cwmafan **SA12** 18 A2
Cwmaman **CF44** 18 D2
Cwmann **SA48** 17 J1
Cwmbach *Carmar.* **SA34** 17 F3
Cwmbach *Carmar.* **SA15** 17 H5
Cwmbach *Powys* **HR3** 28 A5
Cwmbach *Powys* **LD2** 27 K3
Cwmbach *R.C.T.* **CF44** 18 D1
Cwmbelan **SY18** 37 J7
Cwmbrân NP44 19 F2
Cwmbrwyno **SY23** 37 G7
Cwmcarn **NP11** 19 F2
Cwmcarvan **NP25** 19 H1
Cwm-Cewydd **SY20** 37 H4
Cwm-cou **SA38** 17 F1
Cwmcrawnon **LD3** 28 A7
Cwmdare **CF44** 18 C1
Cwmdu *Carmar.* **SA19** 17 K2
Cwmdu *Powys* **NP8** 28 A6
Cwmduad **SA33** 17 G2
Cwmerfyn **SY23** 37 G7
Cwmfelin **CF46** 18 D1
Cwmfelin Boeth **SA34** 16 E4
Cwmfelin Mynach **SA34** 17 F3
Cwmfelinfach **NP11** 18 E2
Cwmffrwd **SA31** 17 H4
Cwmgiedd **SA9** 27 G7
Cwmgors **SA18** 27 G7
Cwmgwili **SA14** 17 J4
Cwmgwrach **SA11** 18 B1
Cwmgwyn **SA2** 17 K6
Cwmhiraeth **SA44** 17 G2
Cwmifor **SA19** 17 K3
Cwmisfael **SA32** 17 H4
Cwm-Llinau **SY20** 37 H5
Cwmllyfri **SA33** 17 G4
Cwmllynfell **SA9** 27 G7
Cwm-mawr **SA14** 17 J4
Cwm-miles **SA34** 16 E3
Cwm-Morgan **SA38** 17 F2
Cwm-parc **CF42** 18 C2
Cwmpengraig **SA44** 17 G2
Cwmpennar **CF45** 18 D1
Cwmsychbant **SA40** 17 H1
Cwmsymlog **SY23** 37 F7
Cwmtillery **NP13** 19 F1
Cwm-twrch Isaf **SA9** 27 G7
Cwm-twrch Uchaf **SA9** 27 G7
Cwm-y-glo **LL55** 46 D6
Cwmyoy **NP7** 28 C6
Cwm-yr-Eglwys **SA42** 16 D1
Cwmyrhaiadr **SY20** 37 G6
Cwmystwyth **SY23** 27 G1
Cwrt **SY20** 37 F5
Cwrt-newydd **SA40** 17 H1
Cwrt-y-cadno **SA19** 17 K1
Cwrt-y-gollen **NP8** 28 B7
Cydweli (Kidwelly) SA17 17 H5
Cyffylliog **LL15** 47 J7
Cyfronydd **SY21** 38 A5
Cymau **LL11** 48 B7
Cymmer *N.P.T.* **SA13** 18 B2
Cymmer *R.C.T.* **CF39** 18 D2
Cyncoed **CF23** 18 E3
Cynghordy **SA20** 27 H5
Cynheidre **SA15** 17 H5
Cynwyd **LL21** 37 K1
Cynwyl Elfed **SA33** 17 G3

D

Dabton **DG3** 68 D4
Daccombe **TQ12** 5 K4
Dacre *Cumb.* **CA11** 61 F4
Dacre *N.Yorks.* **HG3** 57 G3
Dacre Banks **HG3** 57 G3
Daddry Shield **DL13** 61 K3
Dadford **MK18** 31 H5
Dadlington **CV13** 41 G6
Dafen **SA14** 17 J5
Daffy Green **IP25** 44 D5
Dagdale **ST14** 40 C2
Dagenham RM10 23 H3
Daggons **SP6** 10 C3
Daglingworth **GL7** 20 C1
Dagnall **HP4** 32 C7
Dail **PA35** 80 B4
Dail Beag **HS2** 100 E3
Dail Bho Dheas (South Dell) **HS2** 101 G1
Dail Bho Thuath (North Dell) **HS2** 101 G1
Dail Mòr **HS2** 100 E3
Dailly **KA26** 67 G3
Dailnamac **PA35** 80 A4
Dainton **TQ12** 5 J4
Dairsie (Osnaburgh) **KY15** 83 F6
Dairy House **HU12** 59 J7
Daisy Bank **WS5** 40 C6
Daisy Green **IP31** 34 E2
Dalabrog **HS8** 84 C2
Dalavich **PA35** 80 A6
Dalballoch **PH20** 88 D5
Dalbeattie DG5 65 J4
Dalblair **KA18** 68 B2
Dalbog **DD9** 90 D7
Dalbreck **IV28** 104 C7
Dalbury **DE6** 40 E2
Dalby *I.o.M.* **IM5** 54 B6
Dalby *Lincs.* **PE23** 53 H6
Dalby *N.Yorks.* **YO60** 58 C2
Dalcairnie **KA6** 67 J3
Dalchalloch **PH18** 81 J1
Dalchalm **KW9** 97 G1
Dalchenna **PA32** 80 B7
Dalchirach **AB37** 89 J1
Dalchork **IV27** 103 H7
Dalchreichart **IV63** 87 J3
Dalchruin **PH6** 81 J6
Dalcross **IV2** 96 E7
Dalderby **LN9** 53 F6
Dalditch **EX9** 7 J7
Daldownie **AB35** 89 K4
Dale *Cumb.* **CA4** 61 G2
Dale *Gt.Man.* **OL3** 49 J2
Dale *Pembs.* **SA62** 16 B5
Dale Abbey **DE7** 41 G2
Dale End *Derbys.* **DE45** 50 E6
Dale End *N.Yorks.* **BD20** 56 E5
Dale Head **CA10** 60 F5
Dale of Walls **ZE2** 107 K7
Dale Park **BN18** 12 C5
Dalehouse **TS13** 63 J5
Dalelia **PH36** 79 J1
Daless **IV12** 89 F1
Dalestie **AB37** 89 J3
Dalfad **AB35** 90 B4
Dalganachan **KW12** 105 F4
Dalgarven **KA13** 74 A6
Dalgety Bay **KY11** 75 K2
Dalgig **KA18** 67 K2
Dalginross **PH6** 81 J5
Dalgonar **DG3** 68 C3
Dalguise **PH8** 82 A3
Dalhalvaig **KW13** 104 D3
Dalham **CB8** 34 B2
Daligan **G84** 74 B2
Dalinlongart **PA23** 73 K2
Dalivaddy **PA28** 66 A1
Daljarrock **KA26** 67 F5
Dalkeith EH22 76 B4
Dallachulish **PA37** 80 A3
Dallas **IV36** 97 J6
Dallaschyle **IV12** 97 F7
Dallash **DG8** 64 E4
Dalleagles **KA18** 67 K2
Dallinghoo **IP13** 35 G3
Dallington *E.Suss.* **TN21** 13 K5
Dallington *Northants.* **NN5** 31 J2
Dallow **HG4** 57 G2
Dalmadilly **AB51** 91 F3
Dalmally PA33 80 C5
Dalmarnock **PH8** 82 B3
Dalmary **FK8** 74 D1
Dalmellington **KA6** 67 J3
Dalmeny **EH30** 75 K3
Dalmichy **IV27** 103 H7
Dalmigavie **IV13** 88 E3
Dalmore **IV17** 96 D5
Dalmuir **G81** 74 C3
Dalmunzie House Hotel **PH10** 89 H7
Dalnabreck **PH36** 79 J1
Dalnacarn **PH10** 82 B1
Dalnaglar Castle **PH10** 82 C1
Dalnaha **PA63** 79 H5
Dalnahaitnach **PH23** 89 F3
Dalnamain **IV25** 96 E2
Dalnatrat **PA38** 80 A2
Dalnavert **PH21** 89 F4
Dalnavie **IV17** 96 D4
Dalness **PH49** 80 C2
Dalnessie **IV27** 103 J7
Dalnigap **DG8** 64 B3
Dalqueich **KY13** 82 B7
Dalreoch **KA26** 67 F5
Dalriech **PH8** 81 J4
Dalroy **IV2** 96 E7
Dalrulzian **PH10** 82 C2
Dalry KA24 74 A6
Dalrymple **KA6** 67 H2
Dalscote **NN12** 31 H3
Dalserf **ML9** 75 F5
Dalsetter **ZE2** 107 P3
Dalshangan **DG7** 67 K5
Dalskairth **DG2** 65 K3
Dalston **CA5** 60 E1
Dalswinton **DG2** 68 E5
Daltomach **IV13** 88 E2
Dalton *Cumb.* **LA6** 55 J2
Dalton *D. & G.* **DG11** 69 G6
Dalton *Lancs.* **WN8** 48 D2
Dalton *N.Yorks.* **YO7** 57 K2
Dalton *N.Yorks.* **DL11** 62 C6
Dalton *Northumb.* **NE18** 71 G6
Dalton *Northumb.* **NE46** 62 A1
Dalton *S.Yorks.* **S65** 51 G3
Dalton Magna **S65** 51 G3
Dalton Piercy **TS27** 63 F3
Dalton-in-Furness LA15 55 F2
Dalton-le-Dale **SR7** 63 F2
Dalton-on-Tees **DL2** 62 D6
Daltote **PA31** 73 F2
Daltra **IV12** 97 G7
Dalveich **FK19** 81 H5
Dalvennan **KA19** 67 H2
Dalvourn **IV2** 88 D1
Dalwhinnie PH19 88 D6
Dalwood **EX13** 8 B4
Dam Green **NR16** 44 E7
Damask Green **SG4** 33 F6
Damerham **SP6** 10 C3

Dam - Dra

Damgate NR13 45 J5
Damnaglaur DG9 64 B7
Damside PH3 82 A6
Danaway ME9 24 E5
Danbury CM3 24 D1
Danby YO21 63 J6
Danby Wiske DL7 62 E7
Dancers Hill EN5 23 F2
Dandaleith AB38 97 K7
Danderhall EH22 76 B4
Dane Bank M34 49 J3
Dane End SG12 33 G6
Dane Hills LE3 41 H5
Danebridge SK11 49 J6
Danehill RH17 13 H4
Danesmoor S45 51 G6
Danestone AB22 91 H3
Daniel's Water TN26 14 E3
Danskine EH41 76 D4
Danthorpe HU12 59 J6
Danzey Green B94 30 C2
Darby End B64 40 B7
Darby Green GU46 22 B5
Darenth DA2 23 J4
Daresbury WA4 48 E4
Darfield S73 51 G2
Dargate ME13 25 G5
Dargues NE19 70 D4
Darite PL14 4 C4
Darland ME7 24 D5
Darlaston WS10 40 B6
Darley HG3 57 H4
Darley Bridge DE4 50 E6
Darley Dale DE4 50 E6
Darley Head HG3 57 G4
Darley Hillside DE4 50 E6
Darlingscott CV36 30 D4
DARLINGTON DL 62 D5
Darliston SY13 38 E2
Darlton NG22 51 K5
Darnabo AB53 99 F6
Darnall S9 51 F4
Darnconner KA18 67 K1
Darnford AB31 91 F5
Darngarroch DG7 65 G4
Darnick TD6 76 D7
Darowen SY20 37 H5
Darra AB53 99 F6
Darracott EX39 6 A4
Darras Hall NE20 71 G6
Darrington WF8 51 G1
Darrow Green IP20 45 G7
Darsham IP17 35 J2
Dartfield AB43 99 J5
DARTFORD DA 23 J4
Dartington TQ9 5 H4
Dartmeet TQ13 5 G3
Dartmouth TQ6 5 J5
Darton S75 51 F1
Darvel KA17 74 D7
Darvell TN32 14 C5
Darwell Hole TN33 13 K5
Darwen BB3 56 B7
Datchet SL3 22 C4
Datchworth SG3 33 F7
Datchworth Green SG3 33 F7
Daubhill BL3 49 F2
Daugh of Kinermony AB38 97 K7
Dauntsey SN15 20 C3
Dauntsey Green SN15 20 C3
Dauntsey Lock SN15 20 C3
Dava PH26 89 H1
Davaar PA28 66 B2
Davan AB34 90 C4
Davenham CW9 49 F5
Davenport Green WA15 49 H4
Daventry NN11 31 G2
Davidstow PL32 4 B2
Davington DG13 69 H3
Daviot *Aber.* AB51 91 F2
Daviot *High.* IV2 88 E1
Davoch of Grange AB55 98 C5
Davyhulme M41 49 G3
Dawley TF4 39 F5
Dawlish EX7 5 K3
Dawn LL22 47 G5
Daws Heath SS7 24 E3
Daw's House PL15 6 B7
Dawsmere PE12 43 H2
Day Green CW11 49 G7
Dayhills ST15 40 B2
Dayhouse Bank B62 29 J1
Daylesford GL56 30 D6
Ddôl CH7 47 K5
Deadman's Cross MK45 32 E4
Deadwaters ML11 75 F6
Deal CT14 15 J2
Deal Hall CM0 25 G2
Dean *Cumb.* CA14 60 B4
Dean *Devon* TQ11 5 H4
Dean *Dorset* SP5 9 J3
Dean *Hants.* SO32 11 G3
Dean *Oxon.* OX7 30 E6
Dean *Som.* BA4 19 K7
Dean Bank DL17 62 D3
Dean Cross EX34 6 D1
Dean Head S35 50 E2
Dean Prior TQ11 5 H4
Dean Row SK9 49 H4
Dean Street ME15 14 C2
Deanburnhaugh TD9 69 J2
Deane *Gt.Man.* BL3 49 F2
Deane *Hants.* RG25 21 J6
Deanland SP5 9 J3
Deanlane End PO9 11 J3
Deans Bottom ME9 24 E5
Deanscales CA13 60 B4
Deansgreen WA13 49 F4
Deanshanger MK19 31 J5
Deanston FK16 81 J7
Dearham CA15 60 B3
Debach IP13 35 G3
Debate DG11 69 H5
Debden CB11 33 J5
Debden Cross CB11 33 J5
Debden Green *Essex* CB11 33 J5
Debden Green *Essex* IG10 23 H2
Debenham IP14 35 F2
Deblin's Green WR2 29 H4
Dechmont EH52 75 J3
Decker Hill TF11 39 G4
Deddington OX15 31 F5
Dedham CO7 34 E5
Dedham Heath CO7 34 E5
Dedworth SL4 22 C4
Deecastle AB34 90 C5
Deene NN17 42 C6
Deenethorpe NN17 42 C6
Deepcar S36 50 E3
Deepcut GU16 22 C6
Deepdale *Cumb.* LA10 56 C1
Deepdale *N.Yorks.* BD23 56 D2
Deeping Gate PE6 42 E5
Deeping St. James PE6 42 E5
Deeping St. Nicholas PE11 43 F4
Deepweir NP26 19 H3
Deerhill AB55 98 C5
Deerhurst GL19 29 H6
Deerhurst Walton GL19 29 H6
Deerton Street ME9 25 F5
Defford WR8 29 J4
Defynnog LD3 27 J6
Deganwy LL31 47 F5
Degnish PA34 79 J6
Deighton *N.Yorks.* DL6 62 E6
Deighton *W.Yorks.* HD2 50 D1
Deighton *York* YO19 58 C5
Deiniolen LL55 46 D6
Delabole PL33 4 A2
Delamere CW8 48 E6
Delavorar AB37 89 J3
Delfrigs AB23 91 H2
Dell Lodge PH25 89 H3
Dell Quay PO20 12 B6
Delliefure PH26 89 H1
Delly End OX29 30 E7
Delnabo AB37 89 J3
Delny IV18 96 E4
Delph OL3 49 J2
Delphorrie AB33 90 C3
Delves DH8 62 C2
Delvine PH1 82 C3
Dembleby NG34 42 D2
Denaby DN12 51 G3
Denaby Main DN12 51 G3
Denbigh (Dinbych) LL16 47 J6
Denbury TQ12 5 J4
Denby DE5 41 F1
Denby Dale HD8 50 E2
Denchworth OX12 21 G2
Dendron LA12 55 F2
Denend AB54 90 E1
Denford NN14 32 C1
Dengie CM0 25 F1
Denham *Bucks.* UB9 22 D3
Denham *Suff.* IP21 35 F1
Denham *Suff.* IP29 34 B2
Denham Green UB9 22 D3
Denham Street IP23 35 F1
Denhead *Aber.* AB51 91 F3
Denhead *Aber.* AB42 99 J5
Denhead *Fife* KY16 83 F6
Denhead of Arbirlot DD11 83 G3
Denhead of Gray DD2 82 E4
Denholm TD9 70 A2
Denholme BD13 57 F6
Denholme Clough BD13 57 F6
Denio LL53 36 C2
Denmead PO7 11 H3
Denmill AB51 91 G3
Denmoss AB54 98 E6
Dennington IP13 35 G2
Denny FK6 75 G2
Dennyloanhead FK4 75 G2
Denshaw OL3 49 J1
Denside AB31 91 G5
Densole CT18 15 H3
Denston CB8 34 B3
Denstone ST14 40 C1
Dent LA10 56 C1
Denton *Cambs.* PE7 42 E7
Denton *Darl.* DL2 62 D5
Denton *E.Suss.* BN9 13 H6
Denton *Gt.Man.* M34 49 J3
Denton *Kent* CT4 15 H3
Denton *Kent* DA12 24 C4
Denton *Lincs.* NG32 42 B2
Denton *N.Yorks.* LS29 57 G5
Denton *Norf.* IP20 45 G7
Denton *Northants.* NN7 32 B3
Denton *Oxon.* OX44 21 J1
Denton's Green WA10 48 D3
Denver PE38 44 A5
Denvilles PO9 11 J4
Denwick NE66 71 H2
Deopham NR18 44 E5
Deopham Green NR18 44 E6
Depden IP29 34 B3
Depden Green IP29 34 B3
Deptford *Gt.Lon.* SE8 23 G4
Deptford *Wilts.* BA12 10 B1
DERBY DE 41 F2
Derbyhaven IM9 54 B7
Dereham (East Dereham) NR19 44 D4
Dererach PA70 79 G5
Deri CF81 18 E1
Derril EX22 6 B5
Derringstone CT4 15 H3
Derrington ST18 40 A3
Derriton EX22 6 B5
Derry FK19 81 H5
Derry Hill SN11 20 C4
Derrythorpe DN17 52 B2
Dersingham PE31 44 A2
Dervaig PA75 79 F2
Derwen LL21 47 J7
Derwenlas SY20 37 G6
Derwydd SA18 17 K4
Derybruich PA21 73 H3
Desborough NN14 42 B7
Desford LE9 41 G5
Detchant NE70 77 J7
Dethick DE4 51 F7
Detling ME14 14 C2
Deuddwr SY22 38 B4
Deunant LL16 47 H6
Deuxhill WV16 39 F7
Devauden NP16 19 H2
Devil's Bridge (Pontarfynach) SY23 27 G1
Devitts Green CV7 40 E6
Devizes SN10 20 D5
Devonport PL1 4 E5
Devonside FK13 75 H1
Devoran TR3 2 E5
Dewar EH38 76 B6
Dewlish DT2 9 G5
Dewsall Court HR2 28 D5
Dewsbury WF12 57 H7
Dewsbury Moor WF15 57 H7
Dhiseig PA68 79 F4
Dhoon IM7 54 D5
Dhoor IM7 54 D4
Dhowin IM7 54 D3
Dhuhallow IV2 88 C2
Dial Green GU28 12 C4
Dial Post RH13 12 E5
Dibden SO45 11 F4
Dibden Hill HP8 22 C2
Dibden Purlieu SO45 11 F4
Dickleburgh IP21 45 F7
Dickleburgh Moor IP21 45 F7
Didbrook GL54 30 B5
Didcot OX11 21 J2
Diddington PE19 32 E2
Diddlebury SY7 38 E7
Didley HR2 28 D5
Didling GU29 12 B5
Didmarton GL9 20 B3
Didsbury M20 49 H3
Didworthy TQ10 5 G4
Digby LN4 52 D7
Digg IV51 93 K5
Diggle OL3 50 C2
Digmoor WN8 48 D2
Digswell AL6 33 F7
Dihewyd SA48 26 D3
Dildawn DG7 65 H5
Dilham NR28 45 H3
Dilhorne ST10 40 B1
Dillington PE19 32 E2
Dilston NE45 70 E7
Dilton Marsh BA13 20 B7
Dilwyn HR4 28 D3
Dilwyn Common HR4 28 D3
Dimple BL7 49 G1
Dinas *Carmar.* SA33 17 F2
Dinas *Gwyn.* LL54 46 C7
Dinas *Gwyn.* LL53 36 B2
Dinas Cross SA42 16 D2
Dinas Dinlle LL54 46 C7
Dinas Powys CF64 18 E4
Dinas-Mawddwy SY20 37 H4
Dinbych (Denbigh) LL16 47 J6
Dinbych-y-pysgod (Tenby) SA70 16 E5
Dinckley BB6 56 B6
Dinder BA5 19 J7
Dinedor HR2 28 E5
Dingestow NP25 28 D7
Dingley LE16 42 A7
Dingwall IV15 96 C6
Dinlabyre TD9 70 A4
Dinnet AB34 90 C5
Dinnington *S.Yorks.* S25 51 H4
Dinnington *Som.* TA17 8 D3
Dinnington *T. & W.* NE13 71 H6
Dinorwig LL55 46 D6
Dinton *Bucks.* HP17 31 J7
Dinton *Wilts.* SP3 10 B1
Dinvin DG9 64 A5
Dinwoodie Mains DG11 69 G4
Dinworthy EX22 6 B4
Dipford TA3 8 B2
Dippen *Arg. & B.* PA28 73 F7
Dippen *N.Ayr.* KA27 66 E1
Dippenhall GU10 22 B7
Dipple *Moray* IV32 98 B5
Dipple *S.Ayr.* KA26 67 G3
Diptford TQ9 5 H5
Dipton DH9 62 C1
Dirdhu PH26 89 H2
Dirleton EH39 76 D2
Discoed LD8 28 B2
Diseworth DE74 41 G3
Dishes KW17 106 F5
Dishforth YO7 57 J2
Dishley LE11 41 H3
Disley SK12 49 J4
Diss IP22 45 F7
Disserth LD1 27 K3
Distington CA14 60 B4
Ditcheat BA4 9 F1
Ditchingham NR35 45 H6
Ditchley OX29 30 E6
Ditchling BN6 13 G5
Ditteridge SN13 20 B5
Dittisham TQ6 5 J5
Ditton *Halton* WA8 48 D4
Ditton *Kent* ME20 14 C2
Ditton Green CB8 33 K3
Ditton Priors WV16 39 F7
Dixton *Glos.* GL20 29 J5
Dixton *Mon.* NP25 28 E7
Dobcross OL3 49 J2
Dobwalls PL14 4 C4
Doc Penfro (Pembroke Dock) SA72 16 C5
Doccombe TQ13 7 F7
Dochgarroch IV3 88 D1
Dockenfield GU10 22 B7
Docker *Cumb.* LA8 61 G7
Docker *Lancs.* LA6 55 J2
Docking PE31 44 B2
Docklow HR6 28 E3
Dockray *Cumb.* CA11 60 E4
Dockray *Cumb.* CA7 60 D1
Dodbrooke TQ7 5 H6
Doddenham WR6 29 G3
Doddinghurst CM15 23 J2
Doddington *Cambs.* PE15 43 H6
Doddington *Kent* ME9 14 E2
Doddington *Lincs.* LN6 52 C5
Doddington *Northumb.* NE71 77 H7
Doddington *Shrop.* DY14 29 F1
Doddiscombsleigh EX6 7 G7
Doddycross PL14 4 D4
Dodford *Northants.* NN7 31 H2
Dodford *Worcs.* B61 29 J1
Dodington *S.Glos.* BS37 20 A3
Dodington *Som.* TA5 7 K1
Dodington Ash BS37 20 A4
Dodleston CH4 48 C6
Dods Leigh ST10 40 C2
Dodscott EX38 6 D4
Dodworth S75 51 F2
Doehole DE55 51 F7
Doffcocker BL1 49 F1
Dog Village EX5 7 H6
Dogdyke LN4 53 F7
Dogmersfield RG27 22 A6
Dogsthorpe PE1 43 F5
Dol Fawr SY19 37 H5
Dolanog SY21 37 K4
Dolau *Powys* LD1 28 A2
Dolau *R.C.T.* CF72 18 D3
Dolbenmaen LL51 36 E1
Doley ST20 39 G3
Dolfach SY18 27 J1
Dolfor SY16 38 A7
Dolgarreg SA20 27 G5
Dolgarrog LL32 47 F6
Dolgellau LL40 37 G4
Dolgoch LL36 37 F5
Dolgran SA39 17 H2
Doll KW9 97 F1
Dollar FK14 75 H1
Dollarbeg FK14 75 H1
Dolphin CH8 47 K5
Dolphinholme LA2 55 J4
Dolphinton EH46 75 K6
Dolton EX19 6 D4
Dolwen *Conwy* LL22 47 G5
Dolwen *Powys* SY21 37 J5
Dolwyddelan LL25 47 F7
Dôl-y-bont SY24 37 F7
Dol-y-cannau HR5 28 A4
Dolyhir LD8 28 B3
Dolywern LL20 38 B2
Domgay SY22 38 B4
DONCASTER DN 51 H2
Donhead St. Andrew SP7 9 J2
Donhead St. Mary SP7 9 J2
Donibristle KY4 75 K2
Doniford TA23 7 J1
Donington *Lincs.* PE11 43 F2
Donington *Shrop.* WV7 40 A5
Donington le Heath LE67 41 G4
Donington on Bain LN11 53 F4
Donisthorpe DE12 41 F4
Donna Nook LN11 53 H3
Donnington *Glos.* GL56 30 C6
Donnington *Here.* HR8 29 G5
Donnington *Shrop.* SY5 38 E5
Donnington *Tel. & W.* TF2 39 G4
Donnington *W.Berks.* RG14 21 H5
Donnington *W.Suss.* PO20 12 B6
Donyatt TA19 8 C3
DORCHESTER *Dorset* DT 9 F5
Dorchester *Oxon.* OX10 21 J2
Dordon B78 40 E5
Dore S17 51 F4
Dores IV2 88 C1
Dorket Head NG5 41 H1
Dorking RH4 22 E7
Dorley's Corner IP17 35 H2
Dormans Park RH19 23 G7
Dormansland RH7 23 H7
Dormanstown TS10 63 G4
Dormer's Wells UB1 22 E3
Dormington HR1 28 E4
Dormston WR7 29 J3
Dorn GL56 30 D5
Dorney SL4 22 C4
Dorney Reach SL6 22 C4
Dornie IV40 86 E2
Dornoch IV25 96 E3
Dornock DG12 69 H7
Dorrery KW12 105 F3
Dorridge B93 30 C1
Dorrington *Lincs.* LN4 52 D7
Dorrington *Shrop.* SY5 38 D5
Dorsell AB33 90 D3
Dorsington CV37 30 C4
Dorstone HR3 28 C4
Dorton HP18 31 H7
Dorusduain IV40 87 F2
Dosthill B77 40 E6
Dotland NE46 62 A1
Dottery DT6 8 D5
Doublebois PL14 4 B4
Dougalston G62 74 D3
Dougarie KA27 73 G7
Doughton GL8 20 B2
Douglas *I.o.M.* IM1 54 C6
Douglas *S.Lan.* ML11 75 G7
Douglas & Angus DD5 83 F4
Douglas Water ML11 75 G7
Douglastown DD8 83 F3
Doulting BA4 19 K7
Dounby KW17 106 B5
Doune *Arg. & B.* G83 74 B1
Doune *Arg. & B.* FK17 80 E6
Doune *High.* PH22 89 F3
Doune *High.* IV24 96 B1
Doune *Stir.* FK16 81 J7
Dounepark AB45 99 F4
Douneside AB34 90 C4
Dounie *High.* IV19 96 D3
Dounie *High.* IV24 96 C2
Dounreay KW14 104 E2
Dousland PL20 5 F4
Dovaston SY10 38 C3
Dove Holes SK17 50 C5
Dovenby CA13 60 B3
Dovendale LN11 53 G4
Dover CT16 15 J3
Dovercourt CO12 35 G5
Doverdale WR9 29 H2
Doveridge DE6 40 D2
Doversgreen RH2 23 F7
Dowally PH9 82 B3
Dowdeswell GL54 30 B6
Dowhill KA26 67 G3
Dowlais CF48 18 D1
Dowland EX19 6 D4
Dowlands DT7 8 B5
Dowlish Ford TA19 8 C3
Dowlish Wake TA19 8 C3
Down Ampney GL7 20 E2
Down End TA6 19 G7
Down Hatherley GL2 29 H6
Down St. Mary EX17 7 F5
Down Thomas PL9 5 F6
Downderry PL11 4 D5
Downe BR6 23 H5
Downend *I.o.W.* PO30 11 G6
Downend *S.Glos.* BS16 19 K4
Downend *W.Berks.* RG20 21 H4
Downfield DD3 82 E4
Downfields CB7 33 K1
Downgate PL17 4 D3
Downham *Essex* CM11 24 D2
Downham *Lancs.* BB7 56 C5
Downham *Northumb.* TD12 77 G7
Downham Market PE38 44 A5
Downhead *Cornw.* PL15 4 C2
Downhead *Som.* BA4 19 K7
Downhead *Som.* BA22 8 E2
Downholland Cross L39 48 C2
Downholme DL11 62 C7
Downies AB12 91 H5
Downley HP13 22 B2
Downs CF5 18 E4
Downside *N.Som.* BS48 19 H5
Downside *Som.* BA4 19 K7
Downside *Som.* BA3 19 K6
Downside *Surr.* KT11 22 E6
Downton *Devon* EX20 6 D7
Downton *Devon* TQ6 5 J5
Downton *Hants.* SO41 10 D5
Downton *Wilts.* SP5 10 C2
Downton on the Rock SY8 28 D1
Dowsby PE10 42 E3
Dowthwaitehead CA11 60 E4
Doxey ST16 40 B3
Doynton BS30 20 A4
Drabblegate NR11 45 G3
Draethen NP10 19 F3
Draffan ML11 75 F6
Dragley Beck LA12 55 F2
Drakeland Corner PL7 5 F5
Drakelow DY11 40 A7
Drakemyre KA24 74 A5
Drakes Broughton WR10 29 J4
Drakes Cross B47 30 B1
Draughton *N.Yorks.* BD23 57 F4
Draughton *Northants.* NN6 31 J1
Drax YO8 58 C7
Draycot Foliat SN4 20 E4
Draycote CV23 31 F2
Draycott *Derbys.* DE72 41 G2
Draycott *Glos.* GL56 30 C5
Draycott *Shrop.* WV5 40 A6
Draycott *Som.* BS27 19 H6
Draycott *Worcs.* WR5 29 H4
Draycott in the Clay DE6 40 D3
Draycott in the Moors ST10 40 B2
Drayford EX17 7 F4
Draynes PL14 4 C4
Drayton *Leics.* LE16 42 B6
Drayton *Lincs.* PE20 43 F2
Drayton *Norf.* NR8 45 F4
Drayton *Oxon.* OX15 31 F4
Drayton *Oxon.* OX14 21 H2
Drayton *Ports.* PO6 11 H4
Drayton *Som.* TA10 8 D2
Drayton *Warks.* CV37 30 C3
Drayton *Worcs.* DY9 29 J1
Drayton Bassett B78 40 D5
Drayton Beauchamp HP22 32 C7
Drayton Parslow MK17 32 B6
Drayton St. Leonard OX10 21 J2

Dre - Eas

Drebley BD23 57 F4
Dreemskerry IM7 54 D4
Dreenhill SA62 16 C4
Drefach *Carmar.* SA44 17 G2
Drefach *Carmar.* SA14 17 J4
Dre-fach *Cere.* SA40 17 J1
Drefelin SA44 17 G2
Dreghorn KA11 74 B7
Drem EH39 76 D3
Dreumasdal (Drimsdale) HS8 84 C1
Drewsteignton EX6 7 F6
Driby LN13 53 G5
Driffield *E.Riding* YO25 59 G4
Driffield *Glos.* GL7 20 D2
Drigg CA19 60 B7
Drighlington BD11 57 H7
Drimfern PA32 80 B6
Drimlee PA32 80 C6
Drimnin PA34 79 G2
Drimore HS8 92 C7
Drimpton DT8 8 D4
Drimsdale (Dreumasdal) HS8 84 C1
Drimsynie PA24 80 C7
Drimvore PA31 73 G1
Drinan IV49 86 B3
Dringhouses YO24 58 B4
Drinisiader HS3 93 G2
Drinkstone IP30 34 D2
Drinkstone Green IP30 34 D2
Drishaig PA32 80 C6
Drissaig PA35 80 A6
Drointon ST18 40 C3
Droitwich Spa WR9 29 H2
Dron PH2 82 C6
Dronfield S18 51 F5
Dronfield Woodhouse S18 51 F5
Drongan KA6 67 J2
Dronley DD3 82 E4
Droop DT10 9 G4
Dropmore SL1 22 C3
Droxford SO32 11 H3
Droylsden M43 49 J3
Druid LL21 37 K1
Druidston SA62 16 B4
Druimarbin PH33 87 G7
Druimavuic PA38 80 B3
Druimdrishaig PA31 73 F3
Druimindarroch PH39 86 C6
Druimkinnerras IV4 88 B1
Drum *Arg. & B.* PA21 73 H3
Drum *P. & K.* KY13 82 B7
Drumachloy PA20 73 J4
Drumbeg IV27 102 D5
Drumblade AB54 98 D6
Drumblair AB54 98 E6
Drumbuie *D. & G.* DG4 68 C2
Drumbuie *High.* IV40 86 D1
Drumburgh CA7 60 D1
Drumchapel G15 74 D3
Drumchardine IV5 96 C7
Drumchork IV22 94 E3
Drumclog ML10 74 E7
Drumdelgie AB54 98 C6
Drumderfit IV1 96 D6
Drumeldrie KY8 83 F7
Drumelzier ML12 75 K7
Drumfearn IV43 86 C3
Drumfern PH33 87 F7
Drumgarve PA28 66 B1
Drumgley DD8 83 F2
Drumguish PH21 88 E5
Drumhead AB31 90 E5
Drumin AB37 89 J1
Drumine IV2 96 E6
Drumjohn DG7 67 K4
Drumlamford House KA26 64 C3
Drumlasie AB31 90 E4
Drumlemble PA28 66 A2
Drumlithie AB39 91 F6
Drummond *High.* IV16 96 D5
Drummond *Stir.* FK17 81 H7
Drummore DG9 64 B7
Drummuir Castle AB55 98 B6
Drumnadrochit IV63 88 C2
Drumnagorrach AB54 98 D5
Drumnatorran PH36 79 K1
Drumoak AB31 91 F5
Drumore PA28 66 B1
Drumour PH8 82 A4
Drumrash DG7 65 G3
Drumrunie IV26 95 H1
Drums AB41 91 H2
Drumsturdy DD5 83 F4
Drumuie IV51 93 K7
Drumuillie PH24 89 G2
Drumvaich FK17 81 H7
Drumwhindle AB41 91 H1
Drumwhirn DG7 68 C5
Drunkendub DD11 83 H3
Druridge NE61 71 H4
Drury CH7 48 B6
Drws-y-nant LL40 37 H3
Dry Doddington NG23 42 B1
Dry Drayton CB23 33 G2
Dry Harbour IV40 94 C6
Dry Sandford OX13 21 H1
Dry Street SS16 24 C3
Drybeck CA16 61 H5
Drybridge *Moray* AB56 98 C4
Drybridge *N.Ayr.* KA11 74 B7
Drybrook GL17 29 F7
Dryburgh TD6 76 D7
Drygrange TD6 76 D7
Dryhope TD7 69 H1
Drymen G63 74 C2
Drymuir AB42 99 H6
Drynoch IV47 85 K1
Dryslwyn SA32 17 J3
Dryton SY5 38 E5
Duachy PA34 79 K5
Dubford AB45 99 F4
Dubhchladach PA29 73 G4
Dubheads PH7 82 A5
Dublin IP23 35 F2
Dubton DD8 83 G2
Duchal PA13 74 B4
Duchally IV27 103 F7
Duchray FK8 81 F7
Duck Bay G83 74 B2
Duck End *Bed.* MK45 32 D4
Duck End *Cambs.* PE19 33 F2
Duck End *Essex* CM6 33 K6
Duck Street HG3 57 G3
Duckington SY14 48 D7
Ducklington OX29 21 G1
Duckmanton S44 51 G5
Duck's Cross MK44 32 E3
Ducks Island EN5 23 F2
Duddenhoe End CB11 33 H5
Duddingston EH15 76 A3
Duddington PE9 42 C5
Duddlestone TA3 8 B2
Duddleswell TN22 13 H4
Duddo TD15 77 H6
Duddon CW6 48 E6
Duddon Bridge LA18 54 E1
Dudleston SY12 38 C2
Dudleston Heath (Criftins) SY12 38 C2
Dudley *T. & W.* NE23 71 H6
DUDLEY *W.Mid.* DY 40 B6
Dudley Port DY4 40 B6
Dudlow's Green WA4 49 F4
Dudsbury BH22 10 B5
Duffield DE56 41 F1
Duffryn SA13 18 B2
Dufftown AB55 90 B1
Duffus IV30 97 J5
Dufton CA16 61 H4
Duggleby YO17 58 E3
Duiar PH26 89 J1
Duible KW8 104 E6
Duiletter PA33 80 C5
Duinish PH17 81 H1
Duirinish IV40 86 D1
Duisdealmor IV43 86 D3
Duisky PH33 87 G7
Duke End B46 40 E7
Dukestown NP22 28 A7
Dukinfield SK16 49 J3
Dulas LL70 46 C4
Dulax AB36 90 B3
Dulcote BA5 19 J7
Dulford EX15 7 J5
Dull PH15 81 K3
Dullatur G68 75 F3
Dullingham CB8 33 K3
Dullingham Ley CB8 33 K3
Dulnain Bridge PH26 89 G2
Duloe *Bed.* PE19 32 E2
Duloe *Cornw.* PL14 4 C5
Dulsie IV12 97 G7
Dulverton TA22 7 H3
Dulwich SE21 23 G4
Dumbarton G82 74 B3
Dumbleton WR11 30 B5
Dumcrieff DG10 69 G3
Dumeath AB54 90 C1
Dumfin G84 74 B2
DUMFRIES DG 65 K3
Dumgoyne G63 74 D2
Dummer RG25 21 J7
Dun DD10 83 H2
Dunach PA34 79 K5
Dunalastair PH16 81 J2
Dunan *Arg. & B.* PA23 73 K3
Dunan *High.* IV49 86 B2
Dunans PA22 73 J1
Dunball TA6 19 G7
Dunbar EH42 76 E3
Dunbeath KW6 105 G5
Dunbeg PA37 79 K4
Dunblane FK15 81 J7
Dunbog KY14 82 D6
Dunbridge SO51 10 E2
Duncanston *Aber.* AB52 90 D2
Duncanston *High.* IV7 96 C6
Dunchideock EX2 7 G7
Dunchurch CV22 31 F1
Duncote NN12 31 H3
Duncow DG1 68 E5
Duncraggan FK17 81 G7
Duncrievie PH2 82 C7
Duncroist FK21 81 G4
Duncrub PH2 82 B6
Duncryne G83 74 C2
Duncton GU28 12 C5
DUNDEE DD 83 F4
Dundee Airport DD2 82 E5
Dundon TA11 8 D1
Dundon Hayes TA11 8 D1
Dundonald KA2 74 B7
Dundonnell IV23 95 G3
Dundraw CA7 60 D2
Dundreggan IV63 87 K3
Dundrennan DG6 65 H6
Dundridge SO32 11 G3
Dundry BS41 19 J5
Dunearn KY3 76 A2
Dunecht AB32 91 F4
Dunfermline KY12 75 J2
Dunfield GL7 20 E2
Dunford Bridge S36 50 D2
Dungate ME9 14 E2
Dungavel ML10 74 E7
Dungworth S6 50 E3
Dunham NG22 52 B5
Dunham Town WA14 49 G4
Dunham Woodhouses WA14 49 G4
Dunham-on-the-Hill WA6 48 D5
Dunhampton DY13 29 H2
Dunholme LN2 52 D5
Dunino KY16 83 G6
Dunipace FK6 75 G2
Dunira PH6 81 J5
Dunkeld PH8 82 B3
Dunkerton BA2 20 A6
Dunkeswell EX14 7 K5
Dunkeswick LS17 57 J5
Dunkirk *Ches.W. & C.* CH1 48 C5
Dunkirk *Kent* ME13 15 F2
Dunk's Green TN11 23 K6
Dunlappie DD9 83 G1
Dunley *Hants.* RG28 21 H6
Dunley *Worcs.* DY13 29 G2
Dunlop KA3 74 C6
Dunloskin PA23 73 K3
Dunmere PL31 4 A4
Dunmore *Arg. & B.* PA29 73 F4
Dunmore *Falk.* FK2 75 G2
Dunn KW1 105 G3
Dunn Street ME7 24 D5
Dunnabie DG11 69 H5
Dunnet KW14 105 H1
Dunnichen DD8 83 G3
Dunning PH2 82 B6
Dunnington *E.Riding* YO25 59 H4
Dunnington *Warks.* B49 30 B3
Dunnington *York* YO19 58 C4
Dunnockshaw BB11 56 D7
Dunoon PA23 73 K3
Dunragit DG9 64 B5
Dunrostan PA31 73 F2
Duns TD11 77 F5
Duns Tew OX25 31 F6
Dunsa DE45 50 E5
Dunsby PE10 42 E3
Dunscore DG2 68 D5
Dunscroft DN7 51 J2
Dunsdale TS14 63 H5
Dunsden Green RG4 22 A4
Dunsfold GU8 12 D3
Dunsford EX6 7 G7
Dunshalt KY14 82 D6
Dunshill GL19 29 H5
Dunsinnan PH2 82 C4
Dunsland Cross EX22 6 C5
Dunsley *N.Yorks.* YO21 63 K5
Dunsley *Staffs.* DY7 40 A7
Dunsmore HP22 22 B1
Dunsop Bridge BB7 56 B4
Dunstable LU6 32 D6
Dunstall DE13 40 D3
Dunstall Green CB8 34 B2
Dunstan NE66 71 H2
Dunstan Steads NE66 71 H1
Dunster TA24 7 H1
Dunston *Lincs.* LN4 52 D6
Dunston *Norf.* NR14 45 G5
Dunston *Staffs.* ST18 40 B4
Dunston Heath ST18 40 B4
Dunston Hill NE11 71 H7
Dunstone *Devon* TQ13 5 H3
Dunstone *Devon* PL8 5 F5
Dunsville DN7 51 J2
Dunswell HU6 59 G6
Dunsyre ML11 75 J6
Dunterton PL19 4 D3
Duntisbourne Abbots GL7 20 C1
Duntisbourne Leer GL7 20 C1
Duntisbourne Rouse GL7 20 C1
Duntish DT2 9 F4
Duntocher G81 74 C3
Dunton *Bucks.* MK18 32 B6
Dunton *Cen.Beds.* SG18 33 F4
Dunton *Norf.* NR21 44 C2
Dunton Bassett LE17 41 H6
Dunton Green TN13 23 J6
Dunton Wayletts CM13 24 C2
Duntulm IV51 93 K4
Dunure KA7 67 G2
Dunure Mains KA7 67 G2
Dunvant SA2 17 J6
Dunvegan IV55 93 H7
Dunwich IP17 35 J1
Dura KY15 83 F6
Durdar CA2 60 F1
Durgan TR11 2 E6
Durgates TN5 13 K3
DURHAM DH 62 D2
Durham Tees Valley Airport DL2 62 E5
Durinemast PA34 79 H2
Durisdeer DG3 68 D3
Durleigh TA5 8 B1
Durley *Hants.* SO32 11 G3
Durley *Wilts.* SN8 21 F5
Durley Street SO32 11 G3
Durlow Common HR8 29 F5
Durnamuck IV23 95 G2
Durness IV27 103 G2
Durno AB51 91 F2
Duror PA38 80 A2
Durran *Arg. & B.* PA33 80 A7
Durran *High.* KW14 105 G2
Durrants PO9 11 J4
Durrington *W.Suss.* BN13 12 E6
Durrington *Wilts.* SP4 20 E7
Dursley GL11 20 A2
Dursley Cross GL17 29 F7
Durston TA3 8 B2
Durweston DT11 9 H4
Dury ZE2 107 N6
Duston NN5 31 J2
Duthil PH23 89 G2
Dutlas LD7 28 B1
Duton Hill CM6 33 K6
Dutson PL15 6 B7
Dutton WA4 48 E5
Duxford CB22 33 H4
Dwygyfylchi LL34 47 F5
Dwyran LL61 46 C6
Dyce AB21 91 G3
Dyfatty SA16 17 H5
Dyffryn *Bridgend* CF34 18 B2
Dyffryn *Pembs.* SA64 16 C2
Dyffryn *V. of Glam.* CF5 18 D4
Dyffryn Ardudwy LL44 36 E3
Dyffryn Castell SY23 37 G7
Dyffryn Ceidrych SA19 27 G6
Dyffryn Cellwen SA10 27 H7
Dyke *Devon* EX39 6 B3
Dyke *Lincs.* PE10 42 E3
Dyke *Moray* IV36 97 G6
Dykehead *Angus* DD8 82 E1
Dykehead *N.Lan.* ML7 75 G5
Dykehead *Stir.* FK8 74 D1
Dykelands AB30 83 J1
Dykends PH11 82 D2
Dykeside AB53 99 F6
Dylife SY19 37 H6
Dymchurch TN29 15 G5
Dymock GL18 29 G5
Dyrham SN14 20 A4
Dysart KY1 76 B1
Dyserth LL18 47 J5

E

Eachwick NE18 71 G6
Eadar dha Fhadhail HS2 100 C4
Eagland Hill PR3 55 H5
Eagle LN6 52 B6
Eagle Barnsdale LN6 52 B6
Eagle Moor LN6 52 B6
Eaglescliffe TS16 63 F5
Eaglesfield *Cumb.* CA13 60 B4
Eaglesfield *D. & G.* DG11 69 H6
Eaglesham G76 74 D5
Eaglethorpe PE8 42 D6
Eagley BL1 49 G1
Eairy IM4 54 B6
Eakley MK16 32 B3
Eakring NG22 51 J6
Ealand DN17 51 K1
Ealing W5 22 E3
Eamont Bridge CA10 61 G4
Earby BB18 56 E5
Earcroft BB2 56 B7
Eardington WV16 39 G6
Eardisland HR6 28 D3
Eardisley HR3 28 C4
Eardiston *Shrop.* SY11 38 C3
Eardiston *Worcs.* WR15 29 F2
Earith PE28 33 G1
Earl Shilton LE9 41 G6
Earl Soham IP13 35 G2
Earl Sterndale SK17 50 C6
Earl Stonham IP14 35 F3
Earle NE71 70 E1
Earlestown WA12 48 E3
Earley RG6 22 A4
Earlham NR4 45 F5
Earlish IV51 93 J5
Earls Barton NN6 32 B2
Earls Colne CO6 34 C6
Earl's Common WR9 29 J3
Earl's Court SW5 23 F4
Earl's Croome WR8 29 H4
Earl's Green IP14 34 E2
Earlsdon CV5 30 E1
Earlsferry KY9 83 F7
Earlsford AB51 91 G1
Earlsheaton WF12 57 H7
Earlston TD4 76 D7
Earlswood *Mon.* NP16 19 H2
Earlswood *Surr.* RH1 23 F7
Earlswood *Warks.* B94 30 C1
Earnley PO20 12 B7
Earnshaw Bridge PR26 55 J7
Earsairidh HS9 84 C5
Earsdon NE25 71 J6
Earsdon Moor NE61 71 G4
Earsham NR35 45 H7
Earsham Street IP21 35 G1
Earswick YO32 58 C4
Eartham PO18 12 C6
Earthcott Green BS35 19 K3
Easby TS9 63 G6
Easdale PA34 79 J6
Easebourne GU29 12 B4
Easenhall CV23 31 F1
Eashing GU7 22 C7
Easington *Bucks.* HP18 31 H7
Easington *Dur.* SR8 63 F2
Easington *E.Riding* HU12 53 G1
Easington *Northumb.* NE70 77 K7
Easington *Oxon.* OX49 21 K2
Easington *R. & C.* TS13 63 J5
Easington Colliery SR8 63 F2
Easington Lane DH5 62 E2
Easingwold YO61 58 B3
Easole Street CT15 15 H2
Eassie DD8 82 E3
East Aberthaw CF62 18 D5
East Acton W3 23 F3
East Allington TQ9 5 H6
East Anstey EX16 7 G3
East Anton SP11 21 G7
East Appleton DL10 62 D7
East Ardsley WF3 57 J7
East Ashey PO33 11 G6
East Ashling PO18 12 B6
East Auchronie AB32 91 G4
East Ayton YO12 59 F1
East Barkwith LN8 52 E4
East Barming ME16 14 C2
East Barnby YO21 63 K5
East Barnet EN4 23 F2
East Barsham NR22 44 D2
East Beckham NR11 45 F2
East Bedfont TW14 22 D4
East Bergholt CO7 34 E5
East Bierley BD4 57 G7
East Bilney NR20 44 D4
East Blatchington BN25 13 H7
East Boldon NE36 71 J7
East Boldre SO42 10 E4
East Bolton NE66 71 G2
East Bower TA6 8 C1
East Brent TA9 19 G6
East Bridge IP16 35 J2
East Bridgford NG13 41 J1
East Brora KW9 97 G1
East Buckland EX32 6 E2
East Budleigh EX9 7 J7
East Burnham SL2 22 C3
East Burra ZE2 107 M9
East Burrafirth ZE2 107 M7
East Burton BH20 9 H6
East Butsfield DL13 62 C2
East Butterleigh EX15 7 H5
East Butterwick DN17 52 B2
East Cairnbeg AB30 91 F7
East Calder EH53 75 J4
East Carleton NR14 45 F5
East Carlton *Northants.* LE16 42 B7
East Carlton *W.Yorks.* LS19 57 H5
East Chaldon (Chaldon Herring) DT2 9 G6
East Challow OX12 21 G3
East Charleton TQ7 5 H6
East Chelborough DT2 8 E4
East Chiltington BN7 13 G5
East Chinnock BA22 8 D3
East Chisenbury SN9 20 E6
East Clandon GU4 22 D6
East Claydon MK18 31 J6
East Clyne KW9 97 G1
East Clyth KW3 105 H5
East Coker BA22 8 E3
East Compton *Dorset* SP7 9 H3
East Compton *Som.* BA4 19 K7
East Coombe EX17 7 G5
East Cornworthy TQ9 5 J5
East Cottingwith YO42 58 D5
East Cowes PO32 11 G5
East Cowick DN14 58 C7
East Cowton DL7 62 E6
East Cramlington NE23 71 H6
East Cranmore BA4 19 K7
East Creech BH20 9 J6
East Croachy IV2 88 D2
East Darlochan PA28 66 A1
East Davoch AB34 90 C4
East Dean *E.Suss.* BN20 13 J7
East Dean *Hants.* SP5 10 D2
East Dean *W.Suss.* PO18 12 C5
East Dereham (Dereham) NR19 44 D4
East Down EX31 6 D1
East Drayton DN22 51 K5
East Dundry BS41 19 J5
East Ella HU5 59 G7
East End *E.Riding* HU12 59 H6
East End *E.Riding* HU12 59 J7
East End *Essex* CM0 25 G1
East End *Hants.* RG20 21 H5
East End *Hants.* SO41 10 E5
East End *Herts.* SG9 33 H6
East End *Kent* TN17 14 D4
East End *Kent* ME12 25 F4
East End *M.K.* MK16 32 C4
East End *N.Som.* BS48 19 H4
East End *Oxon.* OX29 30 E7
East End *Poole* BH21 9 J5
East End *Som.* BA3 19 J6
East End *Suff.* CO7 35 F5
East End *Suff.* IP14 35 F3
East Farleigh ME15 14 C2
East Farndon LE16 42 A7
East Ferry DN21 52 B3
East Firsby LN8 52 D4
East Fleetham NE68 71 H1
East Fortune EH39 76 D3
East Garston RG17 21 G4
East Ginge OX12 21 H3
East Goscote LE7 41 J4
East Grafton SN8 21 F5
East Green *Suff.* IP17 35 J2
East Green *Suff.* CB8 33 K3
East Grimstead SP5 10 D2
East Grinstead RH19 13 G3
East Guldeford TN31 14 E5
East Haddon NN6 31 H2
East Hagbourne OX11 21 J3
East Halton DN40 52 E1
East Ham E6 23 H3
East Hanney OX12 21 H2
East Hanningfield CM3 24 D1
East Hardwick WF8 51 G1
East Harling NR16 44 D7
East Harlsey DL6 63 F7
East Harnham SP2 10 C2
East Harptree BS40 19 J6
East Hartford NE23 71 H6
East Harting GU31 11 J3
East Hatch SP3 9 J2
East Hatley SG19 33 F3
East Hauxwell DL8 62 C7
East Haven DD7 83 G4
East Heckington PE20 42 E1
East Hedleyhope DL13 62 C2
East Helmsdale KW8 105 F7

Eas - Elt

East Hendred OX12 21 H3
East Herrington SR3 62 E1
East Heslerton YO17 59 F2
East Hewish BS24 19 H5
East Hoathly BN8 13 J5
East Holme BH20 9 H6
East Horndon CM13 24 C3
East Horrington BA5 19 J7
East Horsley KT24 22 D6
East Horton NE71 77 J7
East Howe BH10 10 B5
East Huntspill TA9 19 G7
East Hyde LU2 32 E7
East Ilsley RG20 21 H3
East Keal PE23 53 G6
East Kennett SN8 20 E5
East Keswick LS17 57 J5
East Kilbride G74 74 E5
East Kimber EX20 6 C6
East Kirkby PE23 53 G6
East Knapton YO17 58 E2
East Knighton DT2 9 H6
East Knowstone EX36 7 G3
East Knoyle SP3 9 H1
East Kyloe NE70 77 J7
East Lambrook TA13 8 D3
East Langdon CT15 15 J3
East Langton LE16 42 A6
East Langwell IV28 96 E1
East Lavant PO18 12 B6
East Lavington GU28 12 C5
East Layton DL11 62 C5
East Leake LE12 41 H3
East Learmouth TD12 77 G7
East Learney AB31 90 E4
East Leigh *Devon* EX17 7 F5
East Leigh *Devon* EX18 7 F4
East Leigh *Devon* TQ9 5 H5
East Leigh *Devon* PL21 5 G5
East Lexham PE32 44 C4
East Lilburn NE66 71 F1
East Linton EH40 76 D3
East Liss GU33 11 J2
East Lockinge OX12 21 H3
East Looe PL13 4 C5
East Lound DN9 51 K2
East Lulworth BH20 9 H6
East Lutton YO17 59 F3
East Lydford TA11 8 E1
East Lyn EX35 7 F1
East Lyng TA3 8 C2
East Mains AB31 90 E5
East Malling ME19 14 C2
East Malling Heath ME19 23 K6
East March DD4 83 F4
East Marden PO18 12 B5
East Markham NG22 51 K5
East Martin SP6 10 B3
East Marton BD23 56 E4
East Meon GU32 11 H2
East Mere EX16 7 H4
East Mersea CO5 34 E7
East Mey KW14 105 J1
East Molesey KT8 22 E5
East Moor WF1 57 J7
East Morden BH20 9 J5
East Morriston TD4 76 E6
East Morton BD20 57 G5
East Ness YO62 58 C2
East Newton HU11 59 J6
East Norton LE7 42 A5
East Oakley RG23 21 J6
East Ogwell TQ12 5 J3
East Orchard SP7 9 H3
East Ord TD15 77 H5
East Panson PL15 6 B6
East Parley BH23 10 C5
East Peckham TN12 23 K7
East Pennard BA4 8 E1
East Portlemouth TQ8 5 H7
East Prawle TQ7 5 H7
East Preston BN16 12 D6
East Pulham DT2 9 G4
East Putford EX22 6 B4
East Quantoxhead TA5 7 K1
East Rainton DH5 62 E2
East Ravendale DN37 53 F3
East Raynham NR21 44 C3
East Retford (Retford) DN22 51 K4
East Rigton LS17 57 J5
East Rolstone BS24 19 G5
East Rounton DL6 63 F6
East Row YO21 63 K5
East Rudham PE31 44 C3
East Runton NR27 45 G1
East Ruston NR12 45 H3
East Saltoun EH34 76 C4
East Shefford RG17 21 G4
East Sleekburn NE22 71 H5
East Somerton NR29 45 J4
East Stockwith DN21 51 K3
East Stoke *Dorset* BH20 9 H6
East Stoke *Notts.* NG23 42 A1
East Stour SP8 9 H2
East Stourmouth CT3 25 J5
East Stratton SO21 21 J7
East Street BA6 8 E1
East Studdal CT15 15 J3
East Suisnish IV40 86 B1
East Taphouse PL14 4 B4
East Thirston NE65 71 G4
East Tilbury RM18 24 C4
East Tisted GU34 11 J1
East Torrington LN8 52 E4
East Town BA4 19 K7
East Tuddenham NR20 44 E4
East Tytherley SP5 10 D2
East Tytherton SN15 20 C4
East Village EX17 7 G5
East Wall TF13 38 E6
East Walton PE32 44 B4
East Wellow SO51 10 E2
East Wemyss KY1 76 B1
East Whitburn EH47 75 H4
East Wickham DA16 23 H4
East Williamston SA70 16 D5
East Winch PE32 44 A4
East Winterslow SP5 10 D1
East Wittering PO20 11 J5
East Witton DL8 57 G1
East Woodburn NE48 70 E5
East Woodhay RG20 21 H5
East Woodlands BA11 20 A7
East Worldham GU34 11 J1
East Worlington EX17 7 F4
East Youlstone EX23 6 A4
Eastacott EX37 6 E3
Eastbourne BN21 13 K7
Eastbrook CF64 18 E4
Eastburn *E.Riding* YO25 59 F4
Eastburn *W.Yorks.* BD20 57 F5
Eastbury *Herts.* HA6 22 E2
Eastbury *W.Berks.* RG17 21 G4
Eastby BD23 57 F4
Eastchurch ME12 25 F4
Eastcombe *Glos.* GL6 20 B1
Eastcombe *Som.* TA4 7 K2
Eastcote *Gt.Lon.* HA5 22 E3
Eastcote *Northants.* NN12 31 H3
Eastcote *W.Mid.* B92 30 C1
Eastcott *Cornw.* EX23 6 A4
Eastcott *Wilts.* SN10 20 D6
Eastcourt SN16 20 C2
Eastdown TQ6 5 J6
Eastend OX7 30 E6
Easter Ardross IV17 96 D4
Easter Balgedie KY13 82 C7
Easter Balmoral AB35 89 K5
Easter Boleskine IV2 88 C2
Easter Borland FK8 81 H7
Easter Brae IV7 96 D5
Easter Buckieburn FK6 75 F2
Easter Compton BS35 19 J3
Easter Drummond IV2 88 B3
Easter Dullater FK17 81 H7
Easter Ellister PA48 72 A5
Easter Fearn IV24 96 D3
Easter Galcantray IV12 97 F7
Easter Howlaws TD10 77 F6
Easter Kinkell IV7 96 C6
Easter Knox DD11 83 G3
Easter Lednathie DD8 82 E1
Easter Moniack IV5 96 C7
Easter Ord AB32 91 G4
Easter Poldar FK8 74 E1
Easter Skeld (Skeld) ZE2 107 M8
Easter Suddie IV8 96 D6
Easter Tulloch AB30 91 F7
Easter Whyntie AB45 98 E4
Eastergate PO20 12 C6
Easterhouse G34 74 E4
Easterton SN10 20 D6
Easterton Sands SN10 20 D6
Eastertown BS24 19 G6
Eastfield *Bristol* BS9 19 J4
Eastfield *N.Lan.* ML7 75 G4
Eastfield *N.Yorks.* YO11 59 G1
Eastfield Hall NE65 71 H3
Eastgate *Dur.* DL13 62 A3
Eastgate *Lincs.* PE10 42 E4
Eastgate *Norf.* NR10 45 F3
Easthall SG4 33 F6
Eastham *Mersey.* CH62 48 C4
Eastham *Worcs.* WR15 29 F2
Easthampstead RG12 22 B5
Easthampton HR6 28 D2
Easthaugh NR9 44 E4
Eastheath RG41 22 B5
Easthope TF13 38 E6
Easthorpe *Essex* CO6 34 D6
Easthorpe *Leics.* NG13 42 B2
Easthorpe *Notts.* NG25 51 K7
Easthouses EH22 76 B4
Eastington *Devon* EX17 7 F5
Eastington *Glos.* GL54 30 C7
Eastington *Glos.* GL10 20 A1
Eastleach Martin GL7 21 F1
Eastleach Turville GL7 21 F1
Eastleigh *Devon* EX39 6 C3
Eastleigh *Hants.* SO50 11 F3
Eastling ME13 14 E2
Eastmoor *Derbys.* S42 51 F5
Eastmoor *Norf.* PE33 44 B5
Eastnor HR8 29 G5
Eastoft DN17 52 B1
Eastoke PO11 11 J5
Easton *Cambs.* PE28 32 E1
Easton *Cumb.* CA6 69 K6
Easton *Cumb.* CA7 60 D1
Easton *Devon* TQ13 7 F7
Easton *Dorset* DT5 9 F7
Easton *Hants.* SO21 11 G1
Easton *I.o.W.* PO40 10 E6
Easton *Lincs.* NG33 42 C3
Easton *Norf.* NR9 45 F4
Easton *Som.* BA5 19 J7
Easton *Suff.* IP13 35 G3
Easton *Wilts.* SN13 20 B4
Easton Grey SN16 20 B3
Easton Maudit NN29 32 B3
Easton on the Hill PE9 42 D5
Easton Royal SN9 21 F5
Easton-in-Gordano BS20 19 J4
Eastrea PE7 43 F6
Eastriggs DG12 69 H7
Eastrington DN14 58 D6
Eastry CT13 15 J2
Eastside KW17 106 D8
East-the-Water EX39 6 C3
Eastville BS16 19 K4
Eastwell LE14 42 A3
Eastwick CM20 33 H7
Eastwood *Notts.* NG16 41 G1
Eastwood *S'end* SS9 24 E3
Eastwood *S.Yorks.* S65 51 G3
Eastwood *W.Yorks.* OL14 56 E7
Eastwood End PE15 43 H6
Eathorpe CV33 30 E2
Eaton *Ches.E.* CW12 49 H6
Eaton *Ches.W. & C.* CW6 48 E6
Eaton *Leics.* NG32 42 A3
Eaton *Norf.* NR2 45 G5
Eaton *Norf.* PE36 44 A2
Eaton *Notts.* DN22 51 K5
Eaton *Oxon.* OX13 21 H1
Eaton *Shrop.* SY6 38 E6
Eaton *Shrop.* SY9 38 C7
Eaton Bishop HR2 28 D5
Eaton Bray LU6 32 C6
Eaton Constantine SY5 39 F5
Eaton Ford PE19 32 E3
Eaton Hall CH4 48 D6
Eaton Hastings SN7 21 F2
Eaton Socon PE19 32 E3
Eaton upon Tern TF9 39 F3
Eaves Green CV7 40 E7
Eavestone HG4 57 H3
Ebberston YO13 58 E1
Ebbesborne Wake SP5 9 J2
Ebbw Vale (Glynebwy) NP23 18 E1
Ebchester DH8 62 C1
Ebdon BS22 19 G5
Ebford EX3 7 H7
Ebley GL5 20 B1
Ebnal SY14 38 D1
Ebost IV56 85 J1
Ebrington GL55 30 C4
Ebsworthy Town EX20 6 D6
Ecchinswell RG20 21 H6
Ecclaw TD13 77 F4
Ecclefechan DG11 69 G6
Eccles *Gt.Man.* M30 49 G3
Eccles *Kent* ME20 24 D5
Eccles *Sc.Bord.* TD5 77 F6
Eccles Green HR4 28 C4
Eccles Road NR16 44 E6
Ecclesfield S35 51 F3
Ecclesgreig DD10 83 J1
Eccleshall ST21 40 A3
Eccleshill BD10 57 G6
Ecclesmachan EH52 75 J3
Eccles-on-Sea NR12 45 J3
Eccleston *Ches.W. & C.* CH4 48 D6
Eccleston *Lancs.* PR7 48 E1
Eccleston *Mersey.* WA10 48 D3
Eccup LS16 57 H5
Echt AB32 91 F4
Eckford TD5 70 C1
Eckington *Derbys.* S21 51 G5
Eckington *Worcs.* WR10 29 J4
Ecton *Northants.* NN6 32 B2
Ecton *Staffs.* SK17 50 C7
Edale S33 50 D4
Eday KW17 106 E4
Eday Airfield KW17 106 E4
Edburton BN5 13 F5
Edderside CA15 60 C2
Edderton IV19 96 E3
Eddington RG17 21 G5
Eddleston EH45 76 A6
Eden Park BR3 23 G5
Eden Vale TS27 63 F3
Edenbridge TN8 23 H7
Edendonich PA33 80 C5
Edenfield BL0 49 G1
Edenhall CA11 61 G3
Edenham PE10 42 D3
Edensor DE45 50 E5
Edentaggart G83 74 B1
Edenthorpe DN3 51 J2
Edern LL53 36 B2
Edgarley BA6 8 E1
Edgbaston B15 40 C7
Edgcote OX17 31 G4
Edgcott *Bucks.* HP18 31 H6
Edgcott *Som.* TA24 7 G2
Edgcumbe TR10 2 E5
Edge *Glos.* GL6 20 B1
Edge *Shrop.* SY5 38 C5
Edge End GL16 28 E7
Edge Green *Ches.W. & C.* SY14 48 D7
Edge Green *Gt.Man.* WA3 48 E3
Edge Green *Norf.* NR16 44 E7
Edgebolton SY4 38 E3
Edgefield NR24 44 E2
Edgehead EH37 76 B4
Edgeley SY13 38 E1
Edgerley SY10 38 C4
Edgerton HD2 50 D1
Edgeworth GL6 20 C1
Edginswell TQ2 5 J4
Edgmond TF10 39 G4
Edgmond Marsh TF10 39 G3
Edgton SY7 38 C7
Edgware HA8 23 F2
Edgworth BL7 49 G1
Edinample FK19 81 H5
Edinbanchory AB33 90 C3
Edinbane IV51 93 J6
Edinbarnet G81 74 D3
EDINBURGH EH 76 A3
Edinburgh Airport EH12 75 K3
Edinchip FK19 81 G5
Edingale B79 40 E4
Edingley NG22 51 J7
Edingthorpe NR28 45 H2
Edingthorpe Green NR28 45 H2
Edington *Som.* TA7 8 C1
Edington *Wilts.* BA13 20 C6
Edintore AB55 98 C6
Edinvale IV36 97 J6
Edistone EX39 6 A3
Edith Weston LE15 42 C5
Edithmead TA9 19 G7
Edlaston DE6 40 D1
Edlesborough LU6 32 C7
Edlingham NE66 71 G3
Edlington LN9 53 F5
Edmondsham BH21 10 B3
Edmondsley DH7 62 D2
Edmondstown CF40 18 D2
Edmondthorpe LE14 42 B4
Edmonstone KW17 106 E5
Edmonton *Cornw.* PL27 3 G1
Edmonton *Gt.Lon.* N18 23 G2
Edmundbyers DH8 62 B1
Ednam TD5 77 F7
Ednaston DE6 40 E1
Edney Common CM1 24 C1
Edra FK17 81 F6
Edradynate PH15 81 K2
Edrom TD11 77 G5
Edstaston SY4 38 E2
Edstone B95 30 C2
Edvin Loach HR7 29 F3
Edwalton NG12 41 H2
Edwardstone CO10 34 D4
Edwardsville CF46 18 D2
Edwinsford SA19 17 K2
Edwinstowe NG21 51 J6
Edworth SG18 33 F4
Edwyn Ralph HR7 29 F3
Edzell DD9 83 H1
Efail Isaf CF38 18 D3
Efail-fâch SA12 18 A2
Efailnewydd LL53 36 C2
Efailwen SA66 16 E3
Efenechtyd LL15 47 K7
Effingham KT24 22 E6
Effirth ZE2 107 M7
Efflinch DE13 40 D4
Efford EX17 7 G5
Egbury SP11 21 H6
Egdean RH20 12 C4
Egdon WR7 29 J3
Egerton *Gt.Man.* BL7 49 G1
Egerton *Kent* TN27 14 E3
Egerton Forstal TN27 14 D3
Egerton Green SY14 48 E7
Egg Buckland PL6 4 E5
Eggborough DN14 58 B7
Eggerness DG8 64 E6
Eggesford Barton EX18 6 E4
Eggington LU7 32 C6
Egginton DE65 40 E3
Egglescliffe TS16 63 F5
Eggleston DL12 62 A4
Egham TW20 22 D4
Egham Wick TW20 22 C4
Egilsay KW17 106 D5
Egleton LE15 42 B5
Eglingham NE66 71 G2
Eglinton KA12 74 B6
Egloshayle PL27 4 A3
Egloskerry PL15 4 C2
Eglwys Cross SY13 38 D1
Eglwys Fach SY20 37 F6
Eglwys Nunydd SA13 18 A3
Eglwysbach LL28 47 G5
Eglwys-Brewis CF62 18 D5
Eglwyswrw SA41 16 E2
Egmanton NG22 51 K6
Egmere NR22 44 D2
Egremont CA22 60 B5
Egton YO21 63 K6
Egton Bridge YO21 63 K6
Egypt SO21 21 H7
Eigg PH42 85 K6
Eight Ash Green CO6 34 D6
Eignaig PA34 79 J3
Eil PH22 89 F3
Eilanreach IV40 86 E3
Eildon TD6 76 D7
Eilean Bhearnaraigh (Berneray) HS6 92 E3
Eilean Darach IV23 95 H3
Eilean Iarmain (Isleornsay) IV43 86 C3
Eilean Leodhais (Isle of Lewis) HS 101 F3
Eilean Scalpaigh (Scalpay) HS4 93 H2
Eilean Shona PH36 86 C7
Einacleit HS2 100 D5
Eiriosgaigh (Eriskay) HS8 84 C3
Eisgean HS2 101 F6
Eisingrug LL47 37 F2
Eisteddfa Gurig SY23 37 G7
Elan Village LD6 27 J2
Elberton BS35 19 K3
Elborough BS24 19 G6
Elburton PL9 5 F5
Elcho PH2 82 C5
Elcombe SN4 20 E3
Elder Street CB10 33 J5
Eldernell PE7 43 G6
Eldersfield GL19 29 G5
Elderslie PA5 74 C4
Eldon DL14 62 D4
Eldrick KA26 67 G5
Eldroth LA2 56 C3
Eldwick BD16 57 G5
Elemore Vale DH5 62 E2
Elerch (Bont-goch) SY24 37 F7
Elford *Northumb.* NE68 77 K7
Elford *Staffs.* B79 40 D4
Elford Closes CB6 33 J1
Elgin IV30 97 K5
Elgol IV49 86 B3
Elham CT4 15 G3
Elie KY9 83 F7
Elilaw NE65 70 E3
Elim LL65 46 B4
Eling *Hants.* SO40 10 E3
Eling *W.Berks.* RG18 21 J4
Eliock DG4 68 D3
Elishader IV51 93 K5
Elishaw NE19 70 D4
Elkesley DN22 51 J5
Elkington NN6 31 H1
Elkstone GL53 29 J7
Elland HX5 57 G7
Elland Upper Edge HX5 57 G7
Ellary PA31 73 F3
Ellastone DE6 40 D1
Ellbridge PL12 4 E4
Ellel LA2 55 H4
Ellemford TD11 77 F4
Ellenborough CA15 60 B3
Ellenhall ST21 40 A3
Ellen's Green RH12 12 D3
Ellerbeck DL6 63 F7
Ellerby TS13 63 J5
Ellerdine TF6 39 F3
Ellerdine Heath TF6 39 F3
Elleric PA38 80 B3
Ellerker HU15 59 F7
Ellerton *E.Riding* YO42 58 D5
Ellerton *N.Yorks.* DL10 62 D7
Ellerton *Shrop.* TF9 39 G3
Ellerton Abbey DL11 62 B7
Ellesborough HP17 22 B1
Ellesmere SY12 38 C2
Ellesmere Park M30 49 G3
Ellesmere Port CH65 48 D5
Ellingham *Hants.* BH24 10 C4
Ellingham *Norf.* NR35 45 H6
Ellingham *Northumb.* NE67 71 G1
Ellingstring HG4 57 G1
Ellington *Cambs.* PE28 32 E1
Ellington *Northumb.* NE61 71 H4
Ellington Thorpe PE28 32 E1
Elliot's Green BA11 20 A7
Ellisfield RG25 21 K7
Ellistown LE67 41 G4
Ellon AB41 91 H1
Ellonby CA11 60 F3
Ellough NR34 45 J7
Ellough Moor NR34 45 J7
Elloughton HU15 59 F7
Ellwood GL16 19 J1
Elm PE14 43 H5
Elm Park RM12 23 J3
Elmbridge WR9 29 J2
Elmdon *Essex* CB11 33 H5
Elmdon *W.Mid.* B26 40 D7
Elmdon Heath B92 40 D7
Elmers End BR3 23 G5
Elmer's Green WN8 48 D2
Elmesthorpe LE9 41 G6
Elmhurst WS13 40 D4
Elmley Castle WR10 29 J4
Elmley Lovett WR9 29 H2
Elmore GL2 29 G7
Elmore Back GL2 29 G7
Elmscott EX39 6 A3
Elmsett IP7 34 E4
Elmstead *Essex* CO7 34 E6
Elmstead *Gt.Lon.* BR7 23 H4
Elmstead Market CO7 34 E6
Elmstone CT3 25 J5
Elmstone Hardwicke GL51 29 J6
Elmswell *E.Riding* YO25 59 F4
Elmswell *Suff.* IP30 34 D2
Elmton S80 51 H5
Elphin IV27 102 E7
Elphinstone EH33 76 B3
Elrick *Aber.* AB32 91 G4
Elrick *Moray* AB54 90 C2
Elrig DG8 64 D6
Elrigbeag PA32 80 C6
Elsdon NE19 70 E4
Elsecar S74 51 F2
Elsenham CM22 33 J6
Elsfield OX3 31 G7
Elsham DN20 52 D1
Elsing NR20 44 E4
Elslack BD23 56 E5
Elson *Hants.* PO12 11 H4
Elson *Shrop.* SY12 38 C2
Elsrickle ML12 75 J6
Elstead GU8 22 C7
Elsted GU29 12 B5
Elsthorpe PE10 42 D3
Elstob TS21 62 E4
Elston *Lancs.* PR2 55 J6
Elston *Notts.* NG23 42 A1
Elstone EX18 6 E4
Elstow MK42 32 D4
Elstree WD6 22 E2
Elstronwick HU12 59 J6
Elswick PR4 55 H6
Elsworth CB23 33 G2
Elterwater LA22 60 E6
Eltham SE9 23 H4
Eltisley PE19 33 F3
Elton *Cambs.* PE8 42 D6
Elton *Ches.W. & C.* CH2 48 D5
Elton *Derbys.* DE4 50 E6
Elton *Glos.* GL14 29 G7
Elton *Gt.Man.* BL8 49 G1
Elton *Here.* SY8 28 D1
Elton *Notts.* NG13 42 A2
Elton *Stock.* TS21 63 F5
Elton Green CH2 48 D5

Elv - Fid

Elvanfoot **ML12** 68 E2
Elvaston **DE72** 41 G2
Elveden **IP24** 34 C1
Elvingston **EH33** 76 C3
Elvington *Kent* **CT15** 15 H2
Elvington *York* **YO41** 58 D5
Elwick *Hart.* **TS27** 63 F3
Elwick *Northumb.* **NE70** 77 K7
Elworth **CW11** 49 G6
Elworthy **TA4** 7 J2
Ely *Cambs.* CB7 **33** **J1**
Ely *Cardiff* **CF5** 18 E4
Emberton **MK46** 32 B4
Embleton *Cumb.* **CA13** 60 C3
Embleton *Hart.* **TS22** 63 F4
Embleton *Northumb.* **NE66** 71 H1
Embo **IV25** 97 F2
Embo Street **IV25** 97 F2
Emborough **BA3** 19 K6
Embsay **BD23** 57 F4
Emerson Park **RM11** 23 J3
Emery Down **SO43** 10 D4
Emley **HD8** 50 E1
Emmington **OX39** 22 A1
Emneth **PE14** 43 H5
Emneth Hungate **PE14** 43 J5
Empingham **LE15** 42 C5
Empshott **GU33** 11 J1
Empshott Green **GU33** 11 J1
Emsworth PO10 **11** **J4**
Enborne **RG20** 21 H5
Enborne Row **RG20** 21 H5
Enchmarsh **SY6** 38 E6
Enderby **LE19** 41 H6
Endmoor **LA8** 55 J1
Endon **ST9** 49 J7
Endon Bank **ST9** 49 J7
ENFIELD EN **23** **G2**
Enfield Wash **EN3** 23 G2
Enford **SN9** 20 E6
Engine Common **BS37** 19 K3
Englefield **RG7** 21 K4
Englefield Green **TW20** 22 C4
Englesea-brook **CW2** 49 G7
English Bicknor **GL16** 28 E7
English Frankton **SY12** 38 D3
Englishcombe **BA2** 20 A5
Enham Alamein **SP11** 21 G7
Enmore **TA5** 8 B1
Ennerdale Bridge **CA23** 60 B5
Enniscaven **PL26** 3 G3
Ennochdhu **PH10** 82 B1
Ensay **PA75** 78 E3
Ensdon **SY4** 38 D4
Ensis **EX31** 6 D3
Enson **ST18** 40 B3
Enstone **OX7** 30 E6
Enterkinfoot **DG3** 68 D3
Enterpen **TS15** 63 F6
Enton Green **GU8** 22 C7
Enville **DY7** 40 A7
Eolaigearraidh **HS9** 84 C4
Eorabus **PA67** 78 E5
Eorodal **HS2** 101 H1
Eoropaidh **HS2** 101 H1
Epney **GL2** 29 G7
Epperstone **NG14** 41 J1
Epping CM16 **23** **H1**
Epping Green *Essex* **CM16** 23 H1
Epping Green *Herts.* **SG13** 23 F1
Epping Upland **CM16** 23 H1
Eppleby **DL11** 62 C5
Eppleworth **HU16** 59 G6
Epsom KT17 **23** **F5**
Epwell **OX15** 30 E4
Epworth **DN9** 51 K2
Epworth Turbary **DN9** 51 K2
Erbistock **LL13** 38 C1
Erbusaig **IV40** 86 D2
Erchless Castle **IV4** 96 B7
Erdington **B24** 40 D6
Eredine **PA33** 80 A7
Eriboll **IV27** 103 G3
Ericstane **DG10** 69 F2
Eridge Green **TN3** 13 J3
Eriff **DG7** 67 K3
Erines **PA29** 73 G3
Erisey Barton **TR12** 2 E7
Eriskay (Eiriosgaigh) **HS8** 84 C3
Eriswell **IP28** 34 B1
Erith DA8 **23** **J4**
Erlestoke **SN10** 20 C6
Ermington **PL21** 5 G5
Ernesettle **PL5** 4 E4
Erpingham **NR11** 45 F2
Erringden Grange **HX7** 56 E7
Errogie **IV2** 88 C2
Errol **PH2** 82 D5
Errollston **AB42** 91 J1
Erskine PA8 **74** **C3**
Ervie **DG9** 66 D7
Erwarton **IP9** 35 G5
Erwood **LD2** 27 K4
Eryholme **DL2** 62 E6
Eryrys **CH7** 48 B7
Escart **PA29** 73 G4
Escart Farm **PA29** 73 G5
Escomb **DL14** 62 C4
Escrick **YO19** 58 C5
Esgair **SA33** 17 G3
Esgairgeiliog **SY20** 37 G5
Esgyryn **LL31** 47 G5
Esh **DH7** 62 C2
Esh Winning **DH7** 62 C2
Esher KT10 **22** **E5**
Eshott **NE65** 71 H4
Eshton **BD23** 56 E4
Eskadale **IV4** 88 B1
Eskbank **EH22** 76 B4
Eskdale Green **CA19** 60 C6
Eskdalemuir **DG13** 69 H4
Eskham **DN36** 53 G3
Esknish **PA44** 72 B4
Esperley Lane Ends **DL13** 62 C4
Espley Hall **NE61** 71 G4
Esprick **PR4** 55 H6
Essendine **PE9** 42 D4
Essendon **AL9** 23 F1
Essich **IV2** 88 D1
Essington **WV11** 40 B5
Esslemont **AB41** 91 H2
Eston **TS6** 63 G5
Eswick **ZE2** 107 N7
Etal **TD12** 77 H7
Etchilhampton **SN10** 20 D5
Etchingham TN19 **14** **C5**
Etchinghill *Kent* **CT18** 15 G4
Etchinghill *Staffs.* **WS15** 40 C4
Etherdwick Grange **HU11** 59 J6
Etherley Dene **DL14** 62 C4
Ethie Mains **DD11** 83 H3
Eton **SL4** 22 C4
Eton Wick **SL4** 22 C4
Etteridge **PH20** 88 D5
Ettiley Heath **CW11** 49 G6
Ettington **CV37** 30 D4
Etton *E.Riding* **HU17** 59 F5
Etton *Peter.* **PE6** 42 E5
Ettrick **TD7** 69 H2
Ettrickbridge **TD7** 69 J1
Ettrickhill **TD7** 69 H2
Etwall **DE65** 40 E2
Eudon George **WV16** 39 F7
Eurach **PA31** 79 K7
Euston **IP24** 34 C1
Euxton **PR7** 48 E1
Evanstown **CF39** 18 C3
Evanton **IV16** 96 D5
Evedon **NG34** 42 D1
Evelix **IV25** 96 E2
Evenjobb **LD8** 28 B2
Evenley **NN13** 31 G5
Evenlode **GL56** 30 D6
Evenwood **DL14** 62 C4
Evenwood Gate **DL14** 62 C4
Everbay **KW17** 106 F5
Evercreech **BA4** 9 F1
Everdon **NN11** 31 G3
Everingham **YO42** 58 E5
Everleigh **SN8** 21 F6
Everley *High.* **KW1** 105 J2
Everley *N.Yorks.* **YO13** 59 F1
Evershot **MK17** 32 C5
Evershot **DT2** 8 E4
Eversley **RG27** 22 A5
Eversley Cross **RG27** 22 A5
Everthorpe **HU15** 59 F6
Everton *Cen.Beds.* **SG19** 33 F3
Everton *Hants.* **SO41** 10 D5
Everton *Mersey.* **L5** 48 C3
Everton *Notts.* **DN10** 51 J3
Evertown **DG14** 69 J6
Eves Corner **CM0** 25 F2
Evesbatch **WR6** 29 F4
Evesham WR11 **30** **B4**
Evie **KW17** 106 C5
Evington **LE5** 41 J5
Ewart Newtown **NE71** 77 H7
Ewden Village **S36** 50 E3
Ewell **KT17** 23 F5
Ewell Minnis **CT15** 15 H3
Ewelme **OX10** 21 K2
Ewen **GL7** 20 D2
Ewenny **CF35** 18 C4
Ewerby **NG34** 42 E1
Ewerby Thorpe **NG34** 42 E1
Ewhurst **GU6** 22 D7
Ewhurst Green *E.Suss.* **TN32** 14 C5
Ewhurst Green *Surr.* **GU6** 12 D3
Ewloe **CH5** 48 C6
Ewloe Green **CH5** 48 B6
Ewood **BB2** 56 B7
Ewood Bridge **BB4** 56 C7
Eworthy **EX21** 6 C6
Ewshot **GU10** 22 B7
Ewyas Harold **HR2** 28 C6
Exbourne **EX20** 6 E5
Exbury **SO45** 11 F4
Exceat **BN25** 13 J7
Exebridge **TA22** 7 H3
Exelby **DL8** 57 H1
EXETER EX **7** **H6**
Exeter International Airport **EX5** 7 H6
Exford **TA24** 7 G2
Exfords Green **SY5** 38 D5
Exhall *Warks.* **B49** 30 C3
Exhall *Warks.* **CV7** 41 F7
Exlade Street **RG8** 21 K3
Exminster **EX6** 7 H7
Exmouth EX8 **7** **J7**
Exnaboe **ZE3** 107 M11
Exning **CB8** 33 K2
Exton *Devon* **EX3** 7 H7
Exton *Hants.* **SO32** 11 H2
Exton *Rut.* **LE15** 42 C4
Exton *Som.* **TA22** 7 H2
Exwick **EX4** 7 H6
Eyam **S32** 50 E5
Eydon **NN11** 31 G4
Eye *Here.* **HR6** 28 D2
Eye *Peter.* **PE6** 43 F5
Eye *Suff.* IP23 **35** **F1**
Eye Green **PE6** 43 F5
Eyemouth TD14 **77** **H4**
Eyeworth **SG19** 33 F4
Eyhorne Street **ME17** 14 D2
Eyke **IP12** 35 H3
Eynesbury **PE19** 32 E3
Eynort **IV47** 85 J2
Eynsford **DA4** 23 J5
Eynsham **OX29** 21 H1
Eype **DT6** 8 D5
Eyre **IV51** 93 K6
Eythorne **CT15** 15 H3
Eyton *Here.* **HR6** 28 D2
Eyton *Shrop.* **SY7** 38 C7
Eyton on Severn **SY5** 38 E5
Eyton upon the Weald Moors **TF6** 39 F4
Eywood **HR5** 28 C3

F

Faccombe **SP11** 21 G6
Faceby **TS9** 63 F6
Fachwen **LL55** 46 D6
Facit **OL12** 49 H1
Faddiley **CW5** 48 E7
Fadmoor **YO62** 58 C1
Faebait **IV6** 96 B6
Faifley **G81** 74 D3
Fail **KA5** 67 J1
Failand **BS8** 19 J4
Failford **KA5** 67 J1
Failsworth **M35** 49 H2
Fain **IV23** 95 H4
Fair Isle **ZE2** 107 K2
Fair Isle Airstrip **ZE2** 107 K2
Fair Oak *Devon* **EX16** 7 J4
Fair Oak *Hants.* **SO50** 11 F3
Fair Oak *Hants.* **RG19** 21 J5
Fair Oak Green **RG7** 21 K5
Fairbourne LL38 **37** **F4**
Fairburn **WF11** 57 K7
Fairfield *Derbys.* **SK17** 50 C5
Fairfield *Gt.Man.* **M43** 49 J3
Fairfield *Kent* **TN29** 14 E5
Fairfield *Mersey.* **CH63** 48 B4
Fairfield *Stock.* **TS19** 63 F5
Fairfield *Worcs.* **B61** 29 J1
Fairford GL7 **20** **E1**
Fairgirth **DG5** 65 J5
Fairhaven **FY8** 55 G7
Fairhill **ML3** 75 F5
Fairholm **ML9** 75 F5
Fairley **AB15** 91 G4
Fairlie **KA29** 74 A5
Fairlight **TN35** 14 D6
Fairlight Cove **TN35** 14 D6
Fairmile *Devon* **EX11** 7 J6
Fairmile *Surr.* **KT11** 22 E5
Fairmilehead **EH10** 76 A4
Fairnington **TD5** 70 B1
Fairoak **ST21** 39 G2
Fairseat **TN15** 24 C5
Fairstead **CM3** 34 B7
Fairwarp **TN22** 13 H4
Fairwater **CF5** 18 E4
Fairy Cross **EX39** 6 C3
Fairyhill **SA3** 17 H6
Fakenham NR21 **44** **D3**
Fala **EH37** 76 C4
Fala Dam **EH37** 76 C4
Falahill **EH38** 76 B5
Faldingworth **LN8** 52 D4
Falfield *Fife* **KY15** 83 F7
Falfield *S.Glos.* **GL12** 19 K2
Falin-Wnda **SA44** 17 G1
Falkenham **IP10** 35 G5
FALKIRK FK **75** **G3**
Falkland **KY15** 82 D7
Falla **TD8** 70 C2
Fallgate **S45** 51 F6
Fallin **FK7** 75 G1
Falmer **BN1** 13 G6
Falmouth TR11 **3** **F5**
Falsgrave **YO12** 59 G1
Falstone **NE48** 70 C5
Fanagmore **IV27** 102 D4
Fanans **PA35** 80 B5
Fancott **LU5** 32 D6
Fangdale Beck **TS9** 63 G7
Fangfoss **YO41** 58 D4
Fankerton **FK6** 75 F2
Fanmore **PA73** 79 F3
Fanner's Green **CM3** 33 K7
Fans **TD4** 76 E6
Far Cotton **NN4** 31 J3
Far Forest **DY14** 29 G1
Far Gearstones **LA6** 56 C1
Far Green **GL11** 20 A1
Far Moor **WN5** 48 E2
Far Oakridge **GL6** 20 C1
Far Royds **LS12** 57 H6
Far Sawrey **LA22** 60 E7
Farcet **PE7** 43 F6
Farden **SY8** 28 E1
Fareham PO16 **11** **G4**
Farewell **WS13** 40 C4
Farforth **LN11** 53 G5
Faringdon SN7 **21** **F2**
Farington **PR25** 55 J7
Farlam **CA8** 61 G1
Farlary **IV28** 96 E1
Farleigh *N.Som.* **BS48** 19 J5
Farleigh *Surr.* **CR6** 23 G5
Farleigh Hungerford **BA2** 20 B6
Farleigh Wallop **RG25** 21 K7
Farlesthorpe **LN13** 53 H5
Farleton *Cumb.* **LA6** 55 J1
Farleton *Lancs.* **LA2** 55 J3
Farley *Derbys.* **DE4** 50 E6
Farley *Shrop.* **SY5** 38 C5
Farley *Staffs.* **ST10** 40 C1
Farley *Wilts.* **SP5** 10 D2
Farley Green *Suff.* **CB8** 34 B3
Farley Green *Surr.* **GU5** 22 D7
Farley Hill **RG7** 22 A5
Farleys End **GL2** 29 G7
Farlington **YO61** 58 C3
Farlow **DY14** 39 F7
Farm Town **LE67** 41 F4
Farmborough **BA2** 19 K5
Farmcote **GL54** 30 B6
Farmington **GL54** 30 C7
Farmoor **OX2** 21 H1
Farmtown **AB55** 98 D5
Farnborough *Gt.Lon.* **BR6** 23 H5
Farnborough *Hants.* GU14 **22** **B6**
Farnborough *W.Berks.* **OX12** 21 H3
Farnborough *Warks.* **OX17** 31 F4
Farnborough Street **GU14** 22 B6
Farncombe **GU7** 22 C7
Farndish **NN29** 32 C2
Farndon *Ches.W. & C.* **CH3** 48 D7
Farndon *Notts.* **NG24** 51 K7
Farnell **DD9** 83 H2
Farnham *Dorset* **DT11** 9 J3
Farnham *Essex* **CM23** 33 H6
Farnham *N.Yorks.* **HG5** 57 J3
Farnham *Suff.* **IP17** 35 H2
Farnham *Surr.* GU9 **22** **B7**
Farnham Common **SL2** 22 C3
Farnham Green **CM23** 33 H6
Farnham Royal **SL2** 22 C3
Farningham **DA4** 23 J5
Farnley *N.Yorks.* **LS21** 57 H5
Farnley *W.Yorks.* **LS12** 57 H6
Farnley Tyas **HD4** 50 D1
Farnsfield **NG22** 51 J7
Farnworth *Gt.Man.* **BL4** 49 G2
Farnworth *Halton* **WA8** 48 E4
Farr *High.* **KW14** 104 C2
Farr *High.* **IV2** 88 D1
Farr *High.* **PH21** 89 F4
Farr House **IV2** 88 D1
Farraline **IV2** 88 C2
Farringdon **EX5** 7 J6
Farrington Gurney **BS39** 19 K6
Farsley **LS28** 57 H6
Farthing Corner **ME8** 24 E5
Farthing Green **TN12** 14 D3
Farthinghoe **NN13** 31 G5
Farthingstone **NN12** 31 H3
Farthorpe **LN9** 53 F5
Fartown **HD2** 50 D1
Farway **EX24** 7 K6
Fasag **IV22** 94 E6
Fasagrianach **IV23** 95 H3
Fascadale **PH36** 86 B7
Faslane **G84** 74 A2
Fasnacloich **PA38** 80 B3
Fasnakyle **IV4** 87 K2
Fassfern **PH33** 87 G7
Fatfield **NE38** 62 E1
Fattahead **AB45** 98 E5
Faugh **CA8** 61 G1
Fauldhouse **EH47** 75 H4
Faulkbourne **CM8** 34 B7
Faulkland **BA3** 20 A6
Fauls **SY13** 38 E2
Faulston **SP5** 10 B2
Faversham ME13 **25** **G5**
Favillar **AB55** 89 K1
Fawdington **YO61** 57 K2
Fawdon **NE3** 71 H7
Fawfieldhead **SK17** 50 C6
Fawkham Green **DA3** 23 J5
Fawler **OX7** 30 E7
Fawley *Bucks.* **RG9** 22 A3
Fawley *Hants.* **SO45** 11 F4
Fawley *W.Berks.* **OX12** 21 G3
Fawley Chapel **HR1** 28 E6
Fawsyde **DD10** 91 G7
Faxfleet **DN14** 58 E7
Faxton **NN6** 31 J1
Faygate **RH12** 13 F3
Fazakerley **L9** 48 C3
Fazeley **B78** 40 E5
Fearby **HG4** 57 G1
Fearn **IV20** 97 F4
Fearnan **PH15** 81 J3
Fearnbeg **IV54** 94 D6
Fearnhead **WA2** 49 F3
Fearnmore **IV54** 94 D5
Fearnoch *Arg. & B.* **PA22** 73 J3
Fearnoch *Arg. & B.* **PA21** 73 H2
Featherstone *Staffs.* **WV10** 40 B5
Featherstone *W.Yorks.* **WF7** 57 K7
Featherstone Castle **NE49** 70 B7
Feckenham **B96** 30 B2
Feering **CO5** 34 C6
Feetham **DL11** 62 A7
Feith-hill **AB53** 98 E6
Feizor **LA2** 56 C3
Felbridge **RH19** 13 G3
Felbrigg **NR11** 45 G2
Felcourt **RH19** 23 G7
Felden **HP3** 22 D1
Felhampton **SY6** 38 D7
Felindre *Carmar.* **SA19** 27 G6
Felindre *Carmar.* **SA32** 17 J3
Felindre *Carmar.* **SA44** 17 G2
Felindre *Carmar.* **SA19** 17 K2
Felindre *Cere.* **SA48** 26 E3
Felindre *Powys* **LD7** 38 A7
Felindre *Powys* **NP8** 28 A6
Felindre *Swan.* **SA5** 17 K5
Felinfach *Cere.* **SA48** 26 E3
Felinfach *Powys* **LD3** 27 K5
Felinfoel **SA14** 17 J5
Felingwmisaf **SA32** 17 J3
Felingwmuchaf **SA32** 17 J3
Felixkirk **YO7** 57 K1
Felixstowe IP11 **35** **H5**
Felixstowe Ferry **IP11** 35 H5
Felkington **TD15** 77 H6
Felldownhead **PL19** 6 B7
Felling **NE10** 71 H7
Fellonmore **PA65** 79 H5
Felmersham **MK43** 32 C3
Felmingham **NR28** 45 G3
Felpham **PO22** 12 C7
Felsham **IP30** 34 D3
Felsted **CM6** 33 K6
Feltham TW13 **22** **E4**
Felthamhill **TW16** 22 E4
Felthorpe **NR10** 45 F4
Felton *Here.* **HR1** 28 E4
Felton *N.Som.* **BS40** 19 J5
Felton *Northumb.* **NE65** 71 G3
Felton Butler **SY4** 38 C4
Feltwell **IP26** 44 B6
Fen Ditton **CB5** 33 H2
Fen Drayton **CB24** 33 G2
Fen End **CV8** 30 D1
Fen Street *Norf.* **NR17** 44 D6
Fen Street *Norf.* **IP22** 34 E1
Fen Street *Suff.* **IP22** 34 D1
Fen Street *Suff.* **IP14** 35 F2
Fenay Bridge **HD8** 50 D1
Fence **BB12** 56 D6
Fence Houses **DH4** 62 E1
Fencott **OX5** 31 G7
Fendike Corner **PE24** 53 H6
Fenham **TD15** 77 J6
Fenhouses **PE20** 43 F1
Feniscowles **BB2** 56 B7
Feniton **EX14** 7 K6
Fenn Street **ME3** 24 D4
Fenni-fach **LD3** 27 K6
Fenny Bentley **DE6** 50 D7
Fenny Bridges **EX14** 7 K6
Fenny Compton **CV47** 31 F3
Fenny Drayton **CV13** 41 F6
Fenny Stratford **MK2** 32 B5
Fenrother **NE61** 71 G4
Fenstanton **PE28** 33 G2
Fenton *Cambs.* **PE28** 33 G1
Fenton *Lincs.* **LN1** 52 B5
Fenton *Lincs.* **NG23** 52 B7
Fenton *Northumb.* **NE71** 77 H7
Fenton *Notts.* **DN22** 51 K4
Fenton *Stoke* **ST4** 40 A1
Fenton Barns **EH39** 76 D2
Fenwick *E.Ayr.* **KA3** 74 C6
Fenwick *Northumb.* **NE18** 71 F6
Fenwick *Northumb.* **TD15** 77 J6
Fenwick *S.Yorks.* **DN6** 51 H1
Feochaig **PA28** 66 B2
Feock **TR3** 3 F5
Feolin **PA60** 72 D4
Feolin Ferry **PA60** 72 C4
Feorlan **PA28** 66 A3
Feorlin **PA32** 73 H1
Ferguslie Park **PA3** 74 C4
Feriniquarrie **IV55** 93 G6
Fern **DD8** 83 F1
Ferndale CF43 **18** **C2**
Ferndown BH22 **10** **B4**
Ferness **IV12** 97 G7
Fernham **SN7** 21 F2
Fernhill Heath **WR3** 29 H3
Fernhurst **GU27** 12 B4
Fernie **KY15** 82 E6
Ferniela **IV47** 85 J1
Fernilee **SK23** 50 C5
Fernybank **DD9** 90 D7
Ferrensby **HG5** 57 J3
Ferrindonald **IV44** 86 C4
Ferring **BN12** 12 D6
Ferry Hill **PE16** 43 G7
Ferrybridge **WF11** 57 K7
Ferryden **DD10** 83 J2
Ferryhill DL17 **62** **D3**
Ferryside (Glanyferi) SA17 **17** **G4**
Fersfield **IP22** 44 E7
Fersit **PH31** 87 K7
Ferwig **SA43** 16 E1
Feshiebridge **PH21** 89 F4
Fetcham **KT22** 22 E6
Fetlar **ZE2** 107 Q3
Fetterangus **AB42** 99 H5
Fettercairn **AB30** 90 E7
Fetternear House **AB51** 91 F3
Feus of Caldhame **AB30** 83 H1
Fewcott **OX27** 31 G6
Fewston **HG3** 57 G4
Ffairfach **SA19** 17 K3
Ffair-Rhos **SY25** 27 G2
Ffaldybrenin **SA19** 17 K1
Ffarmers **SA19** 17 K1
Ffawyddog **NP8** 28 B7
Ffestiniog (Llan Ffestiniog) **LL41** 37 G1
Ffordd-las *Denb.* **LL16** 47 K6
Fforddlas *Powys* **HR3** 28 B5
Fforest **SA4** 17 J5
Fforest-fach **SA5** 17 K6
Ffostrasol **SA44** 17 G1
Ffos-y-ffin **SA46** 26 D2
Ffridd Uchaf **LL54** 46 D7
Ffrith *Denb.* **LL19** 47 J4
Ffrith *Flints.* **LL11** 48 B7
Ffrwdgrech **LD3** 27 K6
Ffynnon **SA33** 17 G4
Ffynnon Taf (Taff's Well) **CF15** 18 E3
Ffynnongroyw **CH8** 47 K4
Fibhig **HS2** 100 E3
Fichlie **AB33** 90 C3
Fidden **PA66** 78 E5

Fid - Fro

Fiddington *Glos.* **GL20** 29 J5
Fiddington *Som.* **TA5** 19 F7
Fiddleford **DT10** 9 H3
Fiddler's Green *Glos.* **GL51** 29 J6
Fiddler's Green *Here.* **HR1** 28 E5
Fiddler's Green *Norf.* **PE32** 44 C4
Fiddler's Green *Norf.* **NR17** 44 E6
Fiddlers Hamlet **CM16** 23 H1
Field **ST14** 40 C2
Field Broughton **LA11** 55 G1
Field Dalling **NR25** 44 E2
Field Head **LE67** 41 G5
Fife Keith **AB55** 98 C5
Fifehead Magdalen **SP8** 9 G2
Fifehead Neville **DT10** 9 G3
Fifehead St. Quintin **DT10** 9 G3
Fifield *Oxon.* **OX7** 30 D7
Fifield *W. & M.* **SL6** 22 C4
Fifield Bavant **SP5** 10 B2
Figheldean **SP4** 20 E7
Filby **NR29** 45 J4
Filey YO14 59 **H1**
Filgrave **MK16** 32 B4
Filham **PL21** 5 G5
Filkins **GL7** 21 F1
Filleigh *Devon* **EX32** 6 E3
Filleigh *Devon* **EX17** 7 F4
Fillingham **DN21** 52 C4
Fillongley **CV7** 40 E7
Filmore Hill **GU34** 11 H2
Filton **BS34** 19 K4
Fimber **YO25** 58 E3
Finavon **DD8** 83 F2
Fincham **PE33** 44 A5
Finchampstead **RG40** 22 A5
Finchdean **PO8** 11 J3
Finchingfield **CM7** 33 K5
Finchley **N3** 23 F2
Findern **DE65** 41 F2
Findhorn **IV36** 97 H5
Findhorn Bridge **IV13** 89 F2
Findhuglen **PH6** 81 J6
Findo Gask **PH1** 82 B5
Findochty **AB56** 98 C4
Findon *Aber.* **AB12** 91 H5
Findon *W.Suss.* **BN14** 12 E6
Findon Mains **IV7** 96 D5
Findon Valley **BN14** 12 E6
Findrassie **IV30** 97 J5
Findron **AB37** 89 J3
Finedon **NN9** 32 C1
Finegand **PH10** 82 C1
Fingal Street **IP13** 35 G1
Fingask **AB51** 91 F2
Fingerpost **DY14** 29 G1
Fingest **RG9** 22 A2
Finghall **DL8** 57 G1
Fingland *Cumb.* **CA7** 60 D1
Fingland *D. & G.* **DG13** 69 H3
Fingland *D. & G.* **DG4** 68 C2
Finglesham **CT14** 15 J2
Fingringhoe **CO5** 34 E6
Finkle Street **S35** 51 F3
Finlarig **FK21** 81 G4
Finmere **MK18** 31 H5
Finnart *Arg. & B.* **G84** 74 A1
Finnart *P. & K.* **PH17** 81 G2
Finney Hill **LE12** 41 G4
Finningham **IP14** 34 E2
Finningley **DN9** 51 J3
Finnygaud **AB54** 98 E5
Finsbay (Fionnsbhagh) **HS3** 93 F3
Finsbury **EC1R** 23 G3
Finstall **B60** 29 J1
Finsthwaite **LA12** 55 G1
Finstock **OX7** 30 E7
Finstown **KW17** 106 C6
Fintry *Aber.* **AB53** 99 F5
Fintry *Stir.* **G63** 74 E2
Finwood **CV35** 30 C2
Finzean **AB31** 90 D5
Fionnphort **PA66** 78 D5
Fionnsbhagh (Finsbay) **HS3** 93 F3
Fir Tree **DL15** 62 C3
Firbank **LA10** 61 H7
Firbeck **S81** 51 H4
Firby *N.Yorks.* **DL8** 57 H1
Firby *N.Yorks.* **YO60** 58 D3
Firgrove **OL16** 49 J1
Firs Lane **WN7** 49 F2
Firsby **PE23** 53 H6
Firsdown **SP5** 10 D1
Firth **ZE2** 107 N5
Fishbourne *I.o.W.* **PO33** 11 G5
Fishbourne *W.Suss.* **PO18** 12 B6
Fishburn **TS21** 62 E3
Fishcross **FK10** 75 H1
Fisherford **AB51** 90 E1
Fisher's Pond **SO50** 11 F2
Fisher's Row **PR3** 55 H5
Fishersgate **BN41** 13 F6
Fisherstreet **GU8** 12 C3
Fisherton *High.* **IV2** 96 E6
Fisherton *S.Ayr.* **KA7** 67 G2
Fisherton de la Mere **BA12** 9 J1
Fishguard (Abergwaun) SA65 16 **C2**
Fishlake **DN7** 51 J1
Fishleigh Barton **EX37** 6 D3
Fishley **NR13** 45 J4
Fishnish **PA65** 79 H3
Fishpond Bottom **DT6** 8 C5
Fishponds **BS16** 19 K4
Fishpool **BL9** 49 H2
Fishtoft **PE21** 43 G1
Fishtoft Drove **PE22** 43 G1
Fishtown of Usan **DD10** 83 J2
Fishwick **TD15** 77 H5
Fiskerton *Lincs.* **LN3** 52 D5
Fiskerton *Notts.* **NG25** 51 K7
Fitling **HU11** 59 J6
Fittleton **SP4** 20 E7
Fittleworth **RH20** 12 D5
Fitton End **PE13** 43 H4
Fitz **SY4** 38 D4
Fitzhead **TA4** 7 K3
Fitzroy **TA2** 7 K3
Fitzwilliam **WF9** 51 G1
Fiunary **PA34** 79 H3
Five Acres **GL16** 28 E7
Five Ash Down **TN22** 13 H4
Five Ashes **TN20** 13 J4
Five Bridges **WR6** 29 F4
Five Houses **PO30** 11 F6
Five Lanes **NP26** 19 H2
Five Oak Green **TN12** 23 K7
Five Oaks *Chan.I.* **JE2** 3 K7
Five Oaks *W.Suss.* **RH14** 12 D4
Five Roads **SA15** 17 H5
Five Turnings **LD7** 28 B1
Five Wents **ME17** 14 D2
Fivehead **TA3** 8 C2
Fivelanes **PL15** 4 C2
Flack's Green **CM3** 34 B7
Flackwell Heath **HP10** 22 B3
Fladbury **WR10** 29 J4
Fladdabister **ZE2** 107 N9
Flagg **SK17** 50 D6
Flamborough **YO15** 59 J2
Flamstead **AL3** 32 D7
Flamstead End **EN7** 23 G1
Flansham **PO22** 12 C6
Flanshaw **WF2** 57 J7
Flasby **BD23** 56 E4
Flash **SK17** 50 C6
Flashader **IV51** 93 J6
Flask Inn **YO22** 63 J2
Flaunden **HP3** 22 D1
Flawborough **NG13** 42 A1
Flawith **YO61** 57 K3
Flax Bourton **BS48** 19 J5
Flax Moss **BB4** 56 C7
Flaxby **HG5** 57 J4
Flaxholme **DE56** 41 F1
Flaxlands **NR16** 45 F6
Flaxley **GL14** 29 F7
Flaxpool **TA4** 7 K2
Flaxton **YO60** 58 C3
Fleckney **LE8** 41 J6
Flecknoe **CV23** 31 G2
Fledborough **NG22** 52 B5
Fleet *Hants.* GU51 22 **B6**
Fleet *Hants.* **PO11** 11 J4
Fleet *Lincs.* **PE12** 43 G3
Fleet Hargate **PE12** 43 G3
Fleetville **AL1** 22 E1
Fleetwood FY7 55 **G5**
Fleggburgh (Burgh St. Margaret) **NR29** 45 J4
Flemingston **CF62** 18 D4
Flemington **G72** 74 E5
Flempton **IP28** 34 C2
Fleoideabhagh **HS3** 93 F3
Flesherin **HS2** 101 H4
Fletchersbridge **PL30** 4 B4
Fletchertown **CA7** 60 D2
Fletching **TN22** 13 H4
Fleuchats **AB36** 90 B4
Fleur-de-lis **NP12** 18 E2
Flexbury **EX23** 6 A5
Flexford **GU3** 22 C6
Flimby **CA15** 60 B3
Flimwell **TN5** 14 C4
Flint (Y Fflint) CH6 48 **B5**
Flint Cross **SG8** 33 H4
Flint Mountain **CH6** 48 B5
Flintham **NG23** 42 A1
Flinton **HU11** 59 J6
Flint's Green **CV7** 30 D1
Flishinghurst **TN17** 14 C4
Flitcham **PE31** 44 B3
Flitholme **CA16** 61 J5
Flitton **MK45** 32 D5
Flitwick **MK45** 32 D5
Flixborough **DN15** 52 B1
Flixton *Gt.Man.* **M41** 49 G3
Flixton *N.Yorks.* **YO11** 59 G2
Flixton *Suff.* **NR35** 45 H7
Flockton **WF4** 50 E1
Flockton Green **WF4** 50 E1
Flodden **NE71** 77 H7
Flodigarry **IV51** 93 K4
Flood's Ferry **PE15** 43 G6
Flookburgh **LA11** 55 G2
Floors **AB55** 98 C5
Flordon **NR15** 45 F6
Flore **NN7** 31 H2
Flotta **KW16** 106 C8
Flotterton **NE65** 71 F3
Flowton **IP8** 34 E4
Flushdyke **WF5** 57 H7
Flushing *Aber.* **AB42** 99 J6
Flushing *Cornw.* **TR11** 3 F5
Flushing *Cornw.* **TR12** 2 E6
Fluxton **EX11** 7 J6
Flyford Flavell **WR7** 29 J3
Foals Green **IP21** 35 G1
Fobbing **SS17** 24 D3
Fochabers IV32 98 **B5**
Fochriw **CF81** 18 E1
Fockerby **DN17** 52 B1
Fodderletter **AB37** 89 J2
Fodderty **IV15** 96 C6
Foddington **TA11** 8 E2
Foel **SY21** 37 J4
Foelgastell **SA14** 17 J4
Foggathorpe **YO8** 58 D6
Fogo **TD11** 77 F6
Fogorig **TD11** 77 F6
Fogwatt **IV30** 97 K6
Foindle **IV27** 102 D4
Folda **PH11** 82 C1
Fole **ST14** 40 C2
Foleshill **CV6** 41 F7
Folke **DT9** 9 F3
Folkestone CT20 15 **H4**
Folkingham **NG34** 42 D2
Folkington **BN26** 13 J6
Folksworth **PE7** 42 E7
Folkton **YO11** 59 G2
Folla Rule **AB51** 91 F1
Follifoot **HG3** 57 J4
Folly *Dorset* **DT2** 9 G4
Folly *Pembs.* **SA62** 16 C3
Folly Gate **EX20** 6 D6
Fonmon **CF62** 18 D5
Fonthill Bishop **SP3** 9 J1
Fonthill Gifford **SP3** 9 J1
Fontmell Magna **SP7** 9 H3
Fontmell Parva **DT11** 9 H3
Fontwell **BN18** 12 C6
Font-y-gary **CF62** 18 D5
Foolow **S32** 50 D5
Footherley **WS14** 40 D5
Foots Cray **DA14** 23 H4
Forbestown **AB36** 90 B3
Force Forge **LA12** 60 E7
Force Green **TN16** 23 H6
Forcett **DL11** 62 C5
Forches Cross **EX17** 7 F5
Ford *Arg. & B.* **PA31** 79 K7
Ford *Bucks.* **HP17** 22 A1
Ford *Devon* **PL8** 5 G5
Ford *Devon* **EX39** 6 C3
Ford *Devon* **TQ7** 5 H6
Ford *Glos.* **GL54** 30 B6
Ford *Mersey.* **L30** 48 C3
Ford *Midloth.* **EH37** 76 B4
Ford *Northumb.* **TD15** 77 H7
Ford *Pembs.* **SA62** 16 C3
Ford *Plym.* **PL2** 4 E5
Ford *Shrop.* **SY5** 38 D4
Ford *Som.* **TA4** 7 J3
Ford *Staffs.* **ST13** 50 C7
Ford *W.Suss.* **BN18** 12 C6
Ford *Wilts.* **SN14** 20 B4
Ford End **CM3** 33 K7
Ford Green **PR3** 55 H5
Ford Heath **SY5** 38 D4
Ford Street **TA21** 7 K4
Forda **EX20** 6 D6
Fordbridge **B37** 40 D7
Fordcombe **TN3** 23 J7
Forddun (Forden) **SY21** 38 B5
Fordell **KY4** 75 K2
Forden (Forddun) **SY21** 38 B5
Forder Green **TQ13** 5 H4
Fordgate **TA7** 8 C1
Fordham *Cambs.* **CB7** 33 K1
Fordham *Essex* **CO6** 34 D6
Fordham *Norf.* **PE38** 44 A6
Fordham Abbey **CB7** 33 K2
Fordham Heath **CO3** 34 D6
Fordhouses **WV10** 40 B5
Fordingbridge SP6 10 **C3**
Fordon **YO25** 59 G2
Fordoun **AB30** 91 F7
Ford's Green **IP14** 34 E2
Fordstreet **CO6** 34 D6
Fordwells **OX29** 30 E7
Fordwich **CT2** 15 G2
Fordyce **AB45** 98 D4
Forebrae **PH1** 82 A5
Forebridge **ST17** 40 B3
Foredale **BD24** 56 D3
Foreland **PA49** 72 A4
Foremark **DE65** 41 F3
Forest **DL10** 62 D6
Forest Coal Pit **NP7** 28 B6
Forest Gate **E7** 23 H3
Forest Green **RH5** 22 E7
Forest Hall *Cumb.* **LA8** 61 G6
Forest Hall *T. & W.* **NE12** 71 H7
Forest Head **CA8** 61 G1
Forest Hill *Gt.Lon.* **SE23** 23 G4
Forest Hill *Oxon.* **OX33** 21 J1
Forest Lane Head **HG2** 57 J4
Forest Lodge *Arg. & B.* **PA36** 80 D3
Forest Lodge *P. & K.* **PH18** 89 G7
Forest Row RH18 13 **H3**
Forest Side **PO30** 11 F6
Forest Town **NG19** 51 H6
Forestburn Gate **NE61** 71 F4
Forest-in-Teesdale **DL12** 61 K4
Forestmill **FK10** 75 H1
Forestside **PO9** 11 J3
Forfar DD8 83 **F2**
Forgandenny **PH2** 82 B6
Forge **SY20** 37 G6
Forgie **AB55** 98 B5
Forhill **B38** 30 B1
Formby **L37** 48 B2
Forncett End **NR16** 45 F6
Forncett St. Mary **NR16** 45 F6
Forncett St. Peter **NR16** 45 F6
Forneth **PH10** 82 B3
Fornham All Saints **IP28** 34 C2
Fornham St. Martin **IP31** 34 C2
Fornighty **IV12** 97 G6
Forres IV36 97 **H6**
Forrest **ML6** 75 G4
Forrest Lodge **DG7** 67 K5
Forsbrook **ST11** 40 B1
Forse **KW5** 105 H5
Forsie **KW14** 105 F2
Forsinain **KW13** 104 E4
Forsinard KW13 104 **D4**
Forston **DT2** 9 F5
Fort Augustus PH32 87 **K4**
Fort George **IV2** 96 E6
Fort William PH33 87 **H7**
Forter **PH11** 82 C1
Forteviot **PH2** 82 B6
Forth **ML11** 75 H5
Forthampton **GL19** 29 H5
Fortingall **PH15** 81 J3
Fortis Green **N2** 23 F3
Forton *Hants.* **SP11** 21 H7
Forton *Lancs.* **PR3** 55 H4
Forton *Shrop.* **SY4** 38 D4
Forton *Som.* **TA20** 8 C4
Forton *Staffs.* **TF10** 39 G3
Fortrie **AB53** 98 E6
Fortrose IV10 96 **E6**
Fortuneswell **DT5** 9 F7
Forty Green **HP9** 22 C2
Forty Hill **EN2** 23 G2
Forward Green **IP14** 34 E3
Fosbury **SN8** 21 G6
Foscot **OX7** 30 D6
Fosdyke **PE20** 43 G2
Foss **PH16** 81 J2
Foss Cross **GL54** 20 D1
Fossdale **DL8** 61 K7
Fossebridge **GL54** 30 B7
Foster Street **CM17** 23 H1
Fosterhouses **DN14** 51 J1
Foster's Booth **NN12** 31 H3
Foston *Derbys.* **DE65** 40 D2
Foston *Leics.* **LE8** 41 J6
Foston *Lincs.* **NG32** 42 B1
Foston *N.Yorks.* **YO60** 58 C3
Foston on the Wolds **YO25** 59 H4
Fotherby **LN11** 53 G3
Fotheringhay **PE8** 42 D6
Foubister **KW17** 106 E7
Foul Mile **BN27** 13 K5
Foula **ZE2** 107 L2
Foula Airstrip **ZE2** 107 L2
Foulbog **DG13** 69 H3
Foulden *Norf.* **IP26** 44 B6
Foulden *Sc.Bord.* **TD15** 77 H5
Foulness Island **SS3** 25 G2
Foulridge **BB8** 56 D5
Foulsham **NR20** 44 E3
Foulstone **LA6** 55 J1
Foulzie **AB45** 99 F4
Fountainhall **TD1** 76 C6
Four Ashes *Staffs.* **WV10** 40 B5
Four Ashes *Staffs.* **DY7** 40 A7
Four Ashes *Suff.* **IP31** 34 D1
Four Crosses *Denb.* **LL21** 37 K1
Four Crosses *Powys* **SY22** 38 B4
Four Crosses *Powys* **SY21** 37 K5
Four Crosses *Staffs.* **WS11** 40 B5
Four Elms **TN8** 23 H7
Four Forks **TA5** 8 B1
Four Gotes **PE13** 43 H4
Four Lane Ends *B'burn.* **BB1** 56 B7
Four Lane Ends *Ches.W. & C.* **CW6** 48 E6
Four Lane Ends *York* **YO19** 58 C4
Four Lanes **TR16** 2 D5
Four Marks **GU34** 11 H1
Four Mile Bridge **LL65** 46 A5
Four Oaks *E.Suss.* **TN31** 14 D5
Four Oaks *Glos.* **GL18** 29 F6
Four Oaks *W.Mid.* **CV7** 40 E7
Four Oaks *W.Mid.* **B74** 40 D6
Four Oaks Park **B74** 40 D6
Four Roads **SA17** 17 H5
Four Throws **TN18** 14 C5
Fourlane Ends **DE55** 51 F7
Fourlanes End **CW11** 49 H7
Fourpenny **IV25** 97 F2
Fourstones **NE47** 70 D7
Fovant **SP3** 10 B2
Foveran House **AB41** 91 H2
Fowey PL23 4 **B5**
Fowlis **DD2** 82 E4
Fowlis Wester **PH7** 82 A5
Fowlmere **SG8** 33 H4
Fownhope **HR1** 28 E5
Fox Hatch **CM15** 23 J2
Fox Lane **GU14** 22 B6
Fox Street **CO7** 34 E6
Foxbar **PA2** 74 C4
Foxcombe Hill **OX1** 21 H1
Foxcote *Glos.* **GL54** 30 B7
Foxcote *Som.* **BA3** 20 A6
Foxdale **IM4** 54 B6
Foxearth **CO10** 34 C4
Foxfield **LA20** 55 F1
Foxham **SN15** 20 C4
Foxhole *Cornw.* **PL26** 3 G3
Foxhole *High.* **IV4** 88 C1
Foxholes **YO25** 59 G2
Foxhunt Green **TN21** 13 J5
Foxley *Here.* **HR4** 28 D4
Foxley *Norf.* **NR20** 44 E3
Foxley *Northants.* **NN12** 31 H3
Foxley *Wilts.* **SN16** 20 B3
Foxt **ST10** 40 C1
Foxton *Cambs.* **CB22** 33 H4
Foxton *Dur.* **TS21** 62 E4
Foxton *Leics.* **LE16** 41 J6
Foxup **BD23** 56 D2
Foxwist Green **CW8** 49 F6
Foy **HR9** 28 E6
Foyers **IV2** 88 B2
Frachadil **PA75** 78 E2
Fraddam **TR27** 2 C5
Fraddon **TR9** 3 G3
Fradley **WS13** 40 D4
Fradswell **ST18** 40 B2
Fraisthorpe **YO15** 59 H3
Framfield **TN22** 13 H4
Framingham Earl **NR14** 45 G5
Framingham Pigot **NR14** 45 G5
Framlingham **IP13** 35 G2
Frampton *Dorset* **DT2** 9 F5
Frampton *Lincs.* **PE20** 43 G2
Frampton Cotterell **BS36** 19 K3
Frampton Mansell **GL6** 20 C1
Frampton on Severn **GL2** 20 A1
Frampton West End **PE20** 43 F1
Framsden **IP14** 35 G3
Framwellgate Moor **DH1** 62 D2
France Lynch **GL6** 20 C1
Frances Green **PR3** 56 B6
Franche **DY11** 29 H1
Frandley **CW9** 49 F5
Frankby **CH48** 48 B4
Frankfort **NR12** 45 H3
Frankley **B32** 40 B7
Franksbridge **LD1** 28 A3
Frankton **CV23** 31 F1
Frant **TN3** 13 J3
Fraserburgh AB43 99 **H4**
Frating **CO7** 34 E6
Fratton **PO1** 11 H4
Freasley **B78** 40 E6
Freathy **PL10** 4 D5
Freckenham **IP28** 33 K1
Freckleton **PR4** 55 H7
Freeby **LE14** 42 B3
Freefolk **RG28** 21 H7
Freehay **ST10** 40 C1
Freeland **OX29** 31 F7
Freester **ZE2** 107 N7
Freethorpe **NR13** 45 J5
Freethorpe Common **NR13** 45 J5
Freiston **PE22** 43 G1
Freiston Shore **PE22** 43 G1
Fremington *Devon* **EX31** 6 D2
Fremington *N.Yorks.* **DL11** 62 B7
Frenchay **BS16** 19 K4
Frenchbeer **TQ13** 6 E7
Frendraught **AB54** 98 E6
Frenich **FK8** 81 F7
Frensham **GU10** 22 B7
Fresgoe **KW14** 104 E2
Freshbrook **SN5** 20 E3
Freshfield **L37** 48 B2
Freshford **BA2** 20 A6
Freshwater PO40 10 **E6**
Freshwater Bay **PO40** 10 E6
Freshwater East **SA71** 16 D6
Fressingfield **IP21** 35 G1
Freston **IP9** 35 F5
Freswick **KW1** 105 J2
Fretherne **GL2** 20 A1
Frettenham **NR12** 45 G4
Freuchie **KY15** 82 D7
Freystrop Cross **SA62** 16 C4
Friars Carse **DG2** 68 E5
Friar's Gate **TN6** 13 H3
Friarton **PH2** 82 C5
Friday Bridge **PE14** 43 H5
Friday Street *E.Suss.* **BN23** 13 K6
Friday Street *Suff.* **IP13** 35 G3
Friday Street *Suff.* **IP12** 35 H3
Friday Street *Surr.* **RH5** 22 E7
Fridaythorpe **YO25** 58 E4
Friern Barnet **N11** 23 F2
Friesthorpe **LN3** 52 D4
Frieston **NG32** 42 C1
Frieth **RG9** 22 A2
Frilford **OX13** 21 H2
Frilsham **RG18** 21 J4
Frimley **GU16** 22 B6
Frimley Green **GU16** 22 B6
Frindsbury **ME2** 24 D5
Fring **PE31** 44 B2
Fringford **OX27** 31 H6
Friningham **ME14** 14 D2
Frinsted **ME9** 14 D2
Frinton-on-Sea CO13 35 **G7**
Friockheim **DD11** 83 G3
Friog **LL38** 37 F4
Frisby on the Wreake **LE14** 41 J4
Friskney **PE22** 53 H7
Friskney Eaudyke **PE22** 53 H7
Friston *E.Suss.* **BN20** 13 J7
Friston *Suff.* **IP17** 35 J2
Fritchley **DE56** 51 F7
Frith **ME13** 14 E2
Frith Bank **PE22** 43 G1
Frith Common **WR15** 29 F2
Fritham **SO43** 10 D3
Frithelstock **EX38** 6 C4
Frithelstock Stone **EX38** 6 C4
Frithville **PE22** 53 G7
Frittenden **TN17** 14 D3
Frittiscombe **TQ7** 5 J6
Fritton *Norf.* **NR15** 45 G6
Fritton *Norf.* **NR31** 45 J5
Fritwell **OX27** 31 G6
Frizinghall **BD9** 57 G6
Frizington CA26 60 **B5**
Frocester **GL10** 20 A1
Frochas **SY21** 38 B5
Frodesley **SY5** 38 E5
Frodesley Lane **SY5** 38 E5
Frodingham **DN15** 52 B1
Frodsham WA6 48 **E5**
Frog End **CB23** 33 H3
Frog Pool **WR6** 29 G2
Frogden **TD5** 70 C1
Froggatt **S32** 50 E5
Froghall **ST10** 40 C1
Frogham **SP6** 10 C3
Frogland Cross **BS36** 19 K3
Frogmore *Devon* **TQ7** 5 H6

Fro - Gle

Frogmore *Hants.* GU17 22 B6
Frogmore *Herts.* AL2 22 E1
Frogwell PL17 4 D4
Frolesworth LE17 41 H6
Frome BA11 20 A7
Frome Market BA11 20 B6
Frome St. Quintin DT2 8 E4
Frome Whitfield DT2 9 F5
Fromes Hill HR8 29 F4
Fron *Gwyn.* LL53 36 C2
Fron *Powys* SY21 38 B5
Fron *Powys* LD1 27 K2
Fron *Powys* SY15 38 A6
Fron Isaf LL14 38 B2
Froncysyllte LL20 38 B1
Fron-goch LL23 37 J2
Frostenden NR34 45 J7
Frosterley DL13 62 B3
Froxfield SN8 21 G5
Froxfield Green GU32 11 J2
Fryerning CM4 24 C1
Fugglestone St. Peter SP2 10 C1
Fulbeck NG32 52 C7
Fulbourn CB21 33 J3
Fulbrook OX18 30 D7
Fulflood SO22 11 F2
Fulford *Som.* TA2 8 B2
Fulford *Staffs.* ST11 40 B2
Fulford *York* YO10 58 C5
Fulham SW6 23 F4
Fulking BN5 13 F5
Full Sutton YO41 58 D4
Fullaford EX31 6 E2
Fuller Street CM3 34 B7
Fuller's Moor CH3 48 D7
Fullerton SP11 10 E1
Fulletby LN9 53 F5
Fullwood KA3 74 C5
Fulmer SL3 22 D3
Fulmodeston NR21 44 D2
Fulnetby LN8 52 E5
Fulready CV37 30 D4
Fulstone HD9 50 D2
Fulstow LN11 53 G3
Fulwell *Oxon.* OX7 30 E6
Fulwell *T. & W.* SR5 62 E1
Fulwood *Lancs.* PR2 55 J6
Fulwood *S.Yorks.* S10 51 F4
Fundenhall NR16 45 F6
Fundenhall Street NR16 45 F6
Funtington PO18 12 B6
Funtley PO16 11 G4
Funzie ZE2 107 Q3
Furley EX13 8 B4
Furnace *Arg. & B.* PA32 80 B7
Furnace *Carmar.* SA15 17 J5
Furnace *Cere.* SY20 37 F6
Furnace *High.* IV22 95 F4
Furnace End B46 40 E6
Furner's Green TN22 13 H4
Furness Vale SK23 50 C4
Furneux Pelham SG9 33 H6
Furnham TA20 8 C4
Further Quarter TN26 14 D4
Furtho MK19 31 J4
Furze Green IP21 45 G7
Furze Platt SL6 22 B3
Furzehill *Devon* EX35 7 F1
Furzehill *Dorset* BH21 10 B4
Furzeley Corner PO7 11 H3
Furzey Lodge SO42 10 E4
Furzley SO43 10 D3
Fyfett TA20 8 B3
Fyfield *Essex* CM5 23 J1
Fyfield *Glos.* GL7 21 F1
Fyfield *Hants.* SP11 21 F7
Fyfield *Oxon.* OX13 21 H2
Fyfield *Wilts.* SN8 20 E5
Fyfield *Wilts.* SN9 20 E5
Fylingthorpe YO22 63 J2
Fyning GU31 12 B4
Fyvie AB53 91 F1

G

Gabalfa CF14 18 E4
Gabhsunn Bho Dheas HS2 101 G2
Gabhsunn Bho Thuath HS2 101 G2
Gablon IV25 96 E2
Gabroc Hill KA3 74 C5
Gaddesby LE7 41 J4
Gaddesden Row HP2 32 D7
Gadebridge HP1 22 D1
Gadshill ME3 24 D4
Gaer *Newport* NP20 19 F3
Gaer *Powys* NP8 28 A6
Gaer-fawr NP15 19 H2
Gaerllwyd NP16 19 H2
Gaerwen LL60 46 C5
Gagingwell OX7 31 F6
Gaich *High.* PH26 89 H2
Gaich *High.* IV2 88 D1
Gaick Lodge PH21 88 E6
Gailes KA11 74 B7
Gailey ST19 40 B4
Gainford DL2 62 C5
Gainsborough DN21 52 B3
Gainsford End CO9 34 B5
Gairloch IV21 94 D4
Gairlochy PH34 87 H6
Gairney Bank KY13 75 K1
Gairnshiel Lodge AB35 89 K4
Gaitsgill CA5 60 E2
Galabank TD1 76 C6
GALASHIELS TD 76 C7
Galdenoch DG8 64 B4
Gale OL15 49 J1
Galgate LA2 55 H4
Galhampton BA22 9 F2
Gallanach PA34 79 K5
Gallantry Bank SY14 48 E7
Gallatown KY1 76 A1
Gallchoille PA31 73 F1
Gallery AB30 83 H1
Galley Common CV10 41 F6
Galleyend CM2 24 D1
Galleywood CM2 24 D1
Gallowfauld DD8 83 F3
Gallowhill PA3 74 C4
Gallows Green ST10 40 C1
Gallowstree Common RG4 21 K3
Gallowstree Elm DY7 40 A7
Gallt Melyd (Meliden) LL19 47 J4
Galltair IV40 86 E2
Gallt-y-foel LL55 46 D6
Gallypot Street TN7 13 H3
Galmington TA1 8 B2
Galmisdale PH42 85 K6
Galmpton *Devon* TQ7 5 G6
Galmpton *Torbay* TQ5 5 J5
Galmpton Warborough TQ4 5 J5
Galphay HG4 57 H2
Galston KA4 74 C7
Galtrigill IV55 93 G6
Gamble's Green CM3 34 B7
Gamblesby CA10 61 H3
Gamelsby CA7 60 D1
Gamesley SK13 50 C3
Gamlingay SG19 33 F3
Gamlingay Cinques SG19 33 F3
Gamlingay Great Heath SG19 33 F3
Gammaton EX39 6 C3
Gammaton Moor EX39 6 C3
Gammersgill DL8 57 F1
Gamrie AB45 99 F4
Gamston *Notts.* DN22 51 K5
Gamston *Notts.* NG2 41 J2
Ganarew NP25 28 E7
Gang PL14 4 D4
Ganllwyd LL40 37 G3
Gannochy DD9 90 E7
Ganstead HU11 59 H6
Ganthorpe YO60 58 C2
Ganton YO12 59 F2
Ganwick Corner EN5 23 F2
Gaodhail PA72 79 H4
Gappah TQ13 5 J3
Gara Bridge TQ9 5 H5
Garabal G83 80 E6
Garadheancal IV26 95 F1
Garbat IV23 96 B5
Garbhallt PA27 73 J1
Garboldisham IP22 44 E7
Garden FK8 74 D1
Garden City CH5 48 C6
Garden Village S36 50 E3
Gardeners Green RG40 22 B5
Gardenstown AB45 99 F4
Garderhouse ZE2 107 M8
Gardham HU17 59 F5
Gare Hill BA11 20 A7
Garelochhead G84 74 A1
Garford OX13 21 H2
Garforth LS25 57 K6
Gargrave BD23 56 E4
Gargunnock FK8 75 F1
Gariob PA31 73 F2
Garlic Street IP20 45 G7
Garlies Castle DG8 64 E4
Garlieston DG8 64 E6
Garlinge Green CT4 15 G2
Garlogie AB32 91 F4
Garmelow ST21 40 A3
Garmond AB53 99 G5
Garmony PA65 79 H3
Garmouth IV32 98 B4
Garmston SY5 39 F5
Garnant SA18 17 K4
Garndolbenmaen LL51 36 D1
Garneddwen SY20 37 G5
Garnett Bridge LA8 61 G7
Garnfadryn LL53 36 B2
Garnswllt SA18 17 K5
Garrabost HS2 101 H4
Garrachra PA23 73 J2
Garralburn AB55 98 C5
Garras TR12 2 E6
Garreg LL48 37 F1
Garreg Bank SY21 38 B4
Garrett's Green B33 40 D7
Garrick FK15 81 K6
Garrigill CA9 61 J2
Garriston DL8 62 C7
Garroch DG7 67 K5
Garrochty PA20 73 J5
Garros IV51 93 K5
Garrow PH8 81 K4
Garryhorn DG7 67 K4
Garrynahine (Gearraidh na h-Aibhne) HS2 100 E4
Garsdale LA10 56 C1
Garsdale Head LA10 61 J7
Garsdon SN16 20 C3
Garshall Green ST18 40 B2
Garsington OX44 21 J1
Garstang PR3 55 H5
Garston L19 48 D4
Garswood WN4 48 E3
Gartachoil G63 74 D1
Gartally IV63 88 B1
Gartavaich PA29 73 G5
Gartbreck PA43 72 A5
Gartcosh G69 74 E4
Garth *Bridgend* CF34 18 B2
Garth *Cere.* SY23 37 F7
Garth *Gwyn.* LL57 46 D5
Garth *I.o.M.* IM4 54 C6
Garth *Powys* LD4 27 J4
Garth *Shet.* ZE2 107 L7
Garth *Wrex.* LL20 38 B1
Garth Row LA8 61 G7
Garthbrengy LD3 27 K5
Garthdee AB15 91 H4
Gartheli SA48 26 E3
Garthmyl SY21 38 A6
Garthorpe *Leics.* LE14 42 B3
Garthorpe *N.Lincs.* DN17 52 B1
Garths LA8 61 G7
Garthynty SA20 27 G4
Gartincaber FK16 81 H7
Gartly AB54 90 D1
Gartmore FK8 74 D1
Gartnagrenach PA29 73 F5
Gartnatra PA43 72 B4
Gartness G63 74 D2
Gartocharn G83 74 C2
Garton HU11 59 J6
Garton-on-the-Wolds YO25 59 F3
Gartymore KW8 105 F7
Garvald EH41 76 D3
Garvamore PH20 88 C5
Garvan PH33 87 F7
Garvard PA61 72 B1
Garve IV23 95 K5
Garveld PA28 66 A3
Garvestone NR9 44 E5
Garvie PA22 73 J2
Garvock *Aber.* AB30 91 F7
Garvock *Inclyde* PA16 74 A3
Garvock *P. & K.* PH2 82 B6
Garwald DG13 69 H3
Garwaldwaterfoot DG13 69 H3
Garway HR2 28 D6
Garway Hill HR2 28 D6
Gask *Aber.* AB42 99 J6
Gask *Aber.* AB53 99 F6
Gask *P. & K.* PH3 82 A6
Gaskan PH37 86 E7
Gass KA19 67 J3
Gastard SN13 20 B5
Gasthorpe IP22 44 D7
Gaston Green CM22 33 J7
Gatcombe PO30 11 F6
Gate Burton DN21 52 B4
Gate Helmsley YO41 58 C4
Gate House PA60 72 D3
Gateacre L25 48 D4
Gateford S81 51 H4
Gateforth YO8 58 B7
Gatehead KA2 74 B7
Gatehouse PH15 81 K3
Gatehouse of Fleet DG7 65 G5
Gatelawbridge DG3 68 E4
Gateley NR20 44 D3
Gatenby DL7 57 J1
Gatesgarth CA13 60 C5
Gateshaw TD5 70 C1
Gateshead NE8 71 H7
Gatesheath CH3 48 D6
Gateside *Aber.* AB33 90 E3
Gateside *Angus* DD8 83 F3
Gateside *Fife* KY14 82 C7
Gateside *N.Ayr.* KA15 74 B5
Gateslack DG3 68 D3
Gathurst WN5 48 E2
Gatley SK8 49 H4
Gattonside TD6 76 D7
Gatwick Airport (London Gatwick Airport) RH6 23 F7
Gaufron LD6 27 J2
Gaulby LE7 41 J5
Gauldry DD6 82 E5
Gauntons Bank SY13 38 E1
Gaunt's Common BH21 10 B4
Gaunt's Earthcott BS32 19 K3
Gautby LN8 52 E5
Gavinton TD11 77 F5
Gawber S75 51 F2
Gawcott MK18 31 H5
Gawsworth SK11 49 H5
Gawthrop LA10 56 B1
Gawthwaite LA12 55 F1
Gay Bowers CM3 24 D1
Gay Street RH20 12 D4
Gaydon CV35 30 E3
Gayhurst MK16 32 B4
Gayle DL8 56 D1
Gayles DL11 62 C6
Gayton *Mersey.* CH60 48 B4
Gayton *Norf.* PE32 44 B4
Gayton *Northants.* NN7 31 J3
Gayton *Staffs.* ST18 40 B3
Gayton le Marsh LN13 53 H4
Gayton le Wold LN11 53 F4
Gayton Thorpe PE32 44 B4
Gaywood PE30 44 A3
Gazeley CB8 34 B2
Geanies House IV20 97 F4
Gearach PA48 72 A5
Gearnsary KW11 104 C5
Gearradh PH33 80 A1
Gearraidh Bhailteas HS8 84 C2
Gearraidh Bhaird HS2 101 F5
Gearraidh na h-Aibhne (Garrynahine) HS2 100 E4
Gearraidh na Monadh HS8 84 C3
Gearrannan HS2 100 D3
Geary IV55 93 H5
Gedding IP30 34 D3
Geddington NN14 42 B7
Gedgrave Hall IP12 35 J4
Gedintailor IV51 86 B1
Gedling NG4 41 J1
Gedney PE12 43 H3
Gedney Broadgate PE12 43 H3
Gedney Drove End PE12 43 H3
Gedney Dyke PE12 43 H3
Gedney Hill PE12 43 G4
Gee Cross SK14 49 J3
Geilston G82 74 B3
Geirinis HS8 92 C7
Geisiadar HS2 100 D4
Geldeston NR34 45 H6
Gell *Conwy* LL22 47 G6
Gell *Gwyn.* LL52 36 D2
Gelli CF41 18 C2
Gelli Gynan CH7 47 K7
Gellideg CF48 18 D1
Gellifor LL15 47 K6
Gelligaer CF82 18 E2
Gellilydan LL41 37 F2
Gellioedd LL21 37 J1
Gelly SA66 16 D4
Gellyburn PH1 82 B4
Gellywen SA33 17 F3
Gelston *D. & G.* DG7 65 H5
Gelston *Lincs.* NG32 42 C1
Gembling YO25 59 H4
Gemmil PA31 79 J7
Genoch DG9 64 B5
Genoch Square DG9 64 B5
Gentleshaw WS15 40 C4
Geocrab HS3 93 G2
George Green SL3 22 D3
George Nympton EX36 7 F3
Georgeham EX33 6 C2
Georgetown PA6 74 C4
Gerlan LL57 46 E6
Germansweek EX21 6 C6
Germoe TR20 2 C6
Gerrans TR2 3 F5
Gerrards Cross SL9 22 D3
Gerston KW12 105 G3
Gestingthorpe CO9 34 C5
Geuffordd SY22 38 B4
Geufron SY18 37 H7
Gibbet Hill BA11 20 A7
Gibbshill DG7 65 H3
Gibraltar *Lincs.* PE24 53 J7
Gibraltar *Suff.* IP6 35 F3
Giddeahall SN14 20 B4
Giddy Green BH20 9 H6
Gidea Park RM2 23 J3
Gidleigh TQ13 6 E7
Giffnock G46 74 D5
Gifford EH41 76 D4
Giffordland KA24 74 A6
Giffordtown KY15 82 D6
Giggleswick BD24 56 D3
Gigha PA41 72 E6
Gilberdyke HU15 58 E7
Gilbert's End WR8 29 H4
Gilchriston EH36 76 C4
Gilcrux CA7 60 C3
Gildersome LS27 57 H7
Gildingwells S81 51 H4
Gilesgate Moor DH1 62 D2
Gileston CF62 18 D5
Gilfach CF81 18 E2
Gilfach Goch CF39 18 C3
Gilfachrheda SA45 26 D3
Gilgarran CA14 60 B4
Gill CA11 60 F4
Gillamoor YO62 58 C1
Gillen IV55 93 H5
Gillenbie DG11 69 G5
Gillfoot DG2 65 K3
Gilling East YO62 58 C2
Gilling West DL10 62 C6
Gillingham *Dorset* SP8 9 H2
Gillingham *Med.* ME7 24 D5
Gillingham *Norf.* NR34 45 J6
Gillivoan KW5 105 G5
Gillock KW1 105 H3
Gillow Heath ST8 49 H7
Gills KW1 105 J1
Gill's Green TN18 14 C4
Gilmanscleuch TD7 69 J1
Gilmerton *Edin.* EH17 76 A4
Gilmerton *P. & K.* PH7 81 K5
Gilmilnscroft KA5 67 K1
Gilmonby DL12 62 A5
Gilmorton LE17 41 H7
Gilsland CA8 70 B7
Gilsland Spa CA8 70 B7
Gilson B46 40 D7
Gilstead BD16 57 G6
Gilston EH38 76 C5
Gilston Park CM20 33 H7
Giltbrook NG16 41 G1
Gilwern NP7 28 B7
Gimingham NR11 45 G2
Gin Pit M29 49 F2
Ginclough SK10 49 J5
Ginger's Green BN27 13 K5
Giosla HS2 100 D5
Gipping IP14 34 E2
Gipsey Bridge PE22 43 F1
Girlsta ZE2 107 N7
Girsby DL2 62 E6
Girtford SG19 32 E4
Girthon DG7 65 G5
Girton *Cambs.* CB3 33 H2
Girton *Notts.* NG23 52 B6
Girvan KA26 67 F4
Gisburn BB7 56 D5
Gisburn Cotes BB7 56 D5
Gisleham NR33 45 K7
Gislingham IP23 34 E1
Gissing IP22 45 F7
Gittisham EX14 7 K6
Givons Grove KT22 22 E6
Glackour IV23 95 H3
Gladestry HR5 28 B3
Gladsmuir EH33 76 C3
Glaic PA22 73 J3
Glais SA7 18 A1
Glaisdale YO21 63 J6
Glaister KA27 73 H7
Glame IV40 94 B7
Glamis DD8 82 E3
Glan Conwy LL24 47 G7
Glanaber Terrace LL24 37 G1
Glanaman SA18 17 K4
Glanbran SA20 27 H5
Glan-Denys SA48 26 E3
Glanderston AB52 90 D2
Glandford NR25 44 E1
Glan-Duar SA40 17 J1
Glandwr SA34 16 E3
Glan-Dwyfach LL51 36 D1
Glangrwyney NP8 28 B7
Glanllynfi CF34 18 B2
Glanmule SY16 38 A6
Glan-rhyd *N.P.T.* SA9 18 A1
Glanrhyd *Pembs.* SA43 16 E1
Glanton NE66 71 F2
Glanton Pyke NE66 71 F2
Glantwymyn (Cemmaes Road) SY20 37 H5
Glanvilles Wootton DT9 9 F4
Glanwern SY24 37 F7
Glanwydden LL31 47 G4
Glan-y-don CH8 47 K5
Glanyferi (Ferryside) SA17 17 G4
Glan-y-llyn CF15 18 E3
Glan-y-nant SY18 37 J7
Glan-yr-afon *Gwyn.* LL23 37 J1
Glan-yr-afon *Gwyn.* LL21 37 K1
Glan-yr-afon *I.o.A.* LL58 46 E4
Glan-y-Wern LL47 37 F2
Glapthorn PE8 42 D6
Glapwell S44 51 G6
Glasahoile FK8 81 F7
Glasbury HR3 28 A5
Glaschoil PH26 89 H1
Glascoed *Mon.* NP4 19 G1
Glascoed *Wrex.* LL11 48 B7
Glascorrie AB35 90 C5
Glascote B77 40 E5
Glascwm LD1 28 A3
Glasdrum PA38 80 B3
Glasfryn LL21 47 H7
GLASGOW G 74 D4
Glasgow Airport PA3 74 C4
Glasgow Prestwick Airport KA9 67 H1
Glashmore AB31 91 F4
Glasinfryn LL57 46 D6
Glasnacardoch PH41 86 C5
Glasnakille IV49 86 B3
Glaspant SA38 17 F2
Glaspwll SY20 37 G6
Glassburn IV4 87 K1
Glassel AB31 90 E5
Glassenbury TN17 14 C4
Glasserton DG8 64 E7
Glassford ML10 75 F6
Glasshouse GL17 29 G6
Glasshouses HG3 57 G3
Glassingall FK15 81 J7
Glasslie KY6 82 D7
Glasson *Cumb.* CA7 69 H7
Glasson *Lancs.* LA2 55 H4
Glassonby CA10 61 G3
Glasterlaw DD11 83 G2
Glaston LE15 42 B5
Glastonbury BA6 8 D1
Glatton PE28 42 E7
Glazebrook WA3 49 F3
Glazebury WA3 49 F3
Glazeley WV16 39 G7
Gleadless S12 51 F4
Gleadsmoss SK11 49 H6
Gleann Tholastaidh HS2 101 H3
Gleaston LA12 55 F2
Glecknabae PA20 73 J4
Gledhow LS8 57 J6
Gledrid LL14 38 B2
Glemsford CO10 34 C4
Glen *D. & G.* DG2 65 J3
Glen *D. & G.* DG7 65 F5
Glen Auldyn IM7 54 D4
Glen Mona IM7 54 D5
Glen Parva LE2 41 H6
Glen Trool Lodge DG8 67 J5
Glen Village FK1 75 G3
Glen Vine IM4 54 C6
Glenae DG1 68 E5
Glenaladale PH37 86 E7
Glenald G84 74 A1
Glenamachrie PA34 80 A5
Glenapp Castle KA26 66 E5
Glenarm DD8 82 E1
Glenbarr PA29 72 E7
Glenbatrick PA60 72 D3
Glenbeg *High.* IV24 95 K3
Glenbeg *High.* PH26 89 H2
Glenbeg *High.* PH36 79 G1
Glenbeich FK19 81 H5
Glenbervie *Aber.* AB39 91 F6
Glenbervie *Falk.* FK5 75 G2
Glenboig ML5 75 F4
Glenborrodale PH36 79 H1
Glenbranter PA27 73 K1
Glenbreck ML12 69 F1
Glenbrittle IV47 85 K2
Glenburn PA2 74 C4
Glenbyre PA62 79 G5
Glencaple DG1 65 K4
Glencarse PH2 82 C5
Glencat AB34 90 D5
Glenceitlein PH49 80 C3
Glencloy KA27 73 J7
Glencoe PH49 80 C2
Glenconglass AB37 89 J2
Glencraig KY5 75 K1

Gle - Gre

Glencripesdale PH36 79 H2
Glencrosh DG3 68 C5
Glencruitten PA34 79 K5
Glencuie AB33 90 C3
Glendearg *D. & G.* DG13 69 H3
Glendearg *Sc.Bord.* TD1 76 D7
Glendessary PH34 87 F5
Glendevon FK14 82 A7
Glendoebeg PH32 88 B4
Glendoick PH2 82 D5
Glendoll Lodge DD8 89 K7
Glendoune KA26 67 F4
Glendrissaig KA26 67 F4
Glenduckie KY14 82 D6
Glenduisk KA26 67 G5
Glendye Lodge AB31 90 E6
Gleneagles Hotel PH3 82 A6
Gleneagles House PH3 82 A7
Glenearn PH2 82 C6
Glenegedale PA42 72 B5
Glenelg IV40 86 E3
Glenfarg PH2 82 C6
Glenfeochan PA34 79 K5
Glenfield LE3 41 H5
Glenfinnan PH37 87 F6
Glenfoot PH2 82 C6
Glengalmadale PH33 79 K2
Glengap DG6 65 G5
Glengarnock KA14 74 B5
Glengarrisdale PA60 72 E1
Glengennet KA26 67 G4
Glengolly KW14 105 G2
Glengrasco IV51 93 K7
Glengyle FK8 80 E6
Glenhead DG2 68 D5
Glenhead Farm PH11 82 D1
Glenhurich PH37 79 K1
Glenkerry TD7 69 H2
Glenkiln KA27 73 J7
Glenkin PA23 73 K2
Glenkindie AB33 90 C3
Glenlair DG7 65 H3
Glenlatterach IV30 97 K6
Glenlean PA23 73 J2
Glenlee *Angus* DD9 90 C7
Glenlee *D. & G.* DG7 68 B5
Glenlichorn FK15 81 J6
Glenlivet AB37 89 J2
Glenlochar DG7 65 H4
Glenluce DG8 64 B5
Glenmallan G84 74 A1
Glenmanna DG3 68 C3
Glenmavis ML6 75 F4
Glenmaye IM5 54 B6
Glenmeanie IV6 95 J6
Glenmore *Arg. & B.* PA20 73 J4
Glenmore *High.* IV51 93 K7
Glenmore *High.* PH22 89 G4
Glenmore Lodge PH22 89 G4
Glenmoy DD8 83 F1
Glenmuick IV27 103 F7
Glennoe PA35 80 B4
Glenochar ML12 68 E2
Glenogil DD8 83 F1
Glenprosen Village DD8 82 E1
Glenquiech DD8 83 F1
Glenramskill PA28 66 B2
Glenrazie DG8 64 D4
Glenridding CA11 60 E5
Glenrisdell PA29 73 G5
Glenrossal IV27 96 B1
Glenrothes KY7 82 D7
Glensanda PA34 79 K3
Glensaugh AB30 90 E7
Glensgaich IV14 96 B5
Glenshalg AB31 90 D4
Glenshellish PA27 73 K1
Glensluain PA27 73 J1
Glentaggart ML11 68 D1
Glentham LN8 52 D3
Glenton AB51 90 E2
Glentress EH45 76 A7
Glentrool DG8 64 D3
Glentruan IM7 54 D3
Glentworth DN21 52 C4
Glenuachdarach IV51 93 K6
Glenuig PH38 86 C7
Glenure PA38 80 B3
Glenurquhart IV11 96 E5
Glenwhilly DG8 64 B3
Glespin ML11 68 D1
Gletness ZE2 107 N7
Glewstone HR9 28 E6
Glinton PE6 42 E5
Glooston LE16 42 A6
Glororum NE69 77 K7
Glossop SK13 50 C3
Gloster Hill NE65 71 H3
GLOUCESTER GL 29 H7
Gloup ZE2 107 P2
Gloweth TR1 2 E4
Glusburn BD20 57 F5
Gluss ZE2 107 M5
Glympton OX20 31 F6
Glyn LL24 47 F7
Glyn Ceiriog LL20 38 B2
Glynarthen SA44 17 G1
Glyncoch CF37 18 D2
Glyncorrwg SA13 18 B2
Glyn-Cywarch LL47 37 F2
Glynde BN8 13 H6
Glyndebourne BN8 13 H5
Glyndyfrdwy LL21 38 A1
Glynebwy (Ebbw Vale) NP23 18 E1
Glynneath (Glyn-Nedd) SA11 18 B1
Glyn-Nedd (Glynneath) SA11 18 B1
Glynogwr CF35 18 C3
Glyntaff CF37 18 D3
Gnosall ST20 40 A3
Gnosall Heath ST20 40 A3
Goadby LE7 42 A6
Goadby Marwood LE14 42 A3
Goatacre SN11 20 D4
Goatfield PA32 80 B7
Goathill DT9 9 F3
Goathland YO22 63 K6
Goathurst TA5 8 B1
Gobernuisgeach KW12 104 E5
Gobhaig HS3 100 C7
Gobowen SY11 38 C2
Godalming GU7 22 C7
Goddard's Corner IP13 35 G2
Goddards Green BN6 13 F4
Godden Green TN15 23 J6
Goddington BR6 23 H5
Godford Cross EX14 7 K5
Godington OX27 31 H6
Godleybrook ST10 40 B1
Godmanchester PE29 33 F1
Godmanstone DT2 9 F5
Godmersham CT4 15 F2
Godney BA5 19 H7
Godolphin Cross TR13 2 D5
Godor SY22 38 B4
Godre'r-graig SA9 18 A1
Godshill *Hants.* SP6 10 C3
Godshill *I.o.W.* PO38 11 G6
Godstone RH9 23 G6
Godwick PE32 44 D3
Goetre NP4 19 G1
Goff's Oak EN7 23 G1
Gogar EH12 75 K3
Gogarth LL30 47 F4
Goginan SY23 37 F7
Goirtean a' Chladaich PH33 87 G7
Goirtein PA27 73 H2
Golan LL51 36 E1
Golant PL23 4 B5
Golberdon PL17 4 D3
Golborne WA3 49 F3
Golcar HD7 50 D1
Gold Hill *Cambs.* PE14 43 J6
Gold Hill *Dorset* DT11 9 H3
Goldcliff NP18 19 G3
Golden Cross BN27 13 J5
Golden Green TN11 23 K7
Golden Grove SA32 17 J4
Golden Pot GU34 22 A7
Golden Valley *Derbys.* DE55 51 G7
Golden Valley *Glos.* GL51 29 J6
Goldenhill ST6 49 H7
Golders Green NW11 23 F3
Goldhanger CM9 25 F1
Goldielea DG2 65 K3
Golding SY5 38 E5
Goldington MK41 32 D3
Goldsborough *N.Yorks.* YO21 63 K5
Goldsborough *N.Yorks.* HG5 57 J4
Goldsithney TR20 2 C5
Goldstone TF9 39 G3
Goldthorn Park WV2 40 B6
Goldthorpe S63 51 G2
Goldworthy EX39 6 B3
Golford TN17 14 C4
Gollanfield IV2 97 F6
Gollinglith Foot HG4 57 G1
Golspie KW10 97 F2
Golval KW13 104 D2
Gomeldon SP4 10 C1
Gomersal BD19 57 H7
Gometra PA73 78 E3
Gometra House PA73 78 E3
Gomshall GU5 22 D7
Gonachan Cottage G63 74 E2
Gonalston NG14 41 J1
Gonerby Hill Foot NG31 42 C2
Gonfirth ZE2 107 M6
Good Easter CM1 33 K7
Gooderstone PE33 44 B5
Goodleigh EX32 6 E2
Goodmanham YO43 58 E5
Goodmayes IG3 23 H3
Goodnestone *Kent* CT3 15 H2
Goodnestone *Kent* ME13 25 G5
Goodrich HR9 28 E7
Goodrington TQ4 5 J5
Goodshaw BB4 56 D7
Goodshaw Fold BB4 56 D7
Goodwick (Wdig) SA64 16 C2
Goodworth Clatford SP11 21 G7
Goodyers End CV12 41 F7
Goole DN14 58 D7
Goom's Hill WR7 30 B3
Goonbell TR5 2 E4
Goonhavern TR4 2 E3
Goonvrea TR5 2 E4
Goose Green *Essex* CO11 35 F6
Goose Green *Essex* CO16 35 F6
Goose Green *Gt.Man.* WN3 48 E2
Goose Green *Kent* TN11 23 K6
Goose Green *S.Glos.* BS30 19 K4
Goose Pool HR2 28 D5
Gooseham EX23 6 A4
Goosehill Green WR9 29 J2
Goosewell PL9 5 F5
Goosey SN7 21 G2
Goosnargh PR3 55 J6
Goostrey CW4 49 G5
Gorcott Hill B98 30 B2
Gorddinog LL33 46 E5
Gordon TD3 76 E6
Gordonbush KW9 97 F1
Gordonstoun IV30 97 J5
Gordonstown *Aber.* AB45 98 D5
Gordonstown *Aber.* AB51 91 F1
Gore Cross SN10 20 D6
Gore End RG20 21 H5
Gore Pit CO5 34 C7
Gore Street CT12 25 J5
Gorebridge EH23 76 B4
Gorefield PE13 43 H4
Gorey JE3 3 K7
Gorgie EH11 76 A3
Goring RG8 21 K3
Goring Heath RG8 21 K4
Goring-by-Sea BN12 12 E6
Gorleston-on-Sea NR31 45 K5
Gorllwyn SA33 17 G2
Gornalwood DY3 40 B6
Gorrachie AB45 99 F5
Gorran Churchtown PL26 3 G4
Gorran Haven PL26 4 A6
Gors SY23 27 F1
Gorsedd CH8 47 K5
Gorseinon SA4 17 J6
Gorseness KW17 106 D6
Gorseybank DE4 50 E7
Gorsgoch SA40 26 D3
Gorslas SA14 17 J4
Gorsley HR9 29 F6
Gorsley Common HR9 29 F6
Gorstage CW8 49 F5
Gorstan IV23 95 K5
Gorstanvorran PH37 86 E7
Gorsty Hill ST14 40 D3
Gorten PA64 79 J4
Gortenbuie PA70 79 G4
Gorteneorn PH36 79 H1
Gorton *Arg. & B.* PA78 78 C2
Gorton *Gt.Man.* M18 49 H3
Gosbeck IP6 35 F3
Gosberton PE11 43 F2
Gosberton Clough PE11 42 E3
Goseley Dale DE11 41 F3
Gosfield CO9 34 B6
Gosford *Here.* SY8 28 E2
Gosford *Oxon.* OX5 31 G7
Gosforth *Cumb.* CA20 60 B6
Gosforth *T. & W.* NE3 71 H7
Gosland Green CW6 48 E7
Gosmore SG4 32 E6
Gospel End DY3 40 B6
Gosport PO12 11 H5
Gossabrough ZE2 107 P4
Gossington GL2 20 A1
Gossops Green RH11 13 F3
Goswick TD15 77 J6
Gotham NG11 41 H2
Gotherington GL52 29 J6
Gothers PL26 3 G3
Gott ZE2 107 N8
Gotton TA2 8 B2
Goudhurst TN17 14 C4
Goulceby LN11 53 F5
Gourdas AB53 99 F6
Gourdon DD10 91 G7
Gourock PA19 74 A3
Govan G51 74 D4
Goverton NG14 51 K7
Goveton TQ7 5 H6
Govilon NP7 28 B7
Gowanhill AB43 99 J4
Gowdall DN14 58 C7
Gowerton SA4 17 J6
Gowkhall KY12 75 J2
Gowkthrapple ML2 75 F5
Gowthorpe YO41 58 D4
Goxhill *E.Riding* HU11 59 H5
Goxhill *N.Lincs.* DN19 59 H7
Goytre SA13 18 A3
Gozzard's Ford OX13 21 H2
Grabhair HS2 101 F6
Graby NG34 42 D3
Gradbach SK17 49 J6
Grade TR12 2 E7
Gradeley Green CW5 48 E7
Graffham GU28 12 C5
Grafham *Cambs.* PE28 32 E2
Grafham *Surr.* GU5 22 D7
Grafton *Here.* HR2 28 D5
Grafton *N.Yorks.* YO51 57 K3
Grafton *Oxon.* OX18 21 F1
Grafton *Shrop.* SY4 38 D4
Grafton *Worcs.* HR6 28 E2
Grafton *Worcs.* GL20 29 J5
Grafton Flyford WR7 29 J3
Grafton Regis NN12 31 J4
Grafton Underwood NN14 42 C7
Grafty Green ME17 14 D3
Graianrhyd CH7 48 B7
Graig *Carmar.* SA16 17 H5
Graig *Conwy* LL28 47 G5
Graig *Denb.* LL17 47 J5
Graig-fechan LL15 47 K7
Grain ME3 24 E4
Grainel PA44 72 A4
Grainhow AB53 99 G6
Grains Bar OL4 49 J2
Grainsby DN36 53 F3
Grainthorpe LN11 53 G3
Graiselound DN9 51 K3
Gramisdale (Gramsdale) HS7 92 D6
Grampound TR2 3 G4
Grampound Road TR2 3 G3
Gramsdal (Gramisdale) HS7 92 D6
Granborough MK18 31 J6
Granby NG13 42 A2
Grandborough CV23 31 F2
Grandes Rocques GY5 3 J5
Grandtully PH9 82 A2
Grange *Cumb.* CA12 60 D5
Grange *E.Ayr.* KA1 74 C7
Grange *High.* IV63 87 K1
Grange *Med.* ME7 24 D5
Grange *Mersey.* CH48 48 B4
Grange *P. & K.* PH2 82 D5
Grange Crossroads AB55 98 C5
Grange de Lings LN2 52 C5
Grange Hall IV36 97 H5
Grange Hill IG7 23 H2
Grange Moor WF4 50 E1
Grange of Lindores KY14 82 D6
Grange Villa DH2 62 D1
Grangemill DE4 50 E7
Grangemouth FK3 75 H2
Grangemuir KY10 83 G7
Grange-over-Sands LA11 55 H2
Grangeston KA26 67 G4
Grangetown *Cardiff* CF11 18 E4
Grangetown *R. & C.* TS6 63 G4
Granish PH22 89 G3
Gransmoor YO25 59 H4
Granston SA62 16 B2
Grantchester CB3 33 H3
Grantham NG31 42 C2
Grantley HG4 57 H2
Grantlodge AB51 91 F3
Granton EH5 76 A3
Granton House DG10 69 F3
Grantown-on-Spey PH26 89 H2
Grantsfield HR6 28 E2
Grantshouse TD11 77 G4
Grappenhall WA4 49 F4
Grasby DN38 52 D2
Grasmere LA22 60 E6
Grass Green CO9 34 B5
Grasscroft OL4 49 J2
Grassendale L19 48 C4
Grassgarth LA8 60 F7
Grassholme DL12 62 A4
Grassington BD23 57 F3
Grassmoor S42 51 G6
Grassthorpe NG23 51 K6
Grateley SP11 21 F7
Gratwich ST14 40 C2
Gravel Hill SL9 22 D2
Graveley *Cambs.* PE19 33 F2
Graveley *Herts.* SG4 33 F6
Gravelly Hill B23 40 D6
Gravels SY5 38 C5
Graven ZE2 107 N5
Graveney ME13 25 G5
Gravesend DA11 24 C4
Grayingham DN21 52 C3
Grayrigg LA8 61 G7
Grays RM17 24 C4
Grayshott GU26 12 B3
Grayswood GU27 12 C3
Grazeley RG7 21 K5
Greasbrough S61 51 G3
Greasby CH49 48 B4
Great Abington CB21 33 J4
Great Addington NN14 32 C1
Great Alne B49 30 C3
Great Altcar L37 48 C2
Great Amwell SG12 33 G7
Great Asby CA16 61 H5
Great Ashfield IP31 34 D2
Great Ayton TS9 63 G5
Great Baddow CM2 24 D1
Great Bardfield CM7 33 K5
Great Barford MK44 32 E3
Great Barr B43 40 C6
Great Barrington OX18 30 D7
Great Barrow CH3 48 D6
Great Barton IP31 34 C2
Great Barugh YO17 58 D2
Great Bavington NE19 70 E5
Great Bealings IP13 35 G4
Great Bedwyn SN8 21 F5
Great Bentley CO7 35 F6
Great Bernera HS2 100 D4
Great Billing NN3 32 B2
Great Bircham PE31 44 B2
Great Blakenham IP6 35 F3
Great Bolas TF6 39 F3
Great Bookham KT23 22 E6
Great Bourton OX17 31 F4
Great Bowden LE16 42 A7
Great Bradley CB8 33 K3
Great Braxted CM8 34 C7
Great Bricett IP7 34 E3
Great Brickhill MK17 32 C5
Great Bridgeford ST18 40 A3
Great Brington NN7 31 H2
Great Bromley CO7 34 E6
Great Broughton *Cumb.* CA13 60 B3
Great Broughton *N.Yorks.* TS9 63 G6
Great Buckland DA13 24 C5
Great Budworth CW9 49 F5
Great Burdon DL1 62 E5
Great Burstead CM12 24 C2
Great Busby TS9 63 G6
Great Cambourne CB23 33 G3
Great Canfield CM6 33 J7
Great Canney CM3 24 E1
Great Carlton LN11 53 H4
Great Casterton PE9 42 D5
Great Chalfield SN12 20 B5
Great Chart TN23 14 E3
Great Chatwell TF10 39 G4
Great Chell ST6 49 H7
Great Chesterford CB10 33 J4
Great Cheverell SN10 20 C6
Great Chishill SG8 33 H5
Great Clacton CO15 35 F7
Great Clifton CA14 60 B4
Great Coates DN37 53 F2
Great Comberton WR10 29 J4
Great Corby CA4 61 F1
Great Cornard CO10 34 C4
Great Cowden HU11 59 J5
Great Coxwell SN7 21 F2
Great Crakehall DL8 57 H1
Great Cransley NN14 32 B1
Great Cressingham IP25 44 C5
Great Crosby L23 48 C2
Great Crosthwaite CA12 60 D4
Great Cubley DE6 40 D2
Great Cumbrae KA28 73 K5
Great Dalby LE14 42 A4
Great Doddington NN29 32 B2
Great Doward HR9 28 E7
Great Dunham PE32 44 C4
Great Dunmow CM6 33 K6
Great Durnford SP4 10 C1
Great Easton *Essex* CM6 33 K6
Great Easton *Leics.* LE16 42 B6
Great Eccleston PR3 55 H5
Great Edstone YO62 58 D1
Great Ellingham NR17 44 E6
Great Elm BA11 20 A7
Great Eversden CB23 33 G3
Great Fencote DL7 62 D7
Great Finborough IP14 34 E3
Great Fransham NR19 44 C4
Great Gaddesden HP1 32 D7
Great Gidding PE28 42 E7
Great Givendale YO42 58 E4
Great Glemham IP17 35 H2
Great Glen LE8 41 J6
Great Gonerby NG31 42 B2
Great Gransden SG19 33 F3
Great Green *Cambs.* SG8 33 F4
Great Green *Norf.* IP20 45 G7
Great Green *Suff.* IP30 34 D3
Great Green *Suff.* IP21 35 F1
Great Green *Suff.* IP22 34 E1
Great Habton YO17 58 D2
Great Hale NG34 42 E1
Great Hallingbury CM22 33 J7
Great Hampden HP16 22 B1
Great Harrowden NN9 32 B1
Great Harwood BB6 56 C6
Great Haseley OX44 21 K1
Great Hatfield HU11 59 H5
Great Haywood ST18 40 C3
Great Heath CV6 41 F7
Great Heck DN14 58 B7
Great Henny CO10 34 C5
Great Hinton BA14 20 C6
Great Hockham IP24 44 D6
Great Holland CO13 35 G7
Great Horkesley CO6 34 D5
Great Hormead SG9 33 H5
Great Horton BD7 57 G6
Great Horwood MK17 31 J5
Great Houghton *Northants.* NN4 31 J3
Great Houghton *S.Yorks.* S72 51 G2
Great Hucklow SK17 50 D5
Great Kelk YO25 59 H4
Great Kimble HP17 22 B1
Great Kingshill HP15 22 B2
Great Langton DL7 62 D7
Great Leighs CM3 34 B7
Great Limber DN37 52 E2
Great Linford MK14 32 B4
Great Livermere IP31 34 C1
Great Longstone DE45 50 E5
Great Lumley DH3 62 D2
Great Lyth SY3 38 D5
Great Malvern WR14 29 G4
Great Maplestead CO9 34 C5
Great Marton FY4 55 G6
Great Massingham PE32 44 B3
Great Melton NR9 45 F5
Great Milton OX44 21 K1
Great Missenden HP16 22 B1
Great Mitton BB7 56 C6
Great Mongeham CT14 15 J2
Great Moulton NR15 45 F6
Great Munden SG11 33 G6
Great Musgrave CA17 61 J5
Great Ness SY4 38 D4
Great Notley CM77 34 B6
Great Nurcot TA24 7 H2
Great Oak NP15 19 G1
Great Oakley *Essex* CO12 35 F6
Great Oakley *Northants.* NN18 42 B7
Great Offley SG5 32 E6
Great Ormside CA16 61 J5
Great Orton CA5 60 E1
Great Ouseburn YO26 57 K3
Great Oxendon LE16 42 A7
Great Oxney Green CM1 24 C1
Great Palgrave PE32 44 C4
Great Parndon CM19 23 H1
Great Paxton PE19 33 F2
Great Plumpton PR4 55 G6
Great Plumstead NR13 45 H5
Great Ponton NG33 42 C2
Great Potheridge EX20 6 D4
Great Preston LS26 57 J7
Great Purston NN13 31 G5
Great Raveley PE28 43 F7
Great Rissington GL54 30 C7
Great Rollright OX7 30 E5
Great Ryburgh NR21 44 D3
Great Ryle NE66 71 F2
Great Ryton SY5 38 D5
Great Saling CM7 34 B6
Great Salkeld CA11 61 G3
Great Sampford CB10 33 K5
Great Sankey WA5 48 E4

Gre - Hal

Great Saredon **WV10** 40 B5
Great Saxham **IP29** 34 B2
Great Shefford **RG17** 21 G4
Great Shelford **CB22** 33 H3
Great Smeaton **DL6** 62 E6
Great Snoring **NR21** 44 D2
Great Somerford **SN15** 20 C3
Great Stainton **TS21** 62 E4
Great Stambridge **SS4** 25 F2
Great Staughton **PE19** 32 E2
Great Steeping **PE23** 53 H6
Great Stonar **CT13** 15 J2
Great Strickland **CA10** 61 G4
Great Stukeley **PE28** 33 F1
Great Sturton **LN9** 53 F5
Great Sutton *Ches.W. & C.* **CH66** 48 C5
Great Sutton *Shrop.* **SY8** 38 E7
Great Swinburne **NE48** 70 E6
Great Tew **OX7** 30 E6
Great Tey **CO6** 34 C6
Great Thorness **PO30** 11 F5
Great Thurlow **CB9** 33 K4
Great Torr **TQ7** 5 G6
Great Torrington EX38 6 **C4**
Great Tosson **NE65** 71 F3
Great Totham *Essex* **CM9** 34 C7
Great Totham *Essex* **CM9** 34 C7
Great Tows **LN8** 53 F3
Great Urswick **LA12** 55 F2
Great Wakering **SS3** 25 F3
Great Waldingfield **CO10** 34 D4
Great Walsingham NR22 44 **D2**
Great Waltham **CM3** 33 K7
Great Warley **CM13** 23 J2
Great Washbourne **GL20** 29 J5
Great Weeke **TQ13** 7 F7
Great Welnetham **IP30** 34 C3
Great Wenham **CO7** 34 E5
Great Whittington **NE19** 71 F6
Great Wigborough **CO5** 34 D7
Great Wigsell **TN32** 14 C5
Great Wilbraham **CB21** 33 J3
Great Wilne **DE72** 41 G2
Great Wishford **SP2** 10 B1
Great Witcombe **GL3** 29 J7
Great Witley **WR6** 29 G2
Great Wolford **CV36** 30 D5
Great Wratting **CB9** 33 K4
Great Wymondley **SG4** 33 F6
Great Wyrley **WS6** 40 B5
Great Wytheford **SY4** 38 E4
Great Yarmouth NR30 45 **K5**
Great Yeldham **CO9** 34 B5
Greatford **PE9** 42 D4
Greatgate **ST10** 40 C1
Greatham *Hants.* **GU33** 11 J1
Greatham *Hart.* **TS25** 63 F4
Greatham *W.Suss.* **RH20** 12 D5
Greatness **TN14** 23 J6
Greatstone-on-Sea **TN28** 15 F5
Greatworth **OX17** 31 G4
Green **LL16** 47 J6
Green Cross **GU10** 12 B3
Green End *Bed.* **MK44** 32 E3
Green End *Bucks.* **MK17** 32 C5
Green End *Cambs.* **PE29** 33 F1
Green End *Cambs.* **PE27** 33 G1
Green End *Herts.* **SG12** 33 G6
Green End *Herts.* **SG9** 33 G5
Green End *Warks.* **CV7** 40 E7
Green Hammerton **YO26** 57 K4
Green Hill **SN4** 20 D3
Green Lane **B80** 30 B2
Green Moor **S35** 50 E3
Green Ore **BA5** 19 J6
Green Quarter **LA8** 61 F6
Green Street *E.Suss.* **TN38** 14 C6
Green Street *Herts.* **WD6** 22 E2
Green Street *Herts.* **SG11** 33 H6
Green Street *W.Suss.* **RH13** 12 E4
Green Street *Worcs.* **WR5** 29 H4
Green Street Green *Gt.Lon.* **BR6** 23 H5
Green Street Green *Kent* **DA2** 23 J4
Green Tye **SG10** 33 H7
Greenburn **DD5** 83 F3
Greencroft **DH7** 62 C2
Greendams **AB31** 90 E6
Greendykes **NE66** 71 F1
Greenend **OX7** 30 E6
Greenfaulds **G67** 75 F3
Greenfield *Cen.Beds.* **MK45** 32 D5
Greenfield (Maes-Glas) *Flints.* **CH8** 47 K5
Greenfield *Gt.Man.* **OL3** 50 C2
Greenfield *High.* **PH35** 87 J4
Greenfield *Lincs.* **LN13** 53 H5
Greenfield *Oxon.* **OX49** 22 A2
Greenford UB6 22 **E3**
Greengairs **ML6** 75 F3
Greengates **BD10** 57 G6
Greengill **CA7** 60 C3
Greenhalgh **PR4** 55 H6
Greenhall **AB52** 90 E2
Greenham **RG14** 21 H5
Greenhaugh **NE48** 70 C5
Greenhead **CA8** 70 B7
Greenheads **AB42** 91 J1
Greenheys **BL5** 49 G2
Greenhill *Gt.Lon.* **HA1** 22 E3
Greenhill *High.* **KW9** 97 G1
Greenhill *S.Yorks.* **S8** 51 F4
Greenhithe DA9 23 **J4**
Greenholm **KA16** 74 D7
Greenholme **CA10** 61 G6
Greenhow Hill **HG3** 57 G3
Greenigo **KW15** 106 D7
Greenland **KW14** 105 H2
Greenlands **RG9** 22 A3
Greenlaw *Aber.* **AB45** 98 E5
Greenlaw *Sc.Bord.* **TD10** 77 F6
Greenloaning **FK15** 81 K7
Greenmeadow **NP44** 19 F2
Greenmoor Hill **RG8** 21 K3
Greenmount **BL8** 49 G1
Greenmyre **AB53** 91 G1
Greenock PA16 74 **A3**
Greenodd **LA12** 55 G1
Greens Norton **NN12** 31 H4
Greenscares **FK15** 81 J6
Greenside *T. & W.* **NE40** 71 G7
Greenside *W.Yorks.* **HD5** 50 D1
Greenstead **CO4** 34 E6
Greenstead Green **CO9** 34 C6
Greensted **CM5** 23 J1
Greensted Green **CM5** 23 J1
Greenway *Pembs.* **SA66** 16 D2
Greenway *Som.* **TA3** 8 C2
Greenwell **CA8** 61 G1
Greenwich **SE10** 23 G4
Greet **GL54** 30 B5
Greete **SY8** 28 E1
Greetham *Lincs.* **LN9** 53 G5
Greetham *Rut.* **LE15** 42 C4
Greetland **HX4** 57 F7
Gregson Lane **PR5** 55 J7
Greinetobht (Grenitote) **HS6** 92 D4
Greinton **TA7** 8 D1
Grenaby **IM9** 54 B6
Grendon *Northants.* **NN7** 32 B2
Grendon *Warks.* **CV9** 40 E6
Grendon Common **CV9** 40 E6
Grendon Green **HR6** 28 E3
Grendon Underwood **HP18** 31 H6
Grenitote (Greinetobht) **HS6** 92 D4
Grenofen **PL19** 4 E3
Grenoside **S35** 51 F3
Greosabhagh **HS3** 93 G2
Gresford **LL12** 48 C7
Gresham **NR11** 45 F2
Greshornish **IV51** 93 J6
Gress (Griais) **HS2** 101 G3
Gressenhall **NR20** 44 D4
Gressingham **LA2** 55 J3
Greta Bridge **DL12** 62 B5
Gretna DG16 69 **J7**
Gretna Green **DG16** 69 J7
Gretton *Glos.* **GL54** 30 B5
Gretton *Northants.* **NN17** 42 C6
Gretton *Shrop.* **SY6** 38 E6
Grewelthorpe **HG4** 57 H2
Greygarth **HG4** 57 G2
Greylake **TA7** 8 C1
Greys Green **RG9** 22 A3
Greysouthen **CA13** 60 B4
Greystead **NE48** 70 C5
Greystoke **CA11** 60 F3
Greystone *Aber.* **AB35** 89 K5
Greystone *Angus* **DD11** 83 G3
Greystone *Lancs.* **BB9** 56 D5
Greystones **S11** 51 F4
Greywell **RG29** 22 A6
Griais (Gress) **HS2** 101 G3
Gribthorpe **DN14** 58 D6
Gribton **DG2** 68 E5
Griff **CV10** 41 F7
Griffithstown **NP4** 19 F2
Grigadale **PH36** 79 F1
Grigghall **LA8** 61 F7
Grimeford Village **PR6** 49 F1
Grimesthorpe **S4** 51 F3
Grimethorpe **S72** 51 G2
Griminis (Griminish) **HS7** 92 C6
Griminish (Griminis) **HS7** 92 C6
Grimister **ZE2** 107 N3
Grimley **WR2** 29 H2
Grimmet **KA19** 67 H2
Grimness **KW17** 106 D8
Grimoldby **LN11** 53 G4
Grimpo **SY11** 38 C3
Grimsargh **PR2** 55 J6
Grimsay (Griomsaigh) **HS6** 92 D6
Grimsbury **OX16** 31 F4
Grimsby DN32 53 **F2**
Grimscote **NN12** 31 H3
Grimscott **EX23** 6 A5
Grimshader (Griomsiadar) **HS2** 101 G5
Grimsthorpe **PE10** 42 D3
Grimston *E.Riding* **HU11** 59 J6
Grimston *Leics.* **LE14** 41 J3
Grimston *Norf.* **PE32** 44 B3
Grimstone **DT2** 9 F5
Grimstone End **IP31** 34 D2
Grindale **YO16** 59 H2
Grindiscol **ZE2** 107 N9
Grindle **TF11** 39 G5
Grindleford **S32** 50 E5
Grindleton **BB7** 56 C5
Grindley **ST14** 40 C3
Grindley Brook **SY13** 38 E1
Grindlow **SK17** 50 D5
Grindon *Northumb.* **TD15** 77 H6
Grindon *Staffs.* **ST13** 50 C7
Grindon *Stock.* **TS21** 62 E4
Grindon *T. & W.* **SR4** 62 E1
Gringley on the Hill **DN10** 51 K3
Grinsdale **CA5** 60 E1
Grinshill **SY4** 38 E3
Grinton **DL11** 62 B7
Griomarstaidh **HS2** 100 E4
Griomsaigh (Grimsay) **HS6** 92 D6
Griomsiadar (Grimshader) **HS2** 101 G5
Grisdale **LA10** 61 J7
Grishipoll **PA78** 78 C2
Gristhorpe **YO14** 59 G1
Griston **IP25** 44 D6
Gritley **KW17** 106 E7
Grittenham **SN15** 20 D3
Grittleton **SN14** 20 B4
Grizebeck **LA17** 55 F1
Grizedale **LA22** 60 E7
Grobister **KW17** 106 F5
Groby **LE6** 41 H5
Groes **LL16** 47 J6
Groes-faen **CF72** 18 D3
Groesffordd **LL53** 36 B2
Groesffordd Marli **LL22** 47 J5
Groeslon *Gwyn.* **LL54** 46 C7
Groeslon *Gwyn.* **LL55** 46 D6
Groes-lwyd **SY21** 38 B4
Groes-wen **CF15** 18 E3
Grogport **PA28** 73 G6
Groigearraidh **HS8** 84 C1
Gromford **IP17** 35 H3
Gronant **LL19** 47 J4
Groombridge **TN3** 13 J3
Grosmont *Mon.* **NP7** 28 D6
Grosmont *N.Yorks.* **YO22** 63 K6
Grotaig **IV63** 88 B2
Groton **CO10** 34 D4
Groundistone Heights **TD9** 69 K2
Grouville **JE3** 3 K7
Grove *Bucks.* **LU7** 32 C6
Grove *Dorset* **DT5** 9 F7
Grove *Kent* **CT3** 25 J5
Grove *Notts.* **DN22** 51 K5
Grove *Oxon.* **OX12** 21 H2
Grove End **ME9** 24 E5
Grove Green **ME14** 14 C2
Grove Park **SE12** 23 H4
Grove Town **WF8** 57 K7
Grovehill **HP2** 22 D1
Grovesend *S.Glos.* **BS35** 19 K3
Grovesend *Swan.* **SA4** 17 J5
Gruids **IV27** 96 C1
Grula **IV47** 85 J2
Gruline **PA71** 79 G3
Grumbla **TR20** 2 B6
Grundcruie **PH1** 82 B5
Grundisburgh **IP13** 35 G3
Gruting **ZE2** 107 L8
Grutness **ZE3** 107 N11
Gualachulain **PH49** 80 C3
Guardbridge **KY16** 83 F6
Guarlford **WR13** 29 H4
Guay **PH9** 82 B3
Gubbergill **CA19** 60 B7
Gubblecote **HP23** 32 C7
GUERNSEY GY 3 **J5**
Guernsey Airport **GY8** 3 H5
Guestling Green **TN35** 14 D6
Guestling Thorn **TN35** 14 D6
Guestwick **NR20** 44 E3
Guestwick Green **NR20** 44 E3
Guide **BB1** 56 C7
Guide Post **NE62** 71 H5
Guilden Down **SY7** 38 C7
Guilden Morden **SG8** 33 F4
Guilden Sutton **CH3** 48 D6
GUILDFORD GU 22 **C7**
Guildtown **PH2** 82 C4
Guilsborough **NN6** 31 H1
Guilsfield (Cegidfa) **SY21** 38 B4
Guilthwaite **S60** 51 G4
Guisborough TS14 63 **H5**
Guiseley **LS20** 57 G5
Guist **NR20** 44 E3
Guith **KW17** 106 E4
Guiting Power **GL54** 30 B6
Gulberwick **ZE2** 107 N9
Gullane EH31 76 **C2**
Gulval **TR18** 2 B5
Gulworthy **PL19** 4 E3
Gumfreston **SA70** 16 E5
Gumley **LE16** 41 J6
Gunby *Lincs.* **NG33** 42 C3
Gunby *Lincs.* **PE23** 53 H6
Gundleton **SO24** 11 H1
Gunn **EX32** 6 E2
Gunnersbury **W4** 22 E4
Gunnerside **DL11** 62 A7
Gunnerton **NE48** 70 E6
Gunness **DN15** 52 B1
Gunnislake PL18 4 **E3**
Gunnista **ZE2** 107 P8
Gunnister **ZE2** 107 M5
Gunstone **WV8** 40 A5
Gunter's Bridge **GU28** 12 C4
Gunthorpe *Norf.* **NR24** 44 E2
Gunthorpe *Notts.* **NG14** 41 J1
Gunthorpe *Rut.* **LE15** 42 B5
Gunville **PO30** 11 F6
Gunwalloe **TR12** 2 D6
Gupworthy **TA24** 7 H2
Gurnard **PO31** 11 F5
Gurnett **SK11** 49 J5
Gurney Slade **BA3** 19 K7
Gurnos *M.Tyd.* **CF47** 18 D1
Gurnos *Powys* **SA9** 18 A1
Gushmere **ME13** 15 F2
Gussage All Saints **BH21** 10 B3
Gussage St. Andrew **DT11** 9 J3
Gussage St. Michael **BH21** 9 J3
Guston **CT15** 15 J3
Gutcher **ZE2** 107 P3
Guthram Gowt **PE11** 42 E3
Guthrie **DD8** 83 G2
Guyhirn **PE13** 43 G5
Guynd **DD11** 83 G3
Guy's Head **PE12** 43 H3
Guy's Marsh **SP7** 9 H2
Guyzance **NE65** 71 H3
Gwaelod-y-garth **CF15** 18 E3
Gwaenysgor **LL18** 47 J4
Gwaithla **HR5** 28 B3
Gwalchmai **LL65** 46 B5
Gwastad **SA63** 16 D3
Gwastadnant **LL55** 46 E7
Gwaun-Cae-Gurwen **SA18** 27 G7
Gwaynynog **LL16** 47 J6
Gwbert **SA43** 16 E1
Gweek **TR12** 2 E6
Gwehelog **NP15** 19 G1
Gwenddwr **LD2** 27 K4
Gwendreath **TR12** 2 E7
Gwennap **TR16** 2 E4
Gwenter **TR12** 2 E7
Gwernaffield **CH7** 48 B6
Gwernesney **NP15** 19 H1
Gwernogle **SA32** 17 J2
Gwernymynydd **CH7** 48 B6
Gwern-y-Steeple **CF5** 18 D4
Gwersyllt **LL11** 48 C7
Gwespyr **CH8** 47 K4
Gwinear **TR27** 2 C5
Gwithian **TR27** 2 C4
Gwredog **LL71** 46 C4
Gwrhay **NP12** 18 E2
Gwyddelwern **LL21** 37 K1
Gwyddgrug **SA39** 17 H2
Gwynfryn **LL11** 48 B7
Gwystre **LD1** 27 K2
Gwytherin **LL22** 47 G6
Gyfelia **LL13** 38 C1
Gyre **KW17** 106 C7
Gyrn Goch **LL54** 36 C1

H

Habberley **SY5** 38 C5
Habin **GU31** 12 B4
Habrough **DN40** 52 E1
Haccombe **TQ12** 5 J3
Hacconby **PE10** 42 E3
Haceby **NG34** 42 D2
Hacheston **IP13** 35 H3
Hackbridge **SM6** 23 F5
Hackenthorpe **S12** 51 G4
Hackford **NR18** 44 E5
Hackforth **DL8** 62 D7
Hackland **KW17** 106 C5
Hacklet (Haclait) **HS7** 92 D7
Hacklete (Tacleit) **HS2** 100 D4
Hackleton **NN7** 32 B3
Hacklinge **CT14** 15 J2
Hackness *N.Yorks.* **YO13** 63 J3
Hackness *Ork.* **KW16** 106 C8
Hackney **E9** 23 G3
Hackthorn **LN2** 52 C4
Hackthorpe **CA10** 61 G4
Haclait (Hacklet) **HS7** 92 D7
Hacton **RM14** 23 J3
Hadden **TD5** 77 F7
Haddenham *Bucks.* **HP17** 22 A1
Haddenham *Cambs.* **CB6** 33 H1
Haddington *E.Loth.* EH41 76 **D3**
Haddington *Lincs.* **LN5** 52 C6
Haddiscoe **NR14** 45 J6
Haddon **PE7** 42 E6
Hade Edge **HD9** 50 D2
Hademore **WS14** 40 D5
Hadfield **SK13** 50 C3
Hadham Cross **SG10** 33 H7
Hadham Ford **SG11** 33 H6
Hadleigh *Essex* **SS7** 24 E3
Hadleigh *Suff.* **IP7** 34 E4
Hadleigh Heath **IP7** 34 D4
Hadley *Tel. & W.* **TF1** 39 F4
Hadley *Worcs.* **WR9** 29 H2
Hadley End **DE13** 40 D3
Hadley Wood **EN4** 23 F2
Hadlow **TN11** 23 K7
Hadlow Down **TN22** 13 J4
Hadnall **SY4** 38 E3
Hadspen **BA7** 9 F1
Hadstock **CB21** 33 J4
Hadston **NE65** 71 H4
Hadzor **WR9** 29 J2
Haffenden Quarter **TN27** 14 D3
Hafod Bridge **SA19** 17 K2
Hafod-Dinbych **LL24** 47 G7
Hafodunos **LL22** 47 G6
Hafodyrynys **NP11** 19 F2
Haggate **BB10** 56 D6
Haggbeck **CA6** 69 K6
Haggersta **ZE2** 107 M8
Haggerston *Gt.Lon.* **E2** 23 G3
Haggerston *Northumb.* **TD15** 77 J6
Haggrister **ZE2** 107 M5
Haggs **FK4** 75 F3
Hagley *Here.* **HR1** 28 E4
Hagley *Worcs.* **DY9** 40 B7
Hagnaby *Lincs.* **PE23** 53 G6
Hagnaby *Lincs.* **LN13** 53 H5
Hague Bar **SK22** 49 J4
Hagworthingham **PE23** 53 G6
Haigh **WN2** 49 F2
Haighton Green **PR2** 55 J6
Hail Weston **PE19** 32 E2
Haile **CA22** 60 B6
Hailes **GL54** 30 B5
Hailey *Herts.* **SG13** 33 G7
Hailey *Oxon.* **OX10** 21 K3
Hailey *Oxon.* **OX29** 30 E7
Hailsham BN27 13 **J6**
Haimer **KW14** 105 G2
Hainault **IG6** 23 H2
Haine **CT12** 25 K5
Hainford **NR10** 45 G4
Hainton **LN8** 52 E4
Haisthorpe **YO25** 59 H3
Hakin **SA73** 16 B5
Halam **NG22** 51 J7
Halbeath **KY11** 75 K2
Halberton **EX16** 7 J4
Halcro **KW1** 105 H2
Hale *Cumb.* **LA7** 55 J2
Hale *Gt.Man.* **WA15** 49 G4
Hale *Halton* **L24** 48 D4
Hale *Hants.* **SP6** 10 C3
Hale *Surr.* **GU9** 22 B7
Hale Bank **WA8** 48 D4
Hale Barns **WA15** 49 G4
Hale Nook **PR3** 55 G5
Hale Street **TN12** 23 K7
Hales *Norf.* **NR14** 45 H6
Hales *Staffs.* **TF9** 39 G2
Hales Green **DE6** 40 D1
Hales Place **CT2** 15 G2
Halesgate **PE12** 43 G3
Halesowen B63 40 **B7**
Halesworth IP19 35 **H1**
Halewood **L26** 48 D4
Half Way Inn **EX5** 7 J6
Halford *Devon* **TQ12** 5 J3
Halford *Shrop.* **SY7** 38 D7
Halford *Warks.* **CV36** 30 D4
Halfpenny **LA8** 55 J1
Halfpenny Green **DY7** 40 A6
Halfway *Carmar.* **SA19** 17 K2
Halfway *Carmar.* **SA15** 17 J5
Halfway *Powys* **SA20** 27 H5
Halfway *S.Yorks.* **S20** 51 G4
Halfway *W.Berks.* **RG20** 21 H5
Halfway Bridge **GU28** 12 C4
Halfway House **SY5** 38 C4
Halfway Houses *Kent* **ME12** 25 F4
Halfway Houses *Lincs.* **LN6** 52 B6
Halghton Mill **LL13** 38 D1
HALIFAX HX 57 **F7**
Halistra **IV55** 93 H6
Halket **KA3** 74 C5
Halkirk KW12 105 **G3**
Halkyn **CH8** 48 B5
Hall **G78** 74 C5
Hall Cross **PR4** 55 H7
Hall Dunnerdale **LA20** 60 D7
Hall Green *Ches.E.* **ST7** 49 H7
Hall Green *Lancs.* **PR4** 55 H7
Hall Green *W.Mid.* **B28** 40 D7
Hall Grove **AL7** 33 F7
Hall of the Forest **SY7** 38 B7
Halland **BN8** 13 J5
Hallaton **LE16** 42 A6
Hallatrow **BS39** 19 K6
Hallbankgate **CA8** 61 G1
Hallen **BS10** 19 J3
Hallfield Gate **DE55** 51 F7
Hallglen **FK1** 75 G3
Hallin **IV55** 93 H6
Halling **ME2** 24 D5
Hallington *Lincs.* **LN11** 53 G4
Hallington *Northumb.* **NE19** 70 E6
Halliwell **BL1** 49 F1
Halloughton **NG25** 51 J7
Hallow **WR2** 29 H3
Hallow Heath **WR2** 29 H3
Hallrule **TD9** 70 A2
Halls **EH42** 76 E3
Halls Green *Essex* **CM19** 23 H1
Hall's Green *Herts.* **SG4** 33 F6
Hallsands **TQ7** 5 J7
Hallthwaites **LA18** 54 E1
Hallwood Green **GL18** 29 F5
Hallworthy **PL32** 4 B2
Hallyne **EH45** 75 K6
Halmer End **ST7** 39 G1
Halmond's Frome **WR6** 29 F4
Halmore **GL13** 19 K1
Halmyre Mains **EH46** 75 K6
Halnaker **PO18** 12 C6
Halsall **L39** 48 C1
Halse *Northants.* **NN13** 31 G4
Halse *Som.* **TA4** 7 K3
Halsetown **TR26** 2 C5
Halsham **HU12** 59 J7
Halsinger **EX33** 6 D2
Halstead *Essex* CO9 34 **C5**
Halstead *Kent* **TN14** 23 H5
Halstead *Leics.* **LE7** 42 A5
Halstock **BA22** 8 E4
Halsway **TA4** 7 K2
Haltemprice Farm **HU10** 59 G6
Haltham **LN9** 53 F6
Haltoft End **PE22** 43 G1
Halton *Bucks.* **HP22** 32 B7
Halton *Halton* **WA7** 48 E4
Halton *Lancs.* **LA2** 55 J3
Halton *Northumb.* **NE45** 70 E7
Halton *Wrex.* **LL14** 38 C2
Halton East **BD23** 57 F4
Halton Gill **BD23** 56 D2
Halton Green **LA2** 55 J3
Halton Holegate **PE23** 53 H6
Halton Lea Gate **CA8** 61 H1
Halton Park **LA2** 55 J3
Haltwhistle NE49 70 **C7**
Halvergate **NR13** 45 J5
Halwell **TQ9** 5 H5
Halwill **EX21** 6 C6
Halwill Junction **EX21** 6 C6
Ham *Devon* **EX13** 8 B4
Ham *Glos.* **GL13** 19 K2

Ham - Hea

Ham *Glos.* **GL52** 29 J6
Ham *Gt.Lon.* **TW10** 22 E4
Ham *High.* **KW14** 105 H1
Ham *Kent* **CT14** 15 J2
Ham *Plym.* **PL2** 4 E5
Ham *Shet.* **ZE2** 107 L2
Ham *Som.* **TA3** 8 B2
Ham *Som.* **TA20** 8 B3
Ham *Wilts.* **SN8** 21 G5
Ham Common **SP8** 9 H2
Ham Green *Here.* **WR13** 29 G4
Ham Green *Kent* **ME9** 24 E5
Ham Green *Kent* **TN30** 14 D5
Ham Green *N.Som.* **BS20** 19 J4
Ham Green *Worcs.* **B97** 30 B2
Ham Hill **ME6** 24 C5
Ham Street **BA6** 8 E1
Hambleden **RG9** 22 A3
Hambledon *Hants.* **PO7** 11 H3
Hambledon *Surr.* **GU8** 12 C3
Hamble-le-Rice **SO31** 11 F4
Hambleton *Lancs.* **FY6** 55 G5
Hambleton *N.Yorks.* **YO8** 58 B6
Hambridge **TA10** 8 C2
Hambrook *S.Glos.* **BS16** 19 K4
Hambrook *W.Suss.* **PO18** 11 J4
Hameringham **LN9** 53 G6
Hamerton **PE28** 32 E1
Hamilton ML3 75 F5
Hamlet *Devon* **EX14** 7 K6
Hamlet *Dorset* **DT9** 8 E4
Hammer **GU27** 12 B3
Hammerpot **BN16** 12 D6
Hammersmith **W6** 23 F4
Hammerwich **WS7** 40 C5
Hammerwood **RH19** 13 H3
Hammond Street **EN7** 23 G1
Hammoon **DT10** 9 H3
Hamnavoe *Shet.* **ZE2** 107 M9
Hamnavoe *Shet.* **ZE2** 107 N4
Hamnavoe *Shet.* **ZE2** 107 L4
Hamnavoe *Shet.* **ZE2** 107 N5
Hamnish Clifford **HR6** 28 E3
Hamp **TA6** 8 B1
Hampden Park **BN22** 13 K6
Hamperden End **CB11** 33 J5
Hampnett **GL54** 30 C7
Hampole **DN6** 51 H2
Hampreston **BH21** 10 B5
Hampstead **NW3** 23 F3
Hampstead Norreys **RG18** 21 J4
Hampsthwaite **HG3** 57 H4
Hampton *Devon* **EX13** 8 B5
Hampton *Gt.Lon.* TW12 22 E5
Hampton *Kent* **CT6** 25 H5
Hampton *Peter.* **PE7** 42 E6
Hampton *Shrop.* **WV16** 39 G7
Hampton *Swin.* **SN6** 20 E2
Hampton *Worcs.* **WR11** 30 B4
Hampton Bishop **HR1** 28 E5
Hampton Fields **GL6** 20 B2
Hampton Heath **SY14** 38 D1
Hampton in Arden **B92** 40 E7
Hampton Loade **WV15** 39 G7
Hampton Lovett **WR9** 29 H2
Hampton Lucy **CV35** 30 D3
Hampton on the Hill **CV35** 30 D2
Hampton Poyle **OX5** 31 G7
Hampton Wick **KT1** 22 E5
Hamptworth **SP5** 10 D3
Hamsey **BN8** 13 H5
Hamstall Ridware **WS15** 40 D4
Hamstead **PO41** 10 E5
Hamstead Marshall **RG20** 21 H5
Hamsteels **DH7** 62 C2
Hamsterley *Dur.* **DL13** 62 C3
Hamsterley *Dur.* **NE17** 62 C1
Hamstreet **TN26** 15 F4
Hamworthy **BH15** 9 J5
Hanbury *Staffs.* **DE13** 40 D3
Hanbury *Worcs.* **B60** 29 J2
Hanbury Woodend **DE13** 40 D3
Hanby **NG33** 42 D2
Hanchurch **ST4** 40 A1
Handa Island **IV27** 102 D4
Handale **TS13** 63 J5
Handbridge **CH4** 48 D6
Handcross **RH17** 13 F3
Handforth **SK9** 49 H4
Handley *Ches.W. & C.* **CH3** 48 D7
Handley *Derbys.* **DE55** 51 F6
Handley Green **CM4** 24 C1
Handsacre **WS15** 40 C4
Handside **AL8** 33 F7
Handsworth *S.Yorks.* **S13** 51 G4
Handsworth *W.Mid.* **B21** 40 C6
Handwoodbank **SY5** 38 D4
Handy Cross **SL7** 22 B2
Hanford *Dorset* **DT11** 9 H3
Hanford *Stoke* **ST4** 40 A1
Hanging Bridge **DE6** 40 D1
Hanging Houghton **NN6** 31 J1
Hanging Langford **SP3** 10 B1
Hangingshaw **DG11** 69 G5
Hanham **BS15** 19 K4
Hankelow **CW3** 39 F1
Hankerton **SN16** 20 C2
Hankham **BN24** 13 K6
Hanley **ST1** 40 A1
Hanley Castle **WR8** 29 H4
Hanley Child **WR15** 29 F2
Hanley Swan **WR8** 29 H4
Hanley William **WR15** 29 F2
Hanlith **BD23** 56 E3
Hanmer **SY13** 38 D2
Hannah **LN13** 53 H5
Hannington *Hants.* **RG26** 21 J6
Hannington *Northants.* **NN6** 32 B1
Hannington *Swin.* **SN6** 20 E2
Hannington Wick **SN6** 20 E2
Hanslope **MK19** 32 B4
Hanthorpe **PE10** 42 D3
Hanwell *Gt.Lon.* **W7** 22 E3
Hanwell *Oxon.* **OX17** 31 F4
Hanwood **SY5** 38 D5
Hanworth *Gt.Lon.* **TW13** 22 E4
Hanworth *Norf.* **NR11** 45 F2
Happisburgh **NR12** 45 H2
Happisburgh Common **NR12** 45 H3
Hapsford **WA6** 48 D5
Hapton *Lancs.* **BB11** 56 C6
Hapton *Norf.* **NR15** 45 F6
Harberton **TQ9** 5 H5
Harbertonford **TQ9** 5 H5
Harbledown **CT2** 15 G2
Harborne **B17** 40 C7
Harborough Magna **CV23** 31 F1
Harbost (Tabost) **HS2** 101 H1
Harbottle **NE65** 70 E3
Harbourneford **TQ10** 5 H4
Harbridge **BH24** 10 C3
Harbridge Green **BH24** 10 C3
Harburn **EH55** 75 J4
Harbury **CV33** 30 E3
Harby *Leics.* **LE14** 42 A2
Harby *Notts.* **NG23** 52 B5
Harcombe **EX10** 7 K6
Harcombe Bottom **DT7** 8 C5
Harden *W.Mid.* **WS3** 40 C5
Harden *W.Yorks.* **BD16** 57 F6
Hardendale **CA10** 61 G5
Hardenhuish **SN14** 20 C4
Hardgate *Aber.* **AB31** 91 F4
Hardgate *N.Yorks.* **HG3** 57 H3
Hardham **RH20** 12 D5
Hardhorn **FY6** 55 G6
Hardingham **NR9** 44 E5
Hardingstone **NN4** 31 J3
Hardington **BA11** 20 A6
Hardington Mandeville **BA22** 8 E3
Hardington Marsh **BA22** 8 E4
Hardington Moor **BA22** 8 E3
Hardley **SO45** 11 F4
Hardley Street **NR14** 45 H5
Hardmead **MK16** 32 C4
Hardraw **DL8** 61 K7
Hardstoft **S45** 51 G6
Hardway *Hants.* **PO12** 11 H4
Hardway *Som.* **BA10** 9 G1
Hardwick *Bucks.* **HP22** 32 B7
Hardwick *Cambs.* **CB23** 33 G3
Hardwick *Lincs.* **LN1** 52 B5
Hardwick *Norf.* **NR15** 45 G6
Hardwick *Northants.* **NN9** 32 B2
Hardwick *Oxon.* **OX29** 21 G1
Hardwick *Oxon.* **OX27** 31 G6
Hardwick *S.Yorks.* **S26** 51 G4
Hardwick *W.Mid.* **B74** 40 C6
Hardwick Village **S80** 51 J5
Hardwicke *Glos.* **GL51** 29 J6
Hardwicke *Glos.* **GL2** 29 G7
Hardwicke *Here.* **HR3** 28 B4
Hardy's Green **CO2** 34 D6
Hare Green **CO7** 34 E6
Hare Hatch **RG10** 22 B4
Hare Street *Herts.* **SG2** 33 G6
Hare Street *Herts.* **SG9** 33 G6
Hareby **PE23** 53 G6
Harecroft **BD15** 57 F6
Hareden **BB7** 56 B4
Harefield **UB9** 22 D2
Harehill **DE6** 40 D2
Harehills **LS9** 57 J6
Harehope **NE66** 71 F1
Harelaw **ML11** 75 H6
Hareplain **TN27** 14 D4
Haresceugh **CA10** 61 H2
Harescombe **GL4** 29 H7
Haresfield **GL10** 29 H7
Hareshaw *N.Lan.* **ML1** 75 G4
Hareshaw *S.Lan.* **ML10** 74 E6
Harestock **SO22** 11 F1
Harewood **LS17** 57 J5
Harewood End **HR2** 28 E6
Harford *Devon* **PL21** 5 G5
Harford *Devon* **EX6** 7 G6
Hargate **NR16** 45 F6
Hargatewall **SK17** 50 D5
Hargrave *Ches.W. & C.* **CH3** 48 D6
Hargrave *Northants.* **NN9** 32 D1
Hargrave *Suff.* **IP29** 34 B3
Hargrave Green **IP29** 34 B3
Harker **CA6** 69 J7
Harkstead **IP9** 35 F5
Harlaston **B79** 40 E4
Harlaxton **NG32** 42 B2
Harle Syke **BB10** 56 D6
Harlech LL46 36 E2
Harlequin **NG12** 41 J2
Harlescott **SY1** 38 E4
Harlesden **NW10** 23 F3
Harleston *Devon* **TQ7** 5 H6
Harleston *Norf.* IP20 45 G7
Harleston *Suff.* **IP14** 34 E2
Harlestone **NN7** 31 J2
Harley *S.Yorks.* **S62** 51 F3
Harley *Shrop.* **SY5** 38 E5
Harleyholm **ML12** 75 H7
Harlington *Cen.Beds.* **LU5** 32 D5
Harlington *Gt.Lon.* **UB3** 22 D4
Harlosh **IV55** 93 H7
Harlow CM17 33 H7
Harlow Hill **NE15** 71 F7
Harlthorpe **YO8** 58 D6
Harlton **CB23** 33 G3
Harlyn **PL28** 3 F1
Harman's Cross **BH19** 9 J6
Harmby **DL8** 57 G1
Harmer Green **AL6** 33 F7
Harmer Hill **SY4** 38 D3
Harmondsworth **UB7** 22 D4
Harmston **LN5** 52 C6
Harnage **SY5** 38 E5
Harnham **SP2** 10 C2
Harnhill **GL7** 20 D1
Harold Hill **RM3** 23 J2
Harold Park **RM3** 23 J2
Harold Wood **RM3** 23 J2
Haroldston West **SA62** 16 B4
Haroldswick **ZE2** 107 Q1
Harome **YO62** 58 C1
Harpenden AL5 32 E7
Harpford **EX10** 7 J6
Harpham **YO25** 59 G3
Harpley *Norf.* **PE31** 44 B3
Harpley *Worcs.* **WR6** 29 F2
Harpole **NN7** 31 H2
Harprigg **LA6** 56 B1
Harpsdale **KW12** 105 G3
Harpsden **RG9** 22 A3
Harpswell **DN21** 52 C4
Harpur Hill **SK17** 50 C5
Harpurhey **M9** 49 H2
Harracott **EX31** 6 D3
Harrapool **IV49** 86 C2
Harrietfield **PH1** 82 A5
Harrietsham **ME17** 14 D2
Harringay **N8** 23 G3
Harrington *Cumb.* **CA14** 60 A4
Harrington *Lincs.* **PE23** 53 G5
Harrington *Northants.* **NN6** 31 J1
Harringworth **NN17** 42 C6
Harris **PH43** 85 J5
Harris Green **NR15** 45 G6
Harriseahead **ST7** 49 H7
Harriston **CA7** 60 C2
HARROGATE HG 57 J4
Harrold **MK43** 32 C3
Harrop Fold **BB7** 56 C5
HARROW *Gt.Lon.* HA 22 E3
Harrow *High.* **KW14** 105 H1
Harrow Green **IP29** 34 C3
Harrow on the Hill **HA1** 22 E3
Harrow Weald **HA3** 22 E2
Harrowbarrow **PL17** 4 E3
Harrowden **MK42** 32 D4
Harrowgate Hill **DL3** 62 D5
Harry Stoke **BS34** 19 K4
Harston *Cambs.* **CB22** 33 H3
Harston *Leics.* **NG32** 42 B2
Harswell **YO42** 58 E5
Hart **TS27** 63 F3
Hartburn **NE61** 71 F5
Hartest **IP29** 34 C3
Hartfield *E.Suss.* TN7 13 H3
Hartfield *High.* **IV54** 94 D7
Hartford *Cambs.* **PE29** 33 F1
Hartford *Ches.W. & C.* **CW8** 49 F5
Hartford *Som.* **TA22** 7 H3
Hartford End **CM3** 33 K7
Hartfordbridge **RG27** 22 A6
Hartforth **DL10** 62 C6
Hartgrove **SP7** 9 H3
Harthill *Ches.W. & C.* **CH3** 48 E7
Harthill *N.Lan.* **ML7** 75 H4
Harthill *S.Yorks.* **S26** 51 G4
Hartington **SK17** 50 D6
Hartington Hall **NE61** 71 F5
Hartland **EX39** 6 A3
Hartland Quay **EX39** 6 A3
Hartlebury **DY11** 29 H1
Hartlepool TS24 63 G3
Hartley *Cumb.* **CA17** 61 J6
Hartley *Kent* **DA3** 24 C5
Hartley *Kent* **TN17** 14 C4
Hartley *Northumb.* **NE26** 71 J6
Hartley Green **ST18** 40 B3
Hartley Mauditt **GU34** 11 J1
Hartley Wespall **RG27** 21 K6
Hartley Wintney **RG27** 22 A6
Hartlington **BD23** 57 F3
Hartlip **ME9** 24 E5
Hartoft End **YO18** 63 J7
Harton *N.Yorks.* **YO60** 58 D3
Harton *Shrop.* **SY6** 38 D7
Harton *T. & W.* **NE34** 71 J7
Hartpury **GL19** 29 H6
Hartrigge **TD8** 70 B1
Hartshead **WF15** 57 G7
Hartshill **CV10** 41 F6
Hartshorne **DE11** 41 F3
Hartsop **CA10** 60 F5
Hartwell *Bucks.* **HP17** 31 J7
Hartwell *E.Suss.* **TN7** 13 H3
Hartwell *Northants.* **NN7** 31 J3
Hartwith **HG3** 57 H3
Hartwood **ML7** 75 G5
Harvel **DA13** 24 C5
Harvington *Worcs.* **WR11** 30 B4
Harvington *Worcs.* **DY10** 29 H1
Harwell *Notts.* **DN10** 51 J3
Harwell *Oxon.* **OX11** 21 H3
Harwich CO12 35 G5
Harwood *Dur.* **DL12** 61 K3
Harwood *Gt.Man.* **BL2** 49 G1
Harwood *Northumb.* **NE61** 71 F4
Harwood Dale **YO13** 63 J3
Harwood on Teviot **TD9** 69 K3
Harworth **DN11** 51 J3
Hasbury **B63** 40 B7
Hascombe **GU8** 22 D7
Haselbech **NN6** 31 J1
Haselbury Plucknett **TA18** 8 D3
Haseley **CV35** 30 D2
Haseley Knob **CV35** 30 D1
Haselor **B49** 30 C3
Hasfield **GL19** 29 H6
Hasguard **SA62** 16 B5
Haskayne **L39** 48 C2
Hasketon **IP13** 35 G3
Hasland **S41** 51 F6
Hasland Green **S41** 51 F6
Haslemere GU27 12 C3
Haslingden **BB4** 56 C7
Haslingden Grane **BB4** 56 C7
Haslingfield **CB23** 33 H3
Haslington **CW1** 49 G7
Hassall **CW11** 49 G7
Hassall Green **CW11** 49 G7
Hassell Street **TN25** 15 F3
Hassendean **TD9** 70 A1
Hassingham **NR13** 45 H5
Hassocks BN6 13 G5
Hassop **DE45** 50 E5
Haster **KW1** 105 J3
Hasthorpe **LN13** 53 H6
Hastigrow **KW1** 105 H2
Hastingleigh **TN25** 15 F3
Hastings *E.Suss.* TN34 14 D7
Hastings *Som.* **TA19** 8 C3
Hastingwood **CM17** 23 H1
Hastoe **HP23** 22 C1
Haswell **DH6** 62 E2
Haswell Plough **DH6** 62 E2
Hatch *Cen.Beds.* **SG19** 32 E4
Hatch *Hants.* **RG24** 21 K6
Hatch Beauchamp **TA3** 8 C2
Hatch End **HA5** 22 E2
Hatch Green **TA3** 8 C3
Hatching Green **AL5** 32 E7
Hatchmere **WA6** 48 E5
Hatcliffe **DN37** 53 F2
Hatfield *Here.* **HR6** 28 E3
Hatfield *Herts.* AL10 23 F1
Hatfield *S.Yorks.* **DN7** 51 J2
Hatfield Broad Oak **CM22** 33 J7
Hatfield Heath **CM22** 33 J7
Hatfield Peverel **CM3** 34 B7
Hatfield Woodhouse **DN7** 51 J2
Hatford **SN7** 21 G2
Hatherden **SP11** 21 G6
Hatherleigh **EX20** 6 D5
Hathern **LE12** 41 G3
Hatherop **GL7** 20 E1
Hathersage **S32** 50 E4
Hathersage Booths **S32** 50 E4
Hathershaw **OL8** 49 J2
Hatherton *Ches.E.* **CW5** 39 F1
Hatherton *Staffs.* **WS11** 40 B4
Hatley St. George **SG19** 33 F3
Hatt **PL12** 4 D4
Hattingley **GU34** 11 H1
Hatton *Aber.* **AB42** 91 J1
Hatton *Derbys.* **DE65** 40 E3
Hatton *Gt.Lon.* **TW14** 22 E4
Hatton *Lincs.* **LN8** 52 E5
Hatton *Shrop.* **SY6** 38 D6
Hatton *Warks.* **CV35** 30 D2
Hatton *Warr.* **WA4** 48 E4
Hatton Castle **AB53** 99 F6
Hatton Heath **CH3** 48 D6
Hatton of Fintray **AB21** 91 G3
Hattoncrook **AB21** 91 G2
Haugh **LN13** 53 H5
Haugh Head **NE71** 71 F1
Haugh of Glass **AB54** 90 C1
Haugh of Urr **DG7** 65 J4
Haugham **LN11** 53 G4
Haughhead **G66** 74 E3
Haughley **IP14** 34 E2
Haughley Green **IP14** 34 E2
Haughley New Street **IP14** 34 E2
Haughs **AB54** 98 D6
Haughton *Ches.E.* **CW6** 48 E7
Haughton *Notts.* **DN22** 51 J5
Haughton *Powys* **SY22** 38 C4
Haughton *Shrop.* **WV16** 39 F6
Haughton *Shrop.* **SY4** 38 E4
Haughton *Shrop.* **SY11** 38 C3
Haughton *Staffs.* **ST18** 40 A3
Haughton Green **M34** 49 J3
Haughton Le Skerne **DL1** 62 E5
Haultwick **SG11** 33 G6
Haunn **HS8** 84 C3
Haunton **B79** 40 E4
Hauxton **CB22** 33 H3
Havannah **CW12** 49 H6
Havant PO9 11 J4
Haven **HR4** 28 D3
Havenstreet **PO33** 11 G5
Havercroft **WF4** 51 F1
Haverfordwest (Hwlffordd) SA61 16 C4
Haverhill CB9 33 K4
Haverigg **LA18** 54 E2
Havering Park **RM5** 23 H2
Havering-atte-Bower **RM4** 23 J2
Haversham **MK19** 32 B4
Haverthwaite **LA12** 55 G1
Haverton Hill **TS23** 63 F4
Haviker Street **TN12** 14 C3
Havyat **BA6** 8 E1
Hawarden (Penarlâg) **CH5** 48 C6
Hawbridge **WR8** 29 J4
Hawbush Green **CM77** 34 B6
Hawcoat **LA14** 55 F2
Hawes DL8 56 D1
Hawe's Green **NR15** 45 G6
Hawick TD9 70 A2
Hawkchurch **EX13** 8 C4
Hawkedon **IP29** 34 B3
Hawkenbury *Kent* **TN2** 13 J3
Hawkenbury *Kent* **TN12** 14 D3
Hawkeridge **BA13** 20 B6
Hawkerland **EX10** 7 J7
Hawkes End **CV5** 40 E7
Hawkesbury **GL9** 20 A3
Hawkesbury Upton **GL9** 20 A3
Hawkhill **NE66** 71 H2
Hawkhurst **TN18** 14 C4
Hawkinge **CT18** 15 H3
Hawkley **GU33** 11 J2
Hawkridge **TA22** 7 G2
Hawkshead **LA22** 60 E7
Hawkshead Hill **LA22** 60 E7
Hawksheads **LA5** 55 H3
Hawksland **ML11** 75 G6
Hawkswick **BD23** 56 E2
Hawksworth *Notts.* **NG13** 42 A1
Hawksworth *W.Yorks.* **LS20** 57 G5
Hawksworth *W.Yorks.* **LS18** 57 H6
Hawkwell *Essex* **SS5** 24 E2
Hawkwell *Northumb.* **NE18** 71 F6
Hawley *Hants.* **GU17** 22 B6
Hawley *Kent* **DA2** 23 J4
Hawley's Corner **TN16** 23 H6
Hawling **GL54** 30 B6
Hawnby **YO62** 58 B1
Haworth **BD22** 57 F6
Hawstead **IP29** 34 C3
Hawstead Green **IP29** 34 C3
Hawthorn *Dur.* **SR7** 63 F2
Hawthorn *Hants.* **GU34** 11 H1
Hawthorn *R.C.T.* **CF37** 18 D3
Hawthorn *Wilts.* **SN13** 20 B5
Hawthorn Hill *Brack.F.* **RG42** 22 B4
Hawthorn Hill *Lincs.* **LN4** 53 F7
Hawthorpe **PE10** 42 D3
Hawton **NG24** 51 K7
Haxby **YO32** 58 C4
Haxey **DN9** 51 K2
Haxted **TN8** 23 H7
Haxton **SP4** 20 E7
Hay Green **PE34** 43 J4
Hay Mills **B25** 40 D7
Hay Street **SG11** 33 G6
Haydock **WA11** 48 E3
Haydon *Dorset* **DT9** 9 F3
Haydon *Swin.* **SN25** 20 E3
Haydon Bridge **NE47** 70 D7
Haydon Wick **SN25** 20 E3
Hayes *Gt.Lon.* UB3 22 D3
Hayes *Gt.Lon.* **BR2** 23 H5
Hayes End **UB4** 22 D3
Hayfield *Arg. & B.* **PA35** 80 B5
Hayfield *Derbys.* **SK22** 50 C4
Hayfield *Fife* **KY2** 76 A1
Hayfield *High.* **KW14** 105 G2
Haygrove **TA6** 8 B1
Hayhillock **DD11** 83 G3
Hayle TR27 2 C5
Hayling Island PO11 11 J4
Haymoor Green **CW5** 49 F7
Hayne **EX16** 7 H4
Haynes **MK45** 32 E4
Haynes Church End **MK45** 32 D4
Haynes West End **MK45** 32 D4
Hay-on-Wye (Y Gelli Gandryll) **HR3** 28 B4
Hayscastle **SA62** 16 B3
Hayscastle Cross **SA62** 16 C3
Hayton *Cumb.* **CA8** 61 G1
Hayton *Cumb.* **CA7** 60 C2
Hayton *E.Riding* **YO42** 58 E5
Hayton *Notts.* **DN22** 51 K4
Hayton's Bent **SY8** 38 E7
Haytor Vale **TQ13** 5 H3
Haytown **EX22** 6 B4
Haywards Heath RH16 13 G4
Haywood Oaks **NG21** 51 J7
Hazel End **CM23** 33 H6
Hazel Grove **SK7** 49 J4
Hazel Street **TN12** 13 K3
Hazelbank *Arg. & B.* **PA25** 80 B7
Hazelbank *S.Lan.* **ML11** 75 G6
Hazelbury Bryan **DT10** 9 G4
Hazeleigh **CM3** 24 E1
Hazeley **RG27** 22 A6
Hazelhurst **BL8** 49 G1
Hazelside **ML11** 68 D1
Hazelslack **LA7** 55 H2
Hazelslade **WS12** 40 C4
Hazelton Walls **KY15** 82 E5
Hazelwood *Derbys.* **DE56** 41 F1
Hazelwood *Gt.Lon.* **TN14** 23 H5
Hazlefield **DG7** 65 H6
Hazlehead *Aberdeen* **AB15** 91 G4
Hazlehead *S.Yorks.* **S36** 50 D2
Hazlemere **HP15** 22 B2
Hazlerigg **NE13** 71 H6
Hazleton **GL54** 30 B7
Hazon **NE65** 71 G3
Heacham **PE31** 44 A2
Head Bridge **EX37** 6 E4
Headbourne Worthy **SO23** 11 F1
Headcorn **TN27** 14 D3
Headingley **LS6** 57 H6
Headington **OX3** 21 J1
Headlam **DL2** 62 C5
Headless Cross **B97** 30 B2
Headley *Hants.* **GU35** 12 B3
Headley *Hants.* **RG19** 21 J5
Headley *Surr.* **KT18** 23 F6
Headley Down **GU35** 12 B3
Headley Heath **B38** 30 B1
Headon **DN22** 51 K5
Heads Nook **CA8** 61 F1

Heady Hill OL10 49 H1
Heage DE56 51 F7
Healaugh N.Yorks. DL11 62 B7
Healaugh N.Yorks. LS24 58 B5
Heald Green SK8 49 H4
Heale Devon EX31 6 E1
Heale Som. TA10 8 C2
Healey Lancs. OL12 49 H1
Healey N.Yorks. HG4 57 G1
Healey Northumb. NE44 62 B1
Healey W.Yorks. WF17 57 H7
Healeyfield DH8 62 B2
Healing DN41 53 F1
Heamoor TR18 2 B5
Heaning LA23 60 F7
Heanish PA77 78 B3
Heanor DE75 41 G1
Heanton Punchardon EX31 6 D2
Heanton Satchville EX20 6 D4
Heap Bridge BL9 49 H1
Heapey PR6 56 B7
Heapham DN21 52 B4
Hearn GU35 12 B3
Hearthstane ML12 69 G1
Heasley Mill EX36 7 F2
Heast IV49 86 C3
Heath Cardiff CF14 18 E3
Heath Derbys. S44 51 G6
Heath W.Yorks. WF1 51 F1
Heath & Reach LU7 32 C6
Heath End Derbys. LE65 41 F3
Heath End Hants. RG26 21 J5
Heath End Hants. RG20 21 H5
Heath End Surr. GU9 22 B7
Heath Hayes WS12 40 C4
Heath Hill TF11 39 G4
Heath House BS28 19 H7
Heath Town WV10 40 B6
Heathbrook TF9 39 F3
Heathcot AB12 91 G4
Heathcote Derbys. SK17 50 D6
Heathcote Shrop. TF9 39 F3
Heathencote NN12 31 J4
Heather LE67 41 F4
Heathfield Devon TQ12 5 J3
Heathfield E.Suss. TN21 13 J4
Heathfield N.Yorks. HG3 57 G3
Heathfield Som. TA4 7 K3
Heathrow Airport TW6 22 D4
Heathton WV5 40 A6
Heatley WA13 49 G4
Heaton Lancs. LA3 55 H3
Heaton Staffs. SK11 49 J6
Heaton T. & W. NE6 71 H7
Heaton W.Yorks. BD9 57 G6
Heaton Moor SK4 49 H3
Heaton's Bridge L40 48 D1
Heaverham TN15 23 J6
Heaviley SK2 49 J4
Heavitree EX1 7 H6
Hebburn NE31 71 J7
Hebden BD23 57 F3
Hebden Bridge HX7 56 E7
Hebden Green CW7 49 F6
Hebing End SG2 33 G6
Hebron Carmar. SA34 16 E3
Hebron Northumb. NE61 71 G5
Heck DG11 69 F5
Heckfield RG27 22 A5
Heckfield Green IP21 35 F1
Heckfordbridge CO3 34 D6
Heckingham NR14 45 H6
Heckington NG34 42 E1
Heckmondwike WF16 57 H7
Heddington SN11 20 C5
Heddle KW17 106 C6
Heddon-on-the-Wall NE15 71 G7
Hedenham NR35 45 H6
Hedge End SO30 11 F3
Hedgerley SL2 22 C3
Hedging TA7 8 C2
Hedley on the Hill NE43 62 B1
Hednesford WS12 40 C4
Hedon HU12 59 H7
Hedsor HP10 22 C3
Heeley S8 51 F4
Heglibister ZE2 107 M7
Heighington Darl. DL5 62 D4
Heighington Lincs. LN4 52 D6
Heightington DY12 29 G1
Heights of Brae IV14 96 C5
Heilam IV27 103 G3
Heisker Islands
(Monach Islands) HS6 92 B5
Heithat DG11 69 G5
Heiton TD5 77 F7
Hele Devon EX34 6 D1
Hele Devon EX5 7 H5
Hele Devon PL15 6 B6
Hele Devon TQ13 5 H3
Hele Som. TA4 7 K3
Hele Torbay TQ1 5 K4
Hele Bridge EX20 6 D5
Hele Lane EX17 7 F4
Helebridge EX23 6 A5
Helensburgh G84 74 A2
Helford TR12 2 E6
Helhoughton NR21 44 C3
Helions Bumpstead CB9 33 K4
Hell Corner RG17 21 G5
Hellaby S66 51 H3
Helland Cornw. PL30 4 A3
Helland Som. TA3 8 C2
Hellandbridge PL30 4 A3
Hellesdon NR6 45 G4
Hellidon NN11 31 G3
Hellifield BD23 56 D4
Hellingly BN27 13 J5
Hellington NR14 45 H5
Hellister ZE2 107 M8
Helmdon NN13 31 G4
Helmingham IP14 35 F3
Helmington Row DL15 62 C3
Helmsdale KW8 105 F7
Helmshore BB4 56 C7
Helmsley YO62 58 C1
Helperby YO61 57 K3
Helperthorpe YO17 59 F2
Helpringham NG34 42 E1
Helpston PE6 42 E5
Helsby WA6 48 D5
Helsey PE24 53 J5
Helston TR13 2 D6
Helstone PL32 4 A2
Helton CA10 61 G4
Helwith DL11 62 B6
Helwith Bridge BD24 56 D3
Hem SY21 38 B5
Hemborough Post TQ9 5 J5
HEMEL HEMPSTEAD HP 22 D1
Hemerdon PL7 5 F5
Hemingbrough YO8 58 C6
Hemingby LN9 53 F5
Hemingfield S73 51 F2
Hemingford Abbots PE28 33 F1
Hemingford Grey PE28 33 F1
Hemingstone IP6 35 F3
Hemington Leics. DE74 41 G3
Hemington Northants.
PE8 42 D7
Hemington Som. BA3 20 A6
Hemley IP12 35 G4
Hemlington TS8 63 F5
Hemp Green IP17 35 H2
Hempholme YO25 59 G4
Hempnall NR15 45 G6
Hempnall Green NR15 45 G6
Hempriggs IV36 97 J5
Hempriggs House KW1 105 J4
Hempstead Essex CB10 33 K5
Hempstead Med. ME7 24 D5
Hempstead Norf. NR12 45 J3
Hempstead Norf. NR25 45 F2
Hempsted GL2 29 H7
Hempton Norf. NR21 44 D3
Hempton Oxon. OX15 31 F5
Hemsby NR29 45 J4
Hemswell DN21 52 C3
Hemswell Cliff DN21 52 C4
Hemsworth WF9 51 G1
Hemyock EX15 7 K4
Henbury Bristol BS10 19 J4
Henbury Ches.E. SK10 49 H5
Henderland DG2 65 J3
Hendersyde Park TD5 77 F7
Hendham TQ7 5 H5
Hendon Gt.Lon. NW4 23 F3
Hendon T. & W. SR2 62 E1
Hendraburnick PL32 4 B2
Hendre Bridgend CF35 18 C3
Hendre Gwyn. LL53 36 C2
Hendreforgan CF39 18 C3
Hendy SA4 17 J5
**Hendy-Gwyn (Whitland)
SA34 17 F4**
Heneglwys LL77 46 C5
Henfield S.Glos. BS36 19 K4
Henfield W.Suss. BN5 13 F5
Henford EX21 6 B6
Hengherst TN26 14 E4
Hengoed Caerp. CF82 18 E2
Hengoed Powys HR5 28 B3
Hengoed Shrop. SY10 38 B2
Hengrave IP28 34 C2
Henham CM22 33 J6
Heniarth SY21 38 A5
Henlade TA3 8 B2
Henley Dorset DT2 9 F4
Henley Shrop. SY8 28 E1
Henley Som. TA10 8 D1
Henley Som. TA18 8 D4
Henley Suff. IP6 35 F3
Henley W.Suss. GU27 12 B4
Henley Corner TA10 8 D1
Henley Park GU3 22 C6
Henley-in-Arden B95 30 C2
Henley-on-Thames RG9 22 A3
Henley's Down TN33 14 C6
Henllan Carmar. SA44 17 G1
Henllan Denb. LL16 47 J6
Henllan Amgoed SA34 16 E3
Henllys NP44 19 F2
Henlow SG16 32 E5
Hennock TQ13 7 G7
Henny Street CO10 34 C5
Henryd LL32 47 F5
Henry's Moat SA63 16 D3
Hensall DN14 58 B7
Henshaw NE47 70 C7
Hensingham CA28 60 A5
Henstead NR34 45 J7
Hensting SO21 11 F2
Henstridge BA8 9 G3
Henstridge Ash BA8 9 G3
Henstridge Bowden BA8 9 F2
Henstridge Marsh BA8 9 G2
Henton Oxon. OX39 22 A1
Henton Som. BA5 19 H7
Henwood PL14 4 C3
Heogan ZE2 107 N8
Heol Senni LD3 27 J6
Heolgerrig CF48 18 D1
Heol-y-Cyw CF35 18 C3
Hepburn NE66 71 F1
Hepburn Bell NE66 71 F1
Hepple NE65 70 E3
Hepscott NE61 71 H5
Hepthorne Lane S42 51 G6
Heptonstall HX7 56 E7
Hepworth Suff. IP22 34 D1
Hepworth W.Yorks. HD9 50 D2
Hepworth South Common
IP22 34 D1
Herbrandston SA73 16 B5
HEREFORD HR 28 E4
Heriot EH38 76 B5
Herm GY1 3 J5
Hermiston EH14 75 K3
Hermitage D. & G. DG7 65 H4
Hermitage Dorset DT2 9 F4
Hermitage Sc.Bord. TD9 70 A4
Hermitage W.Berks. RG18 21 J4
Hermitage W.Suss. PO10 11 J4
Hermitage Green WA2 49 F3
Hermon Carmar. SA33 17 G2
Hermon I.o.A. LL62 46 B6
Hermon Pembs. SA36 17 F2
Herne CT6 25 H5
Herne Bay CT6 25 H5
Herne Common CT6 25 H5
Herne Pound ME18 23 K6
Herner EX32 6 D3
Hernhill ME13 25 G5
Herodsfoot PL14 4 C4
Herongate CM13 24 C2
Heron's Ghyll TN22 13 H4
Heronsgate WD3 22 D2
Herriard RG25 21 K7
Herringfleet NR32 45 J6
Herring's Green MK45 32 D4
Herringswell IP28 34 B2
Herringthorpe S65 51 G3
Hersden CT3 25 H5
Hersham Cornw. EX23 6 A5
Hersham Surr. KT12 22 E5
Herstmonceux BN27 13 K5
Herston KW17 106 D8
Hertford SG14 33 G7
Hertford Heath SG13 33 G7
Hertingfordbury SG14 33 G7
Hesket Newmarket CA7 60 E3
Hesketh Bank PR4 55 H7
Hesketh Lane PR3 56 B5
Heskin Green PR7 48 E1
Hesleden TS27 63 F3
Hesleyside NE48 70 D5
Heslington YO10 58 C4
Hessay YO26 58 B4
Hessenford PL11 4 D5
Hessett IP30 34 D2
Hessle HU13 59 G7
Hest Bank LA2 55 H3
Hester's Way GL51 29 J6
Hestley Green IP23 35 F2
Heston TW5 22 E4
Heswall CH60 48 B4
Hethe OX27 31 G6
Hethelpit Cross GL19 29 G6
Hetherington NE48 70 D6
Hethersett NR9 45 F5
Hethersgill CA6 69 K7
Hethpool NE71 70 D1
Hett DH6 62 D3
Hetton BD23 56 E4
Hetton-le-Hole DH5 62 E2
Heugh NE18 71 F6
Heugh-head Aber. AB36 90 B3
Heugh-head Aber. AB34 90 D5
Heveningham IP19 35 H1
Hever TN8 23 H7
Heversham LA7 55 H1
Hevingham NR10 45 F3
Hewas Water PL26 3 G4
Hewell Grange B97 30 B2
Hewell Lane B60 30 B2
Hewelsfield GL15 19 J1
Hewelsfield Common GL15 19 J1
Hewish N.Som. BS24 19 H5
Hewish Som. TA18 8 D4
Hewood TA20 8 C4
Heworth YO31 58 C4
Hewton EX20 6 D6
Hexham NE46 70 E7
Hextable BR8 23 J4
Hexthorpe DN4 51 H2
Hexton SG5 32 E5
Hexworthy PL20 5 G3
Hey BB8 56 D5
Hey Houses FY8 55 G7
Heybridge Essex CM4 24 C2
Heybridge Essex CM9 24 E1
Heybridge Basin CM9 24 E1
Heybrook Bay PL9 4 E6
Heydon Cambs. SG8 33 H4
Heydon Norf. NR11 45 F3
Heydour NG32 42 D2
Heylipoll PA77 78 A3
Heylor ZE2 107 L4
Heyop LD7 28 B1
Heysham LA3 55 H3
Heyshaw HG3 57 G3
Heyshott GU29 12 B5
Heyside OL2 49 J2
Heytesbury BA12 20 C7
Heythrop OX7 30 E6
Heywood Gt.Man. OL10 49 H1
Heywood Wilts. BA13 20 B6
Hibaldstow DN20 52 C2
Hibb's Green IP29 34 C3
Hickleton DN5 51 G2
Hickling Norf. NR12 45 J3
Hickling Notts. LE14 41 J3
Hickling Green NR12 45 J3
Hickling Heath NR12 45 J3
Hickstead RH17 13 F4
Hidcote Bartrim GL55 30 C4
Hidcote Boyce GL55 30 C4
High Ackworth WF7 51 G1
High Angerton NE61 71 F5
High Balantyre PA32 80 B6
High Bankhill CA10 61 G2
High Beach IG10 23 H2
High Bentham (Higher
Bentham) LA2 56 B3
High Bickington EX37 6 D3
High Birkwith BD24 56 C2
High Blantyre G72 74 E5
High Bonnybridge FK4 75 G3
High Borgue DG6 65 G5
High Borve HS2 101 G2
High Bradfield S6 50 E3
High Bradley BD20 57 F5
High Bransholme HU7 59 H6
High Bray EX32 6 E2
High Bridge CA5 60 E2
High Brooms TN4 23 J7
High Bullen EX38 6 D3
High Burton HG4 57 H1
High Buston NE66 71 H3
High Callerton NE20 71 G6
High Casterton LA6 56 B2
High Catton YO41 58 D4
High Close DL11 62 C5
High Cogges OX29 21 G1
High Common IP21 45 F7
High Coniscliffe DL2 62 D5
High Crompton OL2 49 J2
High Cross Hants. GU32 11 J2
High Cross Herts. SG11 33 G7
High Cross W.Suss. BN6 13 F5
High Easter CM1 33 K7
High Ellington HG4 57 G1
High Entercommon DL6 62 E6
High Ercall TF6 38 E4
High Etherley DL14 62 C4
High Ferry PE22 43 G1
High Flatts HD8 50 E2
High Garrett CM7 34 B6
High Gate HX7 56 E7
High Grange DL15 62 C3
High Green Norf. NR9 45 F5
High Green Norf. NR19 44 D4
High Green Norf. IP25 44 D5
High Green S.Yorks. S35 51 F3
High Green Suff. IP29 34 C2
High Green Worcs. WR8 29 H4
High Halden TN26 14 D4
High Halstow ME3 24 D4
High Ham TA10 8 D1
High Harrington CA14 60 B4
High Harrogate HG2 57 J4
High Hatton SY4 39 F3
High Hauxley NE65 71 H3
High Hawsker YO22 63 J2
High Heath Shrop. TF9 39 F3
High Heath W.Mid. WS4 40 C5
High Hesket CA4 61 F2
High Hesleden TS27 63 F3
High Hoyland S75 50 E1
High Hunsley YO43 59 F6
High Hurstwood TN22 13 H4
High Hutton YO60 58 D3
High Ireby CA7 60 D3
High Kelling NR25 44 E2
High Kilburn YO61 58 B2
High Kingthorpe YO18 58 E1
High Knipe CA10 61 G5
High Lane Derbys. DE7 41 G1
High Lane Gt.Man. SK6 49 J4
High Lane Worcs. WR6 29 F2
High Laver CM5 23 J1
High Legh WA16 49 G4
High Leven TS15 63 F5
High Littleton BS39 19 K6
High Lorton CA13 60 C4
High Marishes YO17 58 E2
High Marnham NG23 52 B5
High Melton DN5 51 H2
High Moor S21 51 G4
High Newton LA11 55 H1
High Newton-by-the-Sea
NE66 71 H1
High Nibthwaite LA12 60 D7
High Offley ST20 39 G3
High Ongar CM5 23 J1
High Onn ST20 40 A4
High Park Corner CO5 34 E6
High Roding CM6 33 K7
High Shaw DL8 61 K7
High Spen NE39 71 G7
High Stoop DL13 62 C2
High Street Cornw. PL26 3 G3
High Street Kent TN18 14 C4
High Street Suff. IP12 35 J3
High Street Suff. NR35 45 H7
High Street Suff. IP17 35 J1
High Street Suff. CO10 34 C4
High Street Green IP14 34 E3
High Throston TS26 63 F3
High Town WS11 40 B4
High Toynton LN9 53 F6
High Trewhitt NE65 71 F3
High Wham DL13 62 C4
High Wigsell TN32 14 C5
High Woolaston GL15 19 J2
High Worsall TS15 62 E6
High Wray LA22 60 E7
High Wych CM21 33 H7
High Wycombe HP13 22 B2
Higham Derbys. DE55 51 F7
Higham Kent ME3 24 D4
Higham Lancs. BB12 56 D6
Higham S.Yorks. S75 51 F2
Higham Suff. IP28 34 B2
Higham Suff. CO7 34 E5
Higham Dykes NE20 71 G6
Higham Ferrers NN10 32 C2
Higham Gobion SG5 32 E5
Higham on the Hill CV13 41 F6
Higham Wood TN10 23 K7
Highampton EX21 6 C5
Highams Park E4 23 G2
Highbridge Hants. SO50 11 F2
Highbridge Som. TA9 19 G7
Highbrook RH17 13 G3
Highburton HD8 50 D1
Highbury BA3 19 K7
Highclere RG20 21 H5
Highcliffe BH23 10 D5
Higher Alham BA4 19 K7
Higher Ansty DT2 9 G4
Higher Ashton EX6 7 G7
Higher Ballam FY8 55 G6
Higher Bentham
(High Bentham) LA2 56 B3
Higher Blackley M9 49 H2
Higher Brixham TQ5 5 K5
Higher Cheriton EX14 7 K5
Higher Combe TA22 7 H2
Higher Folds WN7 49 F2
Higher Gabwell TQ1 5 K4
Higher Green M29 49 G3
Higher Halstock Leigh BA22 8 E4
Higher Kingcombe DT2 8 E5
Higher Kinnerton CH4 48 C6
Higher Muddiford EX31 6 D2
Higher Nyland SP8 9 G2
Higher Prestacott EX21 6 B6
Higher Standen BB7 56 C5
Higher Tale EX14 7 J5
Higher Thrushgill LA2 56 B3
Higher Town Cornw. PL26 4 A4
Higher Town I.o.S. TR25 2 C1
Higher Walreddon PL19 4 E3
Higher Walton Lancs. PR5 55 J7
Higher Walton Warr. WA4 48 E4
Higher Wambrook TA20 8 B4
Higher Whatcombe DT11 9 H4
Higher Wheelton PR6 56 B7
Higher Whiteleigh EX22 4 C1
Higher Whitley WA4 49 F4
Higher Wincham CW9 49 F5
Higher Woodhill BL8 49 G1
Higher Woodsford DT2 9 G6
Higher Wraxall DT2 8 E4
Higher Wych SY14 38 D1
Highfield E.Riding YO8 58 D6
Highfield N.Ayr. KA24 74 B5
Highfield Oxon. OX26 31 G6
Highfield S.Yorks. S2 51 F4
Highfield T. & W. NE39 62 C1
Highfields Cambs. CB23 33 G3
Highfields Northumb.
TD15 77 H5
Highgate E.Suss. RH18 13 H3
Highgate Gt.Lon. N6 23 F3
Highgreen Manor NE48 70 D4
Highlane Ches.E. SK11 49 H6
Highlane Derbys. S12 51 G4
Highlaws CA7 60 C2
Highleadon GL18 29 G6
Highleigh Devon TA22 7 H3
Highleigh W.Suss. PO20 12 B7
Highley WV16 39 G7
Highmead SA40 17 J1
Highmoor Cross RG9 21 K3
Highmoor Hill NP26 19 H3
Highnam GL2 29 G7
Highstead CT3 25 J5
Highsted ME9 25 F5
Highstreet ME13 25 G5
Highstreet Green Essex
CO9 34 B5
Highstreet Green Surr.
GU8 12 C3
Hightae DG11 69 F6
Highter's Heath B14 30 B1
Hightown Hants. BH24 10 C4
Hightown Mersey. L38 48 B2
Hightown Green IP30 34 D3
Highway SN11 20 D4
Highweek TQ12 5 J3
Highwood WR15 29 F2
Highwood Hill NW7 23 F2
Highworth SN6 21 F2
Hilborough IP26 44 C5
Hilcote DE55 51 G7
Hilcott SN9 20 E6
Hilden Park TN11 23 J7
Hildenborough TN11 23 J7
Hildenley YO17 58 D2
Hildersham CB21 33 J4
Hilderstone ST15 40 B2
Hilderthorpe YO15 59 H3
Hilfield DT2 9 F4
Hilgay PE38 44 A6
Hill S.Glos. GL13 19 K2
Hill Warks. CV23 31 F2
Hill Worcs. WR10 29 J4
Hill Brow GU33 11 J2
Hill Chorlton ST5 39 G2
Hill Common NR12 45 J3
Hill Cottages YO18 63 J7
Hill Croome WR8 29 H4
Hill Deverill BA12 20 B7
Hill Dyke PE22 43 G1
Hill End Dur. DL13 62 B3
Hill End Fife KY12 75 J1
Hill End Glos. GL20 29 H5
Hill End Gt.Lon. UB9 22 D2
Hill End N.Yorks. BD23 57 F4
Hill Green CB11 33 H5
Hill Head PO14 11 G4
Hill Houses DY14 29 F1
Hill Mountain SA62 16 C5
Hill of Beath KY4 75 K2
Hill of Fearn IV20 97 F4
Hill Ridware WS15 40 C4
Hill Row CB6 33 H1
Hill Side HD5 50 D1

Hil - Hou

Hill Street SO40 10 E3
Hill Top Hants. SO42 11 F4
Hill Top S.Yorks. DN12 51 G3
Hill Top S.Yorks. S6 50 E4
Hill View BH21 9 J5
Hill Wootton CV35 30 E2
Hillam LS25 58 B7
Hillbeck CA17 61 J5
Hillberry IM4 54 C6
Hillborough CT6 25 J5
Hillbrae Aber. AB54 98 E6
Hillbrae Aber. AB51 91 F2
Hillbrae Aber. AB51 91 G1
Hillbutts BH21 9 J4
Hillcliffane DE56 40 E1
Hillend Aber. AB55 98 C6
Hillend Fife KY11 75 K2
Hillend Midloth. EH10 76 A4
Hillend N.Lan. ML6 75 G4
Hillend Swan. SA3 17 H6
Hillend Green GL18 29 G6
Hillersland GL16 28 E7
Hillesden MK18 31 H6
Hillesley GL12 20 A3
Hillfarrance TA4 7 K3
Hillfoot End SG5 32 E5
Hillhead Devon TQ5 5 K5
Hillhead S.Ayr. KA6 67 J1
Hillhead of Auchentumb AB43 99 H5
Hillhead of Cocklaw AB42 99 J6
Hilliard's Cross WS13 40 D4
Hilliclay KW14 105 G2
Hillingdon UB10 22 D3
Hillington Glas. G52 74 D4
Hillington Norf. PE31 44 B3
Hillmorton CV21 31 G1
Hillockhead Aber. AB36 90 B4
Hillockhead Aber. AB33 90 C3
Hillowton DG7 65 H4
Hillpound SO32 11 G3
Hill's End MK17 32 C5
Hills Town S44 51 G6
Hillsborough S6 51 F3
Hillsford Bridge EX35 7 F1
Hillside Aber. AB12 91 H5
Hillside Angus DD10 83 J1
Hillside Moray IV30 97 J5
Hillside Shet. ZE2 107 N6
Hillside Worcs. WR6 29 G2
Hillswick ZE2 107 L5
Hillway PO35 11 H6
Hillwell ZE2 107 M11
Hillyfields SO16 10 E3
Hilmarton SN11 20 D4
Hilperton BA14 20 B6
Hilsea PO3 11 H4
Hilston HU11 59 J6
Hilton Cambs. PE28 33 F2
Hilton Cumb. CA16 61 J4
Hilton Derbys. DE65 40 E2
Hilton Dorset DT11 9 G4
Hilton Dur. DL2 62 C4
Hilton High. IV20 97 G3
Hilton Shrop. WV15 39 G6
Hilton Staffs. WS14 40 C5
Hilton Stock. TS15 63 F5
Hilton Croft AB41 91 H1
Hilton of Cadboll IV20 97 F4
Hilton of Delnies IV12 97 F6
Himbleton WR9 29 J3
Himley DY3 40 A6
Hincaster LA7 55 J1
Hinchley Wood KT10 22 E5
Hinckley LE10 41 G6
Hinderclay IP22 34 E1
Hinderton CH64 48 C5
Hinderwell TS13 63 J5
Hindford SY11 38 C2
Hindhead GU26 12 B3
Hindley Gt.Man. WN2 49 F2
Hindley Northumb. NE43 62 B1
Hindley Green WN2 49 F2
Hindlip WR3 29 H3
Hindolveston NR20 44 E3
Hindon Som. TA24 7 H1
Hindon Wilts. SP3 9 J1
Hindringham NR21 44 D2
Hingham NR9 44 E5
Hinksford DY3 40 A7
Hinstock TF9 39 F3
Hintlesham IP8 34 E4
Hinton Glos. GL13 19 K1
Hinton Hants. BH23 10 D5
Hinton Here. HR2 28 C5
Hinton Northants. NN11 31 G3
Hinton S.Glos. SN14 20 A4
Hinton Shrop. SY5 38 D5
Hinton Admiral BH23 10 D5
Hinton Ampner SO24 11 G2
Hinton Blewett BS39 19 J6
Hinton Charterhouse BA2 20 A6
Hinton Martell BH21 10 B4
Hinton on the Green WR11 30 B4
Hinton Parva Dorset BH21 9 J4
Hinton Parva Swin. SN4 21 F3
Hinton St. George TA17 8 D3
Hinton St. Mary DT10 9 G3
Hinton Waldrist SN7 21 G2
Hinton-in-the-Hedges NN13 31 G5
Hints Shrop. SY8 29 F1
Hints Staffs. B78 40 D5
Hinwick NN29 32 C2
Hinxhill TN25 15 F3
Hinxton CB10 33 H4
Hinxworth SG7 33 F4
Hipperholme HX3 57 G7
Hipsburn NE66 71 H2
Hipswell DL9 62 C7
Hirn AB31 91 F4
Hirnant SY10 37 K3
Hirst NE63 71 H5
Hirst Courtney YO8 58 C7
Hirwaen LL15 47 K6
Hirwaun CF44 18 C1
Hiscott EX31 6 D3
Histon CB24 33 H2
Hitcham Bucks. SL1 22 C3
Hitcham Suff. IP7 34 D3
Hitchin SG5 32 E6
Hither Green SE13 23 G4
Hittisleigh EX6 7 F6
Hittisleigh Barton EX6 7 F6
Hive HU15 58 E6
Hixon ST18 40 C3
Hoaden CT3 15 H2
Hoaldalbert NP7 28 C6
Hoar Cross DE13 40 D3
Hoarwithy HR2 28 E6
Hoath CT3 25 J5
Hobarris SY7 28 C1
Hobbister KW17 106 C7
Hobbles Green CB8 34 B3
Hobbs Cross CM16 23 H2
Hobbs Lots Bridge PE15 43 G5
Hobkirk TD9 70 A2
Hobland Hall NR31 45 K5
Hobson NE16 62 C1
Hoby LE14 41 J4
Hockerill CM23 33 H6
Hockering NR20 44 E4
Hockerton NG25 51 K7
Hockley SS5 24 E2
Hockley Heath B94 30 C1
Hockliffe LU7 32 C6
Hockwold cum Wilton IP26 44 B7
Hockworthy TA21 7 J4
Hoddesdon EN11 23 G1
Hoddlesden BB3 56 C7
Hodgehill SK11 49 H6
Hodgeston SA71 16 D6
Hodnet TF9 39 F3
Hodnetheath TF9 39 F3
Hodsoll Street TN15 24 C5
Hodson SN4 20 E3
Hodthorpe S80 51 H5
Hoe NR20 44 D4
Hoe Gate PO7 11 H3
Hoff CA16 61 H5
Hoffleet Stow PE20 43 F2
Hoggard's Green IP29 34 C3
Hoggeston MK18 32 B6
Hoggie AB56 98 D5
Hoggrill's End B46 40 E6
Hogha Gearraidh HS6 92 C4
Hoghton PR5 56 B7
Hognaston DE6 50 E7
Hogsthorpe PE24 53 J5
Holbeach PE12 43 G3
Holbeach Bank PE12 43 G3
Holbeach Clough PE12 43 G3
Holbeach Drove PE12 43 G4
Holbeach Hurn PE12 43 G3
Holbeach St. Johns PE12 43 G4
Holbeach St. Marks PE12 43 G2
Holbeach St. Matthew PE12 43 H2
Holbeck S80 51 H5
Holbeck Woodhouse S80 51 H5
Holberrow Green B96 30 B3
Holbeton PL8 5 G5
Holborough ME2 24 D5
Holbrook Derbys. DE56 41 F1
Holbrook Suff. IP9 35 F5
Holbrooks CV6 41 F7
Holburn TD15 77 J7
Holbury SO45 11 F4
Holcombe Devon EX7 5 K3
Holcombe Gt.Man. BL8 49 G1
Holcombe Som. BA3 19 K7
Holcombe Burnell Barton EX6 7 G6
Holcombe Rogus TA21 7 J4
Holcot NN6 31 J2
Holden BB7 56 C5
Holden Gate OL14 56 D7
Holdenby NN6 31 H2
Holdenhurst BH8 10 C5
Holder's Green CM6 33 K6
Holders Hill NW4 23 F3
Holdgate TF13 38 E7
Holdingham NG34 42 D1
Holditch TA20 8 C4
Hole EX15 7 K4
Hole Park TN17 14 D4
Hole Street BN44 12 E5
Holehouse SK13 50 C3
Hole-in-the-Wall HR9 29 F6
Holford TA5 7 K1
Holgate YO26 58 B4
Holker LA11 55 G2
Holkham NR23 44 C1
Hollacombe Devon EX22 6 B5
Hollacombe Devon EX17 7 G5
Hollacombe Town EX18 6 E4
Holland Ork. KW17 106 D2
Holland Ork. KW17 106 F5
Holland Surr. RH8 23 H6
Holland Fen LN4 43 F1
Holland-on-Sea CO15 35 F7
Hollandstoun KW17 106 G2
Hollee DG11 69 H7
Hollesley IP12 35 H4
Hollicombe TQ2 5 K4
Hollingbourne ME17 14 D2
Hollingbury BN1 13 G6
Hollingrove TN32 13 K4
Hollington Derbys. DE6 40 E2
Hollington E.Suss. TN38 14 C6
Hollington Staffs. ST10 40 C2
Hollingworth SK14 50 C3
Hollins S42 51 F5
Hollins Green WA3 49 F3
Hollins Lane PR3 55 H4
Hollinsclough SK17 50 C6
Hollinwood Gt.Man. OL9 49 J2
Hollinwood Shrop. SY13 38 E2
Hollocombe EX18 6 E4
Hollow Meadows S6 50 E4
Holloway DE4 51 F7
Hollowell NN6 31 H1
Holly Bush LL13 38 D1
Holly End PE14 43 H5
Holly Green HP27 22 A1
Hollybush Caerp. NP12 18 E1
Hollybush E.Ayr. KA6 67 H2
Hollybush Worcs. HR8 29 G5
Hollyhurst SY13 38 E1
Hollym HU19 59 K7
Hollywater GU35 12 B3
Hollywood B47 30 B1
Holm D. & G. DG13 69 H4
Holm (Tolm) W.Isles HS2 101 G4
Holm of Drumlanrig DG3 68 D4
Holmbridge HD9 50 D2
Holmbury St. Mary RH5 22 E7
Holmbush RH12 13 F3
Holme Cambs. PE7 42 E7
Holme Cumb. LA6 55 J2
Holme N.Lincs. DN16 52 C2
Holme N.Yorks. YO7 57 J1
Holme Notts. NG23 52 B7
Holme W.Yorks. HD9 50 D2
Holme Chapel BB10 56 D7
Holme Hale IP25 44 C5
Holme Lacy HR2 28 E5
Holme Marsh HR5 28 C3
Holme next the Sea PE36 44 B1
Holme on the Wolds HU17 59 F5
Holme Pierrepont NG12 41 J2
Holme St. Cuthbert CA15 60 C2
Holme-on-Spalding-Moor YO43 58 E6
Holmer HR1 28 E4
Holmer Green HP15 22 C2
Holmes PR4 48 D1
Holmes Chapel CW4 49 G6
Holme's Hill BN8 13 J5
Holmesfield S18 51 F5
Holmeswood L40 48 D1
Holmewood S42 51 G6
Holmfield HX2 57 F7
Holmfirth HD9 50 D2
Holmhead D. & G. DG7 68 C5
Holmhead E.Ayr. KA18 67 K1
Holmpton HU19 59 K7
Holmrook CA19 60 B6
Holmsgarth ZE1 107 N8
Holmside DH7 62 D2
Holmsleigh Green EX14 8 B4
Holmston KA7 67 H1
Holmwrangle CA4 61 G2
Holne TQ13 5 H4
Holnest DT9 9 F4
Holnicote TA24 7 H1
Holsworthy EX22 6 B5
Holsworthy Beacon EX22 6 B5
Holt Dorset BH21 10 B4
Holt Norf. NR25 44 E2
Holt Wilts. BA14 20 B5
Holt Worcs. WR6 29 H2
Holt Wrex. LL13 48 D7
Holt End Hants. GU34 11 H1
Holt End Worcs. B98 30 B2
Holt Fleet WR6 29 H2
Holt Heath Dorset BH21 10 B4
Holt Heath Worcs. WR6 29 H2
Holt Wood BH21 10 B4
Holtby YO19 58 C4
Holton Oxon. OX33 21 K1
Holton Som. BA9 9 F2
Holton Suff. IP19 35 J1
Holton cum Beckering LN8 52 E4
Holton Heath BH16 9 J5
Holton le Clay DN36 53 F2
Holton le Moor LN7 52 D3
Holton St. Mary CO7 34 E5
Holtspur HP9 22 C3
Holtye TN8 13 H3
Holtye Common TN8 13 H3
Holway TA1 8 B2
Holwell Dorset DT9 9 F3
Holwell Herts. SG5 32 E5
Holwell Leics. LE14 42 A3
Holwell Oxon. OX18 21 F1
Holwell Som. BA11 20 A7
Holwick DL12 62 A4
Holworth DT2 9 G6
Holy Cross DY9 29 J1
Holy Island I.o.A. LL65 46 A5
Holy Island Northumb. TD15 77 K6
Holybourne GU34 22 A7
Holyfield EN9 23 G1
Holyhead (Caergybi) LL65 46 A4
Holymoorside S42 51 F6
Holyport SL6 22 B4
Holystone NE65 70 E3
Holytown ML1 75 F4
Holywell Cambs. PE27 33 G1
Holywell Cornw. TR8 2 E3
Holywell Dorset DT2 8 E4
Holywell E.Suss. BN20 13 K7
Holywell (Treffynnon) Flints. CH8 47 K5
Holywell Northumb. NE25 71 J6
Holywell Green HX4 50 C1
Holywell Lake TA21 7 K3
Holywell Row IP28 34 B1
Holywood DG2 68 E5
Hom Green HR9 28 E6
Homer TF13 39 F5
Homersfield IP20 45 G7
Homington SP5 10 C2
Homore (Tobha Mòr) HS8 84 C1
Honey Hill CT2 25 H5
Honey Street SN9 20 E5
Honey Tye CO6 34 D5
Honeyborough SA73 16 C5
Honeybourne WR11 30 C4
Honeychurch EX20 6 E5
Honicknowle PL5 4 E5
Honiley CV8 30 D1
Honing NR28 45 H3
Honingham NR9 45 F4
Honington Lincs. NG32 42 C1
Honington Suff. IP31 34 D1
Honington Warks. CV36 30 D4
Honiton EX14 7 K5
Honkley LL12 48 C7
Honley HD9 50 D1
Hoo Med. ME3 24 D4
Hoo Suff. IP13 35 G3
Hoo Green WA16 49 G4
Hoo Meavy PL20 5 F4
Hood Green S75 51 F2
Hood Hill S35 51 F3
Hooe E.Suss. TN33 13 K6
Hooe Plym. PL9 5 F5
Hooe Common TN33 13 K5
Hook Cambs. PE15 43 H6
Hook E.Riding DN14 58 D7
Hook Gt.Lon. KT9 22 E5
Hook Hants. RG27 22 A6
Hook Hants. SO31 11 G4
Hook Pembs. SA62 16 C4
Hook Wilts. SN4 20 D3
Hook Green Kent TN3 13 K3
Hook Green Kent DA13 24 C5
Hook Green Kent DA13 24 C4
Hook Green Kent DA2 23 J4
Hook Norton OX15 30 E5
Hook-a-Gate SY5 38 D5
Hooke DT8 8 E4
Hookgate TF9 39 G2
Hookway EX17 7 G6
Hookwood RH6 23 F7
Hoole CH3 48 D6
Hooley CR5 23 F6
Hoop NP25 19 J1
Hooton CH66 48 C5
Hooton Levitt S66 51 H3
Hooton Pagnell DN5 51 G2
Hooton Roberts S65 51 G3
Hop Pole PE11 42 E4
Hopcrofts Holt OX25 31 F6
Hope Derbys. S33 50 D4
Hope Devon TQ7 5 G7
Hope Flints. LL12 48 C7
Hope Powys SY21 38 B5
Hope Shrop. SY5 38 C5
Hope Staffs. DE6 50 D7
Hope Bagot SY8 28 E1
Hope Bowdler SY6 38 D6
Hope End Green CM22 33 J6
Hope Mansell HR9 29 F7
Hope under Dinmore HR6 28 E3
Hopehouse TD7 69 H2
Hopeman IV30 97 J5
Hope's Green SS7 24 D3
Hopesay SY7 38 C7
Hopkinstown CF37 18 D2
Hopley's Green HR3 28 C3
Hopperton HG5 57 K4
Hopsford CV7 41 G7
Hopstone WV5 39 G6
Hopton Derbys. DE4 50 E7
Hopton Norf. NR31 45 K6
Hopton Shrop. TF9 38 E3
Hopton Shrop. SY4 38 C3
Hopton Staffs. ST18 40 B3
Hopton Suff. IP22 34 D1
Hopton Cangeford SY8 38 E7
Hopton Castle SY7 28 C1
Hopton Wafers DY14 29 F1
Hoptonheath SY7 28 C1
Hopwas B78 40 D5
Hopwood B48 30 B1
Horam TN21 13 J5
Horbling NG34 42 E2
Horbury WF4 50 E1
Horden SR8 63 F2
Horderley SY7 38 D7
Hordle SO41 10 D5
Hordley SY12 38 C2
Horeb Carmar. SA15 17 H5
Horeb Cere. SA44 17 G1
Horeb Flints. LL12 48 B7
Horfield BS7 19 J4
Horham IP21 35 G1
Horkesley Heath CO6 34 D6
Horkstow DN18 52 C1
Horley Oxon. OX15 31 F4
Horley Surr. RH6 23 F7
Horn Hill SL9 22 D2
Hornblotton BA4 8 E1
Hornblotton Green BA4 8 E1
Hornby Lancs. LA2 55 J3
Hornby N.Yorks. DL8 62 D7
Hornby N.Yorks. DL6 62 E6
Horncastle LN9 53 F6
Hornchurch RM11 23 J3
Horncliffe TD15 77 H6
Horndean Hants. PO8 11 J3
Horndean Sc.Bord. TD15 77 H6
Horndon PL19 6 D7
Horndon on the Hill SS17 24 C3
Horne RH6 23 G7
Horne Row CM3 24 D1
Horner TA24 7 G1
Horniehaugh DD8 83 F1
Horning NR12 45 H4
Horninghold LE16 42 B6
Horninglow DE13 40 E3
Horningsea CB25 33 H2
Horningsham BA12 20 B7
Horningtoft NR20 44 D3
Horningtops PL14 4 C4
Horns Cross Devon EX39 6 B3
Horns Cross E.Suss. TN31 14 D5
Horns Green TN14 23 H6
Hornsbury TA20 8 C3
Hornsby CA8 61 G2
Hornsby Gate CA8 61 G1
Hornsea HU18 59 J5
Hornsey N8 23 G3
Hornton OX15 30 E4
Horrabridge PL20 5 F4
Horridge TQ13 5 H3
Horringer IP29 34 C2
Horrocks Fold BL1 49 G1
Horse Bridge ST9 49 J7
Horsebridge Devon PL19 4 E3
Horsebridge Hants. SO20 10 E1
Horsebrook ST19 40 A4
Horsecastle BS49 19 H5
Horsehay TF4 39 F5
Horseheath CB21 33 K4
Horsehouse DL8 57 F1
Horsell GU21 22 C6
Horseman's Green SY13 38 D1
Horsenden HP27 22 A1
Horseshoe Green TN8 23 H7
Horseway PE16 43 H7
Horsey NR29 45 J3
Horsey Corner NR29 45 J3
Horsford NR10 45 F4
Horsforth LS18 57 H6
Horsham W.Suss. RH12 12 E3
Horsham Worcs. WR6 29 G3
Horsham St. Faith NR10 45 G4
Horsington Lincs. LN10 52 E6
Horsington Som. BA8 9 G2
Horsington Marsh BA8 9 G2
Horsley Derbys. DE21 41 F1
Horsley Glos. GL6 20 B2
Horsley Northumb. NE15 71 F7
Horsley Northumb. NE19 70 D4
Horsley Cross CO11 35 F6
Horsley Woodhouse DE7 41 F1
Horsleycross Street CO11 35 F6
Horsleygate S18 51 F5
Horsleyhill TD9 70 A2
Horsmonden TN12 23 K7
Horspath OX33 21 J1
Horstead NR12 45 G4
Horsted Keynes RH17 13 G4
Horton Bucks. LU7 32 C7
Horton Dorset BH21 10 B4
Horton Lancs. BD23 56 D4
Horton Northants. NN7 32 B3
Horton S.Glos. BS37 20 A3
Horton Shrop. SY4 38 D3
Horton Som. TA19 8 C3
Horton Staffs. ST13 49 J7
Horton Swan. SA3 17 H7
Horton Tel. & W. TF6 39 F4
Horton W. & M. SL3 22 D4
Horton Wilts. SN10 20 D5
Horton Cross TA19 8 C3
Horton Grange NE13 71 H6
Horton Green SY14 38 D1
Horton Heath SO50 11 F3
Horton in Ribblesdale BD24 56 D2
Horton Inn BH21 10 B4
Horton Kirby DA4 23 J5
Horton-cum-Studley OX33 31 H7
Horwich BL6 49 F1
Horwich End SK23 50 C4
Horwood EX39 6 D3
Hoscar L40 48 D1
Hose LE14 42 A3
Hoses LA20 60 D7
Hosh PH7 81 K5
Hosta HS6 92 C4
Hoswick ZE2 107 N10
Hotham YO43 58 E6
Hothfield TN26 14 E3
Hoton LE12 41 H3
Houbie ZE2 107 Q3
Houdston KA26 67 F4
Hough CW2 49 G7
Hough Green WA8 48 D4
Hougham NG32 42 B1
Hough-on-the-Hill NG32 42 C1
Houghton Cambs. PE28 33 F1
Houghton Cumb. CA3 60 F1
Houghton Devon TQ7 5 G6
Houghton Hants. SO20 10 E1
Houghton Pembs. SA73 16 C5
Houghton W.Suss. BN18 12 D5
Houghton Bank DL2 62 D4
Houghton Conquest MK45 32 D4
Houghton le Spring DH4 62 E2
Houghton on the Hill LE7 41 J5
Houghton Regis LU5 32 D6
Houghton St. Giles NR22 44 D2
Houghton-le-Side DL2 62 D4
Houlsyke YO21 63 J6
Hound SO31 11 F4
Hound Green RG27 22 A6
Houndslow TD3 76 E6
Houndsmoor TA4 7 K3
Houndwood TD14 77 G4
Hounsdown SO40 10 E3
Hounslow TW3 22 E4
Housebay KW17 106 F5
Househill IV12 97 F6

Hou - Itt

Houses Hill HD5 50 D1
Housetter ZE2 107 M4
Housham Tye CM17 33 J7
Houss ZE2 107 M9
Houston PA6 74 C4
Houstry KW6 105 G5
Houstry of Dunn KW1 105 H3
Houton KW17 106 C7
Hove BN3 13 F6
Hove Edge HD6 57 G7
Hoveringham NG14 41 J1
Hoveton NR12 45 H4
Hovingham YO62 58 C2
How CA8 61 G1
How Caple HR1 29 F5
How End MK45 32 D4
How Green TN8 23 H7
How Man CA22 60 A5
Howbrook S35 51 F3
Howden DN14 58 D7
Howden Clough WF17 57 H7
Howden-le-Wear DL15 62 C3
Howe Cumb. LA8 55 H1
Howe High. KW1 105 J2
Howe N.Yorks. YO7 57 J1
Howe Norf. NR15 45 G6
Howe Green CM2 24 D1
Howe of Teuchar AB53 99 F6
Howe Street Essex CM3 33 K7
Howe Street Essex CM7 33 K5
Howegreen CM3 24 E1
Howell NG34 42 E1
Howey LD1 27 K3
Howgate Cumb. CA28 60 A4
Howgate Midloth. EH26 76 A5
Howgill Lancs. BB7 56 D5
Howgill N.Yorks. BD23 57 F4
Howick NE66 71 H2
Howle TF10 39 F3
Howle Hill HR9 29 F6
Howlett End CB10 33 J5
Howley TA20 8 B4
Hownam TD5 70 C2
Hownam Mains TD5 70 C1
Howpasley TD9 69 J3
Howsham N.Lincs. LN7 52 D2
Howsham N.Yorks. YO60 58 D3
Howt Green ME9 24 E5
Howtel TD12 77 G7
Howton HR2 28 D6
Howwood PA9 74 C4
Hoxa KW17 106 D8
Hoxne IP21 35 F1
Hoy High. KW14 105 H2
Hoy Ork. KW16 106 B8
Hoylake CH47 48 B4
Hoyland S74 51 F2
Hoylandswaine S36 50 E2
Hoyle GU29 12 C5
Hubberholme BD23 56 E2
Hubberston SA73 16 B5
Hubbert's Bridge PE20 43 F1
Huby N.Yorks. YO61 58 B3
Huby N.Yorks. LS17 57 H5
Hucclecote GL3 29 H7
Hucking ME17 14 D2
Hucknall NG15 41 H1
HUDDERSFIELD HD 50 D1
Huddington WR9 29 J3
Huddlesford WS13 40 D5
Hudnall HP4 32 D7
Hudscott EX37 6 E3
Hudswell DL11 62 C7
Huggate YO42 58 E4
Hugglescote LE67 41 G4
Hugh Town TR21 2 C1
Hughenden Valley HP14 22 B2
Hughley SY5 38 E6
Hugmore LL13 48 C7
Hugus TR3 2 E4
Huish Devon EX20 6 D4
Huish Wilts. SN8 20 E5
Huish Champflower TA4 7 J3
Huish Episcopi TA10 8 D2
Huisinis HS3 100 B6
Hulcote MK17 32 C5
Hulcott HP22 32 B7
HULL HU 59 H7
Hulland DE6 40 E1
Hulland Ward DE6 40 E1
Hullavington SN14 20 B3
Hullbridge SS5 24 E2
Hulme ST3 40 B1
Hulme End SK17 50 D7
Hulme Walfield CW12 49 H6
Hulver Street NR34 45 J7
Hulverstone PO30 10 E6
Humber Devon TQ14 5 J3
Humber Here. HR6 28 E3
Humberside Airport DN39 52 D1
Humberston DN36 53 G2
Humberstone LE5 41 J5
Humberton YO61 57 K3
Humbie EH36 76 C4
Humbleton Dur. DL2 62 B5
Humbleton E.Riding HU11 59 J6
Humbleton Northumb. NE71 70 E1
Humby NG33 42 D2
Hume TD5 77 F6
Humehall TD5 77 F6
Hummer DT9 8 E3
Humshaugh NE46 70 E6
Huna KW1 105 J1
Huncoat BB5 56 C6
Huncote LE9 41 H6
Hundalee TD8 70 B2
Hundall S18 51 F5
Hunderthwaite DL12 62 A4
Hundleby PE23 53 G6
Hundleton SA71 16 C5
Hundon CO10 34 B4
Hundred Acres PO17 11 G3
Hundred End PR4 55 H7
Hundred House LD1 28 A3
Hungarton LE7 41 J5
Hungate End MK19 31 J4
Hungerford Hants. SP6 10 C3
Hungerford Shrop. SY7 38 E7
Hungerford W.Berks. RG17 21 G5
Hungerford Newtown RG17 21 G4
Hungerton NG32 42 B3
Hunglader IV51 93 J4
Hunmanby YO14 59 G2
Hunningham CV33 30 E2
Hunningham Hill CV33 30 E2
Hunny Hill PO30 11 F6
Hunsdon SG12 33 H7
Hunsingore LS22 57 K4
Hunslet LS10 57 J6
Hunsonby CA10 61 G3
Hunspow KW14 105 H1
Hunstanton PE36 44 A1
Hunstanworth DH8 62 A2
Hunston Suff. IP31 34 D2
Hunston W.Suss. PO20 12 B6
Hunsworth BD19 57 G7
Hunt End B97 30 B2
Hunt House YO22 63 K7
Huntercombe End RG9 21 K3
Hunters Forstal CT6 25 H5
Hunter's Inn EX31 6 E1
Hunter's Quay PA23 73 K3
Hunterston KA23 73 K5
Huntford TD8 70 B3
Huntham TA3 8 C2
Huntingdon PE29 33 F1
Huntingfield IP19 35 H1
Huntingford SP8 9 H1
Huntington Here. HR5 28 B3
Huntington Here. HR4 28 D4
Huntington Staffs. WS12 40 B4
Huntington Tel. & W. TF6 39 F5
Huntington York YO32 58 C4
Huntingtower PH1 82 B5
Huntley GL19 29 G7
Huntly AB54 90 D1
Huntlywood TD4 76 E6
Hunton Hants. SO21 21 H7
Hunton Kent ME15 14 C3
Hunton N.Yorks. DL8 62 C7
Hunton Bridge WD4 22 D1
Hunt's Cross L25 48 D4
Huntscott TA24 7 H1
Huntsham EX16 7 J3
Huntshaw EX38 6 D3
Huntshaw Cross EX31 6 D3
Huntshaw Water EX38 6 D3
Huntspill TA9 19 G7
Huntworth TA7 8 C1
Hunwick DL15 62 C3
Hunworth NR24 44 E2
Hurcott Som. TA11 8 E2
Hurcott Som. TA19 8 C3
Hurdley SY15 38 B6
Hurdsfield SK10 49 J5
Hurley W. & M. SL6 22 B3
Hurley Warks. CV9 40 E6
Hurley Bottom SL6 22 B3
Hurlford KA1 74 C7
Hurliness KW16 106 B9
Hurlston Green L40 48 C1
Hurn BH23 10 C5
Hursley SO21 11 F2
Hurst N.Yorks. DL11 62 B6
Hurst W'ham RG10 22 A4
Hurst Green E.Suss. TN19 14 C5
Hurst Green Essex CO7 34 E7
Hurst Green Lancs. BB7 56 B6
Hurst Green Surr. RH8 23 G6
Hurst Wickham BN6 13 F5
Hurstbourne Priors RG28 21 H7
Hurstbourne Tarrant SP11 21 G6
Hurstpierpoint BN6 13 F5
Hurstwood BB10 56 D6
Hurtmore GU7 22 C7
Hurworth-on-Tees DL2 62 E5
Hury DL12 62 A4
Husabost IV55 93 G6
Husbands Bosworth LE17 41 J7
Husborne Crawley MK43 32 C5
Husthwaite YO61 58 B2
Hutcherleigh TQ9 5 H5
Huthwaite NG17 51 G7
Huttoft LN13 53 J5
Hutton Cumb. CA11 60 F4
Hutton Essex CM13 24 C2
Hutton Lancs. PR4 55 H7
Hutton N.Som. BS24 19 G6
Hutton Sc.Bord. TD15 77 H5
Hutton Bonville DL7 62 E6
Hutton Buscel YO13 59 F1
Hutton Conyers HG4 57 J2
Hutton Cranswick YO25 59 G4
Hutton End CA11 60 F3
Hutton Hang DL8 57 G1
Hutton Henry TS27 63 F3
Hutton Magna DL11 62 C5
Hutton Mount CM13 24 C2
Hutton Mulgrave YO21 63 K6
Hutton Roof Cumb. LA6 55 J2
Hutton Roof Cumb. CA11 60 E3
Hutton Rudby TS15 63 F6
Hutton Sessay YO7 57 K2
Hutton Wandesley YO26 58 B4
Hutton-le-Hole YO62 63 J7
Huxham EX5 7 H6
Huxham Green BA4 8 E1
Huxley CH3 48 E6
Huxter Shet. ZE2 107 M7
Huxter Shet. ZE2 107 P6
Huyton L36 48 D3
Hwlffordd (Haverfordwest) SA61 16 C4
Hycemoor LA19 54 D1
Hyde Glos. GL6 20 B1
Hyde Gt.Man. SK14 49 J3
Hyde End W'ham RG7 22 A5
Hyde End W.Berks. RG7 21 J5
Hyde Heath HP6 22 C1
Hyde Lea ST18 40 B4
Hydestile GU8 22 C7
Hyndford Bridge ML11 75 H6
Hyndlee TD9 70 A3
Hynish PA77 78 A4
Hyssington SY15 38 C6
Hythe Hants. SO45 11 F4
Hythe Kent CT21 15 G4
Hythe End TW19 22 D4
Hythie AB42 99 J5
Hyton LA19 54 D1

I

Ianstown AB56 98 C4
Iarsiadar HS2 100 D4
Ibberton DT11 9 G4
Ible DE4 50 E7
Ibsley BH24 10 C4
Ibstock LE67 41 G4
Ibstone HP14 22 A2
Ibthorpe SP11 21 G6
Ibworth RG26 21 J6
Icelton BS22 19 G5
Ickburgh IP26 44 C6
Ickenham UB10 22 D3
Ickford HP18 21 K1
Ickham CT3 15 H2
Ickleford SG5 32 E5
Icklesham TN36 14 D6
Ickleton CB10 33 H4
Icklingham IP28 34 B1
Ickwell Green SG18 32 E4
Icomb GL54 30 D6
Idbury OX7 30 D6
Iddesleigh EX19 6 D5
Ide EX2 7 H6
Ide Hill TN14 23 H6
Ideford TQ13 5 J3
Iden TN31 14 E5
Iden Green Kent TN17 14 C4
Iden Green Kent TN17 14 D4
Idle BD10 57 G6
Idless TR4 3 F4
Idlicote CV36 30 D4
Idmiston SP4 10 C1
Idridgehay DE56 40 E1
Idridgehay Green DE56 40 E1
Idrigil IV51 93 J5
Idstone SN6 21 F3
Idvies DD8 83 G3
Iffley OX4 21 J1
Ifield RH11 13 F3
Ifieldwood RH11 13 F3
Ifold RH14 12 D3
Iford Bourne. BH7 10 C5
Iford E.Suss. BN7 13 H6
Ifton NP26 19 H3
Ifton Heath SY11 38 C2
Ightfield SY13 38 E2
Ightham TN15 23 J6
Iken IP12 35 J3
Ilam DE6 50 D7
Ilchester BA22 8 E2
Ilderton NE66 71 F1
ILFORD IG 23 H3
Ilfracombe EX34 6 D1
Ilkeston DE7 41 G1
Ilketshall St. Andrew NR34 45 H7
Ilketshall St. Lawrence NR34 45 H7
Ilketshall St. Margaret NR35 45 H7
Ilkley LS29 57 G5
Illey B62 40 B7
Illidge Green CW11 49 G6
Illington IP24 44 D7
Illingworth HX2 57 F7
Illogan TR16 2 D4
Illston on the Hill LE7 42 A6
Ilmer HP27 22 A1
Ilmington CV36 30 D4
Ilminster TA19 8 C3
Ilsington Devon TQ13 5 H3
Ilsington Dorset DT2 9 G5
Ilston SA2 17 J6
Ilton N.Yorks. HG4 57 G2
Ilton Som. TA19 8 C3
Imachar KA27 73 G6
Imber BA12 20 C7
Immeroin FK19 81 G6
Immingham DN40 52 E1
Immingham Dock DN40 53 F1
Impington CB24 33 H2
Ince CH2 48 D5
Ince Blundell L38 48 C2
Ince-in-Makerfield WN3 48 E2
Inch Kenneth PA68 79 F4
Inch of Arnhall DD9 90 E7
Inchbae Lodge IV23 96 B5
Inchbare DD9 83 H1
Inchberry IV32 98 B5
Inchbraoch DD10 83 J2
Inchgrundle DD9 90 C7
Inchindown IV18 96 D4
Inchinnan PA4 74 C4
Inchkinloch IV27 103 J4
Inchlaggan PH35 87 H4
Inchlumpie IV17 96 C4
Inchmarlo AB31 90 E5
Inchmarnock PA20 73 J5
Inchnabobart AB35 90 B6
Inchnacardoch Hotel PH32 87 K3
Inchnadamph IV27 102 E6
Inchock DD11 83 H3
Inchrory AB37 89 J4
Inchture PH14 82 D5
Inchvuilt IV4 87 J1
Inchyra PH2 82 C5
Indian Queens TR9 3 G3
Inerval PA42 72 B6
Ingatestone CM4 24 C2
Ingbirchworth S36 50 E2
Ingerthorpe HG3 57 H3
Ingestre ST18 40 B3
Ingham Lincs. LN1 52 C4
Ingham Norf. NR12 45 H3
Ingham Suff. IP31 34 C1
Ingham Corner NR12 45 H3
Ingleborough PE14 43 H4
Ingleby Derbys. DE73 41 F3
Ingleby Lincs. LN1 52 B5
Ingleby Arncliffe DL6 63 F6
Ingleby Barwick TS17 63 F5
Ingleby Cross DL6 63 F6
Ingleby Greenhow TS9 63 G6
Ingleigh Green EX19 6 E5
Inglesbatch BA2 20 A5
Inglesham SN6 21 F2
Ingleton Dur. DL2 62 C4
Ingleton N.Yorks. LA6 56 B2
Inglewhite PR3 55 J6
Ingliston EH28 75 K3
Ingmire Hall LA10 61 H7
Ingoe NE20 71 F6
Ingoldisthorpe PE31 44 A2
Ingoldmells PE25 53 J6
Ingoldsby NG33 42 D2
Ingon CV37 30 D3
Ingram NE66 71 F2
Ingrave CM13 24 C2
Ingrow BD21 57 F6
Ings LA8 60 F7
Ingst BS35 19 J3
Ingworth NR11 45 F3
Inhurst RG26 21 J5
Inistrynich PA33 80 C5
Injebreck IM4 54 C5
Inkberrow WR7 30 B3
Inkersall S43 51 G5
Inkersall Green S43 51 G5
Inkhorn AB41 91 H1
Inkpen RG17 21 G5
Inkstack KW14 105 H1
Inmarsh SN12 20 C5
Innellan PA23 73 K4
Innergellie KY10 83 G7
Innerhadden PH16 81 H2
Innerleithen EH44 76 B7
Innerleven KY8 82 E7
Innermessan DG9 64 A4
Innerwick E.Loth. EH42 77 F3
Innerwick P. & K. PH15 81 G3
Inninbeg PA34 79 H3
Innsworth GL3 29 H6
Insch AB52 90 E2
Insh PH21 89 F4
Inshore IV27 103 F2
Inskip PR4 55 H6
Instow EX39 6 C2
Intake DN2 51 H2
Intwood NR4 45 F5
Inver Aber. AB35 89 K5
Inver Arg. & B. PA38 80 A3
Inver High. IV20 97 F3
Inver High. KW6 105 G5
Inver P. & K. PH8 82 B3
Inver Mallie PH34 87 H6
Inverailort PH38 86 D6
Inveralligin IV22 94 E6
Inverallochy AB43 99 J4
Inveran IV27 96 C2
Inveraray PA32 80 B7
Inverardoch Mains FK15 81 J7
Inverardran FK20 80 E5
Inverarish IV40 86 B1
Inverarity DD8 83 F3
Inverarnan G83 80 E6
Inverasdale IV22 94 E3
Inverbain IV54 94 D6
Inverbeg G83 74 B1
Inverbervie DD10 91 G7
Inverbroom IV23 95 H3
Inverbrough IV13 89 F2
Invercassley IV27 96 B1
Inverchaolain PA23 73 J3
Invercharnan PH49 80 C3
Inverchorachan PA26 80 D6
Inverchoran IV6 95 J6
Invercreran PA38 80 B3
Inverdruie PH22 89 G3
Inverebrie AB41 91 H1
Invereen IV13 88 E1
Inverernie IV36 97 H5
Inveresk EH21 76 B3
Inverey AB35 89 H6
Inverfarigaig IV2 88 C2
Invergarry PH35 87 K4
Invergelder AB35 89 K5
Invergeldie PH6 81 J5
Invergloy PH34 87 J6
Invergordon IV18 96 E5
Invergowrie DD2 82 E4
Inverguseran PH41 86 D4
Inverharroch Farm AB54 90 B1
Inverherive FK20 80 E5
Inverhope IV27 103 G2
Inverie PH41 86 D4
Inverinan PA35 80 A6
Inverinate IV40 87 F2
Inverkeilor DD11 83 H3
Inverkeithing KY11 75 K2
Inverkeithny AB54 98 E6
Inverkip PA16 74 A3
Inverkirkaig IV27 102 C7
Inverlael IV23 95 H3
Inverlauren G84 74 B2
Inverliever PA31 79 K7
Inverliver PA35 80 B4
Inverlochlarig FK19 81 F6
Inverlochy PA33 80 C5
Inverlussa PA60 72 E2
Invermay PH2 82 B6
Invermoriston IV63 88 B3
Invernaver KW14 104 C2
Inverneil PA30 73 G2
INVERNESS IV 96 D7
Inverness Airport IV2 96 E6
Invernettie AB42 99 K6
Invernoaden PA27 73 K1
Inveroran Hotel PA36 80 D3
Inverquharity DD8 83 F2
Inverquhomery AB42 99 J6
Inverroy PH31 87 J6
Inversanda PH33 80 A2
Invershiel IV40 87 F3
Invershore KW3 105 H5
Inversnaid Hotel FK8 80 E7
Invertrossachs FK17 81 G7
Inverugie AB42 99 K6
Inveruglas G83 80 E7
Inveruglass PH21 89 F4
Inverurie AB51 91 F2
Invervar PH15 81 H3
Invervegain PA23 73 J3
Invery House AB31 90 E5
Inverythan AB53 99 F6
Inwardleigh EX20 6 D6
Inworth CO5 34 C7
Iochdar HS8 92 C7
Iona PA76 78 D5
Iping GU29 12 B4
Ipplepen TQ12 5 J4
Ipsden OX10 21 K3
Ipstones ST10 40 C1
IPSWICH IP 35 F4
Irby CH61 48 B4
Irby Hill CH61 48 B4
Irby in the Marsh PE24 53 H6
Irby upon Humber DN37 52 E2
Irchester NN29 32 C2
Ireby Cumb. CA7 60 D3
Ireby Lancs. LA6 56 B2
Ireland Ork. KW16 106 C7
Ireland Shet. ZE2 107 M10
Ireland's Cross CW3 39 G1
Ireleth LA16 55 F2
Ireshopeburn DL13 61 K3
Irlam M44 49 G3
Irnham NG33 42 D3
Iron Acton BS37 19 K3
Iron Cross WR11 30 B3
Ironbridge TF8 39 F5
Irons Bottom RH2 23 F7
Ironside AB53 99 G5
Ironville NG16 51 G7
Irstead NR12 45 H3
Irthington CA6 69 K7
Irthlingborough NN9 32 C1
Irton YO12 59 G1
Irvine KA12 74 B7
Isauld KW14 104 E2
Isbister Ork. KW17 106 C6
Isbister Ork. KW17 106 B5
Isbister Shet. ZE2 107 P6
Isbister Shet. ZE2 107 M3
Isfield TN22 13 H5
Isham NN14 32 B1
Ishriff PA65 79 H4
Isington GU34 22 A7
Island of Stroma KW1 105 J1
Islawr-dref LL40 37 F4
Islay PA 72 A4
Islay Airport PA42 72 B5
Islay House PA44 72 B4
Isle Abbotts TA3 8 C2
Isle Brewers TA3 8 C2
Isle of Lewis (Eilean Leodhais) HS 101 F3
ISLE OF MAN IM 54 C5
Isle of Man Airport IM9 54 B7
Isle of May KY10 76 E1
Isle of Noss ZE2 107 P8
Isle of Sheppey ME12 25 F4
Isle of Walney LA14 54 E3
Isle of Whithorn DG8 64 E7
Isle of Wight PO 11 F6
Isleham CB7 33 K1
Isleornsay (Eilean Iarmain) IV43 86 C3
Isles of Scilly (Scilly Isles) TR 2 C1
Islesburgh ZE2 107 M6
Isleworth TW7 22 E4
Isley Walton DE74 41 G3
Islibhig HS2 100 B5
Islip Northants. NN14 32 C1
Islip Oxon. OX5 31 G7
Isombridge TF6 39 F4
Istead Rise DA13 24 C5
Itchen SO19 11 F3
Itchen Abbas SO21 11 G1
Itchen Stoke SO24 11 G1
Itchingfield RH13 12 E4
Itchington BS35 19 K3
Itteringham NR11 45 F2
Itton Devon EX20 6 E6
Itton Mon. NP16 19 H2

Itt - Kin

Itton Common NP16 19 H2
Ivegill CA4 60 F2
Ivelet DL11 62 A7
Iver SL0 22 D3
Iver Heath SL0 22 D3
Iveston DH8 62 C1
Ivetsey Bank ST19 40 A4
Ivinghoe LU7 32 C7
Ivinghoe Aston LU7 32 C7
Ivington HR6 28 D3
Ivington Green HR6 28 D3
Ivy Hatch TN15 23 J6
Ivy Todd PE37 44 C5
Ivybridge PL21 5 G5
Ivychurch TN29 15 F5
Iwade ME9 25 F5
Iwerne Courtney (Shroton) DT11 9 H3
Iwerne Minster DT11 9 H3
Ixworth IP31 34 D1
Ixworth Thorpe IP31 34 D1

J
Jack Hill LS21 57 G4
Jackfield TF8 39 F5
Jacksdale NG16 51 G7
Jackstown AB51 91 F1
Jackton G75 74 D5
Jacobstow EX23 4 B1
Jacobstowe EX20 6 D5
Jacobswell GU4 22 C6
Jameston SA70 16 D6
Jamestown D. & G. DG13 69 J4
Jamestown High. IV14 96 B6
Jamestown W.Dun. G83 74 B2
Janefield IV10 96 E6
Janetstown High. KW14 105 F2
Janetstown High. KW1 105 J3
Jarrow NE32 71 J7
Jarvis Brook TN6 13 J3
Jasper's Green CM7 34 B6
Jawcraig FK1 75 G3
Jayes Park RH5 22 E7
Jaywick CO15 35 F7
Jealott's Hill RG42 22 B4
Jeater Houses DL6 63 F7
Jedburgh TD8 70 B1
Jeffreyston SA68 16 D5
Jemimaville IV7 96 E5
Jericho BL9 49 H1
Jersay ML7 75 G4
JERSEY JE 3 J7
Jersey Airport JE3 3 J7
Jersey Marine SA1 18 A2
Jerviswood ML11 75 G6
Jesmond NE2 71 H7
Jevington BN26 13 J6
Jockey End HP2 32 D7
Jodrell Bank SK11 49 G5
John o' Groats KW1 105 J1
Johnby CA11 60 F3
John's Cross TN32 14 C5
Johnshaven DD10 83 J1
Johnson Street NR29 45 H4
Johnston SA62 16 C4
Johnston Mains AB30 91 F7
Johnstone PA5 74 C4
Johnstone Castle PA5 74 C4
Johnstonebridge DG11 69 F4
Johnstown Carmar. SA31 17 G4
Johnstown Wrex. LL14 38 C1
Joppa KA6 67 J2
Jordans HP9 22 C2
Jordanston SA62 16 C2
Jordanstone PH11 82 D3
Joy's Green GL17 29 F7
Jumpers Common BH23 10 C5
Juniper Hill NN13 31 G5
Jura PA60 72 D2
Jura House PA60 72 C4
Jurby East IM7 54 C4
Jurby West IM7 54 C4

K
Kaber CA17 61 J5
Kaimes EH17 76 A4
Kames Arg. & B. PA21 73 H3
Kames Arg. & B. PA34 79 K6
Kames E.Ayr. KA18 68 B1
Kea TR3 3 F4
Keadby DN17 52 B1
Keal Cotes PE23 53 G6
Kearsley BL4 49 G2
Kearstwick LA6 56 B2
Kearton DL11 62 B7
Kearvaig IV27 102 E1
Keasden LA2 56 C3
Kebholes AB45 98 E5
Keckwick WA4 48 E4
Keddington LN11 53 G4
Keddington Corner LN11 53 G4
Kedington CB9 34 B4
Kedleston DE22 40 E1
Keelby DN41 52 E1
Keele ST5 40 A1
Keeley Green MK43 32 D4
Keelham BD13 57 F6
Keeres Green CM6 33 J7
Keeston SA62 16 B4
Keevil BA14 20 C6
Kegworth DE74 41 G3
Kehelland TR14 2 D4
Keig AB33 90 E3
Keighley BD21 57 F5
Keil Arg. & B. PA28 66 A3
Keil High. PA38 80 A2
Keilhill AB45 99 F5
Keillmore PA31 72 E2
Keillor PH13 82 D3
Keillour PH1 82 A5
Keills PA46 72 C4
Keils PA60 72 D4
Keinton Mandeville TA11 8 E1
Keir House FK15 75 F1
Keir Mill DG3 68 D4
Keisby PE10 42 D3
Keisley CA16 61 J4
Keiss KW1 105 J2
Keith AB55 98 C5
Keithick PH13 82 D4
Keithmore AB55 90 B1
Keithock DD9 83 H1
Kelbrook BB18 56 E5
Kelby NG32 42 D1
Keld Cumb. CA10 61 G5
Keld N.Yorks. DL11 61 K6
Keldholme YO62 58 D1
Keldy Castle YO18 63 J7
Kelfield N.Lincs. DN9 52 B2
Kelfield N.Yorks. YO19 58 B6
Kelham NG23 51 K7
Kella IM7 54 C4
Kellacott PL15 6 C7
Kellan PA72 79 G3
Kellas Angus DD5 83 F4
Kellas Moray IV30 97 J6
Kellaton TQ7 5 H7
Kellaways SN15 20 C4
Kelleth CA10 61 H6
Kelleythorpe YO25 59 G4
Kelling NR25 44 E1
Kellington DN14 58 B7
Kelloe DH6 62 E3
Kelloholm DG4 68 C2
Kelly Cornw. PL27 4 A3
Kelly Devon PL16 6 B7
Kelly Bray PL17 4 D3
Kelmarsh NN6 31 J1
Kelmscott GL7 21 F2
Kelsale IP17 35 H2
Kelsall CW6 48 E6
Kelsay PA47 72 A5
Kelshall SG8 33 G5
Kelsick CA7 60 D1
Kelso TD5 77 F7
Kelstedge S45 51 F6
Kelstern LN11 53 F3
Kelston BA1 20 A5
Keltneyburn PH15 81 J3
Kelton DG1 65 K3
Kelton Hill (Rhonehouse) DG7 65 H5
Kelty KY4 75 K1
Kelvedon CO5 34 C7
Kelvedon Hatch CM15 23 J2
Kelvinside G12 74 D4
Kelynack TR19 2 A6
Kemacott EX31 6 E1
Kemback KY15 83 F6
Kemberton TF11 39 G5
Kemble GL7 20 C2
Kemerton GL20 29 J5
Kemeys Commander NP15 19 G1
Kemeys Inferior NP18 19 G2
Kemnay AB51 91 F3
Kemp Town BN2 13 G6
Kempe's Corner TN25 15 F3
Kempley GL18 29 F6
Kempley Green GL18 29 F6
Kemps Green B94 30 C1
Kempsey WR5 29 H4
Kempsford GL7 20 E2
Kempshott RG22 21 J7
Kempston MK42 32 D4
Kempston Hardwick MK45 32 D4
Kempston West End MK43 32 C4
Kempton SY7 38 C7
Kemsing TN15 23 J6
Kemsley ME10 25 F5
Kenardington TN26 14 E4
Kenchester HR4 28 D4
Kencot GL7 21 F1
Kendal LA9 61 G7
Kenderchurch HR2 28 D6
Kendleshire BS36 19 K4
Kenfig CF33 18 A3
Kenfig Hill CF33 18 B3
Kenidjack TR19 2 A5
Kenilworth CV8 30 D1
Kenknock P. & K. PH15 81 G3
Kenknock Stir. FK21 81 F4
Kenley Gt.Lon. CR8 23 G5
Kenley Shrop. SY5 38 E5
Kenmore Arg. & B. PA32 80 B7
Kenmore High. IV54 94 D6
Kenmore P. & K. PH15 81 J3
Kenmore W.Isles HS2 100 E7
Kenn Devon EX6 7 H7
Kenn N.Som. BS21 19 H5
Kennacley HS3 93 G2
Kennacraig PA29 73 G4
Kennards House PL15 4 C2
Kennavay HS4 93 H2
Kenneggy Downs TR20 2 C6
Kennerleigh EX17 7 G5
Kennerty AB31 90 E5
Kennet FK10 75 H1
Kennethmont AB54 90 D2
Kennett CB8 33 K2
Kennford EX6 7 H7
Kenninghall NR16 44 E7
Kennington Kent TN24 15 F3
Kennington Oxon. OX1 21 J1
Kennoway KY8 82 E7
Kenny TA19 8 C3
Kennyhill IP28 33 K1
Kennythorpe YO17 58 D3
Kenovay PA77 78 A3
Kensaleyre IV51 93 K6
Kensington W8 23 F3
Kenstone TF9 38 E3
Kensworth LU6 32 D7
Kent Street E.Suss. TN33 14 C6
Kent Street Kent ME18 23 K6
Kentallen PA38 80 B2
Kentchurch HR2 28 D6
Kentford CB8 34 B2
Kentisbeare EX15 7 J5
Kentisbury EX31 6 E1
Kentisbury Ford EX31 6 E1
Kentish Town NW5 23 F3
Kentmere LA8 61 F6
Kenton Devon EX6 7 H7
Kenton Suff. IP14 35 F2
Kenton T. & W. NE3 71 H7
Kenton Corner IP14 35 G2
Kentra PH36 79 H1
Kents Bank LA11 55 G2
Kent's Green GL18 29 G6
Kent's Oak SO51 10 E2
Kenwick SY12 38 D2
Kenwyn TR1 3 F4
Kenyon WA3 49 F3
Keoldale IV27 103 F2
Keose (Ceos) HS2 101 F5
Keppanach PH33 80 B1
Keppoch Arg. & B. G82 74 B3
Keppoch High. IV40 86 E2
Keprigan PA28 66 A2
Kepwick YO7 63 F7
Keresley CV6 41 F7
Kernborough TQ7 5 H6
Kerrera PA34 79 K5
Kerridge SK10 49 J5
Kerris TR19 2 B6
Kerry (Ceri) SY16 38 A6
Kerrycroy PA20 73 K4
Kerry's Gate HR2 28 C5
Kerrysdale IV21 94 E4
Kersall NG22 51 K6
Kersey IP7 34 E4
Kersey Vale IP7 34 E4
Kershopefoot TD9 69 K5
Kerswell EX15 7 J5
Kerswell Green WR5 29 H4
Kerthen Wood TR27 2 C5
Kesgrave IP5 35 G4
Kessingland NR33 45 K7
Kessingland Beach NR33 45 K7
Kestle PL26 3 G4
Kestle Mill TR8 3 F3
Keston BR2 23 H5
Keswick Cumb. CA12 60 D4
Keswick Norf. NR4 45 G5
Keswick Norf. NR12 45 H2
Ketley Bank TF2 39 F4
Ketsby LN11 53 G5
Kettering NN16 32 B1
Ketteringham NR18 45 F5
Kettins PH13 82 D4
Kettle Corner ME15 14 C2
Kettlebaston IP7 34 D3
Kettlebridge KY15 82 E7
Kettlebrook B77 40 E5
Kettleburgh IP13 35 G2
Kettlehill KY15 82 E7
Kettleholm DG11 69 G6
Kettleness YO21 63 K5
Kettleshulme SK23 49 J5
Kettlesing HG3 57 H4
Kettlesing Bottom HG3 57 H4
Kettlesing Head HG3 57 H4
Kettlestone NR21 44 D2
Kettlethorpe LN1 52 B5
Kettletoft KW17 106 F4
Kettlewell BD23 56 E2
Ketton PE9 42 C5
Kevingtown BR5 23 H5
Kew TW9 22 E4
Kewstoke BS22 19 G5
Kexbrough S75 51 F2
Kexby Lincs. DN21 52 B4
Kexby York YO41 58 D4
Key Green CW12 49 H6
Keyham LE7 41 J5
Keyhaven SO41 10 E5
Keyingham HU12 59 J7
Keymer BN6 13 G5
Keynsham BS31 19 K5
Key's Toft PE24 53 H7
Keysoe MK44 32 D2
Keysoe Row MK44 32 D2
Keyston PE28 32 D1
Keyworth NG12 41 J2
Kibblesworth NE11 62 D1
Kibworth Beauchamp LE8 41 J6
Kibworth Harcourt LE8 41 J6
Kidbrooke SE3 23 H4
Kiddemore Green ST19 40 A5
Kidderminster DY10 29 H1
Kiddington OX20 31 F6
Kidlington OX5 31 F7
Kidmore End RG4 21 K4
Kidnal SY14 38 D1
Kidsdale DG8 64 E7
Kidsgrove ST7 49 H7
Kidstones DL8 56 E1
Kidwelly (Cydweli) SA17 17 H5
Kiel Crofts PA37 79 K4
Kielder NE48 70 B4
Kilbarchan PA10 74 C4
Kilbeg IV44 86 C4
Kilberry PA29 73 F4
Kilbirnie KA25 74 B5
Kilblaan PA32 80 C6
Kilbraur KW9 104 D7
Kilbrennan PA73 79 F3
Kilbride Arg. & B. PA34 79 K5
Kilbride Arg. & B. PA20 73 J4
Kilbride High. IV49 86 B2
Kilbride Farm PA21 73 H4
Kilbridemore PA22 73 J1
Kilburn Derbys. DE56 41 F1
Kilburn Gt.Lon. NW6 23 F3
Kilburn N.Yorks. YO61 58 B2
Kilby LE18 41 J6
Kilchattan Bay PA20 73 K5
Kilchenzie PA28 66 A1
Kilcheran PA34 79 K4
Kilchiaran PA48 72 A4
Kilchoan Arg. & B. PA34 79 J6
Kilchoan High. PH36 79 F1
Kilchoman PA49 72 A4
Kilchrenan PA35 80 B5
Kilchrist PA28 66 A2
Kilconquhar KY9 83 F7
Kilcot GL18 29 F6
Kilcoy IV6 96 C6
Kilcreggan G84 74 A2
Kildale YO21 63 H6
Kildary IV18 96 E4
Kildavie PA28 66 B2
Kildermorie Lodge IV17 96 C4
Kildonan N.Ayr. KA27 66 E1
Kildonan (Cilldonnain) W.Isles HS8 84 C2
Kildonan Lodge KW8 104 E6
Kildonnan PH42 85 K6
Kildrochet House DG9 64 A5
Kildrummy AB33 90 C3
Kildwick BD20 57 F5
Kilfinan PA21 73 H3
Kilfinnan PH34 87 J5
Kilgetty SA68 16 E5
Kilgwrrwg Common NP16 19 H2
Kilham E.Riding YO25 59 G3
Kilham Northumb. TD12 77 G7
Kilkenneth PA77 78 A3
Kilkenny GL54 30 B7
Kilkerran Arg. & B. PA28 66 B2
Kilkerran S.Ayr. KA19 67 H3
Kilkhampton EX23 6 A4
Killamarsh S21 51 G4
Killay SA2 17 K6
Killbeg PA72 79 H3
Killean Arg. & B. PA29 72 E6
Killean Arg. & B. PA32 80 B7
Killearn G63 74 D2
Killellan PA28 66 A2
Killen IV9 96 D6
Killerby DL2 62 C4
Killerton EX5 7 H5
Killichonan PH17 81 G2
Killiechonate PH34 87 J6
Killiechronan PA72 79 G3
Killiecrankie PH16 82 A1
Killiehuntly PH21 88 E5
Killiemor PA72 79 F4
Killilan IV40 87 F1
Killimster KW1 105 J3
Killin High. KW9 97 F1
Killin Stir. FK21 81 G4
Killinallan PA44 72 B3
Killinghall HG3 57 H4
Killington Cumb. LA6 56 B1
Killington Devon EX31 6 E1
Killingworth NE12 71 H6
Killochyett TD1 76 C6
Killocraw PA28 72 E7
Killunaig PA70 79 F5
Killundine PA34 79 G3
Kilmacolm PA13 74 B4
Kilmaha PA35 80 A7
Kilmahog FK17 81 H7
Kilmalieu PH33 79 K2
Kilmaluag IV51 93 K4
Kilmany KY15 82 E5
Kilmarie IV49 86 B3
KILMARNOCK KA 74 C7
Kilmartin PA31 73 G1
Kilmaurs KA3 74 C6
Kilmelford PA34 79 K6
Kilmeny PA45 72 B4
Kilmersdon BA3 19 K6
Kilmeston SO24 11 G2
Kilmichael PA28 66 A1
Kilmichael Glassary PA31 73 G1
Kilmichael of Inverlussa PA31 73 F2
Kilmington Devon EX13 8 B5
Kilmington Wilts. BA12 9 G1
Kilmington Common BA12 9 G1
Kilmorack IV4 96 B7
Kilmore Arg. & B. PA34 79 K5
Kilmore High. IV44 86 C4
Kilmory Arg. & B. PA31 73 F2
Kilmory Arg. & B. PA31 73 F3
Kilmory High. PH43 85 J4
Kilmory High. PH36 86 B7
Kilmory N.Ayr. KA27 66 D1
Kilmote KW8 104 E7
Kilmuir High. IV55 93 H7
Kilmuir High. IV1 96 D7
Kilmuir High. IV18 96 E4
Kilmuir High. IV51 93 J4
Kilmun PA23 73 K2
Kilmux KY8 82 E7
Kiln Green Here. HR9 29 F7
Kiln Green W'ham RG10 22 B4
Kiln Pit Hill DH8 62 B1
Kilnave PA44 72 A3
Kilncadzow ML8 75 G6
Kilndown TN17 14 C4
Kilnhurst S64 51 G3
Kilninian PA74 79 F3
Kilninver PA34 79 K5
Kilnsea HU12 53 H1
Kilnsey BD23 56 E3
Kilnwick YO25 59 F5
Kilnwick Percy YO42 58 E4
Kiloran PA61 72 B1
Kilpatrick KA27 66 D1
Kilpeck HR2 28 D5
Kilphedir KW8 104 E7
Kilpin DN14 58 D7
Kilpin Pike DN14 58 D7
Kilrenny KY10 83 G7
Kilsby CV23 31 G1
Kilspindie PH2 82 D5
Kilstay DG9 64 B7
Kilsyth G65 75 F3
Kiltarlity IV4 96 C7
Kilton Notts. S81 51 H5
Kilton R. & C. TS13 63 H5
Kilton Som. TA5 7 K1
Kilton Thorpe TS12 63 H5
Kiltyrie FK21 81 H4
Kilvaxter IV51 93 J5
Kilve TA5 7 K1
Kilverstone IP24 44 C7
Kilvington NG13 42 B1
Kilwinning KA13 74 B6
Kimberley Norf. NR18 44 E5
Kimberley Notts. NG16 41 H1
Kimberworth S61 51 G3
Kimble Wick HP17 22 B1
Kimblesworth DH2 62 D2
Kimbolton Cambs. PE28 32 D2
Kimbolton Here. HR6 28 E2
Kimbridge SO51 10 E2
Kimcote LE17 41 H7
Kimmeridge BH20 9 J7
Kimmerston NE71 77 H7
Kimpton Hants. SP11 21 F7
Kimpton Herts. SG4 32 E7
Kinaldy KY16 83 G6
Kinblethmont DD11 83 H3
Kinbrace KW11 104 D5
Kinbreack PH34 87 G5
Kinbuck FK15 81 J7
Kincaldrum DD8 83 F3
Kincaple KY16 83 F6
Kincardine Fife FK10 75 H2
Kincardine High. IV24 96 D3
Kincardine O'Neil AB34 90 D5
Kinclaven PH1 82 C4
Kincorth AB12 91 H4
Kincraig Aber. AB41 91 H2
Kincraig High. PH21 89 F4
Kincraigie PH8 82 A3
Kindallachan PH9 82 A2
Kindrogan Field Centre PH10 82 B1
Kinellar AB21 91 G3
Kineton Glos. GL54 30 B6
Kineton Warks. CV35 30 E3
Kineton Green B92 40 D7
Kinfauns PH2 82 C5
King Sterndale SK17 50 C5
Kingarth PA20 73 J5
Kingcoed NP15 19 H1
Kingerby LN8 52 D3
Kingham OX7 30 D6
Kingholm Quay DG1 65 K3
Kinghorn KY3 76 A2
Kinglassie KY5 76 A1
Kingoodie DD2 82 E5
King's Acre HR4 28 D4
King's Bank TN31 14 D5
King's Bromley DE13 40 D4
Kings Caple HR1 28 E6
King's Cliffe PE8 42 D6
King's Coughton B49 30 B3
King's Green WR13 29 G5
King's Heath B14 40 C7
Kings Hill Kent ME19 23 K6
King's Hill W.Mid. WS10 40 B6
King's Hill Warks. CV3 30 E1
Kings Langley WD4 22 D1
King's Lynn PE30 44 A3
King's Meaburn CA10 61 H4
Kings Mills GY5 3 H5
King's Moss WA11 48 E2
Kings Muir EH45 76 A7
King's Newnham CV23 31 F1
King's Newton DE73 41 F3
King's Norton Leics. LE7 41 J5
King's Norton W.Mid. B30 30 B1
Kings Nympton EX37 6 E4
King's Pyon HR4 28 D3
Kings Ripton PE28 33 F1
King's Somborne SO20 10 E1
King's Stag DT10 9 G3
King's Stanley GL10 20 B1
King's Sutton OX17 31 F5
King's Tamerton PL5 4 E5
King's Walden SG4 32 E6
Kings Worthy SO23 11 F1
Kingsand PL10 4 E5
Kingsbarns KY16 83 G6
Kingsbridge Devon TQ7 5 H6
Kingsbridge Som. TA23 7 H2
Kingsburgh IV51 93 J6
Kingsbury Gt.Lon. HA3 22 E3
Kingsbury Warks. B78 40 E6
Kingsbury Episcopi TA12 8 D2
Kingscavil EH49 75 J3
Kingsclere RG20 21 J6
Kingscote GL8 20 B2
Kingscott EX38 6 D4
Kingscross KA27 66 E1
Kingsdale KY8 82 E7
Kingsdon TA11 8 E2
Kingsdown Kent CT14 15 J3
Kingsdown Swin. SN2 20 E3
Kingsdown Wilts. SN13 20 B5
Kingseat KY12 75 K1
Kingsey HP17 22 A1

Kin - Lak

Kingsfold *Pembs.* **SA71** 16 C6
Kingsfold *W.Suss.* **RH12** 12 E3
Kingsford *Aber.* **AB53** 99 F6
Kingsford *Aber.* **AB33** 90 D3
Kingsford *Aberdeen* **AB15** 91 G4
Kingsford *E.Ayr.* **KA3** 74 C6
Kingsford *Worcs.* **DY11** 40 A7
Kingsgate **CT10** 25 K4
Kingshall Street **IP30** 34 D2
Kingsheanton **EX31** 6 D2
Kingshouse **FK19** 81 G5
Kingshouse Hotel **PH49** 80 D2
Kingshurst **B37** 40 D7
Kingskerswell **TQ12** 5 J4
Kingskettle **KY15** 82 E7
Kingsland *Here.* **HR6** 28 D2
Kingsland *I.o.A.* **LL65** 46 A4
Kingsley *Ches.W. & C.* **WA6** 48 E5
Kingsley *Hants.* **GU35** 11 J1
Kingsley *Staffs.* **ST10** 40 C1
Kingsley Green **GU27** 12 B3
Kingsley Holt **ST10** 40 C1
Kingslow **WV6** 39 G6
Kingsmoor **CM19** 23 H1
Kingsmuir *Angus* **DD8** 83 F3
Kingsmuir *Fife* **KY16** 83 G7
Kingsnorth **TN23** 15 F4
Kingsnorth Power Station **ME3** 24 E4
Kingstanding **B44** 40 C6
Kingsteignton **TQ12** 5 J3
Kingsteps **IV12** 97 G6
Kingsthorne **HR2** 28 D5
Kingsthorpe **NN2** 31 J2
Kingston *Cambs.* **CB23** 33 G3
Kingston *Cornw.* **PL17** 4 D3
Kingston *Devon* **TQ7** 5 G6
Kingston *Devon* **EX10** 7 J7
Kingston *Dorset* **DT10** 9 G4
Kingston *Dorset* **BH20** 9 J7
Kingston *E.Loth.* **EH39** 76 D2
Kingston *Gt.Man.* **SK14** 49 J3
Kingston *Hants.* **BH24** 10 C4
Kingston *I.o.W.* **PO38** 11 F6
Kingston *Kent* **CT4** 15 G2
Kingston *M.K.* **MK10** 32 C5
Kingston *Moray* **IV32** 98 B4
Kingston *W.Suss.* **BN16** 12 D6
Kingston Bagpuize **OX13** 21 G2
Kingston Blount **OX39** 22 A2
Kingston by Sea **BN43** 13 F6
Kingston Deverill **BA12** 9 H1
Kingston Gorse **BN16** 12 D6
Kingston Lisle **OX12** 21 G3
Kingston Maurward **DT2** 9 G5
Kingston near Lewes **BN7** 13 G6
Kingston on Soar **NG11** 41 H3
Kingston Russell **DT2** 8 E5
Kingston St. Mary **TA2** 8 B2
Kingston Seymour **BS21** 19 H5
Kingston Stert **OX39** 22 A1
KINGSTON UPON HULL HU 59 **H7**
KINGSTON UPON THAMES KT 22 **E5**
Kingston Warren **OX12** 21 G3
Kingstone *Here.* **HR2** 28 D5
Kingstone *Here.* **HR9** 29 F6
Kingstone *Som.* **TA19** 8 C3
Kingstone *Staffs.* **ST14** 40 C3
Kingstone Winslow **SN6** 21 F3
Kingstown **CA3** 60 E1
Kingswear **TQ6** 5 J5
Kingswell **KA3** 74 D6
Kingswells **AB15** 91 G4
Kingswinford DY6 40 **A7**
Kingswood *Bucks.* **HP18** 31 H7
Kingswood *Glos.* **GL12** 20 A2
Kingswood *Here.* **HR5** 28 B3
Kingswood *Kent* **ME17** 14 D2
Kingswood *Powys* **SY21** 38 B5
Kingswood *S.Glos.* **BS15** 19 K4
Kingswood *Som.* **TA4** 7 K2
Kingswood *Surr.* **KT20** 23 F6
Kingswood *Warks.* **B94** 30 C1
Kingthorpe **LN8** 52 E5
Kington *Here.* HR5 28 **B3**
Kington *Worcs.* **WR7** 29 J3
Kington Langley **SN15** 20 C4
Kington Magna **SP8** 9 G2
Kington St. Michael **SN14** 20 C4
Kingussie PH21 88 **E4**
Kingweston **TA11** 8 E1
Kinharrachie **AB41** 91 H1
Kinharvie **DG2** 65 K4
Kinkell **G66** 74 E3
Kinkell Bridge **PH3** 82 A6
Kinknockie **AB42** 99 J6
Kinlet **DY12** 39 G7
Kinloch *Fife* **KY15** 82 D6
Kinloch *High.* **IV27** 103 F5
Kinloch *High.* **PA34** 79 H2
Kinloch *High.* **PH43** 85 K5
Kinloch *High.* **IV16** 96 C4
Kinloch *P. & K.* **PH12** 82 D3
Kinloch *P. & K.* **PH10** 82 C3
Kinloch Hourn **PH35** 87 F4
Kinloch Laggan **PH20** 88 C6
Kinloch Rannoch **PH16** 81 H2
Kinlochan **PH37** 79 K1
Kinlochard **FK8** 81 F7
Kinlocharkaig **PH34** 87 F5
Kinlochbeoraid **PH38** 86 E6
Kinlochbervie **IV27** 102 E3
Kinlocheil **PH33** 87 F7
Kinlochetive **PH49** 80 C3
Kinlochewe **IV22** 95 G5
Kinlochlaich **PA38** 80 A3
Kinlochleven PH50 80 **C1**
Kinlochmoidart **PH38** 86 D7
Kinlochmorar **PH41** 86 E5
Kinlochmore **PH50** 80 C1
Kinlochroag (Ceann Lochroag) **HS2** 100 D5
Kinlochspelve **PA63** 79 H5
Kinloss **IV36** 97 H5
Kinmel Bay (Bae Cinmel) **LL18** 47 H4
Kinmuck **AB51** 91 G3
Kinnaber **DD10** 83 J1
Kinnadie **AB41** 99 H6
Kinnaird **PH14** 82 D5
Kinneff **DD10** 91 G7
Kinnelhead **DG10** 69 F3
Kinnell *Angus* **DD11** 83 H2
Kinnell *Stir.* **FK21** 81 G4
Kinnerley **SY10** 38 C3
Kinnersley *Here.* **HR3** 28 C4
Kinnersley *Worcs.* **WR8** 29 H4
Kinnerton **LD8** 28 B2
Kinnerton Green **CH4** 48 C6
Kinnesswood **KY13** 82 C7
Kinnettles **DD8** 83 F3
Kinninvie **DL12** 62 B4
Kinnordy **DD8** 82 E2
Kinoulton **NG12** 41 J2
Kinrara **PH22** 89 F4
Kinross KY13 82 **C7**
Kinrossie **PH2** 82 C4
Kinsbourne Green **AL5** 32 E7
Kinsham *Here.* **LD8** 28 C2
Kinsham *Worcs.* **GL20** 29 J5
Kinsley **WF9** 51 G1
Kinson **BH10** 10 B5
Kintarvie **HS2** 100 E6
Kintbury **RG17** 21 G5
Kintessack **IV36** 97 G5
Kintillo **PH2** 82 C6
Kintocher **AB33** 90 D4
Kinton *Here.* **SY7** 28 D1
Kinton *Shrop.* **SY4** 38 C4
Kintore **AB51** 91 F3
Kintour **PA42** 72 C5
Kintra *Arg. & B.* **PA42** 72 B6
Kintra *Arg. & B.* **PA66** 78 E5
Kintradwell **KW9** 97 G1
Kintraw **PA31** 79 K7
Kinuachdrachd **PA60** 73 F1
Kinveachy **PH24** 89 G3
Kinver **DY7** 40 A7
Kinwarton **B49** 30 C3
Kiplaw Croft **AB42** 91 J1
Kipp **FK18** 81 G6
Kippax **LS25** 57 K6
Kippen *P. & K.* **PH2** 82 B6
Kippen *Stir.* **FK8** 74 E1
Kippenross House **FK15** 81 J7
Kippford (Scaur) **DG5** 65 J5
Kipping's Cross **TN12** 23 K7
Kippington **TN13** 23 J6
Kirbister *Ork.* **KW17** 106 C7
Kirbister *Ork.* **KW17** 106 F5
Kirbuster **KW17** 106 B5
Kirby Bedon **NR14** 45 G5
Kirby Bellars **LE14** 42 A4
Kirby Cane **NR35** 45 H6
Kirby Corner **CV4** 30 D1
Kirby Cross **CO13** 35 G6
Kirby Fields **LE9** 41 H5
Kirby Green **NR35** 45 H6
Kirby Grindalythe **YO17** 59 F3
Kirby Hill *N.Yorks.* **DL11** 62 C6
Kirby Hill *N.Yorks.* **YO51** 57 J3
Kirby Knowle **YO7** 57 K1
Kirby le Soken **CO13** 35 G6
Kirby Misperton **YO17** 58 D2
Kirby Muxloe **LE9** 41 H5
Kirby Row **NR35** 45 H6
Kirby Sigston **DL6** 63 F7
Kirby Underdale **YO41** 58 E4
Kirby Wiske **YO7** 57 J1
Kirdford **RH14** 12 D4
Kirk **KW1** 105 H3
Kirk Bramwith **DN7** 51 J1
Kirk Deighton **LS22** 57 J4
Kirk Ella **HU10** 59 G7
Kirk Hallam **DE7** 41 G1
Kirk Hammerton **YO26** 57 K4
Kirk Ireton **DE6** 50 E7
Kirk Langley **DE6** 40 E2
Kirk Merrington **DL16** 62 D3
Kirk Michael **IM6** 54 C4
Kirk of Shotts **ML7** 75 G4
Kirk Sandall **DN3** 51 J2
Kirk Smeaton **WF8** 51 H1
Kirk Yetholm **TD5** 70 D1
Kirkabister **ZE2** 107 N9
Kirkandrews **DG6** 65 G6
Kirkandrews-upon-Eden **CA5** 60 E1
Kirkbampton **CA5** 60 E1
Kirkbean **DG2** 65 K5
Kirkbride **CA7** 60 D1
Kirkbuddo **DD8** 83 G3
Kirkburn *E.Riding* **YO25** 59 F4
Kirkburn *Sc.Bord.* **EH45** 76 A7
Kirkburton **HD8** 50 D1
Kirkby *Lincs.* **LN8** 52 D3
Kirkby *Mersey.* **L32** 48 D3
Kirkby *N.Yorks.* **TS9** 63 G6
Kirkby Fleetham **DL7** 62 D7
Kirkby Green **LN4** 52 D7
Kirkby in Ashfield **NG17** 51 G7
Kirkby la Thorpe **NG34** 42 D1
Kirkby Lonsdale **LA6** 56 B2
Kirkby Malham **BD23** 56 D3
Kirkby Mallory **LE9** 41 G5
Kirkby Malzeard **HG4** 57 H2
Kirkby on Bain **LN10** 53 F6
Kirkby Overblow **HG3** 57 J5
Kirkby Stephen CA17 61 **J6**
Kirkby Thore **CA10** 61 H4
Kirkby Underwood **PE10** 42 D3
Kirkby Wharfe **LS24** 58 B5
Kirkby Woodhouse **NG17** 51 G7
Kirkby-in-Furness LA17 55 **F1**
Kirkbymoorside **YO62** 58 C1
KIRKCALDY KY 76 **A1**
Kirkcambeck **CA8** 70 A7
Kirkcolm **DG9** 64 A4
Kirkconnel **DG4** 68 C2
Kirkconnell **DG2** 65 K4
Kirkcowan **DG8** 64 D4
Kirkcudbright DG6 65 **G5**
Kirkdale House **DG8** 65 F5
Kirkdean **EH46** 75 K6
Kirkfieldbank **ML11** 75 G6
Kirkgunzeon **DG2** 65 J4
Kirkham *Lancs.* **PR4** 55 H6
Kirkham *N.Yorks.* **YO60** 58 D3
Kirkhamgate **WF2** 57 J7
Kirkharle **NE19** 71 F5
Kirkhaugh **CA9** 61 H2
Kirkheaton *Northumb.* **NE19** 71 F6
Kirkheaton *W.Yorks.* **HD5** 50 D1
Kirkhill *Angus* **DD10** 83 H1
Kirkhill *High.* **IV5** 96 C7
Kirkhill *Moray* **AB38** 98 B5
Kirkhope **TD7** 69 J1
Kirkibost *High.* **IV49** 86 B3
Kirkibost (Circebost) *W.Isles* **HS2** 100 D4
Kirkinch **PH12** 82 E3
Kirkinner **DG8** 64 E5
Kirkintilloch **G66** 74 E3
Kirkland *Cumb.* **CA10** 61 H3
Kirkland *Cumb.* **CA26** 60 B5
Kirkland *D. & G.* **DG3** 68 D4
Kirkland *D. & G.* **DG4** 68 C2
Kirkland *D. & G.* **DG11** 69 F5
Kirkland of Longcastle **DG8** 64 D6
Kirkleatham **TS10** 63 G4
Kirklevington **TS15** 63 F5
Kirkley **NR33** 45 K6
Kirklington *N.Yorks.* **DL8** 57 J1
Kirklington *Notts.* **NG22** 51 J7
Kirklinton **CA6** 69 K7
Kirkliston EH29 75 **K3**
Kirkmaiden **DG9** 64 B7
Kirkmichael *P. & K.* **PH10** 82 B1
Kirkmichael *S.Ayr.* **KA19** 67 H3
Kirkmuirhill **ML11** 75 F6
Kirknewton *Northumb.* **NE71** 77 H7
Kirknewton *W.Loth.* EH27 75 **K4**
Kirkney **AB54** 90 D1
Kirkoswald *Cumb.* **CA10** 61 G2
Kirkoswald *S.Ayr.* **KA19** 67 G3
Kirkpatrick Durham **DG7** 65 H3
Kirkpatrick-Fleming **DG11** 69 H6
Kirksanton **LA18** 54 E1
Kirkstall **LS5** 57 H6
Kirkstead **LN10** 52 E6
Kirkstile *Aber.* **AB54** 90 D1
Kirkstile *D. & G.* **DG13** 69 J4
Kirkstyle **KW1** 105 J1
Kirkthorpe **WF1** 57 J7
Kirkton *Aber.* **AB52** 90 E2
Kirkton *Aber.* **AB53** 98 E5
Kirkton *Aber.* **AB33** 90 E3
Kirkton *Angus* **DD8** 83 F3
Kirkton *Arg. & B.* **PA31** 79 J7
Kirkton *D. & G.* **DG1** 68 E5
Kirkton *Fife* **DD6** 82 E5
Kirkton *High.* **IV3** 88 D1
Kirkton *High.* **KW10** 96 E2
Kirkton *High.* **IV2** 96 E6
Kirkton *High.* **KW13** 104 D2
Kirkton *High.* **IV40** 86 E2
Kirkton *P. & K.* **PH3** 82 A6
Kirkton *Sc.Bord.* **TD9** 70 A2
Kirkton Manor **EH45** 76 A7
Kirkton of Airlie **DD8** 82 E2
Kirkton of Auchterhouse **DD3** 82 E4
Kirkton of Barevan **IV12** 97 F7
Kirkton of Bourtie **AB51** 91 G2
Kirkton of Collace **PH2** 82 C4
Kirkton of Craig **DD10** 83 J2
Kirkton of Culsalmond **AB52** 90 E1
Kirkton of Durris **AB31** 91 F5
Kirkton of Glenbuchat **AB36** 90 B3
Kirkton of Glenisla **PH11** 82 D1
Kirkton of Kingoldrum **DD8** 82 E2
Kirkton of Lethendy **PH2** 82 C3
Kirkton of Logie Buchan **AB41** 91 H2
Kirkton of Maryculter **AB12** 91 G5
Kirkton of Menmuir **DD9** 83 G1
Kirkton of Monikie **DD5** 83 G4
Kirkton of Rayne **AB51** 90 E1
Kirkton of Skene **AB32** 91 G4
Kirkton of Tealing **DD4** 83 F4
Kirktonhill *Aber.* **AB30** 83 H1
Kirktonhill *W.Dun.* **G82** 74 B3
Kirktown **AB42** 99 J5
Kirktown of Alvah **AB45** 98 E4
Kirktown of Auchterless **AB53** 99 F6
Kirktown of Deskford **AB56** 98 D4
Kirktown of Fetteresso **AB39** 91 G6
Kirktown of Slains **AB41** 91 J2
KIRKWALL KW 106 **D6**
Kirkwall Airport **KW15** 106 D7
Kirkwhelpington **NE19** 70 E5
Kirmington **DN39** 52 E1
Kirmond le Mire **LN8** 52 E3
Kirn **PA23** 73 K3
Kirriemuir DD8 82 **E2**
Kirstead Green **NR15** 45 G6
Kirtlebridge **DG11** 69 H6
Kirtleton **DG11** 69 H5
Kirtling **CB8** 33 K3
Kirtling Green **CB8** 33 K3
Kirtlington **OX5** 31 G7
Kirtomy **KW14** 104 C2
Kirton *Lincs.* **PE20** 43 G2
Kirton *Notts.* **NG22** 51 J6
Kirton *Suff.* **IP10** 35 G5
Kirton End **PE20** 43 F1
Kirton Holme **PE20** 43 F1
Kirton in Lindsey **DN21** 52 C3
Kiscadale **KA27** 66 E1
Kislingbury **NN7** 31 H3
Kismeldon Bridge **EX22** 6 B4
Kites Hardwick **CV23** 31 F2
Kitley **PL8** 5 F5
Kittisford **TA21** 7 J3
Kittisford Barton **TA21** 7 J3
Kittle **SA3** 17 J7
Kitt's End **EN5** 23 F2
Kitt's Green **B33** 40 D7
Kitwood **SO24** 11 H1
Kivernoll **HR2** 28 D5
Kiveton Park **S26** 51 G4
Klibreck **IV27** 103 H5
Knabbygates **AB54** 98 D5
Knaith **DN21** 52 B4
Knaith Park **DN21** 52 B4
Knap Corner **SP8** 9 H2
Knaphill **GU21** 22 C6
Knaplock **TA22** 7 G2
Knapp *P. & K.* **PH14** 82 D4
Knapp *Som.* **TA3** 8 C2
Knapthorpe **NG23** 51 K7
Knaptoft **LE17** 41 J7
Knapton *Norf.* **NR28** 45 H2
Knapton *York* **YO26** 58 B4
Knapton Green **HR4** 28 D3
Knapwell **CB23** 33 G2
Knaresborough HG5 57 **J4**
Knarsdale **CA8** 61 H1
Knarston **KW17** 106 C5
Knaven **AB42** 99 G6
Knayton **YO7** 57 K1
Knebworth SG3 33 **F6**
Knedlington **DN14** 58 D7
Kneesall **NG22** 51 K6
Kneesworth **SG8** 33 G4
Kneeton **NG13** 42 A1
Knelston **SA3** 17 H7
Knenhall **ST15** 40 B2
Knettishall **IP22** 44 D7
Knightacott **EX31** 6 E2
Knightcote **CV47** 31 F3
Knightley **ST20** 40 A3
Knightley Dale **ST20** 40 A3
Knighton *Devon* **PL9** 5 F6
Knighton *Dorset* **DT9** 9 F3
Knighton *Leic.* **LE2** 41 J5
Knighton *Poole* **BH21** 10 B5
Knighton (Tref-y-clawdd) *Powys* LD7 28 **B1**
Knighton *Som.* **TA5** 7 K1
Knighton *Staffs.* **ST20** 39 G3
Knighton *Staffs.* **TF9** 39 G1
Knighton *Wilts.* **SN8** 21 F4
Knighton on Teme **WR15** 29 F1
Knightswood **G15** 74 D4
Knightwick **WR6** 29 G3
Knill **LD8** 28 B2
Knipoch **PA34** 79 K5
Knipton **NG32** 42 B2
Knitsley **DH8** 62 C2
Kniveton **DE6** 50 E7
Knock *Arg. & B.* **PA71** 79 G4
Knock *Cumb.* **CA16** 61 H4
Knock *High.* **IV44** 86 C4
Knock *Moray* **AB54** 98 D5
Knock of Auchnahannet **PH26** 89 H1
Knockalava **PA31** 73 H1
Knockally **KW6** 105 G6
Knockaloe Moar **IM5** 54 B5
Knockan **IV27** 102 E7
Knockandhu **AB37** 89 K2
Knockando **AB38** 97 J7
Knockarthur **IV28** 96 E1
Knockbain **IV8** 96 D6
Knockban **IV23** 95 J5
Knockbreck **IV19** 96 E3
Knockbrex **DG6** 65 F6
Knockdamph **IV26** 95 J2
Knockdee **KW12** 105 G2
Knockdow **PA23** 73 K3
Knockdown **GL8** 20 B3
Knockenkelly **KA27** 66 E1
Knockentiber **KA2** 74 B7
Knockfin **IV4** 87 K2
Knockgray **DG7** 67 K4
Knockholt **TN14** 23 H6
Knockholt Pound **TN14** 23 H6
Knockin **SY10** 38 C3
Knockinlaw **KA3** 74 C7
Knockintorran (Cnoc an Torrain) **HS6** 92 C5
Knocklearn **DG7** 65 H3
Knockmill **TN15** 23 J5
Knocknaha **PA28** 66 A2
Knocknain **DG9** 66 D7
Knocknalling **DG7** 67 K5
Knockrome **PA60** 72 D3
Knocksharry **IM5** 54 B5
Knockville **DG8** 64 D3
Knockvologan **PA66** 78 E6
Knodishall **IP17** 35 J2
Knodishall Common **IP17** 35 J2
Knodishall Green **IP17** 35 J2
Knole **TA10** 8 D2
Knolls Green **WA16** 49 H5
Knolton **LL13** 38 C2
Knook **BA12** 20 C7
Knossington **LE15** 42 B5
Knott End-on-Sea **FY6** 55 G5
Knotting **MK44** 32 D2
Knotting Green **MK44** 32 D2
Knottingley WF11 58 **B7**
Knotts **BD23** 56 C4
Knotty Green **HP9** 22 C2
Knowbury **SY8** 28 E1
Knowe **DG8** 64 D3
Knowes of Elrick **AB54** 98 E5
Knowesgate **NE19** 70 E5
Knoweside **KA19** 67 G2
Knowetownhead **TD9** 70 A2
Knowhead **AB43** 99 H5
Knowl Green **CO10** 34 B4
Knowl Hill **RG10** 22 B4
Knowl Wall **ST4** 40 A2
Knowle *Bristol* **BS4** 19 K4
Knowle *Devon* **EX17** 7 F5
Knowle *Devon* **EX33** 6 C2
Knowle *Devon* **EX9** 7 J7
Knowle *Shrop.* **SY8** 28 E1
Knowle *Som.* **TA24** 7 H1
Knowle *W.Mid.* **B93** 30 C1
Knowle Cross **EX5** 7 J6
Knowle Green **PR3** 56 B6
Knowle Hall **TA7** 19 G7
Knowle St. Giles **TA20** 8 C3
Knowlton *Dorset* **BH21** 10 B3
Knowlton *Kent* **CT3** 15 H2
Knowsley **L34** 48 D3
Knowstone **EX36** 7 G3
Knox Bridge **TN17** 14 C3
Knucklas **LD7** 28 B1
Knutsford WA16 49 **G5**
Knypersley **ST8** 49 H7
Krumlin **HX4** 50 C1
Kuggar **TR12** 2 E7
Kyle of Lochalsh **IV40** 86 D2
Kyleakin **IV41** 86 D2
Kylerhea **IV40** 86 D2
Kyles Scalpay (Caolas Scalpaigh) **HS3** 93 H2
Kylesbeg **PH38** 86 C7
Kylesknoydart **PH41** 86 E5
Kylesku **IV27** 102 E5
Kylesmorar **PH41** 86 E5
Kylestrome **IV27** 102 E5
Kyloag **IV24** 96 D2
Kynaston **SY10** 38 C3
Kynnersley **TF6** 39 F4
Kyre Park **WR15** 29 F2

L

Labost **HS2** 100 E3
Lacasaigh **HS2** 101 F5
Lacasdal (Laxdale) **HS2** 101 G4
Laceby **DN37** 53 F2
Lacey Green **HP27** 22 B1
Lach Dennis **CW9** 49 G5
Lacharn (Laugharne) **SA33** 17 G4
Lackford **IP28** 34 B1
Lacklee (Leac a' Li) **HS3** 93 G2
Lacock **SN15** 20 C5
Ladbroke **CV47** 31 F3
Laddingford **ME18** 23 K7
Lade Bank **PE22** 53 G7
Ladies Hill **PR3** 55 H5
Ladock **TR2** 3 F3
Lady Hall **LA18** 54 E1
Ladybank **KY15** 82 E6
Ladycross **PL15** 6 B7
Ladyfield **PA32** 80 B6
Ladykirk **TD15** 77 G6
Ladysford **AB43** 99 H4
Ladywood **WR9** 29 H2
Laga **PH36** 79 H1
Lagalochan **PA35** 79 K6
Lagavulin **PA42** 72 C6
Lagg *Arg. & B.* **PA60** 72 D3
Lagg *N.Ayr.* **KA27** 66 D1
Lagg *S.Ayr.* **KA7** 67 G2
Laggan *Arg. & B.* **PA43** 72 A5
Laggan *High.* **PH34** 87 J5
Laggan *High.* **PH20** 88 D5
Laggan *Moray* **AB55** 90 B1
Laggan *Stir.* **FK18** 81 G6
Lagganulva **PA73** 79 F3
Lagganvoulin **AB37** 89 J3
Laglingarten **PA25** 80 C7
Lagnalean **IV3** 96 D7
Lagrae **DG4** 68 C2
Laguna **PH1** 82 C4
Laid **IV27** 103 G3
Laide **IV22** 95 F2
Laig **PH42** 85 K6
Laight **KA18** 68 B2
Lainchoil **PH25** 89 H3
Laindon **SS15** 24 C3
Lair **PH10** 82 C1
Lairg IV27 96 **C1**
Lairg Lodge **IV27** 96 C1
Lairigmor **PH33** 80 C1
Laisterdyke **BD4** 57 G6
Laithers **AB53** 98 E6
Laithes **CA11** 61 F3
Lake *Devon* **EX31** 6 D2
Lake *Devon* **PL20** 5 F4

Lak - Lew

Lake I.o.W. PO36 11 G6
Lake Wilts. SP4 10 C1
Lakenham NR1 45 G5
Lakenheath IP27 44 B7
Lakesend PE14 43 J6
Lakeside Cumb. LA12 55 G1
Lakeside S.Yorks. DN4 51 H2
Lakeside Thur. RM20 23 J4
Laleham TW18 22 D5
Laleston CF32 18 B4
Lamancha EH46 76 A5
Lamarsh CO8 34 C5
Lamas NR10 45 G3
Lamb Corner CO7 34 E5
Lamb Roe BB7 56 C6
Lambden TD10 77 F6
Lamberhurst TN3 13 K3
Lamberhurst Quarter TN3 13 K3
Lamberton TD15 77 H5
Lambfell Moar IM4 54 B5
Lambley Northumb. CA8 61 H1
Lambley Notts. NG4 41 J1
Lambourn RG17 21 G4
Lambourn Woodlands RG17 21 G4
Lambourne End RM4 23 H2
Lambs Green RH12 13 F3
Lambston SA62 16 C4
Lambton NE38 62 D1
Lamellion PL14 4 C4
Lamerton PL19 4 E3
Lamesley NE11 62 D1
Lamington High. IV18 96 E4
Lamington S.Lan. ML12 75 H7
Lamlash KA27 73 J7
Lamloch DG7 67 K4
Lamonby CA11 60 F3
Lamorna TR19 2 B6
Lamorran TR2 3 F4
Lampert NE48 70 B6
Lampeter (Llanbedr Pont Steffan) SA48 17 J1
Lampeter Velfrey SA67 16 E4
Lamphey SA71 16 D5
Lamplugh CA14 60 B4
Lamport NN6 31 J1
Lamyatt BA4 9 F1
Lana Devon EX22 6 B6
Lana Devon EX22 6 B5
Lanark ML11 75 G6
Lanarth TR12 2 E6
LANCASTER LA 55 H3
Lanchester DH7 62 C2
Lancing BN15 12 E6
Landbeach CB25 33 H2
Landcross EX39 6 C3
Landerberry AB32 91 F4
Landewednack TR12 2 E7
Landford SP5 10 D3
Landhallow KW5 105 G5
Landican CH49 48 B4
Landimore SA3 17 H6
Landkey EX32 6 D2
Landmoth DL6 63 F7
Landore SA1 17 K6
Landrake PL12 4 D4
Landscove TQ13 5 H4
Landshipping SA67 16 D4
Landulph PL12 4 E4
Landwade CB8 33 K2
Landywood WS6 40 B5
Lane Bottom BB10 56 D6
Lane End Bucks. HP14 22 B2
Lane End Cumb. LA19 60 C7
Lane End Derbys. DE55 51 G6
Lane End Dorset BH20 9 H5
Lane End Hants. SO21 11 G2
Lane End Here. HR9 29 F7
Lane End Kent DA2 23 J4
Lane End Wilts. BA12 20 B7
Lane Ends Derbys. DE6 40 E2
Lane Ends Gt.Man. SK6 49 J3
Lane Ends Lancs. BB11 56 C6
Lane Ends N.Yorks. BD22 56 E5
Lane Green WV8 40 A5
Lane Head Dur. DL11 62 C5
Lane Head Dur. DL13 62 B4
Lane Head Gt.Man. WA3 49 F3
Lane Head W.Yorks. HD8 50 D2
Lane Heads PR3 55 H6
Lane Side BB4 56 C7
Laneast PL15 4 C2
Lane-end PL30 4 A4
Laneham DN22 52 B5
Lanehead Dur. DL13 61 K2
Lanehead Northumb. NE48 70 C5
Lanesend SA68 16 D5
Lanesfield WV4 40 B6
Laneshawbridge BB8 56 E5
Langais HS6 92 D5
Langamull PA75 78 E2
Langar NG13 42 A2
Langbank PA14 74 B3
Langbar LS29 57 F4
Langbaurgh TS9 63 G5
Langcliffe BD24 56 D3
Langdale End YO13 63 J3
Langdon Cornw. EX23 4 C1
Langdon Cornw. PL15 6 B7
Langdon Beck DL12 61 K3
Langdon Hills SS16 24 C3
Langdon House EX7 5 K3
Langdyke KY8 82 E7
Langford Cen.Beds. SG18 32 E4
Langford Essex CM9 24 E1
Langford Notts. NG23 52 B7
Langford Oxon. GL7 21 F1
Langford Budville TA21 7 K3
Langham Essex CO4 34 E5
Langham Norf. NR25 44 E1
Langham Rut. LE15 42 B4
Langham Suff. IP31 34 D2
Langham Moor CO4 34 E5
Langho BB6 56 B6
Langholm DG13 69 J5
Langland SA3 17 K7
Langlands DG6 65 G5
Langlee TD8 70 B2
Langleeford NE71 70 E1
Langley Ches.E. SK11 49 J5
Langley Derbys. NG16 41 G1
Langley Essex CB11 33 H5
Langley Glos. GL54 30 B6
Langley Gt.Man. M24 49 H2
Langley Hants. SO45 11 F4
Langley Herts. SG4 33 F6
Langley Kent ME17 14 C2
Langley Northumb. NE47 70 D7
Langley Oxon. OX29 30 E7
Langley Slo. SL3 22 D4
Langley Som. TA4 7 J3
Langley W.Suss. GU33 12 B4
Langley Warks. CV37 30 C2
Langley Burrell SN15 20 C4
Langley Corner SL3 22 D3
Langley Green Derbys. DE6 40 E2
Langley Green W.Suss. RH11 13 F3
Langley Green Warks. CV35 30 D2
Langley Heath ME17 14 D2
Langley Marsh TA4 7 J3
Langley Mill NG16 41 G1
Langley Moor DH7 62 D2
Langley Park DH7 62 D2
Langley Street NR14 45 H5
Langney BN23 13 K6
Langold S81 51 H4
Langore PL15 4 C2
Langport TA10 8 D2
Langrick PE22 43 F1
Langrick Bridge PE22 43 F1
Langridge B. & N.E.Som. BA1 20 A5
Langridge Devon EX37 6 D3
Langridgeford EX37 6 D3
Langrigg CA7 60 C2
Langrish GU32 11 J2
Langsett S36 50 E2
Langshaw TD1 76 D7
Langshawburn DG13 69 H3
Langside Glas. G43 74 D4
Langside P. & K. PH6 81 J6
Langskaill KW17 106 D3
Langstone Hants. PO9 11 J4
Langstone Newport NP18 19 G2
Langthorne DL8 62 D7
Langthorpe YO51 57 J3
Langthwaite DL11 62 B6
Langtoft E.Riding YO25 59 G3
Langtoft Lincs. PE6 42 E4
Langton Dur. DL2 62 C5
Langton Lincs. PE23 53 G5
Langton Lincs. LN9 53 F6
Langton N.Yorks. YO17 58 D3
Langton by Wragby LN8 52 E5
Langton Green Kent TN3 13 J3
Langton Green Suff. IP23 35 F1
Langton Herring DT3 9 F6
Langton Long Blandford DT11 9 H4
Langton Matravers BH19 9 J7
Langtree EX38 6 C4
Langtree Week EX38 6 C4
Langwathby CA10 61 G3
Langwell IV27 96 B1
Langwell House KW7 105 G6
Langwith NG20 51 H5
Langworth LN3 52 D5
Lanivet PL30 4 A4
Lank PL30 4 A3
Lanlivery PL30 4 A5
Lanner TR16 2 E4
Lanoy PL15 4 C3
Lanreath PL13 4 B5
Lansallos PL13 4 B5
Lansdown BA1 20 A5
Lanteglos PL32 4 A2
Lanteglos Highway PL23 4 B5
Lanton Northumb. NE71 77 H7
Lanton Sc.Bord. TD8 70 B1
Lanvean TR8 3 F2
Lapford EX17 7 F5
Laphroaig PA42 72 B6
Lapley ST19 40 A4
Lapworth B94 30 C1
Larach na Gaibhre PA31 73 F3
Larachbeg PA34 79 H3
Larbert FK5 75 G2
Larbreck PR3 55 H5
Larden Green CW5 48 E7
Larg DG8 64 D3
Largie AB52 90 E1
Largiemore PA21 73 H2
Largoward KY9 83 F7
Largs KA30 74 A5
Largue AB54 98 E6
Largybaan PA28 66 A2
Largybeg KA27 66 E1
Largymore KA27 66 E1
Lark Hall CB8 33 J3
Larkfield PA16 74 A3
Larkhall ML9 75 F5
Larkhill SP4 20 E7
Larklands DE7 41 G1
Larling NR16 44 D7
Larriston TD9 70 A4
Lartington DL12 62 B5
Lary AB35 90 B4
Lasborough GL8 20 B2
Lasham GU34 21 K7
Lashbrook EX22 6 C5
Lashenden TN27 14 D3
Lassington GL2 29 G6
Lassintullich PH16 81 J2
Lassodie KY12 75 K1
Lasswade EH18 76 B4
Lastingham YO62 63 J7
Latchford WA4 49 F4
Latchingdon CM3 24 E1
Latchley PL18 4 E3
Lately Common WN7 49 F3
Lathallan Mill KY9 83 F7
Lathbury MK16 32 B4
Latheron KW5 105 G5
Latheronwheel KW5 105 G5
Lathockar KY16 83 F6
Lathones KY15 83 F7
Lathrisk KY15 82 D7
Latimer HP5 22 D2
Latteridge BS37 19 K3
Lattiford BA9 9 F2
Latton SN6 20 D2
Lauchentyre DG7 65 F5
Lauchintilly AB51 91 F3
Lauder TD2 76 D6
Laugharne (Lacharn) SA33 17 G4
Laughterton LN1 52 B5
Laughton E.Suss. BN8 13 J5
Laughton Leics. LE17 41 J7
Laughton Lincs. NG34 42 D2
Laughton Lincs. DN21 52 B3
Laughton en le Morthen S25 51 H4
Launcells EX23 6 A5
Launcells Cross EX23 6 A5
Launceston PL15 6 B7
Launde Abbey LE7 42 A5
Launton OX26 31 H6
Laurencekirk AB30 91 F7
Laurieston D. & G. DG7 65 G4
Laurieston Falk. FK2 75 H3
Lavendon MK46 32 C3
Lavenham CO10 34 D4
Laverhay DG10 69 G4
Lavernock CF64 18 E5
Laversdale CA6 69 K7
Laverstock SP1 10 C1
Laverstoke RG28 21 H7
Laverton Glos. WR12 30 B5
Laverton N.Yorks. HG4 57 H2
Laverton Som. BA2 20 A6
Lavister LL12 48 C7
Law ML8 75 G5
Lawers P. & K. PH6 81 J5
Lawers P. & K. PH15 81 H4
Lawford Essex CO11 34 E5
Lawford Som. TA4 7 K2
Lawhitton PL15 6 B7
Lawkland LA2 56 C3
Lawkland Green LA2 56 C3
Lawley TF4 39 F5
Lawnhead ST20 40 A3
Lawrence Weston BS11 19 J4
Lawrenny SA68 16 D5
Laws DD5 83 F4
Lawshall IP29 34 C3
Lawshall Green IP30 34 C3
Lawton HR6 28 D3
Laxdale (Lacasdal) HS2 101 G4
Laxey IM4 54 D5
Laxfield IP13 35 G1
Laxfirth Shet. ZE2 107 N7
Laxfirth Shet. ZE2 107 N8
Laxford Bridge IV27 102 E4
Laxo ZE2 107 N6
Laxton E.Riding DN14 58 D7
Laxton Northants. NN17 42 C6
Laxton Notts. NG22 51 K6
Laycock BD22 57 F5
Layer Breton CO2 34 D7
Layer de la Haye CO2 34 D7
Layer Marney CO5 34 D7
Layham IP7 34 E4
Laymore TA20 8 C4
Layter's Green SL9 22 C2
Laytham YO42 58 D6
Layton FY3 55 G6
Lazenby TS6 63 G4
Lazonby CA10 61 G3
Lea Derbys. DE4 51 F7
Lea Here. HR9 29 F6
Lea Lincs. DN21 52 B4
Lea Shrop. SY9 38 C7
Lea Shrop. SY5 38 D5
Lea Wilts. SN16 20 C3
Lea Bridge DE4 51 F7
Lea Green WR6 29 F2
Lea Marston B76 40 E6
Lea Town PR4 55 H6
Lea Yeat LA10 56 C1
Leac a' Li (Lacklee) HS3 93 G2
Leachd PA27 73 J1
Leachkin IV3 96 D7
Leadburn EH46 76 A5
Leaden Roding CM6 33 J7
Leadenham LN5 52 C7
Leaderfoot TD6 76 D7
Leadgate Cumb. CA9 61 J2
Leadgate Dur. DH8 62 C1
Leadgate Northumb. NE17 62 C1
Leadhills ML12 68 D2
Leadingcross Green ME17 14 D2
Leafield OX29 30 E7
Leagrave LU4 32 D6
Leake Commonside PE22 53 G7
Leake Hurn's End PE22 43 H1
Lealands BN27 13 J5
Lealholm YO21 63 J6
Lealt Arg. & B. PA60 72 E1
Lealt High. IV51 94 B5
Leam S32 50 E5
Leamington Hastings CV23 31 F2
Leamington Spa CV32 30 E2
Leamoor Common SY7 38 D7
Leanach Arg. & B. PA27 73 J1
Leanach High. IV2 96 E7
Leanaig IV7 96 C6
Leanoch IV30 97 J6
Leargybreck PA60 72 D3
Leasgill LA7 55 H1
Leasingham NG34 42 D1
Leask AB41 91 J1
Leason SA3 17 H6
Leasowe CH46 48 B3
Leat PL15 6 B7
Leatherhead KT22 22 E6
Leathley LS21 57 H5
Leaton Shrop. SY4 38 D4
Leaton Tel. & W. TF6 39 F4
Leaveland ME13 15 F2
Leavenheath CO6 34 D5
Leavening YO17 58 D3
Leaves Green BR2 23 H5
Leavesden Green WD25 22 E1
Lebberston YO11 59 G1
Lechlade GL7 21 F2
Leck LA6 56 B2
Leckford SO20 10 E1
Leckfurin KW14 104 C3
Leckgruinart PA44 72 A4
Leckhampstead Bucks. MK18 31 J5
Leckhampstead W.Berks. RG20 21 H4
Leckhampstead Thicket RG20 21 H4
Leckhampton GL53 29 J7
Leckie High. IV22 95 G5
Leckie Stir. FK8 74 E1
Leckmelm IV23 95 H3
Leckroy PH31 87 K5
Leckuary PA31 73 G1
Leckwith CF11 18 E4
Leconfield HU17 59 G5
Ledaig PA37 80 A4
Ledard FK8 81 F7
Ledbeg IV27 102 E7
Ledburn LU7 32 B6
Ledbury HR8 29 G5
Ledcharrie FK20 81 G5
Ledgemoor HR4 28 D3
Ledicot HR6 28 D2
Ledmore Arg. & B. PA72 79 G3
Ledmore High. IV27 102 E7
Lednagullin KW14 104 D2
Ledsham Ches.W. & C. CH66 48 C5
Ledsham W.Yorks. LS25 57 K7
Ledston WF10 57 K7
Ledstone TQ7 5 H6
Ledwell OX7 31 F6
Lee Arg. & B. PA67 79 F5
Lee Devon EX34 6 C1
Lee Hants. SO51 10 E3
Lee Lancs. LA2 55 J4
Lee Shrop. SY12 38 D2
Lee Brockhurst SY4 38 E3
Lee Chapel SS15 24 C3
Lee Clump HP16 22 C1
Lee Mill Bridge PL21 5 F5
Lee Moor PL7 5 F4
Leebotten ZE2 107 N10
Leebotwood SY6 38 D6
Leece LA12 55 F3
Leeds Kent ME17 14 D2
LEEDS W.Yorks. LS 57 H6
Leeds Bradford International Airport LS19 57 H5
Leedstown TR27 2 D5
Leegomery TF1 39 F4
Leek ST13 49 J7
Leek Wootton CV35 30 D2
Leekbrook ST13 49 J7
Leeming N.Yorks. DL7 57 H1
Leeming W.Yorks. BD22 57 F6
Leeming Bar DL7 57 H1
Lee-on-the-Solent PO13 11 G4
Lees Derbys. DE6 40 E2
Lees Gt.Man. OL4 49 J2
Leeswood CH7 48 B7
Leftwich CW9 49 F5
Legars TD5 77 F6
Legbourne LN11 53 G4
Legerwood TD4 76 D6
Legsby LN8 52 E4
LEICESTER LE 41 H5
Leicester Forest East LE3 41 H5
Leideag HS9 84 B5
Leigh Dorset DT9 9 F4
Leigh Dorset BH21 10 B5
Leigh Gt.Man. WN7 49 F2
Leigh Kent TN11 23 J7
Leigh Shrop. SY5 38 C5
Leigh Surr. RH2 23 F7
Leigh Wilts. SN6 20 D2
Leigh Worcs. WR6 29 G3
Leigh Beck SS8 24 E3
Leigh Common BA9 9 G2
Leigh Delamere SN14 20 B4
Leigh Green TN30 14 E4
Leigh Park PO9 11 J4
Leigh Sinton WR13 29 G3
Leigh upon Mendip BA3 19 K7
Leigh Woods BS8 19 J4
Leigham PL6 5 F5
Leighland Chapel TA23 7 J2
Leigh-on-Sea SS9 24 E3
Leighterton GL8 20 B2
Leighton N.Yorks. HG4 57 G2
Leighton (Tre'r Llai) Powys SY21 38 B5
Leighton Shrop. SY5 39 F5
Leighton Som. BA11 20 A7
Leighton Bromswold PE28 32 E1
Leighton Buzzard LU7 32 C6
Leinthall Earls HR6 28 D2
Leinthall Starkes SY8 28 D2
Leintwardine SY7 28 D1
Leire LE17 41 H7
Leirinmore IV27 103 G2
Leiston IP16 35 J2
Leitfie PH11 82 D3
Leith EH6 76 A3
Leitholm TD12 77 F6
Lelant TR26 2 C5
Lelley HU12 59 J6
Lemington NE15 71 G7
Lemnas AB43 99 G4
Lempitlaw TD5 77 F7
Lemsford AL8 33 F7
Lenchwick WR11 30 B4
Lendalfoot KA26 67 F4
Lendrick Lodge FK17 81 G7
Lenham ME17 14 D2
Lenham Heath ME17 14 E3
Lenie IV63 88 C2
Lenimore KA27 73 G6
Lennel TD12 77 G6
Lennox Plunton DG6 65 G5
Lennoxtown G66 74 E3
Lent Rise SL6 22 C3
Lenton Lincs. NG33 42 D2
Lenton Nott. NG7 41 H2
Lenton Abbey NG7 41 H2
Lenwade NR9 44 E4
Lenzie G66 74 E3
Leoch DD3 82 E4
Leochel-Cushnie AB33 90 D3
Leominster HR6 28 D3
Leonard Stanley GL10 20 B1
Leorin PA42 72 B6
Lepe SO45 11 F5
Lephin IV55 93 G7
Lephinchapel PA27 73 H1
Lephinmore PA27 73 H1
Leppington YO17 58 D3
Lepton HD8 50 E1
Lerags PA34 79 K5
Lerryn PL22 4 B5
Lerwick ZE1 107 N8
Lesbury NE66 71 H2
Leschangie AB51 91 F3
Leslie Aber. AB52 90 D2
Leslie Fife KY6 82 D7
Lesmahagow ML11 75 G7
Lesnewth PL35 4 B1
Lessendrum AB54 98 D6
Lessingham NR12 45 H3
Lessness Heath DA8 23 H4
Lessonhall CA7 60 D1
Leswalt DG9 64 A4
Letchmore Heath WD25 22 E2
Letchworth Garden City SG6 33 F5
Letcombe Bassett OX12 21 G3
Letcombe Regis OX12 21 G3
Leth Meadhanach HS8 84 C3
Letham Angus DD8 83 G3
Letham Falk. FK2 75 G2
Letham Fife KY15 82 E6
Lethanhill KA6 67 J2
Lethenty AB53 99 G6
Letheringham IP13 35 G3
Letheringsett NR25 44 E2
Lettaford TQ13 7 F7
Letter Finlay PH34 87 J5
Letterewe IV22 95 F4
Letterfearn IV40 86 E2
Lettermorar PH40 86 D6
Lettermore Arg. & B. PA72 79 F3
Lettermore High. IV27 103 J4
Letters IV23 95 H3
Lettershaws ML12 68 D1
Letterston SA62 16 C3
Lettoch High. PH25 89 H3
Lettoch High. PH26 89 J1
Letton Here. SY7 28 C1
Letton Here. HR3 28 C4
Letty Green SG14 33 F7
Letwell S81 51 H4
Leuchars KY16 83 F5
Leumrabhagh HS2 101 F6
Leurbost (Liurbost) HS2 101 F5
Leusdon TQ13 5 H3
Levedale ST18 40 A4
Level's Green CM23 33 H6
Leven E.Riding HU17 59 H5
Leven Fife KY8 82 E7
Levencorroch KA27 66 E1
Levenhall EH21 76 B3
Levens LA8 55 H1
Levens Green SG11 33 G6
Levenshulme M12 49 H3
Levenwick ZE2 107 N10
Leverburgh (An T-Òb) HS5 93 F3
Leverington PE13 43 H4
Leverstock Green HP3 22 D1
Leverton PE22 43 G1
Leverton Lucasgate PE22 43 H1
Leverton Outgate PE22 43 H1
Levington IP10 35 G5
Levisham YO18 63 K7
Levishie IV63 88 B3
Lew OX18 21 G1
Lewannick PL15 4 C2

Lew - Lla

Lewcombe DT2 8 E4
Lewdown EX20 6 C7
Lewes BN7 13 H5
Leweston SA62 16 C3
Lewisham SE13 23 G4
Lewiston IV63 88 C2
Lewistown CF32 18 C3
Lewknor OX49 22 A2
Leworthy EX32 6 E2
Lewson Street ME9 25 F5
Lewth PR4 55 H6
Lewtrenchard EX20 6 C7
Ley *Aber.* AB33 90 D3
Ley *Cornw.* PL14 4 B4
Ley Green SG4 32 E6
Leybourne ME19 23 K6
Leyburn DL8 62 C7
Leyland PR25 55 J7
Leylodge AB51 91 F3
Leymoor HD3 50 D1
Leys *Aber.* AB42 99 J5
Leys *Aber.* AB34 90 C4
Leys *P. & K.* PH13 82 D4
Leys of Cossans DD8 82 E3
Leysdown-on-Sea ME12 25 G4
Leysmill DD11 83 H3
Leysters HR6 28 E2
Leyton E10 23 G3
Leytonstone E11 23 G3
Lezant PL15 4 D3
Lezerea TR13 2 D5
Lhanbryde IV30 97 K5
Liatrie IV4 87 J1
Libanus LD3 27 J6
Libberton ML11 75 H6
Libbery WR7 29 J3
Liberton EH16 76 A4
Liceasto HS3 93 G2
Lichfield WS13 40 D5
Lickey B45 29 J1
Lickey End B60 29 J1
Lickfold GU28 12 C4
Liddaton Green EX20 6 C7
Liddel KW17 106 D9
Liddesdale PH33 79 J2
Liddington SN4 21 F3
Lidgate *Derbys.* S18 51 F5
Lidgate *Suff.* CB8 34 B3
Lidgett NG21 51 J6
Lidlington MK43 32 C5
Lidsey PO22 12 C6
Lidsing ME7 24 D5
Lidstone OX7 30 E6
Lienassie IV40 87 F2
Lieurary KW14 105 F2
Liff DD2 82 E4
Lifton PL16 6 B7
Liftondown PL15 6 B7
Lightcliffe HX3 57 G7
Lighthorne CV35 30 E3
Lighthorne Heath CV33 30 E3
Lightwater GU18 22 C5
Lightwood ST3 40 B1
Lightwood Green *Ches.E.* CW3 39 F1
Lightwood Green *Wrex.* LL13 38 C1
Lilbourne CV23 31 G1
Lilburn Tower NE66 71 F1
Lillesdon TA3 8 C2
Lilleshall TF10 39 G4
Lilley *Herts.* LU2 32 E6
Lilley *W.Berks.* RG20 21 H4
Lilliesleaf TD6 70 A1
Lilling Green YO32 58 C3
Lillingstone Dayrell MK18 31 J5
Lillingstone Lovell MK18 31 J4
Lillington *Dorset* DT9 9 F3
Lillington *Warks.* CV32 30 E2
Lilliput BH14 10 B6
Lilly EX32 6 D2
Lilstock TA5 7 K1
Lilyhurst TF11 39 G4
Limbury LU3 32 D6
Lime Side OL8 49 J2
Limefield BL9 49 H1
Limehillock AB54 98 D5
Limehurst OL8 49 J2
Limekilnburn ML3 75 F5
Limekilns KY11 75 J2
Limerigg FK1 75 G3
Limerstone PO30 11 F6
Limington BA22 8 E2
Limpenhoe NR13 45 H5
Limpley Stoke BA2 20 A5
Limpsfield RH8 23 H6
Limpsfield Chart TN8 23 H6
Linaclate (Lionacleit) HS7 92 C7
Linbriggs NE65 70 D3
Linby NG15 51 H7
Linchmere GU27 12 B3
Lincluden DG2 65 K3
LINCOLN LN 52 C5
Lincomb DY13 29 H2
Lincombe *Devon* TQ9 5 H5
Lincombe *Devon* TQ7 5 H6
Lindal in Furness LA12 55 F2
Lindale LA11 55 H1
Lindean TD7 76 C7
Lindertis DD8 82 E2
Lindfield RH16 13 G4
Lindford GU35 12 B3
Lindifferon KY15 82 E6
Lindisfarne (Holy Island) TD15 77 K6
Lindley LS21 57 H5
Lindores KY14 82 D6
Lindow End WA16 49 H5
Lindridge WR15 29 F2
Lindsaig PA21 73 H3
Lindsell CM6 33 K6
Lindsey IP7 34 D4
Lindsey Tye IP7 34 D4
Linfitts OL3 49 J2
Linford *Hants.* BH24 10 C4
Linford *Thur.* SS17 24 C4
Linford Wood MK13 32 B4
Lingague IM9 54 B6
Lingards Wood HD7 50 C1
Lingdale TS12 63 H5
Lingen SY7 28 C2
Lingfield RH7 23 G7
Lingley Green WA5 48 E4
Lingwood NR13 45 H5
Linhead AB45 98 E5
Linhope TD9 69 K3
Linicro IV51 93 J5
Linkend GL19 29 H5
Linkenholt SP11 21 G6
Linkhill TN18 14 D5
Linkinhorne PL17 4 D3
Linklater KW17 106 D9
Linksness *Ork.* KW16 106 B7
Linksness *Ork.* KW17 106 E6
Linktown KY1 76 A1
Linley *Shrop.* SY9 38 C6
Linley *Shrop.* TF12 39 F6
Linley Green WR6 29 F3
Linlithgow EH49 75 J3
Linlithgow Bridge EH49 75 H3
Linn of Muick Cottage AB35 90 B6
Linnels NE46 70 E7
Linney SA71 16 B6
Linshiels NE65 70 D3
Linsiadar HS2 100 E4
Linsidemore IV27 96 C2
Linslade LU7 32 C6
Linstead Parva IP19 35 H1
Linstock CA6 60 F1
Linthwaite HD7 50 D1
Lintlaw TD11 77 G5
Lintmill AB56 98 D4
Linton *Cambs.* CB21 33 J4
Linton *Derbys.* DE12 40 E4
Linton *Here.* HR9 29 F6
Linton *Kent* ME17 14 C2
Linton *N.Yorks.* BD23 56 E3
Linton *Sc.Bord.* TD5 70 C1
Linton *W.Yorks.* LS22 57 J5
Linton-on-Ouse YO30 57 K3
Lintzford NE39 62 C1
Linwood *Hants.* BH24 10 C4
Linwood *Lincs.* LN8 52 E4
Linwood *Renf.* PA3 74 C4
Lionacleit (Linaclate) HS7 92 C7
Lional (Lionel) HS2 101 H1
Lionel (Lional) HS2 101 H1
Liphook GU30 12 B3
Lipley TF9 39 G2
Liscard CH45 48 C3
Liscombe TA22 7 G2
Liskeard PL14 4 C4
L'Islet GY2 3 J5
Lismore PA34 79 K4
Liss GU33 11 J2
Liss Forest GU33 11 J2
Lissett YO25 59 H4
Lissington LN3 52 E4
Liston CO10 34 C4
Lisvane CF14 18 E3
Liswerry NP19 19 G3
Litcham PE32 44 C4
Litchborough NN12 31 H3
Litchfield RG28 21 H6
Litherland L21 48 C3
Litlington *Cambs.* SG8 33 G4
Litlington *E.Suss.* BN26 13 J6
Little Abington CB21 33 J4
Little Addington NN14 32 C1
Little Alne B95 30 C2
Little Altcar L37 48 C2
Little Amwell SG13 33 G7
Little Ann SP11 21 G7
Little Asby CA16 61 H6
Little Assynt IV27 102 D6
Little Aston B74 40 C6
Little Atherfield PO38 11 F7
Little Ayton TS9 63 G5
Little Baddow CM3 24 D1
Little Badminton GL9 20 B3
Little Ballinluig PH15 82 A2
Little Bampton CA7 60 D1
Little Bardfield CM7 33 K5
Little Barford PE19 32 E3
Little Barningham NR11 45 F2
Little Barrington OX18 30 D7
Little Barrow CH3 48 D6
Little Barugh YO17 58 D2
Little Bavington NE19 70 E6
Little Bealings IP13 35 G4
Little Bedwyn SN8 21 F5
Little Beeby LE7 41 J5
Little Bentley CO7 35 F6
Little Berkhamsted SG13 23 F1
Little Billing NN3 32 B2
Little Birch HR2 28 E5
Little Bispham FY5 55 G5
Little Blakenham IP8 35 F4
Little Bloxwich WS3 40 C5
Little Bollington WA14 49 G4
Little Bookham KT23 22 E6
Little Bourton OX17 31 F4
Little Bowden LE16 42 A7
Little Bradley CB9 33 K3
Little Brampton SY7 38 C7
Little Braxted CM8 34 C7
Little Brechin DD9 83 G1
Little Brickhill MK17 32 C5
Little Bridgeford ST18 40 A3
Little Brington NN7 31 H2
Little Bromley CO11 34 E6
Little Broughton CA13 60 B3
Little Budworth CW6 48 E6
Little Burdon DL1 62 E5
Little Burstead CM12 24 C2
Little Burton YO25 59 H5
Little Bytham NG33 42 D4
Little Canford BH21 10 B5
Little Carlton *Lincs.* LN11 53 G4
Little Carlton *Notts.* NG23 51 K7
Little Casterton PE9 42 D4
Little Catwick HU17 59 H5
Little Catworth PE28 32 D1
Little Cawthorpe LN11 53 G4
Little Chalfield SN12 20 B5
Little Chalfont HP6 22 C2
Little Chart TN27 14 E3
Little Chesterford CB10 33 J4
Little Chesterton OX26 31 G6
Little Cheverell SN10 20 C6
Little Clacton CO16 35 F7
Little Clanfield OX18 21 F1
Little Clifton CA14 60 B4
Little Coates DN34 53 F2
Little Comberton WR10 29 J4
Little Common TN39 14 C7
Little Compton GL56 30 D5
Little Corby CA4 61 F1
Little Cornard CO10 34 C5
Little Cowarne HR7 29 F3
Little Coxwell SN7 21 F2
Little Crakehall DL8 62 D7
Little Cransley NN14 32 B1
Little Crawley MK16 32 C4
Little Creaton NN6 31 J1
Little Creich IV24 96 D3
Little Cressingham IP25 44 C6
Little Crosby L23 48 C2
Little Crosthwaite CA12 60 D4
Little Cubley DE6 40 D2
Little Dalby LE14 42 A4
Little Dens AB42 99 J6
Little Dewchurch HR2 28 E5
Little Ditton CB8 33 K3
Little Doward HR9 28 E7
Little Down SP11 21 G6
Little Downham CB6 43 J7
Little Drayton TF9 39 F2
Little Driffield YO25 59 G4
Little Dunham PE32 44 C4
Little Dunkeld PH8 82 B3
Little Dunmow CM6 33 K6
Little Durnford SP4 10 C1
Little Easton CM6 33 K6
Little Eaton DE21 41 F1
Little Eccleston PR3 55 H6
Little Ellingham NR17 44 E6
Little End CM5 23 J1
Little Everdon NN11 31 G3
Little Eversden CB23 33 G3
Little Fakenham IP24 34 D1
Little Faringdon GL7 21 F1
Little Fencote DL7 62 D7
Little Fenton LS25 58 B6
Little Finborough IP14 34 E3
Little Fransham NR19 44 D4
Little Gaddesden HP4 32 C7
Little Garway HR2 28 D6
Little Gidding PE28 42 E7
Little Glemham IP13 35 H3
Little Glenshee PH1 82 A4
Little Gorsley HR9 29 F6
Little Gransden SG19 33 F3
Little Green *Cambs.* SG8 33 F4
Little Green *Notts.* NG13 42 A1
Little Green *Suff.* IP23 34 E1
Little Green *Suff.* IP22 34 E1
Little Green *Wrex.* SY13 38 D1
Little Grimsby LN11 53 G3
Little Gringley DN22 51 K4
Little Gruinard IV22 95 F3
Little Habton YO17 58 D2
Little Hadham SG11 33 H6
Little Hale NG34 42 E1
Little Hallingbury CM22 33 H7
Little Hampden HP16 22 B1
Little Haresfield GL10 20 B1
Little Harrowden NN9 32 B1
Little Haseley OX44 21 K1
Little Hatfield HU11 59 H5
Little Hautbois NR12 45 G3
Little Haven *Pembs.* SA62 16 B4
Little Haven *W.Suss.* RH12 12 E3
Little Hay WS14 40 D5
Little Hayfield SK22 50 C4
Little Haywood ST18 40 C3
Little Heath CV6 41 F7
Little Hereford SY8 28 E2
Little Hockham IP24 44 D6
Little Horkesley CO6 34 D5
Little Hormead SG9 33 H6
Little Horsted TN22 13 H5
Little Horton SN10 20 D5
Little Horwood MK17 31 J5
Little Houghton NN7 32 B3
Little Hucklow SK17 50 D5
Little Hulton M38 49 G2
Little Hungerford RG18 21 J4
Little Hutton YO7 57 K2
Little Irchester NN8 32 C2
Little Keyford BA11 20 A7
Little Kimble HP17 22 B1
Little Kineton CV35 30 E3
Little Kingshill HP16 22 B2
Little Langdale LA22 60 E6
Little Langford SP3 10 B1
Little Laver CM5 23 J1
Little Lawford CV23 31 F1
Little Leigh CW8 49 F5
Little Leighs CM3 34 B7
Little Lever BL3 49 G2
Little Ley AB51 90 E3
Little Linford MK19 32 B4
Little Linton CB21 33 J4
Little London *Bucks.* HP18 31 H7
Little London *E.Suss.* TN21 13 J5
Little London *Essex* CM23 33 H6
Little London *Hants.* SP11 21 G7
Little London *Hants.* RG26 21 K6
Little London *I.o.M.* IM6 54 C5
Little London *Lincs.* PE12 43 H3
Little London *Lincs.* PE11 43 F3
Little London *Lincs.* LN9 53 G5
Little London *Lincs.* LN8 52 E4
Little London *Norf.* PE34 43 J3
Little London *Norf.* IP26 44 B6
Little London *Oxon.* OX14 21 J1
Little London *Powys* SY17 37 K7
Little London *Suff.* IP14 34 E3
Little London *W.Yorks.* LS19 57 H6
Little Longstone DE45 50 D5
Little Lyth SY3 38 D5
Little Malvern WR14 29 G4
Little Maplestead CO9 34 C5
Little Marcle HR8 29 F5
Little Marland EX20 6 D4
Little Marlow SL7 22 B3
Little Marsden BB9 56 D6
Little Massingham PE32 44 B3
Little Melton NR9 45 F5
Little Milford SA62 16 C4
Little Mill NP4 19 G1
Little Milton OX44 21 K1
Little Missenden HP7 22 C2
Little Musgrave CA17 61 J5
Little Ness SY4 38 D4
Little Neston CH64 48 B5
Little Newcastle SA62 16 C3
Little Newsham DL2 62 C5
Little Oakley *Essex* CO12 35 G6
Little Oakley *Northants.* NN18 42 B7
Little Odell MK43 32 C3
Little Offley SG5 32 E6
Little Onn ST20 40 A4
Little Orton *Cumb.* CA5 60 E1
Little Orton *Leics.* CV9 41 F5
Little Ouse CB7 44 A7
Little Ouseburn YO26 57 K3
Little Overton LL13 38 C1
Little Packington CV7 40 E7
Little Parndon CM20 33 H7
Little Paxton PE19 32 E2
Little Petherick PL27 3 G1
Little Plumpton PR4 55 G6
Little Plumstead NR13 45 H4
Little Ponton NG33 42 C2
Little Posbrook PO14 11 G4
Little Potheridge EX20 6 D4
Little Preston NN11 31 G3
Little Raveley PE28 43 F7
Little Ribston LS22 57 J4
Little Rissington GL54 30 C7
Little Rogart IV28 96 E1
Little Rollright OX7 30 D5
Little Ryburgh NR21 44 D3
Little Ryle NE66 71 F2
Little Ryton SY5 38 D5
Little Salkeld CA10 61 G3
Little Sampford CB10 33 K5
Little Saxham IP29 34 C2
Little Scatwell IV14 95 K6
Little Shelford CB22 33 H3
Little Shrawardine SY5 38 C4
Little Silver EX16 7 H5
Little Singleton FY6 55 G6
Little Smeaton *N.Yorks.* WF8 51 H1
Little Smeaton *N.Yorks.* DL6 62 E6
Little Snoring NR21 44 D2
Little Sodbury BS37 20 A3
Little Sodbury End BS37 20 A3
Little Somborne SO20 10 E1
Little Somerford SN15 20 C3
Little Soudley TF9 39 G3
Little Stainforth BD24 56 D3
Little Stainton TS21 62 E5
Little Stanney CH2 48 D5
Little Staughton MK44 32 E2
Little Steeping PE23 53 H6
Little Stoke ST15 40 B2
Little Stonham IP14 35 F3
Little Street CB6 43 J7
Little Stretton *Leics.* LE2 41 J6
Little Stretton *Shrop.* SY6 38 D6
Little Strickland CA10 61 G5
Little Stukeley PE28 33 F1
Little Sugnall ST21 40 A2
Little Sutton CH66 48 C5
Little Swinburne NE46 70 E6
Little Tarrington HR1 29 F4
Little Tew OX7 30 E6
Little Tey CO6 34 C6
Little Thetford CB6 33 J1
Little Thornage NR25 44 E2
Little Thornton FY5 55 G5
Little Thorpe SR8 63 F2
Little Thurlow CB9 33 K3
Little Thurlow Green CB9 33 K3
Little Thurrock RM17 24 C4
Little Torboll IV25 96 E2
Little Torrington EX38 6 C4
Little Tosson NE65 71 F3
Little Totham CM9 34 C7
Little Town *Cumb.* CA12 60 D5
Little Town *Lancs.* PR3 56 B6
Little Town *Warr.* WA3 49 F3
Little Twycross CV9 41 F5
Little Urswick LA12 55 F2
Little Wakering SS3 25 F3
Little Walden CB10 33 J4
Little Waldingfield CO10 34 D4
Little Walsingham NR22 44 D2
Little Waltham CM3 34 B7
Little Warley CM13 24 C2
Little Washbourne GL20 29 J5
Little Weighton HU20 59 F6
Little Welland WR13 29 H5
Little Welnetham IP30 34 C2
Little Wenham CO7 34 E5
Little Wenlock TF6 39 F5
Little Whittington NE19 70 E7
Little Wilbraham CB21 33 J3
Little Wishford SP2 10 B1
Little Witcombe GL3 29 J7
Little Witley WR6 29 G2
Little Wittenham OX14 21 J2
Little Wittingham Green IP21 35 G1
Little Wolford CV36 30 D5
Little Woodcote SM5 23 F5
Little Wratting CB9 33 K4
Little Wymington NN10 32 C2
Little Wymondley SG4 33 F6
Little Wyrley WS3 40 C5
Little Wytheford SY4 38 E4
Little Yeldham CO9 34 B5
Littlebeck YO22 63 K6
Littleborough *Devon* EX17 7 G4
Littleborough *Gt.Man.* OL15 49 J1
Littleborough *Notts.* DN22 52 B4
Littlebourne CT3 15 H2
Littlebredy DT2 8 E6
Littlebury CB11 33 J5
Littlebury Green CB11 33 H5
Littledean GL14 29 F7
Littleferry KW10 97 F2
Littleham *Devon* EX39 6 C3
Littleham *Devon* EX8 7 J7
Littlehampton BN17 12 D6
Littlehempston TQ9 5 J4
Littlehoughton NE66 71 H2
Littlemill *E.Ayr.* KA6 67 J2
Littlemill *High.* IV12 97 G6
Littlemoor *Derbys.* S45 51 F6
Littlemoor *Dorset* DT3 9 F6
Littlemore OX4 21 J1
Littlemoss M43 49 J3
Littleover DE23 41 F2
Littleport CB6 43 J7
Littlestead Green RG4 22 A4
Littlestone-on-Sea TN28 15 F5
Littlethorpe HG4 57 J3
Littleton *Ches.W. & C.* CH3 48 D6
Littleton *Hants.* SO22 11 F1
Littleton *P. & K.* PH14 82 D4
Littleton *Som.* TA11 8 D1
Littleton *Surr.* TW17 22 D5
Littleton Drew SN14 20 B3
Littleton Panell SN10 20 D6
Littleton-on-Severn BS35 19 J2
Littletown *Dur.* DH6 62 E2
Littletown *I.o.W.* PO33 11 G5
Littlewick Green SL6 22 B4
Littlewindsor DT8 8 D4
Littleworth *Glos.* GL55 30 C5
Littleworth *Oxon.* SN7 21 G2
Littleworth *S.Yorks.* DN11 51 J3
Littleworth *Staffs.* WS12 40 C4
Littleworth *Worcs.* WR5 29 H4
Littley Green CM3 33 K7
Litton *Derbys.* SK17 50 D5
Litton *N.Yorks.* BD23 56 E2
Litton *Som.* BA3 19 J6
Litton Cheney DT2 8 E5
Liurbost (Leurbost) HS2 101 F5
LIVERPOOL L 48 C3
Liverpool John Lennon Airport L24 48 D4
Liversedge WF15 57 H7
Liverton *Devon* TQ12 5 J3
Liverton *R. & C.* TS13 63 J5
Liverton Street ME17 14 D3
Livingston EH54 75 J4
Livingston Village EH54 75 J4
Lixwm CH8 47 K5
Lizard TR12 2 E7
Llaingarreglwyd SA47 26 D3
Llaingoch LL65 46 A4
Llaithddu LD1 37 K7
Llampha CF35 18 C4
Llan SY19 37 H5
Llan Ffestiniog (Ffestiniog) LL41 37 G1
Llanaber LL42 37 F4
Llanaelhaearn LL54 36 C1
Llanaeron SA48 26 D2
Llanafan SY23 27 F1
Llanafan-fawr LD2 27 J3
Llanafan-fechan LD4 27 J3
Llanallgo LL72 46 D4
Llanandras (Presteigne) LD8 28 C2
Llanarmon LL53 36 D2
Llanarmon Dyffryn Ceiriog LL20 38 A2
Llanarmon-yn-Ial CH7 47 K7
Llanarth *Cere.* SA47 26 D3
Llanarth *Mon.* NP15 28 C7
Llanarthney SA32 17 J3
Llanasa CH8 47 K4
Llanbabo LL68 46 B4
Llanbadarn Fawr SY23 36 E7
Llanbadarn Fynydd LD1 27 K1
Llanbadarn-y-garreg LD2 28 A4
Llanbadoc NP15 19 G1

Lla - Lon

Llanbadrig LL67 46 B3
Llanbeder NP18 19 G2
Llanbedr Gwyn. LL45 36 E3
Llanbedr Powys NP8 28 B6
Llanbedr Powys LD2 28 A4
Llanbedr Pont Steffan (Lampeter) SA48 17 J1
Llanbedr-Dyffryn-Clwyd LL15 47 K7
Llanbedrgoch LL76 46 D4
Llanbedrog LL53 36 C2
Llanbedr-y-cennin LL32 47 F6
Llanberis LL55 46 D7
Llanbethery CF62 18 D5
Llanbister LD1 28 A1
Llanblethian CF71 18 C4
Llanboidy SA34 17 F3
Llanbradach CF83 18 E2
Llanbryn-mair SY19 37 H5
Llancadle CF62 18 D5
Llancarfan CF62 18 D4
Llancayo NP15 19 G1
Llancynfelyn SY20 37 F6
Llandafal NP13 18 E1
Llandaff CF5 18 E4
Llandaff North CF14 18 E4
Llandanwg LL46 36 E3
Llandawke SA33 17 F4
Llanddaniel Fab LL60 46 C5
Llanddarog SA32 17 H4
Llanddeiniol SY23 26 E1
Llanddeiniolen LL55 46 D6
Llandderfel LL23 37 J2
Llanddeusant Carmar. SA19 27 G6
Llanddeusant I.o.A. LL65 46 B4
Llanddew LD3 27 K5
Llanddewi SA3 17 H7
Llanddewi Rhydderch NP7 28 C7
Llanddewi Skirrid NP7 28 C7
Llanddewi Velfrey SA67 16 E4
Llanddewi Ystradenni LD1 28 A2
Llanddewi-Brefi SY25 27 F3
Llanddewi'r Cwm LD2 27 K4
Llanddoged LL26 47 G6
Llanddona LL58 46 D5
Llanddowror SA33 17 F4
Llanddulas LL22 47 H5
Llanddwywe LL44 36 E3
Llanddyfnan LL78 46 D5
Llandefaelog Fach LD3 27 K5
Llandefaelog-tre'r-graig LD3 28 A6
Llandefalle LD3 28 A5
Llandegfan LL59 46 D5
Llandegla LL11 47 K7
Llandegley LD1 28 A2
Llandegveth NP18 19 G2
Llandegwning LL53 36 B2
Llandeilo SA19 17 K3
Llandeilo Abercywyn SA33 17 G4
Llandeilo Graban LD2 27 K4
Llandeilo'r-Fan LD3 27 H5
Llandeloy SA62 16 B3
Llandenny NP15 19 H1
Llandevaud NP18 19 H2
Llandevenny NP26 19 H3
Llandinabo HR2 28 E6
Llandinam SY17 37 K7
Llandissilio SA66 16 E3
Llandogo NP25 19 J1
Llandough V. of Glam. CF11 18 E4
Llandough V. of Glam. CF71 18 C4
Llandovery (Llanymddyfri) SA20 27 G5
Llandow CF71 18 C4
Llandre Carmar. SA19 17 K1
Llandre Carmar. SA34 16 E3
Llandre Cere. SY24 37 F7
Llandrillo LL21 37 K2
LLANDRINDOD WELLS LD 27 K2
Llandrinio SY22 38 B4
LLANDUDNO LL 47 F4
Llandudno Junction LL31 47 F5
Llandudoch (St. Dogmaels) SA43 16 E1
Llandwrog LL54 46 C7
Llandybie SA18 17 K4
Llandyfaelog SA17 17 H4
Llandyfan SA18 17 K4
Llandyfriog SA38 17 G1
Llandyfrydog LL71 46 C4
Llandygai LL57 46 D5
Llandygwydd SA43 17 F1
Llandyrnog LL16 47 K6
Llandyry SA17 17 H5
Llandysilio SY22 38 B4
Llandyssil SY15 38 A6
Llandysul SA44 17 H1
Llanedeyrn CF23 19 F3
Llanedy SA4 17 J5
Llaneglwys LD2 27 K5
Llanegryn LL36 37 F5
Llanegwad SA32 17 J3
Llaneilian LL68 46 C3
Llanelian-yn-Rhos LL29 47 G5
Llanelidan LL15 47 K7
Llanelieu LD3 28 A5
Llanellen NP7 28 C7
Llanelli SA15 17 J5
Llanelltyd LL40 37 G4
Llanelly NP7 28 B7
Llanelly Hill NP7 28 B7
Llanelwedd LD2 27 K3
Llanelwy (St. Asaph) LL17 47 J5
Llanenddwyn LL44 36 E3
Llanengan LL53 36 B3
Llanerfyl SY21 37 K5
Llaneuddog LL70 46 C4
Llaneurgain (Northop) CH7 48 B6
Llanfachraeth LL65 46 B4
Llanfachreth LL40 37 G3
Llanfaelog LL63 46 B5
Llanfaelrhys LL53 36 B3
Llanfaenor NP25 28 D7
Llan-faes I.o.A. LL58 46 E5
Llanfaes Powys LD3 27 K6
Llanfaethlu LL65 46 B4
Llanfaglan LL54 46 C6
Llanfair LL46 36 E3
Llanfair Caereinion SY21 38 A5
Llanfair Clydogau SA48 27 F3
Llanfair Dyffryn Clwyd LL15 47 K7
Llanfair Talhaiarn LL22 47 H5
Llanfair Waterdine LD7 28 B1
Llanfairfechan LL33 46 E5
Llanfair-Nant-Gwyn SA37 16 E2
Llanfair-Orllwyn SA44 17 G1
Llanfairpwllgwyngyll LL61 46 D5
Llanfair-ym-Muallt (Builth Wells) LD2 27 K3
Llanfairynghornwy LL65 46 B3
Llanfair-yn-neubwll LL65 46 B5
Llanfallteg SA34 16 E4
Llanfaredd LD2 27 K3
Llanfarian SY23 26 E1
Llanfechain SY22 38 A3
Llanfechell LL68 46 B3
Llanfendigaid LL36 36 E5
Llanferres CH7 47 K6
Llanfflewyn LL68 46 B4
Llanfigael LL65 46 B4
Llanfihangel Crucornau (Llanvihangel Crucorney) NP7 28 C6
Llanfihangel Glyn Myfyr LL21 37 J1
Llanfihangel Nant Bran LD3 27 J5
Llanfihangel Rhydithon LD1 28 A2
Llanfihangel Rogiet NP26 19 H3
Llanfihangel Tal-y-llyn LD3 28 A6
Llanfihangel-ar-arth SA39 17 H1
Llanfihangel-nant-Melan LD8 28 A3
Llanfihangel-uwch-Gwili SA32 17 H3
Llanfihangel-y-Creuddyn SY23 27 F1
Llanfihangel-yng-Ngwynfa SY22 37 K4
Llanfihangel-yn-Nhywyn LL65 46 B5
Llanfihangel-y-pennant Gwyn. LL51 36 E1
Llanfihangel-y-pennant Gwyn. LL36 37 F5
Llanfilo LD3 28 A5
Llanfoist NP7 28 B7
Llanfor LL23 37 J2
Llanfrechfa NP44 19 G2
Llanfrothen LL48 37 F1
Llanfrynach LD3 27 K6
Llanfwrog Denb. LL15 47 K7
Llanfwrog I.o.A. LL65 46 A4
Llanfyllin SY22 38 A4
Llanfynydd Carmar. SA32 17 J3
Llanfynydd Flints. LL11 48 B7
Llanfyrnach SA35 17 F2
Llangadfan SY21 37 K4
Llangadog SA19 27 G6
Llangadwaladr I.o.A. LL62 46 B6
Llangadwaladr Powys SY10 38 A2
Llangaffo LL60 46 C6
Llangain SA33 17 G4
Llangammarch Wells LD4 27 J4
Llangan CF35 18 C4
Llangarron HR9 28 E6
Llangasty-Talyllyn LD3 28 A6
Llangathen SA32 17 J3
Llangattock NP8 28 B7
Llangattock Lingoed NP7 28 C7
Llangattock-Vibon-Avel NP25 28 D7
Llangedwyn SY10 38 A3
Llangefni LL77 46 C5
Llangeinor CF32 18 C3
Llangeitho SY25 27 F3
Llangeler SA44 17 G2
Llangelynin LL36 36 E5
Llangendeirne SA17 17 H4
Llangennech SA14 17 J5
Llangennith SA3 17 H6
Llangenny NP8 28 B7
Llangernyw LL22 47 G6
Llangian LL53 36 B3
Llangiwg SA8 18 A1
Llangloffan SA62 16 C2
Llanglydwen SA34 16 E3
Llangoed LL58 46 E5
Llangoedmor SA43 16 E1
Llangollen LL20 38 B1
Llangolman SA66 16 E3
Llangorse LD3 28 A6
Llangorwen SY23 37 F7
Llangovan NP25 19 H1
Llangower LL23 37 J2
Llangrannog SA44 26 C3
Llangristiolus LL62 46 C5
Llangrove HR9 28 E7
Llangua NP7 28 C6
Llangunllo LD7 28 B1
Llangunnor SA31 17 H3
Llangurig SY18 27 J1
Llangwm Conwy LL21 37 J1
Llangwm Mon. NP15 19 H1
Llangwm Pembs. SA62 16 C5
Llangwnnadl LL53 36 B2
Llangwyfan LL16 47 K6
Llangwyllog LL77 46 C5
Llangwyryfon SY23 27 F1
Llangybi Cere. SA48 27 F3
Llangybi Gwyn. LL53 36 D1
Llangybi Mon. NP15 19 G2
Llangyfelach SA5 17 K6
Llangynhafal LL16 47 K6
Llangynidr NP8 28 A7
Llangyniew SY21 38 A5
Llangynin SA33 17 F4
Llangynllo SA44 17 G1
Llangynog Carmar. SA33 17 G4
Llangynog Powys SY10 37 K3
Llangynwyd CF34 18 B3
Llanhamlach LD3 27 K6
Llanharan CF72 18 D3
Llanharry CF72 18 D3
Llanhennock NP18 19 G2
Llanhilleth NP13 19 F1
Llanidloes SY18 37 J7
Llaniestyn LL53 36 B2
Llanigon HR3 28 B5
Llanilar SY23 27 F1
Llanilid CF35 18 C3
Llanishen Cardiff CF14 18 E3
Llanishen Mon. NP16 19 H1
Llanllawddog SA32 17 H3
Llanllechid LL57 46 E6
Llanlleonfel LD4 27 J3
Llanllugan SY21 37 K5
Llanllwch SA31 17 G4
Llanllwchaiarn SY16 38 A6
Llanllwni SA39 17 H1
Llanllyfni LL54 46 C7
Llanllywel NP15 19 G2
Llanmadoc SA3 17 H6
Llanmaes CF61 18 C5
Llanmartin NP18 19 G3
Llanmerewig SY15 38 A6
Llanmihangel CF71 18 C4
Llan-mill SA67 16 E4
Llanmiloe SA33 17 F5
Llanmorlais SA4 17 J6
Llannefydd LL16 47 H5
Llannerch Hall LL17 47 J5
Llannerch-y-medd LL71 46 C4
Llannerch-y-Môr CH8 47 K5
Llannon Carmar. SA14 17 J5
Llan-non Cere. SY23 26 E2
Llannor LL53 36 C2
Llanover NP7 19 G1
Llanpumsaint SA33 17 H3
Llanreithan SA62 16 B3
Llanrhaeadr LL16 47 J6
Llanrhaeadr-ym-Mochnant SY10 38 A3
Llanrhian SA62 16 B2
Llanrhidian SA3 17 H6
Llanrhyddlad LL65 46 B4
Llanrhystud SY23 26 E2
Llanrothal NP25 28 D7
Llanrug LL55 46 D6
Llanrumney CF3 19 F3
Llanrwst LL26 47 F6
Llansadurnen SA33 17 F4
Llansadwrn Carmar. SA19 17 K2
Llansadwrn I.o.A. LL59 46 D5
Llansaint SA17 17 G5
Llansamlet SA7 17 K6
Llansanffraid SY23 26 E2
Llansanffraid Glan Conwy LL28 47 G5
Llansannan LL16 47 H6
Llansannor CF71 18 C4
Llansantffraed LD3 28 A6
Llansantffraed-Cwmdeuddwr LD6 27 J2
Llansantffraed-in-Elwel LD1 27 K3
Llansantffraid-ym-Mechain SY22 38 B3
Llansawel Carmar. SA19 17 K2
Llansawel (Briton Ferry) N.P.T. SA11 18 A2
Llansilin SY10 38 B3
Llansoy NP15 19 H1
Llanspyddid LD3 27 K6
Llanstadwell SA73 16 C5
Llansteffan SA33 17 G4
Llanstephan LD3 28 A4
Llantarnam NP44 19 G2
Llanteg SA67 16 E4
Llanthony NP7 28 B6
Llantilio Crossenny NP7 28 C7
Llantilio Pertholey NP7 28 C7
Llantood SA43 16 E1
Llantrisant I.o.A. LL65 46 B4
Llantrisant Mon. NP15 19 G2
Llantrisant R.C.T. CF72 18 D3
Llantrithyd CF71 18 D4
Llantwit Fardre CF38 18 D3
Llantwit Major CF61 18 C5
Llantysilio LL20 38 A1
Llanuwchllyn LL23 37 H3
Llanvaches NP26 19 H2
Llanvair-Discoed NP16 19 H2
Llanvapley NP7 28 C7
Llanvetherine NP7 28 C7
Llanveynoe HR2 28 C5
Llanvihangel Crucorney (Llanfihangel Crucornau) NP7 28 C6
Llanvihangel Gobion NP7 19 G1
Llanvihangel-Ystern-Llewern NP25 28 D7
Llanvithyn CF62 18 D4
Llanwarne HR2 28 E6
Llanwddyn SY10 37 K4
Llanwenog SA40 17 H1
Llanwern NP18 19 G3
Llanwinio SA34 17 F3
Llanwnda Gwyn. LL54 46 C7
Llanwnda Pembs. SA64 16 C2
Llanwnnen SA48 17 J1
Llanwnog SY17 37 K6
Llanwonno CF37 18 D2
Llanwrda SA19 27 G5
Llanwrin SY20 37 G5
Llanwrthwl LD1 27 J2
Llanwrtyd LD5 27 H4
Llanwrtyd Wells LD5 27 H4
Llanwyddelan SY16 37 K5
Llanyblodwel SY10 38 B3
Llanybri SA33 17 G4
Llanybydder SA40 17 J1
Llanycefn SA66 16 E3
Llanychaer Bridge SA65 16 C2
Llanycil LL23 37 J2
Llanycrwys SA19 17 K1
Llanymawddwy SY20 37 H4
Llanymddyfri (Llandovery) SA20 27 G5
Llanymynech SY22 38 B3
Llanynghenedl LL65 46 B4
Llanynys LL16 47 K6
Llan-y-pwll LL13 48 C7
Llanyre LD1 27 K2
Llanystumdwy LL52 36 D2
Llanywern LD3 28 A6
Llawhaden SA67 16 D4
Llawndy CH8 47 K4
Llawnt SY10 38 B2
Llawr-y-dref LL53 36 B3
Llawryglyn SY17 37 J6
Llay LL12 48 C7
Llechcynfarwy LL71 46 B4
Llecheiddior LL51 36 D1
Llechfaen LD3 27 K6
Llechryd Caerp. NP22 18 E1
Llechryd Cere. SA43 17 F1
Llechrydau SY10 38 B2
Lledrod Cere. SY23 27 F1
Lledrod Powys SY10 38 B3
Llethrid SA2 17 J6
Llidiad-Nenog SA32 17 J2
Llidiardau LL23 37 H2
Llithfaen LL53 36 C1
Lloc CH8 47 K5
Llong CH7 48 B6
Llowes HR3 28 A4
Lloyney LD7 28 B1
Llundain-fach SA48 26 E3
Llwydcoed CF44 18 C1
Llwydiarth SY21 37 K4
Llwyn M.Tyd. CF48 27 K7
Llwyn Shrop. SY7 38 B7
Llwyncelyn SA46 26 D3
Llwyn-croes SA33 17 H3
Llwyndafydd SA44 26 C3
Llwynderw SY21 38 B5
Llwyndyrys LL53 36 C1
Llwyneinion LL14 38 B1
Llwyngwril LL37 36 E5
Llwynhendy SA14 17 J6
Llwyn-Madoc LD5 27 J3
Llwynmawr LL20 38 B2
Llwyn-onn SA45 26 D3
Llwyn-y-brain Carmar. SA34 16 E4
Llwyn-y-brain Carmar. SA20 27 G5
Llwyn-y-groes SY25 26 E3
Llwynypia CF40 18 C2
Llyn Penmaen (Penmaenpool) LL40 37 F4
Llynclys SY10 38 B3
Llynfaes LL65 46 C5
Llysfaen LL29 47 G5
Llyswen LD3 28 A5
Llysworney CF71 18 C4
Llys-y-frân SA63 16 D3
Llywel LD3 27 H5
Load Brook S6 50 E4
Loandhu IV20 97 F4
Loanhead Aber. AB41 91 H1
Loanhead Midloth. EH20 76 A4
Loans KA10 74 B7
Lobb EX33 6 C2
Lobhillcross EX20 6 C7
Loch a Charnain HS8 92 D7
Loch Baghasdail (Lochboisdale) HS8 84 C3
Loch Choire Lodge KW11 103 J5
Loch Eil Outward Bound PH33 87 G7
Loch Head D. & G. DG8 64 D6
Loch Head D. & G. KA6 67 J4
Loch na Madadh (Lochmaddy) HS6 92 E5
Loch Sgioport HS8 84 D1
Lochailort PH38 86 D6
Lochaline PA34 79 H3
Lochans DG9 64 A5
Locharbriggs DG1 68 E5
Lochawe PA33 80 C5
Lochboisdale (Loch Baghasdail) HS8 84 C3
Lochbuie PA62 79 H5
Lochcarron IV54 94 E7
Lochdhu Hotel KW12 105 F4
Lochdon PA64 79 J4
Lochdrum IV23 95 J4
Lochearnhead FK19 81 G5
Lochee DD2 82 E4
Lochend High. KW14 105 H2
Lochend High. IV3 88 C1
Locheport (Locheuphort) HS6 92 D5
Locheuphort (Locheport) HS6 92 D5
Lochfoot DG2 65 K3
Lochgair PA31 73 H1
Lochgarthside IV2 88 C3
Lochgelly KY5 75 K1
Lochgilphead PA31 73 G2
Lochgoilhead PA24 80 D7
Lochgoyn KA3 74 D6
Lochhill E.Ayr. KA18 67 K2
Lochhill Moray IV30 97 K5
Lochinch Castle DG9 64 B4
Lochinver IV27 102 C6
Lochlair DD8 83 G3
Lochlane PH7 81 K5
Lochlea KA1 74 C7
Lochluichart IV23 95 K5
Lochmaben DG11 69 F5
Lochmaddy (Loch na Madadh) HS6 92 E5
Lochore KY5 75 K1
Lochportain HS6 92 E4
Lochranza KA27 73 H5
Lochside Aber. DD10 83 J1
Lochside High. IV27 103 G3
Lochside High. KW11 104 D5
Lochside High. KW14 105 H2
Lochslin IV20 97 F3
Lochton KA26 67 G5
Lochty KY10 83 G7
Lochuisge PH33 79 J2
Lochurr DG3 68 C5
Lochussie IV7 96 B6
Lochwinnoch PA12 74 B5
Lockengate PL26 4 A4
Lockerbie DG11 69 G5
Lockeridge SN8 20 E5
Lockerley SO51 10 D2
Lockhills CA4 61 G2
Locking BS24 19 G6
Lockington E.Riding YO25 59 F5
Lockington Leics. DE74 41 G3
Lockleywood TF9 39 F3
Locks Heath SO31 11 G4
Locksbottom BR6 23 H5
Locksgreen PO30 11 F5
Lockton YO18 58 E1
Loddington Leics. LE7 42 A5
Loddington Northants. NN14 32 B1
Loddiswell TQ7 5 H6
Loddon NR14 45 H6
Lode CB25 33 J2
Loders DT6 8 D5
Lodsworth GU28 12 C4
Lofthouse N.Yorks. HG3 57 G2
Lofthouse W.Yorks. WF3 57 J7
Loftus TS13 63 J5
Logan D. & G. DG9 64 A6
Logan E.Ayr. KA18 67 K1
Loganlea EH55 75 H4
Loggerheads TF9 39 G2
Loggie IV23 95 H2
Logie Angus DD10 83 H1
Logie Angus DD8 82 E2
Logie Fife KY15 83 F5
Logie Moray IV36 97 H6
Logie Coldstone AB34 90 C4
Logie Hill IV18 96 E4
Logie Newton AB54 90 E1
Logie Pert DD10 83 H1
Logierait PH9 82 A2
Login SA34 16 E3
Lolworth CB23 33 G2
Lonbain IV54 94 C6
Londesborough YO43 58 E5
LONDON E, EC, N, NW, SE, SW, W, WC 23 G3
London Apprentice PL26 4 A6
London Ashford Airport TN29 15 F5
London Beach TN30 14 D4
London City Airport E16 23 H3
London Colney AL2 22 E1
London Gatwick Airport (Gatwick Airport) RH6 23 F7
London Heathrow Airport TW6 22 D4
London Luton Airport (Luton Airport) LU2 32 E6
London Minstead SO43 10 D3
London Southend Airport SS2 24 E3
London Stansted Airport (Stansted Airport) CM24 33 J6
Londonderry DL7 57 H1
Londonthorpe NG31 42 C2
Londubh IV22 94 E3
Lonemore IV25 96 E3
Long Ashton BS41 19 J4
Long Bank DY12 29 G1
Long Bennington NG23 42 B1
Long Bredy DT2 8 E5
Long Buckby NN6 31 H2
Long Clawson LE14 42 A3
Long Compton Staffs. ST18 40 A3
Long Compton Warks. CV36 30 D5
Long Crendon HP18 21 K1
Long Crichel BH21 9 J3
Long Dean SN14 20 B4
Long Downs TR10 2 E5
Long Drax YO8 58 C7
Long Duckmanton S44 51 G5

Lon - Lym

Long Eaton NG10 41 G2
Long Gill BD23 56 C4
Long Green Ches.W. & C. CH3 48 D5
Long Green Essex CO6 34 D6
Long Green Worcs. GL19 29 H5
Long Hanborough OX29 31 F7
Long Itchington CV47 31 F2
Long Lane TF6 39 F4
Long Lawford CV23 31 F1
Long Load TA10 8 D2
Long Marston Herts. HP23 32 B7
Long Marston N.Yorks. YO26 58 B4
Long Marston Warks. CV37 30 C4
Long Marton CA16 61 H4
Long Meadowend SY7 38 D7
Long Melford CO10 34 C4
Long Newnton GL8 20 C2
Long Preston BD23 56 D4
Long Riston HU11 59 H5
Long Stratton NR15 45 F6
Long Street MK19 31 J4
Long Sutton Hants. RG29 22 A7
Long Sutton Lincs. PE12 43 H3
Long Sutton Som. TA10 8 D2
Long Thurlow IP31 34 E2
Long Whatton LE12 41 G3
Long Wittenham OX14 21 J2
Longbenton NE7 71 H7
Longborough GL56 30 C6
Longbridge Plym. PL6 5 F5
Longbridge W.Mid. B31 30 B1
Longbridge Warks. CV34 30 D2
Longbridge Deverill BA12 20 B7
Longburgh CA5 60 E1
Longburton DT9 9 F3
Longcliffe DE4 50 E7
Longcombe TQ9 5 J5
Longcot SN7 21 F2
Longcroft FK4 75 F3
Longcross Devon PL19 4 E3
Longcross Surr. KT16 22 C5
Longden SY5 38 D5
Longdon Staffs. WS15 40 C4
Longdon Worcs. GL20 29 H5
Longdon Green WS15 40 C4
Longdon upon Tern TF6 39 F4
Longdown EX6 7 G6
Longdrum AB23 91 H3
Longfield DA3 24 C5
Longfield Hill DA3 24 C5
Longfleet BH15 10 B5
Longford Derbys. DE6 40 E2
Longford Glos. GL2 29 H6
Longford Gt.Lon. UB7 22 D4
Longford Shrop. TF9 39 F2
Longford Tel. & W. TF10 39 G4
Longford W.Mid. CV6 41 F7
Longforgan DD2 82 E5
Longformacus TD11 76 E5
Longframlington NE65 71 G3
Longham Dorset BH22 10 B5
Longham Norf. NR19 44 D4
Longhill AB42 99 H5
Longhirst NE61 71 H5
Longhope Glos. GL17 29 F7
Longhope Ork. KW16 106 C8
Longhorsley NE65 71 G4
Longhoughton NE66 71 H2
Longlands Aber. AB54 90 C2
Longlands Cumb. CA7 60 D3
Longlands Gt.Lon. SE9 23 H4
Longlane Derbys. DE6 40 E2
Longlane W.Berks. RG18 21 H4
Longlevens GL2 29 H7
Longley HD9 50 D2
Longley Green WR6 29 G3
Longmanhill AB45 99 F4
Longmoor Camp GU33 11 J1
Longmorn IV30 97 K6
Longnewton Sc.Bord. TD6 70 A1
Longnewton Stock. TS21 62 E5
Longney GL2 29 G7
Longniddry EH32 76 C3
Longnor Shrop. SY5 38 D5
Longnor Staffs. SK17 50 C6
Longparish SP11 21 H7
Longridge Lancs. PR3 56 B6
Longridge Staffs. ST18 40 B4
Longridge W.Loth. EH47 75 H4
Longridge End GL19 29 H6
Longridge Towers TD15 77 H5
Longriggend ML6 75 G3
Longrock TR20 2 C5
Longsdon ST9 49 J7
Longshaw WN5 48 E2
Longside AB42 99 J6
Longslow TF9 39 F2
Longsowerby CA2 60 E1
Longstanton CB24 33 H2
Longstock SO20 10 E1
Longstone TR26 2 C5
Longstowe CB23 33 G3
Longstreet SN9 20 E6
Longthorpe PE3 42 E6
Longton Lancs. PR4 55 H7
Longton Stoke ST3 40 B1
Longtown Cumb. CA6 69 J7
Longtown Here. HR2 28 C6
Longville in the Dale TF13 38 E6
Longwell Green BS30 19 K4
Longwick HP27 22 A1
Longwitton NE61 71 F5
Longworth OX13 21 G2
Longyester EH41 76 D4
Lonmay AB43 99 J5
Lonmore IV55 93 H7
Looe PL13 4 C5
Loose ME15 14 C2
Loosebeare EX17 7 F5
Loosegate PE12 43 G3
Loosley Row HP27 22 B1
Lopcombe Corner SP5 10 D1
Lopen TA13 8 D3
Loppington SY4 38 D3
Lorbottle NE65 71 F3
Lorbottle Hall NE66 71 F3
Lordington PO18 11 J4
Lord's Hill SO16 10 E3
Lorgill IV55 93 G7
Lorn G83 74 B2
Lornty PH10 82 C3
Loscoe DE75 41 G1
Loscombe DT6 8 D5
Losgaintir HS3 93 F2
Lossiemouth IV31 97 K4
Lossit PA47 72 A5
Lostock Gralam CW9 49 F5
Lostock Green CW9 49 F5
Lostock Junction BL6 49 F2
Lostwithiel PL22 4 B5
Loth KW17 106 F4
Lothbeg KW8 104 E7
Lothersdale BD20 56 E5
Lothmore KW8 104 E7
Loudwater HP10 22 C2
Loughborough LE11 41 H4
Loughor SA4 17 J6
Loughton Essex IG10 23 H2
Loughton M.K. MK5 32 B5
Loughton Shrop. WV16 39 F7
Lound Lincs. PE10 42 D4
Lound Notts. DN22 51 J4
Lound Suff. NR32 45 K6
Lount LE65 41 F4
Lour DD8 83 F3
Louth LN11 53 G4
Love Clough BB4 56 D7
Lovedean PO8 11 H3
Lover SP5 10 D3
Loversall DN11 51 H3
Loves Green CM1 24 C1
Lovesome Hill DL6 62 E7
Loveston SA68 16 D5
Lovington BA7 8 E1
Low Ackworth WF7 51 G1
Low Angerton NE61 71 F5
Low Ballevain PA28 66 A1
Low Barlay DG7 65 F5
Low Barlings LN3 52 D5
Low Bentham (Lower Bentham) LA2 56 B3
Low Bolton DL8 62 B7
Low Bradfield S6 50 E3
Low Bradley (Bradley) BD20 57 F5
Low Braithwaite CA4 60 F2
Low Brunton NE46 70 E6
Low Burnham DN9 51 K2
Low Burton HG4 57 H1
Low Buston NE65 71 H3
Low Catton YO41 58 D4
Low Coniscliffe DL2 62 D5
Low Craighead KA26 67 G3
Low Dinsdale DL2 62 E5
Low Ellington HG4 57 H1
Low Entercommon DL6 62 E6
Low Etherley DL14 62 C4
Low Fell NE11 62 D1
Low Gate NE46 70 E7
Low Grantley HG4 57 H2
Low Green IP29 34 C2
Low Habberley DY11 29 H1
Low Ham TA10 8 D2
Low Hawsker YO22 63 J2
Low Haygarth LA10 61 H7
Low Hesket CA4 61 F2
Low Hesleyhurst NE65 71 F4
Low Hutton YO60 58 D3
Low Kingthorpe YO18 58 E1
Low Laithe HG3 57 G3
Low Langton LN8 52 E5
Low Leighton SK22 50 C4
Low Lorton CA13 60 C4
Low Marishes YO17 58 E2
Low Marnham NG23 52 B6
Low Middleton NE70 77 K7
Low Mill YO62 63 H7
Low Moor Lancs. BB7 56 C5
Low Moor W.Yorks. BD12 57 G7
Low Moorsley DH5 62 E2
Low Moresby CA28 60 A4
Low Newton-by-the-Sea NE66 71 H1
Low Row Cumb. CA8 70 A7
Low Row N.Yorks. DL11 62 A7
Low Stillaig PA21 73 H4
Low Street NR9 44 E5
Low Tharston NR15 45 F6
Low Torry KY12 75 J2
Low Town NE65 71 G3
Low Toynton LN9 53 F5
Low Wood LA12 55 G1
Low Worsall TS15 62 E6
Lowbands GL19 29 G5
Lowdham NG14 41 J1
Lowe SY4 38 E2
Lowe Hill ST13 49 J7
Lower Achachenna PA35 80 B5
Lower Aisholt TA5 8 B1
Lower Apperley GL19 29 H6
Lower Arncott OX25 31 H7
Lower Ashtead KT21 22 E6
Lower Ashton EX6 7 G7
Lower Assendon RG9 22 A3
Lower Auchalick PA21 73 H3
Lower Ballam PR4 55 G6
Lower Barewood HR6 28 C3
Lower Bartle PR4 55 H6
Lower Bayble (Pabail Iarach) HS2 101 H4
Lower Beeding RH13 13 F4
Lower Benefield PE8 42 C7
Lower Bentham (Low Bentham) LA2 56 B3
Lower Bentley B60 29 J2
Lower Berry Hill GL16 28 E7
Lower Birchwood DE55 51 G7
Lower Boddington NN11 31 F3
Lower Boscaswell TR19 2 A5
Lower Bourne GU10 22 B7
Lower Brailes OX15 30 E5
Lower Breakish IV42 86 C2
Lower Bredbury SK6 49 J3
Lower Broadheath WR2 29 H3
Lower Brynamman SA18 27 G7
Lower Bullingham HR2 28 E5
Lower Bullington SO21 21 H7
Lower Burgate SP6 10 C3
Lower Burrow TA12 8 D2
Lower Burton HR6 28 D3
Lower Caldecote SG18 32 E4
Lower Cam GL11 20 A1
Lower Cambourne CB23 33 G3
Lower Camster KW3 105 H4
Lower Chapel LD3 27 K5
Lower Cheriton EX14 7 K5
Lower Chicksgrove SP3 9 J1
Lower Chute SP11 21 G6
Lower Clent DY9 40 B7
Lower Creedy EX17 7 G5
Lower Cumberworth HD8 50 E2
Lower Darwen BB3 56 B7
Lower Dean PE28 32 D2
Lower Diabaig IV22 94 D5
Lower Dicker BN27 13 J5
Lower Dinchope SY7 38 D7
Lower Down SY7 38 C7
Lower Drift TR19 2 B6
Lower Dunsforth YO26 57 K3
Lower Earley RG6 22 A4
Lower Edmonton N9 23 G2
Lower Elkstone SK17 50 C7
Lower End Bucks. HP18 21 K1
Lower End M.K. MK17 32 C5
Lower End Northants. NN7 32 B2
Lower Everleigh SN8 20 E6
Lower Eythorne CT15 15 H3
Lower Failand BS8 19 J4
Lower Farringdon GU34 11 J1
Lower Fittleworth RH20 12 D5
Lower Foxdale IM4 54 B6
Lower Freystrop SA62 16 C4
Lower Froyle GU34 22 A7
Lower Gabwell TQ12 5 K4
Lower Gledfield IV24 96 C2
Lower Godney BA5 19 H7
Lower Gravenhurst MK45 32 E5
Lower Green Essex CB11 33 H5
Lower Green Herts. SG5 32 E5
Lower Green Kent TN2 23 K7
Lower Green Norf. NR21 44 D2
Lower Green Staffs. WV9 40 B5
Lower Green Bank LA2 55 J4
Lower Halstock Leigh BA22 8 E4
Lower Halstow ME9 24 E5
Lower Hardres CT4 15 G2
Lower Harpton LD8 28 B2
Lower Hartshay DE5 51 F7
Lower Hartwell HP17 31 J7
Lower Hawthwaite LA20 55 F1
Lower Haysden TN11 23 J7
Lower Hayton SY8 38 E7
Lower Heath CW12 49 H6
Lower Hergest HR5 28 B3
Lower Heyford OX25 31 F6
Lower Higham ME3 24 D4
Lower Holbrook IP9 35 F5
Lower Hopton WF14 50 D1
Lower Hordley SY12 38 C3
Lower Horncroft RH20 12 D5
Lower Horsebridge BN27 13 J5
Lower Houses HD5 50 D1
Lower Howsell WR14 29 G4
Lower Kersal M25 49 H2
Lower Kilchattan PA61 72 B1
Lower Kilcott GL12 20 A3
Lower Killeyan PA42 72 A6
Lower Kingcombe DT2 8 E5
Lower Kingswood KT20 23 F6
Lower Kinnerton CH4 48 C6
Lower Langford BS40 19 H5
Lower Largo KY8 83 F7
Lower Leigh ST10 40 C2
Lower Lemington GL56 30 D5
Lower Lovacott EX31 6 D3
Lower Loxhore EX31 6 E2
Lower Lydbrook GL17 28 E7
Lower Lye HR6 28 D2
Lower Machen NP10 19 F3
Lower Maes-coed HR2 28 C5
Lower Mannington BH21 10 B4
Lower Middleton Cheney OX17 31 G4
Lower Milton BA5 19 J7
Lower Moor WR10 29 J4
Lower Morton BS35 19 K2
Lower Nash SA72 16 D5
Lower Nazeing EN9 23 G1
Lower Netchwood WV16 39 F6
Lower Nyland SP8 9 G2
Lower Oddington GL56 30 D6
Lower Ollach IV51 86 B1
Lower Penarth CF64 18 E4
Lower Penn WV4 40 A6
Lower Pennington SO41 10 E5
Lower Peover WA16 49 G5
Lower Pollicott HP18 31 J7
Lower Quinton CV37 30 C4
Lower Race NP4 19 F1
Lower Rainham ME8 24 E5
Lower Roadwater TA23 7 J2
Lower Sapey WR6 29 F2
Lower Seagry SN15 20 C3
Lower Shelton MK43 32 C4
Lower Shiplake RG9 22 A4
Lower Shuckburgh NN11 31 F2
Lower Slaughter GL54 30 C6
Lower Soothill WF17 57 H7
Lower Stanton St. Quintin SN14 20 C3
Lower Stoke ME3 24 E4
Lower Stondon SG16 32 E5
Lower Stone GL13 19 K2
Lower Stonnall WS9 40 C5
Lower Stow Bedon NR17 44 D6
Lower Street Dorset DT11 9 H5
Lower Street E.Suss. TN33 14 C6
Lower Street Norf. NR11 45 G2
Lower Street Norf. NR11 45 F2
Lower Street Suff. IP6 35 F3
Lower Stretton WA4 49 F4
Lower Sundon LU3 32 D6
Lower Swanwick SO31 11 F4
Lower Swell GL54 30 C6
Lower Tadmarton OX15 31 F5
Lower Tale EX14 7 J5
Lower Tean ST10 40 C2
Lower Thurlton NR14 45 J6
Lower Thurnham LA2 55 H4
Lower Town Cornw. TR13 2 D6
Lower Town Devon TQ13 5 H3
Lower Town I.o.S. TR25 2 C1
Lower Town Pembs. SA65 16 C2
Lower Trebullett PL15 4 D3
Lower Tysoe CV35 30 E4
Lower Upcott TQ13 7 G7
Lower Upham SO32 11 G3
Lower Upnor ME2 24 D4
Lower Vexford TA4 7 K2
Lower Wallop SY5 38 C5
Lower Walton WA4 49 F4
Lower Waterhay SN6 20 D2
Lower Weald MK19 31 J5
Lower Wear EX2 7 H7
Lower Weare BS26 19 H6
Lower Welson HR3 28 B3
Lower Whatley BA11 20 A7
Lower Whitley WA4 49 F5
Lower Wick WR2 29 H3
Lower Wield SO24 21 K7
Lower Winchendon (Nether Winchendon) HP18 31 J7
Lower Withington SK11 49 H6
Lower Woodend SL7 22 B3
Lower Woodford SP4 10 C1
Lower Wyche WR14 29 G4
Lowerhouse BB12 56 D6
Lowertown KW17 106 D8
Lowesby LE7 42 A5
Lowestoft NR32 45 K6
Loweswater CA13 60 C4
Lowfield Heath RH11 23 F7
Lowgill Cumb. LA8 61 H7
Lowgill Lancs. LA2 56 B3
Lowick Cumb. LA12 55 F1
Lowick Northants. NN14 42 C7
Lowick Northumb. TD15 77 J7
Lowick Bridge LA12 55 F1
Lowick Green LA12 55 F1
Lownie Moor DD8 83 F3
Lowsonford B95 30 C2
Lowther CA10 61 G4
Lowther Castle CA10 61 G4
Lowthorpe YO25 59 G3
Lowton Devon EX20 6 E5
Lowton Gt.Man. WA3 49 F3
Lowton Som. TA3 7 K4
Lowton Common WA3 49 F3
Loxbeare EX16 7 H4
Loxhill GU8 12 D3
Loxhore EX31 6 E2
Loxley CV35 30 D3
Loxley Green ST14 40 C2
Loxton BS26 19 G6
Loxwood RH14 12 D3
Lubachoinnich IV24 96 B2
Lubcroy IV27 95 K1
Lubenham LE16 42 A7
Lubfearn IV23 95 K4
Lubmore IV22 95 G6
Lubreoch PH15 81 F3
Luccombe TA24 7 H1
Luccombe Village PO37 11 G6
Lucker NE70 77 K7
Luckett PL17 4 D3
Luckington SN14 20 B3
Lucklawhill KY16 83 F5
Luckwell Bridge TA24 7 H2
Lucton HR6 28 D2
Lucy Cross DL11 62 D5
Ludag HS8 84 C3
Ludborough DN36 53 F3
Ludbrook PL21 5 G5
Ludchurch SA67 16 E4
Luddenden HX2 57 F7
Luddenden Foot HX2 57 F7
Luddenham Court ME13 25 F5
Luddesdown DA13 24 C5
Luddington N.Lincs. DN17 52 B1
Luddington Warks. CV37 30 C3
Luddington in the Brook PE8 42 E7
Ludford Lincs. LN8 52 E4
Ludford Shrop. SY8 28 E1
Ludgershall Bucks. HP18 31 H7
Ludgershall Wilts. SP11 21 F6
Ludgvan TR20 2 C5
Ludham NR29 45 H4
Ludlow SY8 28 E1
Ludney LN11 53 G3
Ludstock HR8 29 F5
Ludstone WV5 40 A6
Ludwell SP7 9 J2
Ludworth DH6 62 E2
Luffincott EX22 6 B6
Luffness EH32 76 C2
Lufton BA22 8 E3
Lugar KA18 67 K1
Luggate EH41 76 D3
Luggate Burn EH41 76 E3
Luggiebank G67 75 F3
Lugton KA3 74 C5
Lugwardine HR1 28 E4
Luib IV49 86 B2
Luibeilt PH30 80 D1
Luing PA34 79 J6
Lulham HR2 28 D4
Lullington Derbys. DE12 40 E4
Lullington Som. BA11 20 A6
Lulsgate Bottom BS40 19 J5
Lulsley WR6 29 G3
Lulworth Camp BH20 9 H6
Lumb Lancs. BB4 56 D7
Lumb W.Yorks. HX6 57 F7
Lumbutts OL14 56 E7
Lumby LS25 57 K6
Lumphanan AB31 90 D4
Lumphinnans KY4 75 K1
Lumsdaine TD14 77 G4
Lumsdale DE4 51 F6
Lumsden AB54 90 C2
Lunan DD11 83 H2
Lunanhead DD8 83 F2
Luncarty PH1 82 B5
Lund E.Riding YO25 59 F5
Lund N.Yorks. YO8 58 C6
Lund Shet. ZE2 107 P2
Lundale HS2 100 D4
Lundavra PH33 80 B1
Lunderton AB42 99 K6
Lundie Angus DD2 82 D4
Lundie High. IV63 87 H3
Lundin Links KY8 83 F7
Lundwood S71 51 F2
Lundy EX34 6 A1
Lunga PA31 79 K7
Lunna ZE2 107 N6
Lunning ZE2 107 P6
Lunnon SA3 17 J7
Lunsford's Cross TN39 14 C6
Lunt L29 48 C2
Luntley HR6 28 C3
Luppitt EX14 7 K5
Lupset WF2 51 F1
Lupton LA6 55 J1
Lurgashall GU28 12 C4
Lurignich PA38 80 A2
Lusby PE23 53 G6
Luss G83 74 B1
Lussagiven PA60 72 E2
Lusta IV55 93 H6
Lustleigh TQ13 7 F7
Luston HR6 28 D2
Luthermuir AB30 83 H1
Luthrie KY15 82 E6
Luton Devon EX14 7 J5
Luton Devon TQ13 5 K3
LUTON Luton LU 32 D6
Luton Med. ME5 24 D5
Luton Airport (London Luton Airport) LU2 32 E6
Lutterworth LE17 41 H7
Lutton Devon PL21 5 F5
Lutton Dorset BH20 9 J6
Lutton Lincs. PE12 43 H3
Lutton Northants. PE8 42 E7
Luxborough TA23 7 H2
Luxulyan PL30 4 A5
Lybster High. KW14 105 F2
Lybster High. KW3 105 H5
Lydacott EX21 6 C5
Lydbury North SY7 38 C7
Lydcott EX32 6 E2
Lydd TN29 15 F5
Lydden CT15 15 H3
Lyddington LE15 42 B6
Lydd-on-Sea TN29 15 F5
Lyde Green BS16 19 K4
Lydeard St. Lawrence TA4 7 K2
Lydford EX20 6 D7
Lydford-on-Fosse TA11 8 E1
Lydgate Gt.Man. OL15 49 J1
Lydgate Gt.Man. OL4 49 J2
Lydgate W.Yorks. OL14 56 E7
Lydham SY9 38 C6
Lydiard Millicent SN5 20 D3
Lydiard Tregoze SN5 20 E3
Lydiate L31 48 C2
Lydlinch DT10 9 G3
Lydney GL15 19 K1
Lydstep SA70 16 D6
Lye DY9 40 B7
Lye Cross BS40 19 H5
Lye Green Bucks. HP5 22 C1
Lye Green E.Suss. TN6 13 J3
Lye Green Warks. CV35 30 C2
Lye's Green BA12 20 B7
Lyford OX12 21 G2
Lymbridge Green TN25 15 G3
Lyme Regis DT7 8 C5
Lymekilns G74 74 E5
Lyminge CT18 15 G3
Lymington SO41 10 E5
Lyminster BN17 12 D6
Lymm WA13 49 F4
Lymore SO41 10 D5
Lympne CT21 15 G4

Lym - Mea

Lympsham **BS24** 19 G6
Lympstone **EX8** 7 H7
Lynaberack **PH21** 88 E5
Lynch **TA24** 7 G1
Lynch Green **NR9** 45 F5
Lynchat **PH21** 88 E4
Lyndale House **IV51** 93 J6
Lyndhurst SO43 10 D4
Lyndon **LE15** 42 C5
Lyne *Aber.* **AB51** 91 F4
Lyne *Sc.Bord.* **EH45** 76 A6
Lyne *Surr.* **KT16** 22 D5
Lyne Down **HR8** 29 F5
Lyne of Gorthleck **IV2** 88 C2
Lyne of Skene **AB32** 91 F3
Lyne Station **EH45** 76 A6
Lyneal **SY12** 38 D2
Lynegar **KW1** 105 H3
Lyneham *Oxon.* **OX7** 30 D6
Lyneham *Wilts.* **SN15** 20 D4
Lyneholmeford **CA6** 70 A6
Lynemore *High.* **PH26** 89 H2
Lynemore *Moray* **AB37** 89 J1
Lynemouth **NE61** 71 H4
Lyness **KW16** 106 C8
Lynford **IP26** 44 C6
Lyng *Norf.* **NR9** 44 E4
Lyng *Som.* **TA3** 8 C2
Lyngate **NR28** 45 H3
Lynmouth EX35 7 F1
Lynn **TF10** 39 G4
Lynsted **ME9** 25 F5
Lynstone **EX23** 6 A5
Lynton EX35 7 F1
Lyon's Gate **DT2** 9 F4
Lyonshall **HR5** 28 C3
Lyrabus **PA44** 72 A4
Lytchett Matravers **BH16** 9 J5
Lytchett Minster **BH16** 9 J5
Lyth **KW1** 105 H2
Lytham **FY8** 55 G7
Lytham St. Anne's FY8 55 G7
Lythe **YO21** 63 K5
Lythe Hill **GU27** 12 C3
Lythes **KW17** 106 D9
Lythmore **KW14** 105 F2

M

Maaruig (Maraig) **HS3** 100 E7
Mabe Burnthouse **TR10** 2 E5
Mabie **DG2** 65 K3
Mablethorpe LN12 53 J4
Macclesfield SK11 49 J5
Macclesfield Forest **SK11** 49 J5
Macduff AB44 99 F4
Macedonia **KY6** 82 D7
Machan **ML9** 75 F5
Machany **PH3** 81 K6
Macharioch **PA28** 66 B3
Machen **CF83** 19 F3
Machrie *Arg. & B.* **PA49** 72 A4
Machrie *Arg. & B.* **PA42** 72 B6
Machrie *N.Ayr.* **KA27** 73 G7
Machrihanish **PA28** 66 A1
Machrins **PA61** 72 B1
Machynlleth SY20 37 G5
McInroy's Point **PA19** 74 A3
Mackerye End **AL4** 32 E7
Mackworth **DE22** 41 F2
Macmerry **EH33** 76 C3
Macterry **AB53** 99 F6
Madderty **PH7** 82 A5
Maddiston **FK2** 75 H3
Madehurst **BN18** 12 C5
Madeley *Staffs.* **CW3** 39 G1
Madeley *Tel. & W.* **TF7** 39 G5
Madeley Heath **CW3** 39 G1
Maders **PL17** 4 D3
Madford **EX15** 7 K4
Madingley **CB23** 33 G2
Madjeston **SP8** 9 H2
Madley **HR2** 28 D5
Madresfield **WR13** 29 H4
Madron **TR20** 2 B5
Maenaddwyn **LL71** 46 C4
Maenclochog **SA66** 16 D3
Maendy *Cardiff* **CF14** 18 E4
Maendy *V. of Glam.* **CF71** 18 D4
Maenporth **TR11** 2 E6
Maentwrog **LL41** 37 F1
Maen-y-groes **SA45** 26 C3
Maer *Cornw.* **EX23** 6 A5
Maer *Staffs.* **ST5** 39 G2
Maerdy *Carmar.* **SA19** 17 K3
Maerdy *Carmar.* **SA19** 17 K3
Maerdy *Conwy* **LL21** 37 K1
Maerdy *R.C.T.* **CF43** 18 C2
Maesbrook **SY10** 38 B3
Maesbury Marsh **SY10** 38 C3
Maes-Glas (Greenfield) *Flints.* **CH8** 47 K5
Maes-glas *Newport* **NP20** 19 F3
Maesgwynne **SA34** 17 F3
Maeshafn **CH7** 48 B6
Maesllyn **SA44** 17 G1
Maesmynis **LD2** 27 K4
Maesteg CF34 18 B2
Maes-Treylow **LD8** 28 B2
Maesybont **SA14** 17 J4
Maesycrugiau **SA39** 17 H1
Maesycwmmer **CF82** 18 E2
Maesyfed (New Radnor) **LD8** 28 B2
Magdalen Laver **CM5** 23 J1
Maggieknockater **AB38** 98 B6
Maggots End **CM23** 33 H6
Magham Down **BN27** 13 K5
Maghull **L31** 48 C2
Magna Park **LE17** 41 H7
Magor **NP26** 19 H3
Magpie Green **IP22** 34 E1
Maiden Bradley **BA12** 9 G1
Maiden Head **BS41** 19 J5
Maiden Law **DH7** 62 C2
Maiden Newton **DT2** 8 E5
Maiden Wells **SA71** 16 C6
Maidencombe **TQ1** 5 K4
Maidenhayne **EX13** 8 B5
Maidenhead SL6 22 B3
Maidens **KA26** 67 G3
Maiden's Green **RG42** 22 B4
Maidensgrove **RG9** 22 A3
Maidenwell *Cornw.* **PL30** 4 B3
Maidenwell *Lincs.* **LN11** 53 G5
Maidford **NN12** 31 H3
Maids' Moreton **MK18** 31 J5
Maidstone ME14 14 C2
Maidwell **NN6** 31 J1
Mail **ZE2** 107 N10
Maindee **NP19** 19 G3
Mainland *Ork.* **KW** 106 B6
Mainland *Shet.* **ZE** 107 M7
Mains of Ardestie **DD5** 83 G4
Mains of Balgavies **DD8** 83 G2
Mains of Balhall **DD9** 83 G1
Mains of Ballindarg **DD8** 83 F2
Mains of Burgie **IV36** 97 H6
Mains of Culsh **AB53** 99 G6
Mains of Dillavaird **AB30** 91 F6
Mains of Drum **AB31** 91 G4
Mains of Dudwick **AB41** 91 H1
Mains of Faillie **IV2** 88 E1
Mains of Fedderate **AB42** 99 G6
Mains of Glack **AB51** 91 F2
Mains of Glassaugh **AB45** 98 D4
Mains of Glenbuchat **AB36** 90 B3
Mains of Linton **AB51** 91 F3
Mains of Melgund **DD9** 83 G2
Mains of Pitfour **AB42** 99 H6
Mains of Pittrichie **AB21** 91 G2
Mains of Sluie **IV36** 97 H6
Mains of Tannachy **AB56** 98 B4
Mains of Thornton **AB30** 90 E7
Mains of Tig **KA26** 67 F5
Mains of Watten **KW1** 105 H3
Mainsforth **DL17** 62 E3
Mainsriddle **DG2** 65 K5
Mainstone **SY9** 38 B7
Maisemore **GL2** 29 H6
Major's Green **B90** 30 C1
Makendon **NE65** 70 D3
Makeney **DE56** 41 F1
Makerstoun **TD5** 76 E7
Malacleit **HS6** 92 C4
Malborough **TQ7** 5 H7
Malden Rushett **KT9** 22 E5
Maldon CM9 24 E1
Malham **BD23** 56 E3
Maligar **IV51** 93 K5
Malinbridge **S6** 51 F4
Mallaig PH41 86 C5
Mallaigmore **PH41** 86 C5
Mallaigvaig **PH41** 86 C5
Malleny Mills **EH14** 75 K4
Malletsheugh **G77** 74 D5
Malling **FK8** 81 G7
Mallows Green **CM23** 33 H6
Malltraeth **LL62** 46 C6
Mallwyd **SY20** 37 H4
Malmesbury SN16 20 C3
Malmsmead **EX35** 7 F1
Malpas *Ches.W. & C.* SY14 38 D1
Malpas *Cornw.* **TR1** 3 F4
Malpas *Newport* **NP20** 19 G2
Maltby *Lincs.* **LN11** 53 G4
Maltby *S.Yorks.* **S66** 51 H3
Maltby *Stock.* **TS8** 63 F5
Maltby le Marsh **LN13** 53 H4
Malting End **CB8** 34 B3
Malting Green **CO2** 34 D7
Maltman's Hill **TN27** 14 D3
Malton YO17 58 D2
Malvern Link **WR14** 29 G4
Malvern Wells **WR14** 29 G4
Mambeg **G84** 74 A2
Mamble **DY14** 29 F1
Mamhead **EX6** 7 H7
Mamhilad **NP4** 19 G1
Manaccan **TR12** 2 E6
Manadon **PL5** 4 E5
Manafon **SY21** 38 A5
Manais (Manish) **HS3** 93 G3
Manaton **TQ13** 7 F7
Manby **LN11** 53 G4
Mancetter **CV9** 41 F6
MANCHESTER M 49 H3
Manchester Airport **M90** 49 H4
Mancot Royal **CH5** 48 C6
Mandally **PH35** 87 J4
Manea **PE15** 43 H7
Maneight **KA18** 67 K3
Manfield **DL2** 62 D5
Mangaster **ZE2** 107 M5
Mangerton **DT6** 8 D5
Mangotsfield **BS16** 19 K4
Mangrove Green **LU2** 32 E6
Mangurstadh **HS2** 100 C4
Manish (Manais) **HS3** 93 G3
Mankinholes **OL14** 56 E7
Manley **WA6** 48 E5
Manmoel **NP12** 18 E1
Mannal **PA77** 78 A3
Manningford Abbots **SN9** 20 E6
Manningford Bohune **SN9** 20 E6
Manningford Bruce **SN9** 20 E6
Manningham **BD8** 57 G6
Mannings Heath **RH13** 13 F4
Mannington **BH21** 10 B4
Manningtree CO11 35 F5
Mannofield **AB15** 91 H4
Manor Park **SL2** 22 C3
Manorbier **SA70** 16 D6
Manorbier Newton **SA70** 16 D5
Manordeifi **SA43** 17 F1
Manordeilo **SA19** 17 K3
Manorowen **SA65** 16 C2
Mansell Gamage **HR4** 28 C4
Mansell Lacy **HR4** 28 D4
Mansergh **LA6** 56 B1
Mansfield NG18 51 H6
Mansfield Woodhouse **NG19** 51 H6
Manson Green **NR9** 44 E5
Mansriggs **LA12** 55 F1
Manston *Dorset* **DT10** 9 H3
Manston *Kent* **CT12** 25 K5
Manston *W.Yorks.* **LS15** 57 J6
Manswood **BH21** 9 J4
Manthorpe *Lincs.* **PE10** 42 D4
Manthorpe *Lincs.* **NG31** 42 C2
Manton *N.Lincs.* **DN21** 52 C2
Manton *Notts.* **S80** 51 H5
Manton *Rut.* **LE15** 42 B5
Manton *Wilts.* **SN8** 20 E5
Manuden **CM23** 33 H6
Manwood Green **CM5** 33 J7
Maolachy **PA35** 79 K6
Maperton **BA9** 9 F2
Maple Cross **WD3** 22 D2
Maplebeck **NG22** 51 K6
Mapledurham **RG4** 21 K4
Mapledurwell **RG25** 21 K6
Maplehurst **RH13** 12 E4
Maplescombe **DA4** 23 J5
Mapleton **DE6** 40 D1
Mapperley *Derbys.* **DE7** 41 G1
Mapperley *Notts.* **NG5** 41 H1
Mapperton *Dorset* **DT8** 8 E5
Mapperton *Dorset* **DT11** 9 J5
Mappleborough Green **B80** 30 B2
Mappleton **HU18** 59 J5
Mapplewell **S75** 51 F2
Mappowder **DT10** 9 G4
Mar Lodge **AB35** 89 H5
Maraig (Maaruig) **HS3** 100 E7
Marazion TR17 2 C5
Marbhig **HS2** 101 G6
Marbury **SY13** 38 E1
March PE15 43 H6
Marcham **OX13** 21 H2
Marchamley **SY4** 38 E3
Marchamley Wood **SY4** 38 E2
Marchington **ST14** 40 D2
Marchington Woodlands **ST14** 40 D3
Marchwiel **LL13** 38 C1
Marchwood **SO40** 10 E3
Marcross **CF61** 18 C5
Marcus **DD8** 83 G2
Marden *Here.* **HR1** 28 E4
Marden *Kent* **TN12** 14 C3
Marden *T. & W.* **NE30** 71 J6
Marden *Wilts.* **SN10** 20 D6
Marden Ash **CM5** 23 J1
Marden Beech **TN12** 14 C3
Marden Thorn **TN12** 14 C3
Marden's Hill **TN6** 13 H3
Mardon **TD12** 77 H7
Mardy **NP7** 28 C7
Mare Green **TA3** 8 C2
Marefield **LE7** 42 A5
Mareham le Fen **PE22** 53 F6
Mareham on the Hill **LN9** 53 F6
Maresfield **TN22** 13 H4
Marfleet **HU9** 59 H6
Marford **LL12** 48 C7
Margam **SA13** 18 A3
Margaret Marsh **SP7** 9 H3
Margaret Roding **CM6** 33 J7
Margaretting **CM4** 24 C1
Margaretting Tye **CM4** 24 C1
Margate CT9 25 K4
Margnaheglish **KA27** 73 J7
Margreig **DG2** 65 J3
Margrove Park **TS12** 63 H5
Marham **PE33** 44 B5
Marhamchurch **EX23** 6 A5
Marholm **PE6** 42 E5
Marian Cwm **LL18** 47 J5
Mariandyrys **LL58** 46 E4
Marian-glas LL73 46 D4
Mariansleigh **EX36** 7 F3
Marine Town **ME12** 25 F4
Marishader **IV51** 93 K5
Maristow House **PL6** 4 E4
Mark **TA9** 19 G7
Mark Causeway **TA9** 19 G7
Mark Cross **TN6** 13 J3
Markbeech **TN8** 23 H7
Markby **LN13** 53 H5
Markdhu **DG8** 64 B3
Markeaton **DE22** 41 F2
Market Bosworth **CV13** 41 G5
Market Deeping **PE6** 42 E4
Market Drayton TF9 39 F2
Market Harborough LE16 42 A7
Market Lavington **SN10** 20 D6
Market Overton **LE15** 42 B4
Market Rasen LN8 52 E4
Market Stainton **LN8** 53 F5
Market Street **NR12** 45 G3
Market Warsop **NG20** 51 H6
Market Weighton **YO43** 58 E5
Market Weston **IP22** 34 D1
Markethill **PH13** 82 D4
Markfield LE67 41 G4
Markham **NP12** 18 E1
Markham Moor **DN22** 51 K5
Markinch **KY7** 82 D7
Markington **HG3** 57 H3
Marks Gate **RM6** 23 H2
Marks Tey **CO6** 34 D6
Marksbury **BA2** 19 K5
Markwell **PL12** 4 D5
Markyate **AL3** 32 D7
Marl Bank **WR14** 29 G4
Marland **OL11** 49 H1
Marlborough SN8 20 E5
Marlbrook **B60** 29 J1
Marlcliff **B50** 30 B3
Marldon **TQ3** 5 J4
Marle Green **TN21** 13 J5
Marlesford **IP13** 35 H3
Marley Green **SY13** 38 E1
Marley Hill **NE16** 62 D1
Marlingford **NR9** 45 F5
Marloes **SA62** 16 A5
Marlow *Bucks.* SL7 22 B3
Marlow *Here.* **SY7** 28 D1
Marlpit Hill **TN8** 23 H7
Marlpool **DE75** 41 G1
Marnhull **DT10** 9 G3
Marnoch **AB54** 98 D5
Marple **SK6** 49 J4
Marple Bridge **SK6** 49 J4
Marr **DN5** 51 H2
Marrel **KW8** 105 F7
Marrick **DL11** 62 B7
Marrister **ZE2** 107 P6
Marros **SA33** 17 F5
Marsden *T. & W.* **NE34** 71 J7
Marsden *W.Yorks.* **HD7** 50 C1
Marsett **DL8** 56 E1
Marsh **EX14** 8 B3
Marsh Baldon **OX44** 21 J2
Marsh Benham **RG20** 21 H5
Marsh Gibbon **OX27** 31 H6
Marsh Green *Devon* **EX5** 7 J6
Marsh Green *Gt.Man.* **WN5** 48 E2
Marsh Green *Kent* **TN8** 23 H7
Marsh Green *Tel. & W.* **TF6** 39 F4
Marsh Lane **S21** 51 G5
Marsh Street **TA24** 7 H1
Marshall Meadows **TD15** 77 H5
Marshalsea **DT6** 8 C4
Marshalswick **AL1** 22 E1
Marsham **NR10** 45 F3
Marshaw **LA2** 55 J4
Marshborough **CT13** 15 J2
Marshbrook **SY6** 38 D7
Marshchapel **DN36** 53 G3
Marshfield *Newport* **CF3** 19 F3
Marshfield *S.Glos.* **SN14** 20 A4
Marshgate **PL32** 4 B1
Marshland St. James **PE14** 43 J5
Marshside **PR9** 48 C1
Marshwood **DT6** 8 C5
Marske **DL11** 62 C6
Marske-by-the-Sea **TS11** 63 H4
Marsland Green **M29** 49 F3
Marston *Ches.W. & C.* **CW9** 49 F5
Marston *Here.* **HR6** 28 C3
Marston *Lincs.* **NG32** 42 B1
Marston *Oxon.* **OX3** 21 J1
Marston *Staffs.* **ST18** 40 B3
Marston *Staffs.* **ST20** 40 A4
Marston *Warks.* **B76** 40 E6
Marston *Wilts.* **SN10** 20 C6
Marston Doles **CV47** 31 F3
Marston Green **B37** 40 D7
Marston Magna **BA22** 8 E2
Marston Meysey **SN6** 20 E2
Marston Montgomery **DE6** 40 D2
Marston Moretaine **MK43** 32 C4
Marston on Dove **DE65** 40 E3
Marston St. Lawrence **OX17** 31 G4
Marston Stannett **HR6** 28 E3
Marston Trussell **LE16** 41 J7
Marstow **HR9** 28 E7
Marsworth **HP23** 32 C7
Marten **SN8** 21 F5
Marthall **WA16** 49 G5
Martham **NR29** 45 J4
Martin *Hants.* **SP6** 10 B3
Martin *Lincs.* **LN4** 52 E7
Martin *Lincs.* **LN9** 53 F6
Martin Drove End **SP6** 10 B2
Martin Hussingtree **WR3** 29 H2
Martinhoe **EX31** 6 E1
Martinscroft **WA1** 49 F4
Martinstown **DT2** 9 F6
Martlesham **IP12** 35 G4
Martlesham Heath **IP5** 35 G4
Martletwy **SA67** 16 D4
Martley **WR6** 29 G2
Martock TA12 8 D3
Marton *Ches.E.* **SK11** 49 H6
Marton *Cumb.* **LA12** 55 F2
Marton *E.Riding* **HU11** 59 H6
Marton *E.Riding* **YO15** 59 J3
Marton *Lincs.* **DN21** 52 B4
Marton *Middbro.* **TS7** 63 G5
Marton *N.Yorks.* **YO51** 57 K3
Marton *N.Yorks.* **YO62** 58 D1
Marton *Shrop.* **SY21** 38 B5
Marton *Shrop.* **SY4** 38 D3
Marton *Warks.* **CV23** 31 F2
Marton Abbey **YO61** 58 B3
Marton-in-the-Forest **YO61** 58 B3
Marton-le-Moor **HG4** 57 J2
Martyr Worthy **SO21** 11 G1
Martyr's Green **KT11** 22 D6
Marwick **KW17** 106 B5
Marwood **EX31** 6 D2
Mary Tavy **PL19** 5 F3
Marybank *High.* **IV6** 96 B6
Marybank *W.Isles* **HS2** 101 G4
Maryburgh **IV7** 96 C6
Maryfield *Cornw.* **PL11** 4 E5
Maryfield *Shet.* **ZE2** 107 N8
Marygold **TD11** 77 G5
Maryhill *Aber.* **AB53** 99 G6
Maryhill *Glas.* **G20** 74 D4
Marykirk **AB30** 83 H1
Marylebone *Gt.Lon.* **W1G** 23 F3
Marylebone *Gt.Man.* **WN1** 48 E2
Marypark **AB37** 89 J1
Maryport *Cumb.* CA15 60 B3
Maryport *D. & G.* **DG9** 64 B7
Marystow **PL16** 6 C7
Maryton **DD10** 83 H2
Marywell *Aber.* **AB12** 91 H5
Marywell *Aber.* **AB34** 90 D5
Marywell *Angus* **DD11** 83 H3
Masham **HG4** 57 H1
Mashbury **CM1** 33 K7
Masongill **LA6** 56 B2
Mastin Moor **S43** 51 G5
Mastrick **AB16** 91 H4
Matchborough **B98** 30 B2
Matching **CM17** 33 J7
Matching Green **CM17** 33 J7
Matching Tye **CM17** 33 J7
Matfen **NE20** 71 F6
Matfield **TN12** 23 K7
Mathern **NP16** 19 J2
Mathon **WR13** 29 G4
Mathry **SA62** 16 B2
Matlaske **NR11** 45 F2
Matlock DE4 51 F6
Matlock Bank **DE4** 51 F6
Matlock Bath **DE4** 50 E7
Matson **GL4** 29 H7
Matterdale End **CA11** 60 E4
Mattersey **DN10** 51 J4
Mattersey Thorpe **DN10** 51 J4
Mattingley **RG27** 22 A6
Mattishall **NR20** 44 E4
Mattishall Burgh **NR20** 44 E4
Mauchline KA5 67 J1
Maud **AB42** 99 H6
Maufant **JE2** 3 K7
Maugersbury **GL54** 30 C6
Maughold **IM7** 54 D4
Mauld **IV4** 87 K1
Maulden **MK45** 32 D5
Maulds Meaburn **CA10** 61 H5
Maunby **YO7** 57 J1
Maund Bryan **HR1** 28 E3
Maundown **TA4** 7 J3
Mautby **NR29** 45 J4
Mavesyn Ridware **WS15** 40 C4
Mavis Enderby **PE23** 53 G6
Maw Green **CW1** 49 G7
Mawbray **CA15** 60 B2
Mawdesley **L40** 48 D1
Mawdlam **CF33** 18 B3
Mawgan **TR12** 2 E6
Mawgan Porth **TR8** 3 F2
Mawla **TR16** 2 E4
Mawnan **TR11** 2 E6
Mawnan Smith **TR11** 2 E6
Mawsley **NN14** 32 B1
Mawthorpe **LN13** 53 H5
Maxey **PE6** 42 E5
Maxstoke **B46** 40 E7
Maxted Street **CT4** 15 G3
Maxton *Kent* **CT17** 15 J3
Maxton *Sc.Bord.* **TD6** 76 E7
Maxwellheugh **TD5** 77 F7
Maxwelltown **DG2** 65 K3
Maxworthy **PL15** 4 C1
May Hill **GL17** 29 G6
Mayals **SA3** 17 K6
Maybole KA19 67 H3
Maybury **GU22** 22 D6
Mayen **AB54** 98 D6
Mayfair **W1J** 23 F3
Mayfield *E.Suss.* TN20 13 J4
Mayfield *Midloth.* **EH22** 76 B4
Mayfield *Staffs.* **DE6** 40 D1
Mayford **GU22** 22 C6
Mayland **CM3** 25 F1
Maylandsea **CM3** 25 F1
Maynard's Green **TN21** 13 J5
Maypole *I.o.S.* **TR21** 2 C1
Maypole *Kent* **CT3** 25 H5
Maypole *Mon.* **NP25** 28 D7
Maypole Green *Essex* **CO2** 34 D6
Maypole Green *Norf.* **NR14** 45 J6
Maypole Green *Suff.* **IP13** 35 G2
Maypole Green *Suff.* **IP30** 34 D3
May's Green *N.Som.* **BS24** 19 G5
Mays Green *Oxon.* **RG9** 22 A3
Maywick **ZE2** 107 M10
Mead **EX39** 6 A4
Mead End **SP5** 10 B2
Meadgate **BA2** 19 K6
Meadle **HP17** 22 B1
Meadow Green **WR6** 29 G3
Meadowhall **S9** 51 F3
Meadowmill **EH33** 76 C3
Meadowtown **SY5** 38 C5
Meadwell **PL16** 6 C7
Meaford **ST15** 40 A2
Meal Bank **LA9** 61 G7
Mealabost (Melbost Borve) **HS2** 101 G2
Mealasta **HS2** 100 B5
Meals **LN11** 53 H3

Mea - Min

Mealsgate CA7 60 D2
Meanley BB7 56 C5
Meanwood LS6 57 H6
Mearbeck BD23 56 D3
Meare BA6 19 H7
Meare Green TA3 8 B2
Mearns G77 74 D5
Mears Ashby NN6 32 B2
Measham DE12 41 F4
Meathop LA11 55 H1
Meavy PL20 5 F4
Medbourne LE16 42 B6
Meddon EX39 6 A4
Meden Vale NG20 51 H6
Medlar PR4 55 H6
Medmenham SL7 22 B3
Medomsley DH8 62 C1
Medstead GU34 11 H1
Meer Common HR3 28 C3
Meer End CV8 30 D1
Meerbrook ST13 49 J6
Meesden SG9 33 H5
Meeson TF6 39 F3
Meeth EX20 6 D5
Meeting House Hill NR28 45 H3
Meggethead TD7 69 G1
Meidrim SA33 17 F3
Meifod Denb. LL16 47 J7
Meifod Powys SY22 38 A4
Meigle PH12 82 D3
Meikle Earnock ML3 75 F5
Meikle Grenach PA20 73 J4
Meikle Kilmory PA20 73 J4
Meikle Rahane G84 74 A2
Meikle Strath AB30 90 E7
Meikle Tarty AB41 91 H2
Meikle Wartle AB51 91 F1
Meikleour PH2 82 C4
Meikleyard KA4 74 D7
Meinciau SA17 17 H4
Meir ST3 40 B1
Meirheath ST3 40 B2
Melbost HS2 101 G4
Melbost Borve (Mealabost) HS2 101 G2
Melbourn SG8 33 G4
Melbourne Derbys. DE73 41 F3
Melbourne E.Riding YO42 58 D5
Melbury EX39 6 B4
Melbury Abbas SP7 9 H3
Melbury Bubb DT2 8 E4
Melbury Osmond DT2 8 E4
Melbury Sampford DT2 8 E4
Melby ZE2 107 K7
Melchbourne MK44 32 D2
Melcombe Bingham DT2 9 G4
Melcombe Regis DT4 9 F6
Meldon Devon EX20 6 D6
Meldon Northumb. NE61 71 G5
Meldreth SG8 33 G4
Meledor PL26 3 G3
Melfort PA34 79 K6
Melgarve PH20 88 B5
Melgum AB34 90 C4
Meliden (Gallt Melyd) LL19 47 J4
Melincourt SA11 18 B1
Melin-y-coed LL26 47 G6
Melin-y-ddol SY21 37 K5
Melin-y-grug SY21 37 K5
Melin-y-Wig LL21 37 K1
Melkinthorpe CA10 61 G4
Melkridge NE49 70 C7
Melksham SN12 20 C5
Melksham Forest SN12 20 C5
Melldalloch PA21 73 H3
Melling Lancs. LA6 55 J2
Melling Mersey. L31 48 C2
Melling Mount L31 48 D2
Mellis IP23 34 E1
Mellon Charles IV22 94 E2
Mellon Udrigle IV22 94 E2
Mellor Gt.Man. SK6 49 J4
Mellor Lancs. BB2 56 B6
Mellor Brook BB2 56 B6
Mells BA11 20 A7
Melmerby Cumb. CA10 61 H3
Melmerby N.Yorks. DL8 57 F1
Melmerby N.Yorks. HG4 57 J2
Melplash DT6 8 D5
Melrose Aber. AB45 99 F4
Melrose Sc.Bord. TD6 76 D7
Melsetter KW16 106 B9
Melsonby DL10 62 C6
Meltham HD9 50 C1
Melton E.Riding HU14 59 F7
Melton Suff. IP12 35 G3
Melton Constable NR24 44 E2
Melton Mowbray LE13 42 A4
Melton Ross DN38 52 D1
Meltonby YO42 58 D4
Melvaig IV21 94 D3
Melverley SY10 38 C4
Melverley Green SY10 38 C4
Melvich KW14 104 D2
Membury EX13 8 B4
Memsie AB43 99 H4
Memus DD8 83 F2
Menabilly PL24 4 A5
Menai Bridge (Porthaethwy) LL59 46 D5
Mendham IP20 45 G7
Mendlesham IP14 35 F2
Mendlesham Green IP14 34 E2
Menethorpe YO17 58 D3
Menheniot PL14 4 C4
Menie House AB23 91 H2
Menithwood WR6 29 G2
Mennock DG4 68 D3
Menston LS29 57 G5
Menstrie FK11 75 G1
Mentmore LU7 32 C7
Meoble PH40 86 D6
Meole Brace SY3 38 D4
Meon PO14 11 G4
Meonstoke SO32 11 H3
Meopham DA13 24 C5
Meopham Green DA13 24 C5
Mepal CB6 43 H7
Meppershall SG17 32 E5
Merbach HR3 28 C4
Mercaston DE6 40 E1
Mere Ches.E. WA16 49 G4
Mere Wilts. BA12 9 H1
Mere Brow PR9 48 D1
Mere Green B75 40 D6
Mere Heath CW9 49 F5
Mereclough BB10 56 D6
Mereside FY4 55 G6
Meretown TF10 39 G3
Mereworth ME18 23 K6
Mergie AB39 91 F6
Meriden CV7 40 E7
Merkadale IV47 85 J1
Merkinch IV3 96 D7
Merkland DG7 65 H3
Merley BH21 10 B5
Merlin's Bridge SA61 16 C4
Merridge TA5 8 B1
Merrifield TQ7 5 J6
Merrington SY4 38 D3
Merrion SA71 16 C6
Merriott TA16 8 D3
Merrivale PL19 5 F3
Merrow GU1 22 D6
Merry Hill Herts. WD23 22 E2
Merry Hill W.Mid. DY5 40 B7
Merry Hill W.Mid. WV3 40 A6
Merrymeet PL14 4 C4
Mersea Island CO5 34 E7
Mersham TN25 15 F4
Merstham RH1 23 F6
Merston PO20 12 B6
Merstone PO30 11 G6
Merther TR2 3 F4
Merthyr SA33 17 G3
Merthyr Cynog LD3 27 J5
Merthyr Dyfan CF62 18 E5
Merthyr Mawr CF32 18 B4
Merthyr Tydfil CF47 18 D1
Merthyr Vale CF48 18 D2
Merton Devon EX20 6 D4
Merton Norf. IP25 44 D6
Merton Oxon. OX25 31 G7
Mervinslaw TD8 70 B2
Meshaw EX36 7 F4
Messing CO5 34 D7
Messingham DN17 52 B2
Metcombe EX11 7 J6
Metfield IP20 45 G7
Metheringham LN4 52 D6
Metherwell PL17 4 E4
Methil KY8 76 B1
Methlem LL53 36 A2
Methley LS26 57 J7
Methley Junction LS26 57 J7
Methlick AB41 91 G1
Methven PH1 82 B5
Methwold IP26 44 B6
Methwold Hythe IP26 44 B6
MetroCentre NE11 71 H7
Mettingham NR35 45 H7
Metton NR11 45 F2
Mevagissey PL26 4 A6
Mewith Head LA2 56 C3
Mexborough S64 51 G3
Mey KW14 105 H1
Meysey Hampton GL7 20 E1
Miabhag W.Isles HS3 93 G2
Miabhag W.Isles HS3 100 C7
Miabhaig (Miavaig) HS2 100 C4
Mial IV21 94 D4
Miavaig (Miabhaig) HS2 100 C4
Michaelchurch HR2 28 E6
Michaelchurch Escley HR2 28 C5
Michaelchurch-on-Arrow HR5 28 B3
Michaelston-le-Pit CF64 18 E4
Michaelston-super-Ely CF5 18 E4
Michaelston-y-Fedw CF3 19 F3
Michaelstow PL30 4 A3
Michelcombe TQ13 5 G4
Micheldever SO21 11 G1
Michelmersh SO51 10 E2
Mickfield IP14 35 F2
Mickle Trafford CH2 48 D6
Micklebring S66 51 H3
Mickleby TS13 63 K5
Micklefield LS25 57 K6
Micklefield Green WD3 22 D2
Mickleham RH5 22 E6
Micklehurst OL5 49 J2
Mickleover DE3 41 F2
Micklethwaite Cumb. CA7 60 D1
Micklethwaite W.Yorks. BD20 57 F5
Mickleton Dur. DL12 62 A4
Mickleton Glos. GL55 30 C4
Mickletown LS26 57 J7
Mickley Derbys. S18 51 F5
Mickley N.Yorks. HG4 57 H2
Mickley Green IP29 34 C3
Mickley Square NE43 71 F7
Mid Ardlaw AB43 99 H4
Mid Beltie AB31 90 E4
Mid Calder EH53 75 J4
Mid Clyth KW3 105 H5
Mid Lambrook TA13 8 D3
Mid Lavant PO18 12 B6
Mid Letter PA27 80 B7
Mid Lix FK21 81 G4
Mid Mossdale DL8 61 K7
Mid Yell ZE2 107 P3
Midbea KW17 106 D3
Middle Assendon RG9 22 A3
Middle Aston OX25 31 F6
Middle Barton OX7 31 F6
Middle Bickenhill B92 40 E7
Middle Bockhampton BH23 10 C5
Middle Claydon MK18 31 J6
Middle Drift PL14 4 B4
Middle Drums DD9 83 G2
Middle Duntisbourne GL7 20 C1
Middle Handley S21 51 G5
Middle Harling NR16 44 D7
Middle Kames PA31 73 H2
Middle Littleton WR11 30 B4
Middle Maes-coed HR2 28 C5
Middle Marwood EX31 6 D2
Middle Mill SA62 16 B3
Middle Quarter TN26 14 D4
Middle Rasen LN8 52 D4
Middle Rigg PH2 82 B7
Middle Salter LA2 56 B3
Middle Sontley LL13 38 C1
Middle Stoford TA21 7 K3
Middle Taphouse PL14 4 B4
Middle Town TR25 2 C1
Middle Tysoe CV35 30 E4
Middle Wallop SO20 10 D1
Middle Winterslow SP5 10 D1
Middle Woodford SP4 10 C1
Middlebie DG11 69 H6
Middlecliff S72 51 G2
Middlecott EX22 6 C5
Middleham DL8 57 G1
Middlehill Aber. AB53 99 G6
Middlehill Cornw. PL14 4 C4
Middlehope SY7 38 E7
Middlemarsh DT9 9 F4
Middlemoor PL19 4 E3
Middlequarter (Ceathramh Meadhanach) HS6 92 D4
Middlesbrough TS1 63 F4
Middlesceugh CA4 60 F2
Middleshaw LA8 55 J1
Middlesmoor HG3 57 F2
Middlestone DL16 62 D3
Middlestone Moor DL16 62 D3
Middlestown WF4 50 E1
Middleton Aber. AB21 91 G3
Middleton Angus DD11 83 G3
Middleton Cumb. LA6 56 B1
Middleton Derbys. DE4 50 E7
Middleton Derbys. DE45 50 D6
Middleton Essex CO10 34 C4
Middleton Gt.Man. M24 49 H2
Middleton Hants. SP11 21 H7
Middleton Here. SY8 28 E2
Middleton Lancs. LA3 55 H4
Middleton Midloth. EH23 76 B5
Middleton N.Yorks. YO18 58 D1
Middleton Norf. PE32 44 A4
Middleton Northants. LE16 42 B7
Middleton Northumb. NE61 71 F5
Middleton Northumb. NE70 77 J7
Middleton P. & K. KY13 82 C7
Middleton P. & K. PH10 82 C3
Middleton Shrop. SY15 38 B6
Middleton Shrop. SY8 28 E1
Middleton Shrop. SY11 38 C3
Middleton Suff. IP17 35 J2
Middleton Swan. SA3 17 H7
Middleton W.Yorks. LS29 57 G5
Middleton W.Yorks. LS10 57 J7
Middleton Warks. B78 40 D6
Middleton Baggot WV16 39 F6
Middleton Bank Top NE61 71 F5
Middleton Cheney OX17 31 F4
Middleton Green ST10 40 B2
Middleton Hall NE71 70 E1
Middleton Moor IP17 35 J2
Middleton of Potterton AB23 91 H3
Middleton on the Hill SY8 28 E2
Middleton One Row DL2 62 E5
Middleton Park AB22 91 H3
Middleton Priors WV16 39 F7
Middleton Quernhow HG4 57 J2
Middleton St. George DL2 62 E5
Middleton Scriven WV16 39 F7
Middleton Stoney OX25 31 G6
Middleton Tyas DL10 62 D6
Middleton-in-Teesdale DL12 62 A4
Middleton-on-Leven TS15 63 F5
Middleton-on-Sea PO22 12 C6
Middleton-on-the-Wolds YO25 59 F5
Middletown Cumb. CA22 60 A6
Middletown Powys SY21 38 C4
Middlewich CW10 49 G6
Middlewood Ches.E. SK12 49 J4
Middlewood S.Yorks. S6 51 F3
Middlewood Green IP14 34 E2
Middlezoy TA7 8 C1
Middridge DL4 62 D4
Midfield IV27 103 H2
Midford BA2 20 A5
Midge Hall PR26 55 J7
Midgeholme CA8 61 H1
Midgham RG7 21 J5
Midgley W.Yorks. WF4 50 E1
Midgley W.Yorks. HX2 57 F7
Midhopestones S36 50 E3
Midhurst GU29 12 B4
Midlem TD7 70 A1
Midloe Grange PE19 32 E2
Midpark PA20 73 J5
Midsomer Norton BA3 19 K6
Midthorpe LN9 53 F5
Midtown High. IV27 103 H2
Midtown High. IV22 94 E3
Midtown of Barras AB39 91 G6
Midville PE22 53 G7
Midway DE11 41 F3
Migdale IV24 96 D2
Migvie AB34 90 C4
Milarrochy G63 74 C1
Milber TQ12 5 J3
Milbethill AB54 98 E5
Milborne Port DT9 9 F3
Milborne St. Andrew DT11 9 G5
Milborne Wick DT9 9 F2
Milbourne Northumb. NE20 71 G6
Milbourne Wilts. SN16 20 C3
Milburn CA10 61 H4
Milbury Heath GL12 19 K2
Milcombe OX15 31 F5
Milden IP7 34 D4
Mildenhall Suff. IP28 34 B1
Mildenhall Wilts. SN8 21 F5
Mile Elm SN11 20 C5
Mile End Essex CO4 34 D6
Mile End Glos. GL16 28 E7
Mile Oak TN12 23 K7
Mile Town ME12 25 F4
Milebrook LD7 28 C1
Milebush TN12 14 C3
Mileham PE32 44 D4
Miles Green ST7 40 A1
Miles Hope WR15 28 E2
Milesmark KY12 75 J2
Miles's Green RG7 21 J5
Milfield NE71 77 H7
Milford Derbys. DE56 41 F1
Milford Devon EX39 6 A3
Milford Shrop. SY4 38 D3
Milford Staffs. ST17 40 B3
Milford Surr. GU8 22 C7
Milford Haven (Aberdaugleddau) SA73 16 B5
Milford on Sea SO41 10 D5
Milkwall GL16 19 J1
Mill Bank HX6 57 F7
Mill Brow SK6 49 J4
Mill End Bucks. RG9 22 A3
Mill End Cambs. CB8 33 K3
Mill End Herts. SG9 33 G5
Mill End Green CM6 33 K6
Mill Green Cambs. CB21 33 K4
Mill Green Essex CM4 24 C1
Mill Green Herts. AL9 23 F1
Mill Green Norf. IP22 45 F7
Mill Green Shrop. TF9 39 F3
Mill Green Staffs. WS15 40 C3
Mill Green Suff. IP13 35 H2
Mill Green Suff. IP14 35 F3
Mill Green Suff. IP14 34 D3
Mill Green W.Mid. WS9 40 C5
Mill Hill B'burn. BB2 56 B7
Mill Hill Cambs. SG19 33 F3
Mill Hill Gt.Lon. NW7 23 F2
Mill Houses LA2 56 B3
Mill Lane GU10 22 A6
Mill of Camsail G84 74 A2
Mill of Colp AB53 99 F6
Mill of Elrick AB41 99 H6
Mill of Fortune PH6 81 J5
Mill of Kingoodie AB21 91 G2
Mill of Monquich AB39 91 G5
Mill of Uras AB39 91 G6
Mill Side LA11 55 H1
Mill Street Kent ME19 23 K6
Mill Street Norf. NR20 44 E4
Milland GU30 12 B4
Millbank AB42 99 J6
Millbeck CA12 60 D4
Millbounds KW17 106 E4
Millbreck AB42 99 J6
Millbridge GU10 22 B7
Millbrook Cen.Beds. MK45 32 D5
Millbrook Cornw. PL10 4 E5
Millbrook Devon EX13 8 C5
Millbrook S'ham. SO15 10 E3
Millburn Aber. AB33 90 D2
Millburn Aber. AB54 90 E1
Millcombe TQ9 5 J6
Millcorner TN31 14 D5
Milldale DE6 50 D7
Millden AB23 91 H3
Milldens DD8 83 G2
Millearne PH7 82 A6
Millend OX7 30 E6
Millenheath SY13 38 E2
Millerhill EH22 76 B4
Miller's Dale SK17 50 D5
Millers Green Derbys. DE4 50 E7
Miller's Green Essex CM5 23 J1
Millgate OL12 56 D7
Millhalf HR3 28 B4
Millhayes Devon EX15 7 K4
Millhayes Devon EX14 8 B4
Millholme LA8 61 G7
Millhouse Arg. & B. PA21 73 H3
Millhouse Cumb. CA7 60 E3
Millhouse Green S36 50 E2
Millhousebridge DG11 69 G5
Millikenpark PA10 74 C4
Millin Cross SA62 16 C4
Millington YO42 58 E4
Millington Green DE6 40 E1
Millmeece ST21 40 A2
Millness IV63 87 K1
Millom LA18 54 E1
Millow SG18 33 F4
Millpool PL30 4 B3
Millport KA28 73 K5
Millthorpe S18 51 F5
Millthrop LA10 61 H7
Milltimber AB13 91 G4
Milltown Aber. AB36 89 K4
Milltown Cornw. PL22 4 B5
Milltown D. & G. DG14 69 J6
Milltown Derbys. S45 51 F6
Milltown Devon EX31 6 D2
Milltown High. IV12 97 G7
Milltown of Aberdalgie PH2 82 B5
Milltown of Auchindoun AB55 90 B1
Milltown of Craigston AB53 99 F5
Milltown of Edinvillie AB38 97 K7
Milltown of Kildrummy AB33 90 C3
Milltown of Rothiemay AB54 98 D6
Milltown of Towie AB33 90 C3
Milnathort KY13 82 C7
Milners Heath CH3 48 D6
Milngavie G62 74 D3
Milnrow OL16 49 J1
Milnsbridge HD3 50 D1
Milnthorpe LA7 55 H1
Milovaig IV55 93 G6
Milrig KA4 74 D7
Milson DY14 29 F1
Milstead ME9 14 E2
Milston SP4 20 E7
Milton Angus DD8 82 E3
Milton Cambs. CB24 33 H2
Milton Cumb. CA8 70 A7
Milton D. & G. DG2 68 D5
Milton D. & G. DG2 65 J3
Milton D. & G. DG8 64 C5
Milton Derbys. DE65 41 F3
Milton High. IV6 95 K6
Milton High. IV18 96 E4
Milton High. IV12 97 G6
Milton High. IV54 94 D7
Milton High. IV6 96 C7
Milton High. KW1 105 J3
Milton High. IV63 88 B1
Milton Moray AB56 98 D4
Milton N.Som. BS22 19 G5
Milton Newport NP19 19 G3
Milton Notts. NG22 51 K5
Milton Oxon. OX15 31 F5
Milton Oxon. OX14 21 H2
Milton P. & K. PH8 82 A4
Milton Pembs. SA70 16 D5
Milton Ports. PO4 11 H5
Milton Som. TA12 8 D2
Milton Stir. FK8 81 G7
Milton Stoke ST2 49 J7
Milton W.Dun. G82 74 C3
Milton Abbas DT11 9 H4
Milton Abbot PL19 4 E3
Milton Bridge EH26 76 A4
Milton Bryan MK17 32 C5
Milton Clevedon BA4 9 F1
Milton Combe PL20 4 E4
Milton Damerel EX22 6 B4
Milton End GL2 29 G7
Milton Ernest MK44 32 D3
Milton Green CH3 48 D7
Milton Hill OX13 21 H2
MILTON KEYNES MK 32 B5
Milton Keynes Village MK10 32 B5
Milton Lilbourne SN9 20 E5
Milton Malsor NN7 31 J3
Milton Morenish FK21 81 H4
Milton of Auchinhove AB31 90 D4
Milton of Balgonie KY7 82 E7
Milton of Buchanan G63 74 C1
Milton of Cairnborrow AB54 98 C6
Milton of Callander FK17 81 G7
Milton of Campfield AB31 90 E4
Milton of Campsie G66 74 E3
Milton of Coldwells AB41 91 H1
Milton of Cullerlie AB32 91 F4
Milton of Cushnie AB33 90 D3
Milton of Dalcapon PH9 82 A2
Milton of Inveramsay AB51 91 F2
Milton of Noth AB54 90 D2
Milton of Tullich AB35 90 B5
Milton on Stour SP8 9 G2
Milton Regis ME10 24 E5
Milton Street BN26 13 J6
Miltonduff IV30 97 J5
Miltonhill IV36 97 H5
Miltonise DG8 64 B3
Milton-Lockhart ML8 75 G6
Milton-under-Wychwood OX7 30 D7
Milverton Som. TA4 7 K3
Milverton Warks. CV32 30 E2
Milwich ST18 40 B2
Mimbridge GU24 22 C5
Minard PA32 73 H1
Minard Castle PA32 73 H1
Minchington DT11 9 J3
Minchinhampton GL6 20 B1
Mindrum TD12 77 G7
Mindrummill TD12 77 G7
Minehead TA24 7 H1
Minera LL11 48 B7
Minety SN16 20 D2
Minety Lower Moor SN16 20 D2
Minffordd Gwyn. LL40 37 G4
Minffordd Gwyn. LL48 36 E2
Minffordd Gwyn. LL57 46 D5
Miningsby PE22 53 G6

Min - Nai

Minions PL14 4 C3
Minishant KA19 67 H2
Minley Manor GU17 22 B6
Minllyn SY20 37 H4
Minnes AB41 91 H2
Minngearraidh HS8 84 C2
Minnigaff DG8 64 E4
Minnonie AB45 99 F4
Minskip YO51 57 J3
Minstead SO43 10 D3
Minsted GU29 12 B4
Minster *Kent* ME12 25 F4
Minster *Kent* CT12 25 K5
Minster Lovell OX29 30 E7
Minsteracres NE44 62 B1
Minsterley SY5 38 C5
Minsterworth GL2 29 G7
Minterne Magna DT2 9 F4
Minterne Parva DT2 9 F4
Minting LN9 52 E5
Mintlaw AB42 99 J6
Minto TD9 70 A1
Minton SY6 38 D6
Minwear SA67 16 D4
Minworth B76 40 D6
Miodar PA77 78 B2
Mirbister KW17 106 C6
Mirehouse CA28 60 A5
Mireland KW1 105 J2
Mirfield WF14 57 H7
Miserden GL6 20 C1
Miskin *R.C.T.* CF72 18 D3
Miskin *R.C.T.* CF45 18 D2
Misselfore SP5 10 B2
Misson DN10 51 J3
Misterton *Leics.* LE17 41 H7
Misterton *Notts.* DN10 51 K3
Misterton *Som.* TA18 8 D4
Mistley CO11 35 F5
Mitcham CR4 23 F5
Mitchel Troy NP25 28 D7
Mitcheldean GL17 29 F7
Mitchell TR8 3 F3
Mitchelland LA8 60 F7
Mitcheltroy Common NP25 19 H1
Mitford NE61 71 G5
Mithian TR5 2 E3
Mitton ST19 40 A4
Mixbury NN13 31 H5
Mixenden HX2 57 F7
Moar PH15 81 G3
Moat CA6 69 K6
Moats Tye IP14 34 E3
Mobberley *Ches.E.* WA16 49 G5
Mobberley *Staffs.* ST10 40 C1
Moccas HR2 28 C4
Mochdre *Conwy* LL28 47 G5
Mochdre *Powys* SY16 37 K7
Mochrum DG8 64 D6
Mockbeggar *Hants.* BH24 10 C4
Mockbeggar *Kent* TN12 14 C3
Mockerkin CA13 60 B4
Modbury PL21 5 G5
Moddershall ST15 40 B2
Modsarie KW14 103 J2
Moelfre *I.o.A.* LL72 46 D4
Moelfre *Powys* SY10 38 A3
Moffat DG10 69 F3
Mogerhanger MK44 32 E4
Moin'a'choire PA44 72 B4
Moine House IV27 103 H3
Moira DE12 41 F4
Molash CT4 15 F2
Mol-chlach PH41 85 K3
Mold (Yr Wyddgrug) CH7 48 B6
Molehill Green *Essex* CM6 33 J6
Molehill Green *Essex* CM6 34 B6
Molescroft HU17 59 G5
Molesden NE61 71 G5
Molesworth PE28 32 D1
Mollance DG7 65 H4
Molland EX36 7 G3
Mollington *Ches.W. & C.* CH1 48 C5
Mollington *Oxon.* OX17 31 F4
Mollinsburn G67 75 F3
Monach Islands (Heisker Islands) HS6 92 B5
Monachty SY23 26 E2
Monachyle FK19 81 F6
Monevechadan PA24 80 C7
Monewden IP13 35 G3
Moneydie PH1 82 B5
Moneyrow Green SL6 22 B4
Moniaive DG3 68 C4
Monifieth DD5 83 G4
Monikie DD5 83 G4
Monimail KY15 82 D6
Monington SA43 16 E1
Monk Bretton S71 51 F2
Monk Fryston LS25 58 B7
Monk Hesleden TS27 63 F3
Monk Sherborne RG26 21 K6
Monk Soham IP13 35 G2
Monk Soham Green IP13 35 G2
Monk Street CM6 33 K6
Monken Hadley EN5 23 F2
Monkerton EX1 7 H6
Monkhide HR8 29 F4
Monkhill CA5 60 E1
Monkhopton WV16 39 F6
Monkland HR6 28 D3
Monkleigh EX39 6 C3
Monknash CF71 18 C4
Monkokehampton EX19 6 D5
Monks Eleigh IP7 34 D4
Monks Eleigh Tye IP7 34 D4
Monk's Gate RH13 13 F4
Monks' Heath SK10 49 H5
Monk's Hill TN27 14 D3
Monks Horton TN25 15 G4
Monks Kirby CV23 41 G7
Monks Risborough HP27 22 B1
Monkscross PL17 4 D3
Monkseaton NE25 71 J6
Monkshill AB53 99 F6
Monksilver TA4 7 J2
Monkstadt IV51 93 J5
Monkswood NP15 19 G1
Monkton *Devon* EX14 7 K5
Monkton *Kent* CT12 25 J5
Monkton *Pembs.* SA71 16 C5
Monkton *S.Ayr.* KA9 67 H1
Monkton *T. & W.* NE32 71 J7
Monkton *V. of Glam.* CF71 18 C4
Monkton Combe BA2 20 A5
Monkton Deverill BA12 9 H1
Monkton Farleigh BA15 20 B5
Monkton Heathfield TA2 8 B2
Monkton Up Wimborne BH21 10 B3
Monkton Wyld DT6 8 C5
Monkwearmouth SR6 62 E1
Monkwood SO24 11 H1
Monmouth (Trefynwy) NP25 28 E7
Monnington Court HR2 28 C5
Monnington on Wye HR4 28 C4
Monreith DG8 64 D6
Montacute TA15 8 E3
Monteach AB41 99 G6
Montford SY4 38 D4
Montford Bridge SY4 38 D4
Montgarrie AB33 90 D3
Montgomery (Trefaldwyn) SY15 38 B6
Montgreenan KA13 74 B6
Montrave KY8 82 E7
Montrose DD10 83 J2
Monxton SP11 21 G7
Monyash DE45 50 D6
Monymusk AB51 90 E3
Monzie PH7 81 K5
Moodiesburn G69 74 E3
Moons Moat North B98 30 B2
Moonzie KY15 82 E6
Moor Allerton LS17 57 J6
Moor Cock LA2 56 B3
Moor Crichel BH21 9 J4
Moor End *Bed.* MK43 32 D3
Moor End *Cen.Beds.* LU6 32 C6
Moor End *Cumb.* LA6 55 J2
Moor End *E.Riding* YO43 58 E6
Moor End *Lancs.* FY6 55 G5
Moor End *N.Yorks.* YO19 58 B6
Moor End *W.Yorks.* HX2 57 F7
Moor Green *W.Mid.* B13 40 C7
Moor Green *Wilts.* SN13 20 B5
Moor Head BD18 57 G6
Moor Monkton YO26 58 B4
Moor Row CA24 60 B5
Moor Side *Cumb.* LA16 55 F2
Moor Side *Lancs.* PR4 55 H6
Moor Side *Lancs.* PR4 55 H6
Moor Side *Lincs.* PE22 53 F7
Moor Street ME8 24 E5
Moorby PE22 53 F6
Moorcot HR5 28 C3
Moordown BH9 10 B5
Moore WA4 48 E4
Moorend CA5 60 E1
Moorends DN8 51 J1
Moorfield SK13 50 C3
Moorgreen *Hants.* SO30 11 F3
Moorgreen *Notts.* NG16 41 G1
Moorhall S18 51 F5
Moorhampton HR4 28 C4
Moorhouse *Cumb.* CA5 60 E1
Moorhouse *Notts.* NG23 51 K6
Moorland (Northmoor Green) TA7 8 C1
Moorlinch TA7 8 C1
Moorsholm TS12 63 H5
Moorside *Dorset* DT10 9 G3
Moorside *Gt.Man.* OL1 49 J2
Moorside *W.Yorks.* LS13 57 H6
Moorthorpe WF9 51 G1
Moortown *I.o.W.* PO30 11 F6
Moortown *Lincs.* LN7 52 D3
Moortown *Tel. & W.* TF6 39 F4
Morangie IV19 96 E3
Morar PH40 86 C5
Morborne PE7 42 E6
Morchard Bishop EX17 7 F5
Morcombelake DT6 8 D5
Morcott LE15 42 C5
Morda SY10 38 B3
Morden *Dorset* BH20 9 J5
Morden *Gt.Lon.* SM4 23 F5
Morden Park SM4 23 F5
Mordiford HR1 28 E5
Mordington Holdings TD15 77 H5
Mordon TS21 62 E4
More SY9 38 C6
Morebath EX16 7 H3
Morebattle TD5 70 C1
Morecambe LA4 55 H3
Morefield IV26 95 H2
Moreleigh TQ9 5 H5
Morenish FK21 81 H4
Moresby Parks CA28 60 A5
Morestead SO21 11 G2
Moreton *Dorset* DT2 9 H6
Moreton *Essex* CM5 23 J1
Moreton *Here.* HR6 28 E2
Moreton *Mersey.* CH46 48 B4
Moreton *Oxon.* OX9 21 K1
Moreton *Staffs.* DE6 40 D2
Moreton *Staffs.* TF10 39 G4
Moreton Corbet SY4 38 E3
Moreton Jeffries HR1 29 F4
Moreton Mill SY4 38 E3
Moreton Morrell CV35 30 E3
Moreton on Lugg HR4 28 E4
Moreton Paddox CV35 30 E3
Moreton Pinkney NN11 31 G4
Moreton Say TF9 39 F2
Moreton Valence GL2 20 A1
Moretonhampstead TQ13 7 F7
Moreton-in-Marsh GL56 30 D5
Morfa *Carmar.* SA14 17 J4
Morfa *Cere.* SA44 26 C3
Morfa Bychan LL49 36 E2
Morfa Glas SA11 18 B1
Morfa Nefyn LL53 36 B1
Morgan's Vale SP5 10 C2
Morganstown CF15 18 E3
Mork GL15 19 J1
Morland CA10 61 G4
Morley *Derbys.* DE7 41 F1
Morley *Dur.* DL14 62 C4
Morley *W.Yorks.* LS27 57 H7
Morley Green SK9 49 H4
Morley St. Botolph NR18 44 E6
Mornick PL17 4 D3
Morningside *Edin.* EH10 76 A3
Morningside *N.Lan.* ML2 75 G5
Morningthorpe NR15 45 G6
Morpeth NE61 71 H5
Morphie DD10 83 J1
Morrey DE13 40 D4
Morridge Side ST13 50 C7
Morrilow Heath ST10 40 B2
Morriston *S.Ayr.* KA19 67 G3
Morriston *Swan.* SA6 17 K6
Morristown CF64 18 E4
Morroch PH39 86 C6
Morston NR25 44 E1
Mortehoe EX34 6 C1
Morthen S66 51 G4
Mortimer RG7 21 K5
Mortimer West End RG7 21 K5
Mortimer's Cross HR6 28 D2
Mortlake SW14 23 F4
Morton *Derbys.* DE55 51 G6
Morton *Lincs.* PE10 42 D3
Morton *Lincs.* DN21 52 B3
Morton *Lincs.* LN6 52 B6
Morton *Notts.* NG25 51 K7
Morton *S.Glos.* BS35 19 K2
Morton *Shrop.* SY10 38 B3
Morton Bagot B80 30 C2
Morton on the Hill NR9 45 F4
Morton Tinmouth DL2 62 C4
Morton-on-Swale DL7 62 E7
Morvah TR20 2 B5
Morval PL13 4 C5
Morvich *High.* IV40 87 F2
Morvich *High.* IV28 96 E1
Morvil SA66 16 D2
Morville WV16 39 F6
Morwellham PL19 4 E4
Morwenstow EX23 6 A4
Morwick Hall NE65 71 H3
Mosborough S20 51 G4
Moscow KA4 74 C6
Mosedale CA11 60 E3
Moselden Height HD3 50 C1
Moseley *W.Mid.* B13 40 C7
Moseley *W.Mid.* WV1 40 B6
Moseley *Worcs.* WR2 29 H3
Moses Gate BL3 49 G2
Moss *Arg. & B.* PA77 78 A3
Moss *S.Yorks.* DN6 51 H1
Moss *Wrex.* LL11 48 C7
Moss Bank WA11 48 E3
Moss Houses SK11 49 H5
Moss Nook M22 49 H4
Moss of Barmuckity IV30 97 K5
Moss Side *Gt.Man.* M14 49 H3
Moss Side *Lancs.* FY8 55 G6
Moss Side *Mersey.* L31 48 C2
Mossat AB33 90 C3
Mossbank ZE2 107 N5
Mossblown KA6 67 J1
Mossburnford TD8 70 B2
Mossdale DG7 65 G3
Mossend ML4 75 F4
Mossgiel KA5 67 J1
Mosshead AB54 90 D1
Mosside of Ballinshoe DD8 83 F2
Mossley *Ches.E.* CW12 49 H6
Mossley *Gt.Man.* OL5 49 J2
Mossley Hill L18 48 C4
Mosspaul Hotel TD9 69 J4
Moss-side *High.* IV12 97 F6
Moss-side *Moray* AB54 98 D5
Mosstodloch IV32 98 B4
Mosston DD11 83 G3
Mossy Lea WN6 48 E1
Mosterton DT8 8 D4
Moston *Gt.Man.* M40 49 H2
Moston *Shrop.* SY4 38 E3
Moston Green CW11 49 G6
Mostyn CH8 47 K4
Motcombe SP7 9 H2
Mothecombe PL8 5 G6
Motherby CA11 60 F4
MOTHERWELL ML 75 F5
Mottingham SE9 23 H4
Mottisfont SO51 10 E2
Mottistone PO30 11 F6
Mottram in Longdendale SK14 49 J3
Mottram St. Andrew SK10 49 H5
Mouldsworth WA6 48 E5
Moulin PH16 82 A2
Moulsecoomb BN2 13 G6
Moulsford OX10 21 J3
Moulsham CM2 24 D1
Moulsoe MK16 32 C4
Moulton *Ches.W. & C.* CW9 49 F6
Moulton *Lincs.* PE12 43 G3
Moulton *N.Yorks.* DL10 62 D6
Moulton *Northants.* NN3 31 J2
Moulton *Suff.* CB8 33 K2
Moulton *V. of Glam.* CF62 18 D4
Moulton Chapel PE12 43 F4
Moulton St. Mary NR13 45 J5
Moulton Seas End PE12 43 G3
Mounie Castle AB51 91 F2
Mount *Cornw.* TR8 2 E3
Mount *Cornw.* PL30 4 B4
Mount *High.* IV12 97 G7
Mount *Kent* CT4 15 G3
Mount *W.Yorks.* HD3 50 D1
Mount Ambrose TR16 2 E4
Mount Bures CO8 34 D5
Mount Charles PL25 4 A5
Mount Hawke TR4 2 E4
Mount Manisty CH65 48 C5
Mount Oliphant KA6 67 H2
Mount Pleasant *Ches.E.* ST7 49 H7
Mount Pleasant *Derbys.* DE56 41 F1
Mount Pleasant *Derbys.* DE11 40 E4
Mount Pleasant *E.Suss.* BN8 13 H5
Mount Pleasant *Flints.* CH6 48 B5
Mount Pleasant *Gt.Lon.* UB9 22 D2
Mount Pleasant *Hants.* SO41 10 E5
Mount Pleasant *Norf.* NR17 44 D6
Mount Pleasant *Suff.* CO10 34 B4
Mount Sorrel SP5 10 B2
Mount Tabor HX2 57 F7
Mountain HX2 57 F6
Mountain Ash (Aberpennar) CF45 18 D2
Mountain Cross EH46 75 K6
Mountain Water SA62 16 C3
Mountbenger TD7 69 J1
Mountblairy AB53 98 E5
Mountblow G60 74 C3
Mountfield TN32 14 C5
Mountgerald IV15 96 C5
Mountjoy TR8 3 F2
Mountnessing CM15 24 C2
Mounton NP16 19 J2
Mountsorrel LE12 41 H4
Mousa ZE2 107 N10
Mousehole TR19 2 B6
Mouswald DG1 69 F6
Mow Cop ST7 49 H7
Mowden DL3 62 D5
Mowhaugh TD5 70 D1
Mowsley LE17 41 J7
Mowtie AB39 91 G6
Moxley WS10 40 B6
Moy *High.* PH33 87 H6
Moy *High.* PH31 88 B6
Moy *High.* IV13 88 E1
Moy House IV36 97 H6
Moylgrove SA43 16 E1
Muasdale PA29 72 E6
Much Birch HR2 28 E5
Much Cowarne HR7 29 F4
Much Dewchurch HR2 28 D5
Much Hadham SG10 33 H7
Much Hoole PR4 55 H7
Much Hoole Town PR4 55 H7
Much Marcle HR8 29 F5
Much Wenlock TF13 39 F5
Muchalls AB39 91 H5
Muchelney TA10 8 D2
Muchelney Ham TA10 8 D2
Muchlarnick PL13 4 C5
Muchra TD7 69 H2
Muchrachd IV4 87 J1
Muck PH41 85 K7
Mucking SS17 24 C3
Muckle Roe ZE2 107 M6
Muckleford DT2 9 F5
Mucklestone TF9 39 G2
Muckleton TF6 38 E3
Muckletown AB33 90 D2
Muckley WV16 39 F6
Muckley Corner WS14 40 C5
Muckton LN11 53 G4
Mudale IV27 103 H5
Muddiford EX31 6 D2
Muddles Green BN8 13 J5
Muddleswood BN6 13 F5
Mudeford BH23 10 C5
Mudford BA21 8 E3
Mudgley BS28 19 H7
Mugdock G62 74 D3
Mugeary IV51 85 K1
Mugginton DE6 40 E1
Muggintonlane End DE6 40 E1
Muggleswick DH8 62 B2
Mugswell KT20 23 F6
Muie IV28 96 D1
Muir AB35 89 H6
Muir of Fowlis AB33 90 D3
Muir of Lochs IV32 98 B4
Muir of Ord IV6 96 C6
Muirden AB53 99 F5
Muirdrum DD7 83 G4
Muiredge KY8 76 B1
Muirhead *Aber.* AB33 90 D3
Muirhead *Angus* DD2 82 E4
Muirhead *Fife* KY15 82 D7
Muirhead *Moray* IV36 97 H5
Muirhead *N.Lan.* G69 74 E4
Muirhouses EH51 75 J2
Muirkirk KA18 68 B1
Muirmill FK6 75 F2
Muirtack *Aber.* AB41 91 H1
Muirtack *Aber.* AB53 99 G6
Muirton *High.* IV11 96 E5
Muirton *P. & K.* PH3 82 A6
Muirton *P. & K.* PH1 82 C5
Muirton of Ardblair PH10 82 C3
Muirton of Ballochy DD10 83 H1
Muirtown IV36 97 G6
Muiryfold AB53 99 F5
Muker DL11 62 A7
Mulbarton NR14 45 F5
Mulben AB55 98 B5
Mulhagery HS2 101 F7
Mull PA 79 G4
Mullach Charlabhaigh HS2 100 E3
Mullacott Cross EX34 6 D1
Mullion TR12 2 D7
Mullion Cove TR12 2 D7
Mumby LN13 53 J5
Munderfield Row HR7 29 F3
Munderfield Stocks HR7 29 F3
Mundesley NR11 45 H2
Mundford IP26 44 C6
Mundham NR14 45 H6
Mundon CM9 24 E1
Mundurno AB23 91 H3
Munerigie PH35 87 J4
Mungasdale IV22 95 F2
Mungoswells EH39 76 C3
Mungrisdale CA11 60 E3
Munlochy IV8 96 D6
Munnoch KA22 74 A6
Munsley HR8 29 F4
Munslow SY7 38 E7
Murchington TQ13 6 E7
Murcott *Oxon.* OX5 31 G7
Murcott *Wilts.* SN16 20 C2
Murdostoun ML2 75 G5
Murieston EH54 75 J4
Murkle KW14 105 G2
Murlaganmore FK21 81 G4
Murlaggan *High.* PH34 87 G5
Murlaggan *High.* PH31 87 K6
Murra KW16 106 B7
Murrell Green RG27 22 A6
Murroes DD5 83 F4
Murrow PE13 43 G5
Mursley MK17 32 B6
Murston ME10 25 F5
Murthill DD8 83 F2
Murthly PH1 82 B4
Murton *Cumb.* CA16 61 J4
Murton *Dur.* SR7 62 E2
Murton *Northumb.* TD15 77 H6
Murton *Swan.* SA3 17 J7
Murton *York* YO19 58 C4
Musbury EX13 8 B5
Muscliff BH9 10 B5
Musdale PA34 80 A5
Musselburgh EH21 76 B3
Mustard Hyrn NR29 45 J4
Muston *Leics.* NG13 42 B2
Muston *N.Yorks.* YO14 59 G2
Mustow Green DY10 29 H1
Mutford NR34 45 J7
Muthill PH5 81 K6
Mutley PL3 4 E5
Mutterton EX15 7 J5
Muxton TF2 39 G4
Mybster KW1 105 G3
Myddfai SA20 27 G5
Myddle SY4 38 D3
Myddlewood SY4 38 D3
Mydroilyn SA48 26 D3
Myerscough College PR3 55 H6
Myerscough Smithy PR5 56 B6
Mylor TR11 3 F5
Mylor Bridge TR11 3 F5
Mynachdy CF14 18 E4
Mynachlog-ddu SA66 16 E2
Myndtown SY7 38 C7
Mynydd Llandygai LL57 46 E6
Mynydd-bach *Mon.* NP16 19 H2
Mynydd-bach *Swan.* SA5 17 K6
Mynyddmechell LL68 46 B3
Mynyddygarreg SA17 17 H5
Mynytho LL53 36 C2
Myrebird AB31 91 F5
Mytchett GU16 22 B6
Mytholm HX7 56 E7
Mytholmroyd HX7 57 F7
Mythop FY4 55 G6
Myton-on-Swale YO61 57 K3
Mytton SY4 38 D4

N

Naast IV22 94 E3
Nab's Head PR5 56 B7
Na-Buirgh HS3 93 F2
Naburn YO19 58 B5
Nackington CT4 15 G2
Nacton IP10 35 G4
Nadderwater EX4 7 G6
Nafferton YO25 59 G4
Nailbridge GL17 29 F7
Nailsbourne TA2 8 B2
Nailsea BS48 19 H4
Nailstone CV13 41 G5
Nailsworth GL6 20 B2
Nairn IV12 97 F6

Nan - New

Nancegollan TR13 2 D5
Nancekuke TR16 2 D4
Nancledra TR20 2 B5
Nanhoron LL53 36 B2
Nannau LL40 37 G3
Nannerch CH7 47 K6
Nanpantan LE11 41 H4
Nanpean PL26 3 G3
Nanstallon PL30 4 A4
Nant Peris LL55 46 E7
Nant-ddu CF48 27 K7
Nanternis SA45 26 C3
Nantgaredig SA32 17 H3
Nantgarw CF15 18 E3
Nant-glas LD1 27 J2
Nantglyn LL16 47 J6
Nantgwyn LD6 27 J1
Nantlle LL54 46 D7
Nantmawr SY10 38 B3
Nantmel LD1 27 K2
Nantmor LL55 37 F1
Nantwich CW5 49 F7
Nantycaws SA32 17 H4
Nant-y-derry NP7 19 G1
Nant-y-dugoed SY21 37 J4
Nantyffyllon CF34 18 B2
Nantyglo NP23 28 A7
Nant-y-Gollen SY10 38 B3
Nant-y-groes LD1 27 K2
Nant-y-moel CF32 18 C2
Nant-y-Pandy LL33 46 E5
Naphill HP14 22 B2
Napley Heath TF9 39 G2
Nappa BD23 56 D4
Napton on the Hill CV47 31 F2
Narberth (Arberth) SA67 16 E4
Narborough *Leics.* LE19 41 H6
Narborough *Norf.* PE32 44 B4
Narkurs PL11 4 D5
Narrachan PA35 80 A6
Nasareth LL54 36 D1
Naseby NN6 31 H1
Nash *Bucks.* MK17 31 J5
Nash *Here.* LD8 28 C2
Nash *Newport* NP18 19 G3
Nash *Shrop.* SY8 29 F1
Nash *V. of Glam.* CF71 18 C4
Nash Street DA13 24 C5
Nassington PE8 42 D6
Nasty SG11 33 G6
Nateby *Cumb.* CA17 61 J6
Nateby *Lancs.* PR3 55 H5
Nately Scures RG27 22 A6
Natland LA9 55 J1
Naughton IP7 34 E4
Naunton *Glos.* GL54 30 C6
Naunton *Worcs.* WR8 29 H5
Naunton Beauchamp WR10 29 J3
Navenby LN5 52 C7
Navestock RM4 23 J2
Navestock Side CM14 23 J2
Navidale KW8 105 F7
Navity IV11 96 E5
Nawton YO62 58 C1
Nayland CO6 34 D5
Nazeing EN9 23 H1
Neacroft BH23 10 C5
Neal's Green CV7 41 F7
Neap ZE2 107 P7
Neap House DN15 52 B1
Near Sawrey LA22 60 E7
Nearton End MK17 32 B6
Neasden NW10 23 F3
Neasham DL2 62 E5
Neat Enstone OX7 30 E6
Neath (Castell-nedd) SA11 18 A2
Neatham GU34 22 A7
Neatishead NR12 45 H3
Nebo *Cere.* SY23 26 E2
Nebo *Conwy* LL26 47 G7
Nebo *Gwyn.* LL54 46 C7
Nebo *I.o.A.* LL68 46 C3
Necton PE37 44 C5
Nedd IV27 102 D5
Nedderton NE22 71 H5
Nedging IP7 34 D4
Nedging Tye IP7 34 E4
Needham IP20 45 G7
Needham Market IP6 34 E3
Needham Street CB8 34 B2
Needingworth PE27 33 G1
Needwood DE13 40 D3
Neen Savage DY14 29 F1
Neen Sollars DY14 29 F1
Neenton WV16 39 F7
Nefyn LL53 36 C1
Neighbourne BA3 19 K7
Neilston G78 74 C5
Neithrop OX16 31 F4
Nelson *Caerp.* CF46 18 E2
Nelson *Lancs.* BB9 56 D6
Nelson Village NE23 71 H6
Nemphlar ML11 75 G6
Nempnett Thrubwell BS40 19 J5
Nenthall CA9 61 J2
Nenthead CA9 61 J2
Nenthorn TD5 76 E7
Neopardy EX17 7 F6
Nerabus PA48 72 A5
Nercwys CH7 48 B6
Neriby PA44 72 B4
Nerston G74 74 E5
Nesbit NE71 77 H7
Nesfield LS29 57 F5
Ness CH64 48 C5
Ness of Tenston KW16 106 B6
Nesscliffe SY4 38 C4
Neston *Ches.W. & C.* CH64 48 B5
Neston *Wilts.* SN13 20 B5
Nether Alderley SK10 49 H5
Nether Auchendrane KA7 67 H2
Nether Barr DG8 64 E4
Nether Blainslie TD1 76 D6
Nether Broughton LE14 41 J3
Nether Burrow LA6 56 B2
Nether Cerne DT2 9 F5
Nether Compton DT9 8 E3
Nether Crimond AB51 91 G2
Nether Dalgliesh TD7 69 H3
Nether Dallachy IV32 98 B4
Nether Edge S7 51 F4
Nether End DE45 50 E5
Nether Exe EX5 7 H6
Nether Glasslaw AB43 99 G5
Nether Handwick DD8 82 E3
Nether Haugh S62 51 G3
Nether Heage DE56 51 F7
Nether Heselden BD23 56 D2
Nether Heyford NN7 31 H3
Nether Kellet LA6 55 J3
Nether Kinmundy AB42 99 J6
Nether Langwith NG20 51 H5
Nether Lenshie AB51 98 E6
Nether Loads S42 51 F6
Nether Moor S42 51 F6
Nether Padley S32 50 E5
Nether Pitforthie AB30 91 G7
Nether Poppleton YO26 58 B4
Nether Silton YO7 63 F7
Nether Skyborry LD7 28 B1
Nether Stowey TA5 7 K2
Nether Urquhart KY14 82 C7
Nether Wallop SO20 10 E1
Nether Wasdale CA20 60 C6
Nether Wellwood KA18 68 B1
Nether Welton CA5 60 E2
Nether Westcote OX7 30 D6
Nether Whitacre B46 40 E6
Nether Winchendon (Lower Winchendon) HP18 31 J7
Nether Worton OX7 31 F6
Netheravon SP4 20 E7
Netherbrae AB53 99 F5
Netherbrough KW17 106 C6
Netherburn ML9 75 G6
Netherbury DT6 8 D5
Netherby *Cumb.* CA6 69 J6
Netherby *N.Yorks.* HG3 57 J5
Nethercott OX5 31 F6
Netherfield *E.Suss.* TN33 14 C6
Netherfield *Notts.* NG4 41 J1
Netherfield *S.Lan.* ML10 75 F6
Netherhall KA30 74 A4
Netherhampton SP2 10 C2
Netherhay DT8 8 D4
Netherland Green ST14 40 D2
Netherley AB39 91 G5
Nethermill DG1 69 F5
Nethermuir AB42 99 H6
Netherseal DE12 40 E4
Nethershield KA5 67 K1
Netherstreet SN15 20 C5
Netherthird *D. & G.* DG7 65 H5
Netherthird *E.Ayr.* KA18 67 K2
Netherthong HD9 50 D2
Netherthorpe S80 51 H4
Netherton *Angus* DD9 83 G2
Netherton *Ches.W. & C.* WA6 48 E5
Netherton *Devon* TQ12 5 J3
Netherton *Hants.* SP11 21 G6
Netherton *Mersey.* L30 48 C2
Netherton *N.Lan.* ML2 75 F5
Netherton *Northumb.* NE65 70 E3
Netherton *Oxon.* OX13 21 H1
Netherton *P. & K.* PH10 82 C2
Netherton *S.Lan.* ML11 75 H5
Netherton *W.Mid.* DY2 40 B7
Netherton *W.Yorks.* WF4 50 E1
Netherton *W.Yorks.* HD4 50 D1
Netherton *Worcs.* WR10 29 J4
Netherton Burnfoot NE65 70 E3
Netherton Northside NE65 70 E3
Nethertown *Cumb.* CA22 60 A6
Nethertown *Ork.* KW1 105 J1
Nethertown *Staffs.* WS15 40 D4
Netherwitton NE61 71 G4
Netherwood *D. & G.* DG1 65 K3
Netherwood *E.Ayr.* KA18 68 B1
Nethy Bridge PH25 89 H2
Netley Abbey SO31 11 F4
Netley Marsh SO40 10 E3
Nettlebed RG9 21 K3
Nettlebridge BA3 19 K7
Nettlecombe *Dorset* DT6 8 E5
Nettlecombe *I.o.W.* PO38 11 G7
Nettlecombe *Som.* TA4 7 J2
Nettleden HP1 32 D7
Nettleham LN2 52 D5
Nettlestead *Kent* ME18 23 K6
Nettlestead *Suff.* IP8 34 E4
Nettlestead Green ME18 23 K6
Nettlestone PO34 11 H5
Nettlesworth DH2 62 D2
Nettleton *Lincs.* LN7 52 E2
Nettleton *Wilts.* SN14 20 B4
Nettleton Hill HD7 50 C1
Netton *Devon* PL8 5 F6
Netton *Wilts.* SP4 10 C1
Neuadd *Cere.* SA47 26 D3
Neuadd *I.o.A.* LL67 46 B3
Neuadd *Powys* LD2 27 J4
Nevendon SS12 24 D2
Nevern SA42 16 D2
Nevill Holt LE16 42 B6
New Abbey DG2 65 K4
New Aberdour AB43 99 G4
New Addington CR0 23 G5
New Alresford SO24 11 G1
New Alyth PH11 82 D3
New Arley CV7 40 E6
New Arram HU17 59 G5
New Ash Green DA3 24 C5
New Balderton NG24 52 B7
New Barn DA3 24 C5
New Belses TD8 70 A1
New Bewick NE66 71 F1
New Bolingbroke PE22 53 G7
New Boultham LN6 52 C5
New Bradwell MK13 32 B4
New Brancepeth DH7 62 D2
New Bridge *D. & G.* DG2 65 K3
New Bridge *Devon* TQ13 5 H3
New Brighton *Flints.* CH7 48 B6
New Brighton *Hants.* PO10 11 J4
New Brighton *Mersey.* CH45 48 C3
New Brighton *W.Yorks.* LS27 57 H7
New Brighton *Wrex.* LL11 48 B7
New Brinsley NG16 51 G7
New Broughton LL11 48 C7
New Buckenham NR16 44 E6
New Byth AB53 99 G5
New Cheriton SO24 11 G2
New Cross *Cere.* SY23 27 F1
New Cross *Gt.Lon.* SE14 23 G4
New Cumnock KA18 68 B2
New Deer AB53 99 G6
New Duston NN5 31 J2
New Earswick YO31 58 C4
New Edlington DN12 51 H3
New Elgin IV30 97 K5
New Ellerby HU11 59 H6
New Eltham SE9 23 H4
New End B96 30 B2
New England PE1 42 E5
New Farnley LS12 57 H6
New Ferry CH62 48 C4
New Galloway DG7 65 G3
New Gilston KY8 83 F7
New Greens AL3 22 E1
New Grimsby TR24 2 B1
New Hartley NE25 71 J6
New Haw KT15 22 D5
New Heaton TD12 77 G7
New Hedges SA70 16 E5
New Herrington DH4 62 E1
New Hinksey OX1 21 J1
New Holland DN19 59 G7
New Houghton *Derbys.* NG19 51 H6
New Houghton *Norf.* PE31 44 B3
New Houses BD24 56 D2
New Hunwick DL15 62 C3
New Hutton LA8 61 G7
New Hythe ME20 14 C2
New Inn *Carmar.* SA39 17 H2
New Inn *Fife* KY7 82 D7
New Inn *Mon.* NP16 19 H1
New Inn *Torfaen* NP4 19 F2
New Invention *Shrop.* SY7 28 B1
New Invention *W.Mid.* WV12 40 B5
New Kelso IV54 95 F7
New Lanark ML11 75 G6
New Lane L40 48 D1
New Lane End WA3 49 F3
New Leake PE22 53 H7
New Leeds AB42 99 H5
New Leslie AB52 90 D2
New Lodge S75 51 F2
New Longton PR4 55 J7
New Luce DG8 64 B4
New Mains ML11 75 G7
New Mains of Ury AB39 91 G6
New Malden KT3 23 F5
New Marske TS11 63 H4
New Marton SY11 38 C2
New Mill *Cornw.* TR20 2 B5
New Mill *Herts.* HP23 32 C7
New Mill *W.Yorks.* HD9 50 D2
New Mill End LU1 32 E7
New Mills *Cornw.* TR2 3 F3
New Mills *Derbys.* SK22 50 C4
New Mills *Glos.* GL15 19 K1
New Mills *Mon.* NP25 19 J1
New Mills (Y Felin Newydd) *Powys* SY16 37 K5
New Milton BH25 10 D5
New Mistley CO11 35 F5
New Moat SA63 16 D3
New Ollerton NG22 51 J6
New Orleans PA28 66 B2
New Oscott B44 40 C6
New Park *Cornw.* PL15 4 B2
New Park *N.Yorks.* HG1 57 H4
New Pitsligo AB43 99 G5
New Polzeath PL27 3 G1
New Quay (Ceinewydd) SA45 26 C2
New Rackheath NR13 45 G4
New Radnor (Maesyfed) LD8 28 B2
New Rent CA11 61 F3
New Ridley NE43 71 F7
New Road Side BD22 56 E5
New Romney TN28 15 F5
New Rossington DN11 51 J3
New Row *Cere.* SY25 27 G1
New Row *Lancs.* PR3 56 B6
New Sawley NG10 41 G2
New Shoreston NE69 77 K7
New Silksworth SR3 62 E1
New Stevenston ML1 75 F5
New Swannington LE67 41 G4
New Totley S17 51 F5
New Town *Cen.Beds.* SG18 32 E4
New Town *Cere.* SA43 16 E1
New Town *Dorset* SP5 9 J3
New Town *Dorset* BH21 9 J4
New Town *E.Loth.* EH34 76 C3
New Town *E.Suss.* TN22 13 H4
New Town *Glos.* GL54 30 B5
New Tredegar NP24 18 E1
New Tupton S42 51 F6
New Ulva PA31 73 F2
New Valley HS2 101 G4
New Village DN5 51 H2
New Walsoken PE13 43 H5
New Waltham DN36 53 F2
New Winton EH33 76 C3
New World PE15 43 G6
New Yatt OX29 30 E7
New York *Lincs.* LN4 53 F7
New York *T. & W.* NE27 71 J6
Newall LS21 57 H5
Newark *Ork.* KW17 106 G3
Newark *Peter.* PE1 43 F5
Newark-on-Trent NG24 52 B7
Newarthill ML1 75 F5
Newball LN3 52 D5
Newbarn CT18 15 G4
Newbarns LA14 55 F2
Newbattle EH22 76 B4
Newbiggin *Cumb.* CA11 61 F4
Newbiggin *Cumb.* CA10 61 H4
Newbiggin *Cumb.* CA8 61 G2
Newbiggin *Cumb.* LA12 55 F3
Newbiggin *Cumb.* LA19 60 B7
Newbiggin *Dur.* DL12 62 A4
Newbiggin *N.Yorks.* DL8 57 F1
Newbiggin *N.Yorks.* DL8 62 A7
Newbiggin *Northumb.* NE46 70 E7
Newbiggin-by-the-Sea NE64 71 J5
Newbigging *Aber.* AB39 91 G5
Newbigging *Aber.* AB35 89 J6
Newbigging *Angus* DD4 83 F4
Newbigging *Angus* DD5 83 F4
Newbigging *Angus* PH12 82 D3
Newbigging *S.Lan.* ML11 75 J6
Newbiggin-on-Lune CA17 61 J6
Newbold *Derbys.* S41 51 F5
Newbold *Leics.* LE67 41 G4
Newbold on Avon CV21 31 F1
Newbold on Stour CV37 30 D4
Newbold Pacey CV35 30 D3
Newbold Verdon LE9 41 G5
Newborough (Niwbwrch) *I.o.A.* LL61 46 C6
Newborough *Peter.* PE6 43 F5
Newborough *Staffs.* DE13 40 D3
Newbottle *Northants.* OX17 31 G5
Newbottle *T. & W.* DH4 62 E1
Newbourne IP12 35 G4
Newbridge (Cefn Bychan) *Caerp.* NP11 19 F2
Newbridge *Cornw.* TR20 2 B5
Newbridge *Cornw.* PL17 4 D4
Newbridge *E.Suss.* TN7 13 H3
Newbridge *Edin.* EH28 75 K3
Newbridge *Hants.* SO40 10 D3
Newbridge *I.o.W.* PO41 11 F6
Newbridge *N.Yorks.* YO18 58 E1
Newbridge *Oxon.* OX29 21 H1
Newbridge *Pembs.* SA62 16 C2
Newbridge *Wrex.* LL14 38 B1
Newbridge Green WR8 29 H5
Newbridge on Wye LD1 27 K3
Newbridge-on-Usk NP15 19 G2
Newbrough NE47 70 D7
Newbuildings EX17 7 F5
Newburgh *Aber.* AB43 99 H5
Newburgh *Aber.* AB41 91 H2
Newburgh *Fife* KY14 82 D6
Newburgh *Lancs.* WN8 48 D1
Newburgh *Sc.Bord.* TD7 69 J1
Newburn NE15 71 G7
Newbury *Som.* BA11 19 K6
Newbury *W.Berks.* RG14 21 H5
Newbury *Wilts.* BA12 20 B7
Newbury Park IG2 23 H3
Newby *Cumb.* CA10 61 G4
Newby *Lancs.* BB7 56 D5
Newby *N.Yorks.* TS8 63 G5
Newby *N.Yorks.* LA2 56 C3
Newby *N.Yorks.* YO12 59 G1
Newby Bridge LA12 55 G1
Newby Cote LA2 56 C2
Newby Cross CA5 60 E1
Newby East CA4 61 F1
Newby West CA2 60 E1
Newby Wiske DL7 57 J1
Newcastle *Bridgend* CF31 18 B4
Newcastle *Mon.* NP25 28 D7
Newcastle *Shrop.* SY7 38 B7
Newcastle Emlyn (Castell Newydd Emlyn) SA38 17 G1
Newcastle International Airport NE13 71 G6
NEWCASTLE UPON TYNE NE 71 H7
Newcastleton TD9 69 K5
Newcastle-under-Lyme ST5 40 A1
Newchapel *Pembs.* SA37 17 F2
Newchapel *Staffs.* ST7 49 H7
Newchapel *Surr.* RH7 23 G7
Newchurch *Carmar.* SA33 17 G3
Newchurch *I.o.W.* PO36 11 G6
Newchurch *Kent* TN29 15 F4
Newchurch *Lancs.* BB4 56 D7
Newchurch *Lancs.* BB12 56 D6
Newchurch *Mon.* NP16 19 H2
Newchurch *Powys* HR5 28 B3
Newchurch *Staffs.* DE13 40 D3
Newcott EX14 8 B4
Newcraighall EH21 76 B3
Newdigate RH5 22 E7
Newell Green RG42 22 B4
Newenden TN18 14 D5
Newent GL18 29 G6
Newerne GL15 19 K1
Newfield *Dur.* DL14 62 D3
Newfield *Dur.* DH2 62 D1
Newfield *High.* IV19 96 E4
Newfound RG23 21 J6
Newgale SA62 16 B3
Newgate NR25 44 E1
Newgate Street SG13 23 G1
Newgord ZE2 107 P2
Newhall *Ches.E.* CW5 39 F1
Newhall *Derbys.* DE11 40 E3
Newham NE67 71 G1
Newham Hall NE67 71 G1
Newhaven BN9 13 H6
Newhey OL16 49 J1
Newholm YO21 63 K5
Newhouse ML1 75 F4
Newick BN8 13 H4
Newingreen CT21 15 G4
Newington *Edin.* EH9 76 A3
Newington *Kent* CT18 15 G4
Newington *Kent* ME9 24 E5
Newington *Notts.* DN10 51 J3
Newington *Oxon.* OX10 21 K2
Newington Bagpath GL8 20 B2
Newland *Cumb.* LA12 55 G2
Newland *Glos.* GL16 19 J1
Newland *Hull* HU6 59 G6
Newland *N.Yorks.* YO8 58 C7
Newland *Oxon.* OX28 30 E7
Newland *Worcs.* WR13 29 G4
Newlandrig EH23 76 B4
Newlands *Cumb.* CA7 60 E3
Newlands *Essex* SS8 24 E3
Newlands *Northumb.* DH8 62 B1
Newlands *Sc.Bord.* TD9 70 A4
Newland's Corner GU4 22 D7
Newlands of Geise KW14 105 F2
Newlands of Tynet IV32 98 B4
Newlyn TR18 2 B6
Newmachar AB21 91 G3
Newmains ML2 75 G5
Newman's End CM22 33 J7
Newman's Green CO10 34 C4
Newmarket *Suff.* CB8 33 K2
Newmarket *W.Isles* HS2 101 G4
Newmill *Aber.* AB39 91 F6
Newmill *Aber.* AB41 99 G6
Newmill *Aber.* AB51 91 G2
Newmill *Moray* AB55 98 C5
Newmill *Sc.Bord.* TD9 69 K2
Newmill of Inshewan DD8 83 F1
Newmillerdam WF2 51 F1
Newmills IV7 96 D5
Newmiln *P. & K.* PH2 82 C4
Newmiln *P. & K.* PH1 82 B5
Newmilns KA16 74 D7
Newney Green CM1 24 C1
Newnham *Glos.* GL14 29 F7
Newnham *Hants.* RG27 22 A6
Newnham *Herts.* SG7 33 F5
Newnham *Kent* ME9 14 E2
Newnham *Northants.* NN11 31 G3
Newnham Bridge WR15 29 F2
Newnham Paddox CV23 41 G7
Newnoth AB54 90 D1
Newport *Cornw.* PL15 6 B7
Newport *Devon* EX32 6 D2
Newport *E.Riding* HU15 58 E6
Newport *Essex* CB11 33 J5
Newport *Glos.* GL13 19 K2
Newport *High.* KW7 105 G6
Newport *I.o.W.* PO30 11 G6
NEWPORT (CASNEWYDD) *Newport* NP 19 G3
Newport *Norf.* NR29 45 K4
Newport (Trefdraeth) *Pembs.* SA42 16 D2
Newport *Som.* TA3 8 C2
Newport *Tel. & W.* TF10 39 G4
Newport Pagnell MK16 32 B4
Newport-on-Tay DD6 83 F5
Newpound Common RH14 12 D4
Newquay TR7 3 F2
Newquay Cornwall International Airport TR8 3 F2
Newsbank CW12 49 H6
Newseat AB51 91 F1
Newsells SG8 33 G5
Newsham *Lancs.* PR3 55 J6
Newsham *N.Yorks.* DL11 62 C5
Newsham *N.Yorks.* YO7 57 J1
Newsham *Northumb.* NE24 71 H6
Newsholme *E.Riding* DN14 58 D7
Newsholme *Lancs.* BB7 56 D4
Newsome HD4 50 D1
Newstead *Northumb.* NE67 71 G1
Newstead *Notts.* NG15 51 H7
Newstead *Sc.Bord.* TD6 76 D7
Newthorpe *N.Yorks.* LS25 57 K6
Newthorpe *Notts.* NG16 41 G1
Newtoft LN8 52 D4
Newton *Aber.* AB54 98 C6
Newton *Aber.* AB42 99 J6
Newton *Arg. & B.* PA27 73 J1
Newton *Bridgend* CF36 18 B4
Newton *Cambs.* PE13 43 H4
Newton *Cambs.* CB22 33 H4
Newton *Cardiff* CF3 19 F4

New - Nor

Newton *Ches.W. & C.* **CH3** 48 E7
Newton *Ches.W. & C.* **WA6** 48 E5
Newton *Cumb.* **LA13** 55 F2
Newton *D. & G.* **DG10** 69 G4
Newton *Derbys.* **DE55** 51 G7
Newton *Gt.Man.* **SK14** 49 J3
Newton *Here.* **SY7** 28 C2
Newton *Here.* **HR6** 28 E3
Newton *Here.* **HR2** 28 C5
Newton *High.* **KW1** 105 J3
Newton *High.* **IV2** 96 E7
Newton *High.* **KW1** 105 H3
Newton *High.* **IV27** 102 E5
Newton *High.* **IV6** 96 C6
Newton *High.* **IV11** 96 E5
Newton *Lancs.* **BB7** 56 B4
Newton *Lancs.* **LA6** 55 J2
Newton *Lancs.* **FY3** 55 G6
Newton *Lincs.* **NG34** 42 D2
Newton *Moray* **IV32** 98 B4
Newton *N.Ayr.* **KA27** 73 H5
Newton *Norf.* **PE32** 44 C4
Newton *Northants.* **NN14** 42 B7
Newton *Northumb.* **NE43** 71 F7
Newton *Northumb.* **NE65** 70 E3
Newton *Notts.* **NG13** 41 J1
Newton *P. & K.* **PH8** 81 K4
Newton *Pembs.* **SA62** 16 B3
Newton *Pembs.* **SA71** 16 C5
Newton *S.Glos.* **BS35** 19 K2
Newton *S.Lan.* **ML12** 75 H7
Newton *Sc.Bord.* **TD8** 70 B1
Newton *Shrop.* **SY12** 38 D2
Newton *Som.* **TA4** 7 K2
Newton *Staffs.* **WS15** 40 C3
Newton *Suff.* **CO10** 34 D4
Newton *Swan.* **SA3** 17 K7
Newton *W.Loth.* **EH52** 75 J3
Newton *W.Yorks.* **WF10** 57 K7
Newton *Warks.* **CV23** 31 G1
Newton *Wilts.* **SP5** 10 D2
Newton Abbot TQ12 5 J3
Newton Arlosh **CA7** 60 D1
Newton Aycliffe DL5 62 D4
Newton Bewley **TS22** 63 F4
Newton Blossomville **MK43** 32 C3
Newton Bromswold **MK44** 32 C2
Newton Burgoland **LE67** 41 F5
Newton by Toft **LN8** 52 D4
Newton Ferrers **PL8** 5 F6
Newton Flotman **NR15** 45 G6
Newton Green **NP16** 19 J2
Newton Harcourt **LE8** 41 J6
Newton Kyme **LS24** 57 K5
Newton Longville **MK17** 32 B5
Newton Mearns **G77** 74 D5
Newton Morrell *N.Yorks.* **DL10** 62 D6
Newton Morrell *Oxon.* **OX27** 31 H6
Newton Mountain **SA73** 16 C5
Newton Mulgrave **TS13** 63 J5
Newton of Affleck **DD5** 83 F4
Newton of Ardtoe **PH36** 86 C7
Newton of Balcanquhal **PH2** 82 C6
Newton of Dalvey **IV36** 97 H6
Newton of Falkland **KY15** 82 D7
Newton of Leys **IV2** 88 D1
Newton on the Hill **SY4** 38 D3
Newton on Trent **LN1** 52 B5
Newton Poppleford **EX10** 7 J7
Newton Purcell **MK18** 31 H5
Newton Regis **B79** 40 E5
Newton Reigny **CA11** 61 F3
Newton St. Cyres **EX5** 7 G6
Newton St. Faith **NR10** 45 G4
Newton St. Loe **BA2** 20 A5
Newton St. Petrock **EX22** 6 C4
Newton Solney **DE15** 40 E3
Newton Stacey **SO20** 21 H7
Newton Stewart DG8 64 E4
Newton Tony **SP4** 21 F7
Newton Tracey **EX31** 6 D3
Newton under Roseberry **TS9** 63 G5
Newton Underwood **NE61** 71 G5
Newton upon Derwent **YO41** 58 D5
Newton Valence **GU34** 11 J1
Newton with Scales **PR4** 55 H6
Newtonairds **DG2** 68 D5
Newtongrange **EH22** 76 B4
Newtonhill **AB39** 91 H5
Newton-le-Willows *Mersey.* WA12 48 E3
Newton-le-Willows *N.Yorks.* **DL8** 57 H1
Newtonmill **DD9** 83 H1
Newtonmore PH20 88 E5
Newton-on-Ouse **YO30** 58 B4
Newton-on-Rawcliffe **YO18** 63 K7
Newton-on-the-Moor **NE65** 71 G3
Newtown *Bucks.* **HP5** 22 C1
Newtown *Ches.W. & C.* **CH3** 48 E7
Newtown *Cornw.* **PL15** 4 C3
Newtown *Cornw.* **TR20** 2 C6
Newtown *Cumb.* **CA6** 70 A7
Newtown *Derbys.* **SK22** 49 J4
Newtown *Devon* **EX36** 7 F3
Newtown *Devon* **EX5** 7 J6
Newtown *Dorset* **DT8** 8 D4
Newtown *Glos.* **GL13** 19 K1
Newtown *Gt.Man.* **WN5** 48 E2
Newtown *Gt.Man.* **M27** 49 G2
Newtown *Hants.* **PO17** 11 H3
Newtown *Hants.* **RG20** 21 H5
Newtown *Hants.* **SO43** 10 D3
Newtown *Hants.* **SO51** 10 E2
Newtown *Hants.* **SO32** 11 G3
Newtown *Here.* **HR8** 29 F4
Newtown *Here.* **HR6** 28 D3
Newtown *High.* **PH35** 87 K4
Newtown *I.o.M.* **IM4** 54 C6
Newtown *I.o.W.* **PO30** 11 F5
Newtown *Northumb.* **NE65** 71 F3
Newtown *Northumb.* **NE66** 71 F1
Newtown *Oxon.* **RG9** 22 A3
Newtown (Y Drenewydd) *Powys* SY16 38 A6
Newtown *R.C.T.* **CF45** 18 D2
Newtown *Shrop.* **SY4** 38 D2
Newtown *Som.* **TA6** 8 B1
Newtown *Som.* **TA20** 8 B3
Newtown *Staffs.* **ST8** 49 J6
Newtown *Staffs.* **SK17** 50 C6
Newtown *Staffs.* **WS6** 40 B5
Newtown *Wilts.* **SP3** 9 J2
Newtown *Wilts.* **SN8** 21 G5
Newtown Linford **LE6** 41 H5
Newtown St. Boswells **TD6** 76 D7
Newtown Unthank **LE9** 41 G5
Newtown-in-St-Martin **TR12** 2 E6
Newtyle **PH12** 82 D3
Newyears Green **UB9** 22 D3
Neyland **SA73** 16 C5
Nibley *Glos.* **GL15** 19 K1
Nibley *S.Glos.* **BS37** 19 K3
Nibley Green **GL11** 20 A2
Nicholashayne **TA21** 7 K4
Nicholaston **SA3** 17 J7
Nidd **HG3** 57 J3
Nigg *Aberdeen* **AB12** 91 H4
Nigg *High.* **IV19** 97 F4
Nightcott **TA22** 7 G3
Nilig **LL15** 47 J7
Nilston Rigg **NE47** 70 D7
Nimlet **SN14** 20 A4
Nine Ashes **CM4** 23 J1
Nine Elms **SN5** 20 E3
Nine Mile Burn **EH26** 75 K5
Ninebanks **NE47** 61 J1
Ninemile Bar (Crocketford) **DG2** 65 J3
Nineveh **WR15** 29 F2
Ninfield **TN33** 14 C6
Ningwood **PO30** 11 F6
Nisbet **TD8** 70 B1
Niton **PO38** 11 G7
Nitshill **G53** 74 D4
Niwbwrch (Newborough) **LL61** 46 C6
Nizels **TN11** 23 J6
No Man's Heath *Ches.W. & C.* **SY14** 38 E1
No Man's Heath *Warks.* **B79** 40 E5
No Man's Land **PL13** 4 C5
Noah's Ark **TN15** 23 J6
Noak Hill **RM4** 23 J2
Noblehill **DG1** 65 K3
Noblethorpe **S75** 50 E2
Nobottle **NN7** 31 H2
Nocton **LN4** 52 D6
Noddsdale **KA30** 74 A4
Nogdam End **NR14** 45 H5
Noke **OX3** 31 G7
Nolton **SA62** 16 B4
Nolton Haven **SA62** 16 B4
Nomansland *Devon* **EX16** 7 G4
Nomansland *Wilts.* **SP5** 10 D3
Noneley **SY4** 38 D3
Nonington **CT15** 15 H2
Nook *Cumb.* **CA6** 69 K6
Nook *Cumb.* **LA6** 55 J1
Noonsbrough **ZE2** 107 L7
Noranside **DD8** 83 F1
Norbreck **FY5** 55 G5
Norbury *Ches.E.* **SY13** 38 E1
Norbury *Derbys.* **DE6** 40 D1
Norbury *Gt.Lon.* **SW16** 23 G4
Norbury *Shrop.* **SY9** 38 C6
Norbury *Staffs.* **ST20** 39 G3
Norbury Common **SY13** 38 E1
Norbury Junction **ST20** 39 G3
Norchard **SA70** 16 D6
Norcott Brook **WA4** 49 F4
Nordelph **PE38** 43 J5
Norden *Dorset* **BH20** 9 J6
Norden *Gt.Man.* **OL11** 49 H1
Nordley **WV16** 39 F6
Norham **TD15** 77 H6
Norland Town **HX6** 57 F7
Norley **WA6** 48 E5
Norleywood **SO41** 10 E5
Norlington **BN8** 13 H5
Norman Cross **PE7** 42 E6
Normanby *N.Lincs.* **DN15** 52 B1
Normanby *N.Yorks.* **YO62** 58 D1
Normanby *R. & C.* **TS6** 63 G5
Normanby by Stow **DN21** 52 B4
Normanby le Wold **LN7** 52 E3
Normanby-by-Spital **LN8** 52 D4
Normandy **GU3** 22 C6
Normann's Ruh **PA74** 79 F3
Norman's Bay **BN24** 13 K6
Norman's Green **EX15** 7 J5
Normanston **NR32** 45 K6
Normanton *Derby* **DE23** 41 F2
Normanton *Leics.* **NG13** 42 B1
Normanton *Lincs.* **NG32** 42 C1
Normanton *Notts.* **NG25** 51 K7
Normanton *Rut.* **LE15** 42 C5
Normanton *W.Yorks.* WF6 57 J7
Normanton le Heath **LE67** 41 F4
Normanton on Soar **LE12** 41 H3
Normanton on Trent **NG23** 51 K6
Normanton-on-the-Wolds **NG12** 41 J2
Normoss **FY3** 55 G6
Norrington Common **SN12** 20 B5
Norris Green **PL17** 4 E4
Norris Hill **DE12** 41 F4
Norristhorpe **WF15** 57 H7
North Acton **W3** 23 F3
North Anston **S25** 51 H4
North Ascot **SL5** 22 C5
North Aston **OX25** 31 F6
North Baddesley **SO52** 10 E3
North Ballachulish **PH33** 80 B1
North Balloch **KA26** 67 H4
North Barrow **BA22** 9 F2
North Barsham **NR22** 44 D2
North Benfleet **SS12** 24 D3
North Bersted **PO21** 12 C6
North Berwick EH39 76 D2
North Boarhunt **PO17** 11 H3
North Bogbain **AB55** 98 B5
North Bovey **TQ13** 7 F7
North Bradley **BA14** 20 B6
North Brentor **PL19** 6 C7
North Brewham **BA10** 9 G1
North Bridge **GU8** 12 C3
North Buckland **EX33** 6 C1
North Burlingham **NR13** 45 H4
North Cadbury **BA22** 9 F2
North Cairn **DG9** 66 D6
North Camp **GU14** 22 B6
North Carlton *Lincs.* **LN1** 52 C5
North Carlton *Notts.* **S81** 51 H4
North Cave **HU15** 58 E6
North Cerney **GL7** 20 D1
North Chailey **BN8** 13 G4
North Charford **SP6** 10 C3
North Charlton **NE67** 71 G1
North Cheriton **BA8** 9 F2
North Chideock **DT6** 8 D5
North Cliffe **YO43** 58 E6
North Clifton **NG23** 52 B5
North Cockerington **LN11** 53 G3
North Coker **BA22** 8 E3
North Collafirth **ZE2** 107 M4
North Common *S.Glos.* **BS30** 19 K4
North Common *Suff.* **IP22** 34 D1
North Commonty **AB53** 99 G6
North Connel **PA37** 80 A4
North Coombe **EX17** 7 G5
North Cornelly **CF33** 18 B3
North Corner **BS36** 19 K3
North Cotes **DN36** 53 G2
North Cove **NR34** 45 J7
North Cowton **DL7** 62 D6
North Crawley **MK16** 32 C4
North Cray **DA14** 23 H4
North Creake **NR21** 44 C2
North Curry **TA3** 8 C2
North Dallens **PA38** 80 A3
North Dalton **YO25** 59 F4
North Dawn **KW17** 106 D7
North Deighton **LS22** 57 J4
North Dell (Dail Bho Thuath) **HS2** 101 G1
North Duffield **YO8** 58 C6
North Elkington **LN11** 53 F3
North Elmham **NR20** 44 D3
North Elmsall **WF9** 51 G1
North End *Bucks.* **LU7** 32 B6
North End *Dorset* **SP7** 9 H2
North End *E.Riding* **YO25** 59 H4
North End *E.Riding* **HU11** 59 H5
North End *E.Riding* **HU12** 59 J6
North End *Essex* **CM6** 33 K7
North End *Hants.* **SO24** 11 G2
North End *Hants.* **SP6** 10 C3
North End *Leics.* **LE12** 41 H4
North End *N.Som.* **BS49** 19 H5
North End *Norf.* **NR16** 44 E6
North End *Northumb.* **NE65** 71 G3
North End *Ports.* **PO2** 11 H4
North End *W.Suss.* **BN14** 12 E5
North End *W.Suss.* **BN18** 12 C6
North Erradale **IV21** 94 D3
North Essie **AB42** 99 J5
North Fambridge **CM3** 24 E2
North Ferriby HU14 59 F7
North Frodingham **YO25** 59 H4
North Gorley **SP6** 10 C3
North Green *Norf.* **IP21** 45 G7
North Green *Suff.* **IP19** 35 H1
North Green *Suff.* **IP17** 35 H2
North Green *Suff.* **IP17** 35 H2
North Grimston **YO17** 58 E3
North Halling **ME2** 24 D5
North Harby **NG23** 52 B5
North Hayling **PO11** 11 J4
North Hazelrigg **NE66** 77 J7
North Heasley **EX36** 7 F2
North Heath *W.Berks.* **RG20** 21 H4
North Heath *W.Suss.* **RH20** 12 D4
North Hill **PL15** 4 C3
North Hillingdon **UB10** 22 D3
North Hinksey **OX2** 21 H1
North Holmwood **RH5** 22 E7
North Houghton **SO20** 10 E1
North Huish **TQ10** 5 H5
North Hykeham **LN6** 52 C6
North Johnston **SA62** 16 C4
North Kelsey **LN7** 52 D2
North Kessock **IV1** 96 D7
North Killingholme **DN40** 52 E1
North Kilvington **YO7** 57 K1
North Kilworth **LE17** 41 J7
North Kingston **BH24** 10 C4
North Kyme **LN4** 52 E7
North Lancing **BN15** 12 E6
North Lee **HP22** 22 B1
North Lees **HG4** 57 J2
North Leigh **OX29** 30 E7
North Leverton with Habblesthorpe **DN22** 51 K4
North Littleton **WR11** 30 B4
North Lopham **IP22** 44 E7
North Luffenham **LE15** 42 C5
North Marden **PO18** 12 B5
North Marston **MK18** 31 J6
North Middleton *Midloth.* **EH23** 76 B5
North Middleton *Northumb.* **NE71** 71 F1
North Millbrex **AB53** 99 G6
North Molton **EX36** 7 F3
North Moreton **OX11** 21 J3
North Mundham **PO20** 12 B6
North Muskham **NG23** 51 K7
North Newbald **YO43** 59 F6
North Newington **OX15** 31 F5
North Newnton **SN9** 20 E6
North Newton **TA7** 8 B1
North Nibley **GL11** 20 A2
North Oakley **RG26** 21 J6
North Ockendon **RM14** 23 J3
North Ormesby **TS3** 63 G5
North Ormsby **LN11** 53 F3
North Otterington **DL7** 57 J1
North Owersby **LN8** 52 D3
North Perrott **TA18** 8 D4
North Petherton **TA6** 8 B1
North Petherwin **PL15** 4 C2
North Pickenham **PE37** 44 C5
North Piddle **WR7** 29 J3
North Plain **CA7** 69 G7
North Pool **TQ7** 5 H6
North Poorton **DT6** 8 E5
North Quarme **TA24** 7 H2
North Queensferry **KY11** 75 K2
North Radworthy **EX36** 7 F2
North Rauceby **NG34** 42 D1
North Reston **LN11** 53 G4
North Rigton **LS17** 57 H5
North Rode **CW12** 49 H6
North Roe **ZE2** 107 M4
North Ronaldsay **KW17** 106 G2
North Ronaldsay Airfield **KW17** 106 G2
North Runcton **PE33** 44 A4
North Sandwick **ZE2** 107 P3
North Scale **LA14** 54 E3
North Scarle **LN6** 52 B6
North Seaton **NE63** 71 H5
North Shian **PA38** 80 A3
North Shields NE30 71 J7
North Shoebury **SS3** 25 F3
North Side **PE6** 43 F6
North Skelton **TS12** 63 H5
North Somercotes **LN11** 53 H3
North Stainley **HG4** 57 H2
North Stainmore **CA17** 61 K5
North Stifford **RM16** 24 C3
North Stoke *B. & N.E.Som.* **BA1** 20 A5
North Stoke *Oxon.* **OX10** 21 K3
North Stoke *W.Suss.* **BN18** 12 D5
North Stoneham **SO50** 11 F3
North Street *Hants.* **SO24** 11 H1
North Street *Kent* **ME13** 15 F2
North Street *Med.* **ME3** 24 E4
North Street *W.Berks.* **RG7** 21 K4
North Sunderland **NE68** 77 K4
North Tamerton **EX22** 6 B6
North Tarbothill **AB23** 91 H3
North Tawton EX20 6 E5
North Third **FK7** 75 F2
North Thoresby **DN36** 53 F3
North Tidworth **SP9** 21 F7
North Togston **NE65** 71 H3
North Town *Devon* **EX20** 6 D5
North Town *Hants.* **GU12** 22 B6
North Town *W. & M.* **SL6** 22 B3
North Tuddenham **NR20** 44 E4
North Uist (Uibhist a Tuath) HS6 92 D4
North Walsham NR28 45 G2
North Waltham **RG25** 21 J7
North Warnborough **RG29** 22 A6
North Water Bridge **AB30** 83 H1
North Watten **KW1** 105 H3
North Weald Bassett **CM16** 23 H1
North Wembley **HA0** 22 E3
North Wheatley **DN22** 51 K4
North Whilborough **TQ12** 5 J4
North Wick **BS41** 19 J5
North Widcombe **BS40** 19 J6
North Willingham **LN8** 52 E4
North Wingfield **S42** 51 G6
North Witham **NG33** 42 C3
North Wootton *Dorset* **DT9** 9 F3
North Wootton *Norf.* **PE30** 44 A3
North Wootton *Som.* **BA4** 19 J7
North Wraxall **SN14** 20 B4
North Wroughton **SN4** 20 E3
North Yardhope **NE65** 70 E3
Northacre **NR17** 44 D6
Northall **LU6** 32 C6
Northall Green **NR20** 44 D4
Northallerton DL6 62 E7
Northam *Devon* **EX39** 6 C3
Northam *S'ham.* **SO14** 11 F3
NORTHAMPTON NN 31 J2
Northaw **EN6** 23 F1
Northay *Devon* **EX13** 8 C4
Northay *Som.* **TA20** 8 B3
Northbay **HS9** 84 C4
Northbeck **NG34** 42 D1
Northborough **PE6** 42 E5
Northbourne *Kent* **CT14** 15 J2
Northbourne *Oxon.* **OX11** 21 J3
Northbridge Street **TN32** 14 C5
Northbrook *Hants.* **SO21** 11 G1
Northbrook *Oxon.* **OX5** 31 F6
Northburnhill **AB53** 99 G6
Northchapel **GU28** 12 C4
Northchurch **HP4** 22 C1
Northcote Manor **EX37** 6 E4
Northcott **PL15** 6 B6
Northcourt **OX14** 21 J2
Northdyke **KW16** 106 B5
Northedge **S42** 51 F6
Northend *B. & N.E.Som.* **BA1** 20 A5
Northend *Bucks.* **RG9** 22 A2
Northend *Warks.* **CV47** 30 E3
Northfield *Aber.* **AB45** 99 G4
Northfield *Aberdeen* **AB16** 91 H4
Northfield *High.* **KW1** 105 J4
Northfield *Hull* **HU4** 59 G7
Northfield *Sc.Bord.* **TD14** 77 H4
Northfield *Som.* **TA6** 8 B1
Northfield *W.Mid.* **B31** 30 B1
Northfields **PE9** 42 D5
Northfleet **DA11** 24 C4
Northhouse **TD9** 69 K3
Northiam **TN31** 14 D5
Northill **SG18** 32 E4
Northington *Glos.* **GL14** 20 A1
Northington *Hants.* **SO24** 11 G1
Northlands **PE22** 53 G7
Northleach **GL54** 30 C7
Northleigh *Devon* **EX24** 7 K6
Northleigh *Devon* **EX32** 6 E2
Northlew **EX20** 6 D6
Northmoor **OX29** 21 H1
Northmoor Green (Moorland) **TA7** 8 C1
Northmuir **DD8** 82 E2
Northney **PO11** 11 J4
Northolt UB5 22 E3
Northop (Llaneurgain) **CH7** 48 B6
Northop Hall **CH7** 48 B6
Northorpe *Lincs.* **PE11** 43 F2
Northorpe *Lincs.* **DN21** 52 B3
Northorpe *Lincs.* **PE10** 42 D4
Northover *Som.* **BA22** 8 E2
Northover *Som.* **BA6** 8 D1
Northowram **HX3** 57 G7
Northport **BH20** 9 J6
Northpunds **ZE2** 107 N10
Northrepps **NR27** 45 G2
Northton (Taobh Tuath) **HS3** 92 E3
Northtown **KW17** 106 D8
Northway *Glos.* **GL20** 29 J5
Northway *Som.* **TA4** 7 K3
Northwich CW8 49 F5
Northwick *S.Glos.* **BS35** 19 J3
Northwick *Som.* **TA9** 19 G7
Northwick *Worcs.* **WR3** 29 H3
Northwold **IP26** 44 B6
Northwood *Gt.Lon.* HA6 22 D2
Northwood *I.o.W.* **PO31** 11 F5
Northwood *Kent* **CT12** 25 K5
Northwood *Mersey.* **L33** 48 D3
Northwood *Shrop.* **SY4** 38 D2
Northwood Green **GL14** 29 G7
Northwood Hills **HA6** 22 E2
Norton *Glos.* **GL2** 29 H6
Norton *Halton* **WA7** 48 E4
Norton *Herts.* **SG6** 33 F5
Norton *I.o.W.* **PO41** 10 E6
Norton *Mon.* **NP7** 28 D6
Norton *N.Som.* **BS22** 19 G5
Norton *N.Yorks.* **YO17** 58 D2
Norton *Northants.* **NN11** 31 H2
Norton *Notts.* **NG20** 51 H5
Norton *Powys* **LD8** 28 C2
Norton *S.Yorks.* **DN6** 51 H1
Norton *S.Yorks.* **S8** 51 F4
Norton *Shrop.* **SY4** 38 E5
Norton *Shrop.* **TF11** 39 G5
Norton *Shrop.* **SY7** 38 D7
Norton *Stock.* **TS20** 63 F4
Norton *Suff.* **IP31** 34 D2
Norton *Swan.* **SA3** 17 K7
Norton *V. of Glam.* **CF32** 18 B4
Norton *W.Mid.* **DY8** 40 A7
Norton *W.Suss.* **PO20** 12 C6
Norton *W.Suss.* **PO20** 12 B7
Norton *Wilts.* **SN16** 20 B3
Norton *Worcs.* **WR5** 29 H3
Norton *Worcs.* **WR11** 30 B4
Norton Bavant **BA12** 20 C7
Norton Bridge **ST15** 40 A2
Norton Canes **WS11** 40 C5
Norton Canon **HR4** 28 C4
Norton Disney **LN6** 52 B7
Norton Ferris **BA12** 9 G1
Norton Fitzwarren **TA2** 7 K3
Norton Green *Herts.* **SG1** 33 F6
Norton Green *I.o.W.* **PO40** 10 E6
Norton Green *Stoke* **ST6** 49 J7
Norton Hawkfield **BS40** 19 J5
Norton Heath **CM4** 24 C1
Norton in Hales **TF9** 39 G2
Norton in the Moors **ST6** 49 H7
Norton Lindsey **CV35** 30 D2
Norton Little Green **IP31** 34 D2
Norton Malreward **BS39** 19 K5
Norton Mandeville **CM5** 23 J1
Norton St. Philip **BA2** 20 A6
Norton Subcourse **NR14** 45 J6
Norton Wood **HR4** 28 C4
Norton Woodseats **S8** 51 F4

Nor - Owe

Norton-Juxta-Twycross CV9 41 F5
Norton-le-Clay YO61 57 K2
Norton-sub-Hamdon TA14 8 D3
Norwell NG23 51 K6
Norwell Woodhouse NG23 51 K6
NORWICH NR 45 G5
Norwich International Airport NR6 45 G4
Norwick ZE2 107 Q1
Norwood End CM5 23 J1
Norwood Green *Gt.Lon.* UB2 22 E4
Norwood Green *W.Yorks.* HX3 57 G7
Norwood Hill RH6 23 F7
Norwood Park BA6 8 E1
Noseley LE7 42 A6
Noss Mayo PL8 5 F6
Nosterfield DL8 57 H1
Nosterfield End CB21 33 K4
Nostie IV40 86 E2
Notgrove GL54 30 C6
Nottage CF36 18 B4
Notting Hill W11 23 F3
Nottingham *High.* KW5 105 H5
NOTTINGHAM *Nott.* NG 41 H1
Nottingham East Midlands Airport DE74 41 G3
Nottington DT3 9 F6
Notton *W.Yorks.* WF4 51 F1
Notton *Wilts.* SN15 20 C5
Nottswood Hill GL17 29 G7
Nounsley CM3 34 B7
Noutard's Green WR6 29 G2
Nowton IP29 34 C2
Nox SY5 38 D4
Noyadd Trefawr SA43 17 F1
Nuffield RG9 21 K3
Nun Monkton YO26 58 B4
Nunburnholme YO42 58 E5
Nuneaton CV11 41 F6
Nuneham Courtenay OX44 21 J2
Nunney BA11 20 A7
Nunnington *Here.* HR1 28 E4
Nunnington *N.Yorks.* YO62 58 C2
Nunnington Park TA4 7 J3
Nunnykirk NE61 71 F4
Nunsthorpe DN32 53 F2
Nunthorpe *Middbro.* TS7 63 G5
Nunthorpe *York* YO23 58 B4
Nunton SP5 10 C2
Nunwick *N.Yorks.* HG4 57 J2
Nunwick *Northumb.* NE48 70 D6
Nup End SG4 33 F7
Nupend GL10 20 A1
Nursling SO16 10 E3
Nursted GU31 11 J2
Nurton WV6 40 A6
Nutbourne *W.Suss.* RH20 12 D5
Nutbourne *W.Suss.* PO18 11 J4
Nutfield RH1 23 G6
Nuthall NG16 41 H1
Nuthampstead SG8 33 H5
Nuthurst *W.Suss.* RH13 12 E4
Nuthurst *Warks.* B94 30 C1
Nutley *E.Suss.* TN22 13 H4
Nutley *Hants.* RG25 21 K7
Nutwell DN3 51 J2
Nyadd FK9 75 F1
Nybster KW1 105 J2
Nyetimber PO21 12 B7
Nyewood GU31 11 J2
Nymet Rowland EX17 7 F5
Nymet Tracey EX17 7 F5
Nympsfield GL10 20 B1
Nynehead TA21 7 K3
Nythe TA7 8 D1
Nyton PO20 12 C6

O

Oad Street ME9 24 E5
Oadby LE2 41 J5
Oak Cross EX20 6 D6
Oak Tree DL2 62 E5
Oakamoor ST10 40 C1
Oakbank *Arg. & B.* PA64 79 J4
Oakbank *W.Loth.* EH53 75 J4
Oakdale *Caerp.* NP12 18 E2
Oakdale *Poole* BH15 10 B5
Oake TA4 7 K3
Oaken WV8 40 A5
Oakenclough PR3 55 J5
Oakengates TF2 39 G4
Oakenhead IV31 97 K5
Oakenholt CH6 48 B5
Oakenshaw *Dur.* DL15 62 D3
Oakenshaw *W.Yorks.* BD12 57 G7
Oakerthorpe DE55 51 F7
Oakes HD3 50 D1
Oakfield *I.o.W.* PO33 11 G5
Oakfield *Torfaen* NP44 19 F2
Oakford *Cere.* SA47 26 D3
Oakford *Devon* EX16 7 H3
Oakfordbridge EX16 7 H3
Oakgrove SK11 49 J6
Oakham LE15 42 B5
Oakhanger GU35 11 J1
Oakhill BA3 19 K7
Oakington CB24 33 H2
Oaklands *Conwy* LL26 47 G7
Oaklands *Herts.* AL6 33 F7
Oakle Street GL2 29 G7
Oakley *Bed.* MK43 32 D3
Oakley *Bucks.* HP18 31 H7
Oakley *Fife* KY12 75 J2
Oakley *Hants.* RG23 21 J6
Oakley *Oxon.* OX39 22 A1
Oakley *Poole* BH21 10 B5
Oakley *Suff.* IP21 35 F1
Oakley Green SL4 22 C4
Oakley Park SY17 37 J7
Oakridge Lynch GL6 20 C1
Oaks SY5 38 D5
Oaks Green DE6 40 D2
Oaksey SN16 20 C2
Oakshaw Ford CA6 70 A6
Oakshott GU33 11 J2
Oakthorpe DE12 41 F4
Oaktree Hill DL6 62 E7
Oakwoodhill RH5 12 E3
Oakworth BD22 57 F6
Oare *Kent* ME13 25 G5
Oare *Som.* EX35 7 G1
Oare *Wilts.* SN8 20 E5
Oasby NG32 42 D2
Oatfield PA28 66 A2
Oath TA7 8 C2
Oathlaw DD8 83 F2
Oatlands HG2 57 J4
Oban PA34 79 K5
Obley SY7 28 C1
Oborne DT9 9 F3
Obthorpe PE10 42 D4
Occlestone Green CW10 49 F6
Occold IP23 35 F1
Occumster KW3 105 H5
Ochiltree KA18 67 K1
Ochr-y-foel LL18 47 J5
Ochtermuthill PH5 81 K6
Ochtertyre *P. & K.* PH7 81 K5
Ochtertyre *Stir.* FK9 75 F1
Ockbrook DE72 41 G2
Ockeridge WR6 29 G2
Ockham GU23 22 D6
Ockle PH36 86 B7
Ockley RH5 12 E3
Ocle Pychard HR1 28 E4
Octon YO25 59 G2
Odcombe BA22 8 E3
Odd Down BA2 20 A5
Oddendale CA10 61 H5
Oddingley WR9 29 J3
Oddington OX5 31 G7
Oddsta ZE2 107 P3
Odell MK43 32 C3
Odham EX21 6 C5
Odie KW17 106 F5
Odiham RG29 22 A6
Odsey SG7 33 F5
Odstock SP5 10 C2
Odstone CV13 41 F5
Offchurch CV33 30 E2
Offenham WR11 30 B4
Offerton SK2 49 J4
Offham *E.Suss.* BN8 13 G5
Offham *Kent* ME19 23 K6
Offham *W.Suss.* BN18 12 D6
Offley Hoo SG5 32 E6
Offleymarsh ST21 39 G3
Offord Cluny PE19 33 F2
Offord D'Arcy PE19 33 F2
Offton IP8 34 E4
Offwell EX14 7 K6
Ogbourne Maizey SN8 20 E4
Ogbourne St. Andrew SN8 20 E4
Ogbourne St. George SN8 20 E4
Ogil DD8 83 F1
Ogle NE20 71 G6
Oglet L24 48 D4
Ogmore CF32 18 B4
Ogmore Vale CF32 18 C2
Ogmore-by-Sea CF32 18 B4
Oil Terminal KW16 106 C8
Okeford Fitzpaine DT11 9 H3
Okehampton EX20 6 D6
Okehampton Camp EX20 6 D6
Okraquoy ZE2 107 N9
Olchard TQ13 5 J3
Olchfa SA2 17 K6
Old NN6 31 J1
Old Aberdeen AB24 91 H4
Old Alresford SO24 11 G1
Old Arley CV7 40 E6
Old Basford NG6 41 H1
Old Basing RG24 21 K6
Old Belses TD6 70 A1
Old Bewick NE66 71 F1
Old Blair PH18 81 K1
Old Bolingbroke PE23 53 G6
Old Bramhope LS16 57 H5
Old Brampton S42 51 F5
Old Bridge of Urr DG7 65 H4
Old Buckenham NR17 44 E6
Old Burdon SR7 62 E1
Old Burghclere RG20 21 H6
Old Byland YO62 58 B1
Old Cassop DH6 62 E3
Old Church Stoke SY15 38 B6
Old Cleeve TA24 7 J1
Old Clipstone NG21 51 J6
Old Colwyn LL29 47 G5
Old Craig AB41 91 H2
Old Craighall EH21 76 B3
Old Crombie AB54 98 D5
Old Dailly KA26 67 G4
Old Dalby LE14 41 J3
Old Dam SK17 50 D5
Old Deer AB42 99 H6
Old Dilton BA13 20 B7
Old Down *S.Glos.* BS32 19 K3
Old Down *Som.* BA3 19 K6
Old Edlington DN12 51 H3
Old Eldon DL4 62 D4
Old Ellerby HU11 59 H6
Old Felixstowe IP11 35 H5
Old Fletton PE2 42 E6
Old Ford E3 23 G3
Old Glossop SK13 50 C3
Old Goginan SY23 37 F7
Old Goole DN14 58 D7
Old Gore HR9 29 F6
Old Grimsby TR24 2 B1
Old Hall HU12 53 F1
Old Hall Green SG11 33 G6
Old Hall Street NR28 45 H2
Old Harlow CM20 33 H7
Old Heath CO2 34 E6
Old Heathfield TN21 13 J4
Old Hill B64 40 B7
Old Hurst PE28 33 G1
Old Hutton LA8 55 J1
Old Kea TR3 3 F4
Old Kilpatrick G60 74 C3
Old Kinnernie AB32 91 F4
Old Knebworth SG3 33 F6
Old Leake PE22 53 H7
Old Leslie AB52 90 D2
Old Malton YO17 58 E2
Old Milton BH25 10 D5
Old Milverton CV32 30 D2
Old Montsale CM0 25 G2
Old Netley SO31 11 F3
Old Newton IP14 34 E2
Old Philpstoun EH49 75 J3
Old Poltalloch PA31 79 K7
Old Radnor (Pencraig) LD8 28 B3
Old Rattray AB42 99 J5
Old Rayne AB52 90 E2
Old Romney TN29 15 F5
Old Scone PH2 82 C5
Old Shields G67 75 G3
Old Sodbury BS37 20 A3
Old Somerby NG33 42 C2
Old Stratford MK19 31 J4
Old Sunderlandwick YO25 59 G4
Old Swarland NE65 71 G3
Old Swinford DY8 40 B7
Old Thirsk YO7 57 K1
Old Town *Cumb.* LA6 55 J1
Old Town *I.o.S.* TR21 2 C1
Old Town Farm NE19 70 D4
Old Tupton S42 51 F6
Old Warden SG18 32 E4
Old Weston PE28 32 D1
Old Windsor SL4 22 C4
Old Wives Lees CT4 15 F2
Old Woking GU22 22 D6
Old Woodhall LN9 53 F6
Old Woods SY4 38 D3
Oldberrow B95 30 C2
Oldborough EX17 7 F5
Oldbury *Kent* TN15 23 J6
Oldbury *Shrop.* WV16 39 G6
Oldbury *W.Mid.* B69 40 B7
Oldbury *Warks.* CV10 41 F6
Oldbury Naite BS35 19 K2
Oldbury on the Hill GL9 20 B3
Oldbury-on-Severn BS35 19 K2
Oldcastle *Bridgend* CF31 18 C4
Oldcastle *Mon.* NP7 28 C6
Oldcastle Heath SY14 38 D1
Oldcotes S81 51 H4
Oldcroft GL15 19 K1
Oldeamere PE7 43 G6
Oldfield WR9 29 H2
Oldford BA11 20 A6
Oldhall *Aber.* AB34 90 C5
Oldhall *High.* KW1 105 H3
OLDHAM OL 49 J2
Oldham Edge OL1 49 J2
Oldhamstocks TD13 77 F3
Oldland BS30 19 K4
Oldmeldrum AB51 91 G2
Oldmill AB31 90 D4
Oldpark TF3 39 F5
Oldridge EX4 7 G6
Oldshore Beg IV27 102 D3
Oldshoremore IV27 102 E3
Oldstead YO61 58 B2
Oldtown IV24 96 C3
Oldtown of Aigas IV4 96 B7
Oldtown of Ord AB45 98 E5
Oldwalls SA3 17 H6
Oldways End EX16 7 G3
Oldwhat AB53 99 G5
Oldwich Lane B93 30 D1
Olgrinmore KW12 105 G3
Oliver ML12 69 G1
Oliver's Battery SO22 11 F2
Ollaberry ZE2 107 M4
Ollerton *Ches.E.* WA16 49 G5
Ollerton *Notts.* NG22 51 J6
Ollerton *Shrop.* TF9 39 F3
Olmstead Green CB21 33 K4
Olney MK46 32 B3
Olrig House KW14 105 G2
Olton B92 40 D7
Olveston BS35 19 K3
Ombersley WR9 29 H2
Ompton NG22 51 J6
Onchan IM3 54 C6
Onecote ST13 50 C7
Onehouse IP14 34 E3
Ongar Hill PE34 43 J3
Ongar Street HR6 28 C2
Onibury SY7 28 D1
Onich PH33 80 B1
Onllwyn SA10 27 H7
Onneley CW3 39 G1
Onslow Green CM6 33 K7
Onslow Village GU2 22 C7
Opinan *High.* IV22 94 E2
Opinan *High.* IV21 94 D4
Orange Lane TD12 77 F6
Orasaigh HS2 101 F6
Orbliston IV32 98 B5
Orbost IV55 93 H7
Orby PE24 53 H6
Orcadia PA20 73 K4
Orchard PA23 73 K2
Orchard Portman TA3 8 B2
Orcheston SP3 20 D7
Orcop HR2 28 D6
Orcop Hill HR2 28 D6
Ord IV46 86 C3
Ordhead AB51 90 E3
Ordie AB34 90 C4
Ordiequish IV32 98 B5
Ordsall DN22 51 J5
Ore TN35 14 D6
Oreham Common BN5 13 F5
Oreston PL9 5 F5
Oreton DY14 29 F1
Orford *Suff.* IP12 35 J4
Orford *Warr.* WA2 49 F3
Organford BH16 9 J5
Orgreave DE13 40 D4
Orkney Islands KW 106 B6
Orlestone TN26 14 E4
Orleton *Here.* SY8 28 D2
Orleton *Worcs.* WR6 29 F2
Orleton Common SY8 28 D2
Orlingbury NN14 32 B1
Ormacleit HS8 84 C1
Ormesby TS3 63 G5
Ormesby St. Margaret NR29 45 J4
Ormesby St. Michael NR29 45 J4
Ormidale PA22 73 J2
Ormiscaig IV22 94 E2
Ormiston EH35 76 C4
Ormlie KW14 105 G2
Ormsaigmore PH36 79 F1
Ormsary PA31 73 F3
Ormskirk L39 48 D2
Oronsay PA61 72 B2
Orphir KW17 106 C7
Orpington BR6 23 H5
Orrell *Gt.Man.* WN5 48 E2
Orrell *Mersey.* L20 48 C3
Orrisdale IM6 54 C4
Orrok House AB23 91 H3
Orroland DG6 65 H6
Orsett RM16 24 C3
Orsett Heath RM16 24 C3
Orslow TF10 40 A4
Orston NG13 42 A1
Orton *Cumb.* CA10 61 H6
Orton *Northants.* NN14 32 B1
Orton Longueville PE2 42 E6
Orton Rigg CA5 60 E1
Orton Waterville PE2 42 E6
Orton-on-the-Hill CV9 41 F5
Orwell SG8 33 G3
Osbaldeston BB2 56 B6
Osbaldwick YO10 58 C4
Osbaston *Leics.* CV13 41 G5
Osbaston *Shrop.* SY10 38 C3
Osbaston *Tel. & W.* TF6 38 E4
Osbaston Hollow CV13 41 G5
Osborne PO32 11 G5
Osbournby NG34 42 D2
Oscroft CH3 48 E6
Ose IV56 93 J7
Osgathorpe LE12 41 G4
Osgodby *Lincs.* LN8 52 D3
Osgodby *N.Yorks.* YO8 58 C6
Osgodby *N.Yorks.* YO11 59 G1
Oskaig IV40 86 B1
Osleston DE6 40 E2
Osmaston *Derby* DE24 41 F2
Osmaston *Derbys.* DE6 40 D1
Osmington DT3 9 G6
Osmington Mills DT3 9 G6
Osmondthorpe LS9 57 J6
Osmotherley DL6 63 F7
Osnaburgh (Dairsie) KY15 83 F6
Ospringe ME13 25 G5
Ossett WF5 57 H7
Ossett Street Side WF5 57 H7
Ossington NG23 51 K6
Ostend CM0 25 F2
Osterley TW7 22 E4
Oswaldkirk YO62 58 C2
Oswaldtwistle BB5 56 C7
Oswestry SY11 38 B3
Oteley SY12 38 D2
Otford TN14 23 J6
Otham ME15 14 C2
Otherton ST19 40 B4
Othery TA7 8 C1
Otley *Suff.* IP6 35 G3
Otley *W.Yorks.* LS21 57 H5
Otter PA21 73 H3
Otter Ferry PA21 73 H2
Otterbourne SO21 11 F2
Otterburn *N.Yorks.* BD23 56 D4
Otterburn *Northumb.* NE19 70 D4
Otterburn Camp NE19 70 D4
Otterden Place ME13 14 E2
Otterham PL32 4 B1
Otterham Quay ME8 24 E5
Otterhampton TA5 19 F7
Otternish HS6 92 E4
Ottershaw KT16 22 D5
Otterswick ZE2 107 P4
Otterton EX9 7 J7
Otterwood SO42 11 F4
Ottery St. Mary EX11 7 J6
Ottinge CT4 15 G3
Ottringham HU12 59 J7
Oughterby CA5 60 D1
Oughtershaw BD23 56 D1
Oughterside CA7 60 C2
Oughtibridge S35 51 F3
Oulston YO61 58 B2
Oulton *Cumb.* CA7 60 D1
Oulton *Norf.* NR11 45 F3
Oulton *Staffs.* ST15 40 B2
Oulton *Staffs.* ST20 39 G3
Oulton *Suff.* NR32 45 K6
Oulton *W.Yorks.* LS26 57 J7
Oulton Broad NR33 45 K6
Oulton Grange ST15 40 B2
Oulton Street NR11 45 F3
Oultoncross ST15 40 B2
Oundle PE8 42 D7
Ousby CA10 61 H3
Ousdale KW7 105 F6
Ousden CB8 34 B3
Ousefleet DN14 58 E7
Ouston *Dur.* DH2 62 D1
Ouston *Northumb.* NE18 71 F6
Out Newton HU19 59 K7
Out Rawcliffe PR3 55 H5
Out Skerries Airstrip ZE2 107 Q5
Outcast LA12 55 G2
Outchester NE70 77 K7
Outertown KW16 106 B7
Outgate LA22 60 E7
Outhgill CA17 61 J6
Outlands ST20 39 G3
Outlane HD3 50 C1
Outwell PE14 43 J5
Outwood *Surr.* RH1 23 G7
Outwood *W.Yorks.* WF1 57 J7
Outwoods TF10 39 G4
Ouzlewell Green WF3 57 J7
Ovenden HX3 57 F7
Over *Cambs.* CB24 33 G1
Over *Ches.W. & C.* CW7 49 F6
Over *Glos.* GL2 29 H7
Over *S.Glos.* BS32 19 J3
Over Burrows DE6 40 E2
Over Compton DT9 8 E3
Over Dinsdale DL2 62 E5
Over End DE45 50 E5
Over Green B76 40 D6
Over Haddon DE45 50 E6
Over Hulton BL5 49 F2
Over Kellet LA6 55 J3
Over Kiddington OX20 31 F6
Over Monnow NP25 28 D7
Over Norton OX7 30 E6
Over Peover WA16 49 G5
Over Rankeilour KY15 82 E6
Over Silton YO7 63 F7
Over Stowey TA5 7 K2
Over Stratton TA13 8 D3
Over Tabley WA16 49 G4
Over Wallop SO20 10 D1
Over Whitacre B46 40 E6
Over Winchendon (Upper Winchendon) HP18 31 J7
Over Worton OX7 31 F6
Overbister KW17 106 F3
Overbrae AB53 99 G5
Overbury GL20 29 J5
Overcombe DT3 9 F6
Overgreen S42 51 F5
Overleigh BA16 8 D1
Overpool CH66 48 C5
Overscaig Hotel IV27 103 G6
Overseal DE12 40 E4
Overslade CV22 31 F1
Oversland ME13 15 F2
Oversley Green B49 30 B3
Overstone NN6 32 B2
Overstrand NR27 45 G1
Overthorpe OX17 31 F4
Overton *Aber.* AB51 91 F3
Overton *Aberdeen* AB21 91 G3
Overton *Ches.W. & C.* WA6 48 E5
Overton *Hants.* RG25 21 J7
Overton *Lancs.* LA3 55 H4
Overton *N.Yorks.* YO30 58 B4
Overton *Shrop.* SY8 28 E1
Overton *Swan.* SA3 17 H7
Overton *W.Yorks.* WF4 50 E1
Overton (Owrtyn) *Wrex.* LL13 38 C1
Overton Bridge LL13 38 C1
Overtown *Lancs.* LA6 56 B2
Overtown *N.Lan.* ML2 75 G5
Overtown *Swin.* SN4 20 E4
Overy OX10 21 J2
Oving *Bucks.* HP22 31 J6
Oving *W.Suss.* PO20 12 C6
Ovingdean BN2 13 G6
Ovingham NE42 71 F7
Ovington *Dur.* DL11 62 C5
Ovington *Essex* CO10 34 B4
Ovington *Hants.* SO24 11 G1
Ovington *Norf.* IP25 44 D5
Ovington *Northumb.* NE42 71 F7
Ower *Hants.* SO51 10 E3
Ower *Hants.* SO45 11 F4
Owermoigne DT2 9 G6
Owler Bar S17 50 E5
Owlpen GL11 20 A2
Owl's Green IP13 35 G2
Owlswick HP27 22 A1
Owmby DN38 52 D2
Owmby-by-Spital LN8 52 D4
Owrtyn (Overton) LL13 38 C1
Owslebury SO21 11 G2
Owston LE15 42 A5
Owston Ferry DN9 52 B2
Owstwick HU12 59 J6
Owthorpe NG12 41 J2
Oxborough PE33 44 B5
Oxcliffe Hill LA3 55 H3
Oxcombe LN9 53 G5
Oxen End CM7 33 K6

Oxe - Pen

Oxen Park LA12 55 G1
Oxencombe TQ13 7 G7
Oxenhall GL18 29 G6
Oxenholme LA9 61 G7
Oxenhope BD22 57 F6
Oxenpill BA6 19 H7
Oxenton GL52 29 J5
Oxenwood SN8 21 G6
OXFORD OX 21 J1
Oxhey WD17 22 E2
Oxhill CV35 30 E4
Oxley WV10 40 B5
Oxley Green CM9 34 D7
Oxley's Green TN32 13 K4
Oxnam TD8 70 C2
Oxnead NR10 45 G3
Oxnop Ghyll DL8 62 A7
Oxshott KT22 22 E5
Oxspring S36 50 E2
Oxted RH8 23 G6
Oxton *Mersey.* CH43 48 C4
Oxton *Notts.* NG25 51 J7
Oxton *Sc.Bord.* TD2 76 C5
Oxwich SA3 17 H7
Oxwich Green SA3 17 H7
Oxwick NR21 44 D3
Oykel Bridge IV27 95 K1
Oyne AB52 90 E2
Ozleworth GL12 20 A2

P

Pabail Iarach (Lower Bayble) HS2 101 H4
Pabail Uarach (Upper Bayble) HS2 101 H4
Pabbay HS6 92 E3
Packington LE65 41 F4
Packwood B94 30 C1
Padanaram DD8 83 F2
Padbury MK18 31 J5
Paddington W2 23 F3
Paddlesworth CT18 15 G3
Paddock TN25 14 E2
Paddock Wood TN12 23 K7
Paddockhaugh IV30 97 K6
Paddockhole DG11 69 H5
Paddolgreen SY4 38 E2
Padeswood CH7 48 B6
Padfield SK13 50 C3
Padiham BB12 56 C6
Padside HG3 57 G4
Padstow PL28 3 G1
Padworth RG7 21 K5
Paganhill GL5 20 B1
Pagham PO21 12 B7
Paglesham Churchend SS4 25 F2
Paglesham Eastend SS4 25 F2
Paible HS3 93 F2
Paignton TQ3 5 J4
Pailton CV23 41 G7
Paine's Corner TN21 13 K4
Painscastle LD2 28 A4
Painshawfield NE43 71 F7
Painswick GL6 20 B1
Pairc HS2 100 E3
PAISLEY PA 74 C4
Pakefield NR33 45 K6
Pakenham IP31 34 D2
Pale LL23 37 J2
Palehouse Common TN22 13 H5
Palestine SP11 21 F7
Paley Street SL6 22 B4
Palgowan DG8 67 H5
Palgrave IP22 35 F1
Pallinsburn House TD12 77 G7
Palmarsh CT21 15 G4
Palmers Cross GU5 22 D7
Palmers Green N13 23 G2
Palmerscross IV30 97 K5
Palmerstown CF63 18 E4
Palnackie DG7 65 J5
Palnure DG8 64 E4
Palterton S44 51 G6
Pamber End RG26 21 K6
Pamber Green RG26 21 K6
Pamber Heath RG26 21 K5
Pamington GL20 29 J5
Pamphill BH21 9 J4
Pampisford CB22 33 H4
Pan KW16 106 C8
Panborough BA5 19 H7
Panbride DD7 83 G4
Pancrasweek EX22 6 A5
Pancross CF62 18 D5
Pandy *Gwyn.* LL36 37 F5
Pandy *Mon.* NP7 28 C6
Pandy *Powys* SY19 37 J5
Pandy *Wrex.* LL20 38 A2
Pandy Tudur LL22 47 G6
Pandy'r Capel LL21 47 J7
Panfield CM7 34 B6
Pangbourne RG8 21 K4
Pannal HG3 57 J4
Pannal Ash HG3 57 H4
Panshanger AL7 33 F7
Pant SY10 38 B3
Pant Glas LL54 36 D1
Pant Gwyn LL40 37 H3
Pant Mawr SY18 37 H7
Pantasaph CH8 47 K5
Panteg NP4 19 G2
Pantglas SY20 37 G6
Pantgwyn *Carmar.* SA19 17 J3
Pantgwyn *Cere.* SA43 17 F1
Pant-lasau SA6 17 K5
Panton LN8 52 E5
Pant-pastynog LL16 47 J6
Pantperthog SY20 37 G5
Pant-y-dwr LD6 27 J1
Pantyffordd CH7 48 B7
Pant-y-ffridd SY21 38 A5
Pantyffynnon SA18 17 K4
Pantygasseg NP4 19 F2
Pantygelli NP7 28 C7
Pantymwyn CH7 47 K6
Panxworth NR13 45 H4
Papa Stour ZE2 107 K6
Papa Stour Airstrip ZE2 107 K6
Papa Westray KW17 106 D2
Papa Westray Airfield KW17 106 D2
Papcastle CA13 60 C3
Papil ZE2 107 M9
Papple EH41 76 D3
Papplewick NG15 51 H7
Papworth Everard CB23 33 F2
Papworth St. Agnes CB23 33 F2
Par PL24 4 A5
Parbold WN8 48 D1
Parbrook *Som.* BA6 8 E1
Parbrook *W.Suss.* RH14 12 D4
Parc LL23 37 H2
Parcllyn SA43 26 B3
Parcrhydderch SY25 27 F3
Parc-Seymour NP26 19 H2
Parc-y-rhôs SA48 17 J1
Pardshaw CA13 60 B4
Parham IP13 35 H2
Parish Holm ML11 68 C1
Park AB45 98 D5
Park Close BB18 56 D5
Park Corner *E.Suss.* TN3 13 J3
Park Corner *Oxon.* RG9 21 K3
Park End *Northumb.* NE48 70 D6
Park End *Staffs.* ST7 49 G7
Park End *Worcs.* DY12 29 G1
Park Gate *Hants.* SO31 11 G4
Park Gate *W.Yorks.* LS20 57 G5
Park Gate *W.Yorks.* HD8 50 E1
Park Gate *Worcs.* B61 29 J1
Park Green IP14 35 F2
Park Hill S2 51 F4
Park Lane LL13 38 D2
Park Langley BR4 23 G5
Park Street AL2 22 E1
Parkend *Cumb.* CA7 60 E3
Parkend *Glos.* GL15 19 K1
Parker's Green TN10 23 K7
Parkeston CO12 35 G5
Parkfield *Cornw.* PL14 4 D4
Parkfield *S.Glos.* BS16 19 K4
Parkfield *W.Mid.* WV4 40 B6
Parkford DD8 83 F2
Parkgate *Ches.W. & C.* CH64 48 B5
Parkgate *D. & G.* DG1 69 F5
Parkgate *Kent* TN30 14 D4
Parkgate *S.Yorks.* S62 51 G3
Parkgate *Surr.* RH5 23 F7
Parkham EX39 6 B3
Parkham Ash EX39 6 B3
Parkhead G31 74 E4
Parkhill *Angus* DD11 83 H3
Parkhill *P. & K.* PH10 82 C3
Parkhouse NP25 19 H1
Parkhurst PO30 11 F5
Parkmill SA3 17 J7
Parkmore AB55 98 B6
Parkneuk AB30 91 F7
Parkside LL12 48 C7
Parkstone BH14 10 B5
Parkway BA22 8 E2
Parley Cross BH22 10 B5
Parley Green BH23 10 C5
Parlington LS25 57 K6
Parracombe EX31 6 E1
Parrog SA42 16 D2
Parson Cross S5 51 F3
Parson Drove PE13 43 G5
Parsonage Green CM1 33 K7
Parsonby CA7 60 C3
Partick G11 74 D4
Partington M31 49 G3
Partney PE23 53 H6
Parton *Cumb.* CA28 60 A4
Parton *D. & G.* DG7 65 G3
Partridge Green RH13 12 E5
Parwich DE6 50 D7
Paslow Wood Common CM4 23 J1
Passenham MK19 31 J5
Passfield GU30 12 B3
Passingford Bridge RM4 23 J2
Paston NR28 45 H2
Paston Street NR28 45 H2
Pasturefields ST18 40 B3
Patchacott EX21 6 C6
Patcham BN1 13 G6
Patchetts Green WD25 22 E2
Patching BN13 12 D6
Patchole EX31 6 E1
Patchway BS34 19 K3
Pateley Bridge HG3 57 G3
Path of Condie PH2 82 B6
Pathe TA7 8 C1
Pathfinder Village EX6 7 G6
Pathhead *Aber.* DD10 83 J1
Pathhead *E.Ayr.* KA18 68 B2
Pathhead *Fife* KY1 76 A1
Pathhead *Midloth.* EH37 76 B4
Pathlow CV37 30 C3
Patmore Heath SG11 33 H6
Patna KA6 67 J2
Patney SN10 20 D6
Patrick IM5 54 B5
Patrick Brompton DL8 62 D7
Patrington HU12 59 K7
Patrington Haven HU12 59 K7
Patrishow NP7 28 B6
Patrixbourne CT4 15 G2
Patterdale CA11 60 E5
Pattingham WV6 40 A6
Pattishall NN12 31 H3
Pattiswick CM77 34 C6
Paul TR19 2 B6
Paulerspury NN12 31 J4
Paull HU12 59 H7
Paull Holme HU12 59 H7
Paul's Green TR27 2 D5
Paulton BS39 19 K6
Pauperhaugh NE65 71 F4
Pave Lane TF10 39 G4
Pavenham MK43 32 C3
Pawlett TA6 19 G7
Pawston TD12 77 G7
Paxford GL55 30 C5
Paxhill Park RH16 13 G4
Paxton TD15 77 H5
Payden Street ME17 14 E2
Payhembury EX14 7 J5
Paythorne BB7 56 D4
Peacehaven BN10 13 H6
Peacemarsh SP8 9 H2
Peachley WR2 29 H3
Peak Dale SK17 50 C5
Peak Forest SK17 50 D5
Peakirk PE6 42 E5
Pean Hill CT5 25 H5
Pear Tree DE23 41 F2
Pearsie DD8 82 E2
Pearson's Green TN12 23 K7
Peartree AL7 33 F7
Peartree Green *Essex* CM15 23 J2
Peartree Green *Here.* HR1 29 F5
Pease Pottage RH11 13 F3
Peasedown St. John BA2 20 A6
Peasehill DE5 41 G1
Peaseland Green NR20 44 E4
Peasemore RG20 21 H4
Peasenhall IP17 35 H2
Peaslake GU5 22 D7
Peasley Cross WA9 48 E3
Peasmarsh *E.Suss.* TN31 14 D5
Peasmarsh *Surr.* GU3 22 C7
Peaston EH35 76 C4
Peastonbank EH34 76 C4
Peat Inn KY15 83 F7
Peathill AB43 99 H4
Peathrow DL13 62 C4
Peatling Magna LE8 41 H6
Peatling Parva LE17 41 H7
Peaton SY7 38 E7
Pebble Coombe KT20 23 F6
Pebmarsh CO9 34 C5
Pebworth CV37 30 C4
Pecket Well HX7 56 E7
Peckforton CW6 48 E7
Peckham SE15 23 G4
Peckleton LE9 41 G5
Pedham NR13 45 H4
Pedmore DY9 40 B7
Pedwell TA7 8 D1
Peebles EH45 76 A6
Peel *I.o.M.* IM5 54 B5
Peel *Lancs.* FY4 55 G6
Peening Quarter TN30 14 D5
Peggs Green LE67 41 G4
Pegsdon SG5 32 E5
Pegswood NE61 71 H5
Pegwell CT11 25 K5
Peighinn nan Aoireann HS8 84 C1
Peinchorran IV51 86 B1
Peinlich IV51 93 K6
Pelaw NE10 71 H7
Pelcomb SA62 16 C4
Pelcomb Bridge SA62 16 C4
Pelcomb Cross SA62 16 C4
Peldon CO5 34 D7
Pellon HX2 57 F7
Pelsall WS3 40 C5
Pelton DH2 62 D1
Pelutho CA7 60 C2
Pelynt PL13 4 C5
Pemberton WN5 48 E2
Pembrey (Pen-bre) SA16 17 H5
Pembridge HR6 28 C3
Pembroke (Penfro) SA71 16 C5
Pembroke Dock (Doc Penfro) SA72 16 C5
Pembury TN2 23 K7
Penallt NP25 28 E7
Penally SA70 16 E6
Penalt HR1 28 E6
Penare PL26 3 G4
Penarlâg (Hawarden) CH5 48 C6
Penarran SY16 38 A7
Penarth CF64 18 E4
Pen-bont Rhydybeddau SY23 37 F7
Penboyr SA44 17 G2
Pen-bre (Pembrey) SA16 17 H5
Penbryn SA44 26 B3
Pencader SA39 17 H2
Pen-cae SA47 26 D3
Pen-cae-cwm LL16 47 H6
Pencaenewydd LL53 36 D1
Pencaitland EH34 76 C4
Pencarnisiog LL63 46 B5
Pencarreg SA40 17 J1
Pencarrow PL32 4 B2
Pencelli LD3 27 K6
Pen-clawdd SA4 17 J6
Pencoed CF35 18 C3
Pencombe HR7 28 E3
Pencoyd HR2 28 E6
Pencraig *Here.* HR9 28 E6
Pencraig *Powys* SY10 37 K3
Pencraig (Old Radnor) *Powys* LD8 28 B3
Pendeen TR19 2 A5
Penderyn CF44 18 C1
Pendine (Pentywyn) SA33 17 F5
Pendlebury M27 49 G2
Pendleton BB7 56 C6
Pendock GL19 29 G5
Pendoggett PL29 4 A3
Pendomer BA22 8 E3
Pendoylan CF71 18 D4
Penegoes SY20 37 G5
Penelewey TR3 3 F4
Pen-ffordd SA66 16 D3
Penffordd-las (Staylittle) SY19 37 H6
Penfro (Pembroke) SA71 16 C5
Pengam NP12 18 E2
Penge SE20 23 G4
Pengenffordd LD3 28 A6
Pengorffwysfa LL68 46 C3
Pengover Green PL14 4 C4
Pen-groes-oped NP7 19 G1
Pengwern LL18 47 J5
Penhale TR12 2 D7
Penhallow TR4 2 E3
Penhalvean TR16 2 E5
Penhelig LL35 37 F6
Penhill SN2 20 E3
Penhow NP26 19 H2
Penhurst TN33 13 K5
Peniarth LL36 37 F5
Penicuik EH26 76 A4
Peniel SA32 17 H3
Penifiler IV51 93 K7
Peninver PA28 66 B1
Penisa'r Waun LL55 46 D6
Penisarcwm SY10 37 K4
Penishawain LD3 27 K5
Penistone S36 50 E2
Penjerrick TR11 2 E5
Penketh WA5 48 E4
Penkill KA26 67 G4
Penkridge ST19 40 B4
Penlean EX23 4 C1
Penley LL13 38 D2
Penllech LL53 36 B2
Penllergaer SA4 17 K6
Pen-llyn *I.o.A.* LL65 46 B4
Penllyn *V. of Glam.* CF71 18 C4
Pen-lôn LL61 46 C6
Penmachno LL24 47 F7
Penmaen SA3 17 J7
Penmaenan LL34 47 F5
Penmaenmawr LL34 47 F5
Penmaenpool (Llyn Penmaen) LL40 37 F4
Penmaen-Rhôs LL29 47 G5
Penmark CF62 18 D5
Penmon LL58 46 E4
Penmorfa LL49 36 E1
Penmynydd LL61 46 D5
Penn *Bucks.* HP10 22 C2
Penn *W.Mid.* WV4 40 A6
Penn Street HP7 22 C2
Pennal SY20 37 F5
Pennal-isaf SY20 37 G5
Pennan AB43 99 G4
Pennance TR16 2 E4
Pennant *Cere.* SY23 26 E2
Pennant *Powys* SY19 37 H6
Pennant Melangell SY10 37 K3
Pennar SA72 16 C5
Pennard SA3 17 J7
Pennerley SY5 38 C6
Penninghame DG8 64 D4
Pennington *Cumb.* LA12 55 F2
Pennington *Hants.* SO41 10 E5
Pennington Green WN2 49 F2
Pennorth LD3 28 A6
Pennsylvania SN14 20 A4
Penny Bridge LA12 55 G1
Pennycross PL5 4 E5
Pennyfuir PA34 79 K4
Pennygate NR12 45 H3
Pennyghael PA70 79 G5
Pennyglen KA19 67 G2
Pennygown PA72 79 H3
Pennymoor EX16 7 G4
Penny's Green NR16 45 F6
Pennyvenie KA6 67 J3
Penparc *Cere.* SA43 17 F1
Penparc *Pembs.* SA62 16 B2
Penparcau SY23 36 E7
Penpedairheol NP15 19 G1
Penpethy PL34 4 A2
Penpillick PL24 4 A5
Penpol TR3 3 F5
Penpoll PL22 4 B5
Penponds TR14 2 D5
Penpont *D. & G.* DG3 68 D4
Penpont *Powys* LD3 27 J6
Penprysg CF35 18 C3
Penquit PL21 5 G5
Penrherber SA38 17 F2
Penrhiw SA37 17 F1
Penrhiwceiber CF45 18 D2
Penrhiwgoch SA32 17 J4
Penrhiw-llan SA44 17 G1
Penrhiw-pâl SA44 17 G1
Penrhiwtyn SA11 18 A2
Penrhos *Gwyn.* LL53 36 C2
Penrhos *I.o.A.* LL65 46 A4
Penrhos *Mon.* NP15 28 D7
Penrhos *Powys* SA9 27 H7
Penrhos-garnedd LL57 46 D5
Penrhyn Bay (Bae Penrhyn) LL30 47 G4
Penrhyn-coch SY23 37 F7
Penrhyndeudraeth LL48 37 F2
Penrhyn-side LL30 47 G4
Penrhys CF43 18 D2
Penrice SA3 17 H7
Penrith CA11 61 G4
Penrose *Cornw.* PL27 3 F1
Penrose *Cornw.* PL15 4 C2
Penruddock CA11 60 F4
Penryn TR10 2 E5
Pensarn *Carmar.* SA31 17 H4
Pensarn *Conwy* LL22 47 H5
Pen-sarn *Gwyn.* LL45 36 E3
Pen-sarn *Gwyn.* LL54 36 D1
Pensax WR6 29 G2
Pensby CH61 48 B4
Penselwood BA9 9 G1
Pensford BS39 19 K5
Pensham WR10 29 J4
Penshaw DH4 62 E1
Penshurst TN11 23 J7
Pensilva PL14 4 C4
Pensnett DY5 40 B7
Penston EH33 76 C3
Pentewan PL26 4 A6
Pentir LL57 46 D6
Pentire TR7 2 E2
Pentireglaze PL27 3 G1
Pentlepoir SA69 16 E5
Pentlow CO10 34 C4
Pentlow Street CO10 34 C4
Pentney PE32 44 B4
Penton Mewsey SP11 21 G7
Pentonville N1 23 G3
Pentraeth LL75 46 D5
Pentre *Powys* SY10 37 K3
Pentre *Powys* SY16 37 K7
Pentre *Powys* LD8 28 B2
Pentre *Powys* SY15 38 B6
Pentre *Powys* SY16 38 A7
Pentre *R.C.T.* CF41 18 C2
Pentre *Shrop.* SY4 38 C4
Pentre *Shrop.* SY7 28 C1
Pentre *Wrex.* LL14 38 B1
Pentre *Wrex.* LL14 38 C1
Pentre Berw LL60 46 C5
Pentre Ffwrndan CH6 48 B5
Pentre Galar SA41 16 E2
Pentre Gwenlais SA18 17 K4
Pentre Gwynfryn LL45 36 E3
Pentre Halkyn CH8 48 B5
Pentre Isaf LL22 47 G6
Pentre Llanrhaeadr LL16 47 J6
Pentre Maelor LL13 38 C1
Pentre Meyrick CF71 18 C4
Pentre Poeth SA6 17 K6
Pentre Saron LL16 47 J6
Pentre-bach *Cere.* SA48 17 J1
Pentrebach *M.Tyd.* CF48 18 D1
Pentre-bach *Powys* LD3 27 J5
Pentrebach *R.C.T.* CF37 18 D3
Pentrebach *Swan.* SA4 17 J5
Pentre-bont LL25 47 F7
Pentre-bwlch LL11 38 A1
Pentrecagal SA38 17 G1
Pentre-celyn *Denb.* LL15 47 K7
Pentre-celyn *Powys* SY19 37 H5
Pentre-chwyth SA1 17 K6
Pentreclwydau SA11 18 B1
Pentre-cwrt SA44 17 G2
Pentre-Dolau-Honddu LD3 27 J4
Pentredwr *Denb.* LL20 38 A1
Pentre-dwr *Swan.* SA7 17 K6
Pentrefelin *Carmar.* SA19 17 K3
Pentrefelin *Cere.* SA48 17 K1
Pentrefelin *Conwy* LL28 47 G5
Pentrefelin *Gwyn.* LL52 36 E2
Pentrefelin *Powys* SY10 38 A3
Pentrefoelas LL24 47 G7
Pentregat SA44 26 C3
Pentreheyling SY15 38 B6
Pentre-llwyn-llŵyd LD2 27 J3
Pentre-llyn SY23 27 F1
Pentre-llyn-cymmer LL21 47 H7
Pentre-piod LL23 37 H2
Pentre-poeth NP10 19 F3
Pentre'r beirdd SY21 38 A4
Pentre'r Felin LL28 47 G6
Pentre'r-felin LD3 27 J5
Pentre-tafarn-y-fedw LL26 47 G6
Pentre-ty-gwyn SA20 27 H5
Pentrich DE5 51 F7
Pentridge SP5 10 B3
Pentwyn *Caerp.* CF81 18 E1
Pen-twyn *Caerp.* NP13 19 F1
Pentwyn *Cardiff* CF23 19 F3
Pen-twyn *Mon.* NP25 19 J1
Pentwyn-mawr NP11 18 E2
Pentyrch CF15 18 E3
Pentywyn (Pendine) SA33 17 F5
Penuwch SY25 26 E2
Penwithick PL26 4 A5
Penwood RG20 21 H5
Penwortham PR1 55 J7
Penwortham Lane PR1 55 J7
Penwyllt SA9 27 H7
Pen-y-banc SA19 17 K3
Pen-y-bont *Carmar.* SA20 27 G5
Pen-y-bont *Carmar.* SA33 17 G3
Pen-y-bont *Powys* SY21 37 K4
Pen-y-bont *Powys* SY10 38 B3
Penybont *Powys* LD1 28 A2
Pen-y-bont ar Ogwr (Bridgend) CF31 18 C4
Penybontfawr SY10 37 K3
Penybryn *Caerp.* CF82 18 E2
Pen-y-bryn *Gwyn.* LL40 37 F4
Pen-y-bryn *Pembs.* SA43 16 E1
Pen-y-bryn *Wrex.* LL14 38 B1
Pen-y-cae *Powys* SA9 27 H7
Penycae *Wrex.* LL14 38 B1
Pen-y-cae-mawr NP15 19 H2

Pen - Por

Pen-y-cefn **CH7** 47 K5
Pen-y-clawdd **NP25** 19 H1
Pen-y-coedcae **CF37** 18 D3
Penycwm **SA62** 16 B3
Pen-y-Darren **CF47** 18 D1
Pen-y-fai **CF31** 18 B3
Penyffordd *Flints.* **CH4** 48 C6
Pen-y-ffordd *Flints.* **CH8** 47 K4
Penyffridd **LL54** 46 D7
Pen-y-gaer **NP8** 28 A6
Pen-y-garn *Carmar.* **SA32** 17 J2
Pen-y-garn *Cere.* **SY24** 37 F7
Penygarn *Torfaen* **NP4** 19 F1
Penygarnedd **SY10** 38 A3
Pen-y-garreg **LD2** 27 K4
Pen-y-Graig *Gwyn.* **LL53** 36 B2
Penygraig *R.C.T.* **CF40** 18 D2
Penygroes *Carmar.* **SA14** 17 J4
Penygroes *Gwyn.* **LL54** 46 C7
Pen-y-Gwryd Hotel **LL55** 46 E7
Pen-y-lan **CF23** 18 E4
Penymynydd **CH4** 48 C6
Pen-y-parc **CH7** 48 B6
Pen-y-Park **HR3** 28 B4
Pen-yr-englyn **CF42** 18 C2
Pen-yr-heol *Mon.* **NP25** 28 D7
Penyrheol *Swan.* **SA4** 17 J6
Pen-y-sarn LL69 46 C3
Pen-y-stryt **LL11** 47 K7
Penywaun **CF44** 18 C1
Penzance TR18 2 B5
Penzance Heliport **TR18** 2 B5
Peopleton **WR10** 29 J3
Peover Heath **WA16** 49 G5
Peper Harow **GU8** 22 C7
Peplow **TF9** 39 F3
Pepper Arden **DL7** 62 D6
Pepper's Green **CM1** 33 K7
Perceton **KA11** 74 B6
Percie **AB31** 90 D5
Percyhorner **AB43** 99 H4
Perham Down **SP11** 21 F7
Periton **TA24** 7 H1
Perivale **UB6** 22 E3
Perkhill **AB31** 90 D4
Perkins Beach **SY5** 38 C5
Perkin's Village **EX5** 7 J6
Perlethorpe **NG22** 51 J5
Perran Downs **TR20** 2 C5
Perranarworthal **TR3** 2 E5
Perranporth TR6 2 E3
Perranuthnoe **TR20** 2 C6
Perranzabuloe **TR4** 2 E3
Perrott's Brook **GL7** 20 D1
Perry Barr **B42** 40 C6
Perry Crofts **B79** 40 E5
Perry Green *Essex* **CM77** 34 C6
Perry Green *Herts.* **SG10** 33 H7
Perry Green *Wilts.* **SN16** 20 C3
Perry Street **DA11** 24 C4
Perrymead **BA2** 20 A5
Pershall **ST21** 40 A3
Pershore WR10 29 J4
Persie House **PH10** 82 C2
Pert **AB30** 83 H1
Pertenhall **MK44** 32 D2
PERTH PH 82 C5
Perthcelyn **CF45** 18 D2
Perthy **SY12** 38 C2
Perton **WV6** 40 A6
Pestalozzi Children's Village **TN33** 14 C6
Peter Tavy **PL19** 5 F3
PETERBOROUGH PE 42 E6
Peterburn **IV21** 94 D3
Peterchurch **HR2** 28 C5
Peterculter AB14 91 G4
Peterhead AB42 99 K6
Peterlee SR8 63 F2
Peter's Green **LU2** 32 E7
Peters Marland **EX38** 6 C4
Peters Port (Port Pheadair) **HS7** 92 D7
Petersfield GU32 11 J2
Petersfinger **SP5** 10 C2
Peterstone Wentlooge **CF3** 19 F3
Peterston-super-Ely **CF5** 18 D4
Peterstow **HR9** 28 E6
Petham **CT4** 15 G2
Petrockstowe **EX20** 6 D5
Pett **TN35** 14 D6
Pettaugh **IP14** 35 F3
Petteril Green **CA11** 61 F2
Pettinain **ML11** 75 H6
Pettistree **IP13** 35 G3
Petton *Devon* **EX16** 7 J3
Petton *Shrop.* **SY4** 38 D3
Petts Wood **BR5** 23 H5
Petty **AB53** 91 F1
Pettycur **KY3** 76 A2
Pettymuick **AB41** 91 H2
Petworth GU28 12 C4
Pevensey BN24 13 K6
Pevensey Bay **BN24** 13 K6
Peverell **PL2** 4 E5
Pewsey SN9 20 E5
Pheasant's Hill **RG9** 22 A3
Phesdo **AB30** 90 E7
Philham **EX39** 6 A3
Philiphaugh **TD7** 69 K1
Phillack **TR27** 2 C5
Philleigh **TR2** 3 F5
Philpstoun **EH49** 75 J3
Phocle Green **HR9** 29 F6
Phoenix Green **RG27** 22 A6
Phones **PH20** 88 E5
Phorp **IV36** 97 H6
Pibsbury **TA10** 8 D2
Pica **CA14** 60 B4
Piccadilly Corner **IP20** 45 G7
Pickerells **CM5** 23 J1
Pickering YO18 58 D1
Pickering Nook **NE16** 62 C1
Picket Piece **SP11** 21 G7
Picket Post **BH24** 10 C4
Pickford Green **CV5** 40 E7
Pickhill **YO7** 57 J1
Picklescott **SY6** 38 D6
Pickletillem **KY16** 83 F5
Pickmere **WA16** 49 F5
Pickney **TA2** 7 K3
Pickstock **TF10** 39 G3
Pickston **PH1** 82 A5
Pickup Bank **BB3** 56 C7
Pickwell *Devon* **EX33** 6 C1
Pickwell *Leics.* **LE14** 42 A4
Pickworth *Lincs.* **NG34** 42 D2
Pickworth *Rut.* **PE9** 42 C4
Picton *Ches.W. & C.* **CH2** 48 D5
Picton *N.Yorks.* **TS15** 63 F6
Piddinghoe **BN9** 13 H6
Piddington *Bucks.* **HP14** 22 B2
Piddington *Northants.* **NN7** 32 B3
Piddington *Oxon.* **OX25** 31 H7
Piddlehinton **DT2** 9 G5
Piddletrenthide **DT2** 9 G5
Pidley **PE28** 33 G1
Piercebridge **DL2** 62 D5
Pierowall **KW17** 106 D3
Pigdon **NE61** 71 G5
Pike Hill **BB10** 56 D6
Pikehall **DE4** 50 D7
Pikeshill **SO43** 10 D4
Pilgrims Hatch **CM15** 23 J2
Pilham **DN21** 52 B3
Pill **BS20** 19 J4
Pillaton *Cornw.* **PL12** 4 D4
Pillaton *Staffs.* **ST19** 40 B4
Pillerton Hersey **CV35** 30 D4
Pillerton Priors **CV35** 30 D4
Pilleth **LD7** 28 B2
Pilley *Hants.* **SO41** 10 E5
Pilley *S.Yorks.* **S75** 51 F2
Pilling **PR3** 55 H5
Pilling Lane **FY6** 55 G5
Pillowell **GL15** 19 K1
Pilning **BS35** 19 J3
Pilsbury **SK17** 50 D6
Pilsdon **DT6** 8 D5
Pilsgate **PE9** 42 D5
Pilsley *Derbys.* **S45** 51 G6
Pilsley *Derbys.* **DE45** 50 E5
Pilson Green **NR13** 45 H4
Piltdown **TN22** 13 H4
Pilton *Devon* **EX31** 6 D2
Pilton *Northants.* **PE8** 42 D7
Pilton *Rut.* **LE15** 42 C5
Pilton *Som.* **BA4** 19 J7
Pilton *Swan.* **SA3** 17 H7
Pilton Green **SA3** 17 H7
Pimhole **BL9** 49 H1
Pimlico **HP3** 22 D1
Pimperne **DT11** 9 J4
Pin Mill **IP9** 35 G5
Pinchbeck **PE11** 43 F3
Pinchbeck Bars **PE11** 43 F3
Pinchbeck West **PE11** 43 F3
Pincheon Green **DN14** 51 J1
Pinchinthorpe **TS14** 63 G5
Pindon End **MK19** 31 J4
Pinehurst **SN25** 20 E3
Pinfold **L40** 48 C1
Pinged **SA16** 17 H5
Pinhay **DT7** 8 C5
Pinhoe **EX1** 7 H6
Pinkneys Green **SL6** 22 B3
Pinley Green **CV35** 30 D2
Pinminnoch **KA26** 67 F4
Pinmore **KA26** 67 G4
Pinn **EX10** 7 K7
Pinner HA5 22 E3
Pinner Green **HA5** 22 E2
Pinvin **WR10** 29 J4
Pinwherry **KA26** 67 F5
Pinxton **NG16** 51 G7
Pipe & Lyde **HR1** 28 E4
Pipe Gate **TF9** 39 G1
Pipe Ridware **WS15** 40 C4
Pipehill **WS13** 40 C5
Piperhall **PA20** 73 J5
Piperhill **IV12** 97 F6
Pipers Pool **PL15** 4 C2
Pipewell **NN14** 42 B7
Pippacott **EX31** 6 D2
Pipton **LD3** 28 A5
Pirbright **GU24** 22 C6
Pirnmill **KA27** 73 G6
Pirton *Herts.* **SG5** 32 E5
Pirton *Worcs.* **WR8** 29 H4
Pisgah **FK15** 81 J7
Pishill **RG9** 22 A3
Pistyll **LL53** 36 C1
Pitagowan **PH18** 81 K1
Pitblae **AB43** 99 H4
Pitcairngreen **PH1** 82 B5
Pitcairns **PH2** 82 B6
Pitcaple **AB51** 91 F2
Pitch Green **HP27** 22 A1
Pitch Place *Surr.* **GU3** 22 C6
Pitch Place *Surr.* **GU8** 12 B3
Pitchcombe **GL6** 20 B1
Pitchcott **HP22** 31 J6
Pitchford **SY5** 38 E5
Pitcombe **BA10** 9 F1
Pitcot **CF32** 18 B4
Pitcox **EH42** 76 E3
Pitcur **PH13** 82 D4
Pitfichie **AB51** 90 E3
Pitgrudy **IV25** 96 E2
Pitinnan **AB51** 91 F1
Pitkennedy **DD8** 83 G2
Pitkevy **KY6** 82 D7
Pitlessie **KY15** 82 E7
Pitlochry PH16 82 A2
Pitman's Corner **IP14** 35 F2
Pitmedden **AB41** 91 G2
Pitminster **TA3** 8 B3
Pitmuies **DD8** 83 G3
Pitmunie **AB51** 90 E3
Pitnacree **PH9** 82 A2
Pitney **TA10** 8 D2
Pitroddie **PH2** 82 D5
Pitscottie **KY15** 83 F6
Pitsea **SS13** 24 D3
Pitsford **NN6** 31 J2
Pitsford Hill **TA4** 7 J2
Pitstone **LU7** 32 C7
Pitt *Devon* **EX16** 7 J4
Pitt *Hants.* **SO22** 11 F2
Pittendreich **IV30** 97 J5
Pittentrail **IV28** 96 E1
Pittenweem **KY10** 83 G7
Pitteuchar **KY7** 76 A1
Pittington **DH6** 62 E2
Pittodrie House **AB51** 90 E2
Pitton *Swan.* **SA3** 17 H7
Pitton *Wilts.* **SP5** 10 D1
Pittulie **AB43** 99 H4
Pittville **GL52** 29 J6
Pity Me **DH1** 62 D2
Pityme **PL27** 3 G1
Pixey Green **IP21** 35 G1
Pixley **HR8** 29 F5
Place Newton **YO17** 58 E2
Plaidy **AB53** 99 F5
Plain Dealings **SA67** 16 D4
Plainfield **NE65** 70 E3
Plains **ML6** 75 F4
Plainsfield **TA5** 7 K2
Plaish **SY6** 38 E6
Plaistow *Gt.Lon.* **E13** 23 G3
Plaistow *W.Suss.* **RH14** 12 D3
Plaitford **SO51** 10 D3
Plaitford Green **SO51** 10 D2
Plas **SA32** 17 H3
Plas Gwynant **LL55** 46 E7
Plas Isaf **LL21** 37 K1
Plas Llwyd **LL18** 47 H5
Plas Llwyngwern **SY20** 37 G5
Plas Llysyn **SY17** 37 J6
Plas Nantyr **LL20** 38 A2
Plashett **SA33** 17 F5
Plasisaf **LL16** 47 H6
Plas-rhiw-Saeson **SY19** 37 J5
Plastow Green **RG19** 21 J5
Plas-yn-Cefn **LL17** 47 J5
Platt **TN15** 23 K6
Platt Bridge **WN2** 49 F2
Platt Lane **SY13** 38 E2
Platt's Heath **ME17** 14 D2
Plawsworth **DH2** 62 D2
Plaxtol **TN15** 23 K6
Play Hatch **RG4** 22 A4
Playden **TN31** 14 E5
Playford **IP6** 35 G4
Playing Place **TR3** 3 F4
Playley Green **GL19** 29 G5
Plealey **SY5** 38 D5
Plean **FK7** 75 G2
Pleasance **KY14** 82 D6
Pleasant Valley **CB11** 33 J5
Pleasington **BB2** 56 B7
Pleasley **NG19** 51 H6
Pleasleyhill **NG19** 51 H6
Pleck *Dorset* **DT9** 9 G3
Pleck *W.Mid.* **WS2** 40 B6
Pledgdon Green **CM22** 33 J6
Pledwick **WF2** 51 F1
Plemstall **CH2** 48 D5
Plenmeller **NE49** 70 C7
Pleshey **CM3** 33 K7
Plockton IV52 86 E1
Plocropol **HS3** 93 G2
Plomer's Hill **HP13** 22 B2
Plot Gate **TA11** 8 E1
Plough Hill **CV10** 41 F6
Plowden **SY7** 38 C7
Ploxgreen **SY5** 38 C5
Pluckley **TN27** 14 E3
Pluckley Thorne **TN27** 14 E3
Plucks Gutter **CT3** 25 J5
Plumbland **CA7** 60 C3
Plumbley **S20** 51 G4
Plumley **WA16** 49 G5
Plumpton *Cumb.* **CA11** 61 F3
Plumpton *E.Suss.* **BN7** 13 G5
Plumpton *Northants.* **NN12** 31 H4
Plumpton Green **BN7** 13 G5
Plumpton Head **CA11** 61 G3
Plumstead *Gt.Lon.* **SE18** 23 H4
Plumstead *Norf.* **NR11** 45 F2
Plumtree **NG12** 41 J2
Plungar **NG13** 42 A2
Plush **DT2** 9 G4
Plusha **PL15** 4 C2
Plushabridge **PL14** 4 D3
Plwmp **SA44** 26 C3
Plym Bridge **PL7** 5 F5
PLYMOUTH PL 4 E5
Plymouth City Airport **PL6** 5 F4
Plympton **PL7** 5 F5
Plymstock **PL9** 5 F5
Plymtree **EX15** 7 J5
Pockley **YO62** 58 C1
Pocklington **YO42** 58 E5
Pockthorpe **NR20** 44 E3
Pocombe Bridge **EX2** 7 G6
Pode Hole **PE11** 43 F3
Podimore **BA22** 8 E2
Podington **NN29** 32 C2
Podmore **ST21** 39 G2
Podsmead **GL2** 29 H7
Poffley End **OX29** 30 E7
Point Clear **CO16** 34 E7
Pointon **NG34** 42 E2
Polanach **PA38** 80 A2
Polapit Tamar **PL15** 6 B7
Polbae **DG8** 64 C3
Polbain **IV26** 102 B7
Polbathic **PL11** 4 D5
Polbeth **EH55** 75 J4
Poldean **DG10** 69 G4
Pole Moor **HD3** 50 C1
Polebrook **PE8** 42 D7
Polegate BN26 13 J6
Poles **IV25** 96 E2
Polesworth **B78** 40 E5
Polglass **IV26** 95 G1
Polgooth **PL26** 3 G3
Polgown **DG3** 68 C3
Poling **BN18** 12 D6
Poling Corner **BN18** 12 D6
Polkerris **PL24** 4 A5
Poll a' Charra **HS8** 84 C3
Polla **IV27** 103 F3
Pollardras **TR13** 2 D5
Polldubh **PH33** 80 C1
Pollie **IV28** 104 C7
Pollington **DN14** 51 J1
Polloch **PH37** 79 J1
Pollok **G53** 74 D4
Pollokshaws **G43** 74 D4
Pollokshields **G41** 74 D4
Polmassick **PL26** 3 G4
Polmont **FK2** 75 H3
Polnoon **G76** 74 D5
Polperro **PL13** 4 C5
Polruan **PL23** 4 B5
Polsham **BA5** 19 J7
Polstead **CO6** 34 D5
Polstead Heath **CO6** 34 D4
Poltalloch **PA31** 73 G1
Poltimore **EX4** 7 H6
Polton **EH18** 76 A4
Polwarth **TD10** 77 F5
Polyphant **PL15** 4 C2
Polzeath **PL27** 3 G1
Pomphlett **PL9** 5 F5
Pond Street **CB11** 33 H5
Ponders End **EN3** 23 G2
Pondersbridge **PE26** 43 F6
Ponsanooth **TR3** 2 E5
Ponsonby **CA20** 60 B6
Ponsongath **TR12** 2 E7
Ponsworthy **TQ13** 5 H3
Pont Aber **SA19** 27 G6
Pont Aberglaslyn **LL55** 36 E1
Pont ar Hydfer **LD3** 27 H6
Pont Crugnant **SY19** 37 H6
Pont Cyfyng **LL24** 47 F7
Pont Dolgarrog **LL32** 47 F6
Pont Pen-y-benglog **LL57** 46 E6
Pont Rhyd-sarn **LL23** 37 H3
Pont Rhyd-y-cyff **CF34** 18 B3
Pont Walby **SA11** 18 B1
Pont yr Alwen **LL21** 47 H7
Pontamman **SA18** 17 K4
Pontantwn **SA17** 17 H4
Pontardawe **SA8** 18 A1
Pontarddulais **SA4** 17 J5
Pontarfynach (Devil's Bridge) **SY23** 27 G1
Pontargothi **SA32** 17 J3
Pont-ar-llechau **SA19** 27 G6
Pontarsais **SA32** 17 H3
Pontblyddyn **CH7** 48 B6
Pontbren Llwyd **CF44** 18 C1
Pontefract WF8 57 K7
Ponteland **NE20** 71 G6
Ponterwyd **SY23** 37 G7
Pontesbury **SY5** 38 D5
Pontesbury Hill **SY5** 38 C5
Pontesford **SY5** 38 D5
Pontfadog **LL20** 38 B2
Pontfaen *Pembs.* **SA65** 16 D2
Pont-faen *Powys* **LD3** 27 J5
Pontgarreg **SA44** 26 C3
Pont-Henri **SA15** 17 H5
Ponthir **NP18** 19 G2
Ponthirwaun **SA43** 17 F1
Pont-iets (Pontyates) **SA15** 17 H5
Pontllanfraith **NP12** 18 E2
Pontlliw **SA4** 17 K5
Pontllyfni **LL54** 46 C7
Pontlottyn **CF81** 18 E1
Pontneddfechan **SA11** 18 C1
Pontrhydfendigaid **SY25** 27 G2
Pontrhydyfen **SA12** 18 A2
Pont-rhyd-y-groes **SY25** 27 G1
Pontrhydyrun **NP44** 19 F2
Pontrilas **HR2** 28 C6
Pontrobert **SY22** 38 A4
Pont-rug **LL55** 46 D6
Ponts Green **TN33** 13 K5
Pontshill **HR9** 29 F6
Pont-siân **SA44** 17 H1
Pontsticill **CF48** 27 K7
Pontwelly **SA44** 17 H2
Pontyates (Pont-iets) **SA15** 17 H5
Pontyberem **SA15** 17 J4
Pont-y-blew **LL14** 38 C2
Pontybodkin **CH7** 48 B7
Pontyclun CF72 18 D3
Pontycymer **CF32** 18 C2
Pontygwaith **CF43** 18 D2
Pontymister **NP11** 19 F2
Pontymoel **NP4** 19 F1
Pont-y-pant **LL25** 47 F7
Pontypool NP4 19 F1
Pontypridd CF37 18 D2
Pont-y-rhyl **CF32** 18 C3
Pontywaun **NP11** 19 F2
Pooksgreen **SO40** 10 E3
Pool *Cornw.* **TR15** 2 D4
Pool *W.Yorks.* **LS21** 57 H5
Pool Bank **LA11** 55 H1
Pool Green **WS9** 40 C5
Pool Head **HR1** 28 E3
Pool of Muckhart **FK14** 82 B7
Pool Quay **SY21** 38 B4
Pool Street **CO9** 34 B5
Poole BH15 10 B5
Poole Keynes **GL7** 20 C2
Poolend **ST13** 49 J7
Poolewe **IV22** 94 E3
Pooley Bridge **CA10** 61 F4
Pooley Street **IP22** 44 E7
Poolfold **ST8** 49 H7
Poolhill **GL18** 29 G6
Poolsbrook **S43** 51 G5
Poolthorne Farm **DN20** 52 D2
Pope Hill **SA62** 16 C4
Popeswood **RG42** 22 B5
Popham **SO21** 21 J7
Poplar **E14** 23 G3
Porchfield **PO30** 11 F5
Porin **IV6** 95 K6
Poringland **NR14** 45 G5
Porkellis **TR13** 2 D5
Porlock **TA24** 7 G1
Porlock Weir **TA24** 7 G1
Port Allen **PH2** 82 D5
Port Appin **PA38** 80 A3
Port Askaig **PA46** 72 C4
Port Bannatyne **PA20** 73 J4
Port Carlisle **CA7** 69 H7
Port Charlotte **PA48** 72 A5
Port Clarence **TS2** 63 G4
Port Driseach **PA21** 73 H3
Port e Vullen **IM7** 54 D4
Port Ellen **PA42** 72 B6
Port Elphinstone **AB51** 91 F3
Port Erin **IM9** 54 A7
Port Erroll **AB42** 91 J1
Port Eynon **SA3** 17 H7
Port Gaverne **PL29** 4 A2
Port Glasgow PA14 74 B3
Port Henderson **IV21** 94 D4
Port Isaac PL29 4 A2
Port Logan **DG9** 64 A6
Port Mòr **PH41** 85 K7
Port Mulgrave **TS13** 63 J5
Port na Craig **PH16** 82 A2
Port nan Giùran (Portnaguran) **HS2** 101 H4
Port nan Long **HS6** 92 D4
Port Nis (Port of Ness) **HS2** 101 H1
Port of Menteith **FK8** 81 G7
Port of Ness (Port Nis) **HS2** 101 H1
Port o'Warren **DG5** 65 J5
Port Penrhyn **LL57** 46 D5
Port Pheadair (Peters Port) **HS7** 92 D7
Port Quin **PL29** 3 G1
Port Ramsay **PA34** 79 K3
Port St. Mary **IM9** 54 B7
Port Solent **PO6** 11 H4
Port Sunlight **CH62** 48 C4
Port Talbot SA12 18 A2
Port Tennant **SA1** 17 K6
Port Wemyss **PA47** 72 A5
Port William **DG8** 64 D6
Portachoillan **PA29** 73 F5
Portavadie **PA21** 73 H4
Portbury **BS20** 19 J4
Portchester **PO16** 11 H4
Portencross **KA23** 73 K6
Portesham **DT3** 9 F6
Portessie **AB56** 98 C4
Portfield *Arg. & B.* **PA63** 79 J5
Portfield *W.Suss.* **PO19** 12 B6
Portfield Gate **SA62** 16 C4
Portgate **EX20** 6 C7
Portgordon **AB56** 98 B4
Portgower **KW8** 105 F7
Porth *Cornw.* **TR7** 3 F2
Porth *R.C.T.* CF39 18 D2
Porth Colmon **LL53** 36 A2
Porth Llechog (Bull Bay) **LL68** 46 C3
Porth Navas **TR11** 2 E6
Porthaethwy (Menai Bridge) LL59 46 D5
Porthallow *Cornw.* **TR12** 2 E6
Porthallow *Cornw.* **PL13** 4 C5
Porthcawl CF36 18 B4
Porthcothan **PL28** 3 F1
Porthcurno **TR19** 2 A6
Porthgain **SA62** 16 B2
Porthill **ST5** 40 A1
Porthkerry **CF62** 18 D5
Porthleven **TR13** 2 D6
Porthmadog LL49 36 E2
Porthmeor **TR20** 2 B5
Portholland **PL26** 3 G4
Porthoustock **TR12** 3 F6
Porthpean **PL26** 4 A5
Porthtowan **TR4** 2 D4
Porthyrhyd *Carmar.* **SA19** 27 G5
Porthyrhyd *Carmar.* **SA32** 17 J4
Porth-y-waen **SY10** 38 B3
Portincaple **G84** 74 A1
Portington **DN14** 58 D6
Portinnisherrich **PA33** 80 A6
Portinscale **CA12** 60 D4
Portishead **BS20** 19 H4

Por - Red

Portknockie AB56 98 C4
Portlethen AB12 91 H5
Portlethen Village AB12 91 H5
Portloe TR2 3 G5
Portlooe PL13 4 C5
Portmahomack IV20 97 G3
Portmeirion LL48 36 E2
Portmellon PL26 4 A6
Portmore SO41 10 E5
Port-na-Con IV27 103 G2
Portnacroish PA38 80 A3
Portnaguran (Port nan Giùran) HS2 101 H4
Portnahaven PA47 72 A5
Portnalong IV47 85 J1
Portnaluchaig PH39 86 C6
Portobello EH15 76 B3
Porton SP4 10 C1
Portpatrick DG9 64 A5
Portreath TR16 2 D4
Portree IV51 93 K7
Portscatho TR2 3 F5
Portsea PO1 11 H4
Portskerra KW14 104 D2
Portskewett NP26 19 J3
Portslade BN41 13 F6
Portslade-by-Sea BN41 13 F6
Portslogan DG9 64 A5
PORTSMOUTH PO 11 H5
Portsonachan PA33 80 B5
Portsoy AB45 98 D4
Portuairk PH36 79 F1
Portvoller HS2 101 H4
Portway *Here.* HR4 28 C4
Portway *Here.* HR4 28 D4
Portway *Here.* HR2 28 D5
Portway *Worcs.* B48 30 B1
Portwrinkle PL11 4 D5
Portyerrock DG8 64 E7
Posenhall TF12 39 F5
Poslingford CO10 34 B4
Postbridge PL20 5 G3
Postcombe OX9 22 A2
Postling CT21 15 G4
Post-mawr (Synod Inn) SA44 26 D3
Postwick NR13 45 G5
Potarch AB31 90 E5
Potsgrove MK17 32 C6
Pott Row PE32 44 B3
Pott Shrigley SK10 49 J5
Potten End HP4 22 D1
Potter Brompton YO12 59 F2
Potter Heigham NR29 45 J4
Potter Street CM17 23 H1
Pottergate Street NR16 45 F6
Potterhanworth LN4 52 D6
Potterhanworth Booths LN4 52 D6
Potterne SN10 20 C6
Potterne Wick SN10 20 D6
Potternewton LS7 57 J6
Potters Bar EN6 23 F1
Potters Crouch AL2 22 E1
Potter's Green CV2 41 F7
Potters Marston LE9 41 G6
Potterspury NN12 31 J4
Potterton *Aber.* AB23 91 H3
Potterton *W.Yorks.* LS15 57 K6
Pottle Street BA12 20 B7
Potto DL6 63 F6
Potton SG19 33 F4
Pott's Green CO6 34 D6
Poughill *Cornw.* EX23 6 A5
Poughill *Devon* EX17 7 G5
Poulshot SN10 20 C6
Poulton GL7 20 E1
Poulton-le-Fylde FY6 55 G6
Pound Bank WR14 29 G4
Pound Green *E.Suss.* TN22 13 J4
Pound Green *Suff.* CB8 34 B3
Pound Green *Worcs.* DY12 29 G1
Pound Hill RH10 13 F3
Pound Street RG20 21 H5
Poundbury DT1 9 F5
Poundffald SA4 17 J6
Poundfield TN6 13 J3
Poundgate TN6 13 H4
Poundland KA26 67 F5
Poundon OX27 31 H6
Poundsbridge TN11 23 J7
Poundsgate TQ13 5 H3
Poundstock EX23 4 C1
Povey Cross RH6 23 F7
Pow Green HR8 29 G4
Powburn NE66 71 F2
Powderham EX6 7 H7
Powerstock DT6 8 E5
Powfoot DG12 69 G7
Powick WR2 29 H3
Powler's Piece EX22 6 B4
Powmill FK14 75 J1
Poxwell DT2 9 G6
Poyle SL3 22 D4
Poynings BN45 13 F5
Poyntington DT9 9 F2
Poynton *Ches.E.* SK12 49 J4
Poynton *Tel. & W.* TF6 38 E4
Poynton Green SY4 38 E4
Poyntzfield IV7 96 E5
Poys Street IP17 35 H1
Poyston SA62 16 C4
Poyston Cross SA62 16 C4
Poystreet Green IP30 34 D3
Praa Sands TR13 2 C6
Pratis KY8 82 E7
Pratt's Bottom BR6 23 H5
Praze-an-Beeble TR14 2 D5
Predannack Wollas TR12 2 D7
Prees SY13 38 E2
Prees Green SY13 38 E2
Prees Heath SY13 38 E2
Prees Higher Heath SY13 38 E2
Prees Lower Heath SY13 38 E2
Preesall FY6 55 G5
Preesgweene SY10 38 B2
Prenbrigog CH7 48 B6
Prendergast SA61 16 C4
Prendwick NE66 71 F2
Pren-gwyn SA44 17 H1
Prenteg LL49 36 E1
Prenton CH42 48 C4
Prescot L34 48 D3
Prescott *Devon* EX15 7 J4
Prescott *Shrop.* SY4 38 D3
Presley IV36 97 H6
Pressen TD12 77 G7
Prestatyn LL19 47 J4
Prestbury *Ches.E.* SK10 49 J5
Prestbury *Glos.* GL52 29 J6
Presteigne (Llanandras) LD8 28 C2
Presthope TF13 38 E6
Prestleigh BA4 19 K7
Prestolee M26 49 G2
Preston *B. & H.* BN1 13 G6
Preston *Devon* TQ12 5 J3
Preston *Dorset* DT3 9 G6
Preston *E.Loth.* EH40 76 D3
Preston *E.Riding* HU12 59 H6
Preston *Glos.* HR8 29 F5
Preston *Glos.* GL7 20 D1
Preston *Herts.* SG4 32 E6
Preston *Kent* ME13 25 G5
Preston *Kent* CT3 25 J5
PRESTON *Lancs.* PR 55 J7
Preston *Northumb.* NE67 71 G1
Preston *Rut.* LE15 42 B5
Preston *Sc.Bord.* TD11 77 F5
Preston *Shrop.* SY4 38 E4
Preston *Som.* TA4 7 J2
Preston *Suff.* CO10 34 D3
Preston *Torbay* TQ3 5 J4
Preston *Wilts.* SN15 20 D4
Preston Bagot B95 30 C2
Preston Bissett MK18 31 H5
Preston Bowyer TA4 7 K3
Preston Brockhurst SY4 38 E3
Preston Brook WA7 48 E4
Preston Candover RG25 21 K7
Preston Capes NN11 31 G3
Preston Deanery NN7 31 J3
Preston Gubbals SY4 38 D4
Preston on Stour CV37 30 D4
Preston on the Hill WA4 48 E4
Preston on Wye HR2 28 C4
Preston Plucknett BA20 8 E3
Preston upon the Weald Moors TF6 39 F4
Preston Wynne HR1 28 E4
Preston-le-Skerne DL5 62 E4
Prestonpans EH32 76 B3
Preston-under-Scar DL8 62 B7
Prestwich M25 49 H2
Prestwick *Northumb.* NE20 71 G6
Prestwick *S.Ayr.* KA9 67 H1
Prestwold LE12 41 H3
Prestwood *Bucks.* HP16 22 B1
Prestwood *Staffs.* ST14 40 D1
Price Town CF32 18 C2
Prickwillow CB7 43 J7
Priddy BA5 19 J6
Priest Hill PR3 56 B6
Priest Hutton LA6 55 J2
Priest Weston SY15 38 B6
Priestcliffe SK17 50 D5
Priestland KA17 74 D7
Priestwood DA13 24 C5
Primethorpe LE9 41 H6
Primrose Green NR9 44 E4
Primrose Hill NW1 23 F3
Princes End DY4 40 B6
Princes Gate SA67 16 E4
Princes Risborough HP27 22 B1
Princethorpe CV23 31 F1
Princetown *Caerp.* NP22 28 A7
Princetown *Devon* PL20 5 F3
Prior Muir KY16 83 G6
Prior's Frome HR1 28 E5
Priors Halton SY8 28 D1
Priors Hardwick CV47 31 F3
Priors Marston CV47 31 F3
Prior's Norton GL2 29 H6
Priors Park GL20 29 H5
Priorslee TF2 39 G4
Priory Wood HR3 28 B4
Priston BA2 19 K5
Pristow Green NR16 45 F7
Prittlewell SS0 24 E3
Privett GU34 11 H2
Prixford EX31 6 D2
Proaig PA44 72 C5
Probus TR2 3 F4
Protstonhill AB45 99 G4
Prudhoe NE42 71 F7
Prussia Cove TR20 2 C6
Pubil PH15 81 F3
Publow BS39 19 K5
Puckeridge SG11 33 G6
Puckington TA19 8 C3
Pucklechurch BS16 19 K4
Pucknall SO51 10 E2
Puckrup GL20 29 H5
Puddinglake CW10 49 G6
Puddington *Ches.W. & C.* CH64 48 C5
Puddington *Devon* EX16 7 G4
Puddlebrook GL17 29 F7
Puddledock NR17 44 E6
Puddletown DT2 9 G5
Pudleston HR6 28 E3
Pudsey LS28 57 H6
Pulborough RH20 12 D5
Puldagon KW1 105 J4
Puleston TF10 39 G3
Pulford CH4 48 C7
Pulham DT2 9 G4
Pulham Market IP21 45 F7
Pulham St. Mary IP21 45 G7
Pulley SY3 38 D5
Pulloxhill MK45 32 D5
Pulrossie IV25 96 E3
Pulverbatch SY5 38 D5
Pumpherston EH53 75 J4
Pumsaint SA19 17 K1
Puncheston SA62 16 D3
Puncknowle DT2 8 E6
Punnett's Town TN21 13 K4
Purbrook PO7 11 H4
Purewell BH23 10 C5
Purfleet RM19 23 J4
Puriton TA7 19 G7
Purleigh CM3 24 E1
Purley CR8 23 G5
Purley on Thames RG8 21 K4
Purlogue SY7 28 B1
Purlpit SN12 20 B5
Purls Bridge PE15 43 H7
Purse Caundle DT9 9 F3
Purslow SY7 38 C7
Purston Jaglin WF7 51 G1
Purtington TA20 8 C4
Purton *Glos.* GL13 19 K1
Purton *Glos.* GL15 19 K1
Purton *Wilts.* SN5 20 D3
Purton Stoke SN5 20 D2
Purves Hall TD10 77 F6
Pury End NN12 31 J4
Pusey SN7 21 G2
Putley HR8 29 F5
Putley Green HR8 29 F5
Putney SW15 23 F4
Putsborough EX33 6 C1
Puttenham *Herts.* HP23 32 B7
Puttenham *Surr.* GU3 22 C7
Puttock End CO10 34 C4
Putts Corner EX10 7 K6
Puxton BS24 19 H5
Pwll SA15 17 H5
Pwllcrochan SA71 16 C5
Pwlldefaid LL53 36 A3
Pwll-glas LL15 47 K7
Pwllgloyw LD3 27 K5
Pwllheli LL53 36 C2
Pwll-Mawr CF3 19 F4
Pwllmeyric NP16 19 J2
Pwll-trap SA33 17 F4
Pwll-y-glaw SA12 18 A2
Pye Corner *Herts.* CM20 33 H7
Pye Corner *Kent* ME17 14 D3
Pye Corner *Newport* NP18 19 G3
Pye Green WS12 40 B4
Pyecombe BN45 13 F5
Pyle *Bridgend* CF33 18 B3
Pyle *I.o.W.* PO38 11 F7
Pyleigh TA4 7 K2
Pylle BA4 9 F1
Pymoor (Pymore) CB6 43 H7
Pymore (Pymoor) *Cambs.* CB6 43 H7
Pymore *Dorset* DT6 8 D5
Pyrford GU22 22 D6
Pyrford Green GU22 22 D6
Pyrton OX49 21 K2
Pytchley NN14 32 B1
Pyworthy EX22 6 B5

Q

Quabbs LD7 38 B7
Quadring PE11 43 F2
Quadring Eaudike PE11 43 F2
Quainton HP22 31 J6
Quarff ZE2 107 N9
Quarley SP11 21 F7
Quarndon DE22 41 F1
Quarr Hill PO33 11 G5
Quarrier's Village PA11 74 B4
Quarrington NG34 42 D1
Quarrington Hill DH6 62 E3
Quarry Bank DY5 40 B7
Quarrybank CW6 48 E6
Quarrywood IV30 97 J5
Quarter ML3 75 F5
Quatford WV15 39 G6
Quatt WV15 39 G7
Quebec DH7 62 C2
Quedgeley GL2 29 H7
Queen Adelaide CB7 43 J7
Queen Camel BA22 8 E2
Queen Charlton BS31 19 K5
Queen Dart EX16 7 G4
Queen Oak SP8 9 G1
Queen Street TN12 23 K7
Queenborough ME11 25 F4
Queen's Bower PO36 11 G6
Queen's Head SY11 38 C3
Queensbury *Gt.Lon.* HA3 22 E3
Queensbury *W.Yorks.* BD13 57 G6
Queensferry (South Queensferry) *Edin.* EH30 75 K3
Queensferry *Flints.* CH5 48 C6
Queenzieburn G65 74 E3
Quemerford SN11 20 D5
Quendale ZE2 107 M11
Quendon CB11 33 J5
Queniborough LE7 41 J4
Quenington GL7 20 E1
Quernmore LA2 55 J3
Queslett B43 40 C6
Quethiock PL14 4 D4
Quholm KW16 106 B6
Quick's Green RG8 21 J4
Quidenham NR16 44 E7
Quidhampton RG25 21 J6
Quidinish (Cuidhtinis) HS3 93 F3
Quilquox AB41 91 H1
Quina Brook SY4 38 E2
Quindry KW17 106 D8
Quine's Hill IM4 54 C6
Quinhill PA29 73 F5
Quinton *Northants.* NN7 31 J3
Quinton *W.Mid.* B32 40 B7
Quinton Green NN7 31 J3
Quintrell Downs TR8 3 F2
Quixhill ST14 40 D1
Quoditch EX21 6 C6
Quoig PH7 81 K5
Quoiggs House FK15 81 K7
Quoisley SY13 38 E1
Quorn (Quorndon) LE12 41 H4
Quorndon (Quorn) LE12 41 H4
Quothquan ML12 75 H7
Quoyloo KW16 106 B5
Quoys ZE2 107 Q1
Quoys of Reiss KW1 105 J3

R

Raasay IV40 94 B7
Raby CH63 48 C5
Rachan ML12 75 K7
Rachub LL57 46 E6
Rackenford EX16 7 G4
Rackham RH20 12 D5
Rackheath NR13 45 G4
Racks DG1 69 F6
Rackwick *Ork.* KW16 106 B8
Rackwick *Ork.* KW17 106 D3
Radbourne DE6 40 E2
Radcliffe *Gt.Man.* M26 49 G2
Radcliffe *Northumb.* NE65 71 H3
Radcliffe on Trent NG12 41 J2
Radclive MK18 31 H5
Radcot OX18 21 F2
Raddington TA4 7 J3
Radernie KY15 83 F7
Radford *B. & N.E.Som.* BA2 19 K6
Radford *Nott.* NG7 41 H1
Radford *Oxon.* OX7 31 F6
Radford *W.Mid.* CV6 41 F7
Radford Semele CV31 30 E2
Radipole DT3 9 F6
Radlett WD7 22 E1
Radley OX14 21 J2
Radley Green CM4 24 C1
Radmore Green CW6 48 E7
Radnage HP14 22 A2
Radstock BA3 19 K6
Radstone NN13 31 G4
Radway CV35 30 E4
Radway Green CW1 49 G7
Radwell *Bed.* MK43 32 D3
Radwell *Herts.* SG7 33 F5
Radwinter CB10 33 K5
Radyr CF15 18 E3
Raechester NE19 70 E5
Raemoir House AB31 90 E5
Raffin IV27 102 C5
Rafford IV36 97 H6
Ragdale LE14 41 J3
Ragged Appleshaw SP11 21 G7
Raglan NP15 19 H1
Ragnall NG22 52 B5
Rahoy PA34 79 H2
Rain Shore OL12 49 H1
Rainford WA11 48 D2
Rainham *Gt.Lon.* RM13 23 J3
Rainham *Med.* ME8 24 E5
Rainhill L35 48 D3
Rainhill Stoops L35 48 E3
Rainow SK10 49 J5
Rainsough M27 49 G2
Rainton YO7 57 J2
Rainworth NG21 51 H7
Raisbeck CA10 61 H6
Raise CA9 61 J2
Rait PH2 82 D5
Raithby *Lincs.* PE23 53 G6
Raithby *Lincs.* LN11 53 G4
Rake GU33 12 B4
Raleigh's Cross TA23 7 J2
Ram SA48 17 J1
Ram Alley SN8 21 F5
Ram Lane TN26 14 E3
Ramasaig IV55 93 G7
Rame *Cornw.* TR10 2 E5
Rame *Cornw.* PL10 4 E6
Rampisham DT2 8 E4
Rampside LA13 55 F3
Rampton *Cambs.* CB24 33 H2
Rampton *Notts.* DN22 52 B5
Ramsbottom BL0 49 G1
Ramsbury SN8 21 F4
Ramscraigs KW6 105 G6
Ramsdean GU32 11 J2
Ramsdell RG26 21 J6
Ramsden OX7 30 E7
Ramsden Bellhouse CM11 24 D2
Ramsden Heath CM11 24 D2
Ramsey *Cambs.* PE26 43 F7
Ramsey *Essex* CO12 35 G5
Ramsey *I.o.M.* IM8 54 D4
Ramsey Forty Foot PE26 43 G7
Ramsey Heights PE26 43 F7
Ramsey Island *Essex* CM0 25 F1
Ramsey Island *Pembs.* SA62 16 A3
Ramsey Mereside PE26 43 F7
Ramsey St. Mary's PE26 43 F7
Ramsgate CT11 25 K5
Ramsgate Street NR24 44 E2
Ramsgill HG3 57 G2
Ramsholt IP12 35 H4
Ramshorn ST10 40 C1
Ramsnest Common GU8 12 C3
Ranachan PH36 79 J1
Ranais (Ranish) HS2 101 G5
Ranby *Lincs.* LN8 53 F5
Ranby *Notts.* DN22 51 J4
Rand LN8 52 E5
Randwick GL6 20 B1
Rangemore DE13 40 D3
Rangeworthy BS37 19 K3
Ranish (Ranais) HS2 101 G5
Rankinston KA6 67 J2
Rank's Green CM3 34 B7
Ranmoor S10 51 F4
Rannoch School PH17 81 G2
Ranochan PH38 86 E6
Ranscombe TA24 7 H1
Ranskill DN22 51 J4
Ranton ST18 40 A3
Ranton Green ST18 40 A3
Ranworth NR13 45 H4
Rapness KW17 106 E3
Rapps TA19 8 C3
Rascarrel DG7 65 H6
Rash LA10 56 B1
Rashwood WR9 29 J2
Raskelf YO61 57 K2
Rassau NP23 28 A7
Rastrick HD6 57 G7
Ratagan IV40 87 F3
Ratby LE6 41 H5
Ratcliffe Culey CV9 41 F6
Ratcliffe on Soar NG11 41 G3
Ratcliffe on the Wreake LE7 41 J4
Ratford Bridge SA62 16 B4
Ratfyn SP4 20 E7
Rathen AB43 99 H4
Rathillet KY15 82 E5
Rathliesbeag PH34 87 J6
Rathmell BD24 56 D3
Ratho EH28 75 K3
Ratho Station EH28 75 K3
Rathven AB56 98 C4
Ratley OX15 30 E4
Ratling CT3 15 H2
Ratlinghope SY5 38 D6
Ratsloe EX4 7 H6
Rattar KW14 105 H1
Ratten Row *Cumb.* CA5 60 E2
Ratten Row *Lancs.* PR3 55 H5
Rattery TQ10 5 H4
Rattlesden IP30 34 D3
Rattray PH10 82 C3
Raughton Head CA5 60 E2
Raunds NN9 32 C1
Ravenfield S65 51 G3
Ravenglass CA18 60 B7
Raveningham NR14 45 H6
Raven's Green CO7 35 F6
Ravenscar YO13 63 J2
Ravensdale IM7 54 C4
Ravensden MK44 32 D3
Ravenshaw BD23 56 E5
Ravenshayes EX5 7 H5
Ravenshead NG15 51 H7
Ravensmoor CW5 49 F7
Ravensthorpe *Northants.* NN6 31 H1
Ravensthorpe *W.Yorks.* WF13 57 H7
Ravenstone *Leics.* LE67 41 G4
Ravenstone *M.K.* MK46 32 B3
Ravenstonedale CA17 61 J6
Ravenstruther ML11 75 H6
Ravensworth DL11 62 C6
Raw YO22 63 J2
Rawcliffe *E.Riding* DN14 58 C7
Rawcliffe *York* YO30 58 B4
Rawcliffe Bridge DN14 58 C7
Rawdon LS19 57 H6
Rawmarsh S62 51 G3
Rawnsley WS12 40 C4
Rawreth SS11 24 D2
Rawridge EX14 8 B4
Rawson Green DE56 41 F1
Rawtenstall BB4 56 D7
Rawyards ML6 75 F4
Raxton AB41 91 G1
Raydon IP7 34 E5
Raylees NE19 70 E4
Rayleigh SS6 24 E2
Raymond's Hill EX13 8 C5
Rayne CM77 34 B6
Rayners Lane HA5 22 E3
Raynes Park SW20 23 F5
Reach CB25 33 J2
Read BB12 56 C6
READING RG 22 A4
Reading Green IP21 35 F1
Reading Street TN30 14 E4
Reagill CA10 61 H5
Rearquhar IV25 96 E2
Rearsby LE7 41 J4
Rease Heath CW5 49 F7
Reaster KW1 105 H2
Reaveley NE66 71 F2
Reawick ZE2 107 M8
Reay KW14 104 E2
Reculver CT6 25 J5
Red Ball EX16 7 J4
Red Bull ST7 49 H7

Red - Ros

Red Dial CA7 60 D2
Red Hill *Hants.* PO9 11 J3
Red Hill *Warks.* B49 30 C3
Red Lodge IP28 33 K1
Red Lumb OL11 49 H1
Red Oaks Hill CB10 33 J5
Red Point IV21 94 D5
Red Post *Cornw.* EX23 6 A5
Red Post *Devon* TQ9 5 J4
Red Rail HR2 28 E6
Red Rock WN2 48 E2
Red Roses SA34 17 F4
Red Row NE61 71 H4
Red Street ST5 49 H7
Red Wharf Bay (Traeth Coch) LL75 46 D4
Redberth SA70 16 D5
Redbourn AL3 32 E7
Redbourne DN21 52 C3
Redbrook *Glos.* NP25 28 E7
Redbrook *Wrex.* SY13 38 E1
Redbrook Street TN26 14 E4
Redburn *High.* IV16 96 C5
Redburn *High.* IV12 97 G7
Redburn *Northumb.* NE47 70 C7
Redcar TS10 63 H4
Redcastle *Angus* DD11 83 H2
Redcastle *High.* IV6 96 C7
Redcliff Bay BS20 19 H4
Redcloak AB39 91 G6
Reddingmuirhead FK2 75 H3
Reddish SK5 49 H3
Redditch B97 30 B2
Rede IP29 34 C3
Redenhall IP20 45 G7
Redesmouth NE48 70 D5
Redford *Aber.* AB30 91 F7
Redford *Angus* DD11 83 G3
Redford *Dur.* DL13 62 B3
Redford *W.Suss.* GU29 12 B4
Redgrave IP22 34 E1
Redheugh DD8 83 F1
Redhill *Aber.* AB51 90 E1
Redhill *Aber.* AB32 91 F4
Redhill *Moray* AB54 98 D6
Redhill *N.Som.* BS40 19 H5
Redhill *Notts.* NG5 41 H1
REDHILL *Surr.* RH 23 F6
Redhill Aerodrome & Heliport RH1 23 F7
Redhouse *Aber.* AB33 90 D2
Redhouse *Arg. & B.* PA29 73 G4
Redhouses PA44 72 B4
Redisham NR34 45 J7
Redland *Bristol* BS6 19 J4
Redland *Ork.* KW17 106 C5
Redlingfield IP23 35 F1
Redlynch *Som.* BA10 9 G1
Redlynch *Wilts.* SP5 10 D2
Redmarley D'Abitot GL19 29 G5
Redmarshall TS21 62 E4
Redmile NG13 42 A2
Redmire DL8 62 B7
Redmoor PL30 4 A4
Rednal SY11 38 C3
Redpath TD4 76 D7
Redruth TR15 2 D4
Redscarhead EH45 76 A6
Redshaw ML11 68 D1
Redstone Bank SA67 16 E4
Redwick *Newport* NP26 19 H3
Redwick *S.Glos.* BS35 19 J3
Redworth DL5 62 D4
Reed SG8 33 G5
Reed End SG8 33 G5
Reedham NR13 45 J5
Reedley BB10 56 D6
Reedness DN14 58 D7
Reef (Riof) HS2 100 D4
Reepham *Lincs.* LN3 52 D5
Reepham *Norf.* NR10 44 E3
Reeth DL11 62 B7
Regaby IM7 54 D4
Regil BS40 19 J5
Regoul IV12 97 F6
Reiff IV26 102 B7
Reigate RH2 23 F6
Reighton YO14 59 H2
Reinigeadal (Rhenigidale) HS3 100 E7
Reisgill KW3 105 H5
Reiss KW1 105 J3
Rejerrah TR8 3 F3
Releath TR14 2 D5
Relubbus TR20 2 C5
Relugas IV36 97 G7
Remenham RG9 22 A3
Remenham Hill RG9 22 A3
Remony PH15 81 J3
Rempstone LE12 41 H3
Rendcomb GL7 30 B7
Rendham IP17 35 H2
Rendlesham IP12 35 H3
Renfrew PA4 74 D4
Renhold MK41 32 D3
Renishaw S21 51 G5
Rennington NE66 71 H2
Renton G82 74 B3
Renwick CA10 61 G2
Repps NR29 45 J4
Repton DE65 41 F3
Rescobie DD8 83 G2
Rescorla PL26 4 A5
Resipole PH36 79 J1
Resolis IV7 96 D5
Resolven SA11 18 B1
Resourie PH37 86 E7
Respryn PL30 4 B4
Reston TD14 77 G4
Restormel PL22 4 B4
Reswallie DD8 83 G2
Reterth TR9 3 G2
Retew PL26 3 G3
Retford (East Retford) DN22 51 K4
Rettendon CM3 24 D2
Rettendon Place CM3 24 D2
Retyn TR8 3 F3
Revesby PE22 53 F6
Revesby Bridge PE22 53 G6
Rew TQ13 5 H3
Rew Street PO31 11 F5
Rewe *Devon* EX5 7 H6
Rewe *Devon* EX5 7 G6
Reybridge SN15 20 C5
Reydon IP18 35 J1
Reydon Smear IP18 35 K1
Reymerston NR9 44 E5
Reynalton SA68 16 D5
Reynoldston SA3 17 H7
Rezare PL15 4 D3
Rhadyr NP15 19 G1
Rhaeadr Gwy (Rhaeadr) LD6 27 J2
Rhandirmwyn SA20 27 G4
Rhaoine IV28 96 D1
Rhayader (Rhaeadr Gwy) LD6 27 J2
Rhedyn LL53 36 B2
Rhegreanoch IV27 102 C7
Rheindown IV4 96 C7
Rhelonie IV24 96 C2
Rhemore PA34 79 G2
Rhenigidale (Reinigeadal) HS3 100 E7
Rheola SA11 18 B1
Rhes-y-cae CH8 47 K6
Rhewl *Denb.* LL20 38 A1
Rhewl *Denb.* LL15 47 K6
Rhewl *Shrop.* SY10 38 C2
Rhian IV27 103 H7
Rhicarn IV27 102 C6
Rhiconich IV27 102 E3
Rhicullen IV18 96 D4
Rhidorroch IV26 95 H2
Rhifail KW11 104 C4
Rhigos CF44 18 C1
Rhilochan IV28 96 E1
Rhinduie IV3 96 C7
Rhireavach IV23 95 G2
Rhiroy IV23 95 H3
Rhiston SY15 38 B6
Rhiw LL53 36 B3
Rhiwabon (Ruabon) LL14 38 C1
Rhiwargor SY10 37 J3
Rhiwbina CF14 18 E3
Rhiwbryfdir LL41 37 F1
Rhiwderin NP10 19 F3
Rhiwinder CF39 18 D3
Rhiwlas *Gwyn.* LL57 46 D6
Rhiwlas *Gwyn.* LL23 37 J2
Rhiwlas *Powys* SY10 38 B2
Rhode TA5 8 B1
Rhodes Minnis CT4 15 G3
Rhodesia S80 51 H5
Rhodiad-y-brenin SA62 16 A3
Rhodmad SY23 26 E1
Rhonadale PA28 73 F7
Rhonehouse (Kelton Hill) DG7 65 H5
Rhoose CF62 18 D5
Rhos *Carmar.* SA44 17 G2
Rhos *N.P.T.* SA8 18 A1
Rhos Common SY22 38 B4
Rhosaman SA18 27 G7
Rhoscolyn LL65 46 A5
Rhoscrowther SA71 16 C5
Rhosesmor CH7 48 B6
Rhos-fawr LL53 36 C2
Rhosgadfan LL54 46 D7
Rhos-goch *I.o.A.* LL66 46 C4
Rhosgoch *Powys* LD2 28 A4
Rhos-hill SA43 16 E1
Rhoshirwaun LL53 36 A3
Rhoslan LL52 36 D1
Rhoslefain LL36 36 E5
Rhosllanerchrugog LL14 38 B1
Rhoslligwy LL70 46 C4
Rhosmaen SA19 17 K3
Rhosmeirch LL77 46 C5
Rhosneigr LL64 46 B5
Rhosnesni LL13 48 C7
Rhôs-on-Sea LL28 47 G4
Rhossili SA3 17 H7
Rhosson SA62 16 A3
Rhostrehwfa LL77 46 C5
Rhostryfan LL54 46 C7
Rhostyllen LL14 38 C1
Rhos-y-bol LL68 46 C4
Rhos-y-brithdir SY22 38 A3
Rhosycaerau SA64 16 C2
Rhos-y-garth SY23 27 F1
Rhos-y-gwaliau LL23 37 J2
Rhos-y-llan LL53 36 B2
Rhos-y-mawn LL22 47 G6
Rhos-y-Meirch LD7 28 B2
Rhu G84 74 A2
Rhuallt LL17 47 J5
Rhubodach PA20 73 J3
Rhuddall Heath CW6 48 E6
Rhuddlan LL18 47 J5
Rhue IV26 95 H2
Rhulen LD2 28 A4
Rhumach PH39 86 C6
Rhunahaorine PA29 73 F6
Rhuthun (Ruthin) LL15 47 K7
Rhyd *Gwyn.* LL48 37 F1
Rhyd *Powys* SY17 37 J5
Rhydaman (Ammanford) SA18 17 K4
Rhydargaeau SA32 17 H3
Rhydcymerau SA19 17 J2
Rhyd-Ddu LL54 46 D7
Rhydding SA10 18 A2
Rhydgaled LL16 47 H6
Rhydlanfair LL24 47 G7
Rhydlewis SA44 17 G1
Rhydlios LL53 36 A2
Rhydlydan *Conwy* LL24 47 G7
Rhydlydan *Powys* SY16 37 K6
Rhydolion LL53 36 B3
Rhydowen SA44 17 H1
Rhyd-Rosser SY23 26 E2
Rhydspence HR3 28 B4
Rhydtalog CH7 48 B7
Rhyd-uchaf LL23 37 H2
Rhyd-wen LL23 37 J3
Rhyd-wyn LL65 46 B4
Rhyd-y-ceirw CH7 48 B7
Rhyd-y-clafdy LL53 36 C2
Rhydycroesau SY10 38 B2
Rhydyfelin *Cere.* SY23 26 E1
Rhydyfelin *R.C.T.* CF37 18 D3
Rhyd-y-foel LL22 47 H5
Rhyd-y-fro SA8 18 A1
Rhyd-y-groes LL57 46 D6
Rhydymain LL40 37 H3
Rhydymwyn CH7 48 B6
Rhyd-yr-onnen LL36 37 F5
Rhyd-y-sarn LL41 37 F1
Rhydywrach SA34 16 E4
Rhyl LL18 47 J4
Rhymney NP22 18 E1
Rhyn SY11 38 C2
Rhynd PH2 82 C5
Rhynie *Aber.* AB54 90 C2
Rhynie *High.* IV20 97 F4
Ribbesford DY12 29 G1
Ribchester PR3 56 B6
Ribigill IV27 103 H3
Riby DN37 52 E2
Riccall YO19 58 C6
Riccarton KA1 74 C7
Richards Castle SY8 28 D2
Richings Park SL0 22 D4
Richmond *Gt.Lon.* TW9 22 E4
Richmond *N.Yorks.* DL10 62 C6
Richmond *S.Yorks.* S13 51 G4
Rich's Holford TA4 7 K2
Rickarton AB39 91 G6
Rickerscote ST17 40 B3
Rickford BS40 19 H6
Rickinghall IP22 34 E1
Rickleton NE38 62 D1
Rickling CB11 33 H5
Rickling Green CB11 33 J6
Rickmansworth WD3 22 D2
Riddell TD6 70 A1
Riddings DE55 51 G7
Riddlecombe EX18 6 E4
Riddlesden BD20 57 F5
Ridge *Dorset* BH20 9 J6
Ridge *Herts.* EN6 23 F1
Ridge *Wilts.* SP3 9 J1
Ridge Green RH1 23 G7
Ridge Lane CV10 40 E6
Ridgebourne LD1 27 K2
Ridgeway S12 51 G4
Ridgeway Cross WR13 29 G4
Ridgeway Moor S12 51 G4
Ridgewell CO9 34 B4
Ridgewood TN22 13 H4
Ridgmont MK43 32 C5
Riding Gate BA9 9 G2
Riding Mill NE44 71 F7
Ridley TN15 24 C5
Ridleywood LL13 48 C7
Ridlington *Norf.* NR28 45 H2
Ridlington *Rut.* LE15 42 B5
Ridsdale NE48 70 E5
Riechip PH8 82 B3
Rievaulx YO62 58 B1
Rift House TS25 63 F3
Rigg *D. & G.* DG16 69 H7
Rigg *High.* IV51 94 B6
Riggend ML6 75 F4
Rigifa KW1 105 J1
Rigmaden Park LA6 56 B1
Rigsby LN13 53 H5
Rigside ML11 75 G7
Riley Green PR5 56 B7
Rileyhill WS13 40 D4
Rilla Mill PL17 4 C3
Rillaton PL17 4 C3
Rillington YO17 58 E2
Rimington BB7 56 D5
Rimpton BA22 9 F2
Rimswell HU19 59 K7
Rinaston SA62 16 C3
Ring o' Bells L40 48 D1
Ringford DG7 65 G5
Ringinglow S11 50 E4
Ringland NR8 45 F4
Ringles Cross TN22 13 H4
Ringmer BN8 13 H5
Ringmore *Devon* TQ7 5 G6
Ringmore *Devon* TQ14 5 K3
Ringorm AB38 97 K7
Ring's End PE13 43 G5
Ringsfield NR34 45 J7
Ringsfield Corner NR34 45 J7
Ringshall *Herts.* HP4 32 C7
Ringshall *Suff.* IP14 34 E3
Ringshall Stocks IP14 34 E3
Ringstead *Norf.* PE36 44 B1
Ringstead *Northants.* NN14 32 C1
Ringwood BH24 10 C4
Ringwould CT14 15 J3
Rinloan AB35 89 K4
Rinmore AB33 90 C3
Rinnigill KW16 106 C8
Rinsey TR13 2 C6
Riof (Reef) HS2 100 D4
Ripe BN8 13 J6
Ripley *Derbys.* DE5 51 F7
Ripley *Hants.* BH23 10 C5
Ripley *N.Yorks.* HG3 57 H3
Ripley *Surr.* GU23 22 D6
Riplingham HU15 59 F6
Ripon HG4 57 J2
Rippingale PE10 42 E3
Ripple *Kent* CT14 15 J3
Ripple *Worcs.* GL20 29 H5
Ripponden HX6 50 C1
Risabus PA42 72 B6
Risbury HR6 28 E3
Risby *E.Riding* HU17 59 G6
Risby *Suff.* IP28 34 B2
Risca NP11 19 F2
Rise HU11 59 H5
Riseden TN17 14 C4
Risegate PE11 43 F3
Riseholme LN2 52 C5
Riseley *Bed.* MK44 32 D2
Riseley *W'ham* RG7 22 A5
Rishangles IP23 35 F2
Rishton BB1 56 C6
Rishworth HX6 50 C1
Risinghurst OX3 21 J1
Risley *Derbys.* DE72 41 G2
Risley *Warr.* WA3 49 F3
Risplith HG4 57 H3
Rispond IV27 103 G2
Rivar SN8 21 G5
Rivenhall CM8 34 C7
Rivenhall End CM8 34 C7
River *Kent* CT17 15 H3
River *W.Suss.* GU28 12 C4
River Bank CB7 33 J2
River Bridge TA7 19 G7
Riverford Bridge TQ9 5 H4
Riverhead TN13 23 J6
Riverside CF11 18 E4
Riverton EX32 6 E2
Riverview Park DA12 24 C4
Rivington BL6 49 F1
Roa Island LA13 55 F3
Roach Bridge PR5 55 J7
Road Green NR15 45 G6
Road Weedon NN7 31 H3
Roade NN7 31 J3
Roadhead CA6 70 A6
Roadside *High.* KW12 105 G2
Roadside *Ork.* KW17 106 F3
Roadside of Kinneff DD10 91 G7
Roadwater TA23 7 J2
Roag IV55 93 H7
Roast Green CB11 33 H5
Roath CF24 18 E4
Roberton *S.Lan.* ML12 68 E1
Roberton *Sc.Bord.* TD9 69 K2
Robertsbridge TN32 14 C5
Robertstown *Moray* AB38 97 K7
Robertstown *R.C.T.* CF44 18 D1
Roberttown WF15 57 G7
Robeston Cross SA73 16 B5
Robeston Wathen SA67 16 D4
Robeston West SA73 16 B5
Robin Hood *Derbys.* DE45 50 E5
Robin Hood *Lancs.* WN6 48 E1
Robin Hood *W.Yorks.* LS26 57 J7
Robin Hood Doncaster Sheffield Airport DN10 51 J3
Robin Hood's Bay YO22 63 J2
Robinhood End CO9 34 B5
Robins GU29 12 B4
Roborough *Devon* EX19 6 D4
Roborough *Plym.* PL6 5 F4
Roby L36 48 D3
Roby Mill WN8 48 E2
Rocester ST14 40 D2
Roch SA62 16 B3
Roch Bridge SA62 16 B3
Roch Gate SA62 16 B3
Rochallie PH10 82 C2
Rochdale OL16 49 H1
Roche PL26 3 G2
Rochester *Med.* ME1 24 D5
Rochester *Northumb.* NE19 70 D4
Rochford *Essex* SS4 24 E2
Rochford *Worcs.* WR15 29 F2
Rock *Cornw.* PL27 3 G1
Rock *Northumb.* NE66 71 H2
Rock *Worcs.* DY14 29 G1
Rock Ferry CH42 48 C4
Rockbeare EX5 7 J6
Rockbourne SP6 10 C3
Rockcliffe *Cumb.* CA6 69 J7
Rockcliffe *D. & G.* DG5 65 J5
Rockcliffe Cross CA6 69 J7
Rockfield *Arg. & B.* PA29 73 G5
Rockfield *High.* IV20 97 G3
Rockfield *Mon.* NP25 28 D7
Rockford BH24 10 C4
Rockhampton GL13 19 K2
Rockhead PL33 4 A2
Rockingham LE16 42 B6
Rockland All Saints NR17 44 D6
Rockland St. Mary NR14 45 H5
Rockland St. Peter NR17 44 D6
Rockley SN8 20 E4
Rockside PA49 72 A4
Rockwell End RG9 22 A3
Rockwell Green TA21 7 K4
Rodborough GL5 20 B1
Rodbourne SN16 20 C3
Rodbridge Corner CO10 34 C4
Rodd LD8 28 C2
Roddam NE66 71 F1
Rodden DT3 9 F6
Rode BA11 20 B6
Rode Heath ST7 49 H7
Rodeheath SK11 49 H6
Rodel (Roghadal) HS5 93 F3
Roden TF6 38 E4
Rodhuish TA24 7 J2
Rodington SY4 38 E4
Rodington Heath SY4 38 E4
Rodley GL14 29 G7
Rodmarton GL7 20 C2
Rodmell BN7 13 H6
Rodmersham ME9 25 F5
Rodmersham Green ME9 25 F5
Rodney Stoke BS27 19 H6
Rodsley DE6 40 E1
Rodway TA5 19 F7
Roe Cross SK14 49 J3
Roe Green SG9 33 G5
Roecliffe YO51 57 J3
Roehampton SW15 23 F4
Roesound ZE2 107 M6
Roffey RH12 12 E3
Rogart IV28 96 E1
Rogate GU31 12 B4
Rogerstone NP10 19 F3
Roghadal (Rodel) HS5 93 F3
Rogiet NP26 19 H3
Rokemarsh OX10 21 K2
Roker SR6 63 F1
Rollesby NR29 45 J4
Rolleston *Leics.* LE7 42 A5
Rolleston *Notts.* NG23 51 K7
Rollestone SP3 20 D7
Rolleston-on-Dove DE13 40 E3
Rolston HU18 59 J5
Rolstone BS24 19 G5
Rolvenden TN17 14 D4
Rolvenden Layne TN17 14 D4
Romaldkirk DL12 62 A4
Romanby DL7 62 E7
Romanno Bridge EH46 75 K6
Romansleigh EX36 7 F3
Romesdal IV51 93 K6
Romford *Dorset* BH31 10 B4
ROMFORD *Gt.Lon.* RM 23 J3
Romiley SK6 49 J3
Romney Street TN15 23 J5
Romsey SO51 10 E2
Romsley *Shrop.* WV15 39 G7
Romsley *Worcs.* B62 29 J1
Rona IV40 94 C6
Ronachan PA29 73 F5
Ronague IM9 54 B6
Ronnachmore PA43 72 B5
Rood End B67 40 C7
Rookhope DL13 62 A2
Rookley PO38 11 G6
Rookley Green PO38 11 G6
Rooks Bridge BS26 19 G6
Rook's Nest TA4 7 J2
Rookwith HG4 57 H1
Roos HU12 59 J6
Roose LA13 55 F3
Roosebeck LA12 55 F3
Roosecote LA13 55 F3
Rootham's Green MK44 32 D3
Rootpark ML11 75 H5
Ropley SO24 11 H1
Ropley Dean SO24 11 H1
Ropley Soke SO24 11 H1
Ropsley NG33 42 C2
Rora AB42 99 J5
Rorandle AB51 90 E3
Rorrington SY15 38 C5
Rosarie AB55 98 B6
Rose TR4 2 E3
Rose Ash EX36 7 F3
Rose Green *Essex* CO6 34 D6
Rose Green *W.Suss.* PO21 12 C7
Rose Hill TN22 13 H5
Roseacre *Kent* ME14 14 C2
Roseacre *Lancs.* PR4 55 H6
Rosebank ML8 75 G6
Rosebrough NE67 71 G1
Rosebush SA66 16 D3
Rosecare EX23 4 B1
Rosecliston TR8 3 F3
Rosedale Abbey YO18 63 J7
Roseden NE66 71 F1
Rosehall IV27 96 B1
Rosehearty AB43 99 H4
Rosehill *Aber.* AB34 90 D5
Rosehill *Shrop.* SY4 38 D4
Roseisle IV30 97 J5
Roselands BN22 13 K6
Rosemarket SA73 16 C5
Rosemarkie IV10 96 E6
Rosemary Lane EX15 7 K4
Rosemount *P. & K.* PH10 82 C3
Rosemount *S.Ayr.* KA9 67 H1
Rosenannon PL30 3 G2
Rosenithon TR12 3 F6
Rosepool SA62 16 B4
Rosevean PL26 4 A5
Roseville WV14 40 B6
Rosewell EH24 76 A4
Roseworth TS19 63 F4
Rosgill CA10 61 G5
Roshven PH38 86 D7
Roskhill IV55 93 H7
Roskorwell TR12 2 E6
Rosley CA7 60 E2
Roslin EH25 76 A4
Rosliston DE12 40 E4
Rosneath G84 74 A2
Ross *D. & G.* DG6 65 G6
Ross *Northumb.* NE70 77 K7
Ross *P. & K.* PH6 81 J5

Ros - Sal

Ross Priory G83 74 C2
Rossdhu House G83 74 B2
Rossett LL12 48 C7
Rossett Green HG2 57 J4
Rosside LA12 55 F2
Rossie Farm School DD10 83 H2
Rossie Ochill PH2 82 B6
Rossie Priory PH14 82 D4
Rossington DN11 51 J3
Rosskeen IV18 96 D5
Rossmore BH12 10 B5
Ross-on-Wye HR9 29 F6
Roster KW3 105 H5
Rostherne WA16 49 G4
Rosthwaite *Cumb.* CA12 60 D5
Rosthwaite *Cumb.* LA20 55 F1
Roston DE6 40 D1
Rosudgeon TR20 2 C6
Rosyth KY11 75 K2
Rothbury NE65 71 F3
Rotherby LE14 41 J4
Rotherfield TN6 13 J3
Rotherfield Greys RG9 22 A3
Rotherfield Peppard RG9 22 A3
Rotherham S60 51 G3
Rothersthorpe NN7 31 J3
Rotherwick RG27 22 A6
Rothes AB38 97 K7
Rothesay PA20 73 J4
Rothiebrisbane AB53 91 F1
Rothienorman AB51 91 F1
Rothiesholm KW17 106 F5
Rothley *Leics.* LE7 41 H4
Rothley *Northumb.* NE61 71 F5
Rothney AB52 90 E2
Rothwell *Lincs.* LN7 52 E3
Rothwell *Northants.* NN14 42 B7
Rothwell *W.Yorks.* LS26 57 J7
Rotsea YO25 59 G4
Rottal DD8 82 E1
Rotten Row *Bucks.* RG9 22 A3
Rotten Row *W.Mid.* B93 30 C1
Rottingdean BN2 13 G6
Rottington CA28 60 A5
Roud PO38 11 G6
Roudham NR16 44 D7
Rough Close ST3 40 B2
Rough Common CT2 15 G2
Rougham *Norf.* PE32 44 C3
Rougham *Suff.* IP30 34 D2
Rougham Green IP30 34 D2
Roughburn PH31 87 K6
Roughlee BB9 56 D5
Roughley B75 40 D6
Roughton *Lincs.* LN10 53 F6
Roughton *Norf.* NR11 45 G2
Roughton *Shrop.* WV15 39 G6
Round Bush WD25 22 E2
Roundbush Green CM6 33 J7
Roundham TA18 8 D4
Roundhay LS8 57 J6
Roundstreet Common RH14 12 D4
Roundway SN10 20 D5
Rous Lench WR11 30 B3
Rousay KW17 106 C4
Rousdon DT7 8 B5
Rousham OX25 31 F6
Rousham Gap OX25 31 F6
Routenburn KA30 73 K4
Routh HU17 59 G5
Rout's Green HP14 22 A2
Row *Cornw.* PL30 4 A3
Row *Cumb.* LA8 55 H1
Row *Cumb.* CA10 61 H3
Row Heath CO16 35 F7
Row Town KT15 22 D5
Rowanburn DG14 69 K6
Rowardennan Lodge G63 74 B1
Rowarth SK22 50 C4
Rowbarton TA2 8 B2
Rowberrow BS25 19 H6
Rowchoish G63 80 E7
Rowde SN10 20 C5
Rowden EX20 6 E6
Rowen LL32 47 F5
Rowfields DE6 40 D1
Rowfoot NE49 70 B7
Rowhedge CO5 34 E6
Rowhook RH12 12 E3
Rowington CV35 30 D2
Rowland DE45 50 E5
Rowland's Castle PO9 11 J3
Rowlands Gill NE39 62 C1
Rowledge GU10 22 B7
Rowlestone HR2 28 C6
Rowley *Devon* EX36 7 F4
Rowley *Dur.* DH8 62 B2
Rowley *Shrop.* SY5 38 C5
Rowley Park ST17 40 B3
Rowley Regis B65 40 B7
Rowly GU5 22 D7
Rowner PO13 11 G4
Rowney Green B48 30 B1
Rownhams SO16 10 E3
Rowrah CA26 60 B5
Rowsham HP22 32 B7
Rowsley DE4 50 E6
Rowstock OX11 21 H3
Rowston LN4 52 D7
Rowthorne S44 51 G6
Rowton *Ches.W. & C.* CH3 48 D6
Rowton *Shrop.* SY5 38 C4
Rowton *Tel. & W.* TF6 39 F4
Roxburgh TD5 77 F7
Roxby *N.Lincs.* DN15 52 C1
Roxby *N.Yorks.* TS13 63 J5
Roxton MK44 32 E3
Roxwell CM1 24 C1
Royal British Legion Village ME20 14 C2
Royal Leamington Spa CV32 30 E2
Royal Oak L39 48 D2
ROYAL TUNBRIDGE WELLS TN 13 J3
Royal Wootton Bassett SN4 20 D3
Roybridge PH31 87 J6
Roydon *Essex* CM19 33 H7
Roydon *Norf.* IP22 44 E7
Roydon *Norf.* PE32 44 B3
Roydon Hamlet CM19 23 H1
Royston *Herts.* SG8 33 G4
Royston *S.Yorks.* S71 51 F1
Royton OL2 49 J2
Rozel JE3 3 K6
Ruabon (Rhiwabon) LL14 38 C1
Ruaig PA77 78 B3
Ruan Lanihorne TR2 3 F4
Ruan Major TR12 2 D7
Ruan Minor TR12 2 E7
Ruanaich PA76 78 D5
Ruardean GL17 29 F7
Ruardean Hill GL17 29 F7
Ruardean Woodside GL17 29 F7
Rubery B45 29 J1
Ruckcroft CA4 61 G2
Ruckinge TN26 15 F4
Ruckland LN11 53 G5
Rucklers Lane HP3 22 D1
Ruckley SY5 38 E5
Rudbaxton SA62 16 C3
Rudby TS15 63 F6
Rudchester NE15 71 G7
Ruddington NG11 41 H2
Ruddlemoor PL26 4 A5
Rudford GL2 29 G6
Rudge BA11 20 B6
Rudgeway BS35 19 K3
Rudgwick RH12 12 D3
Rudhall HR9 29 F6
Rudheath CW9 49 F5
Rudley Green CM3 24 E1
Rudloe SN13 20 B5
Rudry CF83 18 E3
Rudston YO25 59 G3
Rudyard ST13 49 J7
Rufford L40 48 D1
Rufforth YO23 58 B4
Ruffside DH8 62 A1
Rugby CV21 31 G1
Rugeley WS15 40 C4
Ruilick IV4 96 C7
Ruishton TA3 8 B2
Ruisigearraidh HS6 92 E3
Ruislip HA4 22 D3
Ruislip Gardens HA4 22 D3
Ruislip Manor HA4 22 E3
Rum PH43 85 J5
Rumbling Bridge KY13 75 J1
Rumburgh IP19 45 H7
Rumford PL27 3 F1
Rumleigh PL20 4 E4
Rumney CF3 19 F4
Rumwell TA4 7 K3
Runacraig FK18 81 G6
Runcorn WA7 48 E4
Runcton PO20 12 B6
Runcton Holme PE33 44 A5
Rundlestone PL20 5 F3
Runfold GU10 22 B7
Runhall NR9 44 E5
Runham *Norf.* NR29 45 J4
Runham *Norf.* NR30 45 K5
Runnington TA21 7 K3
Runsell Green CM3 24 D1
Runshaw Moor PR7 48 E1
Runswick Bay TS13 63 K5
Runtaleave DD8 82 D1
Runwell SS11 24 D2
Ruscombe *Glos.* GL6 20 B1
Ruscombe *W'ham* RG10 22 A4
Rush Green *Gt.Lon.* RM7 23 J3
Rush Green *Herts.* SG4 33 F6
Rushall *Here.* HR8 29 F5
Rushall *Norf.* IP21 45 F7
Rushall *W.Mid.* WS4 40 C5
Rushall *Wilts.* SN9 20 E6
Rushbrooke IP30 34 C2
Rushbury SY6 38 E6
Rushden *Herts.* SG9 33 G5
Rushden *Northants.* NN10 32 C2
Rushford *Devon* PL19 4 E3
Rushford *Norf.* IP24 44 D7
Rushgreen WA13 49 F4
Rushlake Green TN21 13 K5
Rushmere NR33 45 J7
Rushmere St. Andrew IP5 35 G4
Rushmoor GU10 22 B7
Rushock WR9 29 H1
Rusholme M13 49 H3
Rushton *Ches.W. & C.* CW6 48 E6
Rushton *Northants.* NN14 42 B7
Rushton *Shrop.* TF6 39 F5
Rushton Spencer SK11 49 J6
Rushwick WR2 29 H3
Rushy Green BN8 13 H5
Rushyford DL17 62 D4
Ruskie FK8 81 H7
Ruskington NG34 52 D7
Rusko DG7 65 F5
Rusland LA12 55 G1
Rusper RH12 13 F3
Ruspidge GL14 29 F7
Russ Hill RH6 23 F7
Russel IV54 94 E7
Russell Green CM3 34 B7
Russell's Green TN33 14 C6
Russell's Water RG9 22 A2
Russel's Green IP21 35 G1
Rusthall TN4 13 J3
Rustington BN16 12 D6
Ruston YO13 59 F1
Ruston Parva YO25 59 G3
Ruswarp YO21 63 K6
Rutherend ML10 74 E5
Rutherford TD5 76 E7
Rutherglen G73 74 E4
Ruthernbridge PL30 4 A4
Ruthers of Howe KW1 105 J2
Ruthin (Rhuthun) *Denb.* LL15 47 K7
Ruthin *V. of Glam.* CF35 18 C4
Ruthrieston AB10 91 H4
Ruthven *Aber.* AB54 98 D6
Ruthven *Angus* PH12 82 D3
Ruthven *High.* IV13 89 F1
Ruthven *High.* PH21 88 E5
Ruthvoes TR9 3 G2
Ruthwaite CA7 60 D3
Ruthwell DG1 69 F7
Ruyton-XI-Towns SY4 38 C3
Ryal NE20 71 F6
Ryal Fold BB3 56 B7
Ryall *Dorset* DT6 8 D5
Ryall *Worcs.* WR8 29 H4
Ryarsh ME19 23 K6
Rydal LA22 60 E6
Ryde PO33 11 G5
Rydon EX22 6 B5
Rye TN31 14 E5
Rye Foreign TN31 14 D5
Rye Harbour TN31 14 E6
Rye Park EN11 23 G1
Rye Street WR13 29 G5
Ryebank SY4 38 E2
Ryeford HR9 29 F6
Ryehill *Aber.* AB52 90 E2
Ryehill *E.Riding* HU12 59 J7
Ryhall PE9 42 D4
Ryhill WF4 51 F1
Ryhope SR2 63 F1
Rylands NG9 41 H2
Rylstone BD23 56 E4
Ryme Intrinseca DT9 8 E3
Ryther LS24 58 B6
Ryton *Glos.* GL18 29 G5
Ryton *N.Yorks.* YO17 58 D2
Ryton *Shrop.* TF11 39 G5
Ryton *T. & W.* NE40 71 G7
Ryton-on-Dunsmore CV8 30 E1

S

Saasaig IV44 86 C4
Sabden BB7 56 C6
Sabden Fold BB12 56 D6
Sackers Green CO10 34 D5
Sacombe SG12 33 G7
Sacombe Green SG12 33 G7
Sacriston DH7 62 D2
Sadberge DL2 62 E5
Saddell PA28 73 F7
Saddington LE8 41 J6
Saddle Bow PE34 44 A4
Sadgill LA8 61 F6
Saffron Walden CB10 33 J5
Sageston SA70 16 D5
Saham Hills IP25 44 D5
Saham Toney IP25 44 C5
Saighdinis HS6 92 D5
Saighton CH3 48 D6
St. Abbs TD14 77 H4
St. Agnes TR5 2 E3
ST. ALBANS AL 22 E1
St. Allen TR4 3 F3
St. Andrews KY16 83 G6
St. Andrews Major CF64 18 E4
St. Anne GY9 3 K4
St. Anne's FY8 55 G7
St. Ann's DG11 69 F4
St. Ann's Chapel *Cornw.* PL18 4 E3
St. Ann's Chapel *Devon* TQ7 5 G6
St. Anthony TR2 3 F5
St. Anthony-in-Meneage TR12 2 E6
St. Anthony's Hill BN23 13 K6
St. Arvans NP16 19 J2
St. Asaph (Llanelwy) LL17 47 J5
St. Athan CF62 18 D5
St. Aubin JE3 3 J7
St. Audries TA4 7 K1
St. Austell PL25 4 A5
St. Bees CA27 60 A5
St. Blazey PL24 4 A5
St. Blazey Gate PL24 4 A5
St. Boswells TD6 76 D7
St. Brelade JE3 3 J7
St. Breock PL27 3 G1
St. Breward PL30 4 A3
St. Briavels GL15 19 J1
St. Brides SA62 16 B4
St. Brides Major CF32 18 B4
St. Bride's Netherwent NP26 19 H3
St. Brides Wentlooge NP10 19 F3
St. Bride's-super-Ely CF5 18 D4
St. Budeaux PL5 4 E5
St. Buryan TR19 2 B6
St. Catherine BA1 20 A5
St. Catherines PA25 80 C7
St. Clears (Sanclêr) SA33 17 F4
St. Cleer PL14 4 C4
St. Clement *Chan.I.* JE2 3 K7
St. Clement *Cornw.* TR1 3 F4
St. Clether PL15 4 C2
St. Colmac PA20 73 J4
St. Columb Major TR9 3 G2
St. Columb Minor TR7 3 F2
St. Columb Road TR9 3 G3
St. Combs AB43 99 J4
St. Cross South Elmham IP20 45 G7
St. Cyrus DD10 83 J1
St. Davids *Fife* KY11 75 K2
St. David's *P. & K.* PH7 82 A5
St. David's (Tyddewi) *Pembs.* SA62 16 A3
St. Day TR16 2 E4
St. Decumans TA23 7 J1
St. Dennis PL26 3 G3
St. Denys SO17 11 F3
St. Dogmaels (Llandudoch) SA43 16 E1
St. Dogwells SA62 16 C3
St. Dominick PL12 4 E4
St. Donats CF61 18 C5
St. Edith's Marsh SN15 20 C5
St. Endellion PL29 3 G1
St. Enoder TR8 3 F3
St. Erme TR4 3 F4
St. Erney PL12 4 D5
St. Erth TR27 2 C5
St. Erth Praze TR27 2 C5
St. Ervan PL27 3 F1
St. Eval PL27 3 F2
St. Ewe PL26 3 G4
St. Fagans CF5 18 E4
St. Fergus AB42 99 J5
St. Fillans PH6 81 H5
St. Florence SA70 16 D5
St. Gennys EX23 4 B1
St. George *Bristol* BS5 19 K4
St. George *Conwy* LL22 47 H5
St. Georges *N.Som.* BS22 19 G5
St. George's *Tel. & W.* TF2 39 G4
St. George's *V. of Glam.* CF5 18 D4
St. Germans PL12 4 D5
St. Giles in the Wood EX38 6 D4
St. Giles on the Heath PL15 6 B6
St. Harmon LD6 27 J1
St. Helen Auckland DL14 62 C4
St. Helena NR10 45 F4
St. Helen's *E.Suss.* TN34 14 D6
St. Helens *I.o.W.* PO33 11 H6
St. Helens *Mersey.* WA10 48 E3
St. Helier *Chan.I.* JE2 3 J7
St. Helier *Gt.Lon.* SM5 23 F5
St. Hilary *Cornw.* TR20 2 C5
St. Hilary *V. of Glam.* CF71 18 D4
St. Hill RH19 13 G3
St. Ibbs SG4 32 E6
St. Illtyd NP13 19 F1
St. Ippollitts SG4 32 E6
St. Ishmael SA17 17 G5
St. Ishmael's SA62 16 B5
St. Issey PL27 3 G1
St. Ive PL14 4 D4
St. Ives *Cambs.* PE27 33 G1
St. Ives *Cornw.* TR26 2 C4
St. Ives *Dorset* BH24 10 C4
St. James South Elmham IP19 45 H7
St. John *Chan.I.* JE3 3 J6
St. John *Cornw.* PL11 4 E5
St. John's *Gt.Lon.* SE4 23 G4
St. John's *I.o.M.* IM4 54 B5
St. John's *Surr.* GU21 22 C6
St. John's *Worcs.* WR2 29 H3
St. John's Chapel *Devon* EX31 6 D3
St. John's Chapel *Dur.* DL13 61 K3
St. John's Fen End PE14 43 J4
St. John's Hall DL13 62 B3
St. John's Highway PE14 43 J4
St. John's Kirk ML12 75 H7
St. John's Town of Dalry DG7 68 B5
St. Judes IM7 54 C4
St. Just TR19 2 A5
St. Just in Roseland TR2 3 F5
St. Katherines AB51 91 F1
St. Keverne TR12 2 E6
St. Kew PL30 4 A3
St. Kew Highway PL30 4 A3
St. Keyne PL14 4 C4
St. Lawrence *Cornw.* PL30 4 A4
St. Lawrence *Essex* CM0 25 F1
St. Lawrence *I.o.W.* PO38 11 G7
St. Leonards *Bucks.* HP23 22 C1
St. Leonards *Dorset* BH24 10 C4
St. Leonards *E.Suss.* TN37 14 D7
St. Leonards Grange SO42 11 F5
St. Leonard's Street ME19 23 K6
St. Levan TR19 2 A6
St. Lythans CF5 18 E4
St. Mabyn PL30 4 A3
St. Madoes PH2 82 C5
St. Margaret South Elmham IP20 45 H7
St. Margarets *Here.* HR2 28 C5
St. Margarets *Herts.* SG12 33 G7
St. Margarets *Wilts.* SN8 21 F5
St. Margaret's at Cliffe CT15 15 J3
St. Margaret's Hope KW17 106 D8
St. Mark's IM9 54 B6
St. Martin *Chan.I.* GY4 3 J5
St. Martin *Chan.I.* JE3 3 K7
St. Martin *Cornw.* PL13 4 C5
St. Martin *Cornw.* TR12 2 E6
St. Martin's *I.o.S.* TR25 2 C1
St. Martins *P. & K.* PH2 82 C4
St. Martin's *Shrop.* SY11 38 C2
St. Mary JE3 3 J6
St. Mary Bourne SP11 21 H6
St. Mary Church CF71 18 D4
St. Mary Cray BR5 23 H5
St. Mary Hill CF35 18 C4
St. Mary Hoo ME3 24 D4
St. Mary in the Marsh TN29 15 F5
St. Marychurch TQ1 5 K4
St. Mary's *I.o.S.* TR21 2 C1
St. Mary's *Ork.* KW17 106 D7
St. Mary's Airport TR21 2 C1
St. Mary's Bay TN29 15 F5
St. Mary's Grove BS48 19 H5
St. Maughans Green NP25 28 D7
St. Mawes TR2 3 F5
St. Mawgan TR8 3 F2
St. Mellion PL12 4 D4
St. Mellons CF3 19 F3
St. Merryn PL28 3 F1
St. Mewan PL26 3 G3
St. Michael Caerhays PL26 3 G4
St. Michael Church TA7 8 C1
St. Michael Penkevil TR2 3 F4
St. Michael South Elmham NR35 45 H7
St. Michaels *Fife* KY16 83 F5
St. Michaels *Kent* TN30 14 D4
St. Michaels *Worcs.* WR15 28 E2
St. Michael's on Wyre PR3 55 H5
St. Minver PL27 3 G1
St. Monans KY10 83 G7
St. Neot PL14 4 B4
St. Neots PE19 32 E2
St. Newlyn East TR8 3 F3
St. Nicholas *Pembs.* SA64 16 C2
St. Nicholas *V. of Glam.* CF5 18 D4
St. Nicholas at Wade CT7 25 J5
St. Ninians FK7 75 F1
St. Osyth CO16 35 F7
St. Ouen JE3 3 J6
St. Owen's Cross HR2 28 E6
St. Paul's Cray BR5 23 H5
St. Paul's Walden SG4 32 E6
St. Peter JE3 3 J7
St. Peter Port GY1 3 J5
St. Peter's CT10 25 K5
St. Petrox SA71 16 C6
St. Pinnock PL14 4 C4
St. Quivox KA6 67 H1
St. Ruan TR12 2 E7
St. Sampson GY2 3 J5
St. Saviour *Chan.I.* GY7 3 H5
St. Saviour *Chan.I.* JE2 3 K7
St. Stephen PL26 3 G3
St. Stephens *Cornw.* PL12 4 E5
St. Stephens *Cornw.* PL15 6 B7
St. Stephens *Herts.* AL1 22 E1
St. Teath PL30 4 A2
St. Thomas EX2 7 H6
St. Tudy PL30 4 A3
St. Twynnells SA71 16 C6
St. Veep PL22 4 B5
St. Vigeans DD11 83 H3
St. Wenn PL30 3 G2
St. Weonards HR2 28 D6
St. Winnow PL22 4 B5
Saintbury WR12 30 C5
Salachail PA38 80 B2
Salcombe TQ8 5 H7
Salcombe Regis EX10 7 K7
Salcott CM9 34 D7
Sale M33 49 G3
Sale Green WR9 29 J3
Saleby LN13 53 H5
Salehurst TN32 14 C5
Salem *Carmar.* SA19 17 K3
Salem *Cere.* SY23 37 F7
Salem *Gwyn.* LL54 46 D7
Salen *Arg. & B.* PA72 79 G3
Salen *High.* PH36 79 H1
Salendine Nook HD3 50 D1
Salesbury BB1 56 B6
Saleway WR9 29 J3
Salford *Cen.Beds.* MK17 32 C5
Salford *Gt.Man.* M5 49 H3
Salford *Oxon.* OX7 30 D6
Salford Priors WR11 30 B3
Salfords RH1 23 F7
Salhouse NR13 45 H4
Saline KY12 75 J1
SALISBURY SP 10 C2
Salkeld Dykes CA11 61 G3
Sallachan PH33 80 A1
Sallachry PA32 80 B6
Sallachy *High.* IV27 96 C1
Sallachy *High.* IV40 87 F1
Salle NR10 45 F3
Salmonby LN9 53 G5
Salmond's Muir DD11 83 G4
Salperton GL54 30 B6
Salph End MK41 32 D3
Salsburgh ML7 75 G4
Salt ST18 40 B3
Salt Hill SL1 22 C3
Salt Holme TS2 63 F4
Saltaire BD18 57 G6
Saltash PL12 4 E5
Saltburn IV18 96 E4
Saltburn-by-the-Sea TS12 63 H4
Saltby LE14 42 B3
Saltcoats *Cumb.* CA19 60 B7
Saltcoats *N.Ayr.* KA21 74 A6
Saltcotes FY8 55 G7
Saltdean BN2 13 G6
Salterbeck CA14 60 A4
Salterforth BB18 56 D5
Saltergate YO18 63 K7
Salterhill IV30 97 K5
Salterswall CW7 49 F6

Sal - She

Saltfleet **LN11** 53 H3
Saltfleetby All Saints **LN11** 53 H3
Saltfleetby St. Clements **LN11** 53 H3
Saltfleetby St. Peter **LN11** 53 H4
Saltford **BS31** 19 K5
Salthaugh Grange **HU12** 59 J7
Salthouse **NR25** 44 E1
Saltley **B8** 40 C7
Saltmarshe **DN14** 58 D7
Saltness **KW16** 106 B8
Saltney **CH4** 48 C6
Salton **YO62** 58 D1
Saltrens **EX39** 6 C3
Saltwick **NE61** 71 G6
Saltwood **CT21** 15 G4
Salum **PA77** 78 B3
Salvington **BN13** 12 E6
Salwarpe **WR9** 29 H2
Salwayash **DT6** 8 D5
Sambourne **B96** 30 B2
Sambrook **TF10** 39 G3
Samhla **HS6** 92 C5
Samlesbury **PR5** 56 B6
Sampford Arundel **TA21** 7 K4
Sampford Brett **TA4** 7 J1
Sampford Courtenay **EX20** 6 E5
Sampford Moor **TA21** 7 K4
Sampford Peverell **EX16** 7 J4
Sampford Spiney **PL20** 5 F3
Samuelston **EH41** 76 C3
Sanaigmore **PA44** 72 A3
Sanclêr (St. Clears) **SA33** 17 F4
Sancreed **TR20** 2 B6
Sancton **YO43** 59 F6
Sand *Shet.* **ZE2** 107 M8
Sand *Som.* **BS28** 19 H7
Sand Hutton **YO41** 58 C4
Sandaig *Arg. & B.* **PA77** 78 A3
Sandaig *High.* **PH41** 86 D4
Sandaig *High.* **IV40** 86 D3
Sandal Magna **WF2** 51 F1
Sanday **KW17** 106 G3
Sanday Airfield **KW17** 106 F3
Sandbach CW11 49 G6
Sandbank **PA23** 73 K2
Sandbanks **BH13** 10 B6
Sandend **AB45** 98 D4
Sanderstead **CR2** 23 G5
Sandford *Cumb.* **CA16** 61 J5
Sandford *Devon* **EX17** 7 G5
Sandford *Dorset* **BH20** 9 J6
Sandford *I.o.W.* **PO38** 11 G6
Sandford *N.Som.* **BS25** 19 H6
Sandford *S.Lan.* **ML10** 75 F6
Sandford *Shrop.* **SY13** 38 E2
Sandford *Shrop.* **SY11** 38 C3
Sandford Orcas **DT9** 9 F2
Sandford St. Martin **OX7** 31 F6
Sandfordhill **AB42** 99 K6
Sandford-on-Thames **OX4** 21 J1
Sandgarth **KW17** 106 E6
Sandgate **CT20** 15 H4
Sandgreen **DG7** 65 F5
Sandhaven **AB43** 99 H4
Sandhead **DG9** 64 A5
Sandhills *Dorset* **DT9** 9 F3
Sandhills *Dorset* **DT2** 8 E4
Sandhills *Surr.* **GU8** 12 C3
Sandhills *W.Yorks.* **LS14** 57 J6
Sandhoe **NE46** 70 E7
Sandholme *E.Riding* **HU15** 58 E6
Sandholme *Lincs.* **PE20** 43 G2
Sandhurst *Brack.F.* GU47 22 B5
Sandhurst *Glos.* **GL2** 29 H6
Sandhurst *Kent* **TN18** 14 C5
Sandhurst Cross **TN18** 14 C5
Sandhutton **YO7** 57 J1
Sandiacre **NG10** 41 G2
Sandilands **LN12** 53 J4
Sandiway **CW8** 49 F5
Sandleheath **SP6** 10 C3
Sandleigh **OX13** 21 H1
Sandling **ME14** 14 C2
Sandlow Green **CW4** 49 G6
Sandness **ZE2** 107 K7
Sandon *Essex* **CM2** 24 D1
Sandon *Herts.* **SG9** 33 G5
Sandon *Staffs.* **ST18** 40 B3
Sandown PO36 11 H6
Sandplace **PL13** 4 C5
Sandquoy **KW17** 106 G3
Sandridge *Devon* **TQ9** 5 J5
Sandridge *Herts.* **AL4** 32 E7
Sandridge *Wilts.* **SN12** 20 C5
Sandringham PE35 44 A3
Sandrocks **RH16** 13 G4
Sandsend **YO21** 63 K5
Sandside **LA7** 55 H1
Sandside House **KW14** 104 E2
Sandsound **ZE2** 107 M8
Sandtoft **DN8** 51 K2
Sanduck **TQ13** 7 F7
Sandway **ME17** 14 D2
Sandwell **B66** 40 C7
Sandwich CT13 15 J2
Sandwick *Cumb.* **CA10** 60 F5
Sandwick *Shet.* **ZE2** 107 N10
Sandwick (Sanndabhaig) *W.Isles* **HS1** 101 G4
Sandwith **CA28** 60 A5
Sandy *Carmar.* **SA15** 17 H5
Sandy *Cen.Beds.* SG19 32 E4
Sandy Bank **LN4** 53 F7
Sandy Haven **SA62** 16 B5
Sandy Lane *W.Yorks.* **BD15** 57 G6
Sandy Lane *Wilts.* **SN15** 20 C5
Sandy Lane *Wrex.* **LL13** 38 D1
Sandy Way **PO30** 11 F6
Sandycroft **CH5** 48 C6
Sandygate *Devon* **TQ12** 5 J3
Sandygate *I.o.M.* **IM7** 54 C4
Sandyhills **DG5** 65 J5
Sandylands **LA3** 55 H3
Sandypark **TQ13** 7 F7
Sandyway **HR2** 28 D6
Sangobeg **IV27** 103 G2
Sankyn's Green **WR6** 29 G2
Sanna **PH36** 79 F1
Sannaig **PA60** 72 D4
Sanndabhaig (Sandwick) **HS1** 101 G4
Sannox **KA27** 73 J6
Sanquhar DG4 68 C3
Santon Bridge **CA19** 60 C6
Santon Downham **IP27** 44 C7
Sant-y-Nyll **CF5** 18 D4
Sapcote **LE9** 41 G6
Sapey Common **WR6** 29 G2
Sapiston **IP31** 34 D1
Sapperton *Derbys.* **DE65** 40 D2
Sapperton *Glos.* **GL7** 20 C1
Sapperton *Lincs.* **NG34** 42 D2
Saracen's Head **PE12** 43 G3
Sarclet **KW1** 105 J4
Sardis **SA73** 16 C5
Sarisbury **SO31** 11 G4
Sark **GY10** 3 K6
Sarn *Bridgend* **CF32** 18 C3
Sarn *Powys* **SY16** 38 B6
Sarn Bach **LL53** 36 C3
Sarn Meyllteyrn **LL53** 36 B2
Sarnau *Carmar.* **SA33** 17 G4
Sarnau *Cere.* **SA44** 26 C3
Sarnau *Gwyn.* **LL23** 37 J2
Sarnau *Powys* **SY22** 38 B4
Sarnau *Powys* **LD3** 27 K5
Sarnesfield **HR4** 28 C3
Saron *Carmar.* **SA18** 17 K4
Saron *Carmar.* **SA44** 17 G2
Saron *Gwyn.* **LL55** 46 D6
Saron *Gwyn.* **LL54** 46 C7
Sarratt **WD3** 22 D2
Sarre **CT7** 25 J5
Sarsden **OX7** 30 D6
Sarsgrum **IV27** 103 F2
Sartfield **IM7** 54 C4
Satley **DL13** 62 C2
Satron **DL11** 62 A7
Satterleigh **EX37** 6 E3
Satterthwaite **LA12** 60 E7
Sauchen **AB51** 90 E3
Saucher **PH2** 82 C4
Sauchie **FK10** 75 G1
Sauchieburn **AB30** 83 H1
Sauchrie **KA19** 67 H2
Saughall **CH1** 48 C5
Saughall Massie **CH46** 48 B4
Saughtree **TD9** 70 A4
Saul **GL2** 20 A1
Saundby **DN22** 51 K4
Saundersfoot SA69 16 E5
Saunderton **HP27** 22 A1
Saunton **EX33** 6 C2
Sausthorpe **PE23** 53 G6
Saval **IV27** 96 C1
Savalbeg **IV27** 96 C1
Saverley Green **ST11** 40 B2
Savile Town **WF12** 57 H7
Sawbridge **CV23** 31 G2
Sawbridgeworth CM21 33 H7
Sawdon **YO13** 59 F1
Sawley *Derbys.* **NG10** 41 G2
Sawley *Lancs.* **BB7** 56 C5
Sawley *N.Yorks.* **HG4** 57 H3
Sawston **CB22** 33 H4
Sawtry **PE28** 42 E7
Saxby *Leics.* **LE14** 42 B4
Saxby *Lincs.* **LN8** 52 D4
Saxby All Saints **DN20** 52 C1
Saxelbye **LE14** 41 J3
Saxham Street **IP14** 34 E2
Saxilby **LN1** 52 B5
Saxlingham **NR25** 44 E2
Saxlingham Green **NR15** 45 G6
Saxlingham Nethergate **NR15** 45 G6
Saxlingham Thorpe **NR15** 45 G6
Saxmundham IP17 35 H2
Saxon Street **CB8** 33 K3
Saxondale **NG13** 41 J2
Saxtead **IP13** 35 G2
Saxtead Green **IP13** 35 G2
Saxtead Little Green **IP13** 35 G2
Saxthorpe **NR11** 45 F2
Saxton **LS24** 57 K6
Sayers Common **BN6** 13 F5
Scackleton **YO62** 58 C2
Scadabhagh **HS3** 93 G2
Scaftworth **DN10** 51 J3
Scagglethorpe **YO17** 58 E2
Scaitcliffe **BB5** 56 C7
Scalasaig **PA61** 72 B1
Scalby *E.Riding* **HU15** 58 E7
Scalby *N.Yorks.* **YO13** 63 K3
Scaldwell **NN6** 31 J1
Scale Houses **CA10** 61 G2
Scaleby **CA6** 69 K7
Scalebyhill **CA6** 69 K7
Scales *Cumb.* **CA12** 60 E4
Scales *Cumb.* **LA12** 55 F2
Scalford **LE14** 42 A3
Scaling **TS13** 63 J5
Scallasaig **IV40** 86 E3
Scallastle **PA65** 79 J4
Scalloway **ZE1** 107 M9
Scalpay (Eilean Scalpaigh) HS4 93 H2
Scamblesby **LN11** 53 F5
Scammadale **PA34** 79 K5
Scamodale **PH37** 86 E7
Scampston **YO17** 58 E2
Scampton **LN1** 52 C5
Scaniport **IV2** 88 D1
Scapa **KW15** 106 D7
Scapegoat Hill **HD7** 50 C1
Scar **KW17** 106 F3
Scarborough YO11 59 G1
Scarcewater **TR2** 3 G3
Scarcliffe **S44** 51 G6
Scarcroft **LS14** 57 J5
Scardroy **IV6** 95 J6
Scarff **ZE2** 107 L4
Scarfskerry **KW14** 105 H1
Scargill **DL12** 62 B5
Scarinish **PA77** 78 B3
Scarisbrick **L40** 48 C1
Scarning **NR19** 44 D4
Scarrington **NG13** 42 A1
Scarrowhill **CA8** 61 G1
Scarth Hill **L39** 48 D2
Scarthingwell **LS24** 57 K6
Scartho **DN33** 53 F2
Scarwell **KW16** 106 B5
Scatraig **IV2** 88 E1
Scaur *D. & G.* **DG2** 65 J3
Scaur (Kippford) *D. & G.* **DG5** 65 J5
Scawby **DN20** 52 C2
Scawby Brook **DN20** 52 C2
Scawton **YO7** 58 B1
Scayne's Hill **RH17** 13 G4
Scealascro **HS2** 100 D5
Scethrog **LD3** 28 A6
Schaw **KA5** 67 J1
Scholar Green **ST7** 49 H7
Scholes *S.Yorks.* **S61** 51 F3
Scholes *W.Yorks.* **LS15** 57 J6
Scholes *W.Yorks.* **HD9** 50 D2
Scholes *W.Yorks.* **BD19** 57 G7
School Green **CW7** 49 F6
School House **TA20** 8 C4
Schoose **CA14** 60 B4
Sciberscross **IV28** 96 E1
Scilly Isles (Isles of Scilly) **TR** 2 C1
Scissett **HD8** 50 E1
Scleddau **SA65** 16 C2
Sco Ruston **NR12** 45 G3
Scofton **S81** 51 J4
Scole **IP21** 35 F1
Scolpaig **HS6** 92 C4
Scone **PH2** 82 C5
Scones Lethendy **PH2** 82 C5
Sconser **IV48** 86 B1
Scoor **PA67** 79 F6
Scopwick **LN4** 52 D7
Scoraig **IV23** 95 G2
Scorborough **YO25** 59 G5
Scorrier **TR16** 2 E4
Scorriton **TQ11** 5 H4
Scorton *Lancs.* **PR3** 55 J5
Scorton *N.Yorks.* **DL10** 62 D6
Scot Hay **ST5** 40 A1
Scotby **CA4** 60 F1
Scotch Corner **DL10** 62 D6
Scotforth **LA1** 55 H4
Scothern **LN2** 52 D5
Scotland **NG33** 42 D2
Scotland End **OX15** 30 E5
Scotland Street **CO6** 34 D5
Scotlandwell **KY13** 82 C7
Scotnish **PA31** 73 F2
Scots' Gap **NE61** 71 F5
Scotsburn **IV18** 96 E4
Scotston *Aber.* **AB30** 91 F7
Scotston *P. & K.* **PH8** 82 A3
Scotstoun **G14** 74 D4
Scotstown **PH36** 79 K1
Scott Willoughby **NG34** 42 D2
Scotter **DN21** 52 B2
Scotterthorpe **DN21** 52 B2
Scottlethorpe **PE10** 42 D3
Scotton *Lincs.* **DN21** 52 B3
Scotton *N.Yorks.* **DL9** 62 C7
Scotton *N.Yorks.* **HG5** 57 J4
Scottow **NR10** 45 G3
Scoughall **EH39** 76 E2
Scoulton **NR9** 44 D5
Scounslow Green **ST14** 40 C3
Scourie **IV27** 102 D4
Scourie More **IV27** 102 D4
Scousburgh **ZE2** 107 M11
Scouthead **OL4** 49 J2
Scrabster **KW14** 105 G1
Scrafield **LN9** 53 G6
Scrainwood **NE66** 70 E3
Scrane End **PE22** 43 G1
Scraptoft **LE7** 41 J5
Scratby **NR29** 45 K4
Scrayingham **YO41** 58 D4
Scredington **NG34** 42 D1
Scremby **PE23** 53 H6
Scremerston **TD15** 77 J6
Screveton **NG13** 42 A1
Scriven **HG5** 57 J4
Scronkey **PR3** 55 H5
Scrooby **DN10** 51 J3
Scropton **DE65** 40 D2
Scrub Hill **LN4** 53 F7
Scruton **DL7** 62 E7
Sculthorpe **NR21** 44 C2
Scunthorpe DN15 52 B1
Scurlage **SA3** 17 H7
Sea **TA19** 8 C3
Sea Mills **BS9** 19 J4
Sea Palling **NR12** 45 J3
Seabank **PA37** 80 A3
Seaborough **DT8** 8 D4
Seaburn **SR6** 71 K7
Seacombe **CH41** 48 C3
Seacroft *Lincs.* **PE25** 53 J6
Seacroft *W.Yorks.* **LS14** 57 J6
Seadyke **PE20** 43 G2
Seafield *Arg. & B.* **PA31** 73 F2
Seafield *S.Ayr.* **KA7** 67 H1
Seafield *W.Loth.* **EH47** 75 J4
Seaford BN25 13 H7
Seaforth **L21** 48 C3
Seagrave **LE12** 41 J4
Seagry Heath **SN15** 20 C3
Seaham SR7 63 F2
Seaham Grange **SR7** 63 F1
Seahouses NE68 77 K4
Seal **TN15** 23 J6
Sealand **CH1** 48 C6
Seale **GU10** 22 B7
Sealyham **SA62** 16 C3
Seamer *N.Yorks.* **TS9** 63 F5
Seamer *N.Yorks.* **YO12** 59 G1
Seamill **KA23** 73 K6
Searby **DN38** 52 D2
Seasalter **CT5** 25 G5
Seascale CA20 60 B6
Seathorne **PE25** 53 J6
Seathwaite *Cumb.* **CA12** 60 D5
Seathwaite *Cumb.* **LA20** 60 D7
Seatle **LA11** 55 G1
Seatoller **CA12** 60 D5
Seaton *Cornw.* **PL11** 4 D5
Seaton *Cumb.* **CA14** 60 B3
Seaton *Devon* EX12 8 B5
Seaton *Dur.* **SR7** 62 E1
Seaton *E.Riding* **HU11** 59 H5
Seaton *Northumb.* **NE26** 71 J6
Seaton *Rut.* **LE15** 42 C6
Seaton Burn **NE13** 71 H6
Seaton Carew **TS25** 63 G4
Seaton Delaval **NE25** 71 J6
Seaton Junction **EX13** 8 B5
Seaton Ross **YO42** 58 D5
Seaton Sluice **NE26** 71 J6
Seatown *Aber.* **AB42** 99 J5
Seatown *Dorset* **DT6** 8 D5
Seatown *Moray* **AB56** 98 D4
Seave Green **TS9** 63 G6
Seaview PO34 11 H5
Seaville **CA7** 60 C1
Seavington St. Mary **TA19** 8 D3
Seavington St. Michael **TA19** 8 D3
Seawick **CO16** 35 F7
Sebastopol **NP4** 19 F2
Sebergham **CA5** 60 E2
Seckington **B79** 40 E5
Second Coast **IV22** 95 F2
Sedbergh LA10 61 H7
Sedbury **NP16** 19 J2
Sedbusk **DL8** 61 K7
Seddington **SG19** 32 E4
Sedgeberrow **WR11** 30 B5
Sedgebrook **NG32** 42 B2
Sedgefield **TS21** 62 E4
Sedgeford **PE36** 44 B2
Sedgehill **SP7** 9 H2
Sedgemere **CV8** 30 D1
Sedgley **DY3** 40 B6
Sedgwick **LA8** 55 J1
Sedlescombe **TN33** 14 C6
Sedlescombe Street **TN33** 14 C6
Seend **SN12** 20 C5
Seend Cleeve **SN12** 20 C5
Seer Green **HP9** 22 C2
Seething **NR15** 45 H6
Sefton **L29** 48 C2
Seghill **NE23** 71 H6
Seifton **SY8** 38 D7
Seighford **ST18** 40 A3
Seil **PA34** 79 J6
Seilebost **HS3** 93 F2
Seion **LL55** 46 D6
Seisdon **WV5** 40 A6
Seisiadar **HS2** 101 H4
Selattyn **SY10** 38 B2
Selborne **GU34** 11 J1
Selby YO8 58 C6
Selham **GU28** 12 C4
Selhurst **SE25** 23 G5
Selkirk TD7 69 K1
Sellack **HR9** 28 E6
Sellafield **CA20** 60 B6
Sellafirth **ZE2** 107 P3
Sellindge **TN25** 15 G4
Selling **ME13** 15 F2
Sells Green **SN12** 20 C5
Selly Oak **B29** 40 C7
Selmeston **BN26** 13 J6
Selsdon **CR2** 23 G5
Selsey **PO20** 12 B7
Selsfield Common **RH19** 13 G3
Selside *Cumb.* **LA8** 61 G7
Selside *N.Yorks.* **BD24** 56 C2
Selsley **GL5** 20 B1
Selstead **CT15** 15 H3
Selston **NG16** 51 G7
Selworthy **TA24** 7 H1
Semblister **ZE2** 107 M7
Semer **IP7** 34 D4
Semington **BA14** 20 B5
Semley **SP7** 9 H2
Send **GU23** 22 D6
Send Marsh **GU23** 22 D6
Senghenydd **CF83** 18 E2
Sennen **TR19** 2 A6
Sennen Cove **TR19** 2 A6
Sennybridge **LD3** 27 J6
Senwick **DG6** 65 G6
Sequer's Bridge **PL21** 5 G5
Serlby **DN10** 51 J4
Serrington **SP3** 10 B1
Sessay **YO7** 57 K2
Setchey **PE33** 44 A4
Setley **SO42** 10 E4
Setter *Shet.* **ZE2** 107 P8
Setter *Shet.* **ZE2** 107 M7
Settiscarth **KW17** 106 C6
Settle BD24 56 D3
Settrington **YO17** 58 E2
Seven Ash **TA4** 7 K2
Seven Bridges **SN6** 20 E2
Seven Kings **IG3** 23 H3
Seven Sisters **SA10** 18 B1
Seven Springs **GL53** 29 J7
Sevenhampton *Glos.* **GL54** 30 B6
Sevenhampton *Swin.* **SN6** 21 F2
Sevenoaks TN13 23 J6
Sevenoaks Weald **TN14** 23 J6
Severn Beach **BS35** 19 J3
Severn Stoke **WR8** 29 H4
Sevick End **MK44** 32 D3
Sevington **TN24** 15 F3
Sewards End **CB10** 33 J5
Sewardstone **E4** 23 G2
Sewerby **YO15** 59 H3
Seworgan **TR11** 2 E5
Sewstern **NG33** 42 B3
Seymour Villas **EX34** 6 C1
Sezincote **GL56** 30 C5
Sgarasta Mhòr **HS3** 93 F2
Sgiogarstaigh **HS2** 101 H1
Sgodachail **IV24** 96 B2
Shabbington **HP18** 21 K1
Shackerley **WV7** 40 A5
Shackerstone **CV13** 41 F5
Shackleford **GU8** 22 C7
Shadfen **NE61** 71 H5
Shadforth **DH6** 62 E2
Shadingfield **NR34** 45 J7
Shadoxhurst **TN26** 14 E4
Shadsworth **BB1** 56 C7
Shadwell *Norf.* **IP24** 44 D7
Shadwell *W.Yorks.* **LS17** 57 J5
Shaftenhoe End **SG8** 33 H5
Shaftesbury SP7 9 H2
Shafton **S72** 51 F1
Shalbourne **SN8** 21 G5
Shalcombe **PO41** 10 E6
Shalden **GU34** 21 K7
Shalden Green **GU34** 22 A7
Shaldon **TQ14** 5 K3
Shalfleet **PO30** 11 F6
Shalford *Essex* **CM7** 34 B6
Shalford *Surr.* **GU4** 22 D7
Shalford Green **CM7** 34 B6
Shallowford *Devon* **EX31** 7 F1
Shallowford *Staffs.* **ST15** 40 A3
Shalmsford Street **CT4** 15 F2
Shalmstry **KW14** 105 G2
Shalstone **MK18** 31 H5
Shalunt **PA20** 73 J3
Shambellie **DG2** 65 K4
Shamley Green **GU5** 22 D7
Shandon **G84** 74 A2
Shandwick **IV20** 97 F4
Shangton **LE8** 42 A6
Shankend **TD9** 70 A3
Shankhouse **NE23** 71 H6
Shanklin PO37 11 G6
Shannochie **KA27** 66 D1
Shantron **G83** 74 B2
Shantullich **IV8** 96 D6
Shanzie **PH11** 82 D2
Shap **CA10** 61 G5
Shapinsay **KW17** 106 E6
Shapwick *Dorset* **DT11** 9 J4
Shapwick *Som.* **TA7** 8 D1
Sharcott **SN9** 20 E6
Shard End **B34** 40 D7
Shardlow **DE72** 41 G2
Shareshill **WV10** 40 B5
Sharlston **WF4** 51 F1
Sharlston Common **WF4** 51 F1
Sharnal Street **ME3** 24 D4
Sharnbrook **MK44** 32 C3
Sharneyford **OL13** 56 D7
Sharnford **LE10** 41 G6
Sharnhill Green **DT2** 9 G4
Sharow **HG4** 57 J2
Sharp Street **NR29** 45 H3
Sharpenhoe **MK45** 32 D5
Sharperton **NE65** 70 E3
Sharpham House **TQ9** 5 J5
Sharpness **GL13** 19 K1
Sharpthorne **RH19** 13 G3
Sharrington **NR24** 44 E2
Shatterford **DY12** 39 G7
Shatterling **CT3** 15 H2
Shaugh Prior **PL7** 5 F4
Shave Cross **DT6** 8 D5
Shavington **CW2** 49 G7
Shaw *Gt.Man.* **OL2** 49 J2
Shaw *Swin.* **SN5** 20 E3
Shaw *W.Berks.* **RG14** 21 H5
Shaw *Wilts.* **SN12** 20 B5
Shaw Green *Herts.* **SG7** 33 F5
Shaw Green *N.Yorks.* **HG3** 57 H4
Shaw Mills **HG3** 57 H3
Shaw Side **OL2** 49 J2
Shawbost (Siabost) **HS2** 100 E3
Shawbury **SY4** 38 E3
Shawell **LE17** 41 H7
Shawfield *Gt.Man.* **OL12** 49 H1
Shawfield *Staffs.* **SK17** 50 C6
Shawford **SO21** 11 F2
Shawforth **OL12** 56 D7
Shawhead **DG2** 65 J3
Shawtonhill **ML10** 74 E6
Sheanachie **PA28** 66 B2
Sheandow **AB38** 89 K1

She - Sma

Shearington DG1 69 F7
Shearsby LE17 41 J6
Shebbear EX21 6 C5
Shebdon ST20 39 G3
Shebster KW14 105 F2
Shedfield SO32 11 G3
Sheen SK17 50 D6
Sheepridge HD2 50 D1
Sheepscombe GL6 29 H7
Sheepstor PL20 5 F4
Sheepwash *Devon* EX21 6 C5
Sheepwash *Northumb.* NE62 71 H5
Sheepway BS20 19 H4
Sheepy Magna CV9 41 F5
Sheepy Parva CV9 41 F5
Sheering CM22 33 J7
Sheerness ME12 25 F4
Sheet GU32 11 J2
SHEFFIELD S 51 F4
Sheffield Bottom RG7 21 K5
Sheffield Green TN22 13 H4
Shefford SG17 32 E5
Shefford Woodlands RG17 21 G4
Sheigra IV27 102 D2
Sheinton SY5 39 F5
Shelderton SY7 28 D1
Sheldon *Derbys.* DE45 50 D6
Sheldon *Devon* EX14 7 K5
Sheldon *W.Mid.* B26 40 D7
Sheldwich ME13 15 F2
Sheldwich Lees ME13 15 F2
Shelf *Bridgend* CF35 18 C3
Shelf *W.Yorks.* HX3 57 G7
Shelfanger IP22 45 F7
Shelfield *W.Mid.* WS4 40 C5
Shelfield *Warks.* B49 30 C2
Shelfield Green B49 30 C2
Shelford NG12 41 J1
Shellachan *Arg. & B.* PA34 79 K5
Shellachan *Arg. & B.* PA35 80 B5
Shellbrook LE65 41 F4
Shellbrook Hill SY12 38 C1
Shelley *Essex* CM5 23 J1
Shelley *Suff.* IP7 34 E5
Shelley *W.Yorks.* HD8 50 E1
Shellingford SN7 21 G2
Shellow Bowells CM5 24 C1
Shelsley Beauchamp WR6 29 G2
Shelsley Walsh WR6 29 G2
Shelswell MK18 31 H5
Shelthorpe LE11 41 H4
Shelton *Bed.* PE28 32 D2
Shelton *Norf.* NR15 45 G6
Shelton *Notts.* NG23 42 A1
Shelton *Shrop.* SY3 38 D4
Shelve SY5 38 C6
Shelwick HR1 28 E4
Shelwick Green HR1 28 E4
Shenfield CM15 24 C2
Shenington OX15 30 E4
Shenley WD7 22 E1
Shenley Brook End MK5 32 B5
Shenley Church End MK5 32 B5
Shenleybury WD7 22 E1
Shenmore HR2 28 C5
Shennanton DG8 64 D4
Shenstone *Staffs.* WS14 40 D5
Shenstone *Worcs.* DY10 29 H1
Shenstone Woodend WS14 40 D5
Shenton CV13 41 F5
Shenval AB37 89 K2
Shepeau Stow PE12 43 G4
Shephall SG2 33 F6
Shepherd's Bush W12 23 F4
Shepherd's Green RG9 22 A3
Shepherd's Patch GL2 20 A1
Shepherdswell (Sibertswold) CT15 15 H3
Shepley HD8 50 D2
Sheppardstown KW3 105 H5
Shepperdine BS35 19 K2
Shepperton TW17 22 D5
Shepreth SG8 33 G4
Shepshed LE12 41 G4
Shepton Beauchamp TA19 8 D3
Shepton Mallet BA4 19 K7
Shepton Montague BA9 9 F1
Shepway ME15 14 C2
Sheraton TS27 63 F3
Sherborne *Dorset* DT9 9 F3
Sherborne *Glos.* GL54 30 C7
Sherborne St. John RG24 21 K6
Sherbourne CV35 30 D2
Sherbourne Street CO10 34 D4
Sherburn *Dur.* DH6 62 E2
Sherburn *N.Yorks.* YO17 59 F2
Sherburn Hill DH6 62 E2
Sherburn in Elmet LS25 57 K6
Shere GU5 22 D7
Shereford NR21 44 C3
Sherfield English SO51 10 D2
Sherfield on Loddon RG27 21 K6
Sherford *Devon* TQ7 5 H6
Sherford *Som.* TA1 8 B2
Sheriff Hutton YO60 58 C3
Sheriffhales TF11 39 G4
Sheringham NR26 45 F1
Sherington MK16 32 B4
Shernal Green WR9 29 J2
Shernborne PE31 44 B2
Sherramore PH20 88 C5
Sherrington BA12 9 J1
Sherston SN16 20 B3
Sherwood NG5 41 H1
Sherwood Green EX31 6 D3
SHETLAND ISLANDS ZE 107 M7
Shettleston G32 74 E4
Shevington WN6 48 E2
Shevington Moor WN6 48 E1
Sheviock PL11 4 D5
Shide PO30 11 G6
Shiel Bridge IV40 87 F3
Shieldaig *High.* IV54 94 E6
Shieldaig *High.* IV21 94 E4
Shieldhill FK1 75 G3
Shielfoot PH36 86 C7
Shielhill DD8 83 F2
Shiels AB51 90 E4
Shifford OX29 21 G1
Shifnal TF11 39 G5
Shilbottle NE66 71 G3
Shildon DL4 62 D4
Shillingford *Devon* EX16 7 H3
Shillingford *Oxon.* OX10 21 J2
Shillingford Abbot EX2 7 H7
Shillingford St. George EX2 7 H7
Shillingstone DT11 9 H3
Shillington SG5 32 E5
Shillmoor NE65 70 D3
Shilstone EX20 6 D5
Shilton *Oxon.* OX18 21 F1
Shilton *Warks.* CV7 41 G7
Shimpling *Norf.* IP21 45 F7
Shimpling *Suff.* IP29 34 C3
Shimpling Street IP29 34 C3
Shincliffe DH1 62 D2
Shiney Row DH4 62 E1
Shinfield RG2 22 A5
Shingay SG8 33 G4
Shingham PE37 44 B5
Shingle Street IP12 35 H4
Shinness Lodge IV27 103 H7
Shipbourne TN11 23 J6
Shipbrookhill CW9 49 F5
Shipdham IP25 44 D5
Shipham BS25 19 H6
Shiphay TQ2 5 J4
Shiplake RG9 22 A4
Shiplake Row RG9 22 A4
Shipley *Northumb.* NE66 71 G2
Shipley *Shrop.* WV6 40 A6
Shipley *W.Suss.* RH13 12 E4
Shipley *W.Yorks.* BD18 57 G6
Shipley Bridge *Devon* TQ10 5 G4
Shipley Bridge *Surr.* RH6 23 G7
Shipley Common DE7 41 G1
Shipmeadow NR34 45 H6
Shippea Hill CB7 44 A7
Shippon OX13 21 H2
Shipston on Stour CV36 30 D4
Shipton *Glos.* GL54 30 B7
Shipton *N.Yorks.* YO30 58 B4
Shipton *Shrop.* TF13 38 E6
Shipton Bellinger SP9 21 F7
Shipton Gorge DT6 8 D5
Shipton Green PO20 12 B6
Shipton Moyne GL8 20 B3
Shipton Oliffe GL54 30 B7
Shipton Solers GL54 30 B7
Shipton-on-Cherwell OX5 31 F7
Shiptonthorpe YO43 58 E5
Shipton-under-Wychwood OX7 30 D7
Shira PA32 80 C6
Shirburn OX49 21 K2
Shirdley Hill L39 48 C1
Shire Oak WS8 40 C5
Shirebrook NG20 51 H6
Shirecliffe S5 51 F3
Shiregreen S5 51 F3
Shirehampton BS11 19 J4
Shiremoor NE27 71 J6
Shirenewton NP16 19 H2
Shireoaks S81 51 H4
Shirl Heath HR6 28 D3
Shirland DE55 51 F7
Shirley *Derbys.* DE6 40 E1
Shirley *Gt.Lon.* CR0 23 G5
Shirley *Hants.* BH23 10 C5
Shirley *S'ham.* SO15 11 F3
Shirley *W.Mid.* B90 30 C1
Shirley Heath B90 30 C1
Shirley Warren SO16 10 E3
Shirleywich ST18 40 B3
Shirrell Heath SO32 11 G3
Shirwell EX31 6 D2
Shirwell Cross EX31 6 D2
Shiskine KA27 66 D1
Shittlehope DL13 62 B3
Shobdon HR6 28 C2
Shobley BH24 10 C4
Shobrooke EX17 7 G5
Shocklach SY14 38 D1
Shocklach Green SY14 38 D1
Shoeburyness SS3 25 F3
Sholden CT14 15 J2
Sholing SO19 11 F3
Shoot Hill SY5 38 D4
Shooter's Hill DA16 23 H4
Shop *Cornw.* PL28 3 F1
Shop *Cornw.* EX23 6 A4
Shop Corner IP9 35 G5
Shopnoller TA4 7 K2
Shore OL15 49 J1
Shoreditch N1 23 G3
Shoreham TN14 23 J5
Shoreham Airport BN15 12 E6
Shoreham-by-Sea BN43 13 F6
Shoremill IV11 96 E5
Shoresdean TD15 77 H6
Shoreswood TD15 77 H6
Shoreton IV7 96 D5
Shorley SO24 11 G2
Shorncote GL7 20 D2
Shorne DA12 24 C4
Shorne Ridgeway DA12 24 C4
Short Cross SY21 38 B5
Short Green IP22 44 E7
Short Heath *Derbys.* DE12 41 F4
Short Heath *W.Mid.* B23 40 C6
Shortacombe EX20 6 D7
Shortbridge TN22 13 H4
Shortfield Common GU10 22 B7
Shortgate BN8 13 H5
Shortgrove CB11 33 J5
Shorthampton OX7 30 E6
Shortlands BR2 23 G5
Shortlanesend TR4 3 F4
Shorton TQ3 5 J4
Shorwell PO30 11 F6
Shoscombe BA2 20 A6
Shotatton SY4 38 C3
Shotesham NR15 45 G6
Shotgate SS11 24 D2
Shotley *Northants.* NN17 42 C6
Shotley *Suff.* IP9 35 G5
Shotley Bridge DH8 62 B1
Shotley Gate IP9 35 G5
Shotleyfield DH8 62 B1
Shottenden CT4 15 F2
Shottermill GU27 12 B3
Shottery CV37 30 C3
Shotteswell OX17 31 F4
Shottisham IP12 35 H4
Shottle DE56 41 F1
Shottlegate DE56 41 F1
Shotton *Dur.* SR8 63 F3
Shotton *Dur.* TS21 62 E4
Shotton *Flints.* CH5 48 C6
Shotton *Northumb.* NE13 71 H6
Shotton Colliery DH6 62 E2
Shotts ML7 75 G4
Shotwick CH1 48 C5
Shouldham PE33 44 A5
Shouldham Thorpe PE33 44 A5
Shoulton WR2 29 H3
Shover's Green TN5 13 K3
Shrawardine SY4 38 C4
Shrawley WR6 29 H2
Shreding Green SL0 22 D3
Shrewley CV35 30 D2
SHREWSBURY SY 38 D4
Shrewton SP3 20 D7
Shripney PO22 12 C6
Shrivenham SN6 21 F3
Shropham NR17 44 D6
Shroton (Iwerne Courtney) DT11 9 H3
Shrub End CO2 34 D6
Shucknall HR1 28 E4
Shudy Camps CB21 33 K4
Shurdington GL51 29 J7
Shurlock Row RG10 22 B4
Shurnock B96 30 B2
Shurrery KW14 105 F3
Shurrery Lodge KW14 105 F3
Shurton TA5 19 F7
Shustoke B46 40 E6
Shut Heath ST18 40 A3
Shute *Devon* EX13 8 B5
Shute *Devon* EX17 7 G5
Shutford OX15 30 E4
Shuthonger GL20 29 H5
Shutlanger NN12 31 J4
Shutt Green ST19 40 A5
Shuttington B79 40 E5
Shuttlewood S44 51 G5
Shuttleworth BL0 49 G1
Siabost (Shawbost) HS2 100 E3
Siabost Bho Dheas HS2 100 E3
Siabost Bho Thuath HS2 100 E3
Siadar Iarach HS2 101 F2
Siadar Uarach HS2 101 F2
Sibbaldbie DG11 69 G5
Sibbertoft LE16 41 J7
Sibdon Carwood SY7 38 D7
Sibertswold (Shepherdswell) CT15 15 H3
Sibford Ferris OX15 30 E5
Sibford Gower OX15 30 E5
Sible Hedingham CO9 34 B5
Sibley's Green CM6 33 K6
Sibsey PE22 53 G7
Sibson *Cambs.* PE8 42 D6
Sibson *Leics.* CV13 41 F5
Sibster KW1 105 J3
Sibthorpe NG23 42 A1
Sibton IP17 35 H2
Sibton Green IP17 35 H1
Sicklesmere IP30 34 C2
Sicklinghall LS22 57 J5
Sidbury *Devon* EX10 7 K6
Sidbury *Shrop.* WV16 39 F7
Sidcot BS25 19 H6
Sidcup DA14 23 H4
Siddal HX3 57 G7
Siddington *Ches.E.* SK11 49 H5
Siddington *Glos.* GL7 20 D2
Sidemoor B61 29 J1
Sidestrand NR27 45 G2
Sidford EX10 7 K6
Sidlesham PO20 12 B7
Sidley TN39 14 C7
Sidlow RH2 23 F7
Sidmouth EX10 7 K7
Sigford TQ12 5 H3
Sigglesthorne HU11 59 H5
Sigingstone CF71 18 C4
Signet OX18 30 D7
Silchester RG7 21 K5
Sildinis HS2 100 E6
Sileby LE12 41 J4
Silecroft LA18 54 E1
Silfield NR18 45 F6
Silian SA48 26 E3
Silk Willoughby NG34 42 D1
Silkstead SO21 11 F2
Silkstone S75 50 E2
Silkstone Common S75 50 E2
Sill Field LA8 55 J1
Silloth CA7 60 C1
Sills NE19 70 D3
Sillyearn AB55 98 D5
Silpho YO13 63 J3
Silsden BD20 57 F5
Silsoe MK45 32 D5
Silver End *Cen.Beds.* MK45 32 E4
Silver End *Essex* CM8 34 C6
Silver Green NR15 45 G6
Silver Street *Kent* ME9 24 E5
Silver Street *Som.* TA11 8 E1
Silverburn EH26 76 A4
Silvercraigs PA31 73 G2
Silverdale *Lancs.* LA5 55 H2
Silverdale *Staffs.* ST5 40 A1
Silvergate NR11 45 F3
Silverhill TN37 14 C6
Silverlace Green IP13 35 H3
Silverley's Green IP19 35 G1
Silvermoss AB51 91 G1
Silverstone NN12 31 H4
Silverton EX5 7 H5
Silvington DY14 29 F1
Silwick ZE2 107 L8
Simister M25 49 H2
Simmondley SK13 50 C3
Simonburn NE48 70 D6
Simonsbath TA24 7 F2
Simonside NE34 71 J7
Simonstone *Bridgend* CF35 18 C3
Simonstone *Lancs.* BB12 56 C6
Simprim TD12 77 G6
Simpson MK6 32 B5
Sinclair's Hill TD11 77 G5
Sinclairston KA18 67 J2
Sinderby YO7 57 J1
Sinderhope NE47 61 K1
Sindlesham RG41 22 A5
Sinfin DE24 41 F2
Singdean TD9 70 A3
Singleton *Lancs.* FY6 55 G6
Singleton *W.Suss.* PO18 12 B5
Singlewell DA12 24 C4
Singret LL12 48 C7
Sinkhurst Green TN12 14 D3
Sinnahard AB33 90 C3
Sinnington YO62 58 D1
Sinton Green WR2 29 H2
Sipson UB7 22 D4
Sirhowy NP22 28 A7
Sisland NR14 45 H6
Sissinghurst TN17 14 C4
Siston BS16 19 K4
Sithney TR13 2 D6
Sittingbourne ME10 25 F5
Siulaisiadar HS2 101 H4
Six Ashes WV15 39 G7
Six Hills LE14 41 J3
Six Mile Bottom CB8 33 J3
Six Roads End DE6 40 D3
Sixhills LN8 52 E4
Sixmile CT4 15 G3
Sixpenny Handley SP5 10 B3
Sizewell IP16 35 J2
Skail KW11 104 C4
Skaill *Ork.* KW16 106 B6
Skaill *Ork.* KW17 106 E7
Skaill *Ork.* KW17 106 D4
Skares *Aber.* AB54 90 E1
Skares *E.Ayr.* KA18 67 K2
Skarpigarth ZE2 107 K7
Skateraw EH42 77 F3
Skaw ZE2 107 P6
Skeabost IV51 93 K7
Skeabrae KW17 106 B5
Skeeby DL10 62 D6
Skeffington LE7 42 A5
Skeffling HU12 53 G1
Skegby NG17 51 H6
Skegness PE25 53 J6
Skelberry *Shet.* ZE2 107 M11
Skelberry *Shet.* ZE2 107 N6
Skelbo IV25 96 E2
Skelbo Street IV25 96 E2
Skelbrooke DN6 51 H1
Skeld (Easter Skeld) ZE2 107 M8
Skeldon KA6 67 H2
Skeldyke PE20 43 G2
Skellingthorpe LN6 52 C5
Skellister ZE2 107 N7
Skellow DN6 51 H1
Skelmanthorpe HD8 50 E1
Skelmersdale WN8 48 D2
Skelmonae AB41 91 G1
Skelmorlie PA17 73 K4
Skelmuir AB42 99 H6
Skelpick KW14 104 C3
Skelton *Cumb.* CA11 60 F3
Skelton *E.Riding* DN14 58 D7
Skelton *N.Yorks.* DL11 62 B6
Skelton (Skelton-in-Cleveland) *R. & C.* TS12 63 H5
Skelton *York* YO30 58 B4
Skelton-in-Cleveland (Skelton) TS12 63 H5
Skelton-on-Ure HG4 57 J3
Skelwick KW17 106 D3
Skelwith Bridge LA22 60 E6
Skendleby PE23 53 H6
Skendleby Psalter LN13 53 H5
Skenfrith NP7 28 D6
Skerne YO25 59 G4
Skeroblingarry PA28 66 B1
Skerray KW14 103 J2
Skerton LA1 55 H3
Sketchley LE10 41 G6
Sketty SA2 17 K6
Skewen SA10 18 A2
Skewsby YO61 58 C2
Skeyton NR10 45 G3
Skeyton Corner NR10 45 G3
Skidbrooke LN11 53 H3
Skidbrooke North End LN11 53 H3
Skidby HU16 59 G6
Skilgate TA4 7 H3
Skillington NG33 42 B3
Skinburness CA7 60 C1
Skinflats FK2 75 H2
Skinidin IV55 93 H7
Skinnet KW12 105 G2
Skinningrove TS13 63 J5
Skipness PA29 73 G5
Skippool FY5 55 G5
Skipsea YO25 59 H4
Skipsea Brough YO25 59 H4
Skipton BD23 56 E4
Skipton-on-Swale YO7 57 J2
Skipwith YO8 58 C6
Skirbeck PE21 43 G1
Skirbeck Quarter PE21 43 G1
Skirethorns BD23 56 E3
Skirlaugh HU11 59 H6
Skirling ML12 75 J7
Skirmett RG9 22 A3
Skirpenbeck YO41 58 D4
Skirwith *Cumb.* CA10 61 H3
Skirwith *N.Yorks.* LA6 56 C2
Skirza KW1 105 J2
Skittle Green HP27 22 A1
Skomer Island SA62 16 A5
Skulamus IV42 86 C2
Skullomie IV27 103 J2
Skyborry Green LD7 28 B1
Skye IV 85 K1
Skye Green CO5 34 C6
Skye of Curr PH26 89 G2
Skyreholme BD23 57 F3
Slack *Aber.* AB52 90 D1
Slack *Derbys.* S45 51 F6
Slack *W.Yorks.* HX7 56 E7
Slackhall SK23 50 C4
Slackhead AB56 98 C4
Slad GL6 20 B1
Slade *Devon* EX34 6 D1
Slade *Devon* EX14 7 K5
Slade *Pembs.* SA61 16 C4
Slade *Swan.* SA3 17 H7
Slade Green DA8 23 J4
Slade Hooton S25 51 H4
Sladesbridge PL27 4 A3
Slaggyford CA8 61 H1
Slaidburn BB7 56 C4
Slains Park DD10 91 G7
Slaithwaite HD7 50 C1
Slaley NE47 62 A1
Slamannan FK1 75 G3
Slapton *Bucks.* LU7 32 C6
Slapton *Devon* TQ7 5 J6
Slapton *Northants.* NN12 31 H4
Slate Haugh AB56 98 C4
Slatepit Dale S42 51 F6
Slattadale IV22 94 E4
Slaugham RH17 13 F4
Slaughden IP15 35 J3
Slaughterford SN14 20 B4
Slawston LE16 42 A6
Sleaford *Hants.* GU35 12 B3
Sleaford *Lincs.* NG34 42 D1
Sleagill CA10 61 G5
Sleap SY4 38 D3
Sledge Green WR13 29 H5
Sledmere YO25 59 F3
Sleights YO22 63 K6
Slepe BH16 9 J5
Slerra EX39 6 B3
Slickly KW1 105 H2
Sliddery KA27 66 D1
Sliemore PH25 89 H2
Sligachan IV47 85 K2
Slimbridge GL2 20 A1
Slindon *Staffs.* ST21 40 A2
Slindon *W.Suss.* BN18 12 C6
Slinfold RH13 12 E3
Sling GL16 19 J1
Slingsby YO62 58 C2
Slioch AB54 90 D1
Slip End *Cen.Beds.* LU1 32 D7
Slip End *Herts.* SG7 33 F5
Slipton NN14 32 C1
Slitting Mill WS15 40 C4
Slochd PH23 89 F2
Slockavullin PA31 73 G1
Slogarie DG7 65 G4
Sloley NR12 45 G3
Slongaber DG2 65 J3
Sloothby LN13 53 H5
SLOUGH SL 22 C3
Slough Green *Som.* TA3 8 B2
Slough Green *W.Suss.* RH17 13 F4
Sluggan PH23 89 F2
Slyne LA2 55 H3
Smailholm TD5 76 E7
Small Dole BN5 13 F5
Small Hythe TN30 14 D4
Smallbridge OL16 49 J1
Smallbrook EX5 7 G6
Smallburgh NR12 45 H3
Smallburn *Aber.* AB42 99 J6
Smallburn *E.Ayr.* KA18 68 B1
Smalldale SK17 50 C5
Smalley DE7 41 G1
Smallfield RH6 23 G7
Smallford AL4 22 E1
Smallridge EX13 8 B4
Smallthorne ST6 49 H7

Sma - Sta

Smallworth IP22 44 E7
Smannell SP11 21 G7
Smardale CA17 61 J6
Smarden TN27 14 D3
Smaull PA44 72 A4
Smeatharpe EX14 7 K4
Smeeth TN25 15 F4
Smeeton Westerby LE8 41 J6
Smerclet HS8 84 C3
Smerral KW5 105 G5
Smestow DY3 40 A6
Smethwick B66 40 C7
Smethwick Green CW11 49 H6
Smirisary PH38 86 C7
Smisby LE65 41 F4
Smith End Green WR13 29 G3
Smithfield CA6 69 K7
Smithies S71 51 F2
Smithincott EX15 7 J4
Smith's End SG8 33 G5
Smith's Green *Essex* CM22 33 J6
Smith's Green *Essex* CB9 33 K4
Smithstown IV21 94 D4
Smithton IV2 96 E7
Smithy Green WA16 49 G5
Smockington LE10 41 G7
Smyrton KA26 67 F5
Smythe's Green CO5 34 D7
Snailbeach SY5 38 C5
Snailwell CB8 33 K2
Snainton YO13 59 F1
Snaith DN14 58 C7
Snape *N.Yorks.* DL8 57 H1
Snape *Suff.* IP17 35 H3
Snape Green PR8 48 C1
Snape Watering IP17 35 H3
Snarestone DE12 41 F5
Snarford LN8 52 D4
Snargate TN29 14 E5
Snave TN29 15 F5
Sneachill WR7 29 J3
Snead SY15 38 C6
Snead's Green DY13 29 H2
Sneath Common NR15 45 F7
Sneaton YO22 63 K6
Sneatonthorpe YO22 63 J2
Snelland LN3 52 D4
Snellings CA22 60 A6
Snelston DE6 40 D1
Snetterton NR16 44 D6
Snettisham PE31 44 A2
Snipeshill ME10 25 F5
Sniseabhal (Snishival) HS8 84 C1
Snishival (Sniseabhal) HS8 84 C1
Snitter NE65 71 F3
Snitterby DN21 52 C3
Snitterfield CV37 30 D3
Snitterton DE4 50 E6
Snittlegarth CA7 60 D3
Snitton SY8 28 E1
Snodhill HR3 28 C4
Snodland ME6 24 D5
Snow End SG9 33 H5
Snow Street IP22 44 E7
Snowden Hill S35 50 E2
Snowshill WR12 30 B5
Soar *Cardiff* CF15 18 D3
Soar *Carmar.* SA19 17 K3
Soar *Devon* TQ7 5 H7
Soay PH41 85 K3
Soberton SO32 11 H3
Soberton Heath SO32 11 H3
Sockbridge CA10 61 G4
Sockburn DL2 62 E6
Sodom LL16 47 J5
Sodylt Bank SY12 38 C2
Softley DL13 62 B4
Soham CB7 33 J1
Soham Cotes CB7 33 J1
Solas (Sollas) HS6 92 D4
Soldon EX22 6 B4
Soldon Cross EX22 6 B4
Soldridge GU34 11 H1
Sole Street *Kent* CT4 15 F3
Sole Street *Kent* DA12 24 C5
Soleburn DG9 64 A4
Solihull B91 30 C1
Solihull Lodge B90 30 B1
Sollas (Solas) HS6 92 D4
Sollers Dilwyn HR4 28 D3
Sollers Hope HR1 29 F5
Sollom PR4 48 D1
Solomon's Tump GL19 29 G7
Solsgirth FK14 75 H1
Solva SA62 16 A3
Solwaybank DG14 69 J6
Somerby *Leics.* LE14 42 A4
Somerby *Lincs.* DN38 52 D2
Somercotes DE55 51 G7
Somerford ST19 40 B5
Somerford Keynes GL7 20 D2
Somerley PO20 12 B7
Somerleyton NR32 45 J6
Somersal Herbert DE6 40 D2
Somersby PE23 53 G5
Somersham *Cambs.* PE28 33 G1
Somersham *Suff.* IP8 34 E4
Somerton *Newport* NP19 19 G3
Somerton *Oxon.* OX25 31 F6
Somerton *Som.* TA11 8 D2
Somerton *Suff.* IP29 34 C3
Sompting BN15 12 E6
Sompting Abbotts BN15 12 E6
Sonning RG4 22 A4
Sonning Common RG4 22 A3
Sonning Eye RG4 22 A4
Sontley LL13 38 C1
Sookholme NG19 51 H6
Sopley BH23 10 C5
Sopworth SN14 20 B3
Sorbie DG8 64 E6
Sordale KW12 105 G2
Sorisdale PA78 78 D1
Sorn KA5 67 K1
Sornhill KA4 74 D7
Soroba PA34 79 K5
Sortat KW1 105 H2
Sotby LN8 53 F5
Sots Hole LN4 52 E6
Sotterley NR34 45 J7
Soudley TF9 39 G3
Soughton CH7 48 B6
Soulbury LU7 32 B6
Soulby CA17 61 J5
Souldern OX27 31 G5
Souldrop MK44 32 C2
Sound *Ches.E.* CW5 39 F1
Sound *Shet.* ZE1 107 N8
Sound *Shet.* ZE2 107 M7
Sourhope TD5 70 D1
Sourin KW17 106 D4
Sourton EX20 6 D6
Soutergate LA17 55 F1
South Acre PE32 44 C4
South Acton W5 22 E4
South Alkham CT15 15 H3
South Allington TQ7 5 H7
South Alloa FK7 75 G1
South Ambersham GU29 12 C4
South Anston S25 51 H4
South Ascot SL5 22 C5
South Baddesley SO41 10 E5
South Ballachulish PH49 80 B2
South Balloch KA26 67 H4
South Bank TS6 63 G4
South Barrow BA22 9 F2
South Bellsdyke FK2 75 H2
South Benfleet SS7 24 D3
South Bersted PO22 12 C6
South Blackbog AB51 91 F1
South Bockhampton BH23 10 C5
South Bowood DT6 8 D5
South Brent TQ10 5 G4
South Brentor PL19 6 C7
South Brewham BA10 9 G1
South Broomhill NE61 71 H4
South Burlingham NR13 45 H5
South Cadbury BA22 9 F2
South Cairn DG9 66 D7
South Carlton LN1 52 C5
South Cave HU15 59 F6
South Cerney GL7 20 D2
South Chard TA20 8 C4
South Charlton NE66 71 G1
South Cheriton BA8 9 F2
South Church DL14 62 D4
South Cliffe YO43 58 E6
South Clifton NG23 52 B5
South Cockerington LN11 53 G4
South Collafirth ZE2 107 M4
South Common BN8 13 G5
South Cornelly CF33 18 B3
South Corriegills KA27 73 J7
South Cove NR34 45 J7
South Creagan PA37 80 A3
South Creake NR21 44 C2
South Crosland HD4 50 D1
South Croxton LE7 41 J4
South Dalton HU17 59 F5
South Darenth DA4 23 J4
South Dell (Dail Bho Dheas) HS2 101 G1
South Duffield YO8 58 C6
South Elkington LN11 53 F4
South Elmsall WF9 51 G1
South End *Bucks.* LU7 32 B6
South End *Cumb.* LA14 55 F3
South End *Hants.* SP6 10 C3
South End *N.Lincs.* DN19 59 H7
South Erradale IV21 94 D4
South Fambridge SS4 24 E2
South Fawley OX12 21 G3
South Ferriby DN18 59 F7
South Field HU13 59 G7
South Flobbets AB51 91 F1
South Garth ZE2 107 P3
South Godstone RH9 23 G7
South Gorley SP6 10 C3
South Green *Essex* CM11 24 C2
South Green *Essex* CO5 34 E7
South Green *Norf.* NR20 44 E4
South Green *Suff.* IP23 35 F1
South Gyle EH12 75 K3
South Hall PA22 73 J3
South Hanningfield CM3 24 D2
South Harefield UB9 22 D3
South Harting GU31 11 J3
South Hayling PO11 11 J5
South Hazelrigg NE66 77 J7
South Heath HP16 22 C1
South Heighton BN9 13 H6
South Hetton DH6 62 E2
South Hiendley S72 51 F1
South Hill PL17 4 D3
South Hinksey OX1 21 J1
South Hole EX39 6 A4
South Holme YO62 58 C2
South Holmwood RH5 22 E7
South Hornchurch RM13 23 J3
South Hourat KA24 74 A5
South Huish TQ7 5 G6
South Hykeham LN6 52 C6
South Hylton SR4 62 E1
South Kelsey LN7 52 D3
South Kessock IV3 96 D7
South Killingholme DN40 52 E1
South Kilvington YO7 57 K1
South Kilworth LE17 41 J7
South Kirkby WF9 51 G1
South Kirkton AB32 91 F4
South Knighton TQ12 5 J3
South Kyme LN4 42 E1
South Lancing BN15 12 E6
South Ledaig PA37 80 A4
South Leigh OX29 21 G1
South Leverton DN22 51 K4
South Littleton WR11 30 B4
South Lopham IP22 44 E7
South Luffenham LE15 42 C5
South Malling BN7 13 H5
South Marston SN3 20 E3
South Middleton NE71 70 E1
South Milford LS25 57 K6
South Milton TQ7 5 G6
South Mimms EN6 23 F1
South Molton EX36 7 F3
South Moor DH9 62 C1
South Moreton OX11 21 J3
South Mundham PO20 12 B6
South Muskham NG23 51 K7
South Newbald YO43 59 F6
South Newington OX15 31 F5
South Newton SP2 10 B1
South Normanton DE55 51 G7
South Norwood SE25 23 G5
South Nutfield RH1 23 G7
South Ockendon RM15 23 J3
South Ormsby LN11 53 G5
South Ossett WF5 50 E1
South Otterington DL7 57 J1
South Owersby LN8 52 D3
South Oxhey WD19 22 E2
South Park RH2 23 F7
South Parks KY6 82 D7
South Perrott DT8 8 D4
South Petherton TA13 8 D3
South Petherwin PL15 6 B7
South Pickenham PE37 44 C5
South Pool TQ7 5 H6
South Queensferry (Queensferry) EH30 75 K3
South Radworthy EX36 7 F2
South Rauceby NG34 42 D1
South Raynham NR21 44 C3
South Redbriggs AB53 99 F6
South Reston LN11 53 H4
South Ronaldsay KW17 106 D9
South Ruislip HA4 22 E3
South Runcton PE33 44 A5
South Scarle NG23 52 B6
South Shian PA37 80 A3
South Shields NE33 71 J7
South Somercotes LN11 53 H3
South Somercotes Fen Houses LN11 53 H3
South Stainley HG3 57 J3
South Stoke *Oxon.* RG8 21 K3
South Stoke *W.Suss.* BN18 12 D6
South Street *E.Suss.* BN8 13 G5
South Street *Gt.Lon.* TN16 23 H6
South Street *Kent* DA13 24 C5
South Street *Kent* CT5 25 H5
South Street *Kent* ME9 24 E5
South Tawton EX20 6 E6
South Thoresby LN13 53 H5
South Tidworth SP9 21 F7
South Tottenham N15 23 G3
South Town *Devon* EX6 7 H7
South Town *Hants.* GU34 11 H1
South Uist (Uibhist a Deas) HS8 84 C1
South Upper Barrack AB41 99 H6
South View RG21 21 K6
South Walsham NR13 45 H4
South Warnborough RG29 22 A7
South Weald CM14 23 J2
South Weston OX9 22 A2
South Wheatley *Cornw.* PL15 4 C1
South Wheatley *Notts.* DN22 51 K4
South Whiteness ZE2 107 M8
South Wigston LE18 41 H6
South Willingham LN8 52 E4
South Wingfield DE55 51 F7
South Witham NG33 42 C4
South Wonston SO21 11 F1
South Woodham Ferrers CM3 24 E2
South Wootton PE30 44 A3
South Wraxall BA15 20 B5
South Yardley B26 40 D7
South Zeal EX20 6 E6
SOUTHALL UB 22 E3
Southam *Glos.* GL52 29 J6
Southam *Warks.* CV47 31 F2
SOUTHAMPTON SO 11 F3
Southampton Airport SO18 11 F3
Southbar PA4 74 C4
Southborough *Gt.Lon.* BR2 23 H5
Southborough *Kent* TN4 23 J7
Southbourne *Bourne.* BH6 10 C5
Southbourne *W.Suss.* PO10 11 J4
Southbrook EX5 7 J6
Southburgh IP25 44 E5
Southburn YO25 59 F4
Southchurch SS1 25 F3
Southcott *Devon* EX20 6 D6
Southcott *Wilts.* SN9 20 E6
Southcourt HP21 32 B7
Southdean TD9 70 B3
Southdene L32 48 D3
Southease BN7 13 H6
Southend *Aber.* AB53 99 F6
Southend *Arg. & B.* PA28 66 A3
Southend *Bucks.* RG9 22 A3
Southend *W.Berks.* RG7 21 J4
Southend *Wilts.* SN8 20 E4
Southend Airport SS2 24 E3
SOUTHEND-ON-SEA SS 24 E3
Southerfield CA7 60 C2
Southerly EX20 6 D7
Southern Green SG9 33 G5
Southerndown CF32 18 B4
Southerness DG2 65 K5
Southery PE38 44 A6
Southfield KY6 76 A1
Southfields SW18 23 F4
Southfleet DA13 24 C4
Southgate *Cere.* SY23 36 E7
Southgate *Gt.Lon.* N14 23 G2
Southgate *Norf.* NR10 45 F3
Southgate *Norf.* PE31 44 A2
Southgate *Swan.* SA3 17 J7
Southhill SG18 32 E4
Southington RG25 21 J7
Southleigh EX24 8 B5
Southmarsh BA9 9 G1
Southminster CM0 25 F2
Southmuir DD8 82 E2
Southoe PE19 32 E2
Southolt IP23 35 F2
Southorpe PE9 42 D5
Southowram HX3 57 G7
Southport PR8 48 C1
Southrepps NR11 45 G2
Southrey LN3 52 E6
Southrop GL7 20 E1
Southrope RG25 21 K7
Southsea *Ports.* PO4 11 H5
Southsea *Wrex.* LL11 48 B7
Southstoke BA2 20 A5
Southtown *Norf.* NR31 45 K5
Southtown *Ork.* KW17 106 D8
Southwaite *Cumb.* CA17 61 J6
Southwaite *Cumb.* CA4 61 F2
Southwater RH13 12 E4
Southwater Street RH13 12 E4
Southway BA5 19 J7
Southwell *Dorset* DT5 9 F7
Southwell *Notts.* NG25 51 J7
Southwick *D. & G.* DG2 65 K5
Southwick *Hants.* PO17 11 H4
Southwick *Northants.* PE8 42 D6
Southwick *Som.* TA9 19 G7
Southwick *T. & W.* SR5 62 E1
Southwick *W.Suss.* BN42 13 F6
Southwick *Wilts.* BA14 20 B6
Southwold IP18 35 K1
Southwood BA6 8 E1
Sowden EX8 7 H7
Sower Carr FY6 55 G5
Sowerby *N.Yorks.* YO7 57 K1
Sowerby *W.Yorks.* HX6 57 F7
Sowerby Bridge HX6 57 F7
Sowerby Row CA4 60 E2
Sowerhill TA22 7 G3
Sowley Green CB9 34 B3
Sowood HX4 50 C1
Sowton EX5 7 H6
Soyal IV24 96 C2
Spa Common NR28 45 G2
Spadeadam CA8 70 A6
Spalding PE11 43 F3
Spaldington DN14 58 D6
Spaldwick PE28 32 E1
Spalefield KY10 83 G7
Spalford NG23 52 B6
Spanby NG34 42 D2
Sparham NR9 44 E4
Spark Bridge LA12 55 G1
Sparkford BA22 9 F2
Sparkhill B11 40 C7
Sparkwell PL7 5 F5
Sparrow Green NR19 44 D4
Sparrowpit SK17 50 C4
Sparrow's Green TN5 13 K3
Sparsholt *Hants.* SO21 11 F1
Sparsholt *Oxon.* OX12 21 G3
Spartylea NE47 61 K2
Spath ST14 40 C2
Spaunton YO62 58 D1
Spaxton TA5 8 B1
Spean Bridge PH34 87 J6
Spear Hill RH20 12 E5
Speddoch DG2 68 D5
Speedwell BS16 19 K4
Speen *Bucks.* HP27 22 B1
Speen *W.Berks.* RG14 21 H5
Speeton YO14 59 H2
Speke L24 48 D4
Speldhurst TN3 23 J7
Spellbrook CM23 33 H7
Spelsbury OX7 30 E6
Spen Green CW11 49 H6
Spencers Wood RG7 22 A5
Spennithorne DL8 57 G1
Spennymoor DL16 62 D3
Spernall B80 30 B2
Spetchley WR5 29 H3
Spetisbury DT11 9 J4
Spexhall IP19 45 H7
Spey Bay IV32 98 B4
Speybridge PH26 89 H2
Speyview AB38 97 K7
Spilsby PE23 53 G6
Spindlestone NE70 77 K7
Spinkhill S21 51 G5
Spinningdale IV24 96 D3
Spirthill SN11 20 C4
Spital *High.* KW1 105 G3
Spital *W. & M.* SL4 22 C4
Spital in the Street LN8 52 C3
Spitalbrook EN11 23 G1
Spithurst BN8 13 H5
Spittal *D. & G.* DG8 64 E4
Spittal *D. & G.* DG8 64 D5
Spittal *E.Loth.* EH32 76 C3
Spittal *Northumb.* TD15 77 J5
Spittal *Pembs.* SA62 16 C3
Spittal of Glenmuick AB35 90 B6
Spittal of Glenshee PH10 82 C1
Spittalfield PH1 82 C3
Spixworth NR10 45 G4
Splayne's Green TN22 13 H4
Splott CF24 19 F4
Spofforth HG3 57 J4
Spondon DE21 41 G2
Spooner Row NR18 44 E6
Spoonley TF9 39 F2
Sporle PE32 44 C4
Sportsman's Arms LL16 47 H7
Spott EH42 76 E3
Spratton NN6 31 J1
Spreakley GU10 22 B7
Spreyton EX17 6 E6
Spriddlestone PL9 5 F5
Spridlington LN8 52 D4
Spring Grove TW7 22 E4
Spring Vale PO34 11 H5
Springburn G21 74 E4
Springfield *Arg. & B.* PA22 73 J3
Springfield *D. & G.* DG16 69 J7
Springfield *Fife* KY15 82 E6
Springfield *Moray* IV36 97 H6
Springfield *P. & K.* PH13 82 C4
Springfield *W.Mid.* B13 40 C7
Springfields Outlet Village PE12 43 F3
Springhill *Staffs.* WS14 40 C5
Springhill *Staffs.* WV11 40 B5
Springholm DG7 65 J4
Springkell DG11 69 H6
Springleys AB51 91 F1
Springside KA11 74 B7
Springthorpe DN21 52 B4
Springwell NE9 62 D1
Sproatley HU11 59 H6
Sproston Green CW4 49 G6
Sprotbrough DN5 51 H2
Sproughton IP8 35 F4
Sprouston TD5 77 F7
Sprowston NR7 45 G4
Sproxton *Leics.* LE14 42 B3
Sproxton *N.Yorks.* YO62 58 C1
Sprytown PL16 6 C7
Spurlands End HP15 22 B2
Spurstow CW6 48 E7
Spyway DT2 8 E5
Square Point DG7 65 H3
Squires Gate FY4 55 G6
Srondoire PA30 73 G3
Sronphadruig Lodge PH18 88 E7
Stableford *Shrop.* WV15 39 G6
Stableford *Staffs.* ST5 40 A2
Stacey Bank S6 50 E3
Stackhouse BD24 56 D3
Stackpole SA71 16 C6
Stacksteads OL13 56 D7
Staddiscombe PL9 5 F5
Staddlethorpe HU15 58 E7
Staden SK17 50 C5
Stadhampton OX44 21 K2
Stadhlaigearraidh (Stilligarry) HS8 84 C1
Staffield CA10 61 G2
Staffin IV51 93 K5
Stafford ST16 40 B3
Stagden Cross CM1 33 K7
Stagsden MK43 32 C4
Stagshaw Bank NE46 70 E7
Stain KW1 105 J2
Stainburn *Cumb.* CA14 60 B4
Stainburn *N.Yorks.* LS21 57 H5
Stainby NG33 42 C3
Staincross S75 51 F1
Staindrop DL2 62 C4
Staines-upon-Thames TW18 22 D4
Stainfield *Lincs.* PE10 42 D3
Stainfield *Lincs.* LN3 52 E5
Stainforth *N.Yorks.* BD24 56 D3
Stainforth *S.Yorks.* DN7 51 J1
Staining FY3 55 G6
Stainland HX4 50 C1
Stainsacre YO22 63 J2
Stainsby *Derbys.* S44 51 G6
Stainsby *Lincs.* LN9 53 G5
Stainton *Cumb.* LA8 55 J1
Stainton *Cumb.* CA11 61 F4
Stainton *Dur.* DL12 62 B5
Stainton *Middbro.* TS8 63 F5
Stainton *N.Yorks.* DL11 62 C7
Stainton *S.Yorks.* S66 51 H3
Stainton by Langworth LN3 52 D5
Stainton le Vale LN8 52 E3
Stainton with Adgarley LA13 55 F2
Staintondale YO13 63 J3
Stair *Cumb.* CA12 60 D4
Stair *E.Ayr.* KA5 67 J1
Stairfoot S70 51 F2
Staithes TS13 63 J5
Stake Pool PR3 55 H5
Stakeford NE62 71 H5
Stakes PO7 11 H4
Stalbridge DT10 9 G3
Stalbridge Weston DT10 9 G3
Stalham NR12 45 H3
Stalham Green NR12 45 H3
Stalisfield Green ME13 14 E2
Stalling Busk DL8 56 E1
Stallingborough DN41 52 E1
Stallington ST11 40 B2
Stalmine FY6 55 G5
Stalybridge SK15 49 J3
Stambourne CO9 34 B5

Sta - Str

Stamford Lincs. PE9 42 D5
Stamford Northumb. NE66 71 H2
Stamford Bridge Ches.W. & C. CH3 48 D6
Stamford Bridge E.Riding YO41 58 D4
Stamfordham NE18 71 F6
Stanah FY5 55 G5
Stanborough AL8 33 F7
Stanbridge Cen.Beds. LU7 32 C6
Stanbridge Dorset BH21 10 B4
Stanbridge Earls SO51 10 E2
Stanbury BD22 57 F6
Stand ML6 75 F4
Standburn FK1 75 H3
Standeford WV10 40 B5
Standen TN27 14 D4
Standen Street TN17 14 D4
Standerwick BA11 20 B6
Standford GU35 12 B3
Standford Bridge TF10 39 G3
Standish Glos. GL10 20 B1
Standish Gt.Man. WN6 48 E1
Standlake OX29 21 G1
Standon Hants. SO21 11 F2
Standon Herts. SG11 33 G6
Standon Staffs. ST21 40 A2
Standon Green End SG11 33 G7
Stane ML7 75 G5
Stanecastle KA11 74 B7
Stanfield NR20 44 D3
Stanford Cen.Beds. SG18 32 E4
Stanford Kent TN25 15 G4
Stanford Shrop. SY5 38 C4
Stanford Bishop WR6 29 F3
Stanford Bridge WR6 29 G2
Stanford Dingley RG7 21 J4
Stanford End RG7 22 A5
Stanford in the Vale SN7 21 G2
Stanford on Avon NN6 31 G1
Stanford on Soar LE12 41 H3
Stanford on Teme WR6 29 G2
Stanford Rivers CM5 23 J1
Stanford-le-Hope SS17 24 C3
Stanfree S44 51 G5
Stanghow TS12 63 H5
Stanground PE2 43 F6
Stanhoe PE31 44 C2
Stanhope Dur. DL13 62 A3
Stanhope Sc.Bord. ML12 69 G1
Stanion NN14 42 C7
Stanklyn DY10 29 H1
Stanley Derbys. DE7 41 G1
Stanley Dur. DH9 62 C1
Stanley Notts. NG17 51 G6
Stanley P. & K. PH1 82 C4
Stanley Staffs. ST9 49 J7
Stanley W.Yorks. WF3 57 J7
Stanley Wilts. SN15 20 C4
Stanley Common DE7 41 G1
Stanley Crook DL15 62 C3
Stanley Gate L39 48 D2
Stanley Green BH15 10 B5
Stanley Hill HR8 29 F4
Stanleygreen SY13 38 E2
Stanlow Ches.W. & C. CH65 48 D5
Stanlow Shrop. WV6 39 G6
Stanmer BN1 13 G5
Stanmore Gt.Lon. HA7 22 E2
Stanmore W.Berks. RG20 21 H4
Stannersburn NE48 70 C5
Stanningfield IP29 34 C3
Stannington Northumb. NE61 71 H6
Stannington S.Yorks. S6 51 F4
Stansbatch HR6 28 C2
Stansfield CO10 34 B3
Stanshope DE6 50 D7
Stanstead CO10 34 C4
Stanstead Abbotts SG12 33 G7
Stansted TN15 24 C5
Stansted Airport (London Stansted Airport) CM24 33 J6
Stansted Mountfitchet CM24 33 J6
Stanton Derbys. DE15 40 E4
Stanton Glos. WR12 30 B5
Stanton Northumb. NE65 71 G4
Stanton Staffs. DE6 40 D1
Stanton Suff. IP31 34 D1
Stanton by Bridge DE73 41 F3
Stanton by Dale DE7 41 G2
Stanton Drew BS39 19 J5
Stanton Fitzwarren SN6 20 E2
Stanton Harcourt OX29 21 H1
Stanton Hill NG17 51 G6
Stanton in Peak DE4 50 E6
Stanton Lacy SY8 28 D1
Stanton Lees DE4 50 E6
Stanton Long TF13 38 E6
Stanton Prior BA2 19 K5
Stanton St. Bernard SN8 20 D5
Stanton St. John OX33 21 J1
Stanton St. Quintin SN14 20 C4
Stanton Street IP31 34 D2
Stanton under Bardon LE67 41 G4
Stanton upon Hine Heath SY4 38 E3
Stanton Wick BS39 19 K5
Stanton-on-the-Wolds NG12 41 J2
Stanwardine in the Fields SY4 38 D3
Stanwardine in the Wood SY12 38 D3
Stanway Essex CO3 34 D6
Stanway Glos. GL54 30 B5
Stanway Green Essex CO3 34 D6
Stanway Green Suff. IP13 35 G1
Stanwell TW19 22 D4
Stanwell Moor TW19 22 D4
Stanwick NN9 32 C1
Stanwix CA3 60 F1
Stanydale ZE2 107 L7
Staoinebrig HS8 84 C1
Stapeley CW5 39 F1
Stapenhill DE15 40 E3
Staple Kent CT3 15 H2
Staple Som. TA4 7 K1
Staple Cross TA21 7 J3
Staple Fitzpaine TA3 8 B3
Staplecross TN32 14 C5
Staplefield RH17 13 F4
Stapleford Cambs. CB22 33 H3
Stapleford Herts. SG14 33 G7
Stapleford Leics. LE14 42 B4
Stapleford Lincs. LN6 52 B7
Stapleford Notts. NG9 41 G2
Stapleford Wilts. SP3 10 B1
Stapleford Abbotts RM4 23 H2
Stapleford Tawney RM4 23 J2
Staplegrove TA2 8 B2
Staplehay TA3 8 B2
Staplehurst TN12 14 C3
Staplers PO30 11 G6
Staplestreet ME13 25 G5
Stapleton Cumb. CA6 70 A6
Stapleton Here. LD8 28 C2
Stapleton Leics. LE9 41 G6
Stapleton N.Yorks. DL2 62 D5
Stapleton Shrop. SY5 38 D5
Stapleton Som. TA12 8 D2
Stapley TA3 7 K4
Staploe PE19 32 E2
Staplow HR8 29 F4
Star Fife KY7 82 E7
Star Pembs. SA35 17 F2
Star Som. BS25 19 H6
Starbotton BD23 56 E2
Starcross EX6 7 H7
Stareton CV8 30 E1
Starkholmes DE4 51 F7
Starling BL8 49 G1
Starling's Green CB11 33 H5
Starr KA6 67 J4
Starston IP20 45 G7
Startforth DL12 62 B5
Startley SN15 20 C3
Statham WA13 49 F4
Stathe TA7 8 C2
Stathern LE14 42 A2
Station Town TS28 63 F3
Staughton Green PE19 32 E2
Staughton Highway PE19 32 E2
Staunton Glos. GL16 28 E7
Staunton Glos. GL19 29 G6
Staunton Harold Hall LE65 41 F3
Staunton in the Vale NG13 42 B1
Staunton on Arrow HR6 28 C2
Staunton on Wye HR4 28 C4
Staveley Cumb. LA8 61 F7
Staveley Derbys. S43 51 G5
Staveley N.Yorks. HG5 57 J3
Staveley-in-Cartmel LA12 55 G1
Staverton Devon TQ9 5 H4
Staverton Glos. GL51 29 H6
Staverton Northants. NN11 31 G2
Staverton Wilts. BA14 20 B5
Staverton Bridge GL51 29 H6
Stawell TA7 8 C1
Stawley TA21 7 J3
Staxigoe KW1 105 J3
Staxton YO12 59 G2
Staylittle (Penffordd-las) SY19 37 H6
Staynall FY6 55 G5
Staythorpe NG23 51 K7
Stean HG3 57 F2
Steane NN13 31 G5
Stearsby YO61 58 C2
Steart TA5 19 F7
Stebbing CM6 33 K6
Stebbing Green CM6 33 K6
Stechford B33 40 D7
Stedham GU29 12 B4
Steel Cross TN6 13 J3
Steel Green LA18 54 E2
Steele Road TD9 70 A4
Steen's Bridge HR6 28 E3
Steep GU32 11 J2
Steep Marsh GU32 11 J2
Steeple Dorset BH20 9 J6
Steeple Essex CM0 25 F1
Steeple Ashton BA14 20 C6
Steeple Aston OX25 31 F6
Steeple Barton OX25 31 F6
Steeple Bumpstead CB9 33 K4
Steeple Claydon MK18 31 H6
Steeple Gidding PE28 42 E7
Steeple Langford SP3 10 B1
Steeple Morden SG8 33 F4
Steeraway TF1 39 F5
Steeton BD20 57 F5
Stein IV55 93 H6
Steinmanhill AB53 99 F6
Stella NE21 71 G7
Stelling Minnis CT4 15 G3
Stembridge TA12 8 D2
Stemster High. KW12 105 G2
Stemster High. KW5 105 G4
Stemster House KW12 105 G2
Stenalees PL26 4 A5
Stenhill EX15 7 J4
Stenhousemuir FK5 75 G2
Stenigot LN11 53 F4
Stenis HS1 101 G4
Stenness ZE2 107 L5
Stenscholl IV51 93 K5
Stenson DE73 41 F3
Stenton E.Loth. EH42 76 E3
Stenton P. & K. PH8 82 B3
Steornabhagh (Stornoway) HS1 101 G4
Stepaside Pembs. SA67 16 E5
Stepaside Powys SY16 37 K7
Stepney E1 23 G3
Steppingley MK45 32 D5
Stepps G33 74 E4
Sternfield IP17 35 H2
Sterridge EX34 6 D1
Stert SN10 20 D6
Stetchworth CB8 33 K3
STEVENAGE SG 33 F6
Stevenston KA20 74 A6
Steventon Hants. RG25 21 J7
Steventon Oxon. OX13 21 H2
Steventon End CB10 33 K4
Stevington MK43 32 C3
Stewartby MK43 32 D4
Stewarton D. & G. DG8 64 E6
Stewarton E.Ayr. KA3 74 C6
Stewkley LU7 32 B6
Stewley TA19 8 C3
Stewton LN11 53 G4
Steyning BN44 12 E5
Steynton SA73 16 C5
Stibb EX23 6 A4
Stibb Cross EX38 6 C4
Stibb Green SN8 21 F5
Stibbard NR21 44 D3
Stibbington PE8 42 D6
Stichill TD5 77 F7
Sticker PL26 3 G3
Stickford PE22 53 G6
Sticklepath Devon EX20 6 E6
Sticklepath Som. TA20 8 C3
Stickling Green CB11 33 H5
Stickney PE22 53 G7
Stiff Street ME9 24 E5
Stiffkey NR23 44 D1
Stifford's Bridge WR13 29 G4
Stileway BA6 19 H7
Stilligarry (Stadhlaigearraidh) HS8 84 C1
Stillingfleet YO19 58 B5
Stillington N.Yorks. YO61 58 B3
Stillington Stock. TS21 62 E4
Stilton PE7 42 E7
Stinchcombe GL11 20 A2
Stinsford DT2 9 G5
Stirchley Tel. & W. TF3 39 G5
Stirchley W.Mid. B30 40 C7
Stirkoke House KW1 105 J3
Stirling Aber. AB42 99 K6
Stirling Stir. FK8 75 F1
Stirton BD23 56 E4
Stisted CM77 34 C6
Stitchcombe SN8 21 F5
Stithians TR3 2 E5
Stittenham IV17 96 D4
Stivichall CV3 30 E1
Stix PH15 81 J3
Stixwould LN10 52 E6
Stoak CH2 48 D5
Stobo EH45 75 K7
Stoborough BH20 9 J6
Stoborough Green BH20 9 J6
Stobwood ML11 75 H5
Stocinis (Stockinish) HS3 93 G2
Stock CM4 24 C2
Stock Green B96 29 J3
Stock Lane SN8 21 F4
Stock Wood B96 30 B3
Stockbridge Hants. SO20 10 E1
Stockbridge Stir. FK15 81 J7
Stockbridge W.Suss. PO19 12 B6
Stockbury ME9 24 E5
Stockcross RG20 21 H5
Stockdale TR11 2 E5
Stockdalewath CA5 60 E2
Stockerston LE15 42 B6
Stocking Green Essex CB10 33 J5
Stocking Green M.K. MK19 32 B4
Stocking Pelham SG9 33 H6
Stockingford CV10 41 F6
Stockinish (Stocinis) HS3 93 G2
Stockland Cardiff CF5 18 E4
Stockland Devon EX14 8 B4
Stockland Bristol TA5 19 F7
Stockleigh English EX17 7 G5
Stockleigh Pomeroy EX17 7 G5
Stockley SN11 20 D5
Stocklinch TA19 8 C3
STOCKPORT SK 49 H3
Stocksbridge S36 50 E3
Stocksfield NE43 71 F7
Stockton Here. HR6 28 E2
Stockton Norf. NR34 45 H6
Stockton Shrop. TF11 39 G6
Stockton Shrop. SY21 38 B5
Stockton Tel. & W. TF10 39 G4
Stockton Warks. CV47 31 F2
Stockton Wilts. BA12 9 J1
Stockton Heath WA4 49 F4
Stockton on Teme WR6 29 G2
Stockton on the Forest YO32 58 C4
Stockton-on-Tees TS19 63 F5
Stockwell GL4 29 J7
Stockwell Heath WS15 40 C3
Stockwood Bristol BS14 19 K5
Stockwood Dorset DT2 8 E4
Stodday LA2 55 H4
Stodmarsh CT3 25 J5
Stody NR24 44 E2
Stoer IV27 102 C6
Stoford Som. BA22 8 E3
Stoford Wilts. SP2 10 B1
Stogumber TA4 7 J2
Stogursey TA5 19 F7
Stoke Devon EX39 6 A3
Stoke Hants. SP11 21 H6
Stoke Hants. PO11 11 J4
Stoke Med. ME3 24 E4
Stoke Plym. PL3 4 E5
Stoke W.Mid. CV2 30 E1
Stoke Abbott DT8 8 D4
Stoke Albany LE16 42 B7
Stoke Ash IP23 35 F1
Stoke Bardolph NG14 41 J1
Stoke Bishop BS9 19 J4
Stoke Bliss WR15 29 F2
Stoke Bruerne NN12 31 J3
Stoke by Clare CO10 34 B4
Stoke Canon EX5 7 H6
Stoke Charity SO21 11 F1
Stoke Climsland PL17 4 D3
Stoke D'Abernon KT11 22 E6
Stoke Doyle PE8 42 D7
Stoke Dry LE15 42 B6
Stoke Edith HR1 29 F4
Stoke Farthing SP5 10 B2
Stoke Ferry PE33 44 B6
Stoke Fleming TQ6 5 J6
Stoke Gabriel TQ9 5 J5
Stoke Gifford BS34 19 K4
Stoke Golding CV13 41 F6
Stoke Goldington MK16 32 B4
Stoke Green SL2 22 C3
Stoke Hammond MK17 32 B6
Stoke Heath Shrop. TF9 39 F3
Stoke Heath Worcs. B60 29 J2
Stoke Holy Cross NR14 45 G5
Stoke Lacy HR7 29 F4
Stoke Lyne OX27 31 G6
Stoke Mandeville HP22 32 B7
Stoke Newington N16 23 G3
Stoke on Tern TF9 39 F3
Stoke Orchard GL52 29 J6
Stoke Pero TA24 7 G1
Stoke Poges SL2 22 C3
Stoke Pound B60 29 J2
Stoke Prior Here. HR6 28 E3
Stoke Prior Worcs. B60 29 J2
Stoke Rivers EX32 6 E2
Stoke Rochford NG33 42 C3
Stoke Row RG9 21 K3
Stoke St. Gregory TA3 8 C2
Stoke St. Mary TA3 8 B2
Stoke St. Michael BA3 19 K7
Stoke St. Milborough SY8 38 E7
Stoke sub Hamdon TA14 8 D3
Stoke Talmage OX9 21 K2
Stoke Trister BA9 9 G2
Stoke Villice BS40 19 J5
Stoke Wake DT11 9 G4
Stoke-by-Nayland CO6 34 D5
Stokeford BH20 9 H6
Stokeham DN22 51 K5
Stokeinteignhead TQ12 5 K3
Stokenchurch HP14 22 A2
Stokenham TQ7 5 J6
STOKE-ON-TRENT ST 40 A1
Stokesay SY7 38 D7
Stokesby NR29 45 J4
Stokesley TS9 63 G6
Stolford TA5 19 F7
Ston Easton BA3 19 K6
Stonar Cut CT13 25 K5
Stondon Massey CM15 23 J1
Stone Bucks. HP17 31 J7
Stone Glos. GL13 19 K2
Stone Kent DA9 23 J4
Stone Kent TN30 14 E5
Stone S.Yorks. S66 51 H4
Stone Som. BA4 8 E1
Stone Staffs. ST15 40 B2
Stone Worcs. DY10 29 H1
Stone Allerton BS26 19 H6
Stone Cross Dur. DL12 62 B5
Stone Cross E.Suss. BN24 13 K6
Stone Cross E.Suss. TN6 13 J4
Stone Cross Kent TN25 15 F4
Stone Cross Kent TN3 13 J3
Stone House LA10 56 C1
Stone Street Kent TN15 23 J6
Stone Street Suff. IP19 45 H7
Stone Street Suff. CO10 34 D5
Stonea PE15 43 H6
Stonebridge E.Suss. TN22 13 J4
Stonebridge N.Som. BS29 19 G6
Stonebridge Warks. CV7 40 E7
Stonebroom DE55 51 G7
Stonecross Green IP29 34 C3
Stonefield Arg. & B. PA29 73 G3
Stonefield Staffs. ST15 40 B2
Stonegate E.Suss. TN5 13 K4
Stonegate N.Yorks. YO21 63 J6
Stonegrave YO62 58 C2
Stonehaugh NE48 70 C6
Stonehaven AB39 91 G6
Stonehill KT16 22 C5
Stonehouse Ches.W. & C. CH3 48 E5
Stonehouse D. & G. DG2 65 J4
Stonehouse Glos. GL10 20 B1
Stonehouse Northumb. NE49 61 H1
Stonehouse Plym. PL1 4 E5
Stonehouse S.Lan. ML9 75 F6
Stoneleigh Surr. KT17 23 F5
Stoneleigh Warks. CV8 30 E1
Stoneley Green CW5 49 F7
Stonely PE19 32 E2
Stoner Hill GU32 11 J2
Stones OL14 56 E7
Stones Green CO12 35 F6
Stonesby LE14 42 B3
Stonesfield OX29 30 E7
Stonestreet Green TN25 15 F4
Stonethwaite CA12 60 D5
Stoney Cross SO43 10 D3
Stoney Middleton S32 50 E5
Stoney Stanton LE9 41 G6
Stoney Stoke BA9 9 G1
Stoney Stratton BA4 9 F1
Stoney Stretton SY5 38 C5
Stoneyburn EH47 75 H4
Stoneyford EX10 7 J7
Stoneygate LE2 41 J5
Stoneyhills CM0 25 F2
Stoneykirk DG9 64 A5
Stoneywood AB21 91 G3
Stonganess ZE2 107 P2
Stonham Aspal IP14 35 F3
Stonnall WS9 40 C5
Stonor RG9 22 A3
Stonton Wyville LE16 42 A6
Stony Houghton NG19 51 G6
Stony Stratford MK11 31 J4
Stonybreck ZE2 107 K2
Stoodleigh Devon EX16 7 H4
Stoodleigh Devon EX32 6 E2
Stopham RH20 12 D5
Stopsley LU2 32 E6
Stoptide PL27 3 G1
Storeton CH63 48 C4
Stormontfield PH2 82 C5
Stornoway (Steornabhagh) HS1 101 G4
Stornoway Airport HS2 101 G4
Storridge WR13 29 G4
Storrington RH20 12 D5
Storrs S6 50 E4
Storth LA7 55 H1
Storwood YO42 58 D5
Stotfield IV31 97 K4
Stotfold SG5 33 F5
Stottesdon DY14 39 F7
Stoughton Leics. LE2 41 J5
Stoughton Surr. GU2 22 C6
Stoughton W.Suss. PO18 11 J3
Stoughton Cross BS28 19 H7
Stoul PH41 86 D5
Stoulton WR7 29 J4
Stour Provost SP8 9 G2
Stour Row SP7 9 H2
Stourbridge DY8 40 A7
Stourpaine DT11 9 H4
Stourport-on-Severn DY13 29 H1
Stourton Staffs. DY7 40 A7
Stourton Warks. CV36 30 D5
Stourton Wilts. BA12 9 G1
Stourton Caundle DT10 9 G3
Stove KW17 106 F4
Stoven NR34 45 J7
Stow Lincs. LN1 52 B4
Stow Sc.Bord. TD1 76 C6
Stow Bardolph PE34 44 A5
Stow Bedon NR17 44 D6
Stow cum Quy CB25 33 J2
Stow Longa PE28 32 E1
Stow Maries CM3 24 E2
Stow Pasture LN1 52 B4
Stowbridge PE34 43 J5
Stowe Glos. GL15 19 J1
Stowe Shrop. LD7 28 C1
Stowe Staffs. WS13 40 D4
Stowe-by-Chartley ST18 40 C3
Stowehill NN7 31 H3
Stowell Glos. GL54 30 B7
Stowell Som. DT9 9 F2
Stowey BS39 19 J6
Stowford Devon EX20 6 C7
Stowford Devon EX37 6 E3
Stowford Devon EX10 7 K7
Stowlangtoft IP31 34 D2
Stowmarket IP14 34 E3
Stow-on-the-Wold GL54 30 C6
Stowting TN25 15 G3
Stowupland IP14 34 E3
Straad PA20 73 J4
Stracathro DD9 83 H1
Strachan AB31 90 E5
Strachur (Clachan Strachur) PA27 80 B7
Stradbroke IP21 35 G1
Stradishall CB8 34 B3
Stradsett PE33 44 A5
Stragglethorpe LN5 52 C7
Straight Soley RG17 21 G4
Straiton Edin. EH20 76 A4
Straiton S.Ayr. KA19 67 H3
Straloch Aber. AB21 91 G2
Straloch P. & K. PH10 82 B1
Stramshall ST14 40 C2
Strands LA18 54 E1
Strang IM4 54 C6
Strangford HR9 28 E6
Strannda HS5 93 F3
Stranraer DG9 64 A4
Strata Florida SY25 27 G2
Stratfield Mortimer RG7 21 K5
Stratfield Saye RG7 21 K5
Stratfield Turgis RG27 21 K6
Stratford Cen.Beds. SG19 32 E4
Stratford Glos. GL20 29 H5
Stratford Gt.Lon. E15 23 G3
Stratford St. Andrew IP17 35 H3
Stratford St. Mary CO7 34 E5

Str - Tan

Stratford sub Castle SP1 10 C1
Stratford Tony SP5 10 B2
Stratford-upon-Avon CV37 30 D3
Strath KW1 105 H3
Strathan *High.* IV27 102 C6
Strathan *High.* PH34 87 F5
Strathaven ML10 75 F6
Strathblane G63 74 D3
Strathcanaird IV26 95 H1
Strathcarron IV54 95 F7
Strathdon AB36 90 B3
Strathgirnock AB35 90 B5
Strathkinness KY16 83 F6
Strathmiglo KY14 82 D6
Strathpeffer IV14 96 B6
Strathrannoch IV23 95 K4
Strathtay PH9 82 A2
Strathwhillan KA27 73 J7
Strathy KW14 104 D2
Strathyre FK18 81 G6
Stratton *Cornw.* EX23 6 A5
Stratton *Dorset* DT2 9 F5
Stratton *Glos.* GL7 20 D1
Stratton Audley OX27 31 H6
Stratton Hall IP10 35 G5
Stratton St. Margaret SN3 20 E3
Stratton St. Michael NR15 45 G6
Stratton Strawless NR10 45 G3
Stratton-on-the-Fosse BA3 19 K6
Stravanan PA20 73 J5
Stravithie KY16 83 G6
Strawberry Hill TW2 22 E4
Stream TA4 7 J2
Streat BN6 13 G5
Streatham SW16 23 F4
Streatham Vale SW16 23 F4
Streatley *Cen.Beds.* LU3 32 D6
Streatley *W.Berks.* RG8 21 J3
Street *Devon* EX12 7 K7
Street *Lancs.* PR3 55 J4
Street *N.Yorks.* YO21 63 J6
Street *Som.* BA16 8 D1
Street *Som.* TA20 8 C4
Street Ashton CV23 41 G7
Street Dinas SY11 38 C2
Street End PO20 12 B7
Street Gate NE16 62 D1
Street Houses LS24 58 B5
Street Lane DE5 41 F1
Street on the Fosse BA4 9 F1
Streethay WS13 40 D4
Streethouse WF7 57 J7
Streetlam DL7 62 E7
Streetly B74 40 C6
Streetly End CB21 33 K4
Strefford SY7 38 D7
Strelley NG8 41 H1
Strensall YO32 58 C3
Strensham WR8 29 J4
Stretcholt TA6 19 F7
Strete TQ6 5 J6
Stretford *Gt.Man.* M32 49 G3
Stretford *Here.* HR6 28 D3
Stretford *Here.* HR6 28 E3
Strethall CB11 33 H5
Stretham CB6 33 J1
Strettington PO18 12 B6
Stretton *Ches.W. & C.* SY14 48 D7
Stretton *Derbys.* DE55 51 F6
Stretton *Rut.* LE15 42 C4
Stretton *Staffs.* ST19 40 A4
Stretton *Staffs.* DE13 40 E3
Stretton *Warr.* WA4 49 F4
Stretton en le Field DE12 41 F4
Stretton Grandison HR8 29 F4
Stretton Heath SY5 38 C4
Stretton Sugwas HR4 28 D4
Stretton under Fosse CV23 41 G7
Stretton Westwood TF13 38 E6
Stretton-on-Dunsmore CV23 31 F1
Stretton-on-Fosse GL56 30 D5
Stribers LA12 55 G1
Strichen AB43 99 H5
Strines SK6 49 J4
Stringston TA5 7 K1
Strixton NN29 32 C2
Stroat NP16 19 J2
Stromeferry IV53 86 E1
Stromemore IV54 86 E1
Stromness KW16 106 B7
Stronaba PH34 87 J6
Stronachlachar FK8 81 F6
Strone *Arg. & B.* PA23 73 K2
Strone *High.* IV63 88 C2
Strone *High.* PH33 87 H6
Strone *Stir.* FK17 81 F6
Stronechrubie IV27 102 E7
Stronlonag PA23 73 K2
Stronmilchan PA33 80 C5
Stronsay KW17 106 F5
Stronsay Airfield KW17 106 F5
Strontian PH36 79 K1
Strontoiller PA34 80 A5
Stronvar FK19 81 G5
Strood ME2 24 D5
Strood Green *Surr.* RH3 23 F7
Strood Green *W.Suss.* RH14 12 D4
Strood Green *W.Suss.* RH12 12 E3
Stroquhan DG2 68 D5
Stroud *Glos.* GL5 20 B1
Stroud *Hants.* GU32 11 J2
Stroud Common GU5 22 D7
Stroud Green *Essex* SS4 24 E2
Stroud Green *Glos.* GL10 20 B1
Stroude GU25 22 D5
Stroul G84 74 A2
Stroxton NG33 42 C2
Struan *High.* IV56 85 J1
Struan *P. & K.* PH18 81 J1
Strubby *Lincs.* LN13 53 H4
Strubby *Lincs.* LN8 52 E5
Strumpshaw NR13 45 H5
Struthers KY15 82 E7
Struy IV4 88 B1
Stryd y Facsen LL65 46 B4
Stryt-cae-rhedyn CH7 48 B6
Stryt-issa LL14 38 B1
Stuartfield AB42 99 H6
Stub Place LA19 60 B7
Stubber's Green WS9 40 C5
Stubbington PO14 11 G4
Stubbins BL0 49 G1
Stubbs Green NR14 45 H6
Stubhampton DT11 9 J3
Stubley S18 51 F5
Stubshaw Cross WN4 48 E2
Stubton NG23 42 B1
Stuck *Arg. & B.* PA20 73 J4
Stuck *Arg. & B.* PA23 73 K1
Stuckbeg PA24 74 A1
Stuckgowan G83 80 E7
Stuckindroin G83 80 E6
Stuckreoch PA27 73 J1
Stuckton SP6 10 C3
Stud Green SL6 22 B4
Studdon NE47 61 K1
Studfold BD24 56 D2
Studham LU6 32 D7
Studholme CA7 60 D1
Studland BH19 10 B6
Studley *Warks.* B80 30 B2
Studley *Wilts.* SN11 20 C4
Studley Common B80 30 B2
Studley Green HP14 22 A2
Studley Roger HG4 57 H2
Stuggadhoo IM4 54 C6
Stump Cross *Essex* CB10 33 J4
Stump Cross *Lancs.* PR3 55 J6
Stuntney CB7 33 J1
Stunts Green BN27 13 K5
Sturbridge ST21 40 A2
Sturgate DN21 52 B4
Sturmer CB9 33 K4
Sturminster Common DT10 9 G3
Sturminster Marshall BH21 9 J4
Sturminster Newton DT10 9 G3
Sturry CT2 25 H5
Sturton by Stow LN1 52 B4
Sturton le Steeple DN22 51 K4
Stuston IP21 35 F1
Stutton *N.Yorks.* LS24 57 K5
Stutton *Suff.* IP9 35 F5
Styal SK9 49 H4
Styrrup DN11 51 J3
Suainebost (Swainbost) HS2 101 H1
Suardail HS2 101 G4
Succoth *Aber.* AB54 90 C1
Succoth *Arg. & B.* G83 80 D7
Succothmore PA27 80 C7
Suckley WR6 29 G3
Suckley Green WR6 29 G3
Suckley Knowl WR6 29 G3
Sudborough NN14 42 C7
Sudbourne IP12 35 J3
Sudbrook *Lincs.* NG32 42 C1
Sudbrook *Mon.* NP26 19 J3
Sudbrooke LN2 52 D5
Sudbury *Derbys.* DE6 40 D2
Sudbury *Gt.Lon.* HA0 22 E3
Sudbury *Suff.* CO10 34 C4
Sudden OL11 49 H1
Sudgrove GL6 20 C1
Suffield *N.Yorks.* YO13 63 J3
Suffield *Norf.* NR11 45 G2
Sugarloaf TN26 14 E4
Sugnall ST21 39 G2
Sugwas Pool HR4 28 D4
Suie Lodge Hotel FK20 81 F5
Suisnish IV49 86 B3
Sulby *I.o.M.* IM7 54 C4
Sulby *I.o.M.* IM4 54 C5
Sulgrave OX17 31 G4
Sulham RG8 21 K4
Sulhamstead RG7 21 K5
Sullington RH20 12 D5
Sullom ZE2 107 M5
Sullom Voe Oil Terminal ZE2 107 M5
Sully CF64 18 E5
Sumburgh ZE3 107 M11
Sumburgh Airport ZE3 107 M11
Summer Bridge HG3 57 H3
Summer Isles IV26 95 F1
Summer Lodge DL8 62 A7
Summercourt TR8 3 F3
Summerfield *Norf.* PE31 44 B2
Summerfield *Worcs.* DY11 29 H1
Summerhill LL11 48 C7
Summerhouse DL2 62 D5
Summerlands LA8 55 J1
Summerleaze NP26 19 H3
Summertown OX2 21 J1
Summit OL15 49 J1
Sun Green SK15 49 J3
Sunadale PA28 73 G6
Sunbiggin CA10 61 H6
Sunbury TW16 22 E5
Sundaywell DG2 68 D5
Sunderland *Cumb.* CA13 60 C3
Sunderland *Lancs.* LA3 55 H4
SUNDERLAND *T. & W.* SR 62 E1
Sunderland Bridge DH6 62 D3
Sundhope TD7 69 J1
Sundon Park LU3 32 D6
Sundridge TN14 23 H6
Sundrum Mains KA6 67 J1
Sunhill GL7 20 E1
Sunipol PA75 78 E2
Sunk Island HU12 53 F1
Sunningdale SL5 22 C5
Sunninghill SL5 22 C5
Sunningwell OX13 21 H1
Sunniside *Dur.* DL13 62 C3
Sunniside *T. & W.* NE16 62 D1
Sunny Bank LA21 60 D7
Sunny Brow DL15 62 C3
Sunnylaw FK9 75 F1
Sunnyside *Aber.* AB12 91 G5
Sunnyside *Northumb.* NE46 70 E7
Sunnyside *S.Yorks.* S65 51 G3
Sunnyside *W.Suss.* RH19 13 G3
Sunton SN8 21 F6
Sunwick TD15 77 G5
Surbiton KT6 22 E5
Surfleet PE11 43 F3
Surfleet Seas End PE11 43 F3
Surlingham NR14 45 H5
Sustead NR11 45 F2
Susworth DN17 52 B2
Sutcombe EX22 6 B4
Sutcombemill EX22 6 B4
Suton NR18 44 E6
Sutors of Cromarty IV11 97 F5
Sutterby LN13 53 G5
Sutterton PE20 43 F2
Sutton *Cambs.* CB6 33 H1
Sutton *Cen.Beds.* SG19 33 F4
Sutton *Devon* EX17 7 F5
Sutton *Devon* TQ7 5 H6
SUTTON *Gt.Lon.* SM 23 F5
Sutton *Kent* CT15 15 J3
Sutton *Lincs.* LN5 52 B7
Sutton *Norf.* NR12 45 H3
Sutton *Notts.* NG13 42 A2
Sutton *Notts.* DN22 51 J4
Sutton *Oxon.* OX29 21 H1
Sutton *Pembs.* SA62 16 C4
Sutton *Peter.* PE5 42 D6
Sutton *S.Yorks.* DN6 51 H1
Sutton *Shrop.* WV16 39 G7
Sutton *Shrop.* TF9 39 F2
Sutton *Shrop.* SY2 38 E4
Sutton *Shrop.* SY11 38 C3
Sutton *Staffs.* TF10 39 G3
Sutton *Suff.* IP12 35 H4
Sutton *W.Suss.* RH20 12 C5
Sutton Abinger RH5 22 E7
Sutton at Hone DA4 23 J4
Sutton Bassett LE16 42 A7
Sutton Benger SN15 20 C4
Sutton Bingham BA22 8 E3
Sutton Bonington LE12 41 H3
Sutton Bridge PE12 43 H3
Sutton Cheney CV13 41 G5
Sutton Coldfield B74 40 D6
Sutton Courtenay OX14 21 J2
Sutton Crosses PE12 43 H3
Sutton Grange HG4 57 H2
Sutton Green *Oxon.* OX29 21 H1
Sutton Green *Surr.* GU4 22 D6
Sutton Green *Wrex.* LL13 38 D1
Sutton Holms BH21 10 B4
Sutton Howgrave DL8 57 J2
Sutton in Ashfield NG17 51 G7
Sutton in the Elms LE9 41 H6
Sutton Ings HU8 59 H6
Sutton Lane Ends SK11 49 J5
Sutton le Marsh LN12 53 J4
Sutton Leach WA9 48 E3
Sutton Maddock TF11 39 G5
Sutton Mallet TA7 8 C1
Sutton Mandeville SP3 9 J2
Sutton Montis BA22 9 F2
Sutton on Sea LN12 53 J4
Sutton on the Hill DE6 40 E2
Sutton on Trent NG23 51 K6
Sutton Poyntz DT3 9 G6
Sutton St. Edmund PE12 43 G4
Sutton St. James PE12 43 H4
Sutton St. Nicholas HR1 28 E4
Sutton Scarsdale S44 51 G6
Sutton Scotney SO21 11 F1
Sutton upon Derwent YO41 58 D5
Sutton Valence ME17 14 D3
Sutton Veny BA12 20 B7
Sutton Waldron DT11 9 H3
Sutton Weaver WA7 48 E5
Sutton Wick *B. & N.E.Som.* BS40 19 J6
Sutton Wick *Oxon.* OX14 21 H2
Sutton-in-Craven BD20 57 F5
Sutton-on-Hull HU7 59 H6
Sutton-on-the-Forest YO61 58 B3
Sutton-under-Brailes CV36 30 E5
Sutton-under-Whitestonecliffe YO7 57 K1
Swaby LN13 53 G5
Swadlincote DE11 41 F4
Swaffham PE37 44 C5
Swaffham Bulbeck CB25 33 J2
Swaffham Prior CB25 33 J2
Swafield NR28 45 G2
Swainbost (Suainebost) HS2 101 H1
Swainby DL6 63 F6
Swainsthorpe NR14 45 G5
Swainswick BA1 20 A5
Swalcliffe OX15 30 E5
Swalecliffe CT5 25 H5
Swallow LN7 52 E2
Swallow Beck LN6 52 C6
Swallowcliffe SP3 9 J2
Swallowfield RG7 22 A5
Swallows Cross CM15 24 C2
Swampton SP11 21 H6
Swan Green *Ches.W. & C.* WA16 49 G5
Swan Green *Suff.* IP19 35 G1
Swan Street CO6 34 C6
Swanage BH19 10 B7
Swanbach CW3 39 F1
Swanbourne MK17 32 B6
Swanbridge CF64 18 E5
Swancote WV15 39 G6
Swanland HU14 59 F7
Swanlaws TD8 70 C2
Swanley BR8 23 J5
Swanley Village BR8 23 J5
Swanmore *Hants.* SO32 11 G3
Swanmore *I.o.W.* PO33 11 G5
Swannington *Leics.* LE67 41 G4
Swannington *Norf.* NR9 45 F4
Swanscombe DA10 24 C4
SWANSEA (ABERTAWE) SA 17 K6
Swanston EH10 76 A4
Swanton Abbot NR10 45 G3
Swanton Morley NR20 44 E4
Swanton Novers NR24 44 E2
Swanton Street ME9 14 D2
Swanwick *Derbys.* DE55 51 G7
Swanwick *Hants.* SO31 11 G4
Swanwick Green SY13 38 E1
Swarby NG34 42 D1
Swardeston NR14 45 G5
Swarkestone DE73 41 F3
Swarland NE65 71 G3
Swarraton SO24 11 G1
Swarthmoor LA12 55 F2
Swaton NG34 42 E2
Swavesey CB24 33 G2
Sway SO41 10 D5
Swayfield NG33 42 C3
Swaythling SO18 11 F3
Swaythorpe YO25 59 G3
Sweetham EX5 7 G6
Sweethay TA3 8 B2
Sweetshouse PL30 4 A4
Sweffling IP17 35 H2
Swell TA3 8 C2
Swepstone LE67 41 F4
Swerford OX7 30 E5
Swettenham CW12 49 H6
Swffryd NP11 19 F2
Swift's Green TN27 14 D3
Swiftsden TN19 14 C5
Swilland IP6 35 F3
Swillington LS26 57 J6
Swimbridge EX32 6 E3
Swimbridge Newland EX32 6 D2
Swinbrook OX18 30 D7
Swincliffe HG3 57 H4
Swincombe EX31 6 E1
Swinden BD23 56 D4
Swinderby LN6 52 B6
Swindon *Staffs.* DY3 40 A6
SWINDON *Swin.* SN 20 E3
Swindon Village GL51 29 J6
Swine HU11 59 H6
Swinefleet DN14 58 D7
Swineford BS30 19 K5
Swineshead *Bed.* MK44 32 D2
Swineshead *Lincs.* PE20 43 F1
Swineshead Bridge PE20 43 F1
Swineside DL8 57 F1
Swiney KW3 105 H5
Swinford *Leics.* LE17 31 G1
Swinford *Oxon.* OX29 21 H1
Swingate NG16 41 H1
Swingfield Minnis CT15 15 H3
Swingleton Green IP7 34 D4
Swinhoe NE67 71 H1
Swinhope LN8 53 F3
Swining ZE2 107 N6
Swinithwaite DL8 57 F1
Swinscoe DE6 40 D1
Swinside Hall TD8 70 C2
Swinstead NG33 42 D3
Swinton *Gt.Man.* M27 49 G2
Swinton *N.Yorks.* YO17 58 D2
Swinton *N.Yorks.* HG4 57 H2
Swinton *S.Yorks.* S64 51 G3
Swinton *Sc.Bord.* TD11 77 G6
Swinton Quarter TD11 77 G6
Swintonmill TD12 77 G6
Swithland LE12 41 H4
Swordale IV16 96 C5
Swordland PH41 86 D5
Swordle PH36 86 B7
Swordly KW14 104 C2
Sworton Heath WA16 49 F4
Swyddffynnon SY25 27 F2
Swyncombe RG9 21 K2
Swynnerton ST15 40 A2
Swyre DT2 8 E6
Sychnant LD6 27 J1
Syde GL53 29 J7
Sydenham *Gt.Lon.* SE26 23 G4
Sydenham *Oxon.* OX39 22 A1
Sydenham Damerel PL19 4 E3
Syderstone PE31 44 C2
Sydling St. Nicholas DT2 9 F5
Sydmonton RG20 21 H6
Sydney CW1 49 G7
Syerston NG23 42 A1
Sykehouse DN14 51 J1
Sykes BB7 56 B4
Sylen SA15 17 J5
Symbister ZE2 107 P6
Symington *S.Ayr.* KA1 74 B7
Symington *S.Lan.* ML12 75 H7
Symonds Yat HR9 28 E7
Symondsbury DT6 8 D5
Synod Inn (Post-mawr) SA44 26 D3
Syre KW11 103 J4
Syreford GL54 30 B6
Syresham NN13 31 H4
Syston *Leics.* LE7 41 J4
Syston *Lincs.* NG32 42 C1
Sytchampton DY13 29 H2
Sywell NN6 32 B2

T

Taagan IV22 95 G5
Tableyhill WA16 49 G5
Tabost *W.Isles* HS2 101 F6
Tabost (Harbost) *W.Isles* HS2 101 H1
Tachbrook Mallory CV33 30 E2
Tacher KW5 105 G4
Tackley OX5 31 F6
Tacleit (Hacklete) HS2 100 D4
Tacolneston NR16 45 F6
Tadcaster LS24 57 K5
Tadden BH21 9 J4
Taddington *Derbys.* SK17 50 D5
Taddington *Glos.* GL54 30 B5
Taddiport EX38 6 C4
Tadley RG26 21 K5
Tadlow SG8 33 F4
Tadmarton OX15 30 E5
Tadpole Bridge SN7 21 G1
Tadworth KT20 23 F6
Tafarnaubach NP22 28 A7
Tafarn-y-bwlch SA41 16 D2
Tafarn-y-Gelyn CH7 47 K6
Taff Merthyr Garden Village CF46 18 E2
Taff's Well (Ffynnon Taf) CF15 18 E3
Tafolwern SY19 37 H5
Taibach *N.P.T.* SA13 18 A3
Tai-bach *Powys* SY10 38 A3
Taicynhaeaf LL40 37 F4
Tain *High.* KW14 105 H2
Tain *High.* IV19 96 E3
Tai'n Lôn LL54 46 C7
Tai'r Bull LD3 27 J6
Tairbeart (Tarbert) HS3 100 D7
Tairgwaith SA18 27 G7
Tai'r-heol CF46 18 E2
Tairlaw KA19 67 J3
Tai'r-ysgol SA7 17 K6
Takeley CM22 33 J6
Takeley Street CM22 33 J6
Talachddu LD3 27 K5
Talacre CH8 47 K4
Talardd LL23 37 H3
Talaton EX5 7 J6
Talbenny SA62 16 B4
Talbot Green CF72 18 D3
Talbot Village BH10 10 B5
Talerddig SY19 37 J5
Talgarreg SA44 26 D3
Talgarth LD3 28 A5
Taliesin SY20 37 F6
Talisker IV47 85 J1
Talke Pits ST7 49 H7
Talkin CA8 61 G1
Talla Linnfoots ML12 69 G1
Talladale IV22 95 F4
Talladh-a-Bheithe PH17 81 G2
Talland PL13 4 C5
Tallarn Green SY14 38 D1
Tallentire CA13 60 C3
Talley (Talyllychau) SA19 17 K2
Tallington PE9 42 D5
Talmine IV27 103 H2
Talog SA33 17 G3
Tal-sarn SA48 26 E3
Talsarnau LL47 37 F2
Talskiddy TR9 3 G2
Talwrn *I.o.A.* LL77 46 C5
Talwrn *Wrex.* LL14 38 B1
Talwrn *Wrex.* LL13 38 C1
Talybont *Cere.* SY24 37 F7
Tal-y-bont *Conwy* LL32 47 F6
Tal-y-bont *Gwyn.* LL43 36 E3
Tal-y-bont *Gwyn.* LL57 46 E5
Talybont-on-Usk LD3 28 A6
Tal-y-Cae LL57 46 E6
Tal-y-cafn LL28 47 F5
Tal-y-coed NP7 28 D7
Talygarn CF72 18 D3
Talyllychau (Talley) SA19 17 K2
Tal-y-llyn *Gwyn.* LL36 37 G5
Talyllyn *Powys* LD3 28 A6
Talysarn LL54 46 C7
Tal-y-wern SY20 37 H5
Tamavoid FK8 74 D1
Tamerton Foliot PL5 4 E4
Tamworth B79 40 E5
Tamworth Green PE22 43 G1
Tan Office Green IP29 34 B3
Tandem HD5 50 D1
Tandridge RH8 23 G6
Tanerdy SA31 17 H3
Tanfield DH9 62 C1
Tanfield Lea DH9 62 C1
Tang HG3 57 H4
Tang Hall YO10 58 C4
Tangiers SA62 16 C4
Tanglandford AB41 91 G1
Tangley SP11 21 G6
Tangmere PO20 12 C6
Tangwick ZE2 107 L5
Tangy PA28 66 A1

Tan - Thu

Tankerness **KW17** 106 E7
Tankersley **S75** 51 F2
Tankerton **CT5** 25 H5
Tan-lan **LL48** 37 F1
Tannach **KW1** 105 J4
Tannachie **AB39** 91 F6
Tannachy **IV28** 96 E1
Tannadice **DD8** 83 F2
Tannington **IP13** 35 G2
Tannochside **G71** 74 E4
Tansley **DE4** 51 F7
Tansley Knoll **DE4** 51 F7
Tansor **PE8** 42 D6
Tantobie **DH9** 62 C1
Tanton **TS9** 63 G5
Tanworth in Arden **B94** 30 C1
Tan-y-fron **LL16** 47 H6
Tan-y-graig **LL53** 36 C2
Tanygrisiau **LL41** 37 F1
Tan-y-groes **SA43** 17 F1
Tan-y-pistyll **SY10** 37 K3
Tan-yr-allt **LL19** 47 J4
Taobh a' Deas Loch Baghasdail **HS8** 84 C3
Taobh Siar **HS3** 100 D7
Taobh Tuath (Northton) **HS3** 92 E3
Tapeley **EX39** 6 C3
Taplow **SL6** 22 C3
Tapton Grove **S43** 51 G5
Taransay (Tarasaigh) **HS3** 100 C7
Taraphocain **PA38** 80 B3
Tarasaigh (Taransay) **HS3** 100 C7
Tarbat House **IV18** 96 E4
Tarbert *Arg. & B.* PA29 73 G4
Tarbert *Arg. & B.* **PA41** 72 E5
Tarbert *Arg. & B.* **PA60** 72 E2
Tarbert *High.* **PH36** 79 H1
Tarbert (Tairbeart) *W.Isles* **HS3** 100 D7
Tarbet *Arg. & B.* **G83** 80 E7
Tarbet *High.* **PH41** 86 D5
Tarbet *High.* **IV27** 102 D4
Tarbock Green **L35** 48 D4
Tarbolton **KA5** 67 J1
Tarbrax **EH55** 75 J5
Tardebigge **B60** 29 J2
Tardy Gate **PR5** 55 J7
Tarfside **DD9** 90 C7
Tarland **AB34** 90 C4
Tarleton **PR4** 55 H7
Tarlscough **L40** 48 D1
Tarlton **GL7** 20 C2
Tarnbrook **LA2** 55 J4
Tarnock **BS26** 19 G6
Tarporley CW6 48 E6
Tarr **TA4** 7 K2
Tarrant Crawford **DT11** 9 J4
Tarrant Gunville **DT11** 9 J3
Tarrant Hinton **DT11** 9 J3
Tarrant Keyneston **DT11** 9 J4
Tarrant Launceston **DT11** 9 J4
Tarrant Monkton **DT11** 9 J4
Tarrant Rawston **DT11** 9 J4
Tarrant Rushton **DT11** 9 J4
Tarrel **IV20** 97 F3
Tarring Neville **BN9** 13 H6
Tarrington **HR1** 29 F4
Tarrnacraig **KA27** 73 H7
Tarsappie **PH2** 82 C5
Tarskavaig **IV46** 86 B4
Tarves **AB41** 91 G1
Tarvie *High.* **IV14** 96 B6
Tarvie *P. & K.* **PH10** 82 B1
Tarvin **CH3** 48 D6
Tarvin Sands **CH3** 48 D6
Tasburgh **NR15** 45 G6
Tasley **WV16** 39 F6
Taston **OX7** 30 E6
Tatenhill **DE13** 40 E3
Tathall End **MK19** 32 B4
Tatham **LA2** 56 B3
Tathwell **LN11** 53 G4
Tatsfield **TN16** 23 H6
Tattenhall **CH3** 48 D7
Tattenhoe **MK4** 32 B5
Tatterford **NR21** 44 C3
Tattersett **PE31** 44 C2
Tattershall **LN4** 53 F7
Tattershall Bridge **LN4** 52 E7
Tattershall Thorpe **LN4** 53 F7
Tattingstone **IP9** 35 F5
Tatworth **TA20** 8 C4
Tauchers **AB55** 98 B6
TAUNTON TA 8 B2
Tavelty **AB51** 91 F3
Taverham **NR8** 45 F4
Tavernspite **SA34** 16 E4
Tavistock PL19 4 E3
Taw Bridge **EX18** 6 E5
Taw Green **EX20** 6 E6
Tawstock **EX31** 6 D3
Taxal **SK23** 50 C5
Tayburn **KA3** 74 D6
Taychreggan **PA35** 80 B5
Tayinloan **PA29** 72 E6
Taylors Cross **EX23** 6 A4
Taynafead **PA33** 80 B6
Taynish **PA31** 73 F2
Taynton *Glos.* **GL19** 29 G6
Taynton *Oxon.* **OX18** 30 D7
Taynuilt PA35 80 B4
Tayock **DD10** 83 H2
Tayovullin **PA44** 72 A3
Tayport DD6 83 F5
Tayvallich **PA31** 73 F2
Tea Green **LU2** 32 E6
Tealby **LN8** 52 E3
Tealing **DD4** 83 F4
Team Valley **NE11** 71 H7
Teanamachar **HS6** 92 C5
Teangue **IV44** 86 C4
Teasses **KY8** 83 F7
Tebay **CA10** 61 H6
Tebworth **LU7** 32 C6
Tedburn St. Mary **EX6** 7 G6
Teddington *Glos.* **GL20** 29 J5
Teddington *Gt.Lon.* TW11 22 E4
Tedstone Delamere **HR7** 29 F3
Tedstone Wafre **HR7** 29 F3
Teeton **NN6** 31 H1
Teffont Evias **SP3** 9 J1
Teffont Magna **SP3** 9 J1
Tegryn **SA35** 17 F2
Teigh **LE15** 42 B4
Teign Village **TQ13** 7 G7
Teigngrace **TQ12** 5 J3
Teignmouth TQ14 5 K3
TELFORD TF 39 F5
Telham **TN33** 14 C6
Tellisford **BA2** 20 B6
Telscombe **BN7** 13 H6
Telscombe Cliffs **BN10** 13 H6
Tempar **PH16** 81 H2
Templand **DG11** 69 F5
Temple *Cornw.* **PL30** 4 B3
Temple *Midloth.* **EH23** 76 B5
Temple Balsall **B93** 30 D1
Temple Bar **SA48** 26 E3
Temple Cloud **BS39** 19 K6
Temple End **CB9** 33 K3
Temple Ewell **CT16** 15 H3
Temple Grafton **B49** 30 C3
Temple Guiting **GL54** 30 B6
Temple Herdewyke **CV47** 30 E3
Temple Hirst **YO8** 58 C7
Temple Normanton **S42** 51 G6
Temple Sowerby **CA10** 61 H4
Templecombe BA8 9 G2
Templeton *Devon* **EX16** 7 G4
Templeton *Pembs.* **SA67** 16 E4
Templeton Bridge **EX16** 7 G4
Templewood **DD9** 83 H1
Tempsford **SG19** 32 E3
Ten Mile Bank **PE38** 44 A6
Tenbury Wells WR15 28 E2
Tenby (Dinbych-y-pysgod) SA70 16 E5
Tendring **CO16** 35 F6
Tendring Green **CO16** 35 F6
Tenga **PA72** 79 G3
Tenterden TN30 14 D4
Tepersie Castle **AB33** 90 D2
Terally **DG9** 64 B6
Terling **CM3** 34 B7
Tern **TF6** 39 F4
Ternhill **TF9** 39 F2
Terregles **DG2** 65 K3
Terriers **HP13** 22 B2
Terrington **YO60** 58 C2
Terrington St. Clement **PE34** 43 J3
Terrington St. John **PE14** 43 J4
Terry's Green **B94** 30 C1
Tervieside **AB37** 89 K1
Teston **ME18** 14 C2
Testwood **SO40** 10 E3
Tetbury GL8 20 B2
Tetbury Upton **GL8** 20 B2
Tetchill **SY12** 38 C2
Tetcott **EX22** 6 B6
Tetford **LN9** 53 G5
Tetney **DN36** 53 G2
Tetney Lock **DN36** 53 G2
Tetsworth **OX9** 21 K1
Tettenhall **WV6** 40 A5
Tettenhall Wood **WV6** 40 A6
Tetworth **SG19** 33 F3
Teuchan **AB42** 91 J1
Teversal **NG17** 51 G6
Teversham **CB1** 33 H3
Teviothead **TD9** 69 K3
Tewel **AB39** 91 G6
Tewin **AL6** 33 F7
Tewkesbury GL20 29 H5
Teynham **ME9** 25 F5
Thackley **BD10** 57 G6
Thainston **AB30** 90 E7
Thainstone **AB51** 91 F3
Thakeham **RH20** 12 E5
Thame OX9 22 A1
Thames Ditton KT7 22 E5
Thames Haven **SS17** 24 D3
Thamesmead **SE28** 23 H3
Thanington **CT1** 15 G2
Thankerton **ML12** 75 H7
Tharston **NR15** 45 F6
Thatcham RG18 21 J5
Thatto Heath **WA9** 48 E3
Thaxted **CM6** 33 K5
The Apes Hall **CB6** 43 J6
The Bage **HR3** 28 B4
The Balloch **PH7** 81 K6
The Banking **AB51** 91 F1
The Bar **RH13** 12 E4
The Birks **AB32** 91 F4
The Bog **SY5** 38 C6
The Bourne **GU9** 22 B7
The Bratch **WV5** 40 A6
The Broad **HR6** 28 D2
The Bryn **NP7** 19 G1
The Burf **DY13** 29 H2
The Burn **DD9** 90 D7
The Butts **BA11** 20 A7
The Camp **GL6** 20 C1
The Chequer **SY13** 38 D1
The City *Bucks.* **HP14** 22 A2
The City *Suff.* **NR34** 45 H7
The Common *Wilts.* **SP5** 10 D1
The Common *Wilts.* **SN15** 20 D3
The Craigs **IV24** 96 B2
The Cronk **IM7** 54 C4
The Delves **WS5** 40 C6
The Den **KA24** 74 B5
The Dicker **BN27** 13 J6
The Down **WV16** 39 F6
The Drums **DD8** 82 E1
The Eaves **GL15** 19 K1
The Flatt **CA6** 70 A6
The Folly **AL4** 32 E7
The Forge **HR5** 28 C3
The Forstal *E.Suss.* **TN3** 13 J3
The Forstal *Kent* **TN25** 15 F4
The Grange *Lincs.* **LN13** 53 J5
The Grange *Shrop.* **SY12** 38 C2
The Grange *Surr.* **RH9** 23 G7
The Green *Arg. & B.* **PA77** 78 A3
The Green *Cumb.* **LA18** 54 E1
The Green *Essex* **CM8** 34 B7
The Green *Flints.* **CH7** 48 B6
The Green *Wilts.* **SP3** 9 H1
The Grove **WR8** 29 H4
The Haven **RH14** 12 D3
The Headland **TS24** 63 G3
The Heath **ST14** 40 C2
The Herberts **CF71** 18 C4
The Hermitage **KT20** 23 F6
The Hill **LA18** 54 E1
The Holme **HG3** 57 H4
The Howe **IM9** 54 A7
The Isle **SY3** 38 D4
The Laurels **NR14** 45 H6
The Leacon **TN26** 14 E4
The Lee **HP16** 22 B1
The Leigh **GL19** 29 H6
The Lhen **IM7** 54 C3
The Lodge **PA24** 73 K1
The Marsh **SY5** 38 C6
The Moor *E.Suss.* **TN35** 14 D6
The Moor *Kent* **TN18** 14 C5
The Mumbles **SA3** 17 K7
The Murray **G75** 74 E5
The Mythe **GL20** 29 H5
The Narth **NP25** 19 J1
The Neuk **AB31** 91 F5
The Node **SG4** 33 F7
The Oval **BA2** 20 A5
The Polchar **PH22** 89 G4
The Quarter **TN27** 14 D3
The Reddings **GL51** 29 J6
The Rhos **SA62** 16 D4
The Rookery **ST7** 49 H7
The Rowe **ST5** 40 A2
The Sale **DE13** 40 D4
The Sands **GU10** 22 B7
The Shoe **SN14** 20 B4
The Slade **RG7** 21 J4
The Smithies **WV16** 39 F6
The Stocks **TN30** 14 E5
The Swillett **WD3** 22 D2
The Thrift **SG8** 33 G5
The Vauld **HR1** 28 E4
The Wern **LL14** 48 B7
The Wyke **TF11** 39 G5
Theakston **DL8** 57 J1
Thealby **DN15** 52 B1
Theale *Som.* **BS28** 19 H7
Theale *W.Berks.* **RG7** 21 K4
Thearne **HU17** 59 G6
Theberton **IP16** 35 J2
Thedden Grange **GU34** 11 H1
Theddingworth **LE17** 41 J7
Theddlethorpe All Saints **LN12** 53 H4
Theddlethorpe St. Helen **LN12** 53 H4
Thelbridge Barton **EX17** 7 F4
Thelbridge Cross **EX17** 7 F4
Thelnetham **IP22** 34 E1
Thelveton **IP21** 45 F7
Thelwall **WA4** 49 F4
Themelthorpe **NR20** 44 E3
Thenford **OX17** 31 G4
Therfield **SG8** 33 G5
Thetford *Lincs.* **PE6** 42 E4
Thetford *Norf.* IP24 44 C7
Thethwaite **CA5** 60 E2
Theydon Bois **CM16** 23 H2
Theydon Garnon **CM16** 23 H2
Theydon Mount **CM16** 23 H2
Thickwood **SN14** 20 B4
Thimbleby *Lincs.* **LN9** 53 F6
Thimbleby *N.Yorks.* **DL6** 63 F7
Thingley **SN13** 20 B5
Thirkleby **YO7** 57 K2
Thirlby **YO7** 57 K1
Thirlestane **TD2** 76 D6
Thirn **HG4** 57 H1
Thirsk YO7 57 K1
Thirston New Houses **NE65** 71 G4
Thirtleby **HU11** 59 H6
Thistleton *Lancs.* **PR4** 55 H6
Thistleton *Rut.* **LE15** 42 C4
Thistley Green **IP28** 33 K1
Thixendale **YO17** 58 E3
Thockrington **NE48** 70 E6
Tholomas Drove **PE13** 43 G5
Tholthorpe **YO61** 57 K3
Thomas Chapel **SA68** 16 E5
Thomas Close **CA11** 60 F2
Thomastown **AB54** 90 D1
Thompson **IP24** 44 D6
Thomshill **IV30** 97 K6
Thong **DA12** 24 C4
Thongsbridge **HD9** 50 D2
Thoralby **DL8** 57 F1
Thoresby **NG22** 51 J5
Thoresthorpe **LN13** 53 H5
Thoresway **LN8** 52 E3
Thorganby *Lincs.* **DN37** 53 F3
Thorganby *N.Yorks.* **YO19** 58 C5
Thorgill **YO18** 63 J7
Thorington **IP19** 35 J1
Thorington Street **CO6** 34 E5
Thorley **CM23** 33 H7
Thorley Houses **CM23** 33 H6
Thorley Street *Herts.* **CM23** 33 H7
Thorley Street *I.o.W.* **PO41** 10 E6
Thormanby **YO61** 57 K2
Thornaby-on-Tees **TS17** 63 F5
Thornage **NR25** 44 E2
Thornborough *Bucks.* **MK18** 31 J5
Thornborough *N.Yorks.* **DL8** 57 H2
Thornbury *Devon* **EX22** 6 B5
Thornbury *Here.* **HR7** 29 F3
Thornbury *S.Glos.* **BS35** 19 K2
Thornbury *W.Yorks.* **BD3** 57 G6
Thornby **NN6** 31 H1
Thorncliff **HD8** 50 E1
Thorncliffe **ST13** 50 C7
Thorncombe **TA20** 8 C4
Thorncombe Street **GU5** 22 D7
Thorncote Green **SG19** 32 E4
Thorncross **PO30** 11 F6
Thorndon **IP23** 35 F2
Thorndon Cross **EX20** 6 D6
Thorne **DN8** 51 J1
Thorne St. Margaret **TA21** 7 J3
Thorner **LS14** 57 J6
Thorney *Bucks.* **SL0** 22 D4
Thorney *Notts.* **NG23** 52 B5
Thorney *Peter.* **PE6** 43 F5
Thorney *Som.* **TA10** 8 D2
Thorney Close **SR4** 62 E1
Thorney Hill **BH23** 10 D5
Thornfalcon **TA3** 8 B2
Thornford **DT9** 9 F3
Thorngrafton **NE47** 70 C7
Thorngrove **TA7** 8 C1
Thorngumbald **HU12** 59 J7
Thornham **PE36** 44 B1
Thornham Magna **IP23** 35 F1
Thornham Parva **IP23** 35 F1
Thornhaugh **PE8** 42 D5
Thornhill *Cardiff* **CF83** 18 E3
Thornhill *Cumb.* **CA22** 60 B6
Thornhill *D. & G.* DG3 68 D4
Thornhill *Derbys.* **S33** 50 D4
Thornhill *S'ham.* **SO19** 11 F3
Thornhill *Stir.* **FK8** 81 H7
Thornhill *W.Yorks.* **WF12** 50 E1
Thornhill Lees **WF12** 50 E1
Thornholme **YO25** 59 H3
Thornicombe **DT11** 9 H4
Thornley *Dur.* **DH6** 62 E3
Thornley *Dur.* **DL13** 62 C3
Thornley Gate **NE47** 61 K1
Thornliebank **G46** 74 D5
Thornroan **AB41** 91 G1
Thorns **CB8** 34 B3
Thorns Green **WA15** 49 G4
Thornsett **SK22** 50 C4
Thornthwaite *Cumb.* **CA12** 60 D4
Thornthwaite *N.Yorks.* **HG3** 57 G4
Thornton *Angus* **DD8** 82 E3
Thornton *Bucks.* **MK17** 31 J5
Thornton *E.Riding* **YO42** 58 D5
Thornton *Fife* **KY1** 76 A1
Thornton *Lancs.* FY5 55 G5
Thornton *Leics.* **LE67** 41 G5
Thornton *Lincs.* **LN9** 53 F6
Thornton *Mersey.* **L23** 48 C2
Thornton *Middbro.* **TS8** 63 F5
Thornton *Northumb.* **TD15** 77 H6
Thornton *P. & K.* **PH8** 82 B3
Thornton *Pembs.* **SA73** 16 C5
Thornton *W.Yorks.* **BD13** 57 G6
Thornton Bridge **YO61** 57 K2
Thornton Curtis **DN39** 52 D1
Thornton Heath CR7 23 G5
Thornton Hough **CH63** 48 C4
Thornton in Lonsdale **LA6** 56 B2
Thornton le Moor **LN7** 52 D3
Thornton Park **TD15** 77 H6
Thornton Rust **DL8** 56 E1
Thornton Steward **HG4** 57 G1
Thornton Watlass **HG4** 57 H1
Thorntonhall **G74** 74 D5
Thornton-in-Craven **BD23** 56 E5
Thornton-le-Beans **DL6** 63 F7
Thornton-le-Clay **YO60** 58 C3
Thornton-le-Dale **YO18** 58 E1
Thornton-le-Moor **DL7** 57 J1
Thornton-le-Moors **CH2** 48 D5
Thornton-le-Street **YO7** 57 K1
Thorntonloch **EH42** 77 F3
Thornwood **CM16** 23 H1
Thornyhill **AB30** 90 E7
Thornylee **TD1** 76 C7
Thoroton **NG13** 42 A1
Thorp Arch **LS23** 57 K5
Thorpe *Derbys.* **DE6** 50 D7
Thorpe *E.Riding* **YO25** 59 F5
Thorpe *Lincs.* **LN12** 53 H4
Thorpe *N.Yorks.* **BD23** 57 F3
Thorpe *Norf.* **NR14** 45 J6
Thorpe *Notts.* **NG23** 51 K7
Thorpe *Surr.* **TW20** 22 D5
Thorpe Abbotts **IP21** 45 F7
Thorpe Acre **LE11** 41 H4
Thorpe Arnold **LE14** 42 A3
Thorpe Audlin **WF8** 51 G1
Thorpe Bassett **YO17** 58 E2
Thorpe Bay **SS1** 25 F3
Thorpe by Water **LE15** 42 B6
Thorpe Constantine **B79** 40 E5
Thorpe Culvert **PE24** 53 H6
Thorpe End **NR13** 45 G4
Thorpe Green *Essex* **CO16** 35 F6
Thorpe Green *Lancs.* **PR6** 55 J7
Thorpe Green *Suff.* **IP30** 34 D3
Thorpe Hall **YO62** 58 B2
Thorpe Hesley **S61** 51 F3
Thorpe in Balne **DN6** 51 H1
Thorpe in the Fallows **LN1** 52 C4
Thorpe Langton **LE16** 42 A6
Thorpe Larches **TS21** 62 E4
Thorpe le Street **YO42** 58 E5
Thorpe Malsor **NN14** 32 B1
Thorpe Mandeville **OX17** 31 G4
Thorpe Market **NR11** 45 G2
Thorpe Morieux **IP30** 34 D3
Thorpe on the Hill *Lincs.* **LN6** 52 C6
Thorpe on the Hill *W.Yorks.* **WF3** 57 J7
Thorpe Row **IP25** 44 D5
Thorpe St. Andrew **NR7** 45 G5
Thorpe St. Peter **PE24** 53 H6
Thorpe Salvin **S80** 51 H4
Thorpe Satchville **LE14** 42 A4
Thorpe Street **IP22** 34 E1
Thorpe Thewles **TS21** 63 F4
Thorpe Tilney Dales **LN4** 52 E7
Thorpe Underwood *N.Yorks.* **YO26** 57 K4
Thorpe Underwood *Northants.* **NN6** 42 A7
Thorpe Waterville **NN14** 42 D7
Thorpe Willoughby **YO8** 58 B6
Thorpefield **YO7** 57 K2
Thorpe-le-Soken **CO16** 35 F6
Thorpeness **IP16** 35 J2
Thorpland **PE33** 44 A5
Thorrington **CO7** 34 E6
Thorverton **EX5** 7 H5
Thrandeston **IP21** 35 F1
Thrapston **NN14** 32 C1
Threapland **BD23** 56 E3
Threapwood **SY14** 38 D1
Threapwood Head **ST10** 40 C1
Three Ashes **BA3** 19 K7
Three Bridges **RH10** 13 F3
Three Burrows **TR4** 2 E4
Three Chimneys **TN27** 14 D4
Three Cocks (Aberllynfi) **LD3** 28 A5
Three Crosses **SA4** 17 J6
Three Cups Corner **TN21** 13 K4
Three Hammers **PL15** 4 C2
Three Holes **PE14** 43 J5
Three Leg Cross **TN5** 13 K3
Three Legged Cross **BH21** 10 B4
Three Mile Cross **RG7** 22 A5
Three Oaks **TN35** 14 D6
Threehammer Common **NR12** 45 H4
Threekingham **NG34** 42 D2
Threemilestone **TR3** 2 E4
Threlkeld **CA12** 60 E4
Threshfield **BD23** 56 E3
Threxton Hill **IP25** 44 C5
Thriepley **DD2** 82 E4
Thrigby **NR29** 45 J4
Thringarth **DL12** 62 A4
Thringstone **LE67** 41 G4
Thrintoft **DL7** 62 E7
Thriplow **SG8** 33 H4
Throapham **S25** 51 H4
Throckenholt **PE12** 43 G5
Throcking **SG9** 33 G5
Throckley **NE15** 71 G7
Throckmorton **WR10** 29 J4
Throop **DT2** 9 H5
Throphill **NE61** 71 G5
Thropton **NE65** 71 F3
Througham **GL6** 20 C1
Throwleigh **EX20** 6 E6
Throwley **ME13** 14 E2
Throws **CM6** 33 K6
Thrumpton *Notts.* **NG11** 41 H2
Thrumpton *Notts.* **DN22** 51 K5
Thrumster **KW1** 105 J4
Thrunton **NE66** 71 F2
Thrupp *Glos.* **GL5** 20 B1
Thrupp *Oxon.* **OX5** 31 F7
Thrupp *Oxon.* **SN7** 21 F2
Thruscross **HG3** 57 G4
Thrushelton **EX20** 6 C7
Thrussington **LE7** 41 J4
Thruxton *Hants.* **SP11** 21 F7
Thruxton *Here.* **HR2** 28 D5
Thrybergh **S65** 51 G3
Thulston **DE72** 41 G2
Thunder Bridge **HD8** 50 D1
Thundergay **KA27** 73 G6
Thundersley **SS7** 24 E3
Thunderton **AB42** 99 J6
Thundridge **SG12** 33 G7
Thurcaston **LE7** 41 H4
Thurcroft **S66** 51 G4
Thurdistoft **KW14** 105 H2
Thurdon **EX23** 6 A4
Thurgarton *Norf.* **NR11** 45 F2
Thurgarton *Notts.* **NG14** 41 J1
Thurgoland **S35** 50 E2
Thurlaston *Leics.* **LE9** 41 H6
Thurlaston *Warks.* **CV23** 31 F1
Thurlbear **TA3** 8 B2
Thurlby *Lincs.* **PE10** 42 E4
Thurlby *Lincs.* **LN5** 52 C6
Thurlby *Lincs.* **LN13** 53 H5
Thurleigh **MK44** 32 D3
Thurlestone **TQ7** 5 G6
Thurloxton **TA2** 8 B1
Thurlstone **S36** 50 E2

Thu - Tre

Thurlton NR14 45 J6
Thurlwood ST7 49 H7
Thurmaston LE4 41 J5
Thurnby LE7 41 J5
Thurne NR29 45 J4
Thurnham ME14 14 D2
Thurning *Norf.* NR20 44 E3
Thurning *Northants.* PE8 42 D7
Thurnscoe S63 51 G2
Thursby CA5 60 E1
Thursden BB10 56 E6
Thursford NR21 44 D2
Thursley GU8 12 C3
Thurso KW14 105 G2
Thurstaston CH61 48 B4
Thurston IP31 34 D2
Thurston Clough OL3 49 J2
Thurstonfield CA5 60 E1
Thurstonland HD4 50 D1
Thurton NR14 45 H5
Thurvaston *Derbys.* DE6 40 E2
Thurvaston *Derbys.* DE6 40 D2
Thuster KW1 105 H3
Thuxton NR9 44 E5
Thwaite *N.Yorks.* DL11 61 K7
Thwaite *Suff.* IP23 35 F2
Thwaite Head LA12 60 E7
Thwaite St. Mary NR35 45 H6
Thwaites BD21 57 F5
Thwaites Brow BD21 57 F5
Thwing YO25 59 G2
Tibbermore PH1 82 B5
Tibberton *Glos.* GL19 29 G6
Tibberton *Tel. & W.* TF10 39 F3
Tibberton *Worcs.* WR9 29 J3
Tibbie Shiels Inn TD7 69 H1
Tibenham NR16 45 F6
Tibertich PA31 79 K7
Tibshelf DE55 51 G6
Tibthorpe YO25 59 F4
Ticehurst TN5 13 K3
Tichborne SO24 11 G1
Tickencote PE9 42 C5
Tickenham BS21 19 H4
Tickford End MK16 32 B4
Tickhill DN11 51 H3
Ticklerton SY6 38 D6
Ticknall DE73 41 F3
Tickton HU17 59 G5
Tidbury Green B90 30 C1
Tidcombe SN8 21 F6
Tiddington *Oxon.* OX9 21 K1
Tiddington *Warks.* CV37 30 D3
Tiddleywink SN14 20 B4
Tidebrook TN5 13 K4
Tideford PL12 4 D5
Tideford Cross PL12 4 D4
Tidenham NP16 19 J2
Tidenham Chase NP16 19 J2
Tideswell SK17 50 D5
Tidmarsh RG8 21 K4
Tidmington CV36 30 D5
Tidpit SP6 10 B3
Tidworth SP9 21 F7
Tiers Cross SA62 16 C4
Tiffield NN12 31 H3
Tifty AB53 99 F6
Tigerton DD9 83 G1
Tigh a' Gearraidh HS6 92 C4
Tighachnoic PA34 79 H3
Tighnablair PH6 81 J6
Tighnabruaich PA21 73 H3
Tighnacomaire PH33 80 A1
Tigley TQ9 5 H4
Tilbrook PE28 32 D2
Tilbury RM18 24 C4
Tilbury Green CO9 34 B4
Tile Hill CV4 30 D1
Tilehurst RG31 21 K4
Tilford GU10 22 B7
Tilgate RH10 13 F3
Tilgate Forest Row RH10 13 F3
Tillathrowie AB54 90 C1
Tillers' Green GL18 29 F5
Tillery AB41 91 H2
Tilley SY4 38 E3
Tillicoultry FK13 75 H1
Tillingham CM0 25 F1
Tillington *Here.* HR4 28 D4
Tillington *W.Suss.* GU28 12 C4
Tillington Common HR4 28 D4
Tillyarblet DD9 83 G1
Tillybirloch AB51 90 E4
Tillycairn Castle AB51 90 E3
Tillycorthie AB41 91 H2
Tillydrine AB34 90 E5
Tillyfar AB53 99 G6
Tillyfour AB33 90 D3
Tillyfourie AB51 90 E3
Tillygreig AB41 91 G2
Tillypronie AB34 90 C4
Tilmanstone CT14 15 J2
Tiln DN22 51 K4
Tilney All Saints PE34 43 J4
Tilney Fen End PE14 43 J4
Tilney High End PE34 43 J4
Tilney St. Lawrence PE34 43 J4
Tilshead SP3 20 D7
Tilstock SY13 38 E2
Tilston SY14 48 D7
Tilstone Fearnall CW6 48 E6
Tilsworth LU7 32 C6
Tilton on the Hill LE7 42 A5
Tiltups End GL6 20 B2
Timberland LN4 52 E7
Timberland Dales LN10 52 E6
Timbersbrook CW12 49 H6
Timberscombe TA24 7 H1
Timble LS21 57 G4
Timewell EX16 7 H3
Timperley WA15 49 G4
Timsbury *B. & N.E.Som.* BA2 19 K6
Timsbury *Hants.* SO51 10 E2
Timsgearraidh HS2 100 C4
Timworth IP31 34 C2
Timworth Green IP31 34 C2
Tincleton DT2 9 G5
Tindale CA8 61 H1
Tindon End CB10 33 K5
Tingewick MK18 31 H5
Tingley WF3 57 H7
Tingrith MK17 32 D5
Tingwall KW17 106 D5
Tingwall (Lerwick) Airport ZE2 107 N8
Tinhay PL16 6 B7
Tinney EX22 4 C1
Tinshill LS16 57 H6
Tinsley S9 51 G3
Tinsley Green RH10 13 F3
Tintagel PL34 4 A2
Tintern Parva NP16 19 J1
Tintinhull BA22 8 D3
Tintwistle SK13 50 C3
Tinwald DG1 69 F5
Tinwell PE9 42 D5
Tippacott EX35 7 F1
Tipperty *Aber.* AB30 91 F6
Tipperty *Aber.* AB41 91 H2
Tipps End PE14 43 H6
Tiptoe SO41 10 D5
Tipton DY4 40 B6
Tipton St. John EX10 7 J6
Tiptree CO5 34 C7
Tiptree Heath CO5 34 C7
Tirabad LD4 27 H4
Tiree PA77 78 A3
Tiree Airport PA77 78 B3
Tirindrish PH34 87 J6
Tirley GL19 29 H6
Tirphil NP24 18 E1
Tirril CA10 61 G4
Tir-y-dail SA18 17 K4
Tisbury SP3 9 J2
Tisman's Common RH12 12 D3
Tissington DE6 50 D7
Tister KW12 105 G2
Titchberry EX39 6 A3
Titchfield PO14 11 G4
Titchmarsh NN14 32 D1
Titchwell PE31 44 B1
Tithby NG13 41 J2
Titley HR5 28 C2
Titlington NE66 71 G2
Titmore Green SG4 33 F6
Titsey RH8 23 H6
Titson EX23 6 A5
Tittensor ST12 40 A2
Tittleshall PE32 44 C3
Tiverton *Ches.W. & C.* CW6 48 E6
Tiverton *Devon* EX16 7 H4
Tivetshall St. Margaret NR15 45 F7
Tivetshall St. Mary NR15 45 F7
Tivington TA24 7 H1
Tixall ST18 40 B3
Tixover PE9 42 C5
Toab *Ork.* KW17 106 E7
Toab *Shet.* ZE3 107 M11
Tobermory PA75 79 G2
Toberonochy PA34 79 J7
Tobha Mòr (Homore) HS8 84 C1
Tobson HS2 100 D4
Tocher AB51 90 E1
Tockenham SN4 20 D4
Tockenham Wick SN4 20 D3
Tockholes BB3 56 B7
Tockington BS32 19 K3
Tockwith YO26 57 K4
Todber DT10 9 G2
Toddington *Cen.Beds.* LU5 32 D6
Toddington *Glos.* GL54 30 B5
Todenham GL56 30 D5
Todhills *Angus* DD4 83 F4
Todhills *Cumb.* CA6 69 J7
Todlachie AB51 90 E3
Todmorden OL14 56 E7
Todwick S26 51 G4
Toft *Cambs.* CB23 33 G3
Toft *Lincs.* PE10 42 D4
Toft *Shet.* ZE2 107 N5
Toft Hill DL14 62 C4
Toft Monks NR34 45 J6
Toft next Newton LN8 52 D4
Toftcarl KW1 105 J4
Toftrees NR21 44 C3
Tofts KW1 105 J2
Toftwood NR19 44 D4
Togston NE65 71 H3
Tokavaig IV46 86 C3
Tokers Green RG4 21 K4
Tolastadh HS2 101 H3
Tolastadh a' Chaolais HS2 100 D4
Tolastadh Ùr HS2 101 H3
Toll Bar DN5 51 H2
Toll of Birness AB41 91 J1
Tolland TA4 7 K2
Tollard Farnham DT11 9 J3
Tollard Royal SP5 9 J3
Tollcross G32 74 E4
Toller Down Gate DT2 8 E4
Toller Fratrum DT2 8 E5
Toller Porcorum DT2 8 E5
Toller Whelme DT8 8 E4
Tollerton *N.Yorks.* YO61 58 B3
Tollerton *Notts.* NG12 41 J2
Tollesbury CM9 34 D7
Tollesby TS4 63 G5
Tolleshunt D'Arcy CM9 34 D7
Tolleshunt Knights CM9 34 D7
Tolleshunt Major CM9 34 C7
Tolm (Holm) HS2 101 G4
Tolmachan HS3 100 C7
Tolpuddle DT2 9 G5
Tolvah PH21 89 F5
Tolworth KT6 22 E5
Tom an Fhuadain HS2 101 F6
Tomatin IV13 89 F2
Tombreck IV2 88 D1
Tomchrasky IV63 87 J3
Tomdoun PH35 87 H4
Tomdow IV36 97 H7
Tomich *High.* IV4 87 K2
Tomich *High.* IV18 96 E4
Tomich *High.* IV27 96 D1
Tomintoul AB37 89 J3
Tomnacross IV4 96 C7
Tomnamoon IV36 97 H6
Tomnaven AB54 90 C1
Tomnavoulin AB37 89 K2
Tomvaich PH26 89 H1
Ton Pentre CF41 18 C2
Tonbridge TN9 23 J7
Tondu CF32 18 B3
Tonedale TA21 7 K3
Tonfanau LL36 36 E5
Tong *Kent* TN27 14 D3
Tong *Shrop.* TF11 39 G5
Tong *W.Yorks.* BD4 57 H6
Tong Norton TF11 39 G5
Tong Street BD4 57 G6
Tonge DE73 41 G3
Tongham GU10 22 B7
Tongland DG6 65 G5
Tongue IV27 103 H3
Tongue House IV27 103 H3
Tongwynlais CF15 18 E3
Tonmawr SA12 18 B2
Tonna SA11 18 A2
Tonwell SG12 33 G7
Tonypandy CF40 18 C2
Tonyrefail CF39 18 D3
Toot Baldon OX44 21 J1
Toot Hill CM5 23 J1
Toothill *Hants.* SO51 10 E3
Toothill *Swin.* SN5 20 E3
Tooting Graveney SW17 23 F4
Top End MK44 32 D2
Top of Hebers M24 49 H2
Topcliffe YO7 57 K2
Topcroft NR35 45 G6
Topcroft Street NR35 45 G6
Toppesfield CO9 34 B5
Toppings BL7 49 G1
Toprow NR16 45 F6
Topsham EX3 7 H7
Topsham Bridge TQ7 5 H5
Torastan PA78 78 D1
Torbain AB37 89 J3
Torbeg *Aber.* AB35 90 B5
Torbeg *N.Ayr.* KA27 66 D1
Torbothie ML7 75 G5
Torbryan TQ12 5 J4
Torcastle PH33 87 H7
Torcross TQ7 5 J6
Tordarroch IV2 88 D1
Tore IV6 96 D6
Toreduff IV36 97 J5
Toremore *High.* PH26 89 J1
Toremore *High.* KW6 105 G5
Torfrey PL23 4 B5
Torgyle IV63 87 K3
Torksey LN1 52 B5
Torlum HS7 92 C6
Torlundy PH33 87 H7
Tormarton GL9 20 A4
Tormisdale PA47 72 A5
Tormore KA27 73 G7
Tormsdale KW12 105 G3
Tornagrain IV2 96 E6
Tornahaish AB36 89 K4
Tornaveen AB31 90 E4
Torness IV2 88 C2
Toronto DL14 62 C3
Torpenhow CA7 60 D3
Torphichen EH48 75 H3
Torphins AB31 90 E4
Torpoint PL11 4 E5
TORQUAY TQ 5 K4
Torquhan TD1 76 C6
Torr PL8 5 F5
Torran *Arg. & B.* PA31 79 K7
Torran *High.* IV18 96 E4
Torran *High.* IV40 94 B7
Torrance G64 74 E3
Torrance House G75 74 E5
Torrancroy AB36 90 B3
Torre *Som.* TA23 7 J1
Torre *Torbay* TQ1 5 K4
Torrich IV12 97 F6
Torridon IV22 94 E6
Torrin IV49 86 B2
Torrisdale *Arg. & B.* PA28 73 F7
Torrisdale *High.* KW14 103 J2
Torrish KW8 104 E7
Torrisholme LA4 55 H3
Torroble IV27 96 C1
Torry *Aber.* AB54 98 C6
Torry *Aberdeen* AB11 91 H4
Torryburn KY12 75 J2
Torsonce TD1 76 C6
Torterston AB42 99 J6
Torthorwald DG1 69 F6
Tortington BN18 12 C6
Torton DY10 29 H1
Tortworth GL12 20 A2
Torvaig IV51 93 K7
Torver LA21 60 D7
Torwood FK5 75 G2
Torworth DN22 51 J4
Tosberry EX39 6 A3
Toscaig IV54 86 D1
Toseland PE19 33 F2
Tosside BD23 56 C4
Tostarie PA74 78 E3
Tostock IP30 34 D2
Totaig IV55 93 G6
Totamore PA78 78 C2
Tote IV51 93 K7
Tote Hill GU29 12 B4
Totegan KW14 104 D2
Totford SO24 11 G1
Totham Hill CM9 34 C7
Tothill LN13 53 H4
Totland PO39 10 E6
Totley S17 51 F5
Totnes TQ9 5 J4
Toton NG9 41 H2
Totronald PA78 78 C2
Totscore IV51 93 J5
Tottenham N17 23 G2
Tottenhill PE33 44 A4
Tottenhill Row PE33 44 A4
Totteridge *Bucks.* HP13 22 B2
Totteridge *Gt.Lon.* N20 23 F2
Totternhoe LU6 32 C6
Tottington *Gt.Man.* BL8 49 G1
Tottington *Norf.* IP24 44 C6
Totton SO40 10 E3
Toulton TA4 7 K2
Tournaig IV22 94 E3
Toux AB42 99 H5
Tovil ME15 14 C2
Tow Law DL13 62 C3
Towan Cross TR4 2 E4
Toward PA23 73 K4
Towcester NN12 31 H4
Towednack TR26 2 B5
Tower End PE32 44 A4
Towersey OX9 22 A1
Towie *Aber.* AB43 99 G4
Towie *Aber.* AB54 90 D2
Towie *Aber.* AB33 90 C3
Towiemore AB55 98 B6
Town End *Cambs.* PE15 43 H6
Town End *Cumb.* LA11 55 H1
Town End *Mersey.* WA8 48 D4
Town Green *Lancs.* L39 48 D2
Town Green *Norf.* NR13 45 H4
Town of Lowton WA3 49 F3
Town Row TN6 13 J3
Town Street IP27 44 B7
Town Yetholm TD5 70 D1
Townfield DH8 62 A2
Townhead *D. & G.* DG6 65 G6
Townhead *S.Yorks.* S36 50 D2
Townhead of Greenlaw DG7 65 H4
Townhill *Fife* KY12 75 K2
Townhill *Swan.* SA1 17 K6
Towns End RG26 21 J6
Towns Green CW6 49 F6
Townshend TR27 2 C5
Towthorpe *E.Riding* YO25 59 F3
Towthorpe *York* YO32 58 C4
Towton LS24 57 K6
Towyn LL22 47 H5
Toynton All Saints PE23 53 G6
Toynton Fen Side PE23 53 G6
Toynton St. Peter PE23 53 H6
Toy's Hill TN16 23 H6
Trabboch KA5 67 J1
Traboe TR12 2 E6
Tradespark *High.* IV12 97 F6
Tradespark *Ork.* KW15 106 D7
Traeth Coch (Red Wharf Bay) LL75 46 D4
Trafford Centre M17 49 G3
Trafford Park M17 49 G3
Trallong LD3 27 J6
Trallwn SA7 17 K6
Tram Inn HR2 28 D5
Tranent EH33 76 C3
Tranmere CH42 48 C4
Trantlebeg KW13 104 D3
Trantlemore KW13 104 D3
Tranwell NE61 71 G5
Trap SA19 17 K4
Trap Street SK11 49 H6
Traprain EH41 76 D3
Trap's Green B94 30 C2
Traquair EH44 76 B7
Trawden BB8 56 E6
Trawsfynydd LL41 37 G2
Trealaw CF40 18 D2
Treales PR4 55 H6
Trearddur LL65 46 A5
Treaslane IV51 93 J6
Tre-Aubrey CF71 18 D4
Trebanog CF39 18 D2
Trebanos SA8 18 A1
Trebarrow EX22 4 C1
Trebartha PL15 4 C3
Trebarvah TR11 2 E5
Trebarwith PL33 4 A2
Trebeath PL15 4 C2
Trebetherick PL27 3 G1
Trebister ZE1 107 N9
Tre-boeth SA5 17 K6
Treborough TA23 7 J2
Trebudannon TR8 3 F2
Trebullett PL15 4 D3
Treburley PL15 4 D3
Treburrick PL27 3 F1
Trebyan PL30 4 A4
Trecastle LD3 27 H6
Trecott EX20 6 E5
Trecrogo PL15 6 B7
Trecwn SA62 16 C2
Trecynon CF44 18 C1
Tredaule PL15 4 C2
Tredavoe TR20 2 B6
Treddiog SA62 16 B3
Tredegar NP22 18 E1
Tredington *Glos.* GL20 29 J6
Tredington *Warks.* CV36 30 D4
Tredinnick *Cornw.* PL27 3 G2
Tredinnick *Cornw.* PL14 4 C5
Tredogan CF62 18 D5
Tredomen LD3 28 A5
Tredrissi SA42 16 D1
Tredunnock NP15 19 G2
Tredustan LD3 28 A5
Tredworth GL1 29 H7
Treen *Cornw.* TR19 2 A6
Treen *Cornw.* TR20 2 B5
Treesmill PL24 4 A5
Treeton S60 51 G4
Trefaldwyn (Montgomery) SY15 38 B6
Trefasser SA64 16 B2
Trefdraeth *I.o.A.* LL62 46 C5
Trefdraeth (Newport) *Pembs.* SA42 16 D2
Trefecca LD3 28 A5
Trefechan CF48 18 D1
Trefeglwys SY17 37 J6
Trefenter SY23 27 F2
Treffgarne SA62 16 C3
Treffynnon (Holywell) *Flints.* CH8 47 K5
Treffynnon *Pembs.* SA62 16 B3
Trefgarn Owen SA62 16 B3
Trefil NP22 28 A7
Trefilan SA48 26 E3
Trefin SA62 16 B2
Treflach SY10 38 B3
Trefnanney SY22 38 B4
Trefnant LL16 47 J5
Trefonen SY10 38 B3
Trefor *Gwyn.* LL54 36 C1
Trefor *I.o.A.* LL65 46 B4
Treforest CF37 18 D3
Treforest Industrial Estate CF37 18 E3
Trefriw LL27 47 F6
Tref-y-clawdd (Knighton) LD7 28 B1
Trefynwy (Monmouth) NP25 28 E7
Tregadillett PL15 4 C2
Tregaian LL77 46 C5
Tregare NP15 28 D7
Tregarland PL13 4 C5
Tregarne TR12 2 E6
Tregaron SY25 27 F3
Tregarth LL57 46 E6
Tregaswith TR8 3 F2
Tregavethan TR4 2 E4
Tregear TR2 3 F3
Tregeare PL15 4 C2
Tregeiriog LL20 38 A2
Tregele LL67 46 B3
Tregidden TR12 2 E6
Tregiskey PL26 4 A6
Treglemais SA62 16 B3
Tregolds PL28 3 F1
Tregole EX23 4 B1
Tregonetha TR9 3 G2
Tregony TR2 3 G4
Tregoodwell PL32 4 B2
Tregoss PL26 3 G2
Tregowris TR12 2 E6
Tregoyd LD3 28 A5
Tregrehan Mills PL25 4 A5
Tre-groes SA44 17 H1
Treguff CF71 18 D4
Tregullon PL30 4 A4
Tregunnon PL15 4 C2
Tregurrian TR8 3 F2
Tregynon SY16 37 K6
Trehafod CF37 18 D2
Trehan PL12 4 E5
Treharris CF46 18 D2
Treherbert CF42 18 C2
Tre-hill CF5 18 D4
Trekenner PL15 4 D3
Treknow PL34 4 A2
Trelan TR12 2 E7
Trelash PL15 4 B1
Trelassick TR2 3 F3
Trelawnyd LL18 47 J5
Trelech SA33 17 F2
Treleddyd-fawr SA62 16 A3
Trelewis CF46 18 E2
Treligga PL33 4 A2
Trelights PL29 3 G1
Trelill PL30 4 A3
Trelissick TR3 3 F5
Trelleck NP25 19 J1
Trelleck Grange NP16 19 H1
Trelogan CH8 47 K4
Trelowla PL13 4 C5
Trelystan SY21 38 B5
Tremadog LL49 36 E1
Tremail PL32 4 B2
Tremain SA43 17 F1
Tremaine PL15 4 C2
Tremar PL14 4 C4
Trematon PL12 4 D5
Tremeirchion LL17 47 J5
Tremethick Cross TR20 2 B5
Tremore PL30 4 A4
Trenance *Cornw.* TR8 3 F2
Trenance *Cornw.* PL27 3 G1
Trenarren PL26 4 A6
Trench *Tel. & W.* TF2 39 F4
Trench *Wrex.* LL13 38 C2

Tre - Upp

Trencreek TR7 3 F2
Trenear TR13 2 D5
Treneglos PL15 4 C2
Trenewan PL13 4 B5
Trengune EX23 4 B1
Trent DT9 8 E3
Trent Port DN21 52 B4
Trent Vale ST4 40 A1
Trentham ST4 40 A1
Trentishoe EX31 6 E1
Trenwheal TR27 2 D5
Treoes CF35 18 C4
Treorchy CF42 18 C2
Treowen NP11 19 F2
Trequite PL30 4 A3
Tre'r Llai (Leighton) SY21 38 B5
Tre'r-ddol SY20 37 F6
Trerhyngyll CF71 18 D4
Tre-Rhys SA43 16 E1
Trerulefoot PL12 4 D5
Tresaith SA43 26 B3
Tresco TR24 2 B1
Trescott WV6 40 A6
Trescowe TR20 2 C5
Tresean TR8 2 E3
Tresham GL12 20 A2
Treshnish PA75 78 E3
Tresillian TR2 3 F4
Tresinney PL32 4 B2
Tresinwen SA64 16 B1
Treskinnick Cross EX23 4 C1
Treslea PL30 4 B4
Tresmeer PL15 4 C2
Tresowes Green TR13 2 D6
Tresparrett PL32 4 B1
Tresparrett Posts PL32 4 B1
Tressait PH16 81 K1
Tresta *Shet.* ZE2 107 Q3
Tresta *Shet.* ZE2 107 M7
Treswell DN22 51 K5
Trethewey TR19 2 A6
Trethomas CF83 18 E3
Trethurgy PL26 4 A5
Tretio SA62 16 A3
Tretire HR2 28 E6
Tretower NP8 28 A6
Treuddyn CH7 48 B7
Trevadlock PL15 4 C3
Trevalga PL35 4 A1
Trevalyn LL12 48 C7
Trevanson PL27 3 G1
Trevarnon TR27 2 C5
Trevarrack TR18 2 B5
Trevarren TR9 3 G2
Trevarrian TR8 3 F2
Trevarrick PL26 3 G4
Tre-vaughan *Carmar.* SA31 17 G3
Trevaughan *Carmar.* SA34 16 E4
Treveighan PL30 4 A3
Trevellas TR5 2 E3
Trevelmond PL14 4 C4
Trevenen TR13 2 D6
Treverva TR10 2 E5
Trevescan TR19 2 A6
Trevethin NP4 19 F1
Trevigro PL17 4 D4
Trevine PA35 80 B5
Treviscoe PL26 3 G3
Trevivian PL32 4 B2
Trevone PL28 3 F1
Trevor LL20 38 B1
Trewalder PL33 4 A2
Trewarmett PL34 4 A2
Trewarthenick TR2 3 G4
Trewassa PL32 4 B2
Trewellard TR19 2 A5
Trewen *Cornw.* PL15 4 C2
Trewen *Here.* HR9 28 E7
Trewennack TR13 2 D6
Trewent SA71 16 D6
Trewern SY21 38 B4
Trewethern PL27 4 A3
Trewidland PL14 4 C5
Trewilym SA41 16 E1
Trewint *Cornw.* EX23 4 B1
Trewint *Cornw.* PL15 4 C2
Trewithian TR2 3 F5
Trewoon PL25 3 G3
Treworga TR2 3 F4
Treworlas TR2 3 F5
Trewornan PL27 3 G1
Treworthal TR2 3 F5
Tre-wyn NP7 28 C6
Treyarnon PL28 3 F1
Treyford GU29 12 B5
Trezaise PL26 3 G3
Triangle HX6 57 F7
Trickett's Cross BH22 10 B4
Triermain CA8 70 A7
Trimdon TS29 62 E3
Trimdon Colliery TS29 62 E3
Trimdon Grange TS29 62 E3
Trimingham NR11 45 G2
Trimley Lower Street IP11 35 G5
Trimley St. Martin IP11 35 G5
Trimley St. Mary IP11 35 G5
Trimpley DY12 29 G1
Trimsaran SA17 17 H5
Trimstone EX34 6 C1
Trinafour PH18 81 J1
Trinant NP11 19 F2
Tring HP23 32 C7
Trinity *Angus* DD9 83 H1
Trinity *Chan.I.* JE3 3 K6
Trinity *Edin.* EH5 76 A3
Trisant SY23 27 G1
Triscombe *Som.* TA24 7 H2
Triscombe *Som.* TA4 7 K2
Trislaig PH33 87 G7
Trispen TR4 3 F3
Tritlington NE61 71 H4
Trochry PH8 82 A4
Troedyraur SA38 17 G1
Troedyrhiw CF48 18 D1
Trofarth LL22 47 G5
Trondavoe ZE2 107 M5
Troon *Cornw.* TR14 2 D5
Troon *S.Ayr.* KA10 74 B7
Trosairidh HS8 84 C3
Troston IP31 34 C1
Troswell PL15 4 C1
Trottick DD4 83 F4
Trottiscliffe ME19 24 C5
Trotton GU31 12 B4
Trough Gate OL13 56 D7
Troughend NE19 70 D4
Troustan PA22 73 J3
Troutbeck *Cumb.* LA23 60 F6
Troutbeck *Cumb.* CA11 60 E4
Troutbeck Bridge LA23 60 F6
Trow Green GL15 19 J1
Troway S21 51 F5
Trowbridge *Cardiff* CF3 19 F3
Trowbridge *Wilts.* BA14 20 B6
Trowell NG9 41 G2
Trowle Common BA14 20 B6
Trowley Bottom AL3 32 D7
Trows TD5 76 E7
Trowse Newton NR14 45 G5
Troy LS18 57 H6
Trudernish PA42 72 C5
Trudoxhill BA11 20 A7
Trull TA3 8 B2
Trumaisgearraidh HS6 92 D4
Trumpan IV55 93 H5
Trumpet HR8 29 F5
Trumpington CB2 33 H3
Trumps Green GU25 22 C5
Trunch NR28 45 G2
Trunnah FY5 55 G5
TRURO TR 3 F4
Truscott PL15 6 B7
Trusham TQ13 7 G7
Trusley DE6 40 E2
Trusthorpe LN12 53 J4
Truthan TR4 3 F3
Trysull WV5 40 A6
Tubney OX13 21 H2
Tuckenhay TQ9 5 J5
Tuckhill WV15 39 G7
Tuckingmill TR14 2 D4
Tuddenham *Suff.* IP6 35 F4
Tuddenham *Suff.* IP28 34 B1
Tudeley TN11 23 K7
Tudeley Hale TN11 23 K7
Tudhoe DL16 62 D3
Tudweiliog LL53 36 B2
Tuesley GU7 22 C7
Tuffley GL4 29 H7
Tufton *Hants.* RG28 21 H7
Tufton *Pembs.* SA63 16 D3
Tugby LE7 42 A5
Tugford SY7 38 E7
Tughall NE67 71 H1
Tulchan PH1 82 A5
Tullibody FK10 75 G1
Tullich *Arg. & B.* PA32 80 B6
Tullich *Arg. & B.* PA34 79 K6
Tullich *High.* IV20 97 F4
Tullich *High.* IV2 88 D2
Tullich *Moray* AB55 98 B6
Tullich *Stir.* FK21 81 G4
Tullich Muir IV18 96 E4
Tulliemet PH9 82 B2
Tulloch *Aber.* AB51 91 G1
Tulloch *High.* IV24 96 D2
Tulloch *Moray* IV36 97 H6
Tullochgorm PA32 73 H1
Tullochgribban High PH26 89 G2
Tullochvenus AB31 90 D4
Tulloes DD8 83 G3
Tullybannocher PH6 81 J5
Tullybelton PH1 82 B4
Tullyfergus PH11 82 D3
Tullymurdoch PH11 82 C2
Tullynessle AB33 90 D3
Tulse Hill SE21 23 G4
Tumble (Y Tymbl) SA14 17 J4
Tumby PE22 53 F7
Tumby Woodside PE22 53 F7
Tummel Bridge PH16 81 J2
TUNBRIDGE WELLS TN 13 J3
Tundergarth Mains DG11 69 G5
Tunga HS2 101 G4
Tungate NR28 45 G3
Tunley BA2 19 K6
Tunstall *E.Riding* HU12 59 K6
Tunstall *Kent* ME10 24 E5
Tunstall *Lancs.* LA6 56 B2
Tunstall *N.Yorks.* DL10 62 D7
Tunstall *Norf.* NR13 45 J5
Tunstall *Stoke* ST6 49 H7
Tunstall *Suff.* IP12 35 H3
Tunstall *T. & W.* SR2 62 E1
Tunstead *Gt.Man.* OL3 50 C2
Tunstead *Norf.* NR12 45 H3
Tunstead Milton SK23 50 C4
Tunworth RG25 21 K7
Tupholme LN3 52 E6
Tupsley HR1 28 E5
Tupton S42 51 F6
Tur Langton LE8 42 A6
Turbiskill PA31 73 F2
Turclossie AB43 99 G5
Turgis Green RG27 21 K6
Turin DD8 83 G2
Turkdean GL54 30 C7
Turleigh BA15 20 B5
Turn BL0 49 H1
Turnastone HR2 28 C5
Turnberry KA26 67 G3
Turnchapel PL9 4 E5
Turnditch DE56 40 E1
Turner's Green CV35 30 C2
Turners Hill RH10 13 G3
Turners Puddle DT2 9 H5
Turnford EN10 23 G1
Turnworth DT11 9 H4
Turret Bridge PH31 87 K5
Turriff AB53 99 F6
Turton Bottoms BL7 49 G1
Turvey MK43 32 C3
Turville RG9 22 A2
Turville Heath RG9 22 A2
Turweston NN13 31 H5
Tutbury DE13 40 E3
Tutnall B60 29 J1
Tutshill NP16 19 J2
Tuttington NR11 45 G3
Tutts Clump RG7 21 J4
Tutwell PL17 4 D3
Tuxford NG22 51 K5
Twatt *Ork.* KW17 106 B5
Twatt *Shet.* ZE2 107 M7
Twechar G65 75 F3
Tweedbank TD1 76 D7
Tweedmouth TD15 77 H5
Tweedsmuir ML12 69 F1
Twelve Oaks TN32 13 K4
Twelveheads TR4 2 E4
Twemlow Green CW4 49 G6
Twenty PE10 42 E3
TWICKENHAM TW 22 E4
Twigworth GL2 29 H6
Twineham RH17 13 F4
Twineham Green RH17 13 F4
Twinhoe BA2 20 A6
Twinstead CO10 34 C5
Twiss Green WA3 49 F3
Twiston BB7 56 D5
Twitchen *Devon* EX36 7 F2
Twitchen *Shrop.* SY7 28 C1
Twitton TN14 23 J6
Twizell House NE70 71 G1
Two Bridges *Devon* PL20 5 G3
Two Bridges *Glos.* GL14 19 K1
Two Dales DE4 50 E6
Two Gates B77 40 E5
Two Mills CH1 48 C5
Twycross CV9 41 F5
Twyford *Bucks.* MK18 31 H6
Twyford *Derbys.* DE73 41 F3
Twyford *Dorset* SP7 9 H3
Twyford *Hants.* SO21 11 F2
Twyford *Leics.* LE14 42 A4
Twyford *Norf.* NR20 44 E3
Twyford *Oxon.* OX17 31 F5
Twyford *W'ham* RG10 22 A4
Twyford Common HR2 28 E5
Twyn Shôn-Ifan CF82 18 E2
Twynholm DG6 65 G5
Twyning GL20 29 H5
Twynllanan SA19 27 G6
Twyn-yr-odyn CF5 18 E4
Twyn-y-Sheriff NP15 19 H1
Twywell NN14 32 C1
Tyberton HR2 28 C5
Tycroes SA18 17 K4
Tycrwyn SY22 38 A4
Tydd Gote PE13 43 H4
Tydd St. Giles PE13 43 H4
Tydd St. Mary PE13 43 H4
Tyddewi (St. David's) SA62 16 A3
Tye Common CM12 24 C2
Tye Green *Essex* CM77 34 B6
Tye Green *Essex* CM22 33 J6
Tye Green *Essex* CM1 33 K7
Tye Green *Essex* CM18 23 H1
Tyersal BD4 57 G6
Ty-hen LL53 36 A2
Tyldesley M29 49 F2
Tyle-garw CF72 18 D3
Tyler Hill CT2 25 H5
Tylers Green *Bucks.* HP10 22 C2
Tyler's Green *Essex* CM16 23 J1
Tylorstown CF43 18 D2
Tylwch LD6 27 J1
Ty-Mawr *Conwy* LL21 37 J1
Ty-mawr *Denb.* LL21 38 A1
Ty-nant *Conwy* LL21 37 J1
Ty-nant *Gwyn.* LL23 37 J3
Tyndrum FK20 80 E4
Tyneham BH20 9 H6
Tynehead EH37 76 B5
Tynemouth NE30 71 J7
Tynewydd CF42 18 C2
Tyninghame EH42 76 E3
Tynron DG3 68 D4
Tyntesfield BS48 19 J4
Ty'n-y-cefn LL21 37 K1
Ty'n-y-coedcae CF83 18 E3
Tyn-y-cwm SY18 37 H7
Tyn-y-ffridd SY10 38 A2
Ty'n-y-garn CF31 18 B3
Tyn-y-gongl LL74 46 D4
Tynygraig *Cere.* SY25 27 F2
Tyn-y-graig *Powys* LD2 27 K4
Ty'n-y-groes LL32 47 F5
Tyrie AB43 99 H4
Tyringham MK16 32 B4
Tyseley B11 40 D7
Tythegston CF32 18 B4
Tytherington *Ches.E.* SK10 49 J5
Tytherington *S.Glos.* GL12 19 K3
Tytherington *Som.* BA11 20 A7
Tytherington *Wilts.* BA12 20 C7
Tytherton Lucas SN15 20 C4
Tyttenhanger AL4 22 E1
Ty-uchaf SY10 37 J3
Tywardreath PL24 4 A5
Tywardreath Highway PL24 4 A5
Tywyn LL36 36 E5

U

Uachdar HS7 92 D6
Uags IV54 86 D1
Ubberley ST2 40 B1
Ubbeston Green IP19 35 H1
Ubley BS40 19 J6
Uckerby DL10 62 D6
Uckfield TN22 13 H4
Uckinghall GL20 29 H5
Uckington GL51 29 J6
Uddingston G71 74 E4
Uddington ML11 75 G7
Udimore TN31 14 D6
Udley BS40 19 H5
Udny Green AB41 91 G2
Udny Station AB41 91 H2
Udston ML3 75 F5
Udstonhead ML10 75 F6
Uffcott SN4 20 E4
Uffculme EX15 7 J4
Uffington *Lincs.* PE9 42 D5
Uffington *Oxon.* SN7 21 G3
Uffington *Shrop.* SY4 38 E4
Ufford *Peter.* PE9 42 D5
Ufford *Suff.* IP13 35 G3
Ufton CV33 30 E2
Ufton Green RG7 21 K5
Ufton Nervet RG7 21 K5
Ugborough PL21 5 G5
Ugford SP2 10 B1
Uggeshall NR34 45 J7
Ugglebarnby YO22 63 K6
Ugley CM22 33 J6
Ugley Green CM22 33 J6
Ugthorpe YO21 63 J5
Uibhist a Deas (South Uist) HS8 84 C1
Uibhist a Tuath (North Uist) HS6 92 D4
Uidh HS9 84 B5
Uig *Arg. & B.* PA78 78 C2
Uig *Arg. & B.* PA23 73 K2
Uig *High.* IV51 93 J5
Uig *High.* IV55 93 G6
Uigen HS2 100 C4
Uiginish IV55 93 H7
Uigshader IV51 93 K7
Uisgebhagh (Uiskevagh) HS7 92 D6
Uisken PA67 78 E6
Uiskevagh (Uisgebhagh) HS7 92 D6
Ulbster KW2 105 J4
Ulcat Row CA11 60 F4
Ulceby *Lincs.* LN13 53 H5
Ulceby *N.Lincs.* DN39 52 E1
Ulceby Cross LN13 53 H5
Ulceby Skitter DN39 52 E1
Ulcombe ME17 14 D3
Uldale CA7 60 D3
Uldale House CA17 61 J7
Uley GL11 20 A2
Ulgham NE61 71 H4
Ullapool IV26 95 H2
Ullenhall B95 30 C2
Ullenwood GL53 29 J7
Ulleskelf LS24 58 B6
Ullesthorpe LE17 41 H7
Ulley S26 51 G4
Ullingswick HR1 28 E4
Ullinish IV56 85 J1
Ullock CA14 60 B4
Ulpha *Cumb.* LA20 60 C7
Ulpha *Cumb.* LA11 55 H1
Ulrome YO25 59 H4
Ulsta ZE2 107 N4
Ulting CM9 24 E1
Uluvalt PA70 79 G4
Ulva PA73 79 F4
Ulverston LA12 55 F2
Ulwell BH19 10 B6
Ulzieside DG4 68 C3
Umberleigh EX37 6 E3
Unapool IV27 102 E5
Underbarrow LA8 61 F7
Undercliffe BD2 57 G6
Underhill EN5 23 F2
Underhoull ZE2 107 P2
Underling Green TN12 14 C3
Underriver TN15 23 J6
Underwood *Newport* NP18 19 G3
Underwood *Notts.* NG16 51 G7
Underwood *Plym.* PL7 5 F5
Undley IP27 44 A7
Undy NP26 19 H3
Ungisiadar HS2 100 D5
Unifirth ZE2 107 L7
Union Croft AB39 91 G5
Union Mills IM4 54 C6
Union Street TN5 14 C4
Unst ZE2 107 Q1
Unst Airport ZE2 107 Q2
Unstone S18 51 F5
Unstone Green S18 51 F5
Unsworth BL9 49 H2
Unthank *Cumb.* CA11 61 F3
Unthank *Derbys.* S18 51 F5
Up Cerne DT2 9 F4
Up Exe EX5 7 H5
Up Hatherley GL51 29 J6
Up Holland WN8 48 E2
Up Marden PO18 11 J3
Up Mudford BA21 8 E3
Up Nately RG27 21 K6
Up Somborne SO20 10 E1
Up Sydling DT2 9 F4
Upavon SN9 20 E6
Upchurch ME9 24 E5
Upcott *Devon* EX21 6 C6
Upcott *Devon* EX31 6 D2
Upcott *Here.* HR3 28 C3
Upcott *Som.* TA22 7 H3
Upend CB8 33 K3
Upgate NR9 45 F4
Upgate Street *Norf.* NR16 44 E6
Upgate Street *Norf.* NR35 45 G6
Uphall *Dorset* DT2 8 E4
Uphall *W.Loth.* EH52 75 J3
Uphall Station EH54 75 J4
Upham *Devon* EX17 7 G5
Upham *Hants.* SO32 11 G2
Uphampton *Here.* HR6 28 C2
Uphampton *Worcs.* WR9 29 H2
Uphempston TQ9 5 J4
Uphill BS23 19 G6
Uplands *Glos.* GL5 20 B1
Uplands *Swan.* SA2 17 K6
Uplawmoor G78 74 C5
Upleadon GL18 29 G6
Upleatham TS11 63 H5
Uplees ME13 25 G5
Uploders DT6 8 E5
Uplowman EX16 7 J4
Uplyme DT7 8 C5
Upminster RM14 23 J3
Upottery EX14 8 B4
Upper Affcot SY6 38 D7
Upper Ardroscadale PA20 73 J4
Upper Arley DY12 39 G7
Upper Arncott OX25 31 H7
Upper Astley SY4 38 E4
Upper Aston WV5 40 A6
Upper Astrop OX17 31 G5
Upper Barvas HS2 101 F2
Upper Basildon RG8 21 K4
Upper Bayble (Pabail Uarach) HS2 101 H4
Upper Beeding BN44 12 E5
Upper Benefield PE8 42 C7
Upper Bentley B97 29 J2
Upper Berwick SY4 38 D4
Upper Bighouse KW13 104 D3
Upper Boat CF37 18 E3
Upper Boddam AB52 90 E1
Upper Boddington NN11 31 F3
Upper Borth SY24 37 F7
Upper Boyndlie AB43 99 H4
Upper Brailes OX15 30 E5
Upper Breakish IV42 86 C2
Upper Breinton HR4 28 D4
Upper Broadheath WR2 29 H3
Upper Broughton LE14 41 J3
Upper Brynamman SA18 27 G7
Upper Bucklebury RG7 21 J5
Upper Burgate SP6 10 C3
Upper Burnhaugh AB39 91 G5
Upper Caldecote SG18 32 E4
Upper Camster KW3 105 H4
Upper Canada BS24 19 G6
Upper Catesby NN11 31 G3
Upper Catshill B61 29 J1
Upper Chapel LD3 27 K4
Upper Cheddon TA2 8 B2
Upper Chicksgrove SP3 9 J1
Upper Chute SP11 21 F6
Upper Clatford SP11 21 G7
Upper Coberley GL53 29 J7
Upper Colwall WR13 29 G4
Upper Cotton ST10 40 C1
Upper Cound SY5 38 E5
Upper Cumberworth HD8 50 E2
Upper Cwmbran NP44 19 F2
Upper Dallachy IV32 98 B4
Upper Dean PE28 32 D2
Upper Denby HD8 50 E2
Upper Denton CA8 70 B7
Upper Derraid PH26 89 H1
Upper Diabaig IV22 94 E5
Upper Dicker BN27 13 J5
Upper Dovercourt CO12 35 G5
Upper Dunsforth YO26 57 K3
Upper Dunsley HP23 32 C7
Upper Eastern Green CV5 40 E7
Upper Eathie IV11 96 E5
Upper Egleton HR8 29 F4
Upper Elkstone SK17 50 C7
Upper End SK17 50 C5
Upper Enham SP11 21 G7
Upper Farringdon GU34 11 J1
Upper Framilode GL2 29 G7
Upper Froyle GU34 22 A7
Upper Gills KW1 105 J1
Upper Glendessarry PH34 87 F5
Upper Godney BA5 19 H7
Upper Gornal DY3 40 B6
Upper Gravenhurst MK45 32 E5
Upper Green *Essex* CB11 33 H5
Upper Green *Essex* CB10 33 J5
Upper Green *Mon.* NP7 28 C7
Upper Green *W.Berks.* RG17 21 G5
Upper Grove Common HR9 28 E6
Upper Gylen PA34 79 K5
Upper Hackney DE4 50 E6
Upper Halliford TW17 22 D5
Upper Halling ME2 24 C5
Upper Hambleton LE15 42 C5
Upper Harbledown CT2 15 G2
Upper Hardres Court CT4 15 G2
Upper Hartfield TN7 13 H3
Upper Hatton ST21 40 A2
Upper Hawkhillock AB42 91 J1
Upper Hayesden TN11 23 J7
Upper Hayton SY8 38 E7
Upper Heath SY7 38 E7

Upp - Wat

Upper Heaton HD5 50 D1
Upper Hellesdon NR3 45 G4
Upper Helmsley YO41 58 C4
Upper Hengoed SY10 38 B2
Upper Hergest HR5 28 B3
Upper Heyford Northants. NN7 31 H3
Upper Heyford Oxon. OX25 31 F6
Upper Hill Here. HR6 28 D3
Upper Hill S.Glos. GL13 19 K2
Upper Horsebridge BN27 13 J5
Upper Howsell WR14 29 G4
Upper Hulme ST13 50 C6
Upper Inglesham SN6 21 F2
Upper Kilchattan PA61 72 B1
Upper Killay SA2 17 J6
Upper Knockando AB38 97 J7
Upper Lambourn RG17 21 G3
Upper Langford BS40 19 H6
Upper Langwith NG20 51 H6
Upper Largo KY8 83 F7
Upper Leigh ST10 40 C2
Upper Ley GL14 29 G7
Upper Llandwrog (Y Fron) LL54 46 D7
Upper Loads S42 51 F6
Upper Lochton AB31 90 E5
Upper Longdon WS15 40 C4
Upper Longwood SY5 39 F5
Upper Ludstone WV5 40 A6
Upper Lybster KW3 105 H5
Upper Lydbrook GL17 29 F7
Upper Lyde HR4 28 D4
Upper Lye HR6 28 C2
Upper Maes-coed HR2 28 C5
Upper Midhope S36 50 E3
Upper Milovaig IV55 93 F7
Upper Milton OX7 30 D7
Upper Minety SN16 20 D2
Upper Moor WR10 29 J4
Upper Morton BS35 19 K2
Upper Muirskie AB12 91 G5
Upper Nash SA71 16 D5
Upper Newbold S41 51 F5
Upper North Dean HP14 22 B2
Upper Norwood SE19 23 G4
Upper Obney PH1 82 B4
Upper Oddington GL56 30 D6
Upper Ollach IV51 86 B1
Upper Padley S32 50 E5
Upper Pennington SO41 10 D5
Upper Pollicott HP18 31 J7
Upper Poppleton YO26 58 B4
Upper Quinton CV37 30 C4
Upper Ratley SO51 10 E2
Upper Ridinghill AB43 99 J5
Upper Rissington GL54 30 D6
Upper Rochford WR15 29 F2
Upper Sanday KW17 106 E7
Upper Sapey WR6 29 F2
Upper Scolton SA62 16 C3
Upper Seagry SN15 20 C3
Upper Shelton MK43 32 C4
Upper Sheringham NR26 45 F1
Upper Shuckburgh NN11 31 F2
Upper Siddington GL7 20 D2
Upper Skelmorlie PA17 74 A4
Upper Slaughter GL54 30 C6
Upper Sonachan PA33 80 B5
Upper Soudley GL14 29 F7
Upper Staploe PE19 32 E3
Upper Stoke NR14 45 G5
Upper Stondon SG16 32 E5
Upper Stowe NN7 31 H3
Upper Street Hants. SP6 10 C3
Upper Street Norf. NR12 45 H4
Upper Street Norf. IP21 35 F1
Upper Street Suff. IP9 35 F5
Upper Street Suff. IP6 35 F3
Upper Strensham WR8 29 J5
Upper Sundon LU5 32 D6
Upper Swanmore SO32 11 G3
Upper Swell GL54 30 C6
Upper Tean ST10 40 C2
Upper Thurnham LA2 55 H4
Upper Tillyrie KY13 82 C7
Upper Tooting SW17 23 F4
Upper Town Derbys. DE4 50 E6
Upper Town Derbys. DE4 50 E7
Upper Town Derbys. DE6 50 E7
Upper Town Here. HR1 28 E4
Upper Town N.Som. BS40 19 J5
Upper Tysoe CV35 30 E4
Upper Upham SN8 21 F4
Upper Upnor ME2 24 D4
Upper Victoria DD7 83 G4
Upper Vobster BA3 20 A7
Upper Wardington OX17 31 F4
Upper Waterhay SN6 20 D2
Upper Weald MK19 32 B5
Upper Weedon NN7 31 H3
Upper Welson HR3 28 B3
Upper Weston BA1 20 A5
Upper Whiston S60 51 G4
Upper Wick WR2 29 H3
Upper Wield SO24 11 H1
Upper Winchendon (Over Winchendon) HP18 31 J7
Upper Witton B23 40 C6
Upper Woodford SP4 10 C1
Upper Woolhampton RG7 21 J5
Upper Wootton RG26 21 J6
Upper Wraxall SN14 20 B4
Upper Wyche WR13 29 G4
Upperby CA2 60 F1
Uppermill OL3 49 J2
Upperthong HD9 50 D2
Upperton GU28 12 C4
Uppertown Derbys. S45 51 F6
Uppertown Ork. KW1 105 J1
Uppingham LE15 42 B5
Uppington TF6 38 E5
Upsall YO7 57 K1
Upsettlington TD15 77 G6
Upshire EN9 23 H1
Upstreet CT3 25 J5
Upthorpe IP31 34 D1
Upton Bucks. HP17 31 J7
Upton Cambs. PE28 32 E1
Upton Ches.W. & C. CH2 48 D6
Upton Cornw. PL14 4 C3
Upton Cornw. EX23 6 A5
Upton Devon EX14 7 J5
Upton Devon TQ7 5 H6
Upton Dorset BH16 9 J5
Upton Dorset DT2 9 G6
Upton E.Riding YO25 59 H4
Upton Hants. SP11 21 G6
Upton Hants. SO16 10 E3
Upton Leics. CV13 41 F6
Upton Lincs. DN21 52 B4
Upton Mersey. CH49 48 B4
Upton Norf. NR13 45 H4
Upton Northants. NN5 31 J2
Upton Notts. DN22 51 K5
Upton Notts. NG23 51 K7
Upton Oxon. OX11 21 J3
Upton Oxon. OX18 30 D7
Upton Pembs. SA72 16 D5
Upton Peter. PE5 42 E5
Upton Slo. SL1 22 C4
Upton Som. TA4 7 H3
Upton Som. TA10 8 D2
Upton W.Yorks. WF9 51 G1
Upton Wilts. SP3 9 H1
Upton Bishop HR9 29 F6
Upton Cheyney BS30 19 K5
Upton Cressett WV16 39 F6
Upton Crews HR9 29 F6
Upton Cross PL14 4 C3
Upton End SG5 32 E5
Upton Grey RG25 21 K7
Upton Hellions EX17 7 G5
Upton Lovell BA12 20 C7
Upton Magna SY4 38 E4
Upton Noble BA4 9 G1
Upton Park E13 23 H3
Upton Pyne EX5 7 H6
Upton St. Leonards GL4 29 H7
Upton Scudamore BA12 20 B7
Upton Snodsbury WR7 29 J3
Upton upon Severn WR8 29 H4
Upton Warren B61 29 J2
Upwaltham GU28 12 C5
Upware CB7 33 J1
Upwell PE14 43 J5
Upwey DT3 9 F6
Upwick Green SG11 33 H6
Upwood PE26 43 F7
Uradale ZE1 107 N9
Urafirth ZE2 107 M5
Urchany IV12 97 F7
Urchfont SN10 20 D6
Urdimarsh HR1 28 E4
Ure ZE2 107 L5
Urgha HS3 93 G2
Urlay Nook TS16 63 F5
Urmston M41 49 G3
Urpeth DH2 62 D1
Urquhart High. IV7 96 C6
Urquhart Moray IV30 97 K5
Urra TS9 63 G6
Urray IV6 96 C6
Ushaw Moor DH7 62 D2
Usk (Brynbuga) NP15 19 G1
Usselby LN8 52 D3
Usworth NE37 62 E1
Utley BD20 57 F5
Uton EX17 7 G6
Utterby LN11 53 G3
Uttoxeter ST14 40 C2
Uwchmynydd LL53 36 A3
Uxbridge UB8 22 D3
Uyeasound ZE2 107 P2
Uzmaston SA62 16 C4

V

Valley (Y Fali) LL65 46 A5
Valley Truckle PL32 4 B2
Valleyfield D. & G. DG6 65 G5
Valleyfield Fife KY12 75 J2
Valsgarth ZE2 107 Q1
Vange SS16 24 D3
Vardre SA6 17 K5
Varteg NP4 19 F1
Vatersay (Bhatarsaigh) HS9 84 B5
Vatsetter ZE2 107 P4
Vatten IV55 93 H7
Vaul PA77 78 B3
Vaynor CF48 27 K7
Vaynor Park SY21 38 A5
Veaullt HR5 28 A3
Veensgarth ZE2 107 N8
Velindre Pembs. SA41 16 D2
Velindre Powys LD3 28 A5
Vellow TA4 7 J2
Veness KW17 106 E5
Venn TQ7 5 H6
Venn Ottery EX11 7 J6
Venngreen EX22 6 B4
Vennington SY5 38 C5
Venny Tedburn EX17 7 G6
Venterdon PL17 4 D3
Ventnor PO38 11 G7
Venton PL7 5 F5
Vernham Dean SP11 21 G6
Vernham Street SP11 21 G6
Vernolds Common SY7 38 D7
Verwood BH31 10 B4
Veryan TR2 3 G5
Veryan Green TR2 3 G4
Vickerstown LA14 54 E3
Victoria PL26 3 G2
Vidlin ZE2 107 N6
Viewfield KW14 105 F2
Viewpark G71 75 F4
Vigo WS9 40 C5
Vigo Village DA13 24 C5
Villavin EX19 6 D4
Vinehall Street TN32 14 C5
Vine's Cross TN21 13 J5
Viney Hill GL15 19 K1
Virginia Water GU25 22 C5
Virginstow EX21 6 B6
Virley CM9 34 D7
Vobster BA3 20 A7
Voe Shet. ZE2 107 N6
Voe Shet. ZE2 107 M4
Vowchurch HR2 28 C5
Voy KW16 106 B6
Vron Gate SY5 38 C5

W

Waberthwaite CA18 60 C7
Wackerfield DL2 62 C4
Wacton NR15 45 F6
Wadbister ZE2 107 N8
Wadborough WR8 29 J4
Waddesdon HP18 31 J7
Waddeton TQ5 5 J5
Waddicar L31 48 C3
Waddingham DN21 52 C3
Waddington Lancs. BB7 56 C5
Waddington Lincs. LN5 52 C6
Waddingworth LN10 52 E5
Waddon Devon TQ13 5 J3
Waddon Gt.Lon. CR0 23 G5
Wadebridge PL27 3 G1
Wadeford TA20 8 C3
Wadenhoe PE8 42 D7
Wadesmill SG12 33 G7
Wadhurst TN5 13 K3
Wadshelf S42 51 F5
Wadworth DN11 51 H3
Wadworth Hill HU12 59 J7
Waen Denb. LL16 47 K6
Waen Denb. LL16 47 H6
Waen Aberwheeler LL16 47 J6
Waen-fâch SY22 38 B4
Waen-wen LL57 46 D6
Wag KW7 105 F6
Wainfleet All Saints PE24 53 H7
Wainfleet Bank PE24 53 H7
Wainfleet St. Mary PE24 53 H7
Wainford NR35 45 H6
Waingroves DE5 41 G1
Wainhouse Corner EX23 4 B1
Wainscott ME2 24 D4
Wainstalls HX2 57 F7
Waitby CA17 61 J6
WAKEFIELD WF 57 J7
Wakerley LE15 42 C6
Wakes Colne CO6 34 C6
Walberswick IP18 35 J1
Walberton BN18 12 C6
Walbottle NE15 71 G7
Walcot Lincs. NG34 42 D2
Walcot Lincs. LN4 52 E7
Walcot N.Lincs. DN15 58 E7
Walcot Shrop. SY7 38 C7
Walcot Tel. & W. TF6 38 E4
Walcot Green IP22 45 F7
Walcote Leics. LE17 41 H7
Walcote Warks. B49 30 C3
Walcott NR12 45 H2
Walcott Dales LN4 52 E7
Walden DL8 57 F1
Walden Head DL8 56 E1
Walden Stubbs DN6 51 H1
Walderslade ME5 24 D5
Walderton PO18 11 J3
Walditch DT6 8 D5
Waldley DE6 40 D2
Waldridge DH2 62 D2
Waldringfield IP12 35 G4
Waldron TN21 13 J5
Wales S26 51 G4
Walesby Lincs. LN8 52 E3
Walesby Notts. NG22 51 J5
Waleswood S26 51 G4
Walford Here. SY7 28 C1
Walford Here. HR9 28 E6
Walford Shrop. SY4 38 D3
Walford Staffs. ST21 40 A2
Walford Heath SY4 38 D4
Walgherton CW5 39 F1
Walgrave NN6 32 B1
Walhampton SO41 10 E5
Walk Mill BB10 56 D6
Walkden M28 49 G2
Walker NE6 71 H7
Walker Fold BB7 56 B5
Walkerburn EH43 76 B7
Walkeringham DN10 51 K3
Walkerith DN21 51 K3
Walkern SG2 33 F6
Walker's Green HR1 28 E4
Walkford BH23 10 D5
Walkhampton PL20 5 F4
Walkingham Hill HG3 57 J3
Walkington HU17 59 F6
Walkwood B97 30 B2
Wall Cornw. TR27 2 D5
Wall Northumb. NE46 70 E7
Wall Staffs. WS14 40 D5
Wall End LA17 55 F1
Wall Heath DY6 40 A7
Wall Houses NE45 71 F7
Wall under Heywood SY6 38 E6
Wallacehall DG11 69 H6
Wallacetown KA19 67 G3
Wallasey CH45 48 B3
Wallaston Green SA71 16 C5
Wallend ME3 24 E4
Waller's Green HR8 29 F5
Wallingford OX10 21 K3
Wallington Gt.Lon. SM6 23 F5
Wallington Hants. PO16 11 G4
Wallington Herts. SG7 33 F5
Wallington Wrex. LL13 38 D1
Wallingwells S81 51 H4
Wallis SA62 16 D3
Wallisdown BH12 10 B5
Walliswood RH5 12 E3
Walls ZE2 107 L8
Wallsend NE28 71 J7
Wallyford EH21 76 B3
Walmer CT14 15 J2
Walmer Bridge PR4 55 H7
Walmersley BL9 49 H1
Walmley B76 40 D6
Walmsgate LN11 53 G5
Walpole IP19 35 H1
Walpole Cross Keys PE34 43 J4
Walpole Highway PE14 43 J4
Walpole Marsh PE14 43 H4
Walpole St. Andrew PE14 43 J4
Walpole St. Peter PE14 43 J4
Walrond's Park TA3 8 C2
Walrow TA9 19 G7
WALSALL WS 40 C6
Walsall Wood WS9 40 C5
Walsden OL14 56 E7
Walsgrave on Sowe CV2 41 F7
Walsham le Willows IP31 34 E1
Walshford LS22 57 K4
Walsoken PE13 43 H4
Walston ML11 75 J6
Walsworth SG4 32 E5
Walter's Ash HP14 22 B2
Walterston CF62 18 D4
Walterstone HR2 28 C6
Waltham Kent CT4 15 G3
Waltham N.E.Lincs. DN37 53 F2
Waltham Abbey EN9 23 G1
Waltham Chase SO32 11 G3
Waltham Cross EN8 23 G1
Waltham on the Wolds LE14 42 A3
Waltham St. Lawrence RG10 22 B4
Walthamstow E17 23 G3
Walton Bucks. HP21 32 B7
Walton Cumb. CA8 70 A7
Walton Derbys. S42 51 F6
Walton Leics. LE17 41 H7
Walton M.K. MK7 32 B5
Walton Mersey. L9 48 C3
Walton Peter. PE4 42 E5
Walton Powys LD8 28 B3
Walton Shrop. SY7 28 D1
Walton Som. BA16 8 D1
Walton Staffs. ST15 40 A2
Walton Suff. IP11 35 H5
Walton Tel. & W. TF6 38 E4
Walton W.Yorks. WF2 51 F1
Walton W.Yorks. LS23 57 K5
Walton Warks. CV35 30 D3
Walton Cardiff GL20 29 J5
Walton East SA63 16 D3
Walton Elm DT10 9 G3
Walton Highway PE14 43 H4
Walton Lower Street IP11 35 G5
Walton on the Hill KT20 23 F6
Walton on the Naze CO14 35 G6
Walton on the Wolds LE12 41 H4
Walton Park D. & G. DG7 65 H3
Walton Park N.Som. BS21 19 H4
Walton West SA62 16 B4
Walton-in-Gordano BS21 19 H4
Walton-le-Dale PR5 55 J7
Walton-on-Thames KT12 22 E5
Walton-on-the-Hill ST17 40 B3
Walton-on-Trent DE12 40 E4
Walwen Flints. CH8 47 K5
Walwen Flints. CH6 48 B5
Walwick NE46 70 E6
Walworth DL2 62 D5
Walworth Gate DL2 62 D4
Walwyn's Castle SA62 16 B4
Wambrook TA20 8 B4
Wanborough Surr. GU3 22 C7
Wanborough Swin. SN4 21 F3
Wandel ML12 68 E1
Wandon NE71 71 F1
Wandon End LU2 32 E6
Wandsworth SW15 23 F4
Wandylaw NE67 71 G1
Wangford Suff. NR34 35 J1
Wangford Suff. IP27 44 B7
Wanlip LE7 41 J4
Wanlockhead ML12 68 D2
Wannock BN26 13 J6
Wansford E.Riding YO25 59 G4
Wansford Peter. PE8 42 D6
Wanshurst Green TN12 14 C3
Wanstrow BA4 20 A7
Wanswell GL13 19 K1
Wantage OX12 21 G3
Wapley BS37 20 A4
Wappenbury CV33 30 E2
Wappenham NN12 31 H4
Warblebank CA7 60 D2
Warbleton TN21 13 K5
Warblington PO9 11 J4
Warborough OX10 21 J2
Warboys PE28 43 G7
Warbreck FY2 55 G6
Warbstow PL15 4 C1
Warburton WA13 49 F4
Warcop CA16 61 J5
Ward End B8 40 D7
Ward Green IP14 34 E2
Warden Kent ME12 25 G4
Warden Northumb. NE46 70 E7
Warden Hill GL51 29 J6
Warden Street SG18 32 E4
Wardhouse AB52 90 D1
Wardington OX17 31 F4
Wardle Ches.E. CW5 49 F7
Wardle Gt.Man. OL12 49 J1
Wardley Gt.Man. M27 49 G2
Wardley Rut. LE15 42 B5
Wardley T. & W. NE10 71 J7
Wardlow SK17 50 D5
Wardsend SK10 49 J4
Wardy Hill CB6 43 H7
Ware Herts. SG12 33 G7
Ware Kent CT3 25 J5
Wareham BH20 9 J6
Warehorne TN26 14 E4
Waren Mill NE70 77 K7
Warenford NE70 71 G1
Warenton NE70 77 K7
Wareside SG12 33 G7
Waresley Cambs. SG19 33 F3
Waresley Worcs. DY11 29 H1
Warfield RG42 22 B4
Wargrave Mersey. WA12 48 E3
Wargrave W'ham RG10 22 A4
Warham Here. HR4 28 D5
Warham Norf. NR23 44 D1
Wark Northumb. NE48 70 D6
Wark Northumb. TD12 77 G7
Warkleigh EX37 6 E3
Warkton NN16 32 B1
Warkworth Northants. OX17 31 F4
Warkworth Northumb. NE65 71 H3
Warland OL14 56 E7
Warleggan PL30 4 B4
Warley Essex CM14 23 J2
Warley W.Mid. B68 40 C7
Warley Town HX2 57 F7
Warlingham CR6 23 G6
Warmfield WF1 57 J7
Warmingham CW11 49 G6
Warminghurst RH20 12 E5
Warmington Northants. PE8 42 D6
Warmington Warks. OX17 31 F4
Warminster BA12 20 B7
Warmlake ME17 14 D3
Warmley BS30 19 K4
Warmley Hill BS15 19 K4
Warmsworth DN4 51 H2
Warmwell DT2 9 G6
Warndon WR4 29 H3
Warners End HP1 22 D1
Warnford SO32 11 H2
Warnham RH12 12 E3
Warningcamp BN18 12 D6
Warninglid RH17 13 F4
Warren Ches.E. SK11 49 H5
Warren Pembs. SA71 16 C6
Warren House PL20 6 E7
Warren Row RG10 22 B3
Warren Street ME17 14 E2
Warrenby TS10 63 G4
Warren's Green SG4 33 F6
Warrington M.K. MK46 32 B3
WARRINGTON Warr. WA 49 F4
Warroch KY13 82 B7
Warsash SO31 11 F4
Warslow SK17 50 C7
Warsop Vale NG20 51 H6
Warter YO42 58 E4
Warthill YO19 58 C4
Wartle AB31 90 D4
Wartling BN27 13 K6
Wartnaby LE14 42 A3
Warton Lancs. LA5 55 J2
Warton Lancs. PR4 55 H7
Warton Northumb. NE65 71 F3
Warton Warks. B79 40 E5
Warton Bank PR4 55 H7
Warwick CV34 30 D2
Warwick Bridge CA4 61 F1
Warwick Wold RH1 23 G6
Warwick-on-Eden CA4 61 F1
Wasbister KW17 106 C4
Wasdale Head CA20 60 C6
Wash SK23 50 C4
Wash Common RG14 21 H5
Washall Green SG9 33 H5
Washaway PL30 4 A4
Washbourne TQ9 5 H5
Washbrook BS28 19 H7
Washfield EX16 7 H4
Washfold DL11 62 B6
Washford Som. TA23 7 J1
Washford Warks. B98 30 B2
Washford Pyne EX17 7 G4
Washingborough LN4 52 D5
Washington T. & W. NE38 62 E1
Washington W.Suss. RH20 12 E5
Washmere Green CO10 34 D4
Wasing RG7 21 J5
Waskerley DH8 62 B2
Wasperton CV35 30 D3
Wasps Nest LN4 52 D6
Wass YO61 58 B2
Watchet TA23 7 J1
Watchfield Oxon. SN6 21 F2
Watchfield Som. TA9 19 G7

Wat - Wes

Watchgate LA8 61 G7
Watcombe TQ1 5 K4
Watendlath CA12 60 D5
Water BB4 56 D7
Water Eaton M.K. MK2 32 B5
Water Eaton Oxon. OX2 31 G7
Water End Bed. MK44 32 E4
Water End Cen.Beds. MK45 32 D5
Water End E.Riding YO43 58 D6
Water End Essex CB10 33 J4
Water End Herts. AL9 23 F1
Water End Herts. HP1 32 D7
Water Newton PE8 42 E6
Water Orton B46 40 D6
Water Stratford MK18 31 H5
Water Yeat LA12 55 F1
Waterbeach CB25 33 H2
Waterbeck DG11 69 H6
Watercombe DT2 9 G6
Waterend HP14 22 A2
Waterfall ST10 50 C7
Waterfoot E.Renf. G76 74 D5
Waterfoot Lancs. BB4 56 D7
Waterford SG14 33 G7
Watergate PL32 4 B2
Waterhead Cumb. LA22 60 E6
Waterhead D. & G. DG7 68 C5
Waterheath NR34 45 J6
Waterhill of Bruxie AB42 99 H6
Waterhouses Dur. DH7 62 C2
Waterhouses Staffs. ST10 50 C7
Wateringbury ME18 23 K6
Waterlane GL6 20 C1
Waterloo Aber. AB42 91 J1
Waterloo Derbys. S45 51 G6
Waterloo Gt.Man. OL7 49 J2
Waterloo High. IV42 86 C2
Waterloo Mersey. L22 48 C3
Waterloo N.Lan. ML2 75 G5
Waterloo Norf. NR10 45 G4
Waterloo P. & K. PH1 82 B4
Waterloo Pembs. SA72 16 C5
Waterloo Poole BH17 10 B5
Waterloo Cross EX15 7 J4
Waterloo Port LL55 46 C6
Waterlooville PO7 11 H3
Watermeetings ML12 68 E2
Watermillock CA11 60 F4
Waterperry OX33 21 K1
Waterrow TA4 7 J3
Waters Upton TF6 39 F4
Watersfield RH20 12 D5
Watersheddings OL4 49 J2
Waterside Aber. AB36 90 B3
Waterside Aber. AB41 91 J2
Waterside B'burn. BB3 56 C7
Waterside Bucks. HP5 22 C1
Waterside E.Ayr. KA6 67 J3
Waterside E.Ayr. KA3 74 C6
Waterside E.Dun. G66 74 E3
Waterstock OX33 21 K1
Waterston SA73 16 C5
Waterthorpe S20 51 G4
WATFORD Herts. WD 22 E2
Watford Northants. NN6 31 H2
Watford Park CF83 18 E3
Wath N.Yorks. HG4 57 J2
Wath N.Yorks. HG3 57 G3
Wath Brow CA25 60 B5
Wath upon Dearne S63 51 G2
Watley's End BS36 19 K3
Watlington Norf. PE33 44 A4
Watlington Oxon. OX49 21 K2
Watnall NG16 41 H1
Watten KW1 105 H3
Wattisfield IP22 34 E1
Wattisham IP7 34 E3
Watton Dorset DT6 8 D5
Watton E.Riding YO25 59 G5
Watton Norf. IP25 44 D5
Watton at Stone SG14 33 G7
Watton Green IP25 44 D5
Watton's Green CM14 23 J2
Wattston ML6 75 F4
Wattstown CF39 18 D2
Wattsville NP11 19 F2
Waughtonhill AB43 99 H5
Waun Fawr SY23 37 F7
Waun y Clyn SA17 17 H5
Waunarlwydd SA5 17 K6
Waunclunda SA19 17 K2
Waunfawr LL55 46 D7
Waun-Lwyd NP23 18 E1
Wavendon MK17 32 C5
Waverbridge CA7 60 D2
Waverton Ches.W. & C. CH3 48 D6
Waverton Cumb. CA7 60 D2
Wavertree L15 48 C4
Wawne HU7 59 G6
Waxham NR12 45 J3
Waxholme HU19 59 K7
Way Gill BD23 56 E3
Way Village EX16 7 G4
Way Wick BS24 19 G5
Wayford TA18 8 D4
Waytown DT6 8 D5
Wdig (Goodwick) SA64 16 C2
Weachyburn AB45 98 E5
Weacombe TA4 7 K1
Weald OX18 21 G1
Wealdstone HA3 22 E2
Weardley LS17 57 H5
Weare BS26 19 H6
Weare Giffard EX39 6 C3
Wearhead DL13 61 K3
Wearne TA10 8 D2
Weasenham All Saints PE32 44 C3
Weasenham St. Peter PE32 44 C3
Weathercote LA6 56 C2
Weatheroak Hill B48 30 B1
Weaverham CW8 49 F5
Weaverthorpe YO17 59 F2
Webheath B97 30 B2
Webton HR2 28 D5
Wedderlairs AB41 91 G1
Weddington CV10 41 F6
Wedhampton SN10 20 D6
Wedmore BS28 19 H7
Wednesbury WS10 40 B6
Wednesfield WV11 40 B6
Weedon HP22 32 B7
Weedon Bec NN7 31 H3
Weedon Lois NN12 31 H4
Weeford WS14 40 D5
Week Devon EX18 7 F4
Week Devon TQ9 5 H4
Week Som. TA22 7 H2
Week Orchard EX23 6 A5
Week St. Mary EX22 4 C1
Weeke SO22 11 F1
Weekley NN16 42 B7
Weel HU17 59 G6
Weeley CO16 35 F6
Weeley Heath CO16 35 F6
Weem PH15 81 K2
Weeping Cross ST17 40 B3
Weethley B49 30 B3
Weeting IP27 44 B7
Weeton E.Riding HU12 59 K7
Weeton Lancs. PR4 55 G6
Weeton N.Yorks. LS17 57 H5
Weetwood LS16 57 H6
Weir Essex SS6 24 E3
Weir Lancs. OL13 56 D7
Weir Quay PL20 4 E4
Weirbrook SY11 38 C3
Weisdale ZE2 107 M7
Welbeck Abbey S80 51 H5
Welborne NR20 44 E5
Welbourn LN5 52 C7
Welburn N.Yorks. YO60 58 D3
Welburn N.Yorks. YO62 58 C1
Welbury DL6 62 E6
Welby NG32 42 C2
Welches Dam PE16 43 H7
Welcombe EX39 6 A4
Weldon NN17 42 C7
Welford Northants. NN6 41 J7
Welford W.Berks. RG20 21 H4
Welford-on-Avon CV37 30 C3
Welham Leics. LE16 42 A6
Welham Notts. DN22 51 K4
Welham Green AL9 23 F1
Well Hants. RG29 22 A7
Well Lincs. LN13 53 H5
Well N.Yorks. DL8 57 H1
Well End Bucks. SL8 22 B3
Well End Herts. WD6 23 F2
Well Hill BR6 23 H5
Well Street ME19 23 K6
Well Town EX16 7 H5
Welland WR13 29 G4
Wellbank DD5 83 F4
Wellesbourne CV35 30 D3
Wellhill IV36 97 G5
Wellhouse W.Berks. RG18 21 J4
Wellhouse W.Yorks. HD7 50 C1
Welling DA16 23 H4
Wellingborough NN8 32 B2
Wellingham PE32 44 C3
Wellingore LN5 52 C7
Wellington Cumb. CA20 60 B6
Wellington Here. HR4 28 D4
Wellington Som. TA21 7 K3
Wellington Tel. & W. TF1 39 F4
Wellington Heath HR8 29 G4
Wellington Marsh HR4 28 D4
Wellow B. & N.E.Som. BA2 20 A6
Wellow I.o.W. PO41 10 E6
Wellow Notts. NG22 51 J6
Wells BA5 19 J7
Wells Green B92 40 D7
Wellsborough CV13 41 F5
Wells-next-the-Sea NR23 44 D1
Wellstyle Green CM6 33 K7
Wellwood KY12 75 J2
Welney PE14 43 J6
Welsh Bicknor HR9 28 E7
Welsh End SY13 38 E2
Welsh Frankton SY11 38 C2
Welsh Hook SA62 16 C3
Welsh Newton NP25 28 D7
Welsh St. Donats CF71 18 D4
Welshampton SY12 38 D2
Welshpool (Y Trallwng) SY21 38 B5
Welton B. & N.E.Som. BA3 19 K6
Welton Cumb. CA5 60 E2
Welton E.Riding HU15 59 F7
Welton Lincs. LN2 52 D4
Welton Northants. NN11 31 G2
Welton le Marsh PE23 53 H6
Welton le Wold LN11 53 F4
Welwick HU12 59 K7
Welwyn AL6 33 F7
Welwyn Garden City AL8 33 F7
Wem SY4 38 E3
Wembdon TA6 8 B1
Wembley HA0 22 E3
Wembley Park HA9 22 E3
Wembury PL9 5 F6
Wembworthy EX18 6 E4
Wemyss Bay PA18 73 K4
Wenallt Cere. SY23 27 F1
Wenallt Gwyn. LL21 37 J1
Wendens Ambo CB11 33 J5
Wendlebury OX25 31 G7
Wendling NR19 44 D4
Wendover HP22 22 B1
Wendover Dean HP22 22 B1
Wendron TR13 2 D5
Wendy SG8 33 G4
Wenfordbridge PL30 4 A3
Wenhaston IP19 35 J1
Wenlli LL22 47 G6
Wennington Cambs. PE28 33 F1
Wennington Gt.Lon. RM13 23 J3
Wennington Lancs. LA2 56 B2
Wensley Derbys. DE4 50 E6
Wensley N.Yorks. DL8 57 F1
Wentbridge WF8 51 G1
Wentnor SY9 38 C6
Wentworth Cambs. CB6 33 H1
Wentworth S.Yorks. S62 51 F3
Wenvoe CF5 18 E4
Weobley HR4 28 C3
Weobley Marsh HR4 28 D3
Weoley Castle B29 40 C7
Wepham BN18 12 D6
Wepre CH5 48 B6
Wereham PE33 44 A5
Wergs WV6 40 A5
Wern Gwyn. LL49 36 E2
Wern Powys SY21 38 B4
Wern Powys NP8 28 A7
Wern Shrop. SY10 38 B2
Wernffrwd SA4 17 J6
Wern-olau SA4 17 J6
Wernrheolydd NP15 28 C7
Wern-y-cwrt NP15 19 G1
Werrington Cornw. PL15 6 B7
Werrington Peter. PE4 42 E5
Werrington Staffs. ST9 40 B1
Wervil Grange SA44 26 C3
Wervin CH2 48 D5
Wesham PR4 55 H6
Wessington DE55 51 F7
West Aberthaw CF62 18 D5
West Acre PE32 44 B4
West Acton W3 22 E3
West Allerdean TD15 77 H6
West Alvington TQ7 5 H6
West Amesbury SP4 20 E7
West Anstey EX36 7 G3
West Ashby LN9 53 F5
West Ashford EX31 6 D2
West Ashling PO18 12 B6
West Ashton BA14 20 B6
West Auckland DL14 62 C4
West Ayton YO13 59 F1
West Bagborough TA4 7 K2
West Barkwith LN8 52 E4
West Barnby YO21 63 K5
West Barns EH42 76 E3
West Barsham NR21 44 D2
West Bay DT6 8 D5
West Beckham NR25 45 F2
West Benhar ML7 75 G4
West Bergholt CO6 34 D6
West Bexington DT2 8 E6
West Bilney PE32 44 B4
West Blatchington BN3 13 F6
West Boldon NE36 71 J7
West Bourton SP8 9 G2
West Bowling BD5 57 G6
West Brabourne TN25 15 F3
West Bradford BB7 56 C5
West Bradley BA6 8 E1
West Bretton WF4 50 E1
West Bridgford NG2 41 H2
West Bromwich B70 40 C6
West Buckland Devon EX32 6 E2
West Buckland Som. TA21 7 K3
West Burrafirth ZE2 107 L7
West Burton N.Yorks. DL8 57 F1
West Burton W.Suss. RH20 12 C5
West Butsfield DL13 62 B2
West Butterwick DN17 52 B2
West Byfleet KT14 22 D5
West Cairncake AB53 99 G6
West Caister NR30 45 K4
West Calder EH55 75 J4
West Camel BA22 8 E2
West Carbeth G63 74 D3
West Carr Houses DN9 51 K2
West Cauldcoats ML10 74 E6
West Chaldon DT2 9 G6
West Challow OX12 21 G3
West Charleton TQ7 5 H6
West Chevington NE61 71 H4
West Chiltington RH20 12 D5
West Chiltington Common RH20 12 D5
West Chinnock TA18 8 D3
West Chisenbury SN9 20 E6
West Clandon GU4 22 D6
West Cliffe CT15 15 J3
West Clyne KW9 97 F1
West Coker BA22 8 E3
West Compton Dorset DT2 8 E5
West Compton Som. BA4 19 J7
West Cowick DN14 58 C7
West Cross SA3 17 K7
West Crudwell SN16 20 C2
West Curry PL15 4 C1
West Curthwaite CA7 60 E2
West Dean W.Suss. PO18 12 B5
West Dean Wilts. SP5 10 D2
West Deeping PE6 42 E5
West Derby L12 48 C3
West Dereham PE33 44 A5
West Ditchburn NE66 71 G1
West Down EX34 6 D1
West Drayton Gt.Lon. UB7 22 D4
West Drayton Notts. DN22 51 K5
West Dullater FK17 81 G7
West Dunnet KW14 105 H1
West Edington NE61 71 G5
West Ella HU10 59 G7
West End Bed. MK43 32 C3
West End Brack.F. RG42 22 B4
West End Caerp. NP11 19 F2
West End Cambs. PE15 43 H6
West End E.Riding YO25 59 G3
West End Hants. SO30 11 F3
West End Herts. AL9 23 F1
West End Kent CT6 25 H5
West End Lancs. LA4 55 H3
West End Lincs. DN36 53 G3
West End N.Som. BS48 19 H5
West End N.Yorks. HG3 57 G4
West End Norf. NR30 45 J4
West End Norf. IP25 44 D5
West End Oxon. OX29 21 H1
West End Oxon. OX10 21 J3
West End S.Lan. ML11 75 H6
West End Suff. NR34 45 J7
West End Surr. KT10 22 E5
West End Surr. GU24 22 C5
West End Wilts. SN15 20 C4
West End Wilts. SP7 9 J2
West End Wilts. SP5 9 J2
West End Green RG7 21 K5
West Farleigh ME15 14 C2
West Farndon NN11 31 G3
West Felton SY11 38 C3
West Firle BN8 13 H6
West Fleetham NE67 71 G1
West Flotmanby YO14 59 G2
West Garforth LS25 57 J6
West Ginge OX12 21 H3
West Glen PA21 73 H3
West Grafton SN8 21 F5
West Green Gt.Lon. N15 23 G3
West Green Hants. RG27 22 A6
West Grimstead SP5 10 D2
West Grinstead RH13 12 E4
West Haddlesey YO8 58 B7
West Haddon NN6 31 H1
West Hagbourne OX11 21 J3
West Hagley DY9 40 B7
West Hall CA8 70 A7
West Hallam DE7 41 G1
West Halton DN15 59 F7
West Ham E15 23 G3
West Handley S21 51 F5
West Hanney OX12 21 H2
West Hanningfield CM2 24 D2
West Hardwick WF4 51 G1
West Harnham SP2 10 C2
West Harptree BS40 19 J6
West Harrow HA1 22 E3
West Harting GU31 11 J2
West Hatch Som. TA3 8 B2
West Hatch Wilts. SP3 9 J2
West Head PE34 43 J5
West Heath Ches.E. CW12 49 H6
West Heath Gt.Lon. SE2 23 H4
West Heath Hants. RG24 22 B6
West Heath Hants. RG26 21 J6
West Heath W.Mid. B31 30 B1
West Helmsdale KW8 105 F7
West Hendon NW9 23 F3
West Hendred OX12 21 H3
West Heslerton YO17 59 F2
West Hewish BS24 19 G5
West Hill Devon EX11 7 J6
West Hill E.Riding YO16 59 H3
West Hill N.Som. BS20 19 H4
West Hoathly RH19 13 G3
West Holme BH20 9 H6
West Horndon CM13 24 C3
West Horrington BA5 19 J7
West Horsley KT24 22 D6
West Horton NE71 77 J7
West Hougham CT15 15 H4
West Howe BH11 10 B5
West Howetown TA24 7 H2
West Huntspill TA9 19 G7
West Hyde WD3 22 D2
West Hythe CT21 15 G4
West Ilsley RG20 21 H3
West Itchenor PO20 11 J4
West Keal PE23 53 G6
West Kennett SN8 20 E5
West Kilbride KA23 74 A6
West Kingsdown TN15 23 J5
West Kington SN14 20 B4
West Kington Wick SN14 20 B4
West Kirby CH48 48 B4
West Knapton YO17 58 E2
West Knighton DT2 9 G6
West Knoyle BA12 9 H1
West Kyloe TD15 77 J6
West Lambrook TA13 8 D3
West Langdon CT15 15 J3
West Langwell IV28 96 D1
West Lavington W.Suss. GU29 12 B4
West Lavington Wilts. SN10 20 D6
West Layton DL11 62 C6
West Leake LE12 41 H3
West Learmouth TD12 77 G7
West Lees DL6 63 F6
West Leigh Devon EX17 6 E5
West Leigh Devon TQ9 5 H5
West Leigh Som. TA4 7 K2
West Leith HP23 32 C7
West Lexham PE32 44 C4
West Lilling YO60 58 C3
West Lingo KY9 83 F7
West Linton EH46 75 K5
West Liss GU33 11 J2
West Littleton SN14 20 A4
West Lockinge OX12 21 H3
West Looe PL13 4 C5
West Lulworth BH20 9 H6
West Lutton YO17 59 F3
West Lydford TA11 8 E1
West Lyn EX35 7 F1
West Lyng TA3 8 C2
West Lynn PE34 44 A4
West Mains TD15 77 J6
West Malling ME19 23 K6
West Malvern WR14 29 G4
West Marden PO18 11 J3
West Markham NG22 51 K5
West Marsh DN31 53 F2
West Marton BD23 56 D4
West Melbury SP7 9 H2
West Melton S63 51 G2
West Meon GU32 11 H2
West Meon Hut GU32 11 H2
West Mersea CO5 34 E7
West Milton DT6 8 D5
West Minster ME12 25 F4
West Molesey KT8 22 E5
West Monkton TA2 8 B2
West Moors BH22 10 B4
West Morden BH20 9 J5
West Morriston TD4 76 E6
West Morton BD20 57 F5
West Mostard LA10 61 J7
West Mudford BA21 8 E2
West Muir DD9 83 G1
West Ness YO62 58 C2
West Newbiggin DL2 62 E5
West Newton E.Riding HU11 59 H6
West Newton Norf. PE31 44 A3
West Norwood SE27 23 G4
West Ogwell TQ12 5 J4
West Orchard SP7 9 H3
West Overton SN8 20 E5
West Panson PL15 6 B6
West Park Aber. AB31 91 F5
West Park Mersey. WA10 48 E3
West Parley BH22 10 B5
West Peckham ME18 23 K6
West Pelton DH9 62 D1
West Pennard BA6 8 E1
West Pentire TR8 2 E2
West Perry PE28 32 E2
West Porlock TA24 7 G1
West Prawle TQ8 5 H7
West Preston BN16 12 D6
West Pulham DT2 9 G4
West Putford EX22 6 B4
West Quantoxhead TA4 7 K1
West Raddon EX17 7 G5
West Rainton DH4 62 E2
West Rasen LN8 52 D4
West Raynham NR21 44 C3
West Retford DN22 51 J4
West Rounton DL6 63 F6
West Row IP28 33 K1
West Rudham PE31 44 C3
West Runton NR27 45 F1
West Saltoun EH34 76 C4
West Sandford EX17 7 G5
West Sandwick ZE2 107 N4
West Scrafton DL8 57 F1
West Shepton BA4 19 K7
West Shinness Lodge IV27 103 H7
West Somerton NR29 45 J4
West Stafford DT2 9 G6
West Stockwith DN10 51 K3
West Stoke PO18 12 B6
West Stonesdale DL11 61 K6
West Stoughton BS28 19 H7
West Stour SP8 9 G2
West Stourmouth CT3 25 J5
West Stow IP28 34 C1
West Stowell SN8 20 E5
West Stratton SO21 21 J7
West Street Kent ME17 14 E2
West Street Med. ME3 24 D4
West Street Suff. IP31 34 D1
West Tanfield HG4 57 H2
West Taphouse PL22 4 B4
West Tarbert PA29 73 G4
West Tarring BN13 12 E6
West Thirston NE65 71 G3
West Thorney PO10 11 J4
West Thurrock RM20 23 J4
West Tilbury RM18 24 C4
West Tisted SO24 11 H2
West Tofts Norf. IP26 44 C6
West Tofts P. & K. PH1 82 C4
West Torrington LN8 52 E4
West Town B. & N.E.Som. BS40 19 J5
West Town Hants. PO11 11 J5
West Town N.Som. BS48 19 H5
West Town Som. BA6 8 E1
West Tytherley SP5 10 D2
West Walton PE14 43 H4
West Wellow SO51 10 D3
West Wembury PL9 5 F6
West Wemyss KY1 76 B1
West Wick BS24 19 G5
West Wickham Cambs. CB21 33 K4
West Wickham Gt.Lon. BR4 23 G5
West Williamston SA68 16 D5
West Winch PE33 44 A4
West Winterslow SP5 10 D1
West Wittering PO20 11 J5
West Witton DL8 57 F1
West Woodburn NE48 70 D5
West Woodhay RG20 21 G5
West Woodlands BA11 20 A7

Wes - Wic

West Worldham GU34 11 J1
West Worlington EX17 7 F4
West Worthing BN11 12 E6
West Wratting CB21 33 K3
West Wycombe HP14 22 B2
West Yatton SN14 20 B4
West Yell ZE2 107 N4
West Youlstone EX23 6 A4
Westbere CT2 25 H5
Westborough NG23 42 B1
Westbourne Bourne. BH4 10 B5
Westbourne W.Suss. PO10 11 J4
Westbourne Green W2 23 F3
Westbrook Kent CT9 25 K4
Westbrook W.Berks. RG20 21 H4
Westbrook Wilts. SN15 20 C5
Westbury Bucks. NN13 31 H5
Westbury Shrop. SY5 38 C5
Westbury Wilts. BA13 20 B6
Westbury Leigh BA13 20 B6
Westbury on Trym BS9 19 J4
Westbury-on-Severn GL14 29 G7
Westbury-sub-Mendip BA5 19 J7
Westby Lancs. PR4 55 G6
Westby Lincs. NG33 42 C3
Westcliff-on-Sea SS0 24 E3
Westcombe BA4 9 F1
Westcot OX12 21 G3
Westcott Bucks. HP18 31 J7
Westcott Devon EX15 7 J5
Westcott Surr. RH4 22 E7
Westcott Barton OX7 31 F6
Westcourt SN8 21 F5
Westcroft MK4 32 B5
Westdean BN25 13 J7
Westdowns PL33 4 A2
Westend Town SN14 20 A4
Wester Aberchalder IV2 88 C2
Wester Badentyre AB53 99 F5
Wester Balgedie KY13 82 C7
Wester Culbeuchly AB45 98 E4
Wester Dechmont EH52 75 J3
Wester Fintray AB51 91 G3
Wester Foffarty DD8 83 F3
Wester Greenskares AB45 99 F4
Wester Gruinards IV24 96 C2
Wester Hailes EH14 76 A4
Wester Lealty IV17 96 D4
Wester Lonvine IV18 96 E4
Wester Newburn KY8 83 F7
Wester Ord AB32 91 G4
Wester Quarff ZE2 107 N9
Wester Skeld ZE2 107 L8
Westerdale High. KW12 105 G3
Westerdale N.Yorks. YO21 63 H6
Westerfield Shet. ZE2 107 M7
Westerfield Suff. IP6 35 F4
Westergate PO20 12 C6
Westerham TN16 23 H6
Westerhope NE5 71 G7
Westerleigh BS37 19 K4
Westerloch KW1 105 J3
Westerton Aber. AB31 91 F5
Westerton Angus DD10 83 H2
Westerton Dur. DL14 62 D3
Westerton P. & K. PH5 81 K6
Westerwick ZE2 107 L8
Westfield Cumb. CA14 60 A4
Westfield E.Suss. TN35 14 D6
Westfield High. KW14 105 F2
Westfield N.Lan. G68 75 F3
Westfield Norf. NR19 44 D5
Westfield W.Loth. EH48 75 H3
Westfield W.Yorks. WF16 57 H7
Westfield Sole ME14 24 D5
Westgate Dur. DL13 62 A3
Westgate N.Lincs. DN9 51 K2
Westgate Norf. NR21 44 D1
Westgate Northumb. NE20 71 G6
Westgate Hill BD4 57 H7
Westgate on Sea CT8 25 K4
Westhall Aber. AB52 90 E2
Westhall Suff. IP19 45 J7
Westham Dorset DT4 9 F7
Westham E.Suss. BN24 13 K6
Westham Som. TA7 19 H7
Westhampnett PO18 12 B6
Westhay Devon EX13 8 C4
Westhay Som. BA6 19 H7
Westhead L40 48 D2
Westhide HR1 28 E4
Westhill Aber. AB32 91 G4
Westhill High. IV2 96 E7
Westhope Here. HR4 28 D3
Westhope Shrop. SY7 38 D7
Westhorp NN11 31 G3
Westhorpe Lincs. PE11 43 F2
Westhorpe Notts. NG25 51 J7
Westhorpe Suff. IP14 34 E2
Westhoughton BL5 49 F2
Westhouse LA6 56 B2
Westhouses DE55 51 G7
Westhumble RH5 22 E6
Westing ZE2 107 P2
Westlake PL21 5 G5
Westlands ST5 40 A1
Westlea SN5 20 E3
Westleigh Devon EX39 6 C3
Westleigh Devon EX16 7 J4
Westleigh Gt.Man. WN7 49 F2
Westleton IP17 35 J2
Westley Shrop. SY5 38 C5
Westley Suff. IP33 34 C2
Westley Heights SS16 24 C3
Westley Waterless CB8 33 K3
Westlington HP17 31 J7
Westlinton CA6 69 J7
Westloch EH45 76 A5
Westmancote GL20 29 J5
Westmarsh CT3 25 J5
Westmeston BN6 13 G5
Westmill SG9 33 G6
Westminster SW1H 23 F4
Westmuir DD8 82 E2
Westness KW17 106 C5
Westnewton Cumb. CA7 60 C2
Westnewton Northumb. NE71 77 H7
Westoe NE33 71 J7
Weston B. & N.E.Som. BA1 20 A5
Weston Ches.E. CW2 49 G7
Weston Devon EX12 7 K7
Weston Devon EX14 7 K5
Weston Dorset DT5 9 F7
Weston Halton WA7 48 E4
Weston Hants. GU32 11 J2
Weston Here. HR6 28 C3
Weston Herts. SG4 33 F5
Weston Lincs. PE12 43 F3
Weston Moray AB56 98 C4
Weston N.Yorks. LS21 57 G5
Weston Northants. NN12 31 G4
Weston Notts. NG23 51 K6
Weston S'ham. SO19 11 F3
Weston Shrop. SY4 38 E3
Weston Shrop. TF13 38 E6
Weston Shrop. SY7 28 C1
Weston Staffs. ST18 40 B3
Weston W.Berks. RG20 21 H4
Weston Bampfylde BA22 9 F2
Weston Beggard HR1 28 E4
Weston by Welland LE16 42 A6
Weston Colville CB21 33 K3
Weston Corbett RG25 21 K7
Weston Coyney ST3 40 B1
Weston Favell NN3 31 J2
Weston Green Cambs. CB21 33 K3
Weston Green Norf. NR9 45 E4
Weston Heath TF11 39 G4
Weston Hills PE12 43 F3
Weston in Arden CV12 41 F7
Weston Jones TF10 39 G3
Weston Longville NR9 45 F4
Weston Lullingfields SY4 38 D3
Weston Patrick RG25 21 K7
Weston Point WA7 48 D4
Weston Rhyn SY10 38 B2
Weston Subedge GL55 30 C4
Weston Town BA4 20 A7
Weston Turville HP22 32 B7
Weston under Penyard HR9 29 F6
Weston under Wetherley CV33 30 E2
Weston Underwood Derbys. DE6 40 E1
Weston Underwood M.K. MK46 32 B3
Westonbirt GL8 20 B3
Westoning MK45 32 D5
Weston-in-Gordano BS20 19 H4
Weston-on-Avon CV37 30 C3
Weston-on-the-Green OX25 31 G7
Weston-on-Trent DE72 41 G3
Weston-super-Mare BS23 19 G5
Weston-under-Lizard TF11 40 A4
Westonzoyland TA7 8 C1
Westow YO60 58 D3
Westport Arg. & B. PA28 66 A1
Westport Som. TA10 8 C2
Westra CF64 18 E4
Westray KW17 106 D3
Westray Airfield KW17 106 D2
Westridge Green RG8 21 J4
Westrigg EH48 75 H4
Westruther TD3 76 E6
Westry PE15 43 G6
Westside AB12 91 G5
Westvale L32 48 D3
Westville NG15 41 H1
Westward CA7 60 D2
Westward Ho! EX39 6 C3
Westwell Kent TN25 14 E3
Westwell Oxon. OX18 21 F1
Westwell Leacon TN27 14 E3
Westwick Cambs. CB24 33 H2
Westwick Dur. DL12 62 B5
Westwick N.Yorks. YO51 57 J3
Westwick Norf. NR10 45 G3
Westwood Devon EX5 7 J6
Westwood Peter. PE3 42 E6
Westwood S.Lan. G75 74 E5
Westwood Wilts. BA15 20 B6
Westwood Heath CV4 30 D1
Westwoodside DN9 51 K3
Wetham Green ME9 24 E5
Wetheral CA4 61 F1
Wetherby LS22 57 K5
Wetherden IP14 34 E2
Wetherden Upper Town IP14 34 E2
Wetheringsett IP14 35 F2
Wethersfield CM7 34 B5
Wethersta ZE2 107 M6
Wetherup Street IP14 35 F2
Wetley Abbey ST9 40 B1
Wetley Rocks ST9 40 B1
Wettenhall CW7 49 F6
Wettenhall Green CW7 49 F6
Wetton DE6 50 D7
Wetwang YO25 59 F4
Wetwood ST21 39 G2
Wexcombe SN8 21 F6
Wexham Street SL3 22 C3
Weybourne Norf. NR25 45 F1
Weybourne Surr. GU9 22 B7
Weybread IP21 45 G7
Weybread Street IP21 35 G1
Weybridge KT13 22 D5
Weycroft EX13 8 C4
Weydale KW14 105 G2
Weyhill SP11 21 G7
Weymouth DT4 9 F7
Whaddon Bucks. MK17 32 B5
Whaddon Cambs. SG8 33 G4
Whaddon Glos. GL4 29 H7
Whaddon Glos. GL52 29 J6
Whaddon Wilts. SP5 10 C2
Whaddon Wilts. BA14 20 B5
Whaddon Gap SG8 33 G4
Whale CA10 61 G4
Whaley NG20 51 H5
Whaley Bridge SK23 50 C4
Whaley Thorns NG20 51 H5
Whaligoe KW2 105 J4
Whalley BB7 56 C6
Whalsay ZE2 107 P6
Whalsay Airport ZE2 107 P6
Whalton NE61 71 G5
Wham BD24 56 C3
Whaplode PE12 43 G3
Whaplode Drove PE12 43 G4
Whaplode St. Catherine PE12 43 G4
Wharfe LA2 56 C3
Wharles PR4 55 H6
Wharley End MK43 32 C4
Wharncliffe Side S35 50 E3
Wharram le Street YO17 58 E3
Wharram Percy YO17 58 E3
Wharton Ches.W. & C. CW7 49 F6
Wharton Here. HR6 28 E3
Whashton DL11 62 C6
Whatcote CV36 30 E4
Whateley B78 40 E6
Whatfield IP7 34 E4
Whatley BA11 20 A7
Whatlington TN33 14 C6
Whatsole Street TN25 15 G3
Whatstandwell DE4 51 F7
Whatton NG13 42 A2
Whauphill DG8 64 E6
Whaw DL11 62 A6
Wheatacre NR34 45 J6
Wheatcroft DE4 51 F7
Wheatenhurst GL2 20 A1
Wheatfield OX9 21 K2
Wheathampstead AL4 32 E7
Wheathill Shrop. WV16 39 F7
Wheathill Som. TA11 8 E1
Wheatley Hants. GU34 11 J1
Wheatley Oxon. OX33 21 K1
Wheatley W.Yorks. HX3 57 F7
Wheatley Hill DH6 62 E3
Wheatley Lane BB12 56 D6
Wheatley Park DN2 51 H2
Wheaton Aston ST19 40 A4
Wheddon Cross TA24 7 H2
Wheedlemont AB54 90 C2
Wheelerstreet GU8 22 C7
Wheelock CW11 49 G7
Wheelock Heath CW11 49 G7
Wheelton PR6 56 B7
Wheen DD8 90 B7
Wheldale WF10 57 K7
Wheldrake YO19 58 C5
Whelford GL7 20 E2
Whelpley Hill HP5 22 C1
Whelpo CA7 60 E3
Whelston CH6 48 B5
Whenby YO61 58 C3
Whepstead IP29 34 C3
Wherstead IP9 35 F4
Wherwell SP11 21 G7
Wheston SK17 50 D5
Whetley Cross DT8 8 D4
Whetsted TN12 23 K7
Whetstone Gt.Lon. N20 23 F2
Whetstone Leics. LE8 41 H6
Whicham LA18 54 E1
Whichford CV36 30 E5
Whickham NE16 71 H7
Whiddon EX21 6 C5
Whiddon Down EX6 6 E6
Whifflet ML5 75 F4
Whigstreet DD8 83 F3
Whilton NN11 31 H2
Whim EH46 76 A5
Whimble EX22 6 B5
Whimple EX5 7 J6
Whimpwell Green NR12 45 H3
Whin Lane End PR3 55 G5
Whinburgh NR19 44 E5
Whinny Hill TS21 62 E5
Whinnyfold AB42 91 J1
Whippingham PO32 11 G5
Whipsnade LU6 32 D7
Whipton EX1 7 H6
Whirlow S11 51 F4
Whisby LN6 52 C6
Whissendine LE15 42 B4
Whissonsett NR20 44 D3
Whisterfield SK11 49 H5
Whistley Green RG10 22 A4
Whiston Mersey. L35 48 D3
Whiston Northants. NN7 32 B2
Whiston S.Yorks. S60 51 G3
Whiston Staffs. ST10 40 C1
Whiston Staffs. ST19 40 A4
Whiston Cross WV7 39 G5
Whiston Eaves ST10 40 C1
Whitacre Fields B46 40 E6
Whitacre Heath B46 40 E6
Whitbeck LA19 54 E1
Whitbourne WR6 29 G3
Whitburn T. & W. SR6 71 K7
Whitburn W.Loth. EH47 75 H4
Whitby Ches.W. & C. CH65 48 C5
Whitby N.Yorks. YO21 63 K5
Whitbyheath CH65 48 C5
Whitchurch B. & N.E.Som. BS14 19 K5
Whitchurch Bucks. HP22 32 B6
Whitchurch Cardiff CF14 18 E3
Whitchurch Devon PL19 4 E3
Whitchurch Hants. RG28 21 H7
Whitchurch Here. HR9 28 E7
Whitchurch Pembs. SA62 16 A3
Whitchurch Shrop. SY13 38 E1
Whitchurch Warks. CV37 30 D4
Whitchurch Canonicorum DT6 8 C5
Whitchurch Hill RG8 21 K4
Whitchurch-on-Thames RG8 21 K4
Whitcombe DT2 9 G6
Whitcott Keysett SY7 38 B7
White Ball TA21 7 J4
White Colne CO6 34 C6
White Coppice PR6 49 F1
White Cross Cornw. TR8 3 F3
White Cross Devon EX5 7 J6
White Cross Here. HR4 28 D4
White Cross Wilts. BA12 9 G1
White End GL19 29 H6
White Hill BA12 9 H1
White Houses DN22 51 K5
White Kirkley DL13 62 B3
White Lackington DT2 9 G5
White Ladies Aston WR7 29 J3
White Lund LA3 55 H3
White Mill SA32 17 H3
White Moor DE56 41 F1
White Notley CM8 34 B7
White Ox Mead BA2 20 A6
White Pit LN13 53 G5
White Rocks HR2 28 D6
White Roding CM6 33 J7
White Waltham SL6 22 B4
Whiteacen AB38 97 K7
Whiteash Green CO9 34 B5
Whitebirk BB1 56 C7
Whitebog AB43 99 H5
Whitebridge High. KW14 105 H1
Whitebridge High. IV2 88 B3
Whitebrook NP25 19 J1
Whiteburn TD2 76 D6
Whitecairn DG8 64 C5
Whitecairns AB23 91 H3
Whitecastle ML12 75 J6
Whitechapel PR3 55 J5
Whitechurch SA41 16 E2
Whitecote LS13 57 H6
Whitecraig EH21 76 B3
Whitecroft GL15 19 K1
Whitecrook DG9 64 B5
Whitecross Cornw. TR20 2 C5
Whitecross Cornw. PL27 3 G1
Whitecross Dorset DT6 8 D5
Whitecross Falk. EH49 75 H3
Whiteface IV25 96 E3
Whitefield Aber. AB51 91 F2
Whitefield Devon EX32 7 F2
Whitefield Dorset BH20 9 J5
Whitefield Gt.Man. M45 49 H2
Whitefield High. IV2 88 C2
Whitefield High. KW1 105 H3
Whitefield P. & K. PH13 82 C4
Whiteford AB51 91 F2
Whitegate CW8 49 F6
Whitehall Devon EX15 7 K4
Whitehall Hants. RG29 22 A6
Whitehall Ork. KW17 106 F5
Whitehall W.Suss. RH13 12 E4
Whitehaven CA28 60 A5
Whitehill Aber. AB42 99 H5
Whitehill Hants. GU35 11 J1
Whitehill Kent ME13 14 E2
Whitehill Midloth. EH22 76 B4
Whitehill N.Ayr. KA24 74 A5
Whitehills AB45 98 E4
Whitehouse Aber. AB33 90 E3
Whitehouse Arg. & B. PA29 73 G4
Whitehouse Common B75 40 D6
Whitekirk EH42 76 D2
Whitelackington TA19 8 C3
Whitelaw TD11 77 G5
Whiteleen KW2 105 J4
Whitelees KA1 74 B7
Whiteley PO15 11 G4
Whiteley Bank PO38 11 G6
Whiteley Green SK10 49 J5
Whiteley Village KT12 22 D5
Whiteleys DG9 64 A5
Whitemans Green RH17 13 G4
Whitemire IV36 97 G6
Whitemoor PL26 3 G3
Whiteness ZE2 107 M8
Whiteoak Green OX29 30 E7
Whiteparish SP5 10 D2
Whiterashes AB21 91 G2
Whiterow KW1 105 J4
Whiteshill GL6 20 B1
Whiteside Northumb. NE49 70 C7
Whiteside W.Loth. EH48 75 H4
Whitesmith BN8 13 J5
Whitestaunton TA20 8 B3
Whitestone Aber. AB31 90 E5
Whitestone Arg. & B. PA28 73 F7
Whitestone Devon EX4 7 G6
Whitestreet Green CO10 34 D5
Whitestripe AB43 99 H5
Whiteway GL6 29 J7
Whitewell Aber. AB43 99 H4
Whitewell Lancs. BB7 56 B5
Whitewell Wrex. SY13 38 D1
Whiteworks PL20 5 G3
Whitewreath IV30 97 K6
Whitfield Here. HR2 28 D5
Whitfield Kent CT16 15 J3
Whitfield Northants. NN13 31 H5
Whitfield Northumb. NE47 61 J1
Whitfield S.Glos. GL12 19 K2
Whitford Devon EX13 8 B5
Whitford (Chwitffordd) Flints. CH8 47 K5
Whitgift DN14 58 E7
Whitgreave ST18 40 A3
Whithorn DG8 64 E6
Whiting Bay KA27 66 E1
Whitkirk LS15 57 J6
Whitlam AB21 91 G2
Whitland (Hendy-Gwyn) SA34 17 F4
Whitland Abbey SA34 17 F4
Whitleigh PL5 4 E4
Whitletts KA8 67 H1
Whitley N.Yorks. DN14 58 B7
Whitley Read. RG2 22 A5
Whitley W.Mid. CV3 30 E1
Whitley Wilts. SN12 20 B5
Whitley Bay NE26 71 J6
Whitley Chapel NE47 62 A1
Whitley Heath ST21 40 A3
Whitley Lower WF12 50 E1
Whitley Row TN14 23 H6
Whitlock's End B90 30 C1
Whitminster GL2 20 A1
Whitmore Dorset BH21 10 B4
Whitmore Staffs. ST5 40 A1
Whitnage EX16 7 J4
Whitnash CV31 30 E2
Whitnell TA5 19 F7
Whitney-on-Wye HR3 28 B4
Whitrigg Cumb. CA7 60 D1
Whitrigg Cumb. CA7 60 D3
Whitsbury SP6 10 C3
Whitsome TD11 77 G5
Whitson NP18 19 G3
Whitstable CT5 25 H5
Whitstone EX22 4 C1
Whittingham NE66 71 F2
Whittingslow SY6 38 D7
Whittington Derbys. S41 51 F5
Whittington Glos. GL54 30 B6
Whittington Lancs. LA6 56 B2
Whittington Norf. PE33 44 B6
Whittington Shrop. SY11 38 C2
Whittington Staffs. DY7 40 A7
Whittington Staffs. WS14 40 D5
Whittington Worcs. WR5 29 H3
Whittlebury NN12 31 H4
Whittle-le-Woods PR6 55 J7
Whittlesey PE7 43 F6
Whittlesford CB22 33 H4
Whittlestone Head BL7 49 G1
Whitton Gt.Lon. TW2 22 E4
Whitton N.Lincs. DN15 59 F7
Whitton Northumb. NE65 71 F3
Whitton Powys LD7 28 B2
Whitton Shrop. SY8 28 E1
Whitton Stock. TS21 62 E4
Whitton Suff. IP1 35 F4
Whittonditch SN8 21 F4
Whittonstall DH8 62 B1
Whitway RG20 21 H6
Whitwell Derbys. S80 51 H5
Whitwell Herts. SG4 32 E6
Whitwell I.o.W. PO38 11 G7
Whitwell N.Yorks. DL10 62 D7
Whitwell Rut. LE15 42 C5
Whitwell Street NR10 45 F3
Whitwell-on-the-Hill YO60 58 D3
Whitwick LE67 41 G4
Whitwood WF10 57 K7
Whitworth OL12 49 H1
Whixall SY13 38 E2
Whixley YO26 57 K4
Whorlton Dur. DL12 62 C5
Whorlton N.Yorks. DL6 63 F6
Whygate NE48 70 C6
Whyle HR6 28 E2
Whyteleafe CR3 23 G6
Wibdon NP16 19 J2
Wibsey BD6 57 G6
Wibtoft LE17 41 G7
Wichenford WR6 29 G2
Wichling ME9 14 E2
Wick Bourne. BH6 10 C5
Wick Devon EX14 7 K5
Wick High. KW1 105 J3
Wick S.Glos. BS30 20 A4
Wick Som. TA5 19 F7
Wick Som. BA6 8 E1
Wick V. of Glam. CF71 18 C4
Wick W.Suss. BN17 12 D6
Wick Wilts. SP5 10 C2
Wick Worcs. WR10 29 J4
Wick John O'Groats Airport KW1 105 J3
Wick Hill Kent TN27 14 D3
Wick Hill W'ham RG40 22 A5
Wick St. Lawrence BS22 19 G5
Wicken Cambs. CB7 33 J1
Wicken Northants. MK19 31 J5
Wicken Bonhunt CB11 33 H5
Wickenby LN3 52 D4
Wicker Street Green CO10 34 D4

Wic - Woo

Wickerslack CA10 61 H5
Wickersley S66 51 G3
Wicketwood Hill NG4 41 J1
Wickford SS12 24 D2
Wickham *Hants.* PO17 11 G3
Wickham *W.Berks.* RG20 21 G4
Wickham Bishops CM8 34 C7
Wickham Heath RG20 21 H5
Wickham Market IP13 35 H3
Wickham St. Paul CO9 34 C5
Wickham Skeith IP23 34 E2
Wickham Street *Suff.* CB8 34 B3
Wickham Street *Suff.* IP23 34 E2
Wickhambreaux CT3 15 H2
Wickhambrook CB8 34 B3
Wickhamford WR11 30 B4
Wickhampton NR13 45 J5
Wicklewood NR18 44 E5
Wickmere NR11 45 F2
Wickstreet BN26 13 J6
Wickwar GL12 20 A3
Widcombe BA2 20 A5
Widdington CB11 33 J5
Widdop HX7 56 E6
Widdrington NE61 71 H4
Widdrington Station NE61 71 H4
Wide Open NE13 71 H6
Widecombe in the Moor TQ13 5 H3
Widegates PL13 4 C5
Widemouth Bay EX23 6 A5
Widewall KW17 106 D8
Widford *Essex* CM2 24 C1
Widford *Herts.* SG12 33 H7
Widford *Oxon.* OX18 30 D7
Widgham Green CB8 33 K3
Widmer End HP15 22 B2
Widmerpool NG12 41 J3
Widnes WA8 48 E4
Widworthy EX14 8 B5
WIGAN WN 48 E2
Wiganthorpe YO60 58 C2
Wigborough TA13 8 D3
Wiggaton EX11 7 K6
Wiggenhall St. Germans PE34 43 J4
Wiggenhall St. Mary Magdalen PE34 43 J4
Wiggenhall St. Mary the Virgin PE34 43 J4
Wiggenhall St. Peter PE34 44 A4
Wiggens Green CB9 33 K4
Wigginton *Herts.* HP23 32 C7
Wigginton *Oxon.* OX15 30 E5
Wigginton *Shrop.* SY11 38 C2
Wigginton *Staffs.* B79 40 E5
Wigginton *York* YO32 58 C4
Wigglesworth BD23 56 D4
Wiggonby CA7 60 E1
Wiggonholt RH20 12 D5
Wighill LS24 57 K5
Wighton NR23 44 D2
Wightwizzle S36 50 E3
Wigley SO51 10 E3
Wigmore *Here.* HR6 28 D2
Wigmore *Med.* ME8 24 E5
Wigsley NG23 52 B5
Wigsthorpe NN14 42 D7
Wigston LE18 41 J6
Wigston Parva LE10 41 G7
Wigthorpe S81 51 H4
Wigtoft PE20 43 F2
Wigton CA7 60 D2
Wigtown DG8 64 E5
Wike LS17 57 J5
Wilbarston LE16 42 B7
Wilberfoss YO41 58 D4
Wilburton CB6 33 H1
Wilby *Norf.* NR16 44 E6
Wilby *Northants.* NN8 32 B2
Wilby *Suff.* IP21 35 G1
Wilcot SN9 20 E5
Wilcott SY4 38 C4
Wilcrick NP26 19 H3
Wilday Green S18 51 F5
Wildboarclough SK11 49 J6
Wilde Street IP28 34 B1
Wilden *Bed.* MK44 32 D3
Wilden *Worcs.* DY13 29 H1
Wildhern SP11 21 G6
Wildhill AL9 23 F1
Wildmoor B61 29 J1
Wildsworth DN21 52 B3
Wilford NG11 41 H2
Wilkesley SY13 39 F1
Wilkhaven IV20 97 G3
Wilkieston EH27 75 K4
Wilksby PE22 53 F6
Willand *Devon* EX15 7 J4
Willand *Som.* TA3 7 K4
Willaston *Ches.E.* CW5 49 F7
Willaston *Ches.W. & C.* CH64 48 C5
Willaston *Shrop.* SY13 38 E2
Willen MK15 32 B4
Willenhall *W.Mid.* WV13 40 B6
Willenhall *W.Mid.* CV3 30 E1
Willerby *E.Riding* HU10 59 G6
Willerby *N.Yorks.* YO12 59 G2
Willersey WR12 30 C5
Willersley HR3 28 C4
Willesborough TN24 15 F3
Willesborough Lees TN24 15 F3
Willesden NW10 23 F3
Willesleigh EX32 6 D2
Willesley GL8 20 B3
Willett TA4 7 K2
Willey *Shrop.* TF12 39 F6
Willey *Warks.* CV23 41 G7
Willey Green GU3 22 C6
William's Green IP7 34 D4
Williamscot OX17 31 F4
Williamthorpe S42 51 G6
Willian SG6 33 F5
Willimontswick NE47 70 C7
Willingale CM5 23 J1
Willingdon BN20 13 J6
Willingham CB24 33 H1
Willingham by Stow DN21 52 B4
Willingham Green CB8 33 K3
Willington *Bed.* MK44 32 E3
Willington *Derbys.* DE65 40 E3
Willington *Dur.* DL15 62 C3
Willington *Kent* ME15 14 C2
Willington *T. & W.* NE28 71 J7
Willington *Warks.* CV36 30 D5
Willington Corner CW6 48 E6
Willisham IP8 34 E3
Willitoft YO8 58 D6
Williton TA4 7 J1
Willoughbridge TF9 39 G1
Willoughby *Lincs.* LN13 53 H5
Willoughby *Warks.* CV23 31 G2
Willoughby Waterleys LE8 41 H6
Willoughby-on-the-Wolds LE12 41 J3
Willoughton DN21 52 C3
Willow Green CW8 49 F5
Willows Green CM3 34 B7
Willsbridge BS30 19 K4
Willslock ST14 40 C2
Willsworthy PL19 6 D7
Willtown TA10 8 C2
Wilmcote CV37 30 C3
Wilmington *B. & N.E.Som.* BA2 19 K5
Wilmington *Devon* EX14 8 B4
Wilmington *E.Suss.* BN26 13 J6
Wilmington *Kent* DA2 23 J4
Wilmslow SK9 49 H4
Wilnecote B77 40 E5
Wilney Green IP22 44 E7
Wilpshire BB1 56 B6
Wilsden BD15 57 F6
Wilsford *Lincs.* NG32 42 D1
Wilsford *Wilts.* SP4 10 C1
Wilsford *Wilts.* SN9 20 E6
Wilsham EX35 7 F1
Wilshaw HD9 50 D2
Wilsill HG3 57 G3
Wilsley Green TN17 14 C4
Wilsley Pound TN17 14 C4
Wilson DE73 41 G3
Wilstead MK45 32 D4
Wilsthorpe *E.Riding* YO15 59 H3
Wilsthorpe *Lincs.* PE9 42 D4
Wilstone HP23 32 C7
Wilton *Cumb.* CA22 60 B5
Wilton *Here.* HR9 28 E6
Wilton *N.Yorks.* YO18 58 E1
Wilton *R. & C.* TS10 63 G5
Wilton *Sc.Bord.* TD9 69 K2
Wilton *Wilts.* SN8 21 F5
Wilton *Wilts.* SP2 10 B1
Wiltown EX15 7 K4
Wimbish CB10 33 J5
Wimbish Green CB10 33 K5
Wimblebury WS12 40 C4
Wimbledon SW19 23 F4
Wimblington PE15 43 H6
Wimborne Minster BH21 10 B4
Wimborne St. Giles BH21 10 B3
Wimbotsham PE34 44 A5
Wimpole SG8 33 G3
Wimpole Lodge SG8 33 G4
Wimpstone CV37 30 D4
Wincanton BA9 9 G2
Winceby LN9 53 G6
Wincham CW9 49 F5
Winchburgh EH52 75 J3
Winchcombe GL54 30 B6
Winchelsea TN36 14 E6
Winchelsea Beach TN36 14 E6
Winchester SO23 11 F2
Winchet Hill TN17 14 C3
Winchfield RG27 22 A6
Winchmore Hill *Bucks.* HP7 22 C2
Winchmore Hill *Gt.Lon.* N21 23 G2
Wincle SK11 49 J6
Wincobank S9 51 F3
Windermere LA23 60 F7
Winderton OX15 30 E4
Windhill IV4 96 C7
Windle Hill CH64 48 C5
Windlehurst SK6 49 J4
Windlesham GU20 22 C5
Windley DE56 41 F1
Windmill SK17 50 D5
Windmill Hill *E.Suss.* BN27 13 K5
Windmill Hill *Som.* TA19 8 C3
Windmill Hill *Worcs.* WR7 29 J4
Windrush OX18 30 C7
Windsor SL4 22 C4
Windsor Green IP30 34 C3
Windy Nook NE10 71 H7
Windygates KY8 82 E7
Windy-Yett KA3 74 C5
Wineham BN5 13 F4
Winestead HU12 59 J7
Winewall BB8 56 E6
Winfarthing IP22 45 F7
Winford *I.o.W.* PO36 11 G6
Winford *N.Som.* BS40 19 J5
Winforton HR3 28 B4
Winfrith Newburgh DT2 9 H6
Wing *Bucks.* LU7 32 B6
Wing *Rut.* LE15 42 B5
Wingate TS28 62 E3
Wingates *Gt.Man.* BL5 49 F2
Wingates *Northumb.* NE65 71 F4
Wingerworth S42 51 F6
Wingfield *Cen.Beds.* LU7 32 D6
Wingfield *Suff.* IP21 35 G1
Wingfield *Wilts.* BA14 20 B6
Wingfield Green IP21 35 G1
Wingham CT3 15 H2
Wingham Well CT3 15 H2
Wingmore CT4 15 G3
Wingrave HP22 32 B7
Winkburn NG22 51 K7
Winkfield SL4 22 C4
Winkfield Row RG42 22 B4
Winkhill ST13 50 C7
Winkleigh EX19 6 E5
Winksley HG4 57 H2
Winkton BH23 10 C5
Winlaton NE21 71 G7
Winlaton Mill NE21 71 G7
Winless KW1 105 J3
Winmarleigh PR3 55 H5
Winnard's Perch TR9 3 G2
Winnersh RG41 22 A4
Winnington CW8 49 F5
Winscombe BS25 19 H6
Winsford *Ches.W. & C.* CW7 49 F6
Winsford *Som.* TA24 7 H2
Winsham *Devon* EX33 6 C2
Winsham *Som.* TA20 8 C4
Winshill DE15 40 E3
Winsh-wen SA7 17 K6
Winskill CA10 61 G3
Winslade RG25 21 K7
Winsley BA15 20 B5
Winslow MK18 31 J6
Winson GL7 20 D1
Winsor SO40 10 E3
Winster *Cumb.* LA23 60 F7
Winster *Derbys.* DE4 50 E6
Winston *Dur.* DL2 62 C5
Winston *Suff.* IP14 35 F2
Winston Green IP14 35 F2
Winstone GL7 20 C1
Winswell EX38 6 C4
Winterborne Came DT2 9 G6
Winterborne Clenston DT11 9 H4
Winterborne Herringston DT2 9 F6
Winterborne Houghton DT11 9 H4
Winterborne Kingston DT11 9 H5
Winterborne Monkton DT2 9 F6
Winterborne Stickland DT11 9 H4
Winterborne Whitechurch DT11 9 H4
Winterborne Zelston DT11 9 H5
Winterbourne *S.Glos.* BS36 19 K3
Winterbourne *W.Berks.* RG20 21 H4
Winterbourne Abbas DT2 9 F5
Winterbourne Bassett SN4 20 E4
Winterbourne Dauntsey SP4 10 C1
Winterbourne Earls SP4 10 C1
Winterbourne Gunner SP4 10 C1
Winterbourne Monkton SN4 20 E4
Winterbourne Steepleton DT2 9 F6
Winterbourne Stoke SP3 20 D7
Winterbrook OX10 21 K3
Winterburn BD23 56 E4
Wintercleugh ML12 68 E2
Winteringham DN15 59 F7
Winterley CW11 49 G7
Wintersett WF4 51 F1
Wintershill SO32 11 G3
Winterslow SP5 10 D1
Winterton DN15 52 C1
Winterton-on-Sea NR29 45 J4
Winthorpe *Lincs.* PE25 53 J6
Winthorpe *Notts.* NG24 52 B7
Winton *Bourne.* BH9 10 B5
Winton *Cumb.* CA17 61 J5
Wintringham YO17 58 E2
Winwick *Cambs.* PE28 42 E7
Winwick *Northants.* NN6 31 H1
Winwick *Warr.* WA2 49 F3
Wirksworth DE4 50 E7
Wirksworth Moor DE4 51 F7
Wirswall SY13 38 E1
Wisbech PE13 43 H5
Wisbech St. Mary PE13 43 H5
Wisborough Green RH14 12 D4
Wiseton DN10 51 K4
Wishaw *N.Lan.* ML2 75 F5
Wishaw *Warks.* B76 40 D6
Wisley GU23 22 D6
Wispington LN9 53 F5
Wissett IP19 35 H1
Wissington CO6 34 D5
Wistanstow SY7 38 D7
Wistanswick TF9 39 F3
Wistaston CW2 49 F7
Wiston *Pembs.* SA62 16 D4
Wiston *S.Lan.* ML12 75 H7
Wistow *Cambs.* PE28 43 F7
Wistow *N.Yorks.* YO8 58 B6
Wiswell BB7 56 C6
Witcham CB6 33 H1
Witchampton BH21 9 J4
Witchburn PA28 66 B1
Witchford CB6 33 J1
Witcombe TA12 8 D2
Witham CM8 34 C7
Witham Friary BA11 20 A7
Witham on the Hill PE10 42 D4
Withcall LN11 53 F4
Withcote LE15 42 A5
Withdean BN1 13 G6
Witherenden Hill TN19 13 K4
Witherhurst TN19 13 K4
Witheridge EX16 7 G4
Witherley CV9 41 F6
Withern LN13 53 H4
Withernsea HU19 59 K7
Withernwick HU11 59 H5
Withersdale Street IP20 45 G7
Withersfield CB9 33 K4
Witherslack LA11 55 H1
Witherslack Hall LA11 55 H1
Withiel PL30 3 G2
Withiel Florey TA24 7 H2
Withielgoose PL30 4 A4
Withington *Glos.* GL54 30 B7
Withington *Gt.Man.* M20 49 H3
Withington *Here.* HR1 28 E4
Withington *Shrop.* SY4 38 E4
Withington *Staffs.* ST10 40 C2
Withington Green SK11 49 H5
Withington Marsh HR1 28 E4
Withleigh EX16 7 H4
Withnell PR6 56 B7
Withnell Fold PR6 56 B7
Withybrook *Som.* BA3 19 K7
Withybrook *Warks.* CV7 41 G7
Withycombe TA24 7 J1
Withycombe Raleigh EX8 7 J7
Withyham TN7 13 H3
Withypool TA24 7 G2
Witley GU8 12 C3
Witnesham IP6 35 F3
Witney OX28 21 G1
Wittering PE8 42 D5
Wittersham TN30 14 D5
Witton *Angus* DD9 90 D7
Witton *Norf.* NR13 45 H5
Witton *Worcs.* WR9 29 H2
Witton Bridge NR28 45 H2
Witton Gilbert DH7 62 D2
Witton Park DL14 62 C3
Witton-le-Wear DL14 62 C3
Wiveliscombe TA4 7 J3
Wivelsfield RH17 13 G4
Wivelsfield Green RH17 13 G5
Wivenhoe CO7 34 E6
Wiveton NR25 44 E1
Wix CO11 35 F6
Wixford B49 30 B3
Wixhill SY4 38 E3
Wixoe CO10 34 B4
Woburn MK17 32 C5
Woburn Sands MK17 32 C5
Wokefield Park RG7 21 K5
Woking GU22 22 D6
Wokingham RG40 22 B5
Wolborough TQ12 5 J3
Wold Newton *E.Riding* YO25 59 G2
Wold Newton *N.E.Lincs.* LN8 53 F3
Woldingham CR3 23 G6
Wolfelee TD9 70 A3
Wolferlow HR7 29 F2
Wolferton PE31 44 A3
Wolfhampcote CV23 31 G2
Wolfhill PH2 82 C4
Wolfpits LD8 28 B3
Wolf's Castle SA62 16 C3
Wolfsdale SA62 16 C3
Woll TD7 69 K1
Wollaston *Northants.* NN29 32 C2
Wollaston *Shrop.* SY5 38 C4
Wollaston *W.Mid.* DY8 40 A7
Wollaton NG8 41 H2
Wollerton TF9 39 F3
Wollescote DY9 40 B7
Wolsingham DL13 62 B3
Wolston CV8 31 F1
Wolsty CA7 60 C1
Wolvercote OX2 21 J1
WOLVERHAMPTON WV 40 B6
Wolverley *Shrop.* SY4 38 D2
Wolverley *Worcs.* DY11 29 H1
Wolvers Hill BS29 19 G5
Wolverton *Hants.* RG26 21 J6
Wolverton *M.K.* MK12 32 B4
Wolverton *Warks.* CV35 30 D2
Wolverton *Wilts.* BA12 9 G1
Wolverton Common RG26 21 J6
Wolvesnewton NP16 19 H2
Wolvey LE10 41 G7
Wolvey Heath LE10 41 G7
Wolviston TS22 63 F4
Womaston LD8 28 B2
Wombleton YO62 58 C1
Wombourne WV5 40 A6
Wombwell S73 51 F2
Womenswold CT4 15 H2
Womersley DN6 51 H1
Wonastow NP25 28 D7
Wonersh GU5 22 D7
Wonford EX2 7 H6
Wonson EX20 6 E7
Wonston SO21 11 F1
Wooburn HP10 22 C3
Wooburn Green HP10 22 C3
Wood Bevington B49 30 B3
Wood Burcote NN12 31 H4
Wood Dalling NR11 44 E3
Wood Eaton ST20 40 A4
Wood End *Bed.* MK43 32 D4
Wood End *Bed.* MK44 32 D2
Wood End *Bucks.* MK17 31 J5
Wood End *Herts.* SG2 33 G6
Wood End *W.Mid.* WV11 40 B5
Wood End *Warks.* CV9 40 E6
Wood End *Warks.* B94 30 C1
Wood End *Warks.* CV7 40 E7
Wood Enderby PE22 53 F6
Wood Green *Essex* EN9 23 H1
Wood Green *Gt.Lon.* N22 23 G2
Wood Green *Norf.* NR15 45 G6
Wood Lane SY12 38 D2
Wood Norton NR20 44 E3
Wood Seats S35 51 F3
Wood Stanway GL54 30 B5
Wood Street NR29 45 H3
Wood Street Village GU3 22 C6
Woodacott EX22 6 B5
Woodale DL8 57 F2
Woodall S26 51 G4
Woodbastwick NR13 45 H4
Woodbeck DN22 51 K5
Woodborough *Notts.* NG14 41 J1
Woodborough *Wilts.* SN9 20 E6
Woodbridge *Devon* EX24 7 K6
Woodbridge *Dorset* DT10 9 G3
Woodbridge *Suff.* IP12 35 G4
Woodbury *Devon* EX5 7 J7
Woodbury *Som.* BA5 19 J7
Woodbury Salterton EX5 7 J7
Woodchester GL5 20 B1
Woodchurch *Kent* TN26 14 E4
Woodchurch *Mersey.* CH49 48 B4
Woodcombe TA24 7 H1
Woodcote *Oxon.* RG8 21 K3
Woodcote *Tel. & W.* TF10 39 G4
Woodcote Green B61 29 J1
Woodcott RG28 21 H6
Woodcroft NP16 19 J2
Woodcutts SP5 9 J3
Woodditton CB8 33 K3
Woodeaton OX3 31 G7
Woodend *Aber.* AB51 90 E3
Woodend *Cumb.* CA18 60 C7
Woodend *High.* IV13 88 E2
Woodend *High.* PH36 79 J1
Woodend *Northants.* NN12 31 H4
Woodend *P. & K.* PH15 81 J3
Woodend *W.Suss.* PO18 12 B6
Woodend Green CM22 33 J6
Woodfalls SP5 10 C2
Woodfield *Oxon.* OX26 31 G6
Woodfield *S.Ayr.* KA8 67 H1
Woodfoot CA10 61 H5
Woodford *Cornw.* EX23 6 A4
Woodford *Devon* TQ9 5 H5
Woodford *Glos.* GL13 19 K2
Woodford *Gt.Lon.* IG8 23 H2
Woodford *Gt.Man.* SK7 49 H4
Woodford *Northants.* NN14 32 C1
Woodford *Som.* TA4 7 J2
Woodford Bridge IG8 23 H2
Woodford Green IG8 23 H2
Woodford Halse NN11 31 G3
Woodgate *Devon* EX15 7 K4
Woodgate *Norf.* NR20 44 E4
Woodgate *W.Mid.* B32 40 B7
Woodgate *W.Suss.* PO20 12 C6
Woodgate *Worcs.* B60 29 J2
Woodgreen SP6 10 C3
Woodhall *Inclyde* PA14 74 B3
Woodhall *N.Yorks.* DL8 62 A7
Woodhall Hills LS28 57 G6
Woodhall Spa LN10 52 E6
Woodham *Bucks.* HP18 31 H7
Woodham *Dur.* DL17 62 D4
Woodham *Surr.* KT15 22 D5
Woodham Ferrers CM3 24 D2
Woodham Mortimer CM9 24 E1
Woodham Walter CM9 24 E1
Woodhaven DD6 83 F5
Woodhead *Aber.* AB53 91 F1
Woodhead *Staffs.* ST10 40 C1
Woodhey CH42 48 C4
Woodhey Green CW5 48 E7
Woodhill *Shrop.* WV16 39 G7
Woodhill *Som.* TA3 8 C2
Woodhorn NE63 71 H5
Woodhouse *Cumb.* LA7 55 J1
Woodhouse *Leics.* LE12 41 H4
Woodhouse *S.Yorks.* S13 51 G4
Woodhouse *W.Yorks.* LS2 57 H6
Woodhouse *W.Yorks.* WF6 57 J7
Woodhouse *W.Yorks.* HD6 57 G7
Woodhouse Down BS32 19 K3
Woodhouse Eaves LE12 41 H4
Woodhouse Green SK11 49 J6
Woodhouses *Gt.Man.* M35 49 H2
Woodhouses *Staffs.* DE13 40 D4
Woodhouses *Staffs.* WS7 40 C5
Woodhuish TQ6 5 K5
Woodhurst PE28 33 G1
Woodingdean BN2 13 G6
Woodington SO51 10 E2
Woodland *Devon* TQ13 5 H4
Woodland *Dur.* DL13 62 B4
Woodland *Kent* CT18 15 G3
Woodland Head EX17 7 F6
Woodlands *Dorset* BH21 10 B4
Woodlands *Hants.* SO40 10 E3
Woodlands *N.Yorks.* HG2 57 J4
Woodlands *Shrop.* WV16 39 G7
Woodlands *Som.* TA5 7 K1
Woodlands Park SL6 22 B4
Woodlands St. Mary RG17 21 G4
Woodlane DE13 40 D3
Woodleigh TQ7 5 H6
Woodlesford LS26 57 J7

Woo - Zou

Woodley *Gt.Man.* **SK6** 49 J3
Woodley *W'ham* **RG5** 22 A4
Woodmancote *Glos.* **GL7** 20 D1
Woodmancote *Glos.* **GL52** 29 J6
Woodmancote *Glos.* **GL11** 20 A2
Woodmancote *W.Suss.* **BN5** 13 F5
Woodmancote *W.Suss.* **PO10** 11 J4
Woodmancott **SO21** 21 J7
Woodmansey **HU17** 59 G6
Woodmansterne **SM7** 23 F6
Woodmanton **EX5** 7 J7
Woodmill **DE13** 40 D3
Woodminton **SP5** 10 B2
Woodmoor **SY15** 38 B5
Woodnesborough **CT13** 15 J2
Woodnewton **PE8** 42 D6
Woodperry **OX33** 31 G7
Woodplumpton **PR4** 55 J6
Woodrising **NR9** 44 D5
Woodrow **DT10** 9 G3
Wood's Corner **TN21** 13 K5
Woods Eaves **HR3** 28 B4
Wood's Green **TN5** 13 K3
Woodseaves *Shrop.* **TF9** 39 F2
Woodseaves *Staffs.* **ST20** 39 G3
Woodsend **SN8** 21 F4
Woodsetts **S81** 51 H4
Woodsford **DT2** 9 G5
Woodside *Aberdeen* **AB24** 91 H4
Woodside *Brack.F.* **SL4** 22 C4
Woodside *Cen.Beds.* **LU1** 32 D7
Woodside *Cumb.* **CA15** 60 B3
Woodside *D. & G.* **DG1** 69 F6
Woodside *Fife* **KY8** 83 F7
Woodside *Fife* **KY7** 82 D7
Woodside *Gt.Lon.* **SE25** 23 G5
Woodside *Hants.* **SO41** 10 E5
Woodside *Herts.* **AL9** 23 F1
Woodside *N.Ayr.* **KA15** 74 B5
Woodside *P. & K.* **PH13** 82 C4
Woodside *Shrop.* **SY7** 38 C7
Woodside *W.Mid.* **DY5** 40 B7
Woodside Green **ME17** 14 E2
Woodstock *Oxon.* OX20 31 **F7**
Woodstock *Pembs.* **SA63** 16 D3
Woodthorpe *Derbys.* **S43** 51 G5
Woodthorpe *Leics.* **LE12** 41 H4
Woodthorpe *Lincs.* **LN13** 53 H4
Woodthorpe *S.Yorks.* **S2** 51 F4
Woodton **NR35** 45 G6
Woodtown **EX39** 6 C3
Woodvale **PR8** 48 C1
Woodville **DE11** 41 F4
Woodwall Green **ST21** 39 G2
Woodwalton **PE28** 43 F7
Woodwick **KW17** 106 C5
Woodworth Green **CW6** 48 E7
Woodyates **SP5** 10 B3
Woofferton **SY8** 28 E2
Wookey **BA5** 19 J7
Wookey Hole **BA5** 19 J7
Wool **BH20** 9 H6
Woolacombe EX34 6 **C1**
Woolage Green **CT4** 15 H3
Woolage Village **CT4** 15 H2
Woolaston **GL15** 19 J1
Woolaston Slade **GL15** 19 J1
Woolavington **TA7** 19 G7
Woolbeding **GU29** 12 B4
Woolcotts **TA22** 7 H2
Wooldale **HD9** 50 D2
Wooler NE71 70 **E1**
Woolfardisworthy *Devon* **EX17** 7 G5
Woolfardisworthy *Devon* **EX39** 6 B3
Woolfold **BL8** 49 G1
Woolfords Cottages **EH55** 75 J5
Woolgarston **BH20** 9 J6
Woolgreaves **WF2** 51 F1
Woolhampton **RG7** 21 J5
Woolhope **HR1** 29 F5
Woolland **DT11** 9 G4
Woollard **BS39** 19 K5
Woollaton **EX38** 6 C4
Woollensbrook **EN11** 23 G1
Woolley *B. & N.E.Som.* **BA1** 20 A5
Woolley *Cambs.* **PE28** 32 E1
Woolley *Cornw.* **EX23** 6 A4
Woolley *Derbys.* **DE55** 51 F6
Woolley *W.Yorks.* **WF4** 51 F1
Woolley Green *W. & M.* **SL6** 22 B4
Woolley Green *Wilts.* **BA15** 20 B5
Woolmer Green **SG3** 33 F7
Woolmere Green **B60** 29 J2
Woolmersdon **TA5** 8 B1
Woolpit **IP30** 34 D2
Woolpit Green **IP30** 34 D2
Woolscott **CV23** 31 F2
Woolsgrove **EX17** 7 F5
Woolstaston **SY6** 38 D6
Woolsthorpe **NG32** 42 B2
Woolsthorpe by Colsterworth **NG33** 42 C3
Woolston *Devon* **TQ7** 5 H6
Woolston *S'ham.* **SO19** 11 F3
Woolston *Shrop.* **SY10** 38 C3
Woolston *Shrop.* **SY6** 38 D7
Woolston *Warr.* **WA1** 49 F4
Woolston Green **TQ13** 5 H4
Woolstone *Glos.* **GL52** 29 J5
Woolstone *M.K.* **MK15** 32 B5
Woolstone *Oxon.* **SN7** 21 F3
Woolton **L25** 48 D4
Woolton Hill **RG20** 21 H5
Woolverstone **IP9** 35 F5
Woolverton **BA2** 20 A6
Woolwich **SE18** 23 H4
Woonton **HR3** 28 C3
Wooperton **NE66** 71 F2
Woore **CW3** 39 G1
Wootten Green **IP21** 35 G1
Wootton *Bed.* **MK43** 32 D4
Wootton *Hants.* **BH25** 10 D5
Wootton *I.o.W.* **PO33** 11 G5
Wootton *Kent* **CT4** 15 H3
Wootton *N.Lincs.* **DN39** 52 D1
Wootton *Northants.* **NN4** 31 J3
Wootton *Oxon.* **OX20** 31 F7
Wootton *Oxon.* **OX1** 21 H1
Wootton *Shrop.* **SY7** 28 D1
Wootton *Shrop.* **SY11** 38 C3
Wootton *Staffs.* **ST21** 40 A3
Wootton *Staffs.* **DE6** 40 D1
Wootton Bridge **PO33** 11 G5
Wootton Common **PO33** 11 G5
Wootton Courtenay **TA24** 7 H1
Wootton Fitzpaine **DT6** 8 C5
Wootton Green **MK43** 32 D4
Wootton Rivers **SN8** 20 E5
Wootton St. Lawrence **RG23** 21 J6
Wootton Wawen **B95** 30 C2
WORCESTER WR 29 **H3**
Worcester Park KT4 23 **F5**
Wordsley **DY8** 40 A7
Wordwell **IP28** 34 C1
Worfield **WV15** 39 G6
Worgret **BH20** 9 J6
Work **KW15** 106 D6
Workhouse End **MK41** 32 E3
Workington CA14 60 **B4**
Worksop S80 51 **H5**
Worlaby *Lincs.* **LN11** 53 G5
Worlaby *N.Lincs.* **DN20** 52 D1
World's End *Bucks.* **HP22** 22 B1
Worlds End *Hants.* **PO7** 11 H3
World's End *W.Berks.* **RG20** 21 H4
Worlds End *W.Mid.* **B91** 40 D7
Worle **BS22** 19 G5
Worleston **CW5** 49 F7
Worlingham **NR34** 45 J7
Worlington **IP28** 33 K1
Worlingworth **IP13** 35 G2
Wormald Green **HG3** 57 J3
Wormbridge **HR2** 28 D5
Wormegay **PE33** 44 A4
Wormelow Tump **HR2** 28 D5
Wormhill **SK17** 50 D5
Wormiehills **DD11** 83 H4
Wormingford **CO6** 34 D5
Worminghall **HP18** 21 K1
Wormington **WR12** 30 B5
Worminster **BA4** 19 J7
Wormiston **KY10** 83 H7
Wormit **DD6** 82 E5
Wormleighton **CV47** 31 F3
Wormley *Herts.* **EN10** 23 G1
Wormley *Surr.* **GU8** 12 C3
Wormley West End **EN10** 23 G1
Wormshill **ME9** 14 D2
Wormsley **HR4** 28 D4
Worplesdon **GU3** 22 C6
Worrall **S35** 51 F3
Worsbrough **S70** 51 F2
Worsley **M28** 49 G2
Worstead **NR28** 45 H3
Worsted Lodge **CB21** 33 J3
Worsthorne **BB10** 56 D6
Worston **BB7** 56 C5
Worswell **PL8** 5 F6
Worth *Kent* **CT14** 15 J2
Worth *W.Suss.* **RH10** 13 F3
Worth Matravers **BH19** 9 J7
Wortham **IP22** 34 E1
Worthen **SY5** 38 C5
Worthenbury **LL13** 38 D1
Worthing *Norf.* **NR20** 44 D4
Worthing *W.Suss.* BN11 12 **E6**
Worthington **LE65** 41 G3
Worting **RG23** 21 K6
Wortley *Glos.* **GL12** 20 A2
Wortley *S.Yorks.* **S35** 51 F3
Wortley *W.Yorks.* **LS12** 57 H6
Worton *N.Yorks.* **DL8** 56 E1
Worton *Wilts.* **SN10** 20 C6
Wortwell **IP20** 45 G7
Wothersome **LS23** 57 K5
Wotherton **SY15** 38 B5
Wotter **PL7** 5 F4
Wotton **RH5** 22 E7
Wotton Underwood **HP18** 31 H7
Wotton-under-Edge GL12 20 **A2**
Woughton on the Green **MK6** 32 B5
Wouldham **ME1** 24 D5
Wrabness **CO11** 35 F5
Wrae **AB53** 99 F5
Wrafton **EX33** 6 C2
Wragby **LN8** 52 E5
Wragholme **LN11** 53 G3
Wramplingham **NR18** 45 F5
Wrangaton **PL21** 5 G5
Wrangham **AB52** 90 E1
Wrangle **PE22** 53 H7
Wrangle Lowgate **PE22** 53 H7
Wrangway **TA21** 7 K4
Wrantage **TA3** 8 C2
Wrawby **DN20** 52 D2
Wraxall *N.Som.* **BS48** 19 H4
Wraxall *Som.* **BA4** 9 F1
Wray **LA2** 56 B3
Wray Castle **LA22** 60 E6
Wrays **RH6** 23 F7
Wraysbury **TW19** 22 D4
Wrayton **LA6** 56 B2
Wrea Green **PR4** 55 G6
Wreay *Cumb.* **CA11** 60 F4
Wreay *Cumb.* **CA4** 60 F2
Wrecclesham **GU10** 22 B7
Wrecsam (Wrexham) LL13 38 **C1**
Wrekenton **NE9** 62 D1
Wrelton **YO18** 58 D1
Wrenbury **CW5** 38 E1
Wrench Green **YO13** 59 F1
Wreningham **NR16** 45 F6
Wrentham **NR34** 45 J7
Wrenthorpe **WF2** 57 J7
Wrentnall **SY5** 38 D5
Wressle *E.Riding* **YO8** 58 D6
Wressle *N.Lincs.* **DN20** 52 C2
Wrestlingworth **SG19** 33 F4
Wretham **IP24** 44 D6
Wretton **PE33** 44 A5
Wrexham (Wrecsam) LL13 38 **C1**
Wrexham Industrial Estate **LL13** 38 C1
Wribbenhall **DY12** 29 G1
Wrightington Bar **WN6** 48 E1
Wrightpark **FK8** 74 E1
Wright's Green **CM22** 33 J7
Wrinehill **CW3** 39 G1
Wrington **BS40** 19 H5
Writhlington **BA3** 19 K6
Writtle **CM1** 24 C1
Wrockwardine **TF6** 39 F4
Wroot **DN9** 51 K2
Wrose **BD18** 57 G6
Wrotham **TN15** 23 K6
Wrotham Heath **TN15** 23 K6
Wrotham Hill Park **TN15** 24 C5
Wrottesley **WV8** 40 A5
Wroughton **SN4** 20 E3
Wroxall *I.o.W.* **PO38** 11 G7
Wroxall *Warks.* **CV35** 30 D1
Wroxeter **SY5** 38 E5
Wroxham **NR12** 45 H4
Wroxton **OX15** 31 F4
Wstrws **SA44** 17 G1
Wyaston **DE6** 40 D1
Wyberton **PE21** 43 G1
Wyboston **MK44** 32 E3
Wybunbury **CW5** 39 F1
Wych Cross **RH18** 13 H3
Wychbold **WR9** 29 J2
Wychnor **DE13** 40 D4
Wychnor Bridges **DE13** 40 D4
Wyck **GU34** 11 J1
Wyck Rissington **GL54** 30 C6
Wycliffe **DL12** 62 C5
Wycoller **BB8** 56 E6
Wycomb **LE14** 42 A3
Wycombe Marsh **HP11** 22 B2
Wyddial **SG9** 33 G5
Wye **TN25** 15 F3
Wyesham **NP25** 28 E7
Wyfordby **LE14** 42 A4
Wyke *Devon* **EX17** 7 G6
Wyke *Dorset* **SP8** 9 G2
Wyke *Shrop.* **TF13** 39 F5
Wyke *Surr.* **GU3** 22 C6
Wyke *W.Yorks.* **BD12** 57 G7
Wyke Champflower **BA10** 9 F1
Wyke Regis **DT4** 9 F7
Wykeham *N.Yorks.* **YO13** 59 F1
Wykeham *N.Yorks.* **YO17** 58 E2
Wyken *Shrop.* **WV15** 39 G6
Wyken *W.Mid.* **CV2** 41 F7
Wykey **SY4** 38 C3
Wylam NE41 71 **G7**
Wylde Green **B72** 40 D6
Wyllie **NP12** 18 E2
Wylye **BA12** 10 B1
Wymering **PO6** 11 H4
Wymeswold **LE12** 41 J3
Wymington **NN10** 32 C2
Wymondham *Leics.* **LE14** 42 B4
Wymondham *Norf.* NR18 45 **F5**
Wyndham **CF32** 18 C2
Wynford Eagle **DT2** 8 E5
Wynnstay Park **LL14** 38 C1
Wynyard **TS22** 63 F4
Wyre Piddle **WR10** 29 J4
Wyresdale Tower **LA2** 56 B4
Wysall **NG12** 41 J3
Wyson **SY8** 28 E2
Wythall **B47** 30 B1
Wytham **OX2** 21 H1
Wythburn **CA12** 60 E5
Wythenshawe **M22** 49 H4
Wyton *Cambs.* **PE28** 33 F1
Wyton *E.Riding* **HU11** 59 H6
Wyverstone **IP14** 34 E2
Wyverstone Street **IP14** 34 E2
Wyville **NG32** 42 B3
Wyvis Lodge **IV16** 96 B4

Y

Y Bala (Bala) LL23 37 **J2**
Y Bryn **LL23** 37 H3
Y Drenewydd (Newtown) SY16 38 **A6**
Y Fali (Valley) **LL65** 46 A5
Y Fan **SY18** 37 J7
Y Felin Newydd (New Mills) **SY16** 37 K5
Y Felinheli LL56 46 **D6**
Y Fenni (Abergavenny) NP7 28 **B7**
Y Fflint (Flint) CH6 48 **B5**
Y Ffôr **LL53** 36 C2
Y Fron (Upper Llandwrog) **LL54** 46 D7
Y Gelli Gandryll (Hay-on-Wye) **HR3** 28 B4
Y Trallwng (Welshpool) SY21 38 **B5**
Y Tymbl (Tumble) **SA14** 17 J4
Y Waun (Chirk) **LL14** 38 B2
Yaddlethorpe **DN17** 52 B2
Yafford **PO30** 11 F6
Yafforth **DL7** 62 E7
Yalberton **TQ4** 5 J5
Yalding **ME18** 23 K7
Yanley **BS41** 19 J5
Yanwath **CA10** 61 G4
Yanworth **GL54** 30 B7
Yapham **YO42** 58 D4
Yapton **BN18** 12 C6
Yarburgh **LN11** 53 G3
Yarcombe **EX14** 8 B4
Yardley **B25** 40 D7
Yardley Gobion **NN12** 31 J4
Yardley Hastings **NN7** 32 B3
Yardro **LD8** 28 B3
Yarford **TA2** 8 B2
Yarkhill **HR1** 29 F4
Yarlet **ST18** 40 B3
Yarley **BA5** 19 J7
Yarlington **BA9** 9 F2
Yarm TS15 63 **F5**
Yarmouth PO41 10 **E6**
Yarnacott **EX32** 6 E2
Yarnbrook **BA14** 20 B6
Yarnfield **ST15** 40 A2
Yarnscombe **EX31** 6 D3
Yarnton **OX5** 31 F7
Yarpole **HR6** 28 D2
Yarrow *Sc.Bord.* **TD7** 69 J1
Yarrow *Som.* **TA9** 19 G7
Yarrow Feus **TD7** 69 J1
Yarrowford **TD7** 69 K1
Yarsop **HR4** 28 D4
Yarwell **PE8** 42 D6
Yate **BS37** 20 A3
Yatehouse Green **CW10** 49 G6
Yateley GU46 22 **B5**
Yatesbury **SN11** 20 D4
Yattendon **RG18** 21 J4
Yatton *Here.* **HR6** 28 D2
Yatton *N.Som.* **BS49** 19 H5
Yatton Keynell **SN14** 20 B4
Yaverland **PO36** 11 H6
Yawl **DT7** 8 C5
Yaxham **NR19** 44 E4
Yaxley *Cambs.* **PE7** 42 E6
Yaxley *Suff.* **IP23** 35 F1
Yazor **HR4** 28 D4
Yeabridge **TA13** 8 D3
Yeading **UB4** 22 E3
Yeadon **LS19** 57 H5
Yealand Conyers **LA5** 55 J2
Yealand Redmayne **LA5** 55 J2
Yealand Storrs **LA5** 55 J2
Yealmbridge **PL8** 5 F5
Yealmpton **PL8** 5 F5
Yearby **TS11** 63 H4
Yearsley **YO61** 58 B2
Yeaton **SY4** 38 D4
Yeaveley **DE6** 40 D1
Yeavering **NE71** 77 H7
Yedingham **YO17** 58 E2
Yelford **OX29** 21 G1
Yell **ZE2** 107 N4
Yelland *Devon* **EX31** 6 C2
Yelland *Devon* **EX20** 6 D6
Yelling **PE19** 33 F2
Yelvertoft **NN6** 31 G1
Yelverton *Devon* PL20 5 **F4**
Yelverton *Norf.* **NR14** 45 G5
Yenston **BA8** 9 G2
Yeo Mill **EX36** 7 G3
Yeo Vale **EX39** 6 C3
Yeoford **EX17** 7 F6
Yeolmbridge **PL15** 6 B7
Yeomadon **EX22** 6 B5
Yeovil BA20 8 **E3**
Yeovil Marsh **BA21** 8 E3
Yeovilton **BA22** 8 E2
Yerbeston **SA68** 16 D5
Yesnaby **KW16** 106 B6
Yetholm Mains **TD5** 70 D1
Yetlington **NE66** 71 F2
Yetminster **DT9** 8 E3
Yettington **EX9** 7 J7
Yetts o'Muckhart **FK14** 82 B7
Yew Green **CV35** 30 D2
Yielden **MK44** 32 D2
Yieldshields **ML8** 75 G5
Ynys **LL47** 36 E2
Ynys Enlli (Bardsey Island) **LL53** 36 A3
Ynys Môn (Anglesey) **LL** 46 B4
Ynys Tachwedd **SY24** 37 F6
Ynysboeth **CF45** 18 D2
Ynysddu **NP11** 18 E2
Ynyshir **CF39** 18 D2
Ynyslas **SY24** 37 F6
Ynysmaerdy **CF72** 18 D3
Ynysmeudwy **SA8** 18 A1
Ynystawe **SA6** 17 K5
Ynyswen **CF42** 18 C2
Ynysybwl **CF37** 18 D2
Yockenthwaite **BD23** 56 E2
Yockleton **SY5** 38 C4
Yokefleet **DN14** 58 E7
Yoker **G14** 74 D4
Yonder Bognie **AB54** 98 E6
YORK YO 58 **C4**
Yorkletts **CT5** 25 G5
Yorkley **GL15** 19 K1
Yorton **SY4** 38 E3
Yorton Heath **SY4** 38 E3
Youldon **EX22** 6 B5
Youldonmoor Cross **EX22** 6 B5
Youlgreave **DE45** 50 E6
Youlthorpe **YO41** 58 D4
Youlton **YO61** 57 K3
Young's End **CM3** 34 B7
Yoxall **DE13** 40 D4
Yoxford **IP17** 35 H2
Yr Wyddgrug (Mold) CH7 48 **B6**
Ysbyty Cynfyn **SY23** 27 G1
Ysbyty Ifan **LL24** 37 H1
Ysbyty Ystwyth **SY25** 27 G1
Ysceifiog **CH8** 47 K5
Ysgubor-y-coed **SY20** 37 F6
Ystalyfera **SA9** 18 A1
Ystrad **CF41** 18 C2
Ystrad Aeron **SA48** 26 E3
Ystrad Meurig SY25 27 **G2**
Ystrad Mynach **CF82** 18 E2
Ystradfellte **CF44** 27 J7
Ystradffin **SA20** 27 G4
Ystradgynlais **SA9** 27 G7
Ystradowen *Carmar.* **SA9** 27 G7
Ystradowen *V. of Glam.* **CF71** 18 D4
Ystumtuen **SY23** 27 G1
Ythanwells **AB54** 90 E1
Ythsie **AB41** 91 G1

Z

Zeal Monachorum **EX17** 7 F5
Zeals **BA12** 9 G1
Zelah **TR4** 3 F3
Zennor **TR26** 2 B5
Zouch **LE12** 41 H3

INDEX TO NORTHERN IRELAND

Administrative area abbreviations

A. & N.	Antrim & Newtownabbey		F. & O.	Fermanagh & Omagh	N., M. & D.	Newry, Mourne & Down
A., B. & C.	Armagh City, Banbridge & Craigavon		M. & E. Ant.	Mid & East Antrim		
Derry & Str.	Derry City & Strabane		M. Ulster	Mid Ulster		

A

Name	Page	Grid
Acton BT35	109	G6
Aghadowey BT51	109	F2
Aghagallon BT67	109	G5
Aghalee BT67	109	G5
Aghanloo BT49	108	E2
Ahoghill BT42	109	G3
Aldergrove BT29	109	G5
Annaclone BT32	109	G6
Annahilt BT26	109	H6
Annalong BT34	109	H8
Annsborough BT31	109	H7
Antrim BT41	109	G4
Ardboe BT80	109	F5
Ardglass BT30	109	J7
Ardmillan BT23	109	J5
Ardstraw BT78	108	C4
Armagh BT60	109	F6
Armoy BT53	109	G2
Arney BT92	108	C7
Articlave BT51	108	E2
Artigarvan BT82	108	C3
Attical BT34	109	H8
Aughamullan BT71	109	F5
Augher BT77	108	D6
Aughnacloy BT69	108	E6

B

Name	Page	Grid
Ballinamallard BT94	108	C6
Ballintoy BT54	109	G1
Ballybogy BT53	109	F2
Ballycarry BT38	109	J4
Ballycassidy BT94	108	C6
Ballycastle BT54	109	G1
Ballyclare BT39	109	H4
Ballyeaston BT39	109	H4
Ballygalley BT40	109	H3
Ballygawley BT70	108	E6
Ballygowan BT23	109	J5
Ballyhalbert BT22	109	K5
Ballyhoe Bridge BT53	109	G2
Ballyholme BT20	109	J4
Ballyhornan BT30	109	J6
Ballykeel BT25	109	H6
Ballykelly BT49	108	E2
Ballykinler BT30	109	J7
Ballyleny BT61	109	F6
Ballyloughbeg BT57	109	F2
Ballymackilroy BT70	108	E6
Ballymagorry BT82	108	C3
Ballymartin BT34	109	H8
Ballymena BT43	109	G3
Ballymoney BT53	109	F2
Ballynahatty BT78	108	D5
Ballynahinch BT24	109	H6
Ballynakilly BT71	109	F5
Ballynamallaght BT82	108	D4
Ballyneaner BT82	108	D3
Ballynoe BT30	109	J7
Ballynure BT39	109	H4
Ballyrobert BT39	109	H4
Ballyronan BT45	109	F4
Ballyroney BT32	109	H7
Ballystrudder BT40	109	J4
Ballyvoy BT54	109	G1
Ballywalter BT22	109	K5
Ballyward BT31	109	H7
Balnamore BT53	109	F2
Banbridge BT32	109	G6
Bangor BT20	109	J4
Bannfoot BT66	109	F5
Belcoo BT93	108	B7
BELFAST BT	109	H5
Belfast City Airport BT3	109	H5
Belfast International Airport BT29	109	G4
Bellaghy BT45	109	F4
Bellanaleck BT92	108	C7
Belleek BT 93	108	A6
Belleek N., M. & D. BT35	109	F7
Belmont BT38	109	J4
Benburb BT71	109	F6
Bendooragh BT53	109	F2
Beragh BT79	108	D5
Bessbrook BT35	109	G7
Blackwatertown BT71	109	F6
Blaney BT93	108	B6
Bleary BT63	109	G6
Boho BT74	108	B6
Bolea BT49	108	E2
Bovedy BT51	109	F3
Boviel BT47	108	E3
Bready BT82	108	C3
Brookeborough BT94	108	C6
Broughshane BT42	109	G3
Bryansford BT33	109	H7
Burnside BT39	109	H4
Burren BT34	109	G7
Bushmills BT57	109	F1
Butterlope BT79	108	D4

C

Name	Page	Grid
Caddy BT41	109	G4
Caledon BT68	108	E6
Camlough BT35	109	G7
Cappagh BT70	108	E5
Cargan BT43	109	G3
Carland BT71	108	E5
Carncastle BT40	109	H3
Carnduff BT54	109	G1
Carnlough BT44	109	H3
Carnteel BT69	108	E6
Carrickfergus BT38	109	J4
Carrickmore BT79	108	E5
Carryduff BT8	109	H5
Castlecaulfield BT70	108	E5
Castledawson BT45	109	F4
Castlederg BT81	108	C4
Castlereagh BT6	109	H5
Castlerock BT51	108	E2
Castleroe BT51	109	F2
Castlewellan BT31	109	H7
Chapeltown BT41	109	G4
Charlemont BT71	109	F6
Church Ballee BT30	109	J6
Church Hill BT93	108	B6
Churchtown BT30	109	J6
City of Derry Airport BT47	108	D2
Clabby BT75	108	D6
Clady M. Ulster BT44	109	F3
Clady Derry & Str. BT82	108	C4
Clady Milltown BT60	109	F7
Clanabogan BT78	108	D5
Clare BT62	109	G6
Claudy BT47	108	D3
Clogh BT44	109	G3
Cloghcor BT82	108	C3
Clogher BT76	108	D6
Cloghy BT22	109	K6
Clonelly BT93	108	B5
Clonoe BT71	109	F5
Clough BT30	109	J6
Cloughmills BT44	109	G3
Cloughreagh BT35	109	G7
Cloyfin BT52	109	F2
Coagh BT80	109	F5
Coalisland BT71	109	F5
Coleraine BT52	109	F2
Comber BT23	109	J5
Conlig BT23	109	J5
Cookstown BT80	109	F5
Cooneen BT75	108	D6
Corkey BT44	109	G2
Cox's Hill BT62	109	F6
Craig BT40	109	D4
Craigantlet BT23	109	J5
Craigavole BT51	109	F3
Craigavon BT65	109	G6
Craigdarragh BT47	108	D3
Craigs BT42	109	G3
Cranagh BT79	108	D4
Creagh BT94	109	F8
Creggan N., M. & D. BT35	109	F8
Creggan F. & O. BT79	108	E5
Crilly BT69	108	E6
Cromkill BT42	109	G4
Crossgar BT30	109	J6
Crossmaglen BT35	109	F8
Crumlin BT29	109	G5
Culcavy BT26	109	H5
Culkey BT92	108	C6
Cullaville BT35	109	F8
Cullybackey BT42	109	G3
Cullyhanna BT35	109	F7
Culmore BT48	108	D2
Culnady BT46	109	F3
Curragh BT23	109	J6
Curran BT45	109	F4
Cushendall BT44	109	H2
Cushendun BT44	109	H2

D

Name	Page	Grid
Damhead BT52	109	F2
Darkley BT60	109	F7
Darragh Cross BT30	109	J6
Derry (Londonderry) BT48	108	D3
Derrychrin BT80	109	F5
Derrygonnelly BT93	108	B6
Derrylin BT92	108	C7
Derrymacash BT66	109	G5
Derrytrasna BT66	109	F5
Dervock BT53	109	F2
Desertmartin BT45	109	F4
Doagh BT39	109	H4
Dollingstown BT66	109	G6
Donagh BT92	108	C7
Donaghadee BT21	109	J5
Donaghcloney BT66	109	G6
Donaghmore BT70	108	E5
Donemana BT82	108	D3
Downhill BT51	108	E2
Downpatrick BT30	109	J6
Draperstown BT45	108	E4
Dromara BT25	109	H6
Dromore A., B. & C. BT25	109	H6
Dromore F. & O. BT78	108	C5
Drumahoe BT47	108	D3
Drumaness BT24	109	H6
Drumaroad BT31	109	H6
Drumbo BT27	109	H5
Drumcard BT92	108	C7
Drumduff BT94	108	C6
Drumlegagh BT78	108	C5
Drummacabranagher BT92	108	C7
Drumnakilly BT79	108	D5
Drumquin BT78	108	C5
Drumsurn BT49	108	E3
Dunaghy BT53	109	G2
Dundonald BT16	109	J5
Dundrod BT29	109	H5
Dundrum BT33	109	J7
Dungannon BT70	108	E5
Dungiven BT47	108	E3
Dunloy BT44	109	G3
Dunmurry BT17	109	H5
Dunnamore BT80	108	E4
Dyan BT68	108	E6

E

Name	Page	Grid
Eden BT38	109	J4
Ederny BT93	108	C5
Eglinton BT47	108	D2
Eglish BT70	108	E6
Enniskillen BT74	108	C6
Ervey Cross Roads BT47	108	D3
Eshnadarragh BT92	108	D7
Eskragh BT78	108	D6

F

Name	Page	Grid
Feeny BT47	108	E3
Fintona BT78	108	D5
Finvoy BT53	109	F3
Fivemiletown BT75	108	D6
Foreglen BT47	108	E3
Forkhill BT35	109	G8

G

Name	Page	Grid
Gamblestown BT66	109	G6
Garrison BT93	108	A6
Garvagh BT51	109	F3
Garvaghy BT70	108	D5
Garvary BT74	108	C6
Gilford BT63	109	G6
Glarryford BT44	109	G3
Glebe BT82	108	C4
Glenanne BT60	108	F7
Glenariff BT44	109	H2
Glenarm BT44	109	H3
Glenavy BT29	109	G5
Glengormley BT36	109	H4
Glenhead BT49	108	E3
Glenhull BT79	108	E4
Glenoe BT40	109	H4
Glynn BT40	109	J4
Gortaclare BT79	108	D5
Gortin BT79	108	D4
Gortnahey BT47	108	E3
Gracehill BT42	109	G3
Grange Corner BT41	109	G4
Granville BT70	108	E5
Greencastle BT79	108	D4
Greenisland BT38	109	H4
Greyabbey BT22	109	J5
Greysteel BT47	108	D2
Groomsport BT19	109	J4
Gulladuff BT45	109	F4

H

Name	Page	Grid
Hamilton's Bawn BT60	109	F6
Hannahstown BT17	109	H5
Helen's Bay BT19	109	J4
Hillhall BT27	109	H5
Hillsborough BT26	109	H6
Hilltown BT34	109	H7
Holywell BT93	108	B7
Holywood BT18	109	J5

I

Name	Page	Grid
Inishrush BT44	109	F3
Irvinestown BT94	108	C6

J

Name	Page	Grid
Jonesborough BT35	109	G8

K

Name	Page	Grid
Katesbridge BT32	109	H6
Keady BT60	109	F7
Kearney BT22	109	K6
Keeran BT93	108	C5
Kells BT42	109	G4
Kesh BT93	108	B5
Kilcoo BT34	109	H7
Kilkeel BT34	109	H8
Killadeas BT94	108	C6
Killagan Bridge BT44	109	G2
Killen BT81	108	C4
Killeter BT81	108	C4
Killinchy BT23	109	J5
Killough BT30	109	J7
Killowen BT34	109	G8
Killyclogher BT79	108	D5
Killylea BT60	108	E6
Killyleagh BT30	109	J6
Kilmood BT23	109	J5
Kilmore BT30	109	J6
Kilraghts BT53	109	G2
Kilrea BT51	109	F3
Kilskeery BT78	108	C5
Kilwaughter BT40	109	H3
Kinallen BT25	109	H6
Kircubbin BT22	109	J5
Kirkistown BT22	109	K6
Knockcloghrim BT45	109	F4
Knocknacarry BT44	109	H2

L

Name	Page	Grid
Lack BT93	108	C5
Lackagh BT49	108	E3
Lagavara BT54	109	G1
Laghy Corner BT71	109	F5
Larne BT40	109	H3
Laurelvale BT62	109	G6
Lawrencetown BT63	109	G6
Leggs BT93	108	B5
Leitrim BT31	109	H7
Letter BT93	108	B5
Letterbreen BT74	108	B6
Lettershendony BT47	108	D3
Ligoniel BT14	109	H5
Limavady BT49	108	E2
Lisbane BT23	109	J5
Lisbellaw BT94	108	C6
Lisburn BT28	109	H5
Liscloon BT82	108	D3
Liscolman BT53	109	F2
Lisnarick BT94	108	B6
Lisnaskea BT92	108	C7
Lissan BT80	108	E4
Listooder BT30	109	J6
Loanends BT29	109	H4
Londonderry (Derry) BT48	108	D3
Loughbrickland BT32	109	G6
Loughgall BT61	109	F6
Loughguile BT44	109	G2
Loughinisland BT30	109	J6
Loughmacrory BT79	108	D5
Lower Ballinderry BT28	109	G5
Lurgan BT66	109	G6
Lurganare BT34	109	G7

M

Name	Page	Grid
McGregor's Corner BT43	109	G3
Mackan BT92	108	C7
Macosquin BT51	109	F2
Maghaberry BT67	109	G5
Maghera N., M. & D. BT31	109	H7
Maghera M. Ulster BT46	109	F3
Magherafelt BT45	109	F4
Magheralin BT67	109	G6
Magheramason BT47	108	C3
Magheraveely BT92	108	D7
Maghery BT71	109	F5
Magilligan BT49	108	E2
Maguiresbridge BT94	108	C7
Mallusk BT36	109	H4
Markethill BT60	109	F7
Martinstown BT43	109	G3
Mayobridge BT34	109	G7
May's Corner BT32	109	G7
Mazetown BT28	109	H5
Meigh BT35	109	G7
Middletown BT60	108	E7
Mill Town BT41	109	F4
Millbay BT40	109	J4
Millbrook BT40	109	H3
Millford BT60	109	F6
Millisle BT22	109	K5
Milltown A., B. & C. BT32	109	G7
Milltown A., B. & C. BT62	109	G7
Minerstown BT30	109	J7
Moira BT67	109	G5
Monea BT74	108	B6
Moneydig BT51	109	F3
Moneyglass BT41	109	G4
Moneymore BT45	109	F4
Moneyneany BT45	108	E4
Moneyreagh BT23	109	J5
Moneyslane BT31	109	H7
Monteith BT32	109	G6
Moorfields BT42	109	G4
Moortown BT79	109	F5
Mossley BT36	109	H4
Moss-side BT53	109	G2
Mount Hamilton BT79	108	E4
Mount Norris BT60	109	F7
Mountfield BT79	108	D5
Mountjoy M. Ulster BT71	109	E5
Mountjoy F. & O. BT78	108	D5
Moy BT71	109	F6
Moyargert BT54	109	G2
Moygashel BT71	109	E5
Mullaghbane BT35	109	F8
Mullaghmassa BT79	108	D5
Mullan BT92	108	B7
Murley BT75	108	D6

N

Name	Page	Grid
New Buildings BT47	108	D3
New Ferry BT45	109	F4
Newcastle BT33	109	H7
Newmill BT70	109	H4
Newmills BT71	109	F5
Newry BT34	109	G7
Newtown Crommelin BT43	109	G3
Newtownabbey BT36	109	H4
Newtownards BT23	109	J5
Newtownbutler BT92	108	D7
Newtownhamilton BT35	109	F7
Newtownstewart BT78	108	D4
Nutt's Corner BT29	109	H5

O

Name	Page	Grid
Omagh BT79	108	D5
Oritor BT80	108	E5

P

Name	Page	Grid
Park BT47	108	D3
Parkgate BT39	109	H4
Pettigo BT93	108	B5
Plumbridge BT79	108	D4
Pomeroy BT70	108	E5
Portadown BT62	109	G6
Portaferry BT22	109	J6
Portavogie BT22	109	K6
Portballintrae BT57	109	F1
Portglenone BT44	109	F3
Portrush BT56	109	F1
Portstewart BT55	109	F2
Poyntz Pass BT35	109	G7
Prehen BT47	108	D3

R

Name	Page	Grid
Raffrey BT30	109	J6
Raholp BT30	109	J6
Randalstown BT41	109	G4
Rathfriland BT34	109	G7
Ravernet BT27	109	H5
Richhill BT61	109	F6
Ringboy BT22	109	K6
Ringsend BT51	109	F2
Riverside BT34	109	H8
Rosscor BT93	108	A6
Rosslea BT92	108	D7
Rostrevor BT34	109	G8
Rousky BT79	108	D4
Roxhill BT41	109	G4
Rubane BT22	109	K5

S

Name	Page	Grid
Saintfield BT24	109	J6
Sandholes BT80	108	E5
Saul BT30	109	J6
Scarva BT63	109	G6
Scollogstown BT30	109	J7
Scotch Street BT62	109	F6
Scribbagh BT93	108	A6
Seaforde BT30	109	J6
Seapatrick BT32	109	G6
Seskinore BT78	108	D5
Shoptown BT42	109	H4
Shrigley BT30	109	J6
Sion Mills BT82	108	C4
Sixmilecross BT79	108	D5
Soldierstown BT67	109	G5
Spamount BT81	108	C4
Springfield BT74	108	B6
Staffordstown BT41	109	G4
Stewartstown BT71	109	F5
Stonyford BT28	109	H5
Strabane BT82	108	C4
Straid M. & E. Ant. BT42	109	G4
Straid A. & N. BT39	109	H4
Strangford BT30	109	J6
Stranocum BT53	109	G2
Swatragh BT46	109	F3

T

Name	Page	Grid
Tamlaght BT74	108	C6
Tamlaght O'Crilly BT46	109	F3
Tamnamore BT71	109	F5
Tandragee BT62	109	G6
Tassagh BT60	109	F7
Teemore BT92	108	C7
Templepatrick BT39	109	H4
Tempo BT94	108	C6
The Bush BT71	109	F5
The Diamond M. Ulster BT71	109	F5
The Diamond F. & O. BT78	108	D5
The Drones BT53	109	G2
The Loup BT45	109	F4
The Rock BT70	108	E5
The Sheddings BT42	109	H3
The Six Towns BT45	108	E4
The Spa BT24	109	H6
The Temple BT27	109	H5
Tobermore BT45	109	F4
Toome BT41	109	G4
Trillick BT78	108	C6
Trory BT94	108	C6
Tully BT92	108	C7
Tullyhogue BT80	109	F5
Tynan BT60	108	E6

U

Name	Page	Grid
Upper Ballinderry BT28	109	G5
Upperlands BT46	109	F3

V

Name	Page	Grid
Victoria Bridge BT82	108	C4
Vow BT53	109	F3

W

Name	Page	Grid
Waringsford BT25	109	H6
Waringstown BT66	109	G6
Warrenpoint BT34	109	G8
Whitecross BT60	109	F7
Whitehead BT38	109	J4